Green Energy and Technology

Climate change, environmental impact and the limited natural resources urge scientific research and novel technical solutions. The monograph series Green Energy and Technology serves as a publishing platform for scientific and technological approaches to "green"—i.e. environmentally friendly and sustainable—technologies. While a focus lies on energy and power supply, it also covers "green" solutions in industrial engineering and engineering design. Green Energy and Technology addresses researchers, advanced students, technical consultants as well as decision makers in industries and politics. Hence, the level of presentation spans from instructional to highly technical.

Indexed in Scopus.

Indexed in Ei Compendex.

Yaşar Demirel

Energy

Production, Conversion, Storage, Conservation, and Coupling

Fourth Edition

Springer

Yaşar Demirel
Department of Chemical and Biomolecular Engineering
University of Nebraska Lincoln
Lincoln, NE, USA

ISSN 1865-3529 ISSN 1865-3537 (electronic)
Green Energy and Technology
ISBN 978-3-031-97110-5 ISBN 978-3-031-97111-2 (eBook)
https://doi.org/10.1007/978-3-031-97111-2

This Springer imprint is published by the registered company Springer Nature Switzerland AG
The registered company address is: Gewerbestrasse 11, 6330 Cham, Switzerland

To Zuhal, Selçuk, and Can

Preface

This is the fourth edition of the textbook "Energy" to fill the need for a comprehensive textbook on energy. Importance of energy and its effect on everyday life are undisputable. Consequently, many institutions today offer either energy minor or energy major programs. And "Energy Engineering" is emerging as one of the recent engineering disciplines. This textbook is an undergraduate textbook for students with diverse backgrounds and interested to know more on energy and pursue a degree on energy.

The fourth edition discusses five major aspects of energy that are: energy production, energy conversion, energy storage, energy conservation, and energy coupling in separate chapters. Basic definitions in Chap. 1 and the primary and the secondary energy sources are discussed in Chap. 2. Chapter 3 discusses mechanical and electrical energies that are types of major energies other than heat and work. Chapter 4 discusses internal energy and enthalpy. Chapter 5 discusses mass and energy balance equations, heat of reaction, and heat transfer. Chapter 6 discusses energy production mainly using closed and open cycles of heat engines. Chapter 7 discusses the energy conversion with an emphasis on the ways to improve the energy conversion efficiency. Chapter 8 discusses the energy storage by various means. Chapter 9 discusses the energy conservation and recovery. Chapter 10 briefly introduces energy coupling with examples from biological systems. Chapter 11 focuses on sustainability and life cycle analysis in energy systems to emphasize the implications of the use of energy on the environment, society, and economy. Chapters 12 discusses renewable energy types and some important production processes. Chapter 13 energy management economics discusses the importance of energy balance within the context of energy management and economics. This edition has two new chapters: Chap. 14 discusses decarbonization within the context of climate change. The global efforts in reducing the greenhouse gas emissions in industry and the tools achieving decarbonizations are discussed. In Chap. 15, energy transition is discussed by means of electrification and the role of renewable energy.

Most of the chapters contain fully solved example problems and practice problems to support the easy understanding and applications of the topics discussed.

It is obvious that the present textbook will mature further in reoccurring editions based on the technological developments and suggestions from the students and colleagues. I want to sincerely thank those who helped me in preparing, developing, and improving this new edition.

I very much want to encourage those who use this textbook to contact me with suggestions and corrections for future editions.

Lincoln, NE, USA Yaşar Demirel
2025 ydemirel2@unl.edu

Competing Interests The author has no competing interests to declare that are relevant to the content of this manuscript.

Contents

Symbols

B Magnetic field
c Cost
c_i Molar concentration of component i
C Coulomb
°C Degrees Celsius
C_p Constant pressure and constant volume heat capacities
D Diameter
e Total specific energy
e_p Specific energy potential
F Faraday constant
°F Degrees Fahrenheit
$\boldsymbol{F}$ Mass force
G Gibbs free energy
h Specific enthalpy, heat transfer coefficient
h_i Partial specific enthalpy of component I
H Enthalpy
I Charge
m Mass
M Molar mass
n Number of moles
P Pressure
Q Heat, volume flow
$\dot{Q}$ Heat flow
R Universal gas constant
J_r Rate of reaction
s Specific entropy
S Entropy
t Time
T Absolute temperature
u Specific internal energy, mobility
U Internal energy
v Specific volume, center-of-mass, or barycentric velocity

V	Volume
z	Stoichiometric parameter, valance
w_i	Mass fraction of component i
W	Work

Abbreviations

ATP	Adenosine triphosphate
AI	Artificial intelligence
BECCS	Bioenergy with carbon capture and sequestration
CCUS	Carbon capture utilization and sequester
CFC	Chlorofluorocarbons
CI	Carbon intensity
CCU	Carbon capture utilization
CHP	Combined heat and power
CLC	Chemical looping combustion
COP	Coefficient of performance
CRF	Capital recovery factor
DAC	Direct air capture
DT	Digital twins
EES	Energy efficiency standard
ET	Energy transition
EROI	Energy return on investments
EV	Electric vehicle
FAME	Fatty acid methyl ester
GDP	Gross domestic product
GHG	Greenhouse gases
GWP	Global warming potential
HCFC	Hydrochlorofluorocarbons
HRSG	Heat recovery steam generation
HVAC	Heating ventilation and air conditioning
HHV	Higher heating value
KPI	Key performance indicators
IGCC	Integrated gasification combined cycle
IPPC	Intergovernmental panel on climate change
LCA	Life cycle assessment
LCOE	Levelized cost of electricity
LHV	Lower heating value
LPG	Liquefied petroleum gas
LNG	Liquefied natural gas
RFS	Renewable fuel standards
SDG	Sustainable development goals
SMR	Steam methane reforming

TEA	Technoeconomic analysis
TES	Thermal energy storage
TPI	Thermal process intensification
ML	Machine learning
WID	Welfare index dimensions

Greek Symbols

α	Thermal diffusivity
β	Electroosmotic permeability, expansion coefficient
ε	Local field intensity
λ_{eq}	Equivalence conductance
μ_i	Chemical potential of component i
ρ	Density
v	Stoichiometric

Subscripts

e	Effective
gen	Generation
i, j, k	Components
max	Maximum
min	Minimum
prod	Production
q	Heat
r	Chemical reaction

Greek Symbols

Introduction: Basic Definitions

1

Introduction and Learning Objectives: This chapter introduces basic definitions, properties, and units, hence builds vocabulary before discussing the concept of energy and its various aspects. For example, definitions of systems and properties with various units are essential to understand the change of states during energy conversions in various processes. The chapter also discusses heat, work, steam tables, flow, isothermal, and adiabatic processes.

The learning objectives of this chapter are:

- Basic concepts in energy related fields.
- Energy, heat, work, and power.
- State of systems and various processes.

1.1 System

System is a quantity of matter or a region with its boundary in space chosen for study as seen in Fig. 1.1a. *Boundary* separates the system from its surroundings. In *simple systems,* matter or region is homogenous with little or no external force fields, such as magnetic and surface tension effects. The boundary may be fixed or moving as seen in Fig. 1.1b. The universe outside the system's boundary is *surroundings. Environment* refers to the region beyond the immediate surroundings whose properties are not affected by the process at any point. *Control volume* is any arbitrary region in space through which mass and energy can pass across the boundary as seen in Fig. 1.1c. A control volume represents a boundary that may involve heat, work, and material flow interactions with its surroundings.

There are various types of systems with the following brief definitions:

Y. Demirel, *Energy*, Green Energy and Technology,
https://doi.org/10.1007/978-3-031-97111-2_1

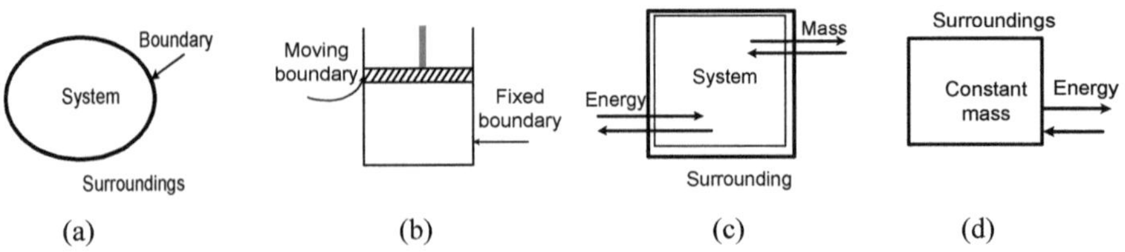

Fig. 1.1 **a** Boundary separates the system from its surroundings, **b** a closed system with moving and fixed boundaries, **c** an open system with a control volume; an open system can exchange mass and energy with its surroundings, **d** a closed system does not exchange material with its surroundings, while it can exchange energy

- *Open systems* exchange mass and energy across the boundary as seen in Fig. 1.1c.
- *Closed systems* exchange energy only through the boundary as seen in Fig. 1.1d.
- *Isolated systems* do not exchange mass and energy with their surroundings.
- *Adiabatic systems* do not exchange heat with their surroundings.
- *Isothermal systems* have a uniform temperature everywhere within the system.
- *Isobaric systems* have a uniform pressure everywhere within the system.
- *Isochoric systems* have a constant volume within the system.
- *Steady-state systems* have properties, which are independent of time; for example, the amount of energy in the systems does not change with time.
- *Unsteady state systems* have properties changing with time.

1.2 Property and Variables

Property is an observable characteristic of a system, such as temperature or pressure. Two types of commonly used properties are *state properties* and *path variables*. The values of *state properties* depend on the state of the system. Temperature, pressure, and volume are examples of state properties. On the other hand, changes in *path variables* depend on the path a system undertakes when it changes from one state to another, such as work and heat. Therefore, the change of heat in a process that proceeds first at constant pressure and then at constant volume will be different from the one that proceeds first at constant volume and then at constant pressure from state 1 to state 2.

- An *extensive property*, such as volume, depends on the size of the system. If the system consists of several parts, the volume of the systems is the sum of the volumes of its parts, and therefore, extensive properties are additive.
- An *intensive property* does not depend on the size of the system, such as temperature, pressure, or density (specific volume). For a homogeneous system, an intensive property can be calculated by dividing an extensive property by the total quantity of the system such as specific volume m^3/kg or specific heat capacity J/g °C.

1.3 Dimensions and Units

Dimensions are basic concepts of measurements such as mass, length, temperature, and time. Units are the means of measuring the dimensions, such as feet or meters for length. Energy has been expressed in several different units. Scientific and engineering quantifications are based on the "International Systems of Units" commonly abbreviated as SI units, which were adopted in 1960 [1]. Table 1.1 shows the seven fundamental measures: length, mass, time, electric current, temperature, amount of substance, and luminous intensity. American Engineering system units, on the other hand, use English system units that are length in foot ft, mass in pound lb_m, and temperature in degree Fahrenheit °F. One kWh is equivalent to 3.6×10^6 J, and one Btu is equivalent to about 1055 J. The candela is the luminous intensity of the surface of a black body at 1 atm and temperature of freezing platinum. Table 1.2 shows the various units of mass and length.

The fundamental measures may be used to derive other quantities and variables called the *derived units*, such as area, volume, density, pressure, force, speed, acceleration, energy, and power. For example, speed is length/time: l/t, acceleration is

Table 1.1 Fundamental measures and units

Measure	Symbol	Unit name (abbreviation)
Length	l	Meter (m), foot (ft)
Mass	m	Kilogram (kg), pound (lb)
Time	t	Second (s)
Current	I	Amper (A)
Temperature	T	Kelvin (K), Rankine (R)
Luminous intensity	I	Candela (cd)
Plane angle	Φ	Radians (rad)

Table 1.2 Mass and length units and their definitions

Measure	Symbol	Unit name
Kilogram	kg	1000 g = 2.204 lb = 32.17 oz
Ounce	oz	28.35 g = 6.25×10^{-2} lb
Pound	lb	0.453 kg = 453 g = 16 oz
Ton, long	ton	2240 lb = 1016.046 kg
Ton, short	sh ton	2000 lb = 907.184 kg
Tonne	t	1000 kg
Ångstrőm	Å	1×10^{-10} m = 0.1 nm
Foot	ft	1/3 yd = 0.3048 m = 12 inches
Inch	in	1/36 yd = 1/12 ft = 0.0254 m
Micron	μ	1×10^{-6} m
Mile	mi	5280 ft = 1760 yd = 1609.344 m
Yard	yd	0.9144 m = 3 ft = 36 in

Table 1.3 Some derived measures and variables

Measure	Symbol	Unit name
Area	A	m^2, ft^2
Volume	V	m^3, ft^3
Density	ρ	kg/m^3, lb/ft^3
Pressure	P	Pascal (Pa), psi, atm
Velocity	v	m/s, ft/s
Acceleration	a	m/s^2 ft/s^2
Volume flow	Q	m^3/s, ft^3/s
Mass flow	m	kg/s, lb/s
Mass flux	m	kg/m^2 s, lb/ft^2 s
Force	F	Newton (N), kg m/s^2
Energy	E	Joule (J), Btu (1055.05 J)
Power	P	Watt (W), (W = J/s), Btu/s
Energy flux	E	J/m^2 s, Btu/ft^2 s
Potential difference	ψ	Volt (V)
Resistance	R	Ohm (Ω)
Electric charge	C	Coulomb (C)
Frequency	f	Hertz (Hz)
Electric flux	e	C/m^2

l/t^2, and force is ml/t^2. Table 1.3 shows some derived variables from fundamental measures, while Table 1.4 shows the prefix system for very small and large magnitudes using metric units.

1.4 Measures of Amounts and Fractions

Common measurements for amount are mass m, number of moles n, or total volume V. Two common fractions are the mass fractions and mole fractions.

- *Mass fraction* is the mass of a substance divided by the total mass of all the substances in the mixture or solution. Mass fraction w_i for a substance i is

$$w_i = \frac{m_i}{m_{\text{total}}} \tag{1.1}$$

- Number of moles n is the ratio of a substance mass m to its molecular weight MW

$$n = \frac{\mathrm{m}}{\mathrm{MW}} \tag{1.2}$$

Table 1.4 Prefixes for SI units

Prefix	Symbol	Factor	Example
Peta	P	10^{15}	Petameter, Pm = 10^{15} m
Tera	T	10^{12}	Terabyte, TB = 10^{12} byte
Giga	G	10^{9}	Gigajoule, GJ = 10^{9} J
Mega	M	10^{6}	Megawatt, MW = 10^{6} W
Kilo	k	10^{3}	Kilojoule, kJ = 10^{3} J
Hecto	h	10^{2}	Hectometer, hm = 10^{2} m
Deka	da	10^{1}	Decaliter, dal = 10^{1} l
Deci	d	10^{-1}	Deciliter, dl = 10^{-1} l
Centi	c	10^{-2}	Centimeter, cm = 10^{-2} m
Milli	m	10^{-3}	Milligram, mg = 10^{-3} g
Micro	µ	10^{-6}	Microvolt, µV = 10^{-6} V
Nano	n	10^{-9}	Nanometer, nm = 10^{-9} m
Pico	p	10^{-12}	Picometer, pm = 10^{-12} m
Femto	f	10^{-15}	Femtometer, fm = 10^{-15} m

- *Mole fraction* x is the ratio of moles of a substance in a mixture or solution divided by the total number of moles in the mixture or solution. Mole fraction x_i for a substance i is

$$x_i = \frac{n_i}{n_{\text{total}}} \tag{1.3}$$

For example, assuming that the mole fraction of nitrogen is 79% and of oxygen is 21% in air, then the mass fractions of nitrogen and oxygen are 76.83% and 23.17%, respectively. Here, the molecular weight of nitrogen is 28.2 g/mol and oxygen 32 g/mol.

1.5 Force

Force F is defined as the product of mass m and acceleration a, according to Newton's second law

$$F = ma \tag{1.4}$$

The SI unit of force is the newton defined by (1 kg)(m/s^2). One kg of mass can be accelerated one m/s^2 by one newton.

The English engineering system of unit of force is the pound force lbf, which can accelerate one pound of mass 32.174 ft/s^2. For the consistency of units, one-pound force is

$$F = \frac{1}{g_c} ma = \frac{1}{g_c} \text{lb}\, 32.174 \frac{\text{ft}}{s^2} \tag{1.5}$$

where g_c is the proportionality constant

$$g_c = 32.174 \frac{\text{lb ft}}{\text{lbf} s^2}$$

A pound force and a pound mass are different quantities. If an equation contains both the pound force and pound mass, the proportionality constant g_c is necessary for the equation to have correct dimensions.

Weight is the force of gravity acting on a body and is expressed in newton or in pound force. Since the acceleration of gravity is $g = 9.8$ m/s^2, the weight in the SI unit is

$$F = mg \tag{1.6}$$

In the English engineering system, the weight is

$$F = m \frac{g}{g_c} \tag{1.7}$$

Since the gravitational force is different at different elevations, weight is a function of elevation. Therefore, the weight for the same mass of a body would vary at different elevations.

1.6 Temperature

Temperature is a physical property that underlies the common notions of hot and cold. Something that feels hotter generally has a higher temperature. Heat flows from a matter with a higher temperature to a matter with a lower temperature. If no net heat flow occurs between two matters, then they have the same temperatures. Statistical physics describes matter as a collection of many particles and defines temperature as a measure of the average energy in each degree of freedom of the particles in the matter. For a matter, this energy is found primarily in the vibrations of its atoms. In an ideal monatomic gas, energy is found in the translational motions resulting in changes in position of the particles; with molecular gases, vibrational and rotational motions also provide thermodynamic degrees of freedom [3, 5].

Temperatures may be calibrated to a variety of scales as seen in Fig. 1.2a. Table 1.5 summarizes the units, definitions, and conversions of temperature to each other. Common temperature scales are:

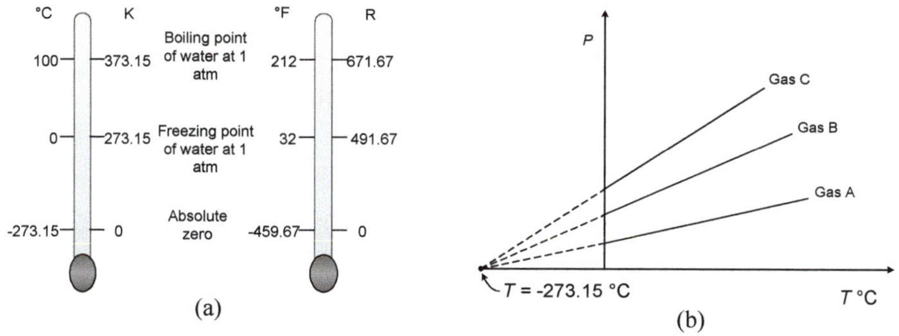

Fig. 1.2 **a** Comparison of different temperature scales; **b** using three different gases in a constant volume gas thermometer, all the straight lines, showing the change of pressure with temperature, will intersect the temperature axis at − 273.15°C after extrapolation of measurements at low pressures. This is the lowest temperature that can be obtained regardless of the nature of gases. So, the absolute gas temperature scale, called the Kelvin scale, is obtained by assigning a value of 0 K to this lowest temperature of − 273.15°C

Table 1.5 Temperature units and their definitions

Unit	Symbol	Definition	Conversion	Equations
Degree Celsius	°C	°C = K − 273.15	°C = (°F − 32)/1.8	(1.8)
Degree Fahrenheit	°F	°F = R − 459.67	°F = °C × 1.8 + 32	(1.9)
Degree Rankine	R	R = °F + 459.67	R = K × 1.8	(1.10)
Degree Kelvin	K	K = °C + 273.15	K = R/1.8	(1.11)

- *Celsius scale* with the freezing point of water is 0°C and the boiling point of water is 100°C (at standard atmospheric pressure). Celsius scale is used in the SI system and named after Anders Celsius.
- *Fahrenheit scale* with the freezing point of water is 32°F and the boiling point is 212°F (at standard atmospheric pressure). The Fahrenheit scale is used in the English engineering unit system and named after Daniel Gabriel Fahrenheit.
- *Thermodynamic temperature scale* is independent of the properties of the substances and is called the *Kelvin scale* or *Rankine scale.* The temperature on this scale is called the *absolute temperature.*
- *The Kelvin scale* is used in the SI units. The lowest temperature on the Kelvin scale is − 273.15°C as seen on Fig. 1.2b. The Kelvin temperature scale matches the Celsius scale: 1 K = 1°C. The Kelvin temperature scale is named after Lord Kelvin.
- *The Rankine scale* is used in the English engineering system units. The lowest temperature on the Rankine scale is − 459.67°F. Rankine and Fahrenheit temperature scales have the same degree intervals: 1R = 1°F. The Rankine temperature scale is named after W. Rankine.

Example 1.1 **Conversion of temperature units**

(a) Convert 27°C to °F, K, and R.
(b) Express a change of 25°C in K and a change of 70°F in R.

Solution:

(a) $T(\text{K}) = T(°\text{C}) + 273.15°\text{C} = 27°\text{C} + 273.15°\text{C} = 300.15\text{K}$

$$T(°\text{F}) = 1.8\text{T}(°\text{C}) + 32 = (1.8)(27)°\text{F} + 32°\text{F} = 80.6°\text{F}$$
$$T(\text{R}) = T(°\text{F}) + 460°\text{F} = 80.6°\text{F} + 460°\text{F} = 540.6\text{R}$$

(b) A change of 25°C in K is: $\Delta T(\text{K}) = \Delta T(°\text{C}) = 25$ K as Kelvin and Celsius scales are identical.

A change 70°F in R is: $\Delta T(\text{R}) = \Delta T(°\text{F}) = 70$ R as Rankine and Fahrenheit scales are identical.

1.7 Pressure

Pressure P is the force per unit cross-sectional area applied in a direction perpendicular to the surface of an object

$$P = \frac{F}{A} \tag{1.8}$$

where F is the component of the force that is perpendicular (normal) and A is the cross-sectional area. The SI unit for pressure is the Pascal (Pa) defined as

$$1\text{Pa} = 1\text{N/m}^2 = \left(1\text{kgm/s}^2\right)/\text{m}^2 \tag{1.9}$$

Two common units for pressure are the pounds force per square inch abbreviated as "psi" and atmosphere abbreviated as "atm," and they are related by 1 atm = 14.659 psi. The atmospheric pressure is variable, while the standard atmosphere is the pressure in a standard gravitational field and is equal to 1 atm, or 101.32 kPa, or 14.69 psi. Pressure may be measured by a manometer. Figure 1.3a shows a basic manometer. Atmospheric pressure is measured by barometer as seen in Fig. 1.3b. The pressure at the bottom of the static column of mercury in Fig. 1.3b is

$$P = \rho g z + P_o \tag{1.10}$$

where P is the pressure at the bottom of the column of the fluid, ρ is the density of the fluid, g is the acceleration of gravity, z is the height of the fluid column, and P_o is the pressure at the top of the column of the fluid.

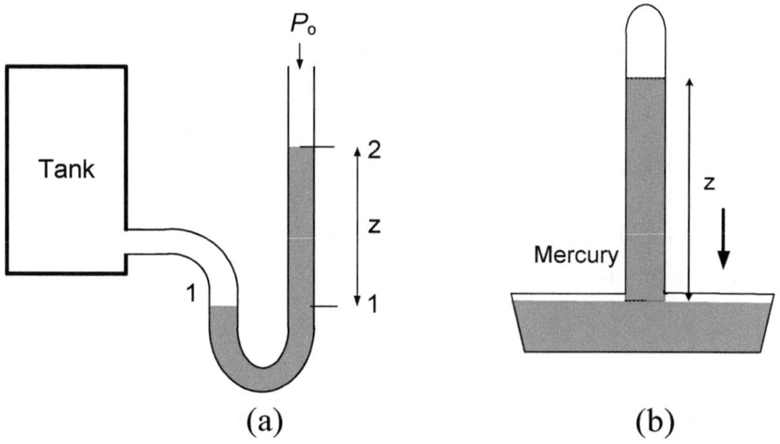

Fig. 1.3 **a** Basic manometer: $P = \rho gz + P_o$, **b** basic mercury barometer to measure atmospheric pressure. The length or the cross-sectional area of the tube has no effect on the height of the fluid in the barometer

- According to the measuring device, we can measure *relative* or *absolute* pressures. *Gauge pressure*, psig, is the pressure relative to the local atmospheric or ambient pressure. Gauge pressure is measured by an open-end manometer as seen in Fig. 1.3a. If the open end of the manometer is closed and a vacuum is created in that end, then we measure *absolute pressure*, psia. For example, the standard atmosphere is absolute pressure. The relationship between absolute and relative pressures is

$$\text{Absolute pressure} = \text{Gauge pressure} + \text{Atmospheric pressure} \tag{1.11}$$

- In common usage, pressure is frequently called *technical atmosphere*, (kilogram-force per square centimeter) (kg_f/cm^2), and $kg_f/cm^2 = 9.8 \times 10^4$ Pa. Table 1.6 shows some conversion factors among the common pressure units.

Example 1.2 Pressure calculations

Consider a column of mercury Hg that has an area of 1 cm^2 and a height of 20 cm. The top of the column is open to atmosphere. The density of mercury at 20°C is 13.55 g/cm^3. Estimate the pressure exerted by the column of mercury on the sealing plate.

Table 1.6 Pressure conversion factors [2]

	kPa	bar	atm	mm Hg	psi
kPa	1	10^{-2}	9.869×10^{-3}	7.50	145.04×10^{-3}
bar	100	1	0.987	750.06	14.503
atm	101.32	1.013	1	760	14.696
mm Hg	0.133	1.333×10^{-3}	1.316×10^{-3}	1	19.337×10^{-3}
psi	6.894×10^3	68.948×10^{-3}	68.046×10^{-3}	51.715	1

1 Pa = 1 N/m^2 = 10^{-5} bar

Solution:

Data: $A = 1$ cm^2, $z = 20$ cm, $g = 980$ cm/s^2, $\rho = 13.55$ g/cm^3 at 20°C, $P_o = 101.32$ kPa.

$$P = \rho g z + P_o = \left(\frac{13.55\ \text{g}}{\text{cm}^3}\right)\left(\frac{980\ \text{cm}}{\text{s}^2}\right)(20\ \text{cm})$$
$$\left(\frac{\text{kg}}{1000\text{g}}\right)\left(\frac{\text{N s}^2}{\text{kg m}}\right)\left(\frac{100\ \text{cm}}{1\ \text{m}}\right)\left(\frac{\text{m}^2\ \text{Pa}}{\text{N}}\right)\left(\frac{\text{kPa}}{1000\ \text{Pa}}\right)$$
$$+ 101.32\ \text{kPa} = 26.56\ \text{kPa} + 101.32\ \text{kPa} = 127.88\ \text{kPa}$$

In English engineering units, we have the density of mercury

$$\rho = \left(\frac{13.55\ \text{g}}{\text{cm}^3}\right)\left(\frac{62.4\ \text{lbm}}{\text{ft}^3}\right) = 845\ \text{lbm/ft}^3$$

The pressure in English units becomes:

$$P = \rho g z + P_o = \left(\frac{845.5\ \text{lbm}}{\text{ft}^3}\right)\left(\frac{32.2\ \text{ft}}{\text{s}^2}\right)(20\ \text{cm})$$
$$\left(\frac{\text{in}}{2.54\ \text{cm}}\right)\left|\frac{\text{ft}}{12\ \text{in.}}\right|\left|\frac{\text{s}^2\ \text{lbf}}{32.174\ \text{ft lbm}}\right|\left(\frac{\text{ft}^2}{144\ \text{in}^2}\right)$$
$$+ 14.69\,\text{psi} = 3.85\ \text{psi} + 14.69\,\text{psi} = 18.54\,\text{psi}$$

In practice, the height of a liquid column is referred to as *head* of liquid. Therefore, the pressure of the column of mercury is expressed as 20 cm Hg, while the pressure on the sealing plate at the bottom becomes 20 cm Hg + 76 cm Hg (1 atm) = 96 cm Hg.

Example 1.3 **Pressure conversions**

Convert the pressure of 10 GPa to (a) psia, (b) atm, and (c) bar.

Solution:

(a) $\frac{10\ \text{GPa}}{1}\left(\frac{10^9\text{Pa}}{\text{GPa}}\right)\left(\frac{14.69\ \text{psia}}{101.32\times10^3\text{Pa}}\right) = 1.45 \times 10^6$ psia

(b) $\frac{10\ \text{GPa}}{1}\left(\frac{10^9\text{Pa}}{\text{GPa}}\right)\left(\frac{\text{atm}}{101.32\times10^3\text{Pa}}\right) = 0.098 \times 10^6$ atm

(c) $\frac{10\ \text{GPa}}{1}\left(\frac{10^9\text{Pa}}{\text{GPa}}\right)\left(\frac{\text{bar}}{100.0\times10^3\text{Pa}}\right) = 0.1 \times 10^6$ bar

Example 1.4 **Absolute pressure estimations**

A gauge connected to a chamber kept under vacuum reads 7.6 psig. The atmospheric pressure at the location is 14.3 psi. Estimate the absolute pressure in the chamber.

Solution:

Absolute pressure shows the actual pressure at a given point. Pressures below atmospheric pressure are called vacuum pressure, and the gauges measuring it indicate the difference between atmospheric pressure and absolute pressure.

$$P_{\text{gauge}} = P_{\text{abs}} - P_{\text{atm}} \quad \text{(For pressures above atmospheric pressure)}$$

$$P_{\text{gauge}} = P_{\text{vacuum}} = P_{\text{atm}} - P_{\text{abs}} \quad \text{(For pressures below atmospheric pressure)}$$

$$P_{\text{abs}} = P_{\text{atm}} - P_{\text{gauge}} = 14.3\ \text{psi} - 7.6\ \text{psig} = 6.7\ \text{psia}$$

1.8 Volume

Volume is how much three-dimensional space occupies or contains a substance. The substance may be one or more solid, liquid, gas, or plasma states. The units of volume depend on the units of length. Table 1.7 shows common units and their definitions for volume. Table 1.8 shows the conversion factor between the various volume units. Volume related derivative properties are:

- Total volume V_t of a system may be divided by the mass to calculate *specific volume* v. Specific volume is the inverse of density, such as m^3/kg of ft^3/lb

$$\text{specific volume } (v) = \frac{1}{\rho} = \frac{V}{m} \tag{1.12}$$

- Total volume may be divided by number of moles n to calculate *molar volume*

$$V_m = \frac{V_t}{n} \tag{1.13}$$

Table 1.7 Volume units and their definitions [1, 4]

Name of unit	Symbol	Definitions
Barrel (Imperial)	bl (Imp)	36 gal (Imp) = 0.163 m^3
Barrel	bl; bbl	42 gal (US) = 0.158 m^3
Cubic foot	cu ft	0.028 m^3
Cubic inch	cu in	16.387×10^{-6} m^3
Cubic meter	m^3	1 m^3 = 1000 L
Cubic yard	cu yd	27 cu ft = 0.764 m^3
Gallon (US)	gal (US)	3.785×10^{-3} m^3 = 3.785 1
Ounce	US fl oz	1/128 gal (US) = 29.573×10^{-6} m^3
Pint	pt (US dry)	1/8 gal (US dry) = 550.610×10^{-6} m^3
Quart	qt (US)	¼ gal (US dry) = 1.101×10^{-3} m^3
Liter	L	1000 cm^3 = 10^{-3} m^3

Table 1.8 Volume conversion factors [2]

	in^3	ft^3	US gal	Liter	m^3
in^3	1	5.787×10^{-4}	4.329×10^{-3}	1.639×10^{-2}	1.639×10^{-5}
ft^3	1.728×10^3	1	7.481	28.32	2.832×10^{-2}
US gal	231	0.133	1	3.785	3.785×10^{-3}
Liters	61.03	3.531×10^{-2}	0.264	1	1.000×10^{-3}
m^3	6.102×10^4	35.31	264.2	1000	1

- *Density* ρ is the ratio of mass per unit volume, such as lb/ft^3, or g/cm^3

$$\rho = \frac{\text{mass}}{\text{volume}} = \frac{m}{V} \tag{1.14}$$

The densities of liquids and solids do not change significantly with pressure in ordinary conditions; however, they change with temperature. The densities of gases change with pressure and temperature. For example, the density of ethanol is 0.790 g/cm^3 and the volume of 1 lb of ethanol is (454 g/lb) (cm^3/0.790 g) = 574.7 cm^3/lb.

- *Specific gravity* (sp. gr.) is the ratio of densities of the substance to density of a reference substance

$$\text{specific gravity (sp. gr.)} = \frac{\rho}{\rho_{H_2O \text{ at } 4°C}} \tag{1.15}$$

In specific gravity, the reference substance usually is water at 4°C, with density of 1 g/cm^3, or 62.43 lb/ft^3. The specific gravity of gases frequently refers to air. For example, sp. gr. = 0.82 at 25°/4° means the substance is at 20°C and the reference substance is at 4°C. As the density of water at 4°C is 1 g/cm^3, the numerical values of the specific gravity and density are the same in the SI unit system.

In the petroleum industry, the specific gravity of petroleum products is often reported in terms of a hydrometer scale called API gravity, and expressed with a standard temperature of 60°F

$$\text{API gravity} = \frac{141.5}{\text{sp. gr.}\frac{60°\text{F}}{60°\text{F}}} - 131.5 \tag{1.16}$$

Concentration refers to the quantity of a substance per unit volume. Some common concentrations are:

- *Mass concentration* is the mass per unit volume such as lb solute/ft^3.
- *Molar concentration* is the moles per unit volume, such as lb mole/ft^3.
- *Parts per million*, ppm, or *parts per billion*, ppb, expresses the concentration of extremely dilute solutions or mixtures.

$$\text{ppm} = \frac{\text{parts}}{10^6\ \text{parts}} \tag{1.17}$$

They are dimensionless and equivalent to a mass fraction for solids and liquids and a mole fraction for gases. For example, about 9 ppm carbon monoxide means that 9 parts per million parts of air in mass or moles.

- *Parts per million by volume*, ppmv, or *parts per billion by volume*, ppbv.

1.9 Energy

Energy is the capacity to do work. Energy comes in various forms, such as motion, heat, light, electrical, chemical, nuclear energy, and gravitation. *Total energy* is the sum of all forms of energy a system possesses. In the absence of magnetic, electrical, and surface tension effects, the total energy of a system consists of the *kinetic, potential, and internal energies.* For example, kinetic energy of $mv^2/2$ in SI unit defines Joule (J) as J = kg (m^2/s^2), where m is the mass and v the average velocity as seen in Table 1.9. The *internal energy* of a system is made up of sensible, latent, chemical, and nuclear energy. The sensible internal energy is due to translational, rotational, and vibrational effects of molecules. *Thermal energy* is the sensible form and latent form of internal energy. The classification of energy into different "types" often follows the boundaries of the fields of study in the natural sciences. For example, *chemical energy* is the kind of potential energy stored in chemical bonds, and *nuclear energy* is the energy stored in interactions between the particles in the atomic nucleus. *Microscopic forms of energy* are related to the molecular structure of a system, and they are independent of outside reference frames [3].

Table 1.9 Some energy units and definitions

Name of unit	Symbol	Definitions
British thermal unit	Btu	1055 J = 5.4039 psia ft^3
Btu/lb	Btu/lb	2.326 kJ/kg
Joule	J	J = m·N = 1 kg·m^2/s^2
Calorie	Cal	4.1868 J
kJ	kJ	kPa m^3 = 1000 J
kJ/kg	kJ/kg	0.43 Btu/lb
Erg	erg	g·$cm^2/s^2 = 10^{-7}$ J
Foot pound force	ft lbf	1.355 J
Horsepower hour	hph	hp × h = 2.684×10^6 J
Kilowatt hour	kWh	kW × h = 3.6×10^6 J
Quad	Quad	1×10^{15} Btu = 1.055×10^{18} J
Atmosphere liter	Atml	atm × l = 101.325 J
kW	kW	3412 Btu/h
Horsepower	hp	2545 Btu/h
Therm	therm	29.3 kWh
Electronvolt	eV	$\approx 1.602\ 17 \times 10^{-19} \pm 4.9 \times 10^{-26}$ J

Hydrogen represents a store of potential energy that can be released by fusion of hydrogen (combination of hydrogen atoms) in the Sun. Some of the *fusion energy* is then transformed into sunlight, which may again be stored as gravitational potential energy after it strikes the earth. Sunlight is also captured by plants as *a potential chemical energy* in photosynthesis, when carbon dioxide and water are converted into carbohydrates, lipids, and proteins. This chemical potential energy is responsible for the growth and development of a biological cell.

British thermal unit (Btu) is the energy unit (in the English system) needed to raise the temperature of 1 lb_m of water at 68°F by 1°F. *Calorie* (cal) is the amount of energy (in the metric system) needed to raise the temperature of 1 g of water at 15°C by 1°C. Table 1.9 displays some of the important energy units and their definitions.

1.10 Work

Energy is a conserved and extensive property of every system in any state and hence its value depends only on the state. *Work*, on the other hand, is not a property of a system. Energy may be transferred in the form of heat or work through the boundary of a system. The mechanical work of expansion or compression proceeds with the observable motion of the coordinates of the particles of matter. *Chemical work*, on the other hand, proceeds with changes in internal energy due to

Table 1.10 Heat, energy, or work conversions

	ft lb	kWh	hph	Btu	Calorie	Joule
ft lb	1	3.766×10^{-7}	5.050×10^{-7}	1.285×10^{-3}	0.324	1.356
kWh	2.655×10^{6}	1	1.341	3.413×10^{3}	8.606×10^{5}	3.6×10^{6}
hph	1.98×10^{6}	0.745	1	2.545×10^{3}	6.416×10^{5}	2.684×10^{6}
Btu	778.16	2.930×10^{-4}	3.930×10^{-4}	1	252	1.055×10^{3}
Calorie	3.086	1.162×10^{-6}	1.558×10^{-6}	3.97×10^{-3}	1	4.184
Joule	0.737	2.773×10^{-7}	3.725×10^{-7}	9.484×10^{-4}	0.2390	1

changes in the chemical composition (mass action). *Potential energy* is the capacity for mechanical work related to the position of a body, while *kinetic energy* is the capacity for mechanical work related to the motion of a body. Potential and kinetic energies are *external energies*, while sensible heat and latent heat are *internal energies*. Table 1.10 lists the conversion coefficients between various units of energy [3].

1.11 Power

The rate of work done is measured by *power*. The instantaneous power created by a force F on an object moving with velocity $v = dx/dt$ becomes: $P = F\ (dx/dt) = Fv$. For example, Watt (W) gives rise to one joule of energy in one second of time. Table 1.11 displays some of the power definitions and their conversions. The rate of work, that is the work per unit time, is called power. In SI units, power is defined as:

$\mathrm{W} = \mathrm{J/s} = \mathrm{Nm/s} = \mathrm{kg \cdot m^2/s^3}$.

In English units' power is:

$$\begin{aligned}\mathrm{Btu/s} &= 778.1\ \mathrm{ft\ lbf/s} = 1.415\ \mathrm{hp} : \mathrm{mechanical\ hp} = 550\,\mathrm{ft\,lbf/s} \\ &= 746\ \mathrm{W} = (\mathrm{electrical\ hp}).\end{aligned}$$

Table 1.11 Power conversions

	hp	kW	ft lbf/s	Btu/s	J/s
hp	1	0.745	550	0.707	745.7
kW	1.341	1	737.56	0.948	1000
ft lbf/s	1.818×10^{-3}	1.356×10^{-3}	1	1.285×10^{-3}	1.356
Btu/s	1.415	1.055	778.16	1	1055
J/s	1.341×10^{-3}	0.001	0.737	9.478×10^{-3}	1

- *Horsepower* is the imperial (British) unit of power. Horsepower is the ability to do work at the rate of 33,000 lbf ft/min or 550 lbf ft/s. For example, the total horsepower developed by water falling from a given height is

$$P_{hp} = \frac{\dot{m}gz}{550} \tag{1.18}$$

Or in terms of volumetric flow rate, horsepower is

$$P_{hp} = \frac{\dot{Q}\rho gz}{550} \tag{1.19}$$

where P_{hp} is the horsepower hp, $\dot{m}$ is the mass flow rate (lb/s), z is the head or height (ft), g is the acceleration of gravity (32 ft/s^2), and $\dot{Q}$ is the volumetric flow rate (ft^3/s).

- *The brake horsepower* is the amount of real horsepower going to, for example, a pump, not the horsepower used by the motor driving the pump. Due to hydraulic, mechanical, and volumetric losses in a pump or turbine, the actual horsepower available for work is less than the total horsepower supplied.

Example 1.5 **Power conversions**

A 20 kW electric motor is used to pump ground water into a storage tank. Estimate the work done by the pump in (a) Btu/hr, (b) hp, (c) J/s, and (d) ft lbf/h.

Solution:

(a) $(20\ \mathrm{kW})\left(\frac{0.948\ \mathrm{Btu/s}}{\mathrm{kW}}\right)\left(\frac{3600\ \mathrm{s}}{\mathrm{h}}\right) = 6.825 \times 10^4\ \mathrm{Btu/h}$

(b) $(20\ \mathrm{kW})\left(\frac{1.341\ \mathrm{hp}}{\mathrm{kW}}\right) = 26.82\ \mathrm{hp}$

(c) $(20\ \mathrm{kW})\left(\frac{1000\ \mathrm{J/s}}{\mathrm{kW}}\right) = 2 \times 10^4\ \mathrm{J/s}$

(d) $(20\ \mathrm{kW})\left(\frac{737.5\ \mathrm{ft\ lbf/s}}{\mathrm{kW}}\right)\left(\frac{3600\ \mathrm{s}}{\mathrm{h}}\right) = 53.1 \times 10^6\ \mathrm{ft\ lbf/h}$

1.12 Heat Capacity

The heat capacity C is the amount of heat necessary to raise the temperature of a matter (system) by one degree. When the temperature change occurs at constant pressure or at constant volume, the heat capacity is shown by C_p or C_v, respectively. For a unit amount, such as gram or pound, it is called the specific heat capacity with the units of J/g °C. It is also expressed in molar units as J/mole °C.

1.13 Heat

Heat is part of the total energy flow across a system boundary that is caused by a temperature difference between the system and its surroundings or between two systems at different temperatures. Heat flows from high temperature region to cold temperature region. Therefore, heat is not stored and defined as thermal energy in transit. The unit for heat in the SI system is the joule, J. When the heat capacity is constant, such as $C_{p,\mathrm{av}}$, the amount of heat changes when a substance changes its temperature from T_1 to another temperature T_2 is

$$q = mC_{p,\mathrm{av}}(T_2 - T_1) \tag{1.20}$$

where q is the heat, and m is the mass. The values of specific heat capacity at constant pressure, C_p, for various substances are shown in Table B3 in Appendix B. The value of heat is positive when transferred to the system and negative when transferred from the system to its surroundings. There is no heat flow through the boundary in an *adiabatic process* as the system and its surroundings have the same temperatures. *Thermal equilibrium* is achieved when two systems in thermal contact with each other cease to exchange energy by heat.

1.14 Internal Energy

Internal energy U of a system or a body with well-defined boundaries is the total of the kinetic energy due to the motion of molecules (translational, rotational, and vibrational) and the potential energy associated with the vibrational motion and electric energy of atoms within molecules. Internal energy also includes energy in all the chemical bonds. For gas, internal energy comprises of translational kinetic energy, rotational kinetic energy, vibrational energy, electronic energy, chemical energy, and nuclear energy. For any material, solid, liquid, or gaseous, it may also consist of the potential energy of attraction or repulsion between the individual molecules [3].

Internal energy is a state function of a system and is an extensive quantity. One can have a corresponding intensive thermodynamic property called *specific internal energy*, commonly symbolized by the lower-case letter u, which is internal energy per mass of the substance in question. As such, the SI unit of specific internal energy would be the J/g. For a closed system, the internal energy is essentially defined by

$$\Delta U = \delta q + \delta W \tag{1.21}$$

where ΔU is the change in internal energy of a system during a process, δq is the path dependent heat change, and δW is the path dependent mechanical work change. The heat absorbed by a system increases its internal energy. If an energy exchange occurs because of temperature difference between a system and its surroundings, this energy appears as *heat* otherwise it appears as *work*.

1.15 Enthalpy

Enthalpy H of a system is defined by

$$H = U + PV \tag{1.22}$$

where U is the internal energy, P is the pressure at the boundary of the system and its environment, and V is the volume of the system. The PV term is equivalent to the energy which would be required to make the pressure of the environment remain constant. For example, a gas changing its volume can maintain a constant pressure P. This may be possible, for example, by a chemical reaction that causes the change in volume of a gas in a cylinder and pushes a piston. The SI unit of enthalpy is the joule. The SI unit for specific enthalpy is joules per kilogram. Enthalpy is a state function of a system and is an extensive quantity.

Whenever the enthalpy of a matter is independent of pressure, regardless of the process, or at constant pressure, the change of enthalpy is

$$\Delta H = H_2 - H_1 = C_{p,\text{av}}(T_2 - T_1) \tag{1.23}$$

where the $C_{p,\text{av}}$, is an average value within the temperature interval $(T_2 - T_1)$. The change of enthalpy (ΔH) of a system is equal to the sum of nonmechanical work done on it and the heat supplied to it. The increase in enthalpy of a system is exactly equal to the energy added through heat, provided that the system is under constant pressure: $\Delta H = \delta q$. Therefore, enthalpy is sometimes described as the "*heat content*" of a system under a given pressure [3].

1.16 Entropy

Entropy change is determined by the following equation

$$\mathrm{d}S = \frac{\delta q_{\text{rev}}}{T} \tag{1.24}$$

where δq_{rev} is the reversible heat flow. The rate of change of entropy when a small amount of energy is added to a system at fixed volume in thermodynamic equilibrium defines temperature as: $\frac{1}{T} = \left.\frac{\mathrm{d}S}{\mathrm{d}U}\right|_V$. In an integrated form, the reversible change of heat flow is estimated by

$$q_{\text{rev}} = T\Delta S \tag{1.25}$$

When a fluid system changes from state A to state B by an irreversible process, then the change of its entropy is be $\Delta S = S_B - S_A$. Entropy is a state function and an extensive property. The total change of entropy is expressed by: $\Delta S = \Delta S_{\text{boundary}} + \Delta S_{\text{internal}}$, where $\Delta S_{\text{boundary}}$ is the change due to the interaction of a system with its surroundings, and $\Delta S_{\text{internal}}$ is the increase due to a natural change within the system, such as a chemical reaction, and is always positive for irreversible changes and zero at equilibrium [1]. For one mole of ideal gases ($PV = RT$, and $dV/dT = R/P$), the change in entropy is

$$dS = C_p\frac{dT}{T} - R\frac{dP}{P} \tag{1.26}$$

For incompressible fluids, entropy change becomes

$$dS = C_{p,\text{av}}\frac{dT}{T} \tag{1.27}$$

Energy exchange accompanied by entropy transfer is heat transfer, and energy exchange that is not accompanied by entropy transfer is the work transfer.

1.17 State

State is a set of determined properties that completely describes the condition of that system. There are various definitions of state:

- *Standard reference state* for the properties of chemical components is chosen as 25°C (77°F) and 1 atm. The property values at the standard reference state are mainly indicated by a superscript (°).
- *In a steady state,* any kind of accumulation, such as mass or energy, within the system is zero. The flows in and out remain constant in time, and the properties of the system do not change in time.
- *Gas* and *vapor states* are often used as synonymous words. The gas state of a substance is above its critical temperature. Vapor, on the other hand, implies a gas state that is not far from its condensation temperature.
- *Thermal equilibrium* means that the temperature is the same throughout the entire system.
- A system is in *mechanical equilibrium* if there is no change in pressure at any point of the system with time.
- In *chemical equilibrium*, the reaction rates in forward and backward directions become equal.
- An ideal gas *equation of state* relates the pressure, temperature and volume of a substance:

$$PV = nRT \tag{1.28}$$

where P is the absolute pressure, T is the absolute temperature, n is the number of moles of a substance and R is called the universal gas constant, which is the same for all substances:

$$R = 8.314\ \text{kJ/kmole K} = 8.314\ \text{kPa m}^3/\text{kmole K}$$

$$R = 1.986\ \text{Btu/lbmole R} = 10.73\ \text{psia ft}^3/\text{lbmole R}$$

The ideal gas relation approximately represents the *PVT* behavior of real gases only at low densities (high volume and low pressure). For a constant mass, the properties of an ideal gas at two different states are related to each other by

$$\frac{P_1 V_1}{T_1} = \frac{P_2 V_2}{T_2} \tag{1.29}$$

1.17.1 Saturated Liquid and Saturated Vapor States

A pure substance will start to boil when its vapor pressure is equal to atmospheric pressure. The boiling temperature is called saturation temperature. *Saturated liquid and saturated vapor states* exist at saturation temperature corresponding to total pressure. Consequently, *superheated vapor* exists at a higher temperature and pressure compared with the saturated temperature and pressure. Figure 1.4a shows a typical phase diagram for a pure substance. Between the solid and liquid phases, there is a solid–liquid equilibrium curve. Between liquid and vapor phases, there is a liquid–vapor equilibrium curve. At critical point (P_c, T_c), the saturated liquid and vapor states become identical. At triple point, the three phases coexist in equilibrium. At the supercritical region: $T > T_c$ and $P > P_c$.

The proportions of vapor and liquids phases in the saturated *liquid–vapor* mixture is called the *quality x,* which is the ratio of mass of vapor to the total mass of the mixture

$$\text{quality } (x) = \frac{\text{mass of vapor}}{\text{total mass of mixture}} \tag{1.30}$$

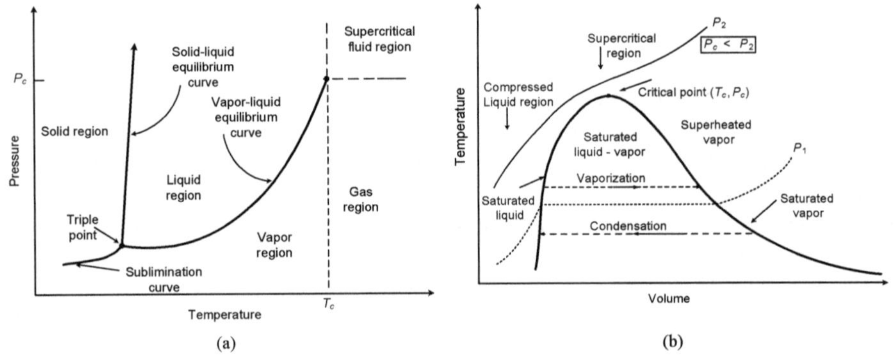

Fig. 1.4 **a** Pressure–temperature diagram of a pure substance; **b** temperature-volume diagram of a pure substance. At a critical point, saturated liquid line and saturated vapor lines merge. Within the saturation curves, a mixture of vapor and liquid phases exist. Super critical region represents a state of matter for temperature and pressure above P_c and T_c that are the critical pressure and temperature

Average values of specific volume, enthalpy, and internal energy of a saturated liquid–vapor mixture are estimated by

$$V = (1 - x)V_{\text{liq sat}} + xV_{\text{vap sat}} \tag{1.31}$$

$$H = (1 - x)H_{\text{liq sat}} + xH_{\text{vap sat}} \tag{1.32}$$

$$U = (1 - x)U_{\text{liq sat}} + xU_{\text{vap sat}} \tag{1.33}$$

1.17.2 Partial Pressure and Saturation Pressure

Partial pressure is the pressure exerted by a single component in a gaseous mixture if it existed alone in the same volume occupied by the mixture and at the same temperature and pressure of the mixture. Partial pressure depends on the pressure and temperature. On the other hand, *saturation pressure* is the pressure at which liquid and vapor phases are at equilibrium at a given temperature. Saturation pressure depends only on the temperature. When the saturation pressure of a pure component reaches the atmospheric pressure, the liquid starts to boil. For example, water boils at 100°C when its saturation pressure equals to the atmospheric pressure of 1 atm. The Antoine equation may be used to estimate the saturation pressure, which is temperature dependent only

$$\ln P_{\text{sat}} = A - \frac{B}{T + C} \tag{1.34}$$

where *A, B*, and *C* are the Antoine constants. Table 1.12 lists the Antoine constants for some species [4].

Example 1.6 **Estimation of saturated vapor pressure**

(a) Estimate the saturation vapor pressure of water and acetone at 50°C, (b) Estimate the boiling point temperature of water and acetone at $P = 101.32$ kPa.

Solution:

The Antoine constants from Table 1.12:

Water: $A = 16.3872$; $B = 3885.70$; $C = 230.17$.

Acetone: $A = 14.3145$; $B = 2756.22$; $C = 228.06$.

(a) $T = 50°\text{C}$ use $\ln P_{\text{sat}} = A - \frac{B}{T+C}$

Table 1.12 Antoine constants [4] for some species: P is in kPa and T is in °C

Species	A	B	C	Range (°C)
Acetone	14.314	2756.22	228.06	– 26 to 77
Acetic acid	15.071	3580.80	224.65	24 to 142
Benzene	13.782	2726.81	217.57	6 to 104
1-Butanol	15.314	3212.43	182.73	37 to 138
Carbon tetrachloride	14.057	2914.23	232.15	– 14 to 101
Chloroform	13.732	2548.74	218.55	– 23 to 84
Ethanol	16.895	3795.17	230.92	3 to 96
Ethyl benzene	13.972	3259.93	212.30	33 to 163
Methanol	16.578	3638.27	239.50	– 11 to 83
Methyl acetate	14.245	2662.78	219.69	– 23 to 78
1-propanol	16.115	3483.67	205.81	20 to 116
2-propanol	16.679	3640.20	219.61	8 to 100
Toluene	13.932	3056.96	217.62	13 to 136
Water	16.387	3885.70	230.17	0 to 200

Plug the Antoine constants into the equation for water and acetone, respectively,

$$P_{w,\mathrm{sat}} = 12.33\,\mathrm{kPa\ at}\ T = 50°\mathrm{C}$$
$$P_{a,\mathrm{sat}} = 82.1\,\mathrm{kPa\ at}\ T = 50°\mathrm{C}$$

(b) Boiling point, T_b, of water at 101.32 kPa

$$T_b = \frac{B}{A - \ln P} - C$$

$$T_{w,b} = 100°\mathrm{C\ and}\ T_{a,b} = 56.4°\mathrm{C\ at}\ 101.33\,\mathrm{kPa}$$

1.18 Steam Tables

Steam tables tabulate the properties of saturated and superheated steam and widely available in English units and SI units (http://www.eisco.co/boilr/steam_tables.pdf). Tables 1.13 and 1.14 show examples of saturated steam tables in English and SI units. Saturated steam is fully defined by either its temperature or pressure. For example, at 140°F liquid water has an enthalpy of 107.95 Btu/lb and water vapor has an enthalpy of 1122.0 Btu/lb. The difference would be the heat of vaporization at 140°F.

Table 1.13 Saturated water in English engineering units

T (°F)	P_{sat} psia	Spec. vol. V ft^3/lb		U Btu/lb		Enthalpy H Btu/lb		Entropy S Btu/lb R	
		V_l	V_g	U_l	U_g	H_l	H_g	S_l	S_g
132	2.345	0.01626	149.66	99,95	1053.7	99.95	1118.6	0.1851	1.9068
134	2.472	0.01626	142.41	101.94	1054.3	101.95	1119.5	0.1884	1.9024
136	2.605	0.01627	135.57	103.94	1055.0	103.95	1120.3	0.1918	1.8980
140	2.889	0.01629	123.00	107.94	1056.2	107.95	1122.0	0.1985	1.8895
142	3.041	0.01630	117.22	109.94	1056.8	109.95	1122.8	0.2018	1.8852
144	3.200	0.01631	111.76	111.94	1057.5	111.95	1123.6	0.2051	1.8810

Table 1.14 Saturated water in SI units

T (K)	P_{sat} kPa	V cm^3/g		U, kJ/kg		Enthalpy H kJ/kg		Entropy S kJ/kg K	
		V_l	V_g	U_l	U_g	H_l	H_g	S_l	S_g
372.15	97.76	1.043	1730.0	414.7	2505.3	414.8	264.4	1.2956	7.3675
373.15	101.33	1.044	1673.0	419.0	2506.5	419.1	2676.0	1.3069	7.3554
375.15	108.78	1.045	1565.5	427.4	2508.8	427.5	2679.1	1.3294	7.3315
377.15	116.68	1.047	1466.2	435.8	2511.1	435.9	2682.2	1.3518	7.3078
379.15	125.04	1.049	1374.2	444.3	2513.4	444.4	2685.3	1.3742	7.2845
381.15	133.90	1.050	1288.9	452.7	2515.7	452.9	2688.3	1.3964	7.2615
383.15	143.27	1.052	1209.9	461.2	2518.0	461.3	2691.3	1.4185	7.2388

Tables 1.15 and 1.16 show examples of superheated steam tables in English and SI units. The properties of superheated steam are determined by the values of temperature and pressure.

Example 1.7 Energy change during evaporation

A mass of 22 kg of saturated liquid water at 101.3 kPa is evaporated completely at constant pressure and produces saturated vapor. Estimate the temperature of the vapor and amount of energy added to the water.

Table 1.15 Superheated steam in English engineering units

T (°F)	V ft^3/lb	U Btu/lb	H Btu/lb	S Btu/lb R	V ft^3/lb	U Btu/lb	H Btu/lb	S Btu/lb R
	$P = 250$ psia, $T_{sat} = 400.97$°F				$P = 255$ psia, $T_{sat} = 402.72$°F			
Sat liq	0.019	375.3	376.1	0.5679	0.019	377.2	378.0	0.5701
Sat vap	1.843	1115.8	1201.1	1.5264	1.808	1116.0	1201.3	1.5247
420	1.907	1125.8	1214.0	1.5413	1.865	1125.1	1213.1	1.5383
440	1.970	1135.9	1227.1	1.5559	1.928	1135.3	1226.3	1.5530
460	2.032	1145.6	1239.6	1.5697	1.989	1145.0	1238.9	1.5669
480	2.092	1154.9	1251.7	1.5827	2.048	1154.5	1251.1	1.5800
500	2.150	1164.0	1263.5	1.5951	2.105	1163.6	1262.9	1.5925

Table 1.16 Superheated steam in SI units

T (°C)	V cm³/g	U kJ/kg	H kJ/kg	S kJ/kg K	V cm³/g	U kJ/kg	H kJ/kg	S kJ/kg K
	$P = 1200$ kPa, $T_{sat} = 187.96$°C				$P = 1250$ kPa, $T_{sat} = 189.81$°C			
Sat liq	1.139	797.1	798.4	2.2161	1.141	805.3	806.7	2.2338
Sat vap	163.20	2586.9	2782.7	6.5194	156.93	2588.0	2784.1	6.5050
200	169.23	2611.3	2814.4	6.5872	161.88	2608.9	2811.2	6.5630
220	178.80	2650.0	2864.5	6.6909	171.17	2648.0	2861.9	6.6680
240	187.95	2686.7	2912.2	6.7858	180.02	2685.1	2910.1	6.7637
260	196.79	2722.1	2958.2	6.8738	188 56	2720.8	2956.5	6.8523
280	205.40	2756.5	3003.0	6.9562	196.88	2755.4	3001.5	6.9353
300	213.85	2790.3	3046.9	7.0342	205.02	2789.3	3045.6	7.0136

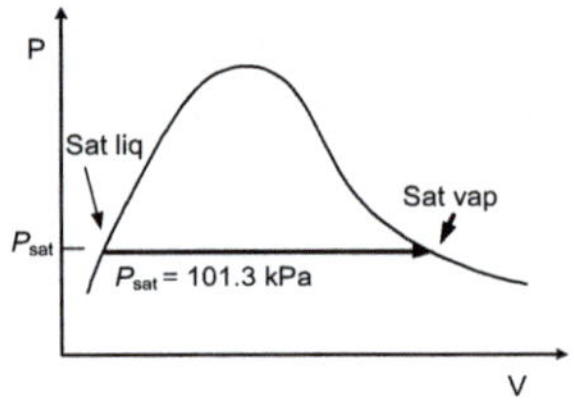

Solution:

From steam table: $P_{\text{sat liq}} = 101.3$ kPa and $T_{\text{sat liq}} = 100$°C.

$$H_{\text{sat liq}} = 419.1\,\text{kJ/kg and} \quad H_{\text{sat vap}} = 2676.0\,\text{kJ/kg}$$

$$\Delta H_{\text{vap}} = H_{\text{sat vap}} - H_{\text{sat liq}} = (2676.0 - 419.1)\,\text{kJ/kg} = 2256.9\,\text{kJ/kg}$$

$$\text{Total amount of heat added } = (2256.9\,\text{kJ/kg})(22\,\text{kg}) = 49{,}651.8\,\text{kJ}.$$

Example 1.8 Energy change during condensation

A mass of 22 kg of superheated steam at 1250 kPa and 220°C condenses completely to saturated liquid state at constant pressure. Estimate the temperature of the saturated vapor and heat removed from the steam.

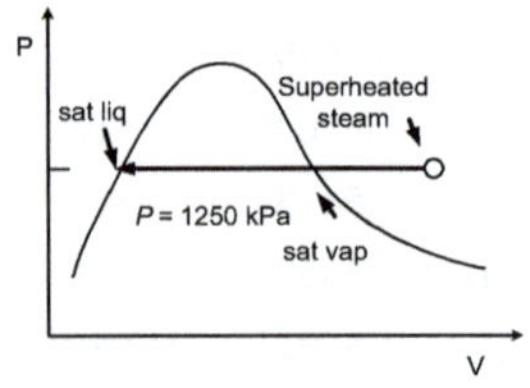

Solution:

From steam tables: Superheated steam:

$P = 1250$ kPa and $T = 220\,^{\circ}$C, $H_{\text{superheat vap}} = 2861.9$ kJ/kg

Saturated steam and liquid at $T_{\text{sat}} = 189.8$°C.

Saturated liquid water at 1250 kPa, and $H_{\text{sat liq}} = 806.7$ kJ/kg , $T_{\text{sat}} = 189.8$°C.

Heat of condensation:

$\Delta H_{\text{cond}} = H_{\text{sat liq}} - H_{\text{superheat vap}} = (806.7 - 2861.9)\text{kJ/kg} = -2055.2\ \text{kJ/kg}.$

Total amount of heat removed $= -(2055.2\ \text{kJ/kg})(22\ \text{kg}) = -45{,}214.4$ kJ.

The removed heat has the negative sign as the heat leaving the system of superheated steam after condensation process as sees in the figure above.

Example 1.9 **Quality of a saturated liquid and vapor mixture of a steam**

A rigid tank contains 15 kg of saturated liquid and vapor water at 85°C. Only 2 kg of the water is in liquid state. Estimate the enthalpy of the saturated mixture and the volume of the tank.

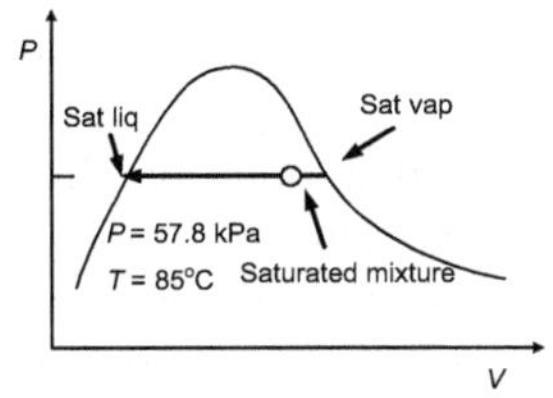

Solution:

From steam tables:

Saturated liquid–vapor water mixture at 57.8 kPa and $T_{\text{sat}} = 85$°C,

Saturated vapor:

$P = 57.8\,\text{kPa}, \quad H_{\text{sat-vap}} = 2652.0\,\text{kJ/kg}, \quad V_{\text{vap}} = 2828\,\text{cm}^3/\text{g}.$

Saturated liquid water at $P = 57.8$ kPa, $H_{\text{sat-liq}} = 355.9$ kJ/kg, $V_{\text{liq}} = 1.003\,\text{cm}^3/\text{g}$.

Amount of liquid water = 2 kg.

Amount of vapor = 15 – 2 = 13 kg.

Quality x of the saturated mixture: $x = \dfrac{m_{\text{vap}}}{m_{\text{total}}} = \dfrac{13\ \text{kg}}{15\ \text{kg}} = 0.87.$

Enthalpy of the mixture, Eq. (1.32):

$$H_{\text{mix}} = (1 - x)H_{\text{sat liq}} + xH_{\text{sat vap}} = (1 - 0.87)355.9\ \text{kJ/kg} + (0.87)2652.0\ \text{kJ/kg} = 2353.5\ \text{kJ/k}$$

Volume of the mixture, Eq. (1.31):

$$V_{\text{mix}} = (1 - x)V_{\text{sat liq}} + xV_{\text{sat vap}} = (1 - 0.87)0.001003\ \text{m}^3/\text{kg} + (0.87)2.828\ \text{m}^3/\text{kg} = 2.46\ \text{m}^3/\text{kg}$$

$$V_{\text{tank}} = 15\text{kg}\left(2.46\text{m}^3/\text{kg}\right) = 36.9\,\text{m}^3$$

Enthalpy of the saturated mixture depends on the quality of the mixture x and will be

$$H_{\text{sat.vap}} > H_{\text{mix}} > H_{\text{sat}\cdot\text{liq}}$$

1.19 Process

Process is any change that a system undergoes from one equilibrium state to another. To describe a process completely, one should specify the initial and final states of the process, as well as the *path* it follows during its progress, and the interactions with the surroundings. There are several types of processes:

- *Adiabatic process* occurs in either open or closed systems without exchanging heat with the surroundings. This is an ideal concept as it implies perfect insulation against the flow of heat, which is impossible in the strict sense.
- In *isothermal* process, the temperature remains constant.
- *Constant pressure processes* are isobaric processes ($P =$ constant) processes).
- *Constant volume processes* are isochoric processes ($V =$ constant) processes).
- *Polytropic process* exists in which the pressure–volume relation is $PV^x =$ constant where x is a constant called the *polytropic index.*
- In a *batch process,* material is neither added to nor removed from the process during its operation.
- In a *semi-batch process,* material may be added but the product is not removed during the operation.
- In a *steady-flow (uniform-flow) process,* a fluid flows through a control volume steadily and no intensive or extensive properties within the control volume change with time. The fluid properties can change from point to point within the control volume, but at any point, they remain constant during the entire process. Figure 1.5 shows a steady-flow system with mass and energy interactions with the surroundings.
- In a *transient-flow* process, properties change within a control volume with time.

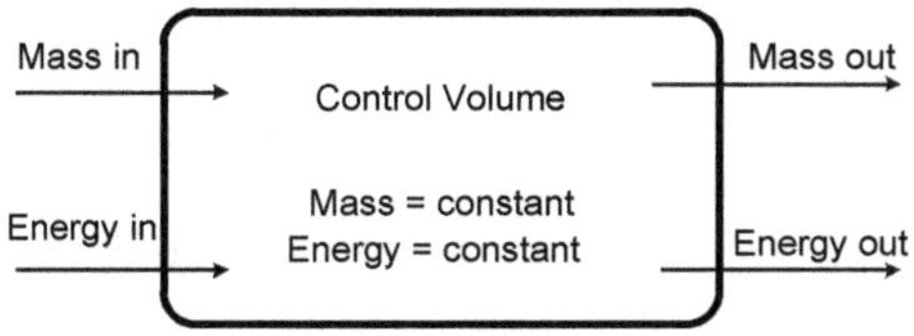

Fig. 1.5 Steady-flow system with mass and energy flow interactions with the surroundings. Mass and energy flows do not change with time in a steady-flow system

- *Irreversible processes* cannot spontaneously reverse themselves and restore the system to its initial state. *Irreversibility* can be viewed as wasted work potential or the lost opportunity to do work. Some of the factors causing *irreversibility* are friction, unrestrained expansion, mixing of two species in different states, heat transfer across a finite temperature difference, electric resistance, inelastic deformation of solids, and chemical reactions. *Friction* is a familiar form of irreversibility associated with bodies in motion which results from the force that opposes the motion developed at the interface of the two bodies in contact.
- *Cyclic process*, or series of processes, allows a system to undergo various state changes and return the system to the initial state at the end of the processes. For a cycle, the initial and final states are identical. *The ideal cycle* consists of internally reversible processes.
- *The isentropic process* is an internally reversible and adiabatic process. In such a process the entropy remains constant.
- *Quasi-static, or quasi-equilibrium, process* is a process which proceeds in such a manner that the system always remains infinitesimally close to an equilibrium state. A quasi-equilibrium process can be viewed as a sufficiently slow process that allows the system to adjust itself internally so that properties in one part of the system do not change any faster than those at other parts.

Problems

1.1. Temperature units are Celsius and Kelvin (K) in the SI units, and Fahrenheit and Rankine (R) in the English engineering system. At what temperature do the values of the temperatures in Celsius (°C) and in Fahrenheit (°F) scales become equal?

1.2. In a cooling process, the temperature of a system drops by 36°F. Express this temperature decrease in Celsius (°C), Kelvin, (K), and Rankine (R).

1.3. In a heating process, the temperature of a system increases by 110°C. Express this temperature increase in Fahrenheit (°F), Kelvin, (K), and Rankine (R).

1.4. The temperature of steam is 420°C. Express the temperature of steam in Fahrenheit (°F), Kelvin, (K), and Rankine (R).

1.5. Convert the following units: (a) atm to bar, (b) Btu to J, (c) hp to kW; (d) quad to kJ, (e) lb_m to ton, (f) quart to liter

1.6. Convert the following units: (a) 4 km to miles, (b) 100 ft^3/day to cm^3/h, (c) 10 miles to ft, (d) 1000 liters to gallons.
1.7. Convert the following units: (a) 40 km to miles, (b) 10 ft^3/day to cm^3/h, (c) 100 miles to ft, (d) 10 liters to gal.
1.8. Convert the following units: (a) 1000 miles to km, (b) 2000 ft^3/day to gal/h, (c) 50000 ft to miles, (d) 1000 gallons to m^3.
1.9. Convert the following units: (a) 550 miles to km, (b) 500 ft^3/day to gal/min, (c) 10000 ft to miles, (d) 150 gallons to liters.
1.10. Convert the following units: (a) 1200 miles to meters, (b) 1500 ft^3/day to gal/s, (c) 2000 ft to meters, (d) 500 gallons to liters.
1.11. Convert the following units: (a) 200 miles to inches, (b) 2500 ft^3/h to gal/s, (c) 600 nanometers to inches, (d) 300 gallons to in^3.
1.12. A heat capacity equation for acetylene is given by

$$C_p = 9.89 + 0.827 \times 10^{-2} T - 0.378 \times 10^{-5} T^2$$

where T is in °F and C_p is in Btu/lbmole °F. Convert the equation for C_p in J/mole °C.
1.13. Parameters in Table B1 require use of temperatures in Kelvin and from that table the molar heat capacity of ethylene in ideal gas state is

$$\frac{C_p^{ig}}{R} = 1.424 + 14.394 \times 10^{-3} T - 4.392 \times 10^{-6} T^2$$

Develop an equation for C_p^{ig} in J/mole °C.
1.14. Parameters in Table B1 require use of temperatures in Kelvin and from that table the molar heat capacity of propylene in ideal gas state is

$$\frac{C_p^{ig}}{R} = 1.637 + 22.706 \times 10^{-3} T - 6.915 \times 10^{-6} T^2$$

Develop an equation for C_p^{ig} in cal/mole °C.
1.15. Parameters in Table B1 require use of temperatures in Kelvin and from that table the molar heat capacity of methane in ideal gas state is

$$\frac{C_p^{ig}}{R} = 1.702 + 9.081 \times 10^{-3} T - 2.164 \times 10^{-6} T^2$$

Develop an equation for C_p^{ig} in cal/mole °C.
1.16. Parameters in Table B2 require use of temperatures in Kelvin and from that table the molar heat capacity of liquid ethanol is

$$\frac{C_p}{R} = 33.866 - 172.60 \times 10^{-3} T + 349.17 \times 10^{-6} T^2$$

Develop an equation for C_p in J/mole °C.

1.17. Parameters in Table B2 require use of temperatures in Kelvin and the molar heat capacity of liquid methanol is

$$\frac{C_p}{R} = 13.431 - 51.28 \times 10^{-3}T + 131.13 \times 10^{-6}\,T^2$$

Develop an equation for C_p in J/mole °C.

1.18. Parameters in Table B3 require use of temperatures in Kelvin and from that table the molar heat capacity of solid $NaHCO_3$ is

$$\frac{C_p}{R} = 5.128 + 18.148 \times 10^{-3}T$$

Develop an equation for C_p in Btu/lbmole R.

1.19. Experimental values of for the heat capacity, C_p, in J/mole °C of air from temperature, T, interval of 0 to 900°C are:

T	0	25	100	200	300	400	500	600	700	800	900
C_p	29.062	29.075	29.142	29.292	29.514	29.782	30.083	30.401	30.711	31.020	31.317

Use the least square method to estimate the values of the coefficients in the following form:

$$C_p = A + BT + C\,T^2$$

1.20 The heat capacity of ethanol is expressed by $C_p = 61.34 + 0.1572T$ where T is in °C and C_p is in J/mol °C. Modify the expression so that the unit for C_p would be Btu/lbmole R.

1.21 Express 30 GPa in: (a) atmosphere, (b) psia, (c) inches of Hg, (d) mm of Hg.

1.22 A gauge connected to a chamber kept under vacuum reads 4.5 psi. The atmospheric pressure at the location is 14.6 psi. Estimate the absolute pressure in the chamber.

1.23 A gauge connected to a chamber kept under vacuum reads 75 kPa. The atmospheric pressure at the location is 101.2 kPa psi. Estimate the absolute pressure in the chamber.

1.24 A gauge connected to a chamber reads 24.8 psi. The atmospheric pressure at the location is 14.2 psi. Estimate the absolute pressure in the chamber.

1.25 Estimate the atmospheric pressure at a location where the temperature is 25°C and the barometric reading is 755 mmHg.

1.26 A dead-weight gauge uses a piston with diameter, D, of 3 mm. If the piston is pressed by a 50 kg of weight, determine the pressure reading on the gauge.

1.27 The pressure reading by a dead-weight gauge is 30 bar. The piston of the gauge has a diameter of 4 mm. Determine the approximate mass in kg of the weights required for pressure reading.

1.28 The pressure reading on a dead-weight gauge is 3 atm. The piston diameter of the gauge is 0.2 in. Determine the approximate mass in lb_m of the weights required.

1.29 The reading on a mercury open manometer (open to the atmosphere at one end) is 15.5 cm. The local acceleration of gravity is 9.832 m/s^2. The atmospheric pressure is 101.75 kPa. Determine the absolute pressure in kPa being measured. The density of mercury at 25°C is 13.534 g/cm^3.

1.30 The reading on a mercury open manometer at 70°F (open to the atmosphere at one end) is 10.25 in. The local acceleration of gravity is 32.243 ft/s^2. Atmospheric pressure is 29.86 in. Hg. Estimate the measured absolute pressure in psia. The density of mercury at 70°F is 13.543 g/cm^3.

1.31 At 300 K the reading on a manometer filled with mercury is 65.7 cm. The local acceleration of gravity is 9.785 m/s^2. What is the pressure in the atmosphere that corresponds to this height of mercury?

1.32 The pressure gauge on a tank filled with carbon dioxide reads 45 psi. The barometer reading is 28.5 in Hg. Estimate the absolute pressure in the tank in psia and atmosphere.

1.33 Complete the following table if $P_{atm} = 100$ kPa and $\rho_{Hg} = 13.6$ g/cm^3

P_{Gauge} (kPa)	P_{abs} (kPa)	P_{abs} (mm Hg)	P_{Gauge} (m H_2O)
5			
	150		
		30	
			30

1.34. Estimate the mass of the air in a room with dimensions of $(4)(5)(5)m^3$ at 20°C and 1 atm.

1.35. Estimate the temperature of 68.2 kg of carbon dioxide gas in a tank with a volume of 20 m^3 and pressure of 250 kPa.

1.36. Estimate the temperature of 12 kg of propane gas in a tank with a volume of 1.6 m^3 and pressure of 400 kPa.

1.37. Estimate the temperature of 6.8 kg of hydrogen gas in a tank with a volume of 40 m^3 and pressure of 300 kPa.

1.38. Estimate the temperature of 12.6 lb of nitrogen gas in a tank with a volume of 150 ft^3 and pressure of 27 psia.

1.39. Estimate the pressure of 2.2 kg of carbon dioxide (CO_2) gas at $T = 350$ K and $V = 0.5$ m^3.

1.40. Estimate the pressure of 0.6 kg of hydrogen gas at $T = 400$ K and $V = 1.0$ m^3.

1.41. A tank has been filled with 240 lb of propane (C_3H_8). The volume of the tank is 150 ft^3 and the temperature is 80°F. Estimate the pressure of propane using the ideal gas equation of state.

1.42. Estimate the molar volume and specific volume of nitrogen (N_2) at 20 psia and 180°F using the ideal gas equation of state.

1.43. Estimate the molar volume and specific volume of propane C_3H_8 at 15 psia and 100°F using the ideal gas equation of state.

1.44. Convert 2.5 μg mole/ml min to lb mole/ft^3 hr.

1.45. The current Operational Safety and Healy Administration (OSHA) 8-hour limit for carbon monoxide (CO) in air is 9 ppm. How many grams carbon monoxide/kg air is 9 ppm?

1.46. (a) Estimate the saturation vapor pressure of water and acetone at 40°C, (b) Estimate the boiling point temperature of water and acetone at 98 kPa.

1.47. (a) Estimate saturation pressures for methanol and ethanol at 40°C, (b) Estimate boiling points for methanol and ethanol at 101.33 kPa.

1.48. (a) Estimate saturation pressures for 1-propanol at 50°C, (b) Estimate boiling points for 1-propanol at 101.33 kPa.

1.49. A tank filled with 10 lb of air is heated. The air originally is at 40 psia and 100°F. After heating, the air pressure becomes 60 psia. Estimate the final temperature of the air.

1.50. A 2.0 m^3 tank filled with hydrogen (H_2) at 200 kPa and 500 K. The hydrogen is cooled until its temperature drops to 300 K. Estimate the amount of hydrogen and the final pressure in the tank after cooling.

1.51. Determine the number of grams of ethanol vapor within a volume of 500 cm^3 at a pressure of 500 kPa and a temperature of 100°C.

1.52. A cylinder of a fixed volume of 15 m^3 containing CO_2 is initially at a temperature of 100°C and a pressure of 101 kPa. When the temperature rises to 150°C find the final pressure.

1.53. A piston cylinder system is under isothermal conditions at a temperature of 15°C with a volume of 37 cm^3 and a pressure of 120 kPa. The piston moves creating a pressure of 260 kPa within the system. What is the change of volume after the piston has reached its final position?

1.54. Determine the volume occupied by 10 kg of water at a pressure of 10 MPa and the following temperatures: 400 and 600°C.

1.55. Determine the volume occupied by 3.5 kg of water vapor at a pressure of 260 psia and the following temperatures of 480 and 1100°F.

1.56. A mass of 10 kg of saturated liquid water at 101.3 kPa is evaporated completely at constant pressure and produces saturated vapor. Estimate the temperature of the vapor and amount of energy added to the water.

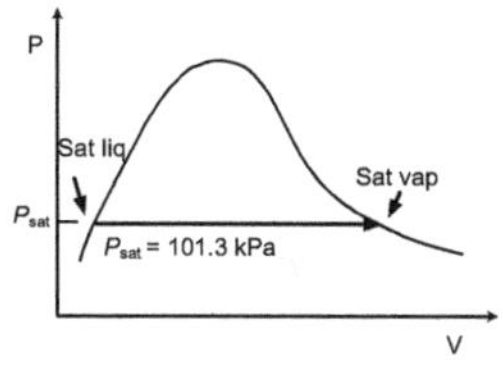

1.57. A mass of 25 lb of saturated liquid water at 3.2 psia is evaporated completely at constant pressure and produces saturated vapor. Estimate the temperature of the vapor and amount of energy added to the water.

1.58. A mass of 100 lb of saturated liquid water at 132°F is evaporated completely at constant pressure and produces saturated vapor. Estimate the temperature of the vapor and amount of energy added to the water.

1.59. A mass of 10 kg of superheated steam at 1250 kPa and 220°C condenses completely to saturated liquid state at constant pressure. Estimate the temperature of the saturated vapor and heat removed from the steam.
1.60. A mass of 150 kg of superheated steam at 500 kPa and 200°C condenses completely to saturated liquid state at constant pressure. Estimate the temperature of the saturated vapor and heat removed from the steam.
1.61. A mass of 25 kg of superheated steam at 470 kPa and 240°C condenses completely saturated liquid state at constant pressure. Estimate the temperature of the saturated vapor and heat removed from the steam.
1.62. A mass of 150 lb of superheated steam at 250 psia and 440°F condenses completely to saturated liquid state at constant pressure. Estimate the temperature of the saturated vapor and heat removed from the steam.
1.63. A mass of 50 lb of superheated steam at 255 psia and 500°F condenses completely to saturated liquid state at constant pressure. Estimate the temperature of the saturated vapor and heat removed from the steam.
1.64. A rigid tank contains 30 kg of saturated liquid and vapor water at 85°C. Only 4 kg of water is in liquid state. Estimate the enthalpy of the saturated mixture and the volume of the tank.

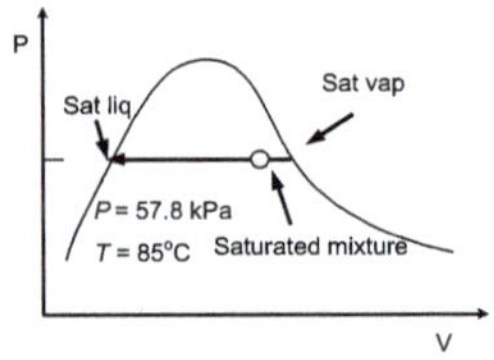

1.65. A rigid tank contains 10 kg of saturated liquid and vapor water at 326.15K. Only 1.5 kg of the water is in liquid state. Estimate the enthalpy of the mixture and the volume of the tank.
1.66. A rigid tank contains 100 lb of saturated liquid and vapor water at 140°F. Only 8.0 lb of the water is in liquid state. Estimate the enthalpy of the saturated mixture and the volume of the tank.

References

1. Çengel YA, Boles MA (2014) Thermodynamics: an engineering approach, 8th edn. McGraw-Hill, New York
2. Himmelblau DM, Riggs JB (2012) Basic principles and calculations in chemical engineering, 8th edn. Prentice Hall, Upper Saddle River
3. Jaffe RL, Taylor W (2018) The physics of energy. Cambridge Univ. Press, Cambridge
4. Poling BE, Prausnitz JM, O'Connell JP (2001) The properties of gases and liquids, 5th edn. McGraw-Hill, New York
5. Demirel Y, Gerbaud V (2025) Nonequilibrium thermodynamics: transport and rate processes in physical, chemical and biological systems, 5th edn. Elsevier, Amsterdam

Energy Sources

2

Introduction and Learning Objectives: Energy has a far reaching impact in everyday life, technology, and development. Primary energy sources are extracted or captured directly from the environment, while the secondary energy sources are derived from the primary energy sources, for example, in the form of electricity or fuel. Primary energies are nonrenewable energy (fossil fuels), renewable energy, and waste. Nonrenewable energy resources are coal, petroleum, natural gas and nuclear, while renewable energy resources are solar, bioenergy, wind, and geothermal. Energy density and the impact of fossil fuels on global warming are discussed briefly.

The learning objectives of this chapter are:

- Sources of energy.
- Energy density, heating value of various fuels.
- Energy usage on the environment.

2.1 Energy Sources

Primary and secondary sources of energy are the two main sources of energy as shown in Fig. 2.1. Primary energy is extracted or captured directly from the environment, while secondary energy is converted from primary energy in the form of electricity or fuel. Differences between primary and secondary energy sources are important in the energy balances to count and record energy supply, convergence, and losses.

Y. Demirel, *Energy*, Green Energy and Technology,
https://doi.org/10.1007/978-3-031-97111-2_2

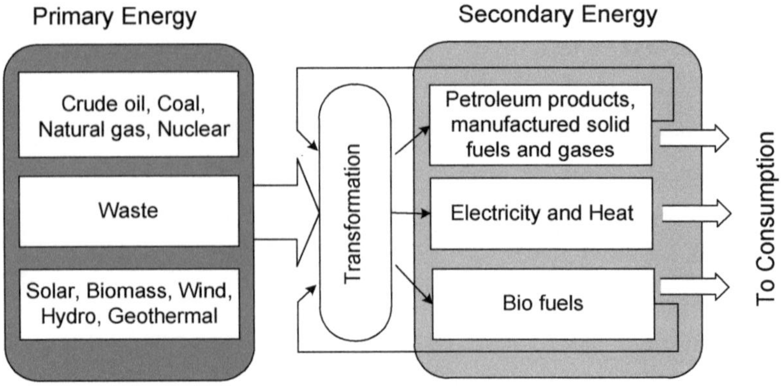

Fig. 2.1 Primary and secondary energy sources. Separating primary and secondary energy is important in energy balances for energy supply, transformation, and losses [13]

2.1.1 Primary Energy Sources

Primary energy is the energy extracted or captured directly from the environment. Three distinctive groups of primary energy are:

- Nonrenewable energy (fossil fuels): Coal, crude oil, natural gas, nuclear fuel.
- Renewable energy: Hydropower, biomass, solar energy, wind, geothermal, and ocean energy.
- Waste.

The principle of supply and demand suggests that as fossil fuels diminish, their prices would rise and renewable energy supplies, particularly biomass, solar, and wind energy resources, would become sufficiently economical to exploit.

2.1.2 Secondary Energy Sources

The primary energy is transformed into *secondary energy* in the form of electrical energy or fuel, such as gasoline, fuel oil, methanol, ethanol, and hydrogen. The primary energy of *renewable energy* sources, such as sun, wind, biomass, geothermal energy, and flowing water is usually equated with either electrical or thermal energy produced from them. Final energy is often electrical energy and fuel, which is referred to as *useful energy*. The selected four forms of final energy are electrical, thermal, mechanical, and chemical energy. These forms of final energy set a boundary between the energy production and the consumption sectors.

2.2 Nonrenewable Energy Sources

It is generally accepted that nonrenewable energy sources or *fossil fuels* are formed from the remains of dead plants and animals by exposure to heat and pressure in the earth's crust over millions of years. Major nonrenewable energy sources are:

- Coal
- Petroleum
- Natural gas
- Nuclear

Fossil fuels contain high percentages of carbon and include mainly coal, petroleum, and natural gas. Natural gas, for example, contains only very low boiling point and gaseous components, while gasoline contains much higher boiling point components. The specific mixture of hydrocarbons gives fuel its characteristic properties, such as boiling point, melting point, density, and viscosity. These types of fuels are known as nonrenewable energy sources.

2.2.1 Coal

Coals are sedimentary rocks containing combustible and incombustible matters as well as water. Coal comes in various compositions and energy content depending on the source and type. Table 2.1 shows some typical properties of various coals. Lignite coals have less than 50% carbon and an energy density lower than wood. Anthracites have more than 90% carbon, while bituminous coals are mostly between 70 and 75%. Bituminous coal ignites easily and burns with a relatively long flame. If improperly fired, bituminous coal is characterized by excess smoke and soot. Anthracite coal is very hard and shiny and creates a steady and clean flame and is preferred for domestic heating. Coal has played a major role as a source of energy in the industrial revolution.

Coal has impurities like sulfur and nitrogen and when it burns the released impurities can combine with water vapor in the air to form droplets that fall to earth as weak forms of sulfuric and nitric acid as acid rain. Coal also contains minerals, which do not burn and make up the ash left behind in a coal combustor.

Table 2.1 Typical properties of various coals [14]

Composition range %	Anthracite coal	Bituminous coal	Lignite coal
Fixed carbon	80.5–85.7	44.9–78.2	31.4
Moisture	2.8–16.3	2.2–15.9	39
Ash	9.7–20.2	3.3–11.7	3.3–11.7
Sulfur	0.6–0.77	0.7–4.0	0.4
Bulk density, lb/ft^3	50–58	42–57	40–54

Because of high carbon content, coal generates more CO_2 per unit of energy than any other fossil fuel such as crude oil. Carbon dioxide is one of several gases that can help trap the earth's heat and, as many scientists believe, cause the earth's temperature to rise and alter the earth's climate. Many new plants are required to have flue gas desulfurization, decarbonization, and denitrification units [14].

2.2.2 Petroleum Fractions

Oil is refined and separated into many products, from gasoline and kerosene to asphalt and chemical reagents used to make plastics and pharmaceuticals. Tables 2.2, 2.3 and 2.4 show some properties of petroleum oil and its fractions. Figure 2.2 shows a part of a typical refinery processing crude oil to produce various fuels. 84% of the volume of the hydrocarbons present in petroleum is converted into energy-rich fuels, including gasoline, diesel, jet fuel, heating, and other fuel oil and liquefied petroleum gases. The remaining oil is converted to pharmaceuticals, solvents, fertilizers, pesticides, and plastics. Therefore, petroleum is vital to many industries and thus is a critical concern to many nations.

Some common fractions from petroleum refining are [13, 14]:

- *Liquefied petroleum gas* (LPG) is a flammable mixture of propane (C_3H_8) (about 38% by volume and more in winter) and butane (C_4H_{10}) (about 60% by volume and more in summer) used as a fuel in heating appliances and vehicles. Energy content of liquefied petroleum gas per kilogram is higher than for gasoline because of higher hydrogen to carbon ratio. Liquefied petroleum gas emits 81% of the CO_2 per kWh produced by oil and 70% of that of coal. Liquefied petroleum gas has a typical specific heat of 46.1 MJ/kg compared with

Table 2.2 Typical elemental composition by weight of crude oil [14]

Element	Percent range %
Carbon	83–87
Hydrogen	10–14
Nitrogen	0.1–2
Oxygen	0.1–1.5
Sulfur	0.5–6
Metals	<0.1

Table 2.3 Composition by weight of hydrocarbons in petroleum [14]

Hydrocarbon	Average %	Range %
Paraffins (alkanes)	30	15–60
Naphtanes (cycloalkanes)	49	30–60
Aromatics	15	3–30
Asphaltics	6	Remainder

Table 2.4 Typical heating values of various fuel oils [14] with permission

Type	Btu/gallon
No. 1 oil	137,400
No. 2 oil	139,600
No. 3 oil	141,800
No. 4 oil	145,100
No. 5 oil	148,800
No. 6 oil	152,400

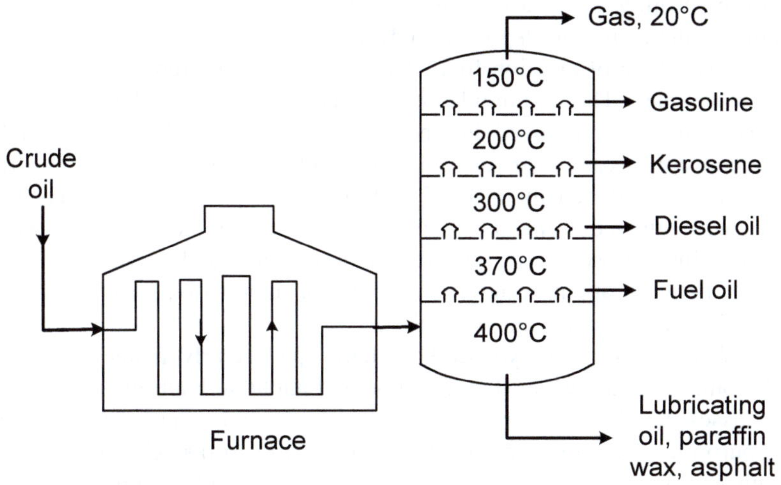

Fig. 2.2 Distillation tower showing the differing weights of various products produced from petroleum

43.5 MJ/kg for gasoline. However, its energy density of 26 MJ/l is lower than that of gasoline. Pure *n*-butane is liquefied at around 220 kPa (2.2 bar), while pure propane (C_3H_8) is liquefied at 2200 kPa (22 bar). At liquid state, the vapor pressure of liquefied petroleum gas is about 550 kPa (5.5 bar).

- *Gasoline* is primarily used as a fuel in internal combustion engines. A typical gasoline consists of hydrocarbons with between 4 and 12 carbon atoms per molecule. It consists mostly of aliphatic hydrocarbons obtained by the fractional distillation of petroleum, enhanced with isooctane or the aromatic hydrocarbons toluene and benzene to increase its octane rating. The specific density of gasoline ranges from 0.71 to 0.77 (6.175 lb/US gal) higher densities having a greater volume of aromatics. Gasoline contains about 132 MJ/US gal (higher heating value), while its blends differ by up to 4% more or less than the average. The emission of CO_2 from gasoline is around 73.38 g/MJ.
- *Petroleum diesel* contains 8–21 carbon atoms per molecule with a boiling point in the range of 180–360°C (360–680°F). The density of petroleum diesel is

about 6.943 lb/gal. About 86.1% of the fuel mass is carbon and it offers a net heating value of around 43.1 MJ/kg. However, due to the higher density, diesel offers a higher volumetric energy density at 128,700 Btu/gal vs. 115,500 Btu/gal for gasoline, some 11% higher. The CO_2 emissions from diesel are 73.25 g/MJ, (like gasoline). Because of quality regulations, additional refining is required to remove sulfur which may contribute to a higher cost.

- *Kerosene* is a thin, clear liquid formed containing between 6 and 16 carbon atoms per molecule, with density of 0.78–0.81 g/cm^3. The flash point of kerosene is between 37 and 65 C (100 and 150°F) and its autoignition temperature is 220 C (428 F). The heat of combustion of kerosene is like that of diesel: its lower heating value is around 18,500 Btu/lb, (43.1 MJ/kg), and its higher heating value is 46.2 MJ/kg (19,861 Btu/lb).
- *Jet fuel* is a type of aviation fuel designed for use in aircraft powered by gas-turbine engines. The commonly used fuels are Jet A and Jet A-1 which are produced to a standardized international specification. Jet B is used for its enhanced cold-weather performance. Jet fuel is a mixture of many different hydrocarbons with density of 0.775 kg/l–0.840 kg/l at 15°C (59°F). The range is restricted by the requirements for the product, for example, the freezing point or smoke point. Kerosene-type jet fuel (including Jet A and Jet A-1) has a carbon number between about 8 and 16; wide-cut or naphtha-type jet fuel (including Jet B), between about 5 and 15.
- *Fuel oil* is made of long hydrocarbon chains, particularly alkanes, cycloalkanes and aromatics and heavier than gasoline and naphtha. Fuel oil is classified into six classes, numbered 1 through 6, according to its boiling point, composition and purpose. The boiling point ranging from 175 to 600°C, and carbon chain length, 9–70 atoms. Viscosity also increases with number, and the heaviest oil must be heated to get it to flow. Price usually decreases as the fuel number increases. Number 1 is similar to kerosene, number 2 is the diesel fuel that trucks and some cars run on, leading to the name "road diesel." Number 4 fuel oil is usually a blend of heavy distillate and residual fuel oils. Number 5 and 6 fuel oils are called residual fuel oils or heavy fuel oils. Table 2.5 shows the heating values of various fuel oils per gallon.

Fossil fuels contain sulfur and other impurities. Combustion of such fuels eventually leads to producing sulfur monoxides (SO) and sulfur dioxide (SO_2) in the exhaust which promotes acid rain. Ozone (O_3) is not emitted directly but made in the air by the action of sunlight on other pollutants to form *ground level ozone.* However, the *ozone layer* in the high atmosphere is useful in blocking the harmful rays from the sun. Ozone is broken down by nitrogen oxides.

2.2.3 Natural Gas

Natural gas is a naturally occurring mixture, consisting mainly of methane. Table 2.5 shows the typical components of natural gas. Natural gas is becoming

Table 2.5 Typical composition in mole % of a natural gas [13, 14]

Component	Composition %	Range %
Methane	95.2	87.0–96.0
Ethane	2.5	1.5–5.1
Propane	0.2	0.1–1.5
Butane	0.03	0.01–0.3
Iso–pentane, n-pentane, hexane plus	0.01	Trace – 0.14
Nitrogen	1.3	0.7–5.6
Carbon dioxide	0.7	0.1–1.0
Oxygen	0.02	0.01–0.1
Hydrogen	Trace	Trace – 0.02
Specific gravity	0.58	0.57–0.62

increasingly popular as an alternative power generation fuel. The typical theoretical flame temperature of natural gas is 1960°C (3562°F), while ignition point is 593°C. The value of gross heating in dry basis is around 37.8 MJ/m^3 and varies between 36.0 MJ/m^3 and 40.2 MJ/m^3, while the typical caloric value is roughly 1000 Btu per cubic foot, depending on gas composition.

Natural gas is a major source of electricity production using gas turbines and steam turbines. It burns more cleanly and produces about 30% less carbon dioxide than burning petroleum and about 45% less than burning coal for an equivalent amount of heat produced. Combined cycle power generation using natural gas is thus the cleanest source of power available using fossil fuels. Liquefied natural gas exists at − 161°C (− 258°F). Impurities and heavy hydrocarbons from the natural gas are removed before the cooling process. The density of liquefied natural gas is in the range of 410–500 kg/m^3 [13].

2.2.4 Nuclear Energy

Nuclear energy plants produce electricity through the fission of nuclear fuel, such as uranium, so they do not pollute the air with harmful gases. *Nuclear fission* is a nuclear reaction in which the nucleus of an atom splits into smaller parts, often producing free neutrons and photons in the form of gamma rays and releasing large amounts of energy. Nuclear fuels undergo fission when struck by free neutrons and generate neutrons leading to a self-sustaining chain reaction that releases energy at a controlled rate in a nuclear reactor [10]. This heat is used to produce steam to be used in a turbine to produce electricity. This is like most coal, oil, and gas-fired power plants.

Typical fission releases about two hundred million eV (200 MeV) of energy, which is much higher than most chemical oxidation reactions. For example, the complete fission energy of uranium-235 isotope is 6.73×10^{10} kJ/kg [10]. The energy of nuclear fission is released as kinetic energy of the fission products and fragments, and as electromagnetic radiation in the form of gamma rays in a nuclear reactor. The energy is converted to heat as the particles and gamma rays collide with the atoms that make up the reactor and its working fluid, usually water or occasionally heavy water. The nuclear fission products, however, are far more radioactive than the heavy elements that fission as fuel, and remain so for a significant amount of time, giving rise to a nuclear waste problem. More than 400 nuclear power plants operating in 25 countries supply almost 17% of the world's electricity.

Nuclear power is essentially carbon-free. However, the electricity from new nuclear power plants would be relatively expensive, and nuclear energy faces several significant obstacles. The biggest challenges are the disposal of radioactive waste and the threat of nuclear proliferation.

Fission and fusion processes both release large amounts of the binding energy of nuclei and generate radioactive waste. Nuclear fission is a spontaneous process on earth and slow-neutron-induced fission can be controlled in nuclear power reactors, while fusion power is a relatively distant possibility. One fission reaction is

$$^{235}_{92}\mathrm{U} + n_{\mathrm{th}} \rightarrow ^{144}_{56}\mathrm{Ba} + ^{90}_{36}\mathrm{Kr} + 2n \tag{2.1}$$

The energy appears as kinetic energy beside the fission fragments, neutrons, and γ-rays.

$$\begin{aligned} q &= \Delta(^{235}\mathrm{U}) + \Delta n \rightarrow \left(\Delta\left(^{144}\mathrm{Ba}\right) + \Delta\left(^{90}\mathrm{Kr}\right) + 2\Delta n\right) \\ q &= 40.922 + 8.071 - (-73.937 - 74.959 + 2(8.071)) = 181.747\,\mathrm{MeV} \end{aligned} \tag{2.2}$$

The core of a thermal-neutron reactor is fuel, moderator, and coolant. The coolant transfers the nuclear energy produced by nuclear fission to the power generator system, while the moderator slows the neutrons. Most reactors use water as a coolant and moderator. Nuclear reactor designs have active and passive safety systems, such as control rods and coolant pumps as active, while pressure release valves or gravity-driven coolant circulation. Nuclear reactors mainly operate at lower temperatures compared with those operated in fossil fuel power plants. Therefore, turbines in nuclear power stations use mixed water/steam phase and light-water reactors can reach efficiencies of 33–37% that are comparable to coal-fired power plants.

Nuclear fusion in the sun combines four protons into helium nucleus and generates heat that produces solar radiation that drives photosynthesis and creates dynamics of atmosphere and oceans. Contrary this, in the nuclear fission large uranium nuclei become unstable as the electromagnetic repulsion between protons opposes the nuclear binding force and decompose into smaller particles. Nuclear fission in a nuclear reactor releases large amount carbon-free energy to produce stream and eventually electric power [10].

2.3 Heating Value of Fuels

The heating value of a fuel is the quantity of heat produced by its combustion at constant pressure and under "normal" conditions (i.e., to 25°C and under pressure of 1 atm). The combustion process generates water. Various definitions of heating values are:

- *Higher heating value* (HHV) consists of the combustion product of water condensed and the heat of vaporization contained in the water vapor is recovered. So, all the water produced in the combustion is in liquid state.
- *Lower heating value* (LHV) assumes that the water product of combustion is at vapor state and the heat of vaporization is not recovered.
- *Net heating value* is the same as lower heating value and is obtained by subtracting the latent heat of vaporization of the water vapor formed by the combustion from the gross or higher heating value.
- *Gross heating value* is the total heat obtained by complete combustion at constant pressure including the heat released by condensing the water vapor in the combustion products. Gross heating accounts for liquid water in the fuel prior to combustion and is valuable for fuels containing water, such as wood and coal. If fuel has no water prior to combustion, then the gross heating value is equal to a higher heating value. A common method of relating HHV to LHV per unit mass of a fuel is [4, 15]

$$\mathrm{HHV} = \mathrm{LHV} + \Delta H_{\mathrm{vap}} \left[\left(\mathrm{MW}_{\mathrm{H_2O}} n_{\mathrm{H_2O,out}} \right) / \left(\mathrm{MW}_{\mathrm{Fuel}} n_{\mathrm{Fuel,in}} \right) \right] \tag{2.3}$$

where ΔH_{vap} is the heat of vaporization per mole of water (kJ/kg or Btu/lb), $n_{\mathrm{H_2O,out}}$ is the moles of water vaporized, $n_{\mathrm{fuel,in}}$ is the number of moles of fuel combusted, and *MW* is the molecular weight. Tables 2.6 and 2.7 show the properties and heating values of some common fuels. The heating value of fossil fuels may vary depending on the source and composition.

2.3.1 Energy Density

Energy density is the amount of energy per unit volume. *Specific energy* is the amount of energy per unit amount. Comparing, for example, the effectiveness of hydrogen fuel to gasoline, hydrogen has a higher specific energy than gasoline but a much lower energy density even in liquid form. Energy per unit volume has the same physical units as pressure. Table 2.8 lists energy densities of some fuel and fuel mixtures.

Table 2.6 Properties heating values of some common fuels and hydrocarbons at 1 atm and 20°C; at 25°C for liquid fuels, and 1 atm and normal boiling temperature for gaseous fuels [4, 13–15]

Fuel (phase)	Formula	MW kg/kmole	ρ kg/l	ΔH_v kJ/kg	T_b °F	C_p kJ/kg °C	HHV[a] kJ/kg	LHV[a] kJ/kg
Carbon (s)	C	12.01	2.000	–		0.71	32,800	32,800
Hydrogen (g)	H_2	2.01	–	–		14.40	141,800	120,000
Methane (g)	CH_4	16.04	–	509	–258.7	2.20	55,530	50,050
Methanol (l)	CH_3OH	32.04	0.790	1168	149.0	2.53	22,660	19,920
Ethane (g)	C_2H_6	30.07	–	172	– 127.5	1.75	51,900	47,520
Ethanol (l)	C_2H_5OH	46.07	0.790	919	172.0	2.44	29,670	26,810
Propane (g)	C_3H_8	44.09	0.500	420	– 43.8	2.77	50,330	46,340
Butane (l)	C_4H_{10}	58.12	0.579	362	31.1	2.42	49,150	45,370
Isopentane (l)	C_5H_{12}	72.15	0.626	–	82.2	2.32	48,570	44,910
Benzene (l)	C_6H_6	78.11	0.877	433	176.2	1.72	41,800	40,100
Hexane (l)	C_6H_{14}	86.18	0.660	366	155.7	2.27	48,310	44,740
Toluene (l)	C_7H_8	92.14	0.867	412	231.1	1.71	42,400	40,500
Heptane (l)	C_7H_{16}	100.204	0.684	365	209.1	2.24	48,100	44,600
Octane (l)	C_8H_{18}	114.23	0.703	363	258.3	2.23	47,890	44,430
Decane (l)	$C_{10}H_{22}$	142.28	0.730	361		2.21	47,640	44,240
Gasoline (l)	$C_nH_{1.87n}$	100–110	0.72–0.78	350		2.40	47,300	44,000
Light diesel	$C_nH_{1.8n}$	170.00	0.78–0.84	270		2.20	46,100	43,200
Heavy diesel	$C_nH_{1.7n}$	200.00	0.82–0.88	230		1.90	45,500	42,800
Natural gas		~18.00	–	–		2.00	50,000	45,000

[a]HHV, LHV: higher heating value and lower heating value, respectively; (s): solid; (l): liquid; (g): gas

Table 2.7 Lower heating values of some common fuels [4, 13, 14]

Fuel	Lower heating value	
	kJ/kg	Btu/lb
Butane	45,510	19,500
Charcoal	28,600	12,300
Coal (anthracite)	32,200	14,000
Coal (bituminous)	30,000	13,000
Coal (semi anthracite)	29,200	12,500
Coal (semi bituminous)	24,500	10,500
Coke	26,000	11,200
Diesel	42,800	18,300
Gasoline	43,300	18,600
Hydrogen	120,790	11,500
Natural gas	47,100	20,200
Propane	46,400	20,000

1 kJ/kg = 1 J/g = 0.43 Btu/lbm = 0.239 kcal/kg
1 Btu/lb_m = 2.326 kJ/kg = 0.55 kcal/kg
1 kcal/kg = 4.187 kJ/kg = 1.8 Btu/lbm

Example 2.1 **Energy consumption by a car**

An average car consumes 50 gallons of gasoline per month. Estimate the energy consumed by the car per year.

Solution:

Assume that gasoline has an average density of 0.72 g/cm^3 and a heating value of 47.3 MJ/kg (Table 2.6).

$$V = 50\,\text{gal/month} = 189.25\,\text{l/month},\ 2271.0\,\text{l/year}\,(3.785\text{l} = 1\text{ gal})$$

$$\rho_{\text{gas}} = 0.72\,\text{g/cm}^3 = 0.72\,\text{kg/l}$$

Mass of gasoline: $m_{gas} = \rho V = 1635.1\,\text{kg/year}$.

Energy consumed per year:
$$\begin{aligned} E_{\text{gas}} &= 1635.1\,\text{kg/year}\,(47{,}300\,\text{kJ/kg}) \\ &= 77{,}340,230\,\text{kJ/year} = 77{,}340.2\,\text{MJ/year}. \end{aligned}$$

Example 2.2 **Fuel consumption by a low and a high-mileage car**

The average daily traveling distance is about 40 miles/day. A car has a city-mileage of 20 miles/gal. If the car is replaced with a new car with a city-mileage of 30 miles/gal and the average cost of gasoline is $3.50/gal, estimate the amount of energy conserved with the new car per year.

Table 2.8 Energy densities of some fuels; [13, 14]

Fuel type	Higher heating value (HHV)		Lower heating value (LHV)
	MJ/l	Btu/gal	Btu/gal
Conventional gasoline	34.8	125,000	115,400
High octane gasoline	33.5	120,200	112,000
LPG (60%Pr. + 40%Bu.)	26.8		
Ethanol	24.0	84,600	75,700
Methanol	17.9	64,600	56,600
Butanol	29.2		
Gasohol E10 (ethanol 10% vol.)	33.2	120,900	112,400
Gasohol E85 (ethanol 85% vol.)	25.6		
Gasoline (petrol)	34.2		115,500
Diesel	38.6	138,700	128,700
Biodiesel	33.5	126,200	117,100
Jet fuel (kerosene based)	35.1	125,935	
Jet fuel (naphtha)	42.8	127,500	118,700
Liquefied petroleum gas	26.8	91,300	83,500
Hydrogen (liquid at 20 K)	10.1		130
Hydrogen gas	0.0108		
Methane (1 atm, 1°C)	0.0378		
Natural gas	0.0364		
LPG propane	25.3		
LPG butane	27.7		
Crude oil	37.0		
Coal, anthracite	72.4		
Coal, lignite			
Coal, bituminous	20.0		
Wood			

Solution:

Assume: The gasoline is incompressible with $\rho_{av} = 0.75\,\text{kg/l}$.

Lower heating value (LHV) = 44,000 kJ/kg; 44,000 kJ of heat is released when 1 kg of gasoline is completely burned, and the produced water is in vapor state (Table 2.6).

Fuel needed for the old car: $(40\,\text{miles/day})/(20\,\text{miles/gal}) = 2\,\text{gal/day}$.

Fuel needed for the new car: $(40\,\text{miles/day})/(30\,\text{miles/gal}) = 1.34\,\text{gal/day}$.

Old car

Mass of gasoline:

$$m_{\text{gas}} = \rho_{\text{av}}(\text{Volume}) = (0.75\,\text{kg/l})(2.0\,\text{gal/day})(3.7851/\text{gal})$$
$$= 5.7\,\text{kg/day}$$

Energy of gasoline:

$$E_{\text{gas}}(\text{LHV}) = (5.7\,\text{kg/day})(44{,}000\,\text{kJ/kg}) = 250{,}800\,\text{kJ/day}\ (365\,\text{day/year})$$
$$= 91{,}542{,}000\,\text{kJ/year} = 91{,}542\,\text{MJ/year}.$$

New car

Mass of gasoline:

$$m_{\text{gas}} = \rho_{\text{av}}(\text{Volume}) = (0.75\,\text{kg/l})(1.34\,\text{gal/day})(3.7851/\text{gal}) = 3.8\,\text{kg/day}$$

Energy of gasoline:

$$E_{\text{gas}}(\text{LHV}) = (3.8\,\text{kg/day})(44{,}000\,\text{kJ/kg}) = 167{,}200\,\text{kJ/day}\ (365\,\text{day/year})$$
$$= 61{,}028{,}000\,\text{kJ/year} = 61{,}028\,\text{MJ/year}.$$

The new car reduces fuel consumption by around 33%:

[(91,542 MJ/year-61,028 MJ/year)/ 91,542 MJ/year], which is significant.

Example 2.3 **Daily consumption of natural gas by a city**

A city consumes natural gas at a rate of 500×10^6 ft^3/day. The volumetric flow is at standard conditions of 60°F and 1 atm = 14.7 psia. If natural gas is costing $6/GJ of higher heating value what is the amount of energy per year for the city.

Solution:

$Q = 500 \times 10^6\,\text{ft}^3/\text{day}$ at 60°F and 1 atm = 14.7 psia.

The higher heating value is the heat of combustion of natural gas when the water product is at liquid state. From Table 2.6, the value of HHV is:

50,000 kJ/kg (~ 1030 Btu/ft^3).

Heating value: $\left(1030\,\mathrm{Btu/ft^3}\right)\left(500\times 10^6\,\mathrm{ft^3/day}\right) = 515.0\times 10^9\mathrm{Btu/day}$

$$\left(515.0\times 10^9\,\mathrm{Btu/day}\right)(1055\,\mathrm{J/Btu}) = 543{,}325\,\mathrm{GJ/day}$$

Consumed energy per year: $198{,}313{,}625\,\mathrm{GJ/year}$.

Example 2.4 **Energy consumed by a car**

An average car consumes about 2 gallons (US gallon = 3.785 l) a day, and the capacity of the fuel tank is about 15 gallons. Therefore, a car needs to be refueled once every week. The density of gasoline ranges from 0.72 to 0.78 kg/l (Table 2.6). The lower heating value of gasoline is about 44,000 kJ/kg. Assume that the average density of gasoline is 0.75 kg/l. If the car was able to use 0.2 kg of nuclear fuel of uranium-235, estimate the time in years for refueling.

Solution:

Assume: The gasoline is incompressible with $\rho_{av} = 0.75\,\mathrm{kg/l}$.

Lower heating value (LHV) = $44{,}000\,\mathrm{kJ/kg}$; 44,000 kJ of heat is released when 1 kg of gasoline is completely burned, and the produced water is in vapor state.

Complete fission energy of U - $235 = 6.73\times 10^{10}\,\mathrm{kJ/kg}$.

Mass of gasoline per day:

$m_{gas} = \rho_{av}V = (0.75\,\mathrm{kg/l})(2\,\mathrm{gal/day})(3.785\,\mathrm{l/gal}) = 5.67\,\mathrm{kg/day}$.
Energy of gasoline per day:

$E_{gas} = m_{gas}(\mathrm{LHV}) = (5.67\,\mathrm{kg/day})(44{,}000\,\mathrm{kJ/kg}) = 249{,}480\,\mathrm{kJ/day}$.

Energy released by the complete fission of 0.2 kg U-235:

$$E_{\text{U-235}} = \left(6.73\times 10^{10}\,\mathrm{kJ/kg}\right)(0.2\,\mathrm{kg}) = 1.346\times 10^{10}\,\mathrm{kJ}$$

Time for refueling:

$\left(1.346\times 10^{10}\,\mathrm{kJ}\right)/(249{,}480\,\mathrm{kJ/day}) = 53{,}952\,\mathrm{days} = 148\,\mathrm{years}$.

Therefore, the car will not need refueling for about 148 years.

2.4 Renewable Energy Resources

Renewable energy comes from natural resources and is naturally replenished. Major renewable energy sources are:

- Hydroelectric
- Solar energy
- Biomass
- Wind energy
- Geothermal heat
- Ocean

In its various forms, renewable energy comes directly from the sun, or from heat generated deep within the earth. The generation of electricity from renewables, such as biomass, wind, solar, geothermal, and biofuels are growing steadily. Climate change concerns and government support are leading to an increase in renewable energy usage and commercialization. Renewable energy replaces conventional fuels in four distinct areas: power generation, hot water/space heating, transport fuels, and rural (off-grid) energy services [2, 3, 9]:

- *Renewable power generation* provides considerable electricity generation worldwide.
- *Solar hot water* contributes a portion of the water heating needs in many countries.
- *Renewable biofuels* have contributed to a decline in oil consumption in many countries.

Renewable energy generally gets cheaper in the long term, while fossil fuels generally get more expensive. Fossil fuel technologies are more mature, while renewable energy technologies are being constantly improved to increase the efficiency of renewable energy and reduce its cost [3].

2.4.1 Hydroenergy

Hydroenergy is derived from the energy of moving water. Most hydroelectric energy comes from the potential energy of dammed water driving a water turbine and generator. The power extracted from the water depends on the volume and on the difference in height between the source and the water's outflow. This height difference is called the head. The amount of potential energy in water is proportional to the head. To deliver water to a turbine while maintaining pressure arising from the head, a large pipe called a penstock may be used.

One of the major advantages of hydroelectricity is the elimination of fossil fuels and hence the emissions and air pollution. Hydroelectric plants also tend to have longer economic lives with lower operating labor costs and maintenance. Micro-hydro-systems typically produce up to 100 kW of power. Hydro-systems without dams derive kinetic energy from rivers and oceans. Ocean energy includes marine current power, ocean thermal energy conversion, and tidal power [6, 9].

2.4.2 Solar Energy

Solar energy is derived from the sun through the form of solar radiation. Solar powered electrical generation relies on photovoltaics and heat engines. Other solar applications include space heating and cooling through solar architecture, daylighting, solar hot water, solar cooking, and high temperature process heat for industrial purposes. Solar technologies are broadly characterized as either passive solar or active solar depending on the way they capture, convert and distribute solar energy [5, 6, 9]:

- *Active solar techniques* include the use of solar thermal collectors to harness energy. The sun is the source of heat and solar collectors collect solar radiation. A part of the incident energy is lost, and the rest is absorbed by the heat transfer fluid (solar collector fluid). Some active solar techniques include s*olar processing heat* by commercial and industrial buildings, *space heating/cooling*, and water heating. A typical water heating system includes solar collectors that work along with a pump, heat exchanger, and one or more large storage tanks. The most common collector is called a *flat-plate collector*. Mounted on a roof, it consists of a thin, flat, rectangular box with a transparent cover that faces the sun. Small tubes run through the box and carry the heat transfer fluid, mainly water or air to be heated. The tubes are attached to an absorber plate, which is painted black to absorb the heat. As heat builds up in the collector, it heats the fluid passing through the tubes. The storage tank then holds the hot liquid. It can be just a modified water heater, but it is usually larger and very well-insulated. Systems that use fluids other than water usually heat the water by passing it through a coil of tubing in the tank, which is full of hot fluid.
- *Passive solar systems* rely on gravity and the tendency for water to naturally circulate as it is heated. *Passive solar techniques* orient buildings to the sun, select materials with favorable thermal mass or light dispersing properties, and design spaces that naturally circulate air.

Nonresidential Solar Collectors

The two main types of solar collectors used for nonresidential buildings are an *evacuated-tube collector* and a *linear concentrator. They* can operate at high temperatures with high efficiency. An evacuated-tube collector is a set of many double-walled, glass tubes, and reflectors to heat the fluid inside the tubes. A vacuum between the two walls insulates the inner tube, retaining the heat. Linear concentrators use long, rectangular, U-shaped mirrors tilted to focus sunlight on tubes that run along the length of the mirrors. The concentrated sunlight heats the fluid within the tubes. Solar absorption systems use thermal energy to evaporate refrigerant fluid to cool the air. In contrast, solar desiccant systems use thermal energy to regenerate desiccants that dry the air, thereby cooling the air [2, 6].

Solar Electric Generating Systems—Solar electric generating system uses parabolic trough collectors to collect the sun's energy to generate steam to drive a conventional steam turbine [2, 6]. The parabolic mirrors automatically track the sun throughout the day. The sunlight is directed to a central tube carrying synthetic oil, which heats around 400°C. Heat is used to convert water to steam to drive a steam turbine and produce electricity.

Photovoltaic—*Solar photovoltaic* (PV) converts light into electricity using semiconductor materials with three types (i) crystalline silicon, (ii) thin film, and (iii) concentrator. A photovoltaic cell is a solid-state electrical device producing direct current, which must be converted to alternating current via an inverter to be used in the grid electricity. Assemblies of cells are known as *solar modules*, which are typically deployed as an array of individual modules on rooftops, building facades, or in large-scale ground-based arrays. A module consists of many jointly connected solar cells. Most crystalline modules usually consist of 60–72 cells. Solar photovoltaic efficiency needs to be increased, and its costs should be decreased [2, 5, 9].

2.4.3 Biomass and Bioenergy

Biomass is an organic material made from plants including microorganisms and animals. Plants absorb the sun's energy in photosynthesis and store the energy as biomass. Therefore, biomass is a renewable energy source based on the carbon cycle. Some examples of biomass fuels include wood, crops, and algae. When burned, the chemical energy in biomass is released as heat. Biomass can be converted to other biofuels, such as ethanol and biodiesel. Biomass grown for biofuel includes corn, soybeans, willow switch grass, rapeseed, sugar beet, palm oil, and sorghum. Cellulosic biomasses such as corn stover, straw, timber, and rice husks can also be used for biofuel production. Anaerobic digestion of biomass produces biogas, while gasification produces syngas, which is the mixture of hydrogen and carbon dioxide to be converted to liquid fuels. Cellulosic ethanol can also be produced from lignocellulosic biomass. Table 2.9 shows lower heating values, moisture, and ash content of some biomass [3].

2.4.4 Carbon Cycle

Carbon in various forms is transported between the various components of the Earth's biosphere, between the atmosphere, hydrosphere (seas and oceans), lithosphere (rocks, soils, and mineral deposits, including fossil fuels) and biological material including plants and animals. Carbon cycle maintains a state of dynamic equilibrium. Other forms, most notably fossil fuels, can potentially store carbon indefinitely. However, if they are burned the carbon is released and makes a net addition to the carbon cycle and raises the total atmospheric carbon. If biomass

Table 2.9 Lower heating values (LHV) for selected biomass [4]

Product	Moisture %	Ash content[a] %	LHV MJ/kg
Bagasse sugarcane	18	4	17–18
Coconut husks	5–10	6	16–17
Coffee husks	13	8–10	16–17
Corn stover	5–6	8	16–17
Corncobs	15	1–2	15–16
Cotton husks	5–10	3	16–18
Oil-palm fibers	55	10	7–8
l-palm husks	55	5	7–8
Poplar wood	5–15	1.2	17–19
Rice hulls	9–11	15–20	13–15
Rice straw and husk	15–30	15–20	17–18
Switchgrass	8–15	6	18–20
Wheat straw and husk	7–15	8–9	17–19
Willow wood	12	1–5	17–19

[a]Approximate

is used without replacement, for example, in the case of forest clearance, this too can make a net addition to the carbon cycle. As growing plants absorb the carbon released by the harvested biomass, sustainable use of biomass makes no direct net contribution [9].

2.4.5 Gross Heating Values of Biomass Fuels

Biomass fuels are usually characterized by *proximate*, *ultimate*, and *sulfate analyses*:

- P*roximate analysis* gives moisture content, volatile content (when heated to 950°C), the free carbon remaining at that point, the ash (mineral) in the sample, and the higher heating value based on the complete combustion of the sample to carbon dioxide and liquid water.
- U*ltimate analysis* is the elemental analysis and provides the composition of the biomass in wt% of carbon, hydrogen, oxygen, and nitrogen.
- Sulfate *analysis* is the elemental analysis and provides the composition of the biomass in wt% of various types of sulfur.

Table 2.10 shows measured and estimated gross heating values as well as the proximate and ultimate analyses of some selected fuels, including biomass components, natural biomass (woods, agricultural products), processed biomass, and other solid and liquid fuels [15].

Table 2.10 Proximate and ultimate analyses of biomass fuels in weight percentage [4, 15]

Biomass name	Fixed Carbon	Volatiles %	Ash %	C %	H %	O %	N %	S %	HHV_m MJ/kg	HHV_{est} MJ/kg
Douglas fir	17.70	81.50	0.80	52.30	6.30	40.50	0.10	0.00	21.05	21.48
Hickory	–	–	0.73	47.67	6.49	43.11	0.00	0.00	20.17	19.82
Maple	–	–	1.35	50.64	6.02	41.74	0.25	0.00	19.96	20.42
Ponderosa pine	17.17	82.54	0.29	49.25	5.99	44.36	0.06	0.03	20.02	19.66
Poplar	–	–	0.65	51.64	6.26	41.45	0.00	0.00	20.75	21.10
Redwood	16.10	83.50	0.40	53.50	5.90	40.30	0.10	0.00	21.03	21.45
Western hemlock	15.20	84.80	2.20	50.40	5.80	41.10	0.10	0.10	20.05	20.14
Yellow pine	–	–	1.31	52.60	7.00	40.10	0.00	0.00	22.30	22.44
White fir	16.58	83.17	0.25	49.00	5.98	44.75	0.05	0.01	19.95	19.52
White oak	17.20	81.28	1.52	49.48	5.38	43.13	0.35	0.01	19.42	19.12
Douglas fir bark	25.80	73.00	1.20	56.20	5.90	36.70	0.00	0.00	22.10	22.75
Loblolly pine bark	33.90	54.70	0.40	56.30	5.60	37.70	0.00	0.00	21.78	22.35
Peach pits	19.85	79.12	1.03	53.00	5.90	39.14	0.32	0.05	20.82	21.39
Walnut shells	21.16	78.28	0.56	49.98	5.71	43.35	0.21	0.01	20.18	19.68
Almond pruning	21.54	76.83	1.63	51.30	5.29	40.90	0.66	0.01	20.01	19.87
Black walnut pruning	18.56	80.69	0.78	49.80	5.82	43.25	0.22	0.01	19.83	19.75
Corncobs	18.54	80.10	1.36	46.58	5.87	45.46	0.47	0.01	18.77	18.44
Wheat straw	19.80	71.30	8.90	43.20	5.00	39.40	0.61	0.11	17.51	16.71
Cotton stalk	22.43	70.89	6.68	43.64	5.81	43.87	0.00	0.00	18.26	17.40
Corn stover	19.25	75.17	5.58	43.65	5.56	43.31	0.61	0.01	17.65	17.19
Sugarcane bagasse	14.95	73.78	11.27	44.80	5.35	39.55	0.38	0.01	17.33	17.61
Rice hulls	15.80	63.60	20.60	38.30	4.36	35.45	0.83	0.06	14.89	14.40
Pine needles	26.12	72.38	1.50	48.21	6.57	43.72			20.12	20.02
Cellulose; $C_6H_{10}O_5$	–	–	162	44.44	6.17	49.38		–	–	17.68

(continued)

Table 2.10 (continued)

Biomass name	Fixed Carbon	Volatiles %	Ash %	C %	H %	O %	N %	S %	HHV_m MJ/kg	HHV_{est} MJ/kg
Lignin (softwood)	–	–	–	63.8	6.30	29.90		–	–	26.60
Lignin (hardwood)	–	–	–	59.8	6.40	33.70		–	–	24.93

A relationship between the high heating value, HHV and the elemental composition is given by

$$\text{HHV (MJ/kg)} = 0.3491\,\text{C} + 1.1783\,\text{H} - 0.1034\text{O} - 0.0211\,\text{A} + 0.1005\,\text{S} - 0.0151\,\text{N} \tag{2.4}$$

where C is the weight fraction of carbon; H of hydrogen; O of oxygen; A of ash; S of sulfur, and N of nitrogen appearing in the ultimate analysis. This equation represents experimental data with an average error of 1.45% and can be used in estimating heat values and modeling of biomass processes [5, 18].

Based on chemical functional groups of the fuels, the heating values may vary. When the oxygen percentage is higher in a fuel, the percentages of carbon and hydrogen available for combustion are reduced. This leads to lower heating values. By using the values of fixed carbon (FC, wt%), the higher heating value of the biomass samples can be estimated by

$$\text{HHV(MJ/kg)} = 0.196(\text{FC}) + 14.119 \tag{2.5}$$

The heating values calculated from Eq. (2.3) shows a mean difference of 2.2% between estimated and measured values [10]. Another correlation between the HHV and dry ash content from proximate analysis of biomass (in weight percent) is estimated by

$$\text{HHV(MJ/kg)} = 19.914 - 0.2324\,\text{Ash} \tag{2.6}$$

Based on the composition of main elements (in wt%) C, H, and O, the heating value is estimated by

$$\text{HHV(MJ/kg)} = 0.3137\,\text{C} + 0.7009\,\text{H} + 0.0318\,\text{O} - 1.3675 \tag{2.7}$$

with more than 90% predictions in the range of ±5% error.

Example 2.5 **Gross heating value estimations**

Using data in Table 2.10, estimate the gross heating values in kJ/kg for the biomass redwood from: (a) ultimate analysis, (b) fixed carbon, (c) dry ash content, and (d) carbon (C), hydrogen (H), and oxygen (O) compositions.

Name	Fixed Carbon	Volatiles %	Ash %	C %	H %	O %	N %	S %	HHV_m MJ/kg	HHV_{est} MJ/kg
Redwood	16.10	83.50	0.40	53.50	5.90	40.30	0.10	0.00	21.03	21.45

Solution:

(a) From ultimate analysis

$$\mathrm{HHV(in\,MJ/kg)} = 0.3491\,\mathrm{C} + 1.1783\,\mathrm{H} - 0.10340 - 0.0211\,\mathrm{A} + 0.1005\,\mathrm{S} - 0.0151\,\mathrm{N}$$

$$\mathrm{HHV(in\,MJ/kg)} = 0.3491(53.50) + 1.1783(5.90) - 0.1034(40.3) - 0.0211(0.0040)$$
$$+ 0.1005(0.0) - 0.0151(0.0010)21.44\,\mathrm{MJ/kg} = 21{,}440\,\mathrm{kJ/kg}$$

(b) From fixed carbon percentage

$$\mathrm{HHV(MJ/kg)} = 0.196(\mathrm{FC}) + 14.119$$

$$\mathrm{HHV(MJ/kg)} = 0.196(16.10) + 14.119 = 17.3\,\mathrm{MJ/kg} = 17{,}300\,\mathrm{kJ/kg}$$

(c) From dry ash conten

$$\mathrm{HHV(MJ/kg)} = 19.914 - 0.2324\,\mathrm{Ash}$$

$$\mathrm{HHV(MJ/kg)} = 19.914 - 0.2324(0.0040) = 19.914\,\mathrm{MJ/kg} = 19{,}914\,\mathrm{kJ/kg}$$

(d) From the main elements (in wt%) C, H, and O

$$\mathrm{HHV(MJ/kg)} = 0.3137\,\mathrm{C} + 0.7009\,\mathrm{H} + 0.0318\,\mathrm{O} - 1.3675$$

$$\mathrm{HHV(MJ/kg)} = 0.3137(53.50) + 0.7009(5.9) + 0.0318(40.3)$$
$$- 1.3675 = 20.83\,\mathrm{MJ/kg} = 20{,}830\,\mathrm{kJ/kg}$$

Estimation from equation above is the closest to the measured value of 21.03 MJ/kg (21,030 kJ/kg).

2.4.6 Biofuels

Biological fuels produced from photosynthesis can be categorized in three groups [11]:

- *Carbohydrates* represent a mixture of mono-di-and polysaccharides (17 kJ/g).
- *Fats* are unsaturated and saturated fatty acids (triglyceride) (39 kJ/g).
- *Proteins* are used partly as fuel source, (17 kJ/g).

Carbohydrates are straight-chain aldehydes or ketones with many hydroxyl groups that can exist as straight chains or rings. Carbohydrates such as starch are the most abundant biological molecules, and play numerous roles, such as the storage and transport of energy, and structural components such as cellulose in plants. Triglycerides and fatty free acids both contain long, linear aliphatic hydrocarbon chains, which are partially unsaturated and have a carbon number range. The fuel value is equal to the heat of combustion (oxidation) of fuel. Carbohydrates and fats can be completely oxidized while proteins can only be partially oxidized and hence have lower fuel values [3].

Some synthetic biofuels are [3]:

- *Bioethanol*—Corn-based ethanol is currently the largest source of biofuel as a gasoline additive. The gasoline is mixed with 10 percent ethanol, a mix known as E10 (or gasohol). Only specific types of vehicles named as flexible fuel vehicles can use mixtures with greater than 10% ethanol. E85 is an alternative fuel that contains up to 85% ethanol.
- *Biodiesel*—Biodiesel is most often blended with petroleum diesel in ratios of 2% (B2), 5% (B5), or 20% (B20). It can also be used as pure biodiesel (B100). Biodiesel can be produced from various feedstock and used in regular diesel vehicles without making any changes to the engines [3, 12].
- *Green diesel*—Green diesel is produced by removing the oxygen by catalytic reaction and saturation of double bond of carbon with hydrogen from renewable feedstock containing triglycerides and fatty acids. Green diesel is a paraffin-rich product and has a heating value close to conventional diesel. It is fully compatible for blending with the standard mix of petroleum-derived diesel fuels. Biodiesel has around 11% oxygen, whereas petroleum-based diesel and green diesel have no oxygen.

Using biomass as a feedstock for liquid fuels production may cut back on waste and greenhouse gas emissions and can offset the use of fossil fuels in heat and power generation [3].

2.4.7 Wind Energy

The earth is unevenly heated by the sun and the differential heating drives a global atmospheric convection system reaching from the earth's surface to the stratosphere. Most of the energy stored in these wind movements can be found at high altitudes where continuous high wind speeds occur. To assess the frequency of wind speeds at a location, a probability distribution function is often fitted to the observed data. Wind power is a totally renewable energy source with no greenhouse gas emissions, but due to its unpredictability, has problems integrating with national grids. The potential for wind to supply a significant quantity of energy is considerable. Availability of transmission capacity helps large-scale deployment by reducing the cost of delivered wind energy [2, 5, 17].

2.4.8 Geothermal Energy

Geothermal energy is the heat originating from the original formation of the planet, from radioactive decay of minerals, from volcanic activity, and from solar energy absorbed at the surface. The geothermal gradient, which is the difference in temperature between the core of the planet and its surface, drives a continuous conduction of thermal energy in the form of heat from the core to the surface. Geothermal power is cost effective, reliable, sustainable, and environmentally friendly. Worldwide, a considerable amount of geothermal power is produced for district heating, space heating, spas, industrial processes, desalination and agricultural applications.

Hot water or steam reservoirs deep in the earth are accessed by drilling. Geothermal reservoirs located near the earth's surface maintain a relatively constant temperature of 50°–60°F. The hot water and steam from reservoirs can be used to drive generators and produce electricity. In other applications, the heat produced from geothermal is used directly in heating buildings and industrial plants. As in the case of biomass electricity, a geothermal plant runs 24 h per day, seven days per week and can provide base load power, thus competing against coal plants [2, 5].

2.4.9 Ocean Energy

Systems to harvest electrical power from ocean waves have been gaining momentum as a viable technology. The potential for this technology is considered promising with the generator powerful enough to power a thousand homes. The turbine has minimal environmental impact, as it is almost entirely submerged, and the rotors pose no danger to wildlife as they turn quite slowly. Ocean thermal energy conversion uses the temperature difference that exists between deep and shallow waters to run a heat engine [2, 5].

2.5 Thermal Energy

Thermal energy is the sum of energy contained in the relative motion of many microscopic particles of a macroscopic whole system. The particles randomly move and collide with each other and the wall of container. The kinetic energy of the motion (including rotation and vibration except for monatomic substances) of the particles relative to the fixed wall is thermal energy. Temperature is a relative measure of the amount of thermal energy. Heat is the transfer of thermal energy from a system with higher temperature to a system with lower temperature. At absolute zero temperature, no further energy can be removed from a system without reorganization of chemical or nuclear binding. Therefore, thermal energy only includes the kinetic and potential energies associated with the relative motion, their rotation and vibrations as well as the vaporization or condensation energies of molecules [10]. Around 90% of current primary energy consumption at some stage involves thermal energy. Heat added to a system at constant volume without work done internal energy $\delta q = \mathrm{d}U = C_\mathrm{v}\mathrm{d}T$, heat added to a system at constant pressure without work done, we have $\delta q = \mathrm{d}H = C_\mathrm{p}\mathrm{d}T$, where $H = U + PV$. Phase changing by melting or vaporization takes place at constant temperatures.

Energy used in a shower-A shower runs around 10 L/min. If a shower takes about 10 min and the water in a shower is heated from 20 to 50°C, estimate the thermal energy used in the shower.

Assume: Density of water is 1 kg/l;

$$q = \dot{m}c_\mathrm{p}\Delta T = (10\,\mathrm{l/min})(10\,\mathrm{min}/60\,\mathrm{s})(1\,\mathrm{kg/l})(4.18\,\mathrm{kJ/kgK})(30\,\mathrm{K}) = 209\,\mathrm{kJ}.$$

2.6 Hydrogen

Hydrogen is the simplest element. In the sun's core, hydrogen atoms combine to form helium atoms (called fusion process) and gives off *radiant energy*. This radiant energy sustains life on earth as it drives photosynthesis in plants and other living systems and is stored as chemical energy in fossil fuels.

Hydrogen does not exist on earth as a gas and is found only in compound form with other elements, such as water H_2O and methane CH_4. Hydrogen is produced from other sources including natural gas, coal, biomass, and water. The two most common production methods are steam reforming natural gas and water electrolysis in which the water is split into oxygen and hydrogen [5, 9]. Hydrogen has the highest energy content of any common fuel by weight (about three times more than gasoline), but the lowest energy content by volume (see Table 2.6). Hydrogen burns cleanly, producing water (H_2O).

About half of hydrogen is used to produce ammonia (NH_3) via the Haber process. Ammonia, in turn, is used directly or indirectly as fertilizer. The other half of current hydrogen production is used in hydrocracking process to convert heavy petroleum sources into lighter fractions suitable for use as fuels. Hydrogen fuel cells make electricity. They are very efficient, but expensive to build. Small fuel cells can power electric cars, while large fuel cells can provide electricity in remote places with no power lines.

2.7 Electric Energy

The protons and electrons of an atom carry an electrical charge. Protons have a positive charge (+) and electrons have a negative charge (−). Opposite charges attract each other. The electrons in an atom's outermost shells do not attract strongly to the protons and can move from one atom to another and create electricity. The amount of electricity a power plant generates, or a customer uses over a period is measured in kilowatt hours (kWh), which is equal to the energy of 1,000 watts working for one hour. For example, if you use a 100-W light bulb for 7 h, you have used 0.7 kWh of electrical energy.

Most of the electricity used in the residential sector is for air conditioning, refrigerators, space and water heating, lighting, and powering appliances and equipment. Electricity is the most well-known energy carrier to transfer energy in coal, natural gas, uranium, wind power, and other energy sources to homes, businesses, and industry. For many energy needs, it is much easier to use electricity than the energy sources themselves [6, 9].

If the current passes through an electric appliance, some of the electric energy will be converted into other forms of energy (although some will always be lost as heat). The amount of electric energy, E_e, due to an electric current can be expressed in several different ways:

$$E_e = VIt = I^2Rt \tag{2.8}$$

where V is the electric potential difference (in volts), I is the current (in amperes), t is the time for which the current flows (in seconds), and R is the electric resistance (in ohms).

In *alternating current* (AC), the direction of the flow of electrons switches back and forth at regular intervals or cycles. The current flowing in power lines and normal household electricity that comes from a wall outlet is alternating current. The standard current used in the USA is 60 cycles per second (i.e., a frequency of 60 Hz); in Europe and most other parts of the world, it is 50 cycles per second (i.e., a frequency of 50 Hz.). In direct current (DC), on the other hand, electrical current flows consistently in one direction. The current that flows in a flashlight is direct current. One advantage of alternating current is that it is relatively cheap to change the voltage of the current. Furthermore, the inevitable loss of energy that occurs when current is carried over long distances is far smaller with alternating current than with direct current [10].

Example 2.6 **Electricity consumption of a laptop computer**

A laptop consuming 90 Watt is used on average 10 h per day. The laptop costs $500 and will be used for four years. Electricity cost is $0.15/kWh. Estimate the total electricity cost in four years for the laptop.

Solution:

$$\begin{aligned}
\mathrm{Cost}_{\mathrm{laptop}} &= \frac{\$500}{4\,\mathrm{year}} = \$125/\mathrm{year} \\
\mathrm{Cost}_{\mathrm{electricity}} &= \left(\frac{\$0.15}{\mathrm{kWh}}\right)\left(\frac{10\,\mathrm{h}}{\mathrm{day}}\right)\left(\frac{365\,\mathrm{days}}{\mathrm{year}}\right)(90\,\mathrm{W})\left(\frac{\mathrm{kW}}{1000\,\mathrm{W}}\right) = \$49.3/\mathrm{year} \\
\mathrm{Cost}_{\mathrm{total}} &= \mathrm{Cost}_{\mathrm{laptop}} + \mathrm{Cost}_{\mathrm{electricity}} \\
&= (\$125/\mathrm{year} + \$49.3/\mathrm{year})(4\,\mathrm{years}) = \$697.2
\end{aligned}$$

2.8 Magnetic Energy

There is no fundamental difference between magnetic energy and electric energy: the two phenomena are related by Maxwell's equations. The potential energy of a magnet of magnetic moment m in a magnetic field B is defined as the work of magnetic force (magnetic torque) is estimated by

$$E_{\mathrm{m}} = -mB \tag{2.9}$$

Calculating work needed to create an electric or magnetic field in unit volume results in the electric and magnetic fields energy densities. Electromagnetic radiation, such as microwaves, visible light, or gamma rays, represents a flow of electromagnetic energy. The energy of electromagnetic radiation has discrete energy levels. The spacing between these levels is equal to $E = hv$ where h is the Planck constant, 6.626×10^{-34} Js, and v is the frequency of the radiation. This quantity of electromagnetic energy is usually called a photon. The photons which make up visible light have energies of 160–310 kJ/mol [10].

2.9 Chemical Energy

Chemical energy results from the associations of atoms in molecules and various other kinds of aggregates of matter. It may be defined as a work done by electric forces that is electrostatic potential energy of electric charges. If the chemical energy of a system decreases during a chemical reaction, the difference is transferred to the surroundings in the form of heat or light. On the other hand, if the chemical energy of a system increases because of a chemical reaction, the difference then is supplied by the surroundings in form of heat or light. Typical values for the change in molar chemical energy during a chemical reaction range from tens to hundreds of kilojoules per mole. For example, 2,2,4-trimethylpentane (isooctane), widely used in petrol, has a chemical formula of C_8H_{18} and it reacts with oxygen exothermically and produces 10.86 MJ per mole of isooctane

$$C_8H_{18}(l) + 25/2O_2(g) \rightarrow 8CO_2(g) + 9H_2O(g) + 10.86\,\text{MJ/mole}$$

When two hydrogen atoms react to form a hydrogen molecule, the chemical energy decreases by the bond energy of the H–H. When the electron is completely removed from a hydrogen atom, forming a hydrogen ion, the chemical energy called the ionization energy increases.

Ionization energy measures the energy required to remove electrons from an atom. This energy would be in the range of 5–100 keV and is the least for the outermost electron and highest for the last electrons. To remove a proton or neutron from a nucleus is approximately 8 meV [10].

2.10 Mass Energy

Mass is a form of energy. Each particle is an excitation of quantum field similar to a single photon of light as a quantum excitation of electromagnetic field. Energy equivalent of a mass by Einstein's' formula is $E = mc^2$, where c is the speed of light in a vacuum, $v = 2.997 \times 10^8$ m/s. Converting mass into energy is only possible by bringing a particle in contact with its antiparticle of the same type. Since the solar system does not contain antimatters, naturally mass energy is important only in nuclear reactions and nuclear power [10].

2.11 Chemical Energy

Energy originating from electromagnetic interactions between atoms at molecular levels is stored in chemical bonds and called the chemical energy of a material. The energy stored in a typical chemical bond is in a magnitude of electron volt ($\text{eV} = 1.602 \times 10^{-19}$ J) that is the energy to move an electron across a one-volt electric potential difference [10].

Problems

2.1 Why is electrical energy so useful?
2.2 How can the energy in the wind be used?
2.3 How can wind power help conserve our oil supplies?
2.4 How might using wind energy help reduce air pollution?
2.5 What is the best energy source to convert to electricity?
2.6 Do the white-colored roof tiles keep houses cool?
2.7 How can energy from the sun be used to heat water?

2.8 With the clear advantages of nuclear power, why is it not more commonly used?

2.9 How can using solar energy help reduce pollution in the atmosphere and help conserve our oil supplies?

2.10 Why is the process of photosynthesis so valuable?

2.11 Name some foods that are known to be high energy foods.

2.12 Why are battery-powered vehicles considered to be the means of transport of the future?

2.13 Why is chemical energy useful to us?

2.14 What other forms of energy can be produced from chemical energy?

2.15 Name three examples of other fuels that contain chemical energy.

2.16 An overused car may consume around 250 gallons of gasoline per month. Estimate the energy consumed by the car per year.

2.17 An overused car may consume around 150 gallons of gasoline per month. Estimate the energy consumed by the car per year.

2.18 A city consumes natural gas at a rate of 500×10^6 ft^3/day. The volumetric flow is at standard conditions of 60°F and 1 atm = 14.7 psia. If the natural is costing $12/GJ of higher heating value what is the daily cost of the gas for the city.

2.19 A city consumes natural gas at a rate of 800×10^6 ft^3/day. The volumetric flow is at standard conditions of 60°F and 1 atm = 14.7 psia. If the natural is costing $10/GJ of higher heating value what is the daily cost of the gas for the city.

2.20 A car consumes about 6 gallons a day, and the capacity of a full tank is about 15 gallons. The density of gasoline ranges from 0.72 to 0.78 kg/l (Table 2.6). The lower heating value of gasoline is about 44,000 kJ/kg. Assume that the average density of gasoline is 0.75 kg/l. If the car was able to use 0.2 kg of nuclear fuel of uranium-235, estimate the time in years for refueling.

2.21 A car consumes about 3 gallons a day, and the capacity of the full tank is about 11 gallons. The density of gasoline ranges from 0.72 to 0.78 kg/l (Table 2.6). The lower heating value of gasoline is about 44,000 kJ/kg. Assume that the average density of gasoline is 0.75 kg/l. If the car was able to use 0.1 kg of nuclear fuel of uranium-235, estimate the time in years for refueling.

2.22 Using data in Table 2.10 and ultimate analysis, fixed carbon, dry ash content, C, H, and O compositions estimate the gross heating values in kJ/kg for the biomass white oak.

2.23 Using data in Table 2.10 and ultimate analysis, fixed carbon, dry ash content, and C, H, and O compositions only estimate the gross heating values in kJ/kg for the biomass corn stover and wheat straw.

2.24 When a hydrocarbon fuel is burned, almost all the carbon in the fuel burns completely to form CO_2 (carbon dioxide), which is the principal gas causing the greenhouse effect and thus global climate change. On average, 0.59 kg of CO_2 is produced for each kWh of electricity generated from a power plant that burns natural gas. A typical new household uses about 7000 kWh of electricity per year. Determine the amount of CO_2 production in a city with 100,000 households.

2.25 When a hydrocarbon fuel is burned, almost all the carbon in the fuel burns completely to form CO_2 (carbon dioxide), which is the principal gas causing the greenhouse effect and thus global climate change. On average, 0.59 kg of CO_2 is produced for each kWh of electricity generated from a power plant that burns natural gas. A typical new household uses about 10,000 kWh of electricity per year. Determine the amount of CO_2 production in a city with 250,000 households.

2.26 A large public computer lab operates Monday through Saturday. There the computers are either being used constantly or remain on until the next user comes. Each computer needs around 240 W. If the computer lab contains 53 computers and each is on for 12 h a day, during the year how much CO_2 will the local coal power plant have to release to the atmosphere to keep these computers running?

2.27 The average university will have a large public computer lab open Monday through Saturday. There the computers are either being used constantly or remain on until the next user comes. Each computer needs around 240 W. If the computer lab contains 53 computers and each is on for 12 h a day, during the year how much coal will the local coal power plant have to consume to keep these computers running?

2.28 A large public computer lab runs six days per week from Monday through Saturday. Each computer uses a power of around 120 W. If the computer lab contains 45 computers and each is on for 12 h a day, during the year how much CO_2 will the local coal power plant have to release into the atmosphere to keep these computers running?

2.29 If a car consumes 60 gallons gasoline per month. Estimate the energy consumed by the car per year.

2.30 A car having an average of 22 miles/gal is used 32 miles every day. If the cost of a gallon fuel is \$3.8 estimate the energy consumed per year.

2.31 A car having an average of 22 miles/gal is used 32 miles every day. Estimate yearly energy usage.

2.32 A 150-Watt electric light bulb is used on average 10 h per day. A new bulb costs \$2.0 and lasts about 5000 h. If electricity cost is \$0.15/kWh, estimate the consumed energy and its cost per year.

2.33 A laptop consuming 90 Watt is used on average 5 h per day. If a laptop is used for five years estimate the total energy.

2.34 A laptop consuming 90 Watt is used on average 7 h per day. If a laptop is used for four years estimate the total electricity needed.

2.35 A 20-hP electric motor is used to pump ground water into a storage tank for four hours every day. Estimate the work done by the pump in kW every year.

2.36 A city consumes natural gas at a rate of 250×10^6 ft^3/day. The volumetric flow is at standard conditions of 60°F and 1 atm = 14.7 psia. If natural gas is costing \$6/GJ of higher heating value what is the yearly energy consumed and its cost of the gas for the city.

2.37 A home consumes natural gas at a rate of 4.3 ft^3/day to heat the home. The volumetric flow is at standard conditions of 60°F and 1 atm = 14.7 psia. If natural gas is costing $0.67/MJ of higher heating value what is the yearly use of energy and its cost of the gas for the home?

2.38 A water heater consumes propane, which provides 80% of the standard heat of combustion. If the price of propane is $2.2/gal measured at 25°C. What is the amount of energy for heating and its cost in $ per million Btu and in $ per MJ/year?

2.39 An average video games system consumes 170 W of power during gameplay. If a person were to play an hour a day for 80% of the year how many liters of gasoline would the person have burned? (Evaluated at HHV)

2.40 A competitive road cyclist can hold an average of 300 W of power during a four-hour race. During long races they must do this each day for a three weeklong race. How many protein bars will the cyclist have to eat at 184 cal per bar to just make up the calories lost during the race?

2.41 Describe the process of how natural gas goes from its natural state to the market?

2.42 Some people like to have background noise when they are falling asleep. Many choose to listen to their television. The television will usually run on about 340 W and will run during the 8 h that you are asleep. With electricity costing $0.20/kWh, calculate how much this will cost you if you do this for five days a week for an entire year.

2.43 What are the advantages and disadvantages of electrical energy in an alternating current?

2.44 What are the advantages and disadvantages of electrical energy flowing in direct current?

2.45 In the search for new sources of energy that are renewable and emit less greenhouse gases, carbon-based biofuels are of major interest. These fuels are still carbon based and must undergo combustion to release chemical energy. Why is this process being looked at as a reasonable energy source?

2.46 While fixing wiring in a house, an electrician aims to deliver the same amount of electric energy to a devise at the same rate it is currently coming in. The wiring he is replacing runs at has 2 ohms of resistance and has a 20-amp current. To deliver the same amount of electric energy how many amps will be needed if the resistance is changed to 4 ohms?

2.47 Calculate the yearly dollar savings if you cut down from a daily nine-minute shower to a six-minute shower. The shower volumetric flow rate is 3.2 gpm and the amount of energy used per gallon is 440 Btu and energy costs $0.13/kWh.

2.48 Rank the following carbon-based fuels in order of lowest to highest gross energy density: diesel, ethanol, conventional gasoline, and kerosene-based jet fuel.

References

1. Alhajji M, Demirel Y (2015) Energy and environmental sustainability assessment of crude oil refinery by thermodynamic analysis. Int J Energy Res 39:1925–1941
2. Blackburn JO (2014) Renewable energy. In: Anwar S (ed) Encyclopedia of energy engineering and technology, 2nd edn. CRC Press, Boca Raton
3. Demirel Y (2025) Biofuels. In: Dincer I (ed) Comprehensive energy systems, vol 2, 2nd edn. Elsevier, Amsterdam, pp 344–384
4. Demirbas A (1997) Calculation of higher heating values of biomass fuels. Fuel 76:431–434
5. Dincer I, Midilli A (2014) Green energy. In: Anwar S (ed) Encyclopedia of energy engineering and technology, 2nd edn. CRC Press, Boca Raton
6. IEA (2024) International energy agency. World Energy Outlook 2024. https://www.iea.org/reports/world-energy-outlook-2024
7. IEO (2023) International Energy Outlook 2023. https://www.eia.gov/outlooks/ieo/
8. IRENA (2023) International renewable energy agency, renewable power generation costs in 2023. https://www.irena.org/Publications/2024/Sep/Renewable-Power-Generation-Costs-in-2023
9. IRENA (2024) International renewable energy agency. World Energy Transition Outlook 2024 . https://www.irena.org/Publications/2024/Nov/World-Energy-Transitions-Outlook-2024
10. Jaffe RL, Taylor W (2018) The physics of energy. Cambridge University Press, Cambridge
11. Marks DB (1999) Biochemistry, 3rd edn. Kluwer, New York
12. Nguyen N, Demirel Y (2013) Economic analysis of Biodiesel and glycerol carbonate production plant by glycerolysis. J Sustain Bioenergy Sys 3:209–216
13. Øvergaard S (2008) Issue paper: definition of primary and secondary energy. www.og.ssb.no
14. Parker G (2014) Coal-to-liquid fuels. In: Anwar S (ed) Encyclopedia of energy engineering and technology, 2nd edn. CRC Press, Boca Raton
15. Sheng C, Azevedo JLT (2005) Estimating the higher heating value of biomass fuels from basic analysis data. Biomass Bioenergy 28:499–507
16. Vimeux F, Cuffey KM, Jouzel J (2002) New insights into southern hemisphere temperature changes from Vostok ice cores using deuterium excess correction. Earth Planet Sci Let 203:829–843
17. Wiser R, Bolinger M (2010) DOE EERE. Wind technologies market report. http://www.osti.gov/bridge

References

Mechanical Energy and Electrical Energy

3

Introduction and Learning Objectives: This chapter introduces mechanical and electrical energy. Mechanical energy can be broadly classified into potential energy and kinetic energy. Potential energy refers to the energy any object has because of its position in a force field. Kinetic energy is the work required to accelerate an object to a given speed. Mechanical energy due to the pressure of the fluid is known as pressure energy. Electric energy is a kind of kinetic energy created by moving electrons through a conductor. This chapter discusses various forms of mechanical energy as well as electric energy.

The learning objectives of this chapter are:

- Forms of mechanical energy
- Electric energy and work

3.1 Mechanical Energy

There are two forces that we can experience: gravitational forces and electromagnetic forces. Both forces act through space. *Mechanical energy* describes the sum of potential energy and kinetic energy present in the components of a *mechanical system.* Mechanical energy is associated with the motion or position of an object under gravitational force. Mass causes gravitational attraction. Charge and mass obey an inverse square law, which states that a specified physical quantity or strength is inversely proportional to the square of the distance from the source of that physical quantity. The gravitational force law depends inversely upon the square of the distance between two masses, so mass plays a role somewhat like the role charge plays in the force law. Charged particles exhibit nongravitational forces between them. The force of attraction or repulsion between two electrically

Y. Demirel, *Energy*, Green Energy and Technology,
https://doi.org/10.1007/978-3-031-97111-2_3

charged particles is directly proportional to the product of the electric charges and is inversely proportional to the square of the distance between them. This is known as Coulomb's law. The law of conservation of mechanical energy states that if a body or system is subjected only to conservative forces the total mechanical energy of that body or system remains constant [1, 2].

Mechanical energy is the form of energy that can be converted to mechanical work completely and directly by a mechanical device such as a turbine. Thermal energy cannot be converted to work directly and completely. The familiar forms of mechanical energy are kinetic and potential energies. Pressure energy is also another form of mechanical energy due to the pressure of a fluid. Therefore, the mechanical energy in metric units may be defined in the following energy balance equation:

$$\dot{m}\left(\frac{\Delta P}{\rho}+\frac{\Delta v^2}{2}+g\Delta z\right)=\dot{W}_{\mathrm{shaft}}+\dot{W}_{\mathrm{loss}} \tag{3.1}$$

Pressure energy + kinetic energy + potential energy
= mechanical energy + work loss.

where $\dot{m}$ is the mas flow rate, P is the pressure, ρ is the density, v is the flow velocity, g is the acceleration of gravity, z is the elevation height, $\dot{W}_{\mathrm{shaft}}$ is net shaft work in per unit mass for a pump, fan, or similar equipment, and $\dot{W}_{\mathrm{loss}}$ represents the work loss due to friction and other irreversibility sources. The mechanical energy equation for a pump, a fan, or a turbine can be written in terms of energy per unit mass [1].

3.2 Kinetic Energy

Kinetic energy KE is the energy of a system, or a material possesses because of its velocity relative to the surroundings. The kinetic energy of a flowing fluid relative to stationary surroundings is estimated by

$$KE=\frac{1}{2}mv^2 \tag{3.2}$$

where m is the mass and v is the average velocity. The value of a change in the kinetic energy occurs in a specified time interval and depends only on the mass and the initial and final values of the average velocities of the material.

$$\Delta KE=\frac{1}{2}m\left(v_2^2-v_1^2\right) \tag{3.3}$$

Mass flow rate $\dot{m}$ is related to the density ρ, cross-sectional area A, and average velocity v by

$$\dot{m}=\rho A v \tag{3.4}$$

in the SI units, kinetic energy unit is Joule: J = kg m^2/s^2. In the English engineering units, for the unit consistency g_c (g_c = 32.174 lbm ft/lbf s^2) is included and the unit of kinetic energy becomes

$$KE = \frac{mv^2}{2g_c} = \frac{\text{lbm ft}^2\ \text{s}^{-2}}{\text{lbm ft (lbf)}^{-1}\text{s}^{-2}} = \text{ft lbf} \tag{3.5}$$

Momentum is related to kinetic energy in mechanics and estimated for a particle of mass moving with velocity $(\mathrm{d}x/\mathrm{d}t)$ by $P = m(\mathrm{d}x/\mathrm{d}t)$. Kinetic energy in terms of momentum is $E = \frac{p^2}{2m}$. Like energy, the total momentum is conserved for an isolated system.

Newton's laws of motion are:

(i) An object in motion experiences no change in its velocity unless an external force acts.
(ii) An acceleration a of an object with a mass m under the influence of a force is: $F = ma$.
(iii) For every action there is an equal and opposite reaction.

Air resistance—Fluid (air/liquid) resistance causes energy loss of a moving object with a speed v

$$\frac{\mathrm{d}E_{\text{loss}}}{\mathrm{d}t} = c_d \frac{A\rho v^3}{2} \tag{3.6}$$

where ρ is the mass density of the fluid, A is the cross-sectional area of the object, and c_d is the drag coefficient, a dimensionless quantity of order one. The product $c_d A$ is often used and called effective area [3].

Example 3.1 **Calculation of the kinetic energy for a flowing fluid**

Water with a flow rate of 2.0 kg/s is pumped from a storage tank through a tube of 2.5 cm inner diameter. Calculate the kinetic energy of the water in the tube.

Solution:

Assume the density of the water as $\rho = 1000$ kg/m^3.

Basis is 2.0 kg/s.

Radius of the tube $(r) = 2.5/2 = 1.25$ cm.

Cross-sectional area of the tube:
$A = \pi r^2 = 3.14(1.25\ \text{cm})^2(1\ \text{m}/100\ \text{cm})^2 = 4.906 \times 10^{-4}\text{m}^2$.

The average velocity of the water: $v = \frac{\dot{m}}{\rho A} = \frac{2.0\ \text{kg/s}}{(1000\ \text{kg/m}^3)(4.906\times10^{-4}\ \text{m}^2)} = 4.076$ m/s .

The kinetic energy of the water in the tube in SI units (J = kg m^2/s^2):

$$KE = \frac{1}{2}\dot{m}v^2 = \frac{1}{2}(2.0\ \text{kg/s})(4.076\ \text{m/s})^2 = 0.0166\ \text{kJ/s} = 16.6\ \text{W}$$

J = N m, where N is the Newton.

Example 3.2 **Kinetic energy of a car**

A car having a mass of 2750 lb is traveling at 55 miles/h. Estimate: (a) the kinetic energy of the car in kJ, (b) the work necessary to stop the car.

Solution:

Equation: $EK = \frac{1}{2}mv^2$.

Data: $m = 2750$ lb (kg/2.2 lb) = 1250 kg;

$$v = 55\ \text{miles}/h \rightarrow v = \left(\frac{55\ \text{miles}}{\text{h}}\right)\left(\frac{1609.34\ \text{m}}{\text{mile}}\right)\left(\frac{\text{h}}{3600\ \text{s}}\right) = 24.58\ \text{m/s}$$

(a) $EK = \frac{1}{2}(1250\ \text{kg})(24.58\ \text{m/s})^2 = 377.6\ \text{kJ}$
(b) $W = EK = 377.6$ kJ

3.3 Potential Energy

Potential energy exists whenever an object has a position within a force field. The gravitational force near the Earth's surface varies very little with the height h and is equal to the mass m multiplied by the gravitational acceleration $g = 9.81\ \text{m/s}^2$. When the force field is the earth's gravitational field, then the potential energy of an object is:

$$PE = mgz \tag{3.7}$$

where z is the height above earth's surface or a reference surface. Potential energy is stored within a system and activated when a restore force tends to pull an object back toward some lower energy position. For example, when a mass is lifted, the force of gravity will act to bring it back down. The energy that went into lifting the mass is stored in its position in the gravitational field. Therefore, the potential energy is the energy difference between the energy of an object in each position and its energy at a reference position Δz.

$$\Delta PE = mg\Delta z \tag{3.8}$$

In the SI units, potential energy is Joule: J = kg (m/s^2) m. In the English units, potential energy is expressed as

$$\Delta PE = \frac{mg\Delta z}{g_c} = \frac{\text{lbm ft ft s}^{-2}}{\text{lbm ft (lbf)}^{-1}\text{ s}^{-2}} = \text{lbf ft} \tag{3.9}$$

For the unit consistency, g_c ($g_c = 32.174$ lbm ft/lbf s^2) is included and the unit of potential energy becomes lbf ft. Various types of potential energy are associated with a force [5]:

- *Gravitational potential energy* is the work of the gravitational force. If an object falls from one point to another inside a gravitational field, the force of gravity will do positive work on the object, such as the production of hydroelectricity by falling water.
- *Elastic potential energy* is the work of an elastic force. Elastic potential energy arises because of a force that tries to restore the object to its original shape, which is most often the electromagnetic force between the atoms and molecules that constitute the object.
- *Chemical potential energy* is related to the structural arrangement of atoms or molecules. For example, chemical potential energy is the energy stored in fossil fuels and when a fuel is burned the chemical energy is converted to heat. Green plants convert solar energy to chemical energy through the process called photosynthesis.
- *Nuclear potential energy* is the potential energy of the particles inside an atomic nucleus. The nuclear particles are bound together by the strong nuclear force.

Example 3.3 **Potential energy change of water**

Water is pumped from one reservoir to another. The water level in the second reservoir is 30 ft above the water level of the first reservoir. What is the increase in specific potential energy of one pound of water?

Solution:

The problem requires the change of potential energy of 1 lbm of water. The difference between the water levels of the two reservoirs is $\Delta z = 30$ ft. Equation (3.8) yields

$$\Delta PE = \left(\frac{32.2\text{ ft}}{\text{s}^2}\right)\left(\frac{\text{lbf s}^2}{32.2\text{ lbm ft}}\right)(30\text{ ft})\left(\frac{\text{Btu}}{778.2\text{ ft lbf}}\right) = 0.0385\text{ Btu/lb}.$$

Example 3.4 **Energy of an elevator**

An elevator with a mass of 1500 kg rests at a level of 3 m above the base of an elevator shaft. The elevator has traveled to 30 m above the base. The elevator falls from this height freely to the base and strikes a strong spring. Assume that the entire process is frictionless. Estimate:

(a) The potential energy of the elevator in its original position relative to the base of the shaft.
(b) The work done traveling the elevator.
(c) The potential energy of the elevator at its highest position relative to the base of the shaft.
(d) The velocity and kinetic energy of the elevator just before it strikes the spring.

Solution:

Assume $g = 9.8\ \text{m/s}^2$

Data: $\dot{m} = 1500\ \text{kg/s}$, $z_1 = 3\ \text{m}$, $z_2 = 30\ \text{m}$

(a) $PE_1 = mgz_1 = (1500)(9.8)(3) = 44{,}100\ \text{J} = 44.1\ \text{kJ}$
(b) $W = \int_{z_1}^{z_2} Fdz = \int_{z_1}^{z_2} mg\text{d}z = mg(z_1 - z_2) = (1500)(9.8)(30 - 3) = 396.9\ \text{kJ}$
(c) $PE_2 = mgz_2 = (1500)(9.8)(30) = 441.0\ \text{kJ}$

So $W = PE_2 - PE_1 = (441.0 - 44.1)\ \text{kJ} = 396.9\ \text{kJ}$

(d) $KE = PE_2 = 441.0\ \text{kJ}$

$$KE = \frac{mv^2}{2} \rightarrow v = \sqrt{\frac{2KE}{m}} = \sqrt{\frac{2(441{,}000)}{1500}} = 24.25\ \text{m/s}.$$

Example 3.5 **Mechanical energy of a plane**

Consider a Boeing 777 airplane with approximately 350 tons at a cruising speed of 850 km/h. Estimate the kinetic and potential energy.

Solution:

This plane has a kinetic energy:

$$E = \frac{1}{2}mv^2 = (3{,}500{,}000\ \text{kg})(850\ \text{km/h})^2 = 126.4\ \text{GJ}$$

The same plane flying at an attitude of 10,000 m has the potential energy:

$$E = mgz = (3{,}500{,}000\ \text{kg})\left(9.8\ \text{mm/s}^2\right)(10{,}000\ \text{m}) = 34.3\ \text{GJ}$$

3.4 Pressure Energy

The pressure unit Pa is defined as: Pa = $\text{N/m}^2 = \text{N m/m}^3 = \text{J/m}^3$, which becomes energy per unit volume. The product of pressure with volume $P(1/\rho)$ leads to J/kg that is the energy per unit mass. Therefore, the mechanical energy due

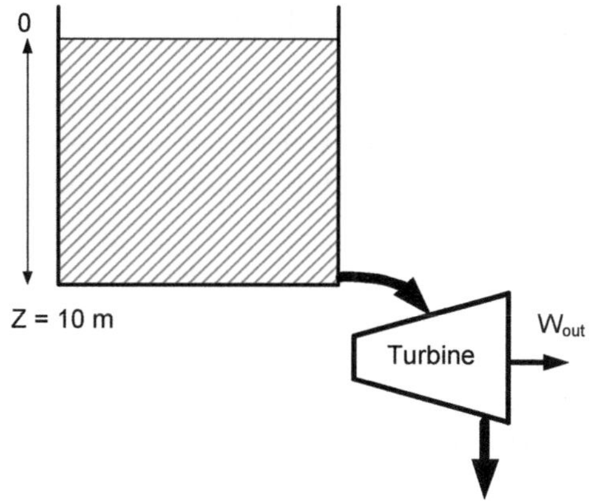

Fig. 3.1 Pressure energy P/ρ can be converted to mechanical energy completely by a reversible turbine

to the pressure of the fluid is known as pressure energy *PV* and expressed by $PV = P/\rho$, where ρ is the density. Pressure energy can be converted to mechanical energy completely, for example, by a reversible turbine. Transferring of mechanical energy is usually accomplished through a rotating shaft, and mechanical work is usually referred to as shaft work. A turbine, for example, converts the mechanical energy of a fluid to shaft work. Only the frictional effects cause the loss of mechanical energy in the form of heat. Consider a tank filled with water, as shown in Fig. 3.1.

$$\dot{W}_{\max} = \dot{m}gz \tag{3.10}$$

For example, for a water mass flow rate of 10 kg/s, maximum work would be

$$\dot{W}_{\max} = \dot{m}gz = (10\ \text{kg/s})\left(9.81\ \text{m/s}^2\right)(10\ \text{m}) = 981\ \text{W}$$

For a flowing fluid, pressure energy is the same as the flow energy using the gauge pressure instead of absolute pressure [1].

Example 3.6 **Pressure energy of a hydraulic turbine**

Electricity is produced by a hydraulic turbine installed near a large lake. Average depth of the water in the lake is 45 m. The mass flow rate of water is 600 kg/s. Determine the work output of the turbine.

Solution:

Assume: The mechanical energy of water at the turbine exit is small and negligible. The density of the water is 1000 kg/m^3.

$$\dot{W}_{\text{out}} = \dot{m}gz = (600\ \text{kg/s})\left(9.81\ \text{m/s}^2\right)(45\ \text{m}) = 264.9\ \text{kW}\ \left(\text{kJ/kg} = 1000\ \text{m}^2/\text{s}^2\right)$$

The lake supplies 264.9 kW of mechanical energy to the turbine.

3.4.1 Pressure Head

For an incompressible fluid, such as liquids, the pressure difference between two elevations can be expressed as *static pressure difference*: $P_2 - P_1 = -\rho g\,(z_2 - z_1)$. The *pressure head* h is estimated by the static pressure difference [5].

$$h = \frac{P_2 - P_1}{\rho g}\ \text{(in m or ft)} \tag{3.11}$$

The pressure head is related to the height of a column of fluid required to give a pressure difference of $\Delta P = P_2 - P_1$. The mechanical energy equation (Eq. 3.1) can also be written in terms of head and per unit mass

$$\left(\frac{\Delta P}{\rho g} + \frac{\Delta v^2}{2g} + \Delta h\right) - h_{\text{shaft}} = h_{\text{loss}} \tag{3.12}$$

where $h_{\text{shaft}} = W_{\text{shaft}}/g$ is the net shaft energy head in per unit mass and the head loss due to friction is estimated by

$$h_{\text{loss}} = W_{\text{loss}}/g = \text{ loss head due to friction} \tag{3.13}$$

Example 3.7 **Pumping water**

Water is pumped from an open tank at level of zero ft to an open tank at level of 12 ft. The pump adds 6 hp to the water when pumping a volumetric flow rate of 3 ft^3/s. Estimate the loss energy in head.

Solution:

Data: $Q = 3$ ft^3/s, $\dot{W}_{\text{shaft}} = 6$ hp.

Since $v_{\text{in}} = v_{\text{out}} = 0$, $P_{\text{in}} = P_{\text{out}} = 0$ and $h_{\text{in}} = 0$, $g = 32.2$ ft/s^2.

Density of water $= 62.4$ lb/ft^3

The special conversion factor with symbol $g_c = 32.174$ ft lbm/(s^2 lbf).

From Eq. (3.12): $h_{\text{shaft}} = h_{\text{loss}} + h_{\text{out}}$

$$h_{\text{shaft}} = \frac{W_{\text{shaft}}}{\rho g Q} = \frac{(6\ \text{hp})\left(\frac{550\ \text{ft lb}}{\text{hp}}\right)}{\left(62.4\ \text{lb/ft}^3\right)\left(3\text{ft}^3/\text{s}\right)} = 17.6\,\text{ft}$$

where specific weight of water is 62.4 lb/ft^3

and 1 hp (English horsepower) $= 550$ ft lb/s

$$h_{\text{loss}} = (17.6 - 12)\text{ft} = 5.6\ \text{ft}$$

Example 3.8 **Calculation of the power needed to pump water**

A pump draws water from 10 feet deep well and discharges it into a tank, which is 2 feet above the ground level. The discharge rate of water is 0.25 ft^3/s. Calculate the power needed by the pump.

Solution:

Assume that negligible heat dissipation due to friction occurs within the pipe and the pump.

The power needed can be calculated from: $\dot{W} = \dot{m}g(z_{\text{out}} - z_{\text{in}})$.

where g is the standard gravity: $g = 9.806$ m/s^2 = 32.2 ft/s^2.

The special conversion factor with symbol $g_c = 32.174$ ft lbm/(s^2 lbf).

Density of water = 62.4 lb/ft^3

With a basis of lbm/s, the mass flow rate of water is

$$\dot{m} = \left(0.25\ \text{ft}^3/\text{s}\right)\left(62.4\ \text{lb/ft}^3\right) = 15.6\ \text{lbm/s}$$

Using the conversion of 1.0 kW = 737.56 lbf ft/s.

$$\text{Power: } \dot{W} = \frac{(15.6\ \text{lbm/s})(32.2\ \text{ft}^2/\text{s})(12\ \text{ft})}{32.2\ \text{ft lbm}/(\text{s}^2\ \text{lbf})}\left(\frac{1.0\,\text{kW}}{737.56\ \text{lbf ft/s}}\right) = 0.253\,\text{kW} = 0.340\,\text{hp}.$$

3.5 Surface Energy

Surface energy is a measure of intermolecular forces that occur when a surface is created. The surface energy is the excess energy at the surface of a material compared to the bulk. For a liquid, the surface tension (surface energy density) is defined as the force per unit length. Water has a surface energy density of 0.072 J/m^2 and a surface tension of 0.072 N/m. Raindrops are spherically shaped because of surface tension or surface energy. Surface energy is a characteristic property of a liquid and directly related to the liquid's surface area. High surface area geometries contain high surface energy, while low surface area geometries contain low surface energy levels. Liquids form themselves into spherical shapes to lower their overall energy state since the sphere has the smallest surface area and has the lowest energy state. The driving force behind spherically shaped liquids involves the cohesive forces, which are the attractive forces that a liquid's molecules have to one another [1, 4].

3.6 Sound Energy

Sound energy is the energy produced by an object's vibrations. Sound energy E results from the integral of the acoustic pressure P times the particle velocity v over a surface A, and is given by the integral

$$E = \int (Pv) \cdot \mathrm{d}A \tag{3.14}$$

The sound energy flux is the average rate of flow of sound energy through any specified area A and is usually referred to as acoustic intensity. Sound energy is, therefore, a form of mechanical energy and related to the pressure of sound vibrations produced. Sound energy is typically not used for electrical power or for other human energy needs because the amount of energy that can be gained from sound is quite small.

Ultrasound is cyclic sound pressure with a frequency greater than the upper limit of human hearing, which is approximately 20 kHz. For a sound wave, the hertz (Hz) is equal to cycles of the sound wave per second. Ultrasound is used typically to penetrate a medium and measure the reflection signals or supply focused energy. The reflection signals can reveal details about the inner structure of the medium. In ultrasonic welding of plastics, high-frequency (15–40 kHz) low-amplitude vibration is used to create heat by way of friction between the materials to be joined. Also, ultrasonic testing is used to find flaws in materials and to measure the thickness of objects [4].

3.7 Electric Energy

Electricity starts with *a charge* that produces electrical forces. When, for example, a light bulb is turned on, charge flows from the wall plug through the bulb and heats up the filament in the bulb generating light. There are two large forces: gravitational forces and electromagnetic forces. Both forces act through space. Mass causes gravitational forces and two masses always attract each other. Some particles possess a charge and may be subject to electromagnetic forces. Charge may be positive or negative and two positive charges or two negative charges will repel each other, whereas the positive and negative charges attract each other. Charge is measured in Coulombs (C) named after Charles Augustin Coulomb who was the first scientist to formulate the force law for charges. A farad (F) is the charge in Coulombs. The Coulomb's law is given by

$$F = k_\mathrm{e} \frac{Q_1 Q_2}{r^2} \tag{3.15}$$

where k_e is the Coulomb's constant and given by $k_\mathrm{e} = 1/(4\pi\varepsilon_0) = 8.987 \times 10^9\,\mathrm{Nm^2/C^2}$, ε_0 is the electric constant and in the SI system $\varepsilon_0 \sim 8.885419 \times 10^{-12}$ F/m. For the two charges, $\boldsymbol{Q_1}$ and $\boldsymbol{Q_2}$, the force between them is proportional to the product of the two charges and inversely proportional to the square of the distance r (in meters) between them. The *electric charge* Coulomb (C) is the amount of electricity carried in one second of time by one ampere of current ($C = As$). One Faraday is 96,485 C. A positive force implies it is repulsive, while a negative force implies it is attractive. Charge comes in discrete sizes. Electrons and protons have the charge of magnitude 1.6×10^{-19} Coulombs that are negative for electrons and positive for protons. The charge on an electron is a fundamental constant of nature.

When charge flows in a wire, it is called current (I), which is measured in coulombs/second and called amperes (A). For example, for a current of 4 amperes and time interval of 15 s, the total charge is: (current)(time) = $(4A)(15\ s) = 60$ coulombs. Current can also flow through an ionic solution and through the ground. Current is measured by a device called *ammeter* [4].

3.7.1 Electric Potential Energy

Electric potential energy is the work of the Coulomb force. The work of intermolecular forces is called *intermolecular potential energy*. There are two main types of electric potential energy: electrostatic potential energy and electrodynamics potential energy (or magnetic potential energy). The *electrostatic potential energy* is the energy of an electrically charged particle in an electric field. It is defined as the work that must be done to move it from an infinite distance away to its present location, in the absence of any nonelectrical forces. Consider two point-like objects A_1 and A_2 with electrical charges Q_1 and Q_2. The work W required to move A_1 from an infinite distance to a distance r away from A_2 is given by:

$$W = \frac{1}{4\pi\varepsilon_0}\frac{Q_1 Q_2}{r} \tag{3.16}$$

Electric potential known as *voltage* is equal to the electric potential energy per unit charge. Voltage is the driving force behind current. When the charged particles are in motion, they generate *magnetic potential energy*. This kind of potential energy is a result of the magnetism having the potential to move other similar objects. Magnetic objects are said to have some *magnetic moment* [4].

Figure 3.2 shows a simple circuit with a battery and two other components. A battery will supply charge at some specified voltage to an applied electrical load. When the charge flows through the load, it gives up the potential energy to the load. If the load is a motor, that energy might be transformed into mechanical energy. If the load is a light bulb, the energy is transformed into light and heat.

Here, V_B represents the voltage across the battery. Considering the circuit in Fig. 3.2, the energy added to a charge Q would be:

- QV_B: When charge moves from the bottom of the battery to the top of the battery, the charge gains an amount of energy.

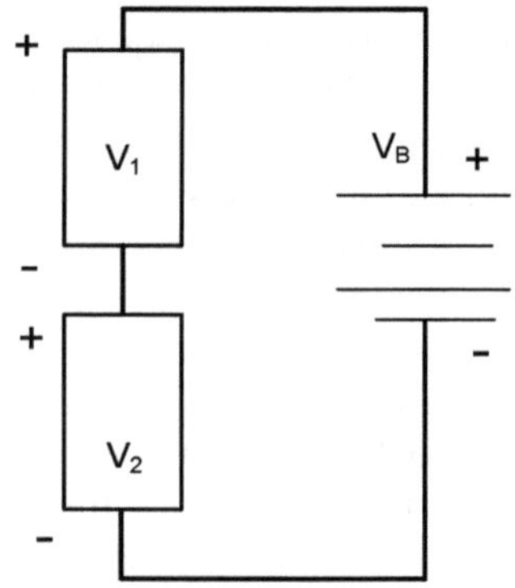

Fig. 3.2 Electrical circuit; V_B represents the voltage across the battery

- QV_1: As a charge Q moves from the top of element 1 to the bottom of element 1, the charge loses an amount of energy.
- QV_2: As a charge Q moves from the top of element 2 to the bottom of element 2, the charge loses an amount of energy.

Like mechanical potential energy, electrical potential energy and voltage are measured from a reference usually chosen as ground. That means the reference is the earth or ground itself. A voltage level of 120 V means that the voltage difference between that point and ground is 120 V. To measure voltage differences in circuits a *Voltmeter* is used.

3.7.2 Estimation of Electrical Energy

Figure 3.3 shows a circuit with a resistance R, which determines how much current will flow through a component. Resistors control voltage and current levels. A high resistance allows a small amount of current to flow. A low resistance allows a large amount of current to flow. Resistance is measured in ohms Ω. One ohm is the resistance through which one volt will maintain a current of one ampere. In a resistor, the voltage and the current are related to each other by the Ohm's law

$$I = V/R \text{ or } V_r = RI_r \tag{3.17}$$

Ohm's law defines the relationships between voltage V, current I, and resistance R. If both V and I remain constant during the time t then electrical energy becomes

$$W_e = VIt \tag{3.18}$$

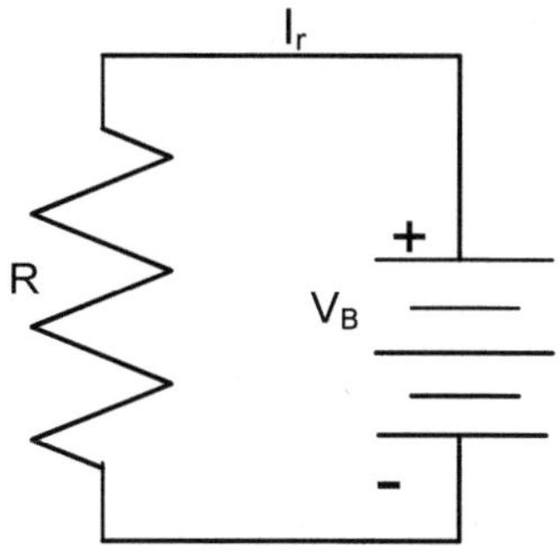

Fig. 3.3 Circuit diagram; battery voltage, V_B, current I_r through a resistance R

3.7.3 Electric Power

Electric power P is the rate at which electrical energy is transferred by an electric circuit

$$P = VI \tag{3.19}$$

where P is the electric power, V is the potential difference, and I is the electric current. Electric power is measured in watts in SI units. The power flowing into the battery is negative, and the power flowing out of the battery is positive. The power delivered to the resistor is the product of the voltage across the resistor and the current through the resistor: VI_r.

In the case of resistive loads (see Fig. 3.3), using the Ohm's law, the following alternative expression for the dissipated power is given by:

$$P = I^2R = V^2/R \tag{3.20}$$

where R is the electrical resistance.

3.7.4 Capacitance

A capacitor's ability to store charge is called the capacitance, which is measured by *Farad* (F). One Farad is the capacitance between two parallel plates that results in one volt of potential difference when charged by one coulomb of electricity ($F = C/V$).

In *alternating current* circuits, energy storage elements such as inductance and capacitance may result in periodic reversals of the direction of energy flow. The portion of power flow that averaged over a complete cycle of the alternating current waveform results in net transfer of energy in one direction is known as real power or active power. That portion of power flow due to stored energy, which returns to the source in each cycle, is known as reactive power as seen in Fig. 3.4.

The relationship between real power, reactive power, and apparent power can be expressed by representing the quantities as vectors. Real power is represented

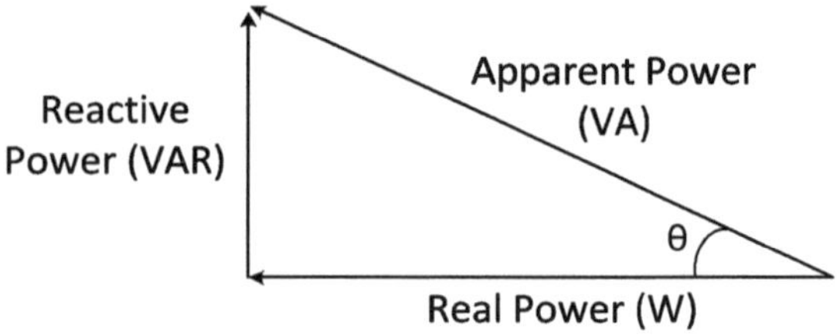

Fig. 3.4 Power triangle and the components of alternating current power

as a horizontal vector and reactive power is represented as a vertical vector. The apparent power vector is the hypotenuse of a right triangle formed by connecting the real and reactive power vectors. This representation is often called the *power triangle*. Using the Pythagorean theorem, the relationship among real, reactive, and apparent powers is:

$$(\text{Apparent power})^2 = (\text{Real power})^2 + (\text{Reactive power})^2 \tag{3.21}$$

Real and reactive powers can also be calculated directly from the apparent power, when the current and voltage are both sinusoids with a known phase angle between them:

$$(\text{Real power}) = (\text{Apparent power})\cos(\theta) \tag{3.22}$$

$$(\text{Reactive power}) = (\text{Apparent power})\sin(\theta) \tag{3.23}$$

The ratio of real power to apparent power is called power factor and is a number always between 0 and 1. Where the currents and voltages have nonsinusoidal forms, power factor is generalized to include the effects of distortion [2, 4].

3.8 Electromagnetic Energy

Electromagnetic force is one of the four fundamental natural forces that are gravity and the strong and weak nuclear forces [4]. Electromagnetic interactions between charge particles and gravitation interactions between bodies are similar to each other. Coulomb' law shows that two charges create large forces equal in magnitude but opposite in direction and same charges repel while opposite-sign charges attract.

$$\boldsymbol{F}_{12} = -\boldsymbol{F}_{21} = \frac{1}{4\pi\varepsilon_0}\frac{Q_1 Q_2}{r_{12}^2}\boldsymbol{r}_{12} \tag{3.24}$$

where Q_1 and Q_2 are the charges (C, coulombs in SI units), r_{12} is the vector separating the two charges, $\boldsymbol{F}_{12}$ is the force exerted by Q_1 on Q_2, and ε_0 is the permittivity of vacuum ($1/4\pi\varepsilon_0 = 8.988 \times 10^9$ Nm^2/C^2).

Electric and magnetic fields that result from charged particles contain electromagnetic energy, which propagates through space as electromagnetic radiation such as visible light (wavelength λ in the range of 400–750 nm). The configuration of electrons and protons can store electromagnetic energy, which can be transmitted through electric circuits. Electromagnetic waves transmit solar energy to earth and become the source of all terrestrial energy sources including fossil fuels, biomass, wind, and waves. Quantum mechanics estimates the electromagnetic energy radiated when a system makes a transition from one quantum energy level to another by Planck's equation $E = h\nu$ where h is the Planck constant ($h = 4.1357 \times 10^{-15}$ eV s or 6.626176×10^{-34} J s) and ν is the frequency. This relation is fundamental to quantum physics. Thermodynamics relates energy and temperature through Boltzmann constant k_B (8.6173×10^{-5} eV/K) [4]

$$E = k_B T = \frac{hc}{\lambda} \tag{3.25}$$

where c is the speed of light. This equation relates the electromagnetic radiation to quantum spectra and thermal processes and defines a thermal wavelength

$$\lambda_{th} = \frac{hc}{k_B T} \tag{3.26}$$

Thermal wavelength sets the scale for the wavelength of thermal radiation from a source at temperature T at which a system has a characteristic magnitude energy of $k_B T$.

Mechanical and electrical energy can be efficiently transformed to one another. However, it is highly inefficient to transform thermal or chemical energy into electromagnetic energy. Storage of electromagnetic energy is challenging yet very important [5]. There is no thermodynamic limit for conversions between mechanical and electromagnetic energy. Solar energy can be converted directly into electromagnetic energy by photovoltaic cells. A 100 W light emitting diode (LED) bulb draws only 16 W from the electric grid and radiates visible light at around 2.6 W [4].

Excitation of some medium can produce waves that can propagate energy over distances without transporting matter. Oscillation periodic in both space and time is the simplest form of wave, such as sine wave $f(x, t) = A\sin(kx - \omega t - \phi)$ where A is the amplitude, $k = 2\pi/1$ is the wave number, λ is the wavelength, $\omega = 2\pi\nu$ is the angular frequency, ν is the frequency, and ϕ is the phase shift. The speed of propagation v is related to λ and ν by

$$v = \lambda\nu = \frac{\omega}{k} \tag{3.27}$$

Excitations at a point may be coupled locally to other close points and waves occur and propagate through the medium in time and space. The energy density in a wave depends on its amplitude quadratically. The one-dimensional wave equation becomes

$$\frac{\partial^2 f(x,t)}{\partial t^2} = v^2 \frac{\partial^2 f(x,t)}{\partial x^2} \tag{3.28}$$

as the wave is a function of $(x - vt)$. The wave equation governs propagation of electromagnetic waves and other typical waves.

A current I and a voltage difference $V(t)$ consumes power $P(t) = IV(t)$. Electromagnetic energy is converted into thermal energy in a resistor R: $P(t) = V^2(t)/R$ (Joule's law and the dissipation of power is called Joule heating) where $V = IR$ (Ohm's law) [4].

3.8.1 Magnetic Force

Moving charges create magnetic fields, which in turn exert a force on moving charged particles. A magnetic field is described by vector $\mathbf{B}(\mathbf{x},t)$. The magnetic force acting on a moving charged (q) particle with velocity v in magnetic field $\mathbf{B}$ becomes

$$F = qv \times \mathbf{B} \tag{3.29}$$

Magnetic force on moving charges may describe the principles of motors and generators. In a motor, an external magnetic field exerts force on a loop carrying current producing a torque on the loop driving a rotational motion. In generator moving the loop in the external magnetic field produces forces that drive an external current. Therefore, generators transform mechanical to electric power [4].

3.9 Various Forms of Work

There may be following forms of work [2]:

- Mechanical work
- Boundary work
- Fluid flow work
- Isentropic process work
- Polytropic process work
- Shaft work
- Spring work
- Stirrer work
- Acceleration work
- Gravitational work
- Electrical work

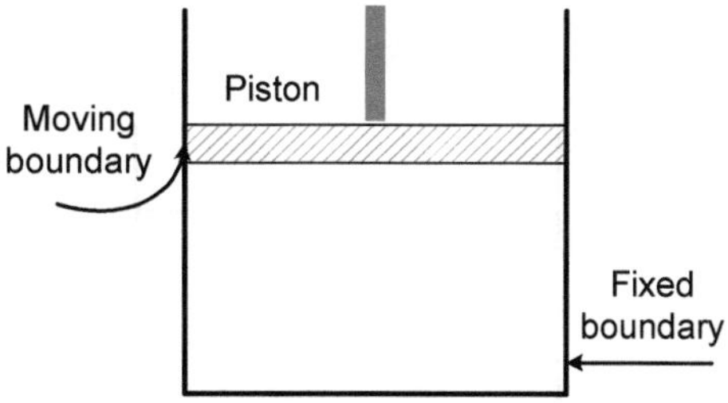

Fig. 3.5 Mechanical work for a mechanical force moving the boundary of a system

3.9.1 Mechanical Work

When a force acts on a system through a distance, then energy is transferred as work. Mechanical work is performed when a mechanical force moves the boundary of a system. For example, in compression work, a piston representing surroundings performs work on the gas (see Fig. 3.5) so that the initial boundary changes and final volume of the gas is less than the initial volume of the gas. The net amount of mechanical work done over a period is

$$W = -\int_{\text{state1}}^{\text{state2}} \mathbf{F} \cdot \mathrm{d}l \tag{3.30}$$

where **F** is an external force vector acting on the system boundary, and **l** is the displacement or pathway vector. Some common types of work are summarized within the next sections.

3.9.2 Boundary Work

The most common form of mechanical work is boundary work. Boundary work is a mechanism for energy interaction between a system and its surroundings. *Boundary work at constant pressure* is the total boundary work done during the entire constant-pressure process is obtained by adding all the differential works from initial state to final state

$$W = -\int_{\text{state1}}^{\text{state2}} P\mathrm{d}V \tag{3.31}$$

For example, during an expansion process, the boundary work represents the energy transferred from the system to its surroundings. Therefore, for the sign convention adapted here, the expansion work is negative. For a compression process, the boundary work represents the energy transferred from surroundings to the system as seen in Figs. 3.5 and 3.6. Therefore, the compression work is positive according to the sign convention. If the boundary is fixed between states 1 and 2,

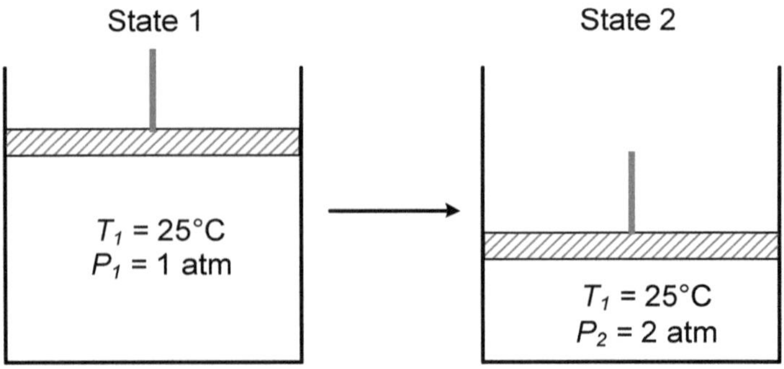

Fig. 3.6 Example of boundary work: compression at constant temperature

there would be no work performed on the system. For example, if a gas in a fixed volume container is heated and its initial temperature changed, there would be no work performed associated with the gas because the boundary of the system has not changed. Work calculated by Eq. (3.31) may be of any value depending on the pathway taken between states 1 and 2. This means that work is a path-dependent function [1, 3].

At constant pressure, Eq. (3.31) becomes

$$W = -P(V_2 - V_1) \tag{3.32}$$

For an ideal gas, *boundary work* at constant temperature for a unit mass is formulated by using

$PV = RT = \text{constant} = C$ and $P = C/V$, which may be substituted to Eq. (3.32), and the work is

$$W = -\int_{\text{state1}}^{\text{state2}} P\mathrm{d}V = -\int_{\text{state1}}^{\text{state2}} \frac{C}{V}\mathrm{d}V = -C\ln\frac{V_2}{V_1} = -P_1V_1\ln\frac{V_2}{V_1} \tag{3.33}$$

The above equation may be modified by using: $P_1V_1 = P_2V_2$ as well as $V_2/V_1 = P_1/P_2$. For example, if a piston cylinder device contains 0.5 m^3 of air at 120 kPa and 100°C, and the air is compressed to 0.2 m^3 at constant temperature, then the work done would be

$$W = -P_1V_1\ln\frac{V_2}{V_1} = -(120\ \text{kPa})\left(0.5\ \text{m}^3\right)\ln\left(\frac{0.2}{0.5}\right)\left(\frac{\text{kJ}}{\text{kPa m}^3}\right) = 54.97\ \text{kJ}$$

As the work done by the surrounding on the system, the work is positive according to the sign convention adopted in this book.

By using $PV = nRT$ in Eq. 3.31, the isothermal boundary work is

$$W = -\int_1^2 P\mathrm{d}V = -\int_1^2 \left(\frac{nRT}{V}\right)\mathrm{d}V = -nRT\ln\left(\frac{V_2}{V_1}\right) = -nRT\ln\left(\frac{P_1}{P_2}\right). \quad (3.34)$$

Example 3.9 **Expansion and compression work of an ideal gas**

Estimate the work during: (a) an expansion of a gas from 1 to 7 m^3 at constant pressure of 120 kPa, (b) a compression of a gas when pressure changes from 100 to 80 kPa and volume changes from 7 to 5 m^3.

Solution:

Assume that the work done by, or on, an ideal gas is computed by taking the product of the pressure and the change in volume.

Data: $P = 120$ kPa, $V_1 = 1\ \mathrm{m}^3$, $V_2 = 7\ \mathrm{m}^3$

(a) Expansion work:

$$W = -P(V_2 - V_1) = -720\ \mathrm{kJ}$$

Notice that the work is negative and done by gas.

(b) When pressure and volume changes at the same time, we calculate the work underneath the P versus V diagram with two parts: a rectangle and a triangle:

$$P_1 = 100\,\mathrm{kPa},\ P_2 = 80\,\mathrm{kPa},\ V_1 = 7\,\mathrm{m}^3,\ V_2 = 5\,\mathrm{m}^3$$

The "area" of the rectangular region: $W_a = -P_2(V_2 - V_1) = 160$ kJ.

The "area" of the triangular region: $W_b = (1/2)(P_1 - P_2)(V_2 - V_1) = 20$ kJ.
Total work done on the gas $= (160 + 20)$ kJ $= 180$ kJ.

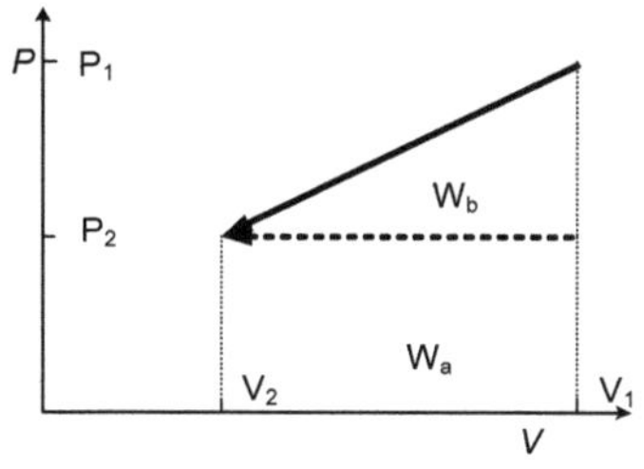

Example 3.10 **Isothermal compression work**

Estimate the boundary work when $V_1 = 0.5\ \text{m}^3$ of carbon dioxide at $P_1 = 110$ kPa and $T_1 = 27°\text{C}$ is compressed isothermally to $P_2 = 750$ kPa.

Solution:

Assume that carbon dioxide is the ideal gas.

Data: $V_1 = 0.5\ \text{m}^3$, $P_1 = 110$ kPa, $T_1 = 27°\text{C}$, $P_2 = 750$ kPa.

Isothermal system: $\Delta H = \Delta U = 0$.

Ideal gas: $PV = nRT$ and $P = \frac{nRT}{V}$

$$W = -\int_1^2 P\text{d}V = -\int_1^2 \left(\frac{nRT}{V}\right)\text{d}V = -P_1V_1 \ln\left(\frac{V_2}{V_1}\right)$$

$$W = -P_1V_1 \ln\frac{V_2}{V_1} = -P_1V_1 \ln\frac{P_1}{P_2} = -(110\ \text{kPa})(0.5\ \text{m}^3)\ln\left(\frac{110}{750}\right)\left(\frac{\text{kJ}}{\text{kPa m}^3}\right)$$
$$= 105.6\ \text{kJ}$$

The positive sign indicates that the work is done on the system.

3.9.3 Fluid Flow Work

Open systems involve mass flows across their boundaries. The work required to maintain a continuous flow into or out of the control volume is called the fluid flow work. It is displacement work done at the moving system boundary. If the fluid pressure is P and the cross-sectional area is A as seen in Fig. 3.7, the force applied on the fluid element becomes

$$F = PA \tag{3.35}$$

When the force acts through a distance L, the fluid flow work becomes

$$W_\text{f} = FL = PAL = PV \tag{3.36}$$

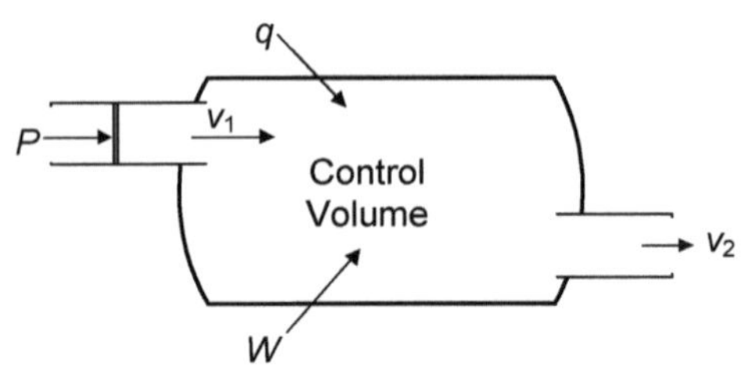

Fig. 3.7 Fluid flow with heat and work interactions between the control volume and surroundings at constant pressure

In most practical processes, flow can be approximated as steady state and one-dimensional, then the total energy of a fluid, E_{fluid}, entering or leaving a system at steady state is

$$\dot{E}_{\text{fluid}} = \dot{m}\left(\Delta U + PV + \frac{\Delta v^2}{2} + g\Delta z\right) \tag{3.37}$$

The dot over a symbol indicates a quantity per unit time. After the combination of $U + PV$ as enthalpy H ($\Delta H = \Delta U + PV$), Eq. (3.37) becomes

$$\dot{E}_{\text{fluid}} = \dot{m}\left(\Delta H + \frac{\Delta v^2}{2} + g\Delta z\right) \tag{3.38}$$

For fluid flow systems, enthalpy rather than the internal energy is important in engineering since the fluid flow work is considered within the enthalpy [2, 3]. The general relation between the heat and work for a fluid flow is

$$\dot{m}\left(\Delta H + \frac{\Delta v^2}{2} + g\Delta z\right) = \dot{q} + \dot{W}_s \tag{3.39}$$

3.9.4 Isentropic Process Work

Isentropic process work occurs at isentropic entropy: $S_1 = S_2$. For isentropic process of ideal gas with constant C_v and C_p, we have the following relations:

$$S_2 - S_1 = C_p \ln\left(\frac{T_2}{T_1}\right) - R\ln\left(\frac{P_2}{P_1}\right) = 0 \tag{3.40}$$

Using the following relationships from the ideal gas equation:

$$\frac{T_2}{T_1} = \frac{P_2V_2}{P_1V_1} \text{ and } C_p - C_v = R \text{ or } R/C_p = (\gamma - 1)/\gamma \tag{3.41}$$

in Eq. (3.41) yields

$$P_1V_1^\gamma = P_2V_2^\gamma \text{ and } \frac{T_2}{T_1} = \left(\frac{P_2}{P_1}\right)^{\frac{\gamma-1}{\gamma}} \tag{3.42}$$

where γ is the ratio of heat capacities $(\gamma = C_p/C_v)$. For an ideal gas, the heat capacity depends only on temperature. The values of R at various units are:

$R = 10.73$ psia ft^3/lbmole R $= 0.7302$ atm ft^3/lbmole R $= 83.14$ bar cm^3/mole K.

$R = 1.987$ cal/mole K $= 1.986$ Btu/lbmol R.

$R = 8.314$ J/mole K $= 8.314$ Pa m^3/mole K $= 8314$ kPa cm^3/mole K.

For the isentropic process (reversible, adiabatic); the heat transfer is zero, and using the ideal gas relations, the work is estimated by

$$\dot{W}_s = \dot{m}C_v(T_2 - T_1) = \dot{m}(U_2 - U_1) \tag{3.43}$$

Example 3.11 **Isentropic compression of air**

A tank filled with air at 1 bar and 20°C is compressed adiabatically (isentropic) to 4 bars. Estimate the compression work in kJ/mole. Constant heat capacities for the air are $C_{v,av} = 20.8$ J/mole K and $C_{p,av} = 29.1$ J/mole K.

Solution:

Assume that the air is ideal gas.

Basis: 1 mol of ideal gas.

Data: $T_1 = 20$°C (293 K); $P_1 = 1$ bar; $P_2 = 4$ bar; $R = 8.314$ J/mole K.

$$C_{v,av} = 20.8\,\text{J/mole K and } C_{p,av} = 29.1\ \text{J/mole K},\ \gamma = C_{p,av}/C_{v,av} = 1.4$$

For adiabatic expansion: $T_2 = T_1\left(\frac{P_2}{P_1}\right)^{\frac{\gamma-1}{\gamma}} = 435$ K

$$W = C_{v,av}(T_2 - T_1) = 2953.9\,\text{J/mole}$$

The positive sign of the work shows that the work is done on air.

3.9.5 Polytropic Process Work

In *polytropic process work*, pressure and volume are related by

$$P_1V_1^\gamma = P_2V_2^\gamma = C \tag{3.44}$$

and the pressure is

$$P = CV^{-\gamma} \tag{3.45}$$

where γ and C are constants. By using this newly derived pressure we obtain the polytropic work by [1]

$$W = -\int_{\text{state1}}^{\text{state2}} PdV = -\int_{\text{state1}}^{\text{state2}} CV^{-\gamma}dV = -C\frac{V_2^{-\gamma+1} - V_1^{-\gamma+1}}{1-\gamma} = -\frac{P_2V_2 - P_1V_1}{1-\gamma} \tag{3.46}$$

Example 3.12 **Calculation of work done by a piston on an ideal gas**

A piston does compression work on an ideal gas. The ideal gas is initially at 400 K and 300 kPa. The volume of the gas is compressed to 0.1 m^3 from 0.3 m^3. Calculate the work done by the piston on the gas by two different paths:

Path A: Isobaric compression at constant pressure $P = 300$ kPa.

Path B: Isothermal compression at constant temperature $T = 400$ K.

Solution:

Assume that the piston is frictionless, compression occurs very slowly, the ideal gas is the system.

From ideal gas law: $PV = nRT$, where $R = 8.314$ kPa m^3/kmole K.

Moles of the gas:

$$n = PV/(RT) = (300\,\text{kPa})\,(0.3\,\text{m}^3)/\left[(8.314\,\text{kPa m}^3/\text{kmole K})(400\,\text{K}))\right] \\ = 0.027\,\text{kmole}$$

Path A: Isobaric compression at $P = 300$ kPa:

$$(V_2 - V_1) = (0.1 - 0.3)\ \text{m}^3 = -0.2\ \text{m}^3;$$

$$W = -\int_{V_1}^{V_2} P\text{d}V = -P(V_2 - V_1) = -300 \times 10^3\ \text{Pa}\frac{N}{m^2\ \text{Pa}}\left(-0.2\ \text{m}^3\right)\frac{J}{\text{N m}} = 60\,\text{kJ}$$

Path B: Isothermal compression at $T = 400$ K:

$$W = -\int_{V_1}^{V_2} P\text{d}V = -\int_{V_1}^{V_2} \frac{nRT}{V}\text{d}V = -nRT\ln\left(\frac{V_2}{V_1}\right)$$

$$= -\ (0.027\,\text{kmole})\left(\frac{8.314\ \text{kJ}}{\text{kmole K}}\right)(400\,\text{K})\ln\left(\frac{0.1\ \text{m}^3}{0.3\ \text{m}^3}\right) = 98.65\,\text{kJ}$$

As the work is path-dependent variable, the work done by the following different paths is different from each other.

Example 3.13 **Polytropic expansion of air**

Air in a cylinder is at 110 kPa and expands polytropically from 0.01 to 0.05 m^3. The exponent $\gamma = 1.3$ in the pressure volume relation (PV^γ = constant) in polytropic process. Estimate the work done during the process.

Solution:

Assume that the air as ideal gas. $PV = RT$ (for one mole of gas).

Data: $V_1 = 0.01\ \text{m}^3$, $V_2 = 0.05\ \text{m}^3$, $P_1 = 110\ \text{kPa}$, and $\gamma = 1.3$

$$P_1 V_1^{\gamma} = P_2 V_2^{\gamma} = C = \text{constant}$$

$$P_2 = P_1 \frac{V_1^{\gamma}}{V_2^{\gamma}} = (110\,\text{kPa}) \left(\frac{0.01}{0.05}\right)^{1.3} = 13.6\,\text{kPa}$$

$$W_\text{b} = \int_1^2 CV^{-\gamma}\text{d}V = \frac{R\Delta T}{\gamma - 1} = \frac{P_2V_2 - P_1V_1}{\gamma - 1}\left(\frac{\text{kJ}}{\text{kPa}\,\text{m}^3}\right) = -1.4\,\text{kJ}$$

The negative sign indicates the work is done on the surroundings.

3.9.6 Shaft Work

Shaft work is the energy transmitted by a rotating shaft and is related to the torque **T** applied to the shaft and the number of revolutions of the shaft per unit time. Shaft work occurs by force acting on a shaft to turn the shaft against a mechanical resistance. A shaft work performed by a pump transferring a liquid body over a distance is positive. On the other hand, the shaft work is negative if a fluid in the system turns a shaft that performs work on the surroundings [4].

A force F acting through a moment arm r produces a torque **T**.

$$\mathbf{T} = Fr \text{ and } F = \frac{\mathbf{T}}{r} \tag{3.47}$$

This force acts through *$2\pi rN$, where N is the number of revolutions. The work done during N revolution becomes

$$W_s = F(2\pi rN) = \frac{\mathbf{T}}{r}2\pi rN = 2\pi N\mathbf{T} \tag{3.48}$$

A rotating shaft can transmit power. The power transmitted through the shaft is

$$\dot{W}_s = 2\pi \dot{N}\mathbf{T} \tag{3.49}$$

Example 3.14 **Estimation of shaft power**

Estimate the power transmitted through the shaft of a car when the torque applied is 100 N m. The shaft rotates at a rate of 2000 revolution per minute.

Solution:

$$\dot{N} = 2000/\text{min}, \boldsymbol{T} = 100\,\text{Nm}$$

$$\dot{W}_s = 2\pi\dot{N}\boldsymbol{T} = 2\pi\left(2000\frac{1}{\text{min}}\left(\frac{\text{min}}{60\,\text{s}}\right)100\,\text{Nm}\left(\frac{\text{kJ}}{1000\,\text{Nm}}\right)\right) = 20.9\,\text{kW} = 28.1\,\text{hp}$$

3.9.7 Spring Work

The length of a spring changes when a force is applied. For a change of length of dx after applying a force **F**, the work done becomes

$$\delta W_{sp} = \mathbf{F}\text{d}x \tag{3.50}$$

For linear elastic springs, the change in length is proportional to the force applied: $F = kx$, where k is the spring constant with unit kN/m. Therefore, for an initial x_1 and final x_2 displacements the work done becomes

$$W_{sp} = \frac{1}{2}k\left(x_2^2 - x_1^2\right)\ (\text{kJ}) \tag{3.51}$$

Example 3.15 **Estimation of spring work**

A piston cylinder system is attached to a spring with a spring constant of 110 kN/m. At state 1 there is no displacement in the spring. At state 2, the piston cylinder system performs compression work and due to the decrease in volume underneath the piston, spring changes its length by 0.12 m. Estimate the spring work.

Solution:

$k = 110$ kN/m, $x_1 = 0$, $x_2 = 0.12$ m

$$W_{sp} = \frac{1}{2}k\left(x_2^2 - x_1^2\right) = 792\,\text{J} = 0.792\,\text{kJ}$$

3.9.8 Stirrer Work

When a shaft induces stirring of the fluid, there occurs a work transfer to the fluid, which is system in this case, as the load, W_{load}. The work transfer W_{sw} is

$$W_{sw} = \int_1^2 mg\text{d}z = \int_1^2 W_{\text{load}}\text{d}z = \int_1^2 \mathbf{T}\text{d}\theta \tag{3.52}$$

where dz is the vertical drop of the load W_{load} when the induced torque **T** has rotated the shaft through an angular displacement by $d\theta$.

3.9.9 Acceleration Work

Acceleration work occurs due to change of velocity of a system under the action of an accelerating force, F_α, which is defined by Newton's Second Law of Motion as F_α = ma, where a is the acceleration $a = \mathrm{d}v/\mathrm{d}t$ and $\mathrm{d}l = v\mathrm{d}t$; hence the acceleration work is

$$W_a = \int_1^2 F_a \mathrm{d}l = \int_1^2 m(\mathrm{d}v/\mathrm{d}t)(v\mathrm{d}t) = m\int_1^2 v\mathrm{d}v = (1/2)m\left(v_2^2 - v_1^2\right) \quad (3.53)$$

where v_1 and v_2 represent the initial and final velocities. The acceleration work represents the change is kinetic energy.

3.9.10 Gravitational Work

Gravitational work is done against gravity. This is simply the change in potential energy. If F_g is the gravitational force acting on the body of mass *m*, then $F_g = mg$, where g is acceleration due to gravity = 9.81 m s^{-2}. Then the gravitational work done to raise a mass m from elevation z_1 to z_2 is

$$W_g = \int_1^2 F_g \mathrm{d}z = mg\int_1^2 \mathrm{d}z = mg(z_2 - z_1) \quad (3.54)$$

3.9.11 Electrical Work

When an electrical current flows through a resistor, it implies that there is work transfer to the system (resistor). For example, the current that flows can drive a motor and hence mechanical work is performed. *Electrical work* occurs when an electrical current passes through an electrical resistance in the circuit. Electrical work in the rate form is the same as power

$$\dot{W}_e = VI = I^2R \quad (3.55)$$

where V is the potential difference, I is the electric current, and R is the resistance.

Example 3.16 **Estimation of electrical work**

A resistance heater within a well-insulated tank is used to heat 12 kg of water. The heater passes a current of 0.6 ampere (A) for 4 h from a 120-V source. Estimate the electric energy used after 4 h of heating.

Solution:

$$V = 120\,\mathrm{V}, I = 0.6\,\mathrm{A}, t = 4\,\mathrm{h}$$

Electrical energy received by the water:

$$W_e = VIt = 120V(0.6A)(4\ \mathrm{h})(3600\ \mathrm{s/h})\left(\frac{\mathrm{kJ/s}}{1000VA}\right) = 1036.8\ \mathrm{kJ}$$

3.10 Other Forms of Work

Nonmechanical forms of work are formulated in terms of generalized force, **F**, and generalized displacement, x

$$W_{\mathrm{other}} = \mathbf{F}x \tag{3.56}$$

- *Magnetic work*: (magnetic field strength) (total magnetic dipole moment).
- *Electric polarization work*: (electric field strength) (polarization of the medium).

Polarization of the medium is the sum of the electric dipole rotation moments of the molecules.

Problems

3.1 Determine the SI units and English units of kinetic and potential energy.
3.2 A one $\mathrm{lb_m}$ glass cup is at rest initially, and dropped onto a solid surface: (a) What is the kinetic energy change, (b) What is the sign of potential energy change? (c) What is the work?
3.3 A 50-hp electric motor is used to pump ground water into a storage tank. Estimate the work done by the pump in Btu/h, kW, and J/s.
3.4 A water pumping process consumes 200 hp. Estimate the pump work in Btu/h and in kW.
3.5 A 25-hP electric motor is used to pump ground water into a storage tank. Estimate the work done by the pump in Btu/h, MW, and J/s.
3.6 A 85-hP electric motor is used to pump ground water into a storage tank. Estimate the work done by the pump in Btu/h, kW, and kcal/s.
3.7 A 40-kW electric motor is used to pump ground water into a storage tank. Estimate the work done by the pump in Btu/h, hp, J/s, and ft lbf/h.
3.8 A car having a mass of 1500 kg is initially traveling with a speed of 80 km/h. It slows down at a constant rate, coming to a stop in a distance of 90 m.
   (a) What is the change in the car's kinetic energy over the 90 m distance it travels while coming to a stop?
   (b) What is the net force on the car while it's coming to a stop?

3.9 A truck weighing 5000 lb is traveling at 55 miles/h. The brakes are suddenly applied to stop the car completely. Estimate the energy transfer through the brakes.

3.10 A car having a mass of 1000 kg is initially at rest. A constant 1,000 N net force acts on the car over 50 m, causing the car to speed up. (a) Estimate the total work done on the car over the 50 m distance it travels, (b) Estimate the speed of the car after 50 m.

3.11 Calculate the kinetic energy (KE) of a 1500-kg automobile with a speed of 30 m/s. If it accelerates to this speed in 20 s, what average power has been developed?

3.12 A load with mass of 200 kg is pushed a horizontal distance of 20.0 m by a force of 60.0 N. For a frictionless surface, estimate the kinetic energy of the load when the work is finished, (b) estimate the speed of the load.

3.13 Estimate the power needed to accelerate a 1400-kg car from rest to a speed of 60 km/h in 10 s on a level road.

3.14 Estimate the power needed to accelerate a 2000-kg car from rest to a speed of 70 km/h in 5 s on a level road.

3.15 A car having a mass of 1300 kg is traveling at 55 miles/h. Estimate: (a) the kinetic energy of the car in kJ, (b) the work necessary to stop the car.

3.16 A car having a mass of 1450 kg is traveling at 65 miles/h. Estimate: (a) the kinetic energy of the car in kWh, (b) the work necessary to stop the car.

3.17 A car having a mass of 1800 kg is traveling at 40 miles/h. Estimate: (a) the kinetic energy of the car in Btu, (b) the work necessary to stop the car.

3.18 A car having a mass of 2250 kg is traveling at 65 miles/h. Estimate (a) the kinetic energy of the car in kJ, (b) the work necessary to stop the car.

3.19 A car weighing 2700 lb is traveling at 60 miles/h. Estimate (a) the kinetic energy of the car in Btu, (b) the work necessary to stop the car.

3.20 A car having a mass of 3000 lb is traveling at 50 miles/h. Estimate: (a) the kinetic energy of the car in kJ, (b) the work necessary to stop the car.

3.21 A car having a mass of 4200 lb is traveling at 70 miles/h. Estimate: (a) the kinetic energy of the car in ft lbf, (b) the work necessary to stop the car.

3.22 A box has a mass of 6 kg. The box is lifted from the garage floor and placed on a shelf. If the box gains 200 J of potential energy estimate the height of the shelf.

3.23 A man climbs on to a wall that is 3.4 m high and gains 2400 J of potential energy. Estimate the mass of the man.

3.24 A 1 kg ball is dropped from the top of a cliff and falls with a constant acceleration due to gravity (9.8 m/s^2). Assume that effects of air resistance can be ignored. (a) Estimate the ball's gravitational potential energy change, (b) How much work has been done on the ball?

3.25 Estimate the potential energy of 1000 kg of water at the surface of a lake that is 50 m above a water turbine, which is used to generate electricity.

3.26 Consider a 1400-kg car traveling at 55 km/h. If the car has to climb a hill with a slope of 20° from the horizontal road, estimate the additional power needed for the speed of the car to remain the same.

3.27 Consider a 1200-kg car traveling at 25 km/h. If the car has to climb a hill with a slope of 30° from the horizontal road, estimate the additional power needed for the speed of the car to remain the same.

3.28 Estimate the potential energy of 10 kg of water at the surface of a lake that is 50 m above a water turbine, which is used to generate electricity.

3.29 Estimate the potential energy of 350 lb of water at the surface of a lake that is 200 ft above a water turbine, which is used to generate electricity.

3.30 Water is flowing over a waterfall 80 m in height. Assume 1 kg of the water as the system. Also assume that the system does not exchange energy with its surroundings. (a) What is the potential energy of the water at the top of the falls with respect to the base of the falls? (b) What is the kinetic energy of the water just before it strikes bottom?

3.31 Estimate the potential energy of 1000 lb of water at the surface of a lake that is 100 ft above a water turbine, which is used to generate electricity.

3.32 An elevator with a mass of 3000 kg rests at a level of 3 m above the base of an elevator shaft. The elevator has traveled to 30 m above the base. The elevator falls from this height freely to the base and strikes a strong spring. Assume that the entire process is frictionless. Estimate:

(a) The potential energy of the elevator in its original position relative to the base of the shaft.
(b) The work done traveling the elevator.
(c) The potential energy of the elevator in its highest position relative to the base of the shaft.
(d) The velocity and kinetic energy of the elevator just before it strikes the spring.

3.33 An elevator with a mass of 2000 kg rests at a level of 2 m above the base of an elevator shaft. The elevator has traveled to 20 m above the base. The elevator falls from this height freely to the base and strikes a strong spring. Assume that the entire process is frictionless. Estimate:

(a) The potential energy of the elevator in its original position relative to the base of the shaft.
(b) The work done traveling the elevator.
(c) The potential energy of the elevator in its highest position relative to the base of the shaft.
(d) The velocity and kinetic energy of the elevator just before it strikes the spring.

3.34 An elevator with a mass of 4000 lb rests at a level of 5 ft above the base of an elevator shaft. The elevator has traveled to 100 ft above the base. The elevator falls from this height freely to the base and strikes a strong spring. Assume that the entire process is frictionless. Estimate:

(a) The potential energy of the elevator in its original position relative to the base of the shaft.
(b) The work done traveling the elevator.
(c) The potential energy of the elevator in its highest position relative to the base of the shaft.

(d) The velocity and kinetic energy of the elevator just before it strikes the spring.

3.35 An elevator with a mass of 5000 lb rests at a level of 10 ft above the base of an elevator shaft. The elevator has traveled to 150 ft above the base. The elevator falls from this height freely to the base and strikes a strong spring. Assume that the entire process is frictionless. Estimate:
(a) The potential energy of the elevator in its original position relative to the base of the shaft.
(b) The work done traveling the elevator.
(c) The potential energy of the elevator in its highest position relative to the base of the shaft.
(d) The velocity and kinetic energy of the elevator just before it strikes the spring.

3.36 Water is pumped from an open tank at level zero to an open tank at level 14 ft. The pump adds 7.5 hp to the water when pumping a volumetric flow rate of 3.6 ft^3/s. Estimate the loss energy in head.

3.37 Water is pumped from an open tank at level zero to an open tank at a level of 24 ft. The pump adds 16 hp to the water when pumping a volumetric flow rate of 5 ft^3/s. Estimate the loss energy in head.

3.38 Estimate the head equivalent of a pressure difference of 12 psi (lbf/in^2) in ft of water and in ft of mercury.

3.39 Estimate the head equivalent of a pressure difference of 20 psi (lbf/in^2) in ft of water and in ft of mercury.

3.40 Estimate the compression work done by a gas during a process with the following recorded data obtained that relates pressure and volume:

P kPa	100	200	300	400	500
V L	1.1	0.8	0.55	0.4	0.3

3.41 A saturated steam is expanded inside a cylinder. The steam is initially at 151.8°C and 500 kPa. The final pressure is 200 kPa. Determine the work produced by the steam.

P kPa	500	450	400	350	300	250	200
V cm^3	374	413	462	524	605	718	885

3.42 Find the work during (a) an expansion of a gas from 1 to 8 m^3 at constant pressure of 120 kPa, (b) a compression of a gas when pressure changes from 140 to 80 kPa and volume changes from 7 to 2 m^3.

3.43 Calculate the boundary work when $V_1 = 1.5\ m^3$ of carbon dioxide at $P_1 = 110$ kPa and $T_1 = 27°C$ is compressed isothermally to $P_2 = 750$ kPa.

3.44 A tank filled with an ideal gas at 200°C and 100 kPa is compressed isothermally to 500 kPa. Estimate the compression work in kJ/mol.

3.45 Air undergoes a constant temperature compression from 100 kPa and 100°C to 200 kPa. Find the net work done for 4 kg of air.

3.46 A tank filled with air at 6 bar and 20°C is expanded isothermally to 2 bar. Estimate the expansion work in kJ/mol.

3.47 A tank filled with 10 mol of air at 2 bar and 20°C is compressed adiabatically (isentropic) to 8 bar. Estimate the compression work in kJ/mole. Constant heat capacities for the air are $C_{v,av} = 20.8$ J/mole K and $C_{p,av} = 29.1$ J/mole K.

3.48 A piston cylinder device contains 2 kg steam at 2 MPa and 450°C. (a) If the steam is cooled to 225°C at constant pressure, estimate the compression work the system performs, (b) if the volume of the cylinder at the final state has the saturated vapor, estimate the work and the temperature at the final state.

3.49 Saturated ammonia vapor enters a compressor at − 10°C and leaves at 76.2°C. Determine the work requirement per 0.4 kg of ammonia for the compressor if the process is adiabatic and reversible. Constant heat capacity is $C_{p,av} = 2.64$ kJ/kg K.

3.50 Air in a cylinder is at 110 kPa and undergoes polytropic expansion from 0.02 to 0.1 m^3. The exponent $\gamma = 1.3$ in the pressure volume relation (PV^{γ} = constant) in polytropic process. Estimate the work done during the process.

3.51 1 kg of air in a cylinder is at 100 kPa and 30°C. After a polytropic compression the temperature of air reaches 62°C. The constant $\gamma = 1.25$ in the pressure volume relation (PV^{γ} = constant) in polytropic process. Estimate the work done during the process.

3.52 Calculate the boundary work when 1.0 m^3 of air at 25°C is compressed to a final volume of 0.8 m^3 of air at a constant temperature process.

3.53 Calculate the boundary work when 60 ft^3 of nitrogen gas at 80 psia and 150°F is expanded to 40 psia at a constant temperature process.

3.54 Estimate the boundary work when 500 ft^3 of hydrogen gas at 30 psia and 150°F is compressed to 150 psia at a constant temperature process.

3.55 Superheated vapor at 45 psia and 500°F in a piston cylinder system is cooled at constant pressure until 80% of it condenses. Calculate the boundary work per pound of steam.

3.56 Calculate the boundary work when 150 ft^3 of carbon dioxide at 20 psia and 100°F is compressed to 70 psia at a constant temperature process.

3.57 57.A piston cylinder device has a 10 kg of saturated vapor at 350 kPa. The vapor is heated to 150°C at constant pressure. Estimate the work done by the steam.

3.58 Superheated vapor at 800 kPa and 200°C in a piston cylinder system is cooled at constant pressure until 70% of it condenses. Calculate the boundary work per kg of steam.

3.59 A piston cylinder device contains 0.5 kg of air at 1000 kPa and 400 K. The air undergoes an isothermal expansion to 300 kPa. Estimate the boundary work for this expansion process.

3.60 A piston cylinder device contains 10 lb of air at 20 psia and 500 R. The air undergoes a polytropic compression to 100 psia. Estimate the boundary work for the compression process.

3.61 A piston cylinder device contains 10 lb of air at 100 psia and 500 R. The air undergoes an isothermal expansion to 20 psia. Estimate the boundary work for the expansion process.

3.62 A piston cylinder device contains 0.5 kg steam at 1 MPa and 400°C. (a) If the steam is cooled to 250°C at constant pressure, estimate the compression work the system performs, (b) if the volume of the cylinder at the final state is 50% of the initial volume and the pressure in the final state is 0.5 MPa, estimate the work and the temperature at the final state.

3.63 A piston cylinder device has a volume of 0.5 m^3 and contains carbon dioxide (CO_2) at 140 kPa and 100°C. At the final state pressure is 90 kPa. Estimate the isentropic expansion work and value of temperature at the final state.

3.64 A piston cylinder device has a volume of 0.5 m^3 and contains carbon dioxide (CO_2) at 140 kPa and 400 K. At the final state pressure is 90 kPa. Estimate the isentropic expansion work and value of temperature at the final state.

3.65 A piston cylinder device has a volume of 1.5 m^3 and contains carbon dioxide (CO_2) at 100 kPa and 300 K. At the final state pressure is 200 kPa. Estimate the isentropic compression work and value of temperature at the final state.

3.66 A car creates a torque of 400 Nm. This torque is transmitted through a shaft which rotates at a rate of 3000 revolution per minute. Estimate the power transmitted through the shaft.

3.67 A car creates a torque of 500 Nm. This torque is transmitted through a shaft which rotates at a rate of 3200 revolution per minute. Estimate the power transmitted through the shaft.

3.68 A 180-hp car creates a rotation rate of 2600 revolution/min. Estimate the torque the car creates at this revolution rate.

3.69 A 210-hp car creates a rotation rate of 3000 revolution/min. Estimate the torque the car creates at this revolution rate.

3.70 Determine the SI system units of (a) electric power; (b) electric charge; (c) electric potential difference, and (d) electric resistance.

3.71 A 2-kW resistance heater is used within a water tank of a shower. If you use the shower during half an hour every day, estimate the cost of electricity per month. The unit cost of electricity is $0.10/kWh.

3.72 A 5-kW resistance heater is used within a room with floor heating systems. If you use the heating system during an average of 8 h every day, estimate the cost of electricity per year. The unit cost of electricity is $0.10/kWh.

3.73 A 120-W laptop is used an average of 12 h every day. Estimate the cost of electricity per year. The unit cost of electricity is $0.15/kWh.

3.74 A resistance heater is used within a well-insulated water tank. The heater passes a current of 1.2 ampere from a 120-V source. Estimate the electric energy after 4 h of heating.

3.75 A resistance heater is used within a room. The heater passes a current of 0.2 ampere from a 120-V source. Estimate the electric energy used after 30 min of heating.

3.76 A resistance heater passes a current of 0.4 ampere (A) from a 120-V source. Estimate the electric energy if the heater is used 10 h.

References

1. Çengel YA, Boles MA (2014) Thermodynamics: an engineering approach, 8th edn. McGraw-Hill, New York
2. Chattopadhyay P (2015) Engineering thermodynamics, 2nd edn. Oxford University Press, Oxford
3. Himmelblau DM, Riggs JB (2012) Basic principles and calculations in chemical engineering, 8th edn. Prentice Hall, Upper Saddle River
4. Jaffe RL, Taylor W (2018) The physics of energy. Cambridge University Press, Cambridge
5. Moran MJ, Shapiro HN, Boettner DD, Bailey MB (2014) Fundamentals of engineering thermodynamics, 8th edn. Wiley, New York

Internal Energy and Enthalpy 4

Introduction and Learning Objectives: Internal energy is associated with the microscopic and collective random motion of particles constituting the media. The energy of the collective random motion of particles is heat. Heat flows across the boundary of a system. The net energy transfer by heat, work, and mass is equal to the sum of the change in internal, kinetic, potential energies. Enthalpy is used in many fluid flow systems. Only changes in internal energy and enthalpy can be calculated. Measurement of heat change by using calorimeters is briefly discussed. Estimations of energy and temperature changes of air using the psychometric chart are briefly discussed. Heat transfer is briefly covered. The concept of exergy in fluid flow systems is discussed briefly.

The learning objectives of this chapter are:

- Internal energy and enthalpy in various closed and open systems
- Sensible and latent heats
- Heat of reaction using the heat of formation
- Heat of combustion
- Psychometric chart
- Heat transfer

4.1 Internal Energy

Internal energy U of a system or a body with well-defined boundaries is the total of the kinetic energy due to the motion of molecules (translational, rotational, and vibrational) and the potential energy associated with the vibrational motion and electric energy of atoms within molecules. Internal energy also includes energy in all the chemical bonds [1]. From a microscopic point of view, internal energy

Y. Demirel, *Energy*, Green Energy and Technology,
https://doi.org/10.1007/978-3-031-97111-2_4

may be found in many different forms. For gas, internal energy comprises of translational kinetic energy, rotational kinetic energy, vibrational energy, electronic energy, chemical energy, and nuclear energy. For any material, solid, liquid, or gaseous, it may also consist of the potential energy of attraction or repulsion between the individual molecules. For a gaseous system,

Internal energy is a state function and an extensive quantity. One can have a corresponding intensive thermodynamic property called *specific internal energy*, *u*, which is internal energy per mass of the substance (J/g). Table 4.1 shows the different components of internal energy of a system, while Table 4.2 shows the changes in internal energy. Nonmechanical work, such as external electric field, or adding energy through stirring, can change the internal energy.

For a closed system, the internal energy is

$$\Delta U = \delta q + \delta W \tag{4.1}$$

Table 4.1 Components of internal energy [1, 4]

Thermal energy	Sensible heat	Energy change of a system associated with: (a) Molecular translation, rotation, vibration (b) Electron translation and spin (c) Nuclear spin of molecules
	Latent heat	Energy required or released for phase change; change from liquid to vapor phase requires heat of vaporization
Chemical energy	Energy associated with the chemical bonds in a molecule	
Nuclear energy	The large amount of energy associated with the bonds within the nucleus of the atom	

Table 4.2 Physical and chemical processes that can change the internal energy of a system

Transferring energy across the system boundary	Heat transfer	Energy transfer from a high temperature to low temperature state
	Work transfer	Energy transfer driven by changes in macroscopic physical properties of a system, such as compression or expansion work
	Mass transfer	Energy transfer by mass flowing across a system boundary
Change through internal processes	Mixing	Heat releases upon mixing components that may lead to lower internal energy
	Chemical reaction	Heat required or released during a chemical reaction that changes chemical energy
	Nuclear reaction	Heat released during a nuclear reaction that changes nuclear energy

where ΔU is the change in internal energy of a system during a process, δq is the heat, and δW is the mechanical work. Both heat and work are not state functions and are path dependent. The heat absorbed by a system increases its internal energy. If an energy exchange occurs because of temperature difference between a system and its surroundings, this energy appears as *heat* otherwise it appears as *work*. When a force acts on a system over a distance, then energy is transferred as work. Equation (4.1) shows that energy is conserved. According to the sign convention adopted here, the value of heat is positive when transferred to the system and negative when it is transferred from the system to its surroundings. Similarly, the value of work is positive when transferred to the system and negative when transferred from the system to its surroundings. Equation (4.1) in infinitesimal terms is

$$dU = \delta q + \delta W \tag{4.2}$$

where the operator *d* before the internal energy function indicates that it is an exact differential as it is a state function. On the other hand, the δ's show the increments of energy which are not state functions; both the heat and work depend on the path taken between the initial and final states of a system.

The change in internal energy in terms of temperature and volume is

$$dU = C_V dT + [T(\beta/\kappa) - P]dV \tag{4.3}$$

where C_V is the heat capacity at constant volume, β is the thermal expansion coefficient, $\beta = \frac{1}{V}\left(\frac{\partial V}{\partial T}\right)_P$, and κ is the isothermal compressibility, $\kappa = -\frac{1}{V}\left(\frac{\partial V}{\partial P}\right)_T$. The value of C_V is the amount of heat required to raise the temperature of one gram of a substance by one degree Celsius at a constant volume process. C_V is a temperature dependent quantity. A gas becomes ideal in the limit as pressure approaches zero, where the interactions between the gas particles will be small enough to neglect. The second term on the right-hand side of Eq. (4.3) may be zero for a constant volume process, for ideal gases, and for incompressible fluids. Whenever internal energy is independent of volume or at constant volume, the change of internal energy is

$$\Delta U = U_2 - U_1 = \int_{T_1}^{T_2} C_V dT \tag{4.4}$$

If the value of the C_V is constant, such as an average value, $C_{V,av}$, within the temperature interval $(T_2 - T_1)$ considered, then integration in Eq. (4.4) yields

$$\Delta U = U_2 - U_1 = C_{V,av}(T_2 - T_1) \tag{4.5}$$

However, if the C_v is expressed by $C_v = A + BT + CT^2$, we have the change of internal energy in terms of temperature estimated by

$$\Delta U = \int_{T_1}^{T_2} C_v \mathrm{d}T = A(T_2 - T_1) + \frac{1}{2}B\left(T_2^2 - T_1^2\right) + \frac{1}{3}C\left(T_2^3 - T_1^3\right) \tag{4.6}$$

Instead of the absolute value of internal energy, the change in internal energy, $\Delta U = U_{\text{final}} - U_{\text{initial}}$, is a more useful in energy estimations. U_{final} is the final value of the internal energy, while U_{initial} is the initial value of the internal energy of the system.

4.2 Enthalpy

Enthalpy H of a system is

$$H = U + PV \tag{4.7}$$

where U is the internal energy, P is the pressure, and V is the volume of the system. The SI unit of enthalpy is the joule with specific enthalpy in J/g. Enthalpy is a state function and an extensive quantity.

Instead of the absolute value of enthalpy, the *change* of a system's enthalpy is commonly used in energy estimations. Enthalpy change is defined by: $\Delta H = H_{\text{final}} - H_{\text{initial}}$, where H_{final} is the final value and H_{initial} is the initial value of the enthalpy of the system. When enthalpy is expressed in terms of T and P, we have the change of enthalpy given by

$$\mathrm{d}H = C_p \mathrm{d}T + (1 - \beta T) V \mathrm{d}P \tag{4.8}$$

where C_p is the heat capacity at constant pressure and β is the thermal expansion coefficient. The value of C_p is the amount of heat required to raise the temperature of one gram of a substance by one degree Celsius at a constant-pressure process. The value of C_p is a temperature dependent quantity. The second term on the right-hand side of Eq. (4.8) may be zero for a constant-pressure process and approximately zero for low-pressure gases. For ideal gases, the enthalpy of a matter is independent of pressure and a function of temperature only. For low-pressure gases and incompressible fluids, the enthalpy of a matter is approximately independent of pressure. Whenever the enthalpy of a matter is independent of pressure, regardless of the process, or at constant pressure, the change of enthalpy is

$$\Delta H = H_2 - H_1 = \int_{T_1}^{T_2} C_p \mathrm{d}T \tag{4.9}$$

If the value of the C_p is constant, such as an average value, $C_{p,av}$, within the temperature interval $(T_2 - T_1)$ considered, then integration in Eq. (4.9) yields

$$\Delta H = H_2 - H_1 = C_{p,av}(T_2 - T_1) \tag{4.10}$$

ΔH of a system is equal to the sum of nonmechanical work done on it and the heat supplied to it. The increase in enthalpy of a system is exactly equal to the energy added through heat, provided that the system is under constant pressure: $\Delta H = q$. Therefore, enthalpy is sometimes described as the "*heat content*" of a system under a given pressure.

Estimation of the enthalpy by the integral in Eq. (4.9) requires an expression for the temperature dependence of heat capacity

$$C_p = A + BT + CT^2 \tag{4.11}$$

where A,B, and C are the parameters. Ideal gas-state heat capacities are functions of temperature only designated by C_p^{ig} and C_v^{ig}. The temperature dependence of an ideal gas may be expressed by

$$C_p^{ig}/R = A + BT + CT^2 \tag{4.12}$$

The ratio C_p/R is dimensionless and the units of C_p are governed by the units of R chosen. In Appendix B, Table B1 lists the parameters used in Eq. (4.12) for various chemical species. The two ideal gas heat capacities, C_p^{ig} and C_v^{ig}, are related by:

$$C_p^{ig} - C_v^{ig} = R \tag{4.13}$$

The departure of real gases from ideal behavior is seldom significant at pressures of several bars, and C_p^{ig} and C_v^{ig} are usually good approximations to their true heat capacities [1].

Figure 4.1 shows the change of enthalpy with pressure and temperature for carbon dioxide. At high pressures, the interactions between the molecules will be considerable and the behavior of carbon dioxide will be nonideal. Figure 4.2 shows the values of enthalpy for water at various temperatures and pressures. The values of enthalpy at as high as 1000 basr can be obtained from these pressure-enthalpy diagrams as they represent the nonideal behavior of carbon dioxide and water. Therefore, they are useful in industrial design of compression and expansion processes [4].

Example 4.1 Unit conversions of heat capacity

The parameters in Table B1 require the use of temperatures in Kelvin. Dimensionless heat capacity of propane in ideal gas state in Table B1 is given by

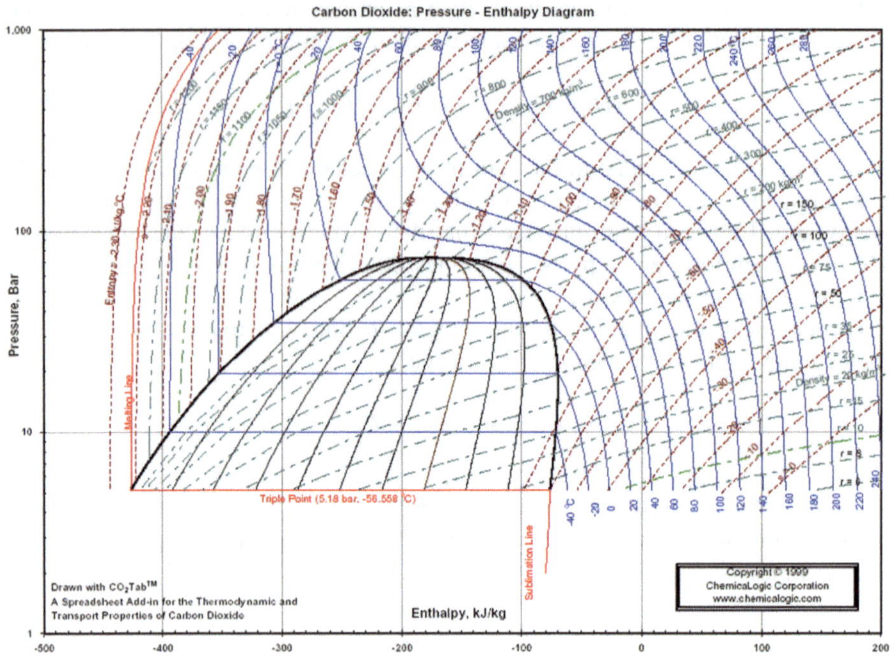

Fig. 4.1 Pressure-enthalpy diagram for carbon dioxide (With permission from Chemicalogic Corporation)

$C_{\mathrm{p}}^{\mathrm{ig}}/R = 1.213 + 28.785 \times 10^{-3}T - 8.824 \times 10^{-6}T^2$. Develop an equation for $C_{\mathrm{p}}^{\mathrm{ig}}$ in J/mole °C.

Solution:

The relation between the two temperature scales: T K $= T$°C + 273.15.

The gas constant: $R = 8.314$ J/mole K = 8.314 J/mole °C

$$C_{\mathrm{p}}^{\mathrm{ig}} = R\left(1.213 + 28.785 \times 10^{-3}(T + 273.15) - 8.824 \times 10^{-6}(T + 273.15)^2\right)$$

$$C_{\mathrm{p}}^{\mathrm{ig}} = 69.981 + 199.24 \times 10^{-3}T - 73.362 \times 10^{-6}T^2 \text{in J/mole°C}$$

Example 4.2 **Calculation of internal energy change**

Calculate the change in internal energy when 0.5 m^3 of air at 200 kPa is heated from 25 to 120°C in a constant volume process.

Solution:

Assume that the system will remain as nearly ideal gas during the process.

Data: $V = 0.5$ m^3, $P = 200$ kPa, $T_1 = 25$°C (298.15 K), $T_2 = 120$°C (393.15 K),

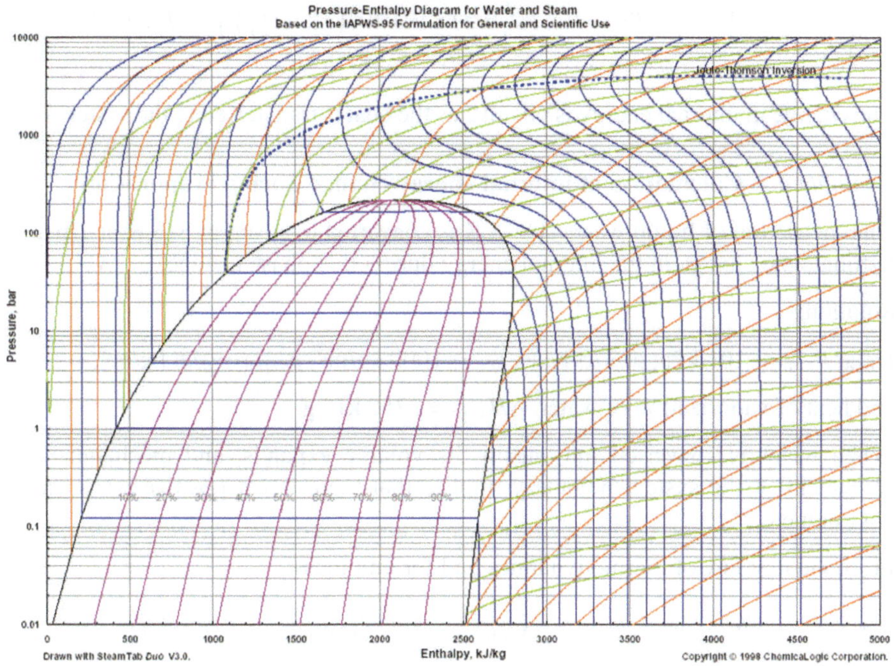

Fig. 4.2 Pressure-enthalpy diagram for water (With permission from Chemicalogic Corporation)

The relation between the two temperature scales:

$T(\mathrm{K}) = T°\mathrm{C} + 273.15$, $R = 8.314$ Pa m^3/mole K.

Moles of air: $n_{\mathrm{air}} = \frac{PV}{RT} = \frac{(200\times10^3\ \mathrm{Pa})(0.5\ \mathrm{m}^3)}{8.314(\mathrm{Pa\ m^3/mole\ K})(298.15\ \mathrm{K})} = 40.3\,\mathrm{mole}.$

The heat capacity parameters in Table B1 require the use of temperatures in Kelvin.

The heat capacity at constant pressure of air in ideal gas state: (Table B1)

$$\frac{C_{\mathrm{p}}^{\mathrm{ig}}}{R} = 3.355 + 0.575\times10^{-3}T \qquad (T \text{ in K})$$

From Eq. (4.13), we have $\frac{C_v^{ig}}{R} = \frac{C_p^{ig}}{R} - 1$ or $C_v^{ig} = C_p^{ig} - R$

$$\frac{C_{\mathrm{V}}^{\mathrm{ig}}}{R} = \left(3.355 + 0.575\times10^{-3}T\right) - 1$$

The gas constant: $R = 8.314$ J/mole K $= 8.314$ J/mole °C

$$\Delta U = n_{\mathrm{air}}R\int_{298\,\mathrm{K}}^{393\,\mathrm{K}} C_{\mathrm{V}}^{\mathrm{ig}}\mathrm{d}T = (40.3\,\mathrm{mole})(8.314\,\mathrm{J/mole\ K})$$

$$\left(2.355(T_2 - T_1) + \frac{1}{2}0.575 \times 10^{-3}\left(T_2^2 - T_1^2\right)\right) = 78.10\,\text{kJ}$$

The change in internal energy is positive as the heat is transferred to the system (air).

Example 4.3 **Calculation of internal energy change during evaporation**

One metric ton of liquid water at 373 K is converted to steam at 373 K at 101 kPa. Estimate the change in the internal energy during this phase change.

Solution:

Data: $P_{\text{atm}} = 101$ kPa; at 373 K, $\rho_{\text{water}} = 1000$ kg/m^3, $\rho_{\text{vapor}} = 0.6$ kg/m^3,

and heat of vaporization of water, $\Delta H_{\text{vap}} = 2250$ kJ/kg,

When water is converted into steam, its volume expands, and mechanical displacement of the system boundary implies that there is net positive work on the environment.

From the first law of thermodynamics and for unit mass:

$$\begin{aligned} W &= P_{\text{atm}}(V_{\text{steam}} - V_{\text{water}}) = P_{\text{atm}}(1/\rho_{\text{steam}} - 1/\rho_{\text{water}}) \\ &= 168.323\,\text{kJ/kg} \end{aligned}$$

The heat added to the system is $q = \Delta H_{\text{vap}} = 2250$ kJ/kg

$$\Delta U = q - W = 2250 - 168.323 = 2081.677\ \text{kJ/kg}$$

One metric ton = 1000 kg, and the total change in internal energy becomes:

$$2081.677\ \text{kJ/kg} \times (1000\ \text{kg}) = 2{,}081{,}677\ \text{kJ} = 2081.677\ \text{MJ}.$$

Example 4.4 **Determination of state properties**

Complete the following table for water with the missing state properties.

T °F	P psia	ΔU Btu/lb	State
250	27	550	
213.03		181.16	Saturated liquid
213.03	15	1077.9	
350	30	1125.5	
298	100		Compressed liquid

Solution:

Using the properties from steam tables.

T °F	P psia	ΔU Btu/lb	State
250	27	550.0	Saturated mixture
213.03	15	181.16	Saturated liquid
213.03	15	1077.90	Saturated vapor
350	30	1125.50	Superheated vapor
298	100	267.42	Compressed liquid

Example 4.5 Heat value of a saturated liquid and vapor mixture of a steam

A rigid tank contains 15 kg of saturated vapor water at 85°C. If the vapor partially condenses and 2 kg of the liquid water is formed, estimate the enthalpy of the saturated liquid–vapor mixture and the amount of heat lost through the condensation.

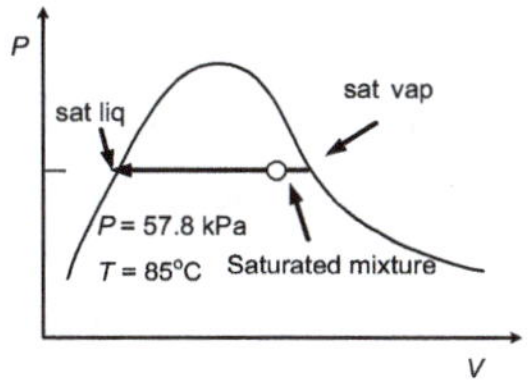

Solution:

Saturated liquid–vapor water mixture at $T_{sat} = 85°C$ at (57.8 kPa).

Saturated vapor:
$H_{sat\ vap} = 2652.0$ kJ/kg, saturated liquid water: $H_{sat\ liq} = 355.9$ kJ/kg.

Amount of liquid water = 2 kg.

Amount of vapor = 15 – 2 = 13 kg.

Quality x of the saturated mixture: $x = \frac{m_{vap}}{m_{total}} = \frac{13\,\text{kg}}{15\,\text{kg}} = 0.87$.

Enthalpy of the liquid–vapor mixture:

$$H_{mix} = (1 - x)H_{sat\ liq} + xH_{sat\ vap} = (1 - 0.87)355.9\,\text{kJ/kg}$$
$$+(0.87)2652.0\,\text{kJ/kg} = 2353.5\,\text{kJ/kg}$$

Heat of condensation:

$\Delta H_{cond} = H_{sat\ vap} - H_{sat\ liq} = (2652.0 - 355.9)$ kJ/kg = 2296.1 kJ/kg.

Heat lost after forming 2 kg of liquid water: (2 kg)(ΔH_{cond}) = 4592.2 kJ.

Enthalpy of the saturated mixture depends on the quality x and will be:

$H_{\text{sat vap}} > H_{\text{mix}} > H_{\text{sat liq}}$.

4.3 Heat

Heat is part of the total energy flow across a system boundary that is caused by a temperature difference between the system and its surroundings or between two systems. Heat flows from high temperature region to cold temperature region. Therefore, heat is not stored and defined as thermal energy in transit. The unit for heat in the SI system is the joule, J, though the Btu and the calorie are still used. When heat capacity is constant, such as $C_{\text{p,av}}$, the amount of heat changes when a substance changes its temperature from T_1 to another temperature T_2 is

$$q = mC_{\text{p,av}}(T_2 - T_1) \tag{4.14}$$

where q is the heat, and m is the mass. For processes taking place at constant volume, the specific heat capacity is denoted by C_{v}. The values of specific heat capacity at constant pressure, C_{p}, for various substances are shown in Table B3 in Appendix B. For a fluid flow, the rate of heat becomes

$$\dot{q} = \dot{m}C_{\text{p,av}}(T_2 - T_1) \tag{4.15}$$

where $\dot{q}$ is the rate of heat transfer (J/s or Btu/s), and $\dot{m}$ is the rate of mass flow (kg/s or lb/s). The value of heat is positive when transferred to the system and negative when transferred from the system to its surroundings. There is no heat flow through the boundary in an *adiabatic process* as the system and its surroundings have the same temperatures [1, 4].

If an amount of mass, *m*, is heated from T_1 to T_2, it will increase the heat. For example, the energy needed to heat 1.0 kg of water from 0 to 100°C when the specific heat of water is 4.18 kJ/kg K becomes q = (1.0 kg) (4.18 kJ/kg K) (100 − 0)°C = 418 kJ. However, if an amount of mass, *m*, is cooled from T_1 to T_2, it will lose the heat. The value of heat is a path dependent function. For example, the value of heat at the constant-pressure process will be different from the value required at a constant volume process.

When there is no work interaction between a closed system and its surroundings the amount of heat is estimated by

$$q = \int_{T_1}^{T_2} C_{\text{p}} \mathrm{d}T \tag{4.16}$$

Using the expression for the temperature dependence of the heat capacity, given in Eq. (4.11), the change of heat between T_1 and T_2 is estimated by

$$q = \int_{T_1}^{T_2} C_p \mathrm{d}T = A(T_2 - T_1) + \frac{1}{2}B\left(T_2^2 - T_1^2\right) + \frac{1}{3}C\left(T_2^3 - T_1^3\right) \quad (4.17)$$

Thermal equilibrium is achieved when two systems in thermal contact with each other cease to exchange energy by heat.

- If two systems are in thermal equilibrium, then their temperatures are the same.
- For an *isolated system,* no energy is transferred to or from the system.
- A *perfectly insulated* system experiences no heat exchange through the boundary.

4.3.1 Sensible Heat

Sensible heat effects refer to the quantity of heat transfer causing the temperature change of a system in which there is no phase transition, no chemical reaction, and no change in composition. For mechanically reversible, constant pressure, and closed systems, the amount of total heat transferred would be

$$q = \Delta H = \int_{T_1}^{T_2} C_p \mathrm{d}T \quad (4.18)$$

When $C_{p,av}$ = constant, we have

$$q = \Delta H = C_{p,av}(T_2 - T_1) \quad (4.19)$$

According to the sign convention used here, the sign of heat is negative when heat is transferred from the system and positive when heat is transferred to the system.

In a wide variety of situations, it is possible to raise the temperature of a system by the energy released from another system. The calorie was originally defined as the amount of energy required to raise the temperature of one gram of water by 1°C (approximately 4.1855 J), and the British thermal unit was defined as the energy required to heat one pound of water by 1°F (later fixed as 1055.06 J) [1, 4].

4.3.2 Latent Heat

Consider the heat needed for liquefaction of a pure substance from the solid state or vaporization from the liquid state at constant pressure. These heat effects are called the *latent heat of fusion* and *latent heat of vaporization*, respectively. In either case, no change in temperature occurs but the transfer of a finite amount of heat into the system is required. Similarly, a substance state may change from liquid to solid state if a finite amount of heat is transferred from the liquid state to the surroundings. This heat is called the *latent heat of freezing*. The state of a substance changes from vapor to liquid state when a finite amount of heat is transferred from the vapor state to the surroundings. This heat is called the *latent heat of condensation.* There are also heats of transitions accompanying the change of a substance from one solid state to another; for example, when rhombic crystalline sulfur changes to monoclinic structure at 95°C and 1 bar, the heat of transition is 360 J per gram-atom [1, 4].

The latent heat of vaporization ΔH_{vap} is a function of temperature only, and is related to other system properties of an ideal gas at temperature T by the *Clausius–Clapeyron* equation

$$\frac{d \ln P_{\text{sat}}}{dT} = \frac{\Delta H_{\text{vap}}}{RT^2} \tag{4.20}$$

where P_{sat} is the saturation pressure of the pure substance. The value of saturation pressure P_{sat} can be calculated from the Antoine equation, given in Eq. (1.37). The Antoine parameters *A*, *B*, and *C* are listed in Table 1.12. When ΔH_{vap} can be assumed to be constant within the temperature interval of $(T_2 - T_1)$, Eq. (4.20) can be integrated as follows

$$\ln\left(\frac{P_{2\text{sat}}}{P_{1\text{sat}}}\right) = -\frac{\Delta H_{\text{vap}}}{R}\left(\frac{1}{T_2} - \frac{1}{T_1}\right) \tag{4.21}$$

Equation (4.21) relates the fluid saturation pressures at two different temperatures to the heat of vaporization. Here the heat of vaporization is assumed to be temperature independent when the temperature difference of $(T_2 - T_1)$ is small [1].

Latent heats may also be measured using a calorimeter, and experimental values are available at selected temperatures for many substances (see Table A1). *Trouton's rule* predicts the heat of vaporization at normal boiling point, ΔH_{n}, that is at a pressure of 1 standard atm (101.32 kPa)

$$\frac{\Delta H_n}{RT_n} \simeq 10 \tag{4.22}$$

where T_n is the absolute temperature of the normal boiling point and R is the gas constant. The units of ΔH_n, T_n, and R must be chosen to yield a dimensionless value for the ratio in Eq. (4.22). Another approximate model is [1]

$$\frac{\Delta H_n}{RT_n} \simeq \frac{1.092(\ln P_{\text{c}} - 1.013)}{0.930 - T_n/T_{\text{c}}} \tag{4.23}$$

where P_c and T_c are the critical pressure and critical temperature, respectively. Predictions by this equation are quite satisfactory. For example, the application of Eq. (4.23) for water is

$$\frac{\Delta H_n}{RT_n} \simeq \frac{1.092(\ln 220.55 - 1.013)}{0.930 - 0.577} = 13.56$$

$$\Delta H_n = 13.56(8.314)(373.15) = 42.065 \text{ kJ/mole}$$

where $R = 8.134$ J/mole K, $T_n = 373.15$ K, $P_c = 220.55$ bar, and $T_c = 646.71$ K (Table A3). The experimental value from Table A1 is 40.65 kJ, which is lower by only 3.4%.

The method proposed by Watson (1943) estimates the latent heat of vaporization of a pure substance at any temperature from a known value of the latent heat at a single temperature

$$\frac{\Delta H_2}{\Delta H_1} \simeq \left(\frac{1 - T_{r_2}}{1 - T_{r_1}}\right)^{0.38} \tag{4.24}$$

where T_{ri} is the reduced temperature obtained from the ratio of T_i/T_c. This equation is fairly accurate [1].

4.3.3 Heating with Phase Change

The heating processes for the water between 25 and 200°C are illustrated on a temperature-volume diagram in Fig. 4.3. The heating with phase change has the following sub processes of sensible and latent heating:

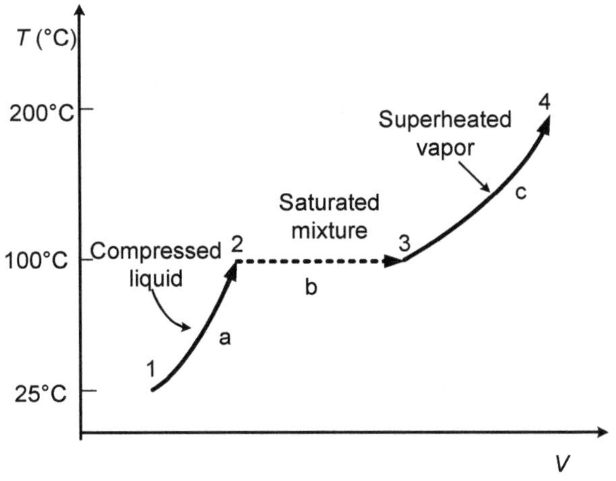

Fig. 4.3 Temperature-volume diagram for heating water with phase change between 25 and 200°C

Process a: Sensible heating of liquid water between 25°C (298 K) and 100°C (373 K):

Process b: Latent heating of vaporization at 100°C (373 K).

Process c: Sensible heating of water vapor between 10°C (373 K) and 200°C (473 K).

The total heat per unit amount of water is calculated by

$$\Delta H_{\text{total}} = \Delta H_{\text{a}} + \Delta H_{\text{b}} + \Delta H_{\text{c}} = \left(\int_{298\,\text{K}}^{373\,\text{K}} C_{\text{p,liq}} \text{d}T + \Delta H_{\text{vap}} + \int_{373\,\text{K}}^{473\,\text{K}} C_{\text{p,vap}}^{\text{ig}} \text{d}T \right)$$

The equations of temperature dependent heat capacities are needed within the integrals to estimate the sensible heating at liquid and vapor states.

Example 4.6 **Calculation of heat of vaporization using antoine equation and Clausius-Clapeyron equation**

Determine the heat of vaporization of benzene at its normal boiling point T_{b} by using the Antoine equation and compare with the experimental value.

Solution:

$T = T_{\text{b}} = 80.1°\text{C} = 353.26$ K (from Table A1), $R = 8.314$ J/mole K.

Vapor pressure of benzene from the Antoine equation with the constants from Table 1.12:

A = 13.7819, B = 2726.81, C = 217.57.

$\ln P_{\text{sat}} = 13.7819 - \frac{2726.81}{T+217.57}$ (P_{sat} is in kPa and T in °C) (Eq. 1.27).

The Clausius–Clapeyron equation: $\frac{d \ln P_{\text{sat}}}{\text{d}T} \simeq \frac{\Delta H_{\text{vap}}}{RT^2}$.

Differentiation of the Antoine equation with respect to temperature T yields

$$\frac{d \ln P_{\text{sat}}}{\text{d}T} = -\frac{2726.81}{(80 + 217.57)^2} \simeq \frac{\Delta H_{\text{vap}}}{8.314(353.15)^2}$$

$\Delta H_{\text{vap}} = 31.95$ kJ/mole or 2.49 kJ/kg (with the molecular weight of benzene = 78 g/mole).

The experimental data from Table A1 is 30.76 kJ/mole and the deviation:

$\frac{31.95-30.76}{30.76} = 0.038 = 3.8\%$ (small enough for many engineering calculations).

Example 4.7 **Estimation of change of enthalpy with sensible and latent heats**

Estimate the change of enthalpy of water flowing at a rate of 4.5 kg/s and heated from 25 to 200°C at atmospheric pressure.

Solution:

Assume that the water will remain in ideal state from the initial and the final temperatures.

The heating processes are shown below on a temperature versus volume diagram:

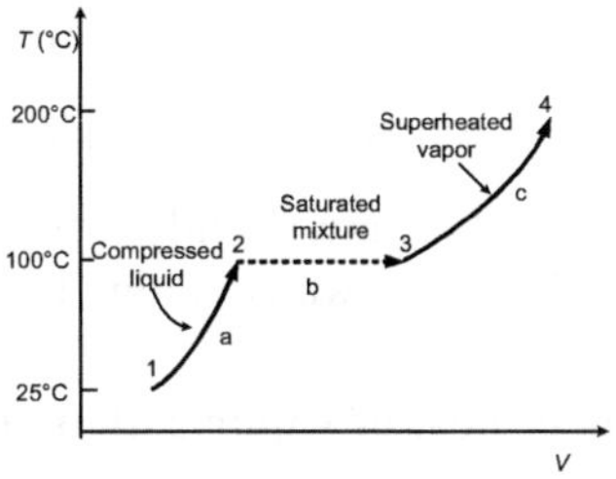

$T_1 = 298$ K, $T_{sat} = 373$ K, $T_2 = 473$ K

The total heat:

$\Delta H_{total} = \Delta H_a + \Delta H_b + \Delta H_c = \dot{m}\left(\int_{298\text{ K}}^{373\text{ K}} C_{p,liq} dT + \Delta H_{vap} + \int_{373\text{ K}}^{473\text{ K}} C_{p,vap} dT\right)$.
where $\dot{m}$ is mass flow rate of water: 4.5 kg/s, $R = 8.314$ J/mole K, $MW = 18.02$ g/mole, $R = 0.4615$ kJ/kg K.

At $T_{sat} = 373$ K: $\Delta H_{vap} = \Delta H_{sat\ vap} - \Delta H_{sat\ liq} = 2257$ kJ/kg (Steam table).

From Table B1 the dimensionless heat capacities at constant pressure are:

$$\frac{C_{p,liq}}{R} = A + BT = 8.712 + 1.25\times10^{-3}T \quad \text{(T in K)}$$

$$\frac{C_p^{ig}}{R} = A + BT = 3.47 + 1.45\times10^{-3}T \quad \text{(T in K)}$$

$$\text{Process a: } \dot{m}\left(\int_{298K}^{373K} C_{p,liq} dT\right)$$

$$= \dot{m}R\left[8.712(373 - 298) + \frac{1}{2}1.25 \times 10^{-3}\left(373^2 - 298^2\right)\right]$$

$$= 1422.3 \text{ kJ/s} = 1422.3 \text{ kW}$$

Process b: $\dot{m}\left(\Delta H_{vap}\right) = 10{,}256.3\,\text{kJ/s} = 10{,}256.3\,\text{kW}$.

$$\text{Process c: } \dot{m}\left(\int_{373\text{ K}}^{573\text{ K}} C_{p,vap}^{ig} dT\right)$$

$$= \dot{m}R\left[3.47(473 - 373) + \frac{1}{2}1.45 \times 10^{-3}\left(473^2 - 373^2\right)\right]^{\cdot}$$

$$= 848.0\,\text{kJ/s} = 848\,\text{kW}$$

Total amount of energy (power) needed = 12,426.8 kJ/s = 12,426.8 kW.

Using the steam tables:

Process a: $\dot{m}(H_2 - H_1) = (4.5 \cdot \text{kg/s})(419.06 - 104.8)\text{kJ/kg} = 1417.5\,\text{kW}$.

Process b: $\dot{m}\Delta H_{\text{vap}} = (4.5\text{kg/s})(2676.0 - 419.06)\text{kJ/kg} = 10156.5\,\text{kW}$.

Process c: $\dot{m}(H_4 - H_3)(4.5\text{kg/s})(2875.3 - 2676.0)\text{kJ/kg} = 898.8\,\text{kW}$.

Total = 1417.5 + 10,156.5 + 898.8 = 12,470.8 kW.

The results from the steam tables are slightly different from the estimated results because of the nonideal behavior of water at high temperatures and pressures.

Example 4.8 **Estimation of heat of vaporization at another temperature**

The latent heat of vaporization of water at 100°C is 2253.0 kJ/kg. Estimate the latent heat of water at 150°C.

Solution:

For water:
T_c = 646.71 K; ΔH_1 = 2253.0 kJ/kg (Steam Tables), T_1 = 373.15 K, T_2 = 423.15 K.

Reduced temperatures are:

$T_{r1} = 373.15/647.1 = 0.577$ and $T_{r2} = 423.15/647.1 = 0.654$

$$\Delta H_2 \simeq (2253.0\,\text{kJ/kg})\left(\frac{1 - 0.654}{1 - 0.577}\right)^{0.38} = 2087.4\,\text{kJ/kg}$$

The value given in the steam table is 2113.2 kJ/kg (From steam tables).

Deviation: $\frac{2113.2 - 2087.4}{2113.2} = 0.012$ or 1.2%

Deviation of the estimated value from the tabulated value is low.

Example 4.9 **Energy exchange in expansion of air**

A closed 5 L (l) cylinder contains air at 100°C and 2 atm. The cylinder is in boiling water at normal atmospheric pressure. A piston is released very slowly so the air expands, while absorbing heat from the boiling water until its pressure reaches 1.5 atm. Estimate the final volume of the air and heat absorbed.

(a) Isothermal expansion
(b) Adiabatic expansion

Solution:
As the temperature remains constant and assuming the air as ideal gas ($PV = RT$, for unit mole), we use

$P_1V_1 = P_2V_2 = RT =$ constant,

$$V_2 = (P_1/P_2)\ \ V_1 = (2\,\text{atm}/1.5\,\text{atm})51 = 6.661$$

For an ideal gas expansion $q = P_1V_1\ \ln(V_2/V_1) = (2\ \text{atm})\ (5\ \text{l})\ \ln\ (6.66\ \text{L}/5\ \text{l}) =$ 280.6 J

(b) Adiabatic expansion with $\gamma = 1.4$

During the expansion, no heat is added into the air and the work done is $\text{W} = (P_1V_1 - P_2V_2)/(1 - \gamma)$

$$(V_2)^\gamma = (P_1/P_2)(V_1)^\gamma$$

$$V_2 = \text{V}_1(\text{P}_1/\text{P}_2)^{1/\gamma} = (51)(2\text{atm}/1.5\text{atm})^{1/\gamma} = 6.11$$

$$q = (P_1V_1 - P_2V_2)/(1 - \gamma) = (25 - 1.56.1)/(1 - 1.4) = 212.7\,\text{J}.$$

4.3.4 Heat of Reaction

Systems with chemical reactions exchange heat with the surroundings and the temperature may change. These effects result because of the differences in molecular structure and therefore in energy of the products and reactants. The heat associated with a specific chemical reaction depends on the temperatures of reactants and products. A general chemical reaction may be written as follows

$$\nu_A A + \nu_B B \rightarrow \nu_C C + \nu_D D \tag{4.25}$$

where v_i is the stoichiometric coefficient for the substances A, B, C, and D, which is as positive for a product and negative for a reactant.

A standard basis for estimation of heat of reaction effects results when the reactants and products are at the same temperature. For the reaction given in Eq. (4.25), the standard heat of reaction is the enthalpy change when ν_A moles of A and ν_B moles of B in their standard states at temperature T react to form ν_C moles of C and ν_D moles of D in their standard states at the same temperature T. Standard state for gases is the pure substance in the ideal gas state at 1 atm. For liquids and solids, the standard state is the real pure liquid or solid at 1 atm. Standard-state properties are therefore functions of temperature. Based on this sign convention, a standard heat of reaction is expressed by [3, 4]

$$\Delta H_r^o = \sum_i v_i \Delta H_{fi}^o \tag{4.26}$$

where H^{o}_{fi} is the *standard heat of formation*. Some values of standard heats of formation are tabulated at 298.15 K or 25°C in Table C1 and are shown as ΔH^{o}_{f298}.

A formation reaction is defined as a reaction which forms 1 mole of a single compound from its constituent elements with each substance in its standard state at 298.15 K (25°C). Standard states are: (i) gases (g); pure ideal gas at 1 atm and 25°C. (b) liquids (l) and solids (s); (ii) pure substance at 1 atm and 25°C; (iii) solutes in aqueous 1-molal ideal solution at 1 atm and 25°C. Standard heats of formation for stable elements in their standard states are zero since they are naturally existing. For example, the formation reaction for ammonia is $N_2 + 3H_2 \rightarrow NH_3$ since all the reactants are elements. However, the reaction $H_2O + SO_3 \rightarrow H_2SO_4$ is not a formation reaction since it forms sulfuric acid not from elements but from other compounds.

The summation in Eq. (4.26) is over all products and reactants. For example, for the reaction

$$4HCl(g) + O_2(g) \rightarrow 2H_2O(g) + 2Cl_2(g) \tag{4.27}$$

the stoichiometric coefficients are $\nu_{HCl} = -4,\ \nu_{O_2} = -1,\ \nu_{H_2O} = 2,\ \nu_{Cl_2} = 2$ and the standard heat of reaction using Eq. (4.26) becomes

$$\Delta H^{o}_{r} = \sum_{i} \nu_i \Delta H^{o}_{fi} = \nu_{H_2O}\Delta H^{o}_{fH_2O} + \nu_{Cl_2}\Delta H^{o}_{fCl_2} + \nu_{HCl}\Delta H^{o}_{fHCl} + \nu_{O_2}\Delta H^{o}_{fO_2} \tag{4.28}$$

where $\Delta H^{o}_{fCl_2} = 0$ and $\Delta H^{o}_{fO_2} = 0$ as they are naturally occurring stable elements.

Heat effects change according to the nature of chemical reactions. Reactions that release heat are called *exothermic reactions* and reactions that require heat are called *endothermic reactions.* For an exothermic reaction at constant pressure, the system's change in enthalpy equals the energy released in the reaction, including the energy retained in the system and lost through expansion against its surroundings. In a similar manner, for an endothermic reaction, the system's change in enthalpy is equal to the energy *absorbed* in the reaction. If ΔH_r is positive, the reaction is endothermic, that is heat is absorbed by the system due to the products of the reaction having a greater enthalpy than the reactants. On the other hand, if ΔH_r is negative, the reaction is exothermic, that is the overall decrease in enthalpy is achieved by the generation of heat [4] (Fig. 4.4).

When a heat of reaction is given for a reaction, this applies for the current stoichiometric coefficients. If the stoichiometric coefficients are doubled, the heat of reaction is doubled. For example, consider the following reactions

$$\frac{1}{2}N_2 + \frac{3}{2}H_2 \rightarrow NH_3 \quad \Delta H^{o}_{r} = -46.11\,\text{kJ}$$

$$N_2 + 3H_2 \rightarrow 2NH_3 \quad \Delta H^{o}_{r} = -92.22\,\text{kJ}$$

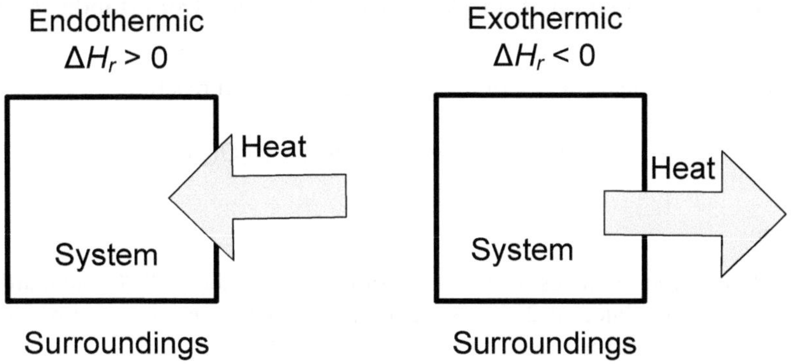

Fig. 4.4 Heat interactions in chemical reactions: endothermic reaction needs energy from outside and exothermic reactions release energy to the surroundings

When chemical reactions are combined by addition, the standard heats of reaction may also be added to estimate the standard heats of the resulting reaction [1, 4]. This is possible since enthalpy is a state function and, for a given initial and final states, its value is independent of path. Chemical reaction equations considered for this purpose often include an indication of the physical state of each reactant and product; the symbol g, l, or s indicates gas, liquid, or solid states, respectively.

Example 4.10 **Estimation of standard heat of reaction**

Estimate the heat of reaction at 298.15 K for the reaction:
$4HCl(g) + O_2(g) \rightarrow 2H_2O(g) + 2Cl_2(g)$.

Solution:

$$4HCl(g) + O_2(g) \rightarrow 2H_2O(g) + 2Cl_2(g)$$

Stoichiometric coefficients ν_i are positive for a product and negative for a reactant. The stoichiometric coefficients for the reaction are $\nu_{HCl} = -4$, $\nu_{O_2} = -1$, $\nu_{H_2O} = 2$, $\nu_{Cl_2} = 2$.
Based on this sign convention, a standard heat of reaction is expressed by $\Delta H_r^o = \sum_i \nu_i \Delta H_{fi}^o$ where H_{fi}^o is the standard heat of formation. The summation is over all the products and the reactants using the standard heat of formations from Table B3.

$$\Delta H_r^o = 2\Delta H_{fH_2O}^o - 4\Delta H_{fHCL}^o = 2(-241.82) - 4(-92.31) = -114.41 \text{ kJ/mole}$$

The heats of formations for the stable and naturally occurring elements of O_2 and Cl_2 are zero.

Example 4.11 **Estimation of standard heats of reaction from standard heats of formation**

Estimate standard heats of reaction for the water-gas shift reaction:

$$CO_2(g) + H_2(g) \rightarrow CO(g) + H_2O(g)$$

Solution:

Although this reaction takes place at temperatures well above 25°C, standard heats of reaction is calculated at 25°C. Stoichiometric coefficients v_i are positive for a product and negative for a reactant.

The pertinent formation reactions and their heats of formation from Table C1 are:

CO_2:	$C(s) + O_2(g) \rightarrow CO_2(g)$	$\Delta H^o_{f298} = -393.51$ kJ/mole, $v = -1$
H_2:	$H_2(g)$	$\Delta H^o_{f298} = 0$ because hydrogen is a stable element
CO:	$C(s) + \frac{1}{2}O_2(g) \rightarrow CO(g)$	$\Delta H^o_{f298} = -110.52$ kJ/mole, $v = 1$
H_2O:	$H_2(g) + \frac{1}{2}O_2(g) \rightarrow H_2O(g)$	$\Delta H^o_{f298} = -241.82$ kJ/mole, $v = 1$

Here all the products and reactants are at their standard states: ideal gas state at 1 atm and 25°C.

Based on this sign convention, the standard heat of reaction is expressed by Eq. (4.26)

$$\Delta H_r^o = \sum_i v_i \Delta H_{fi}^o = -110.52 - 241.82 - (-393.51) = 41.16\,\mathrm{kJ/mole}$$

This result shows that the enthalpy of 1 mole of CO plus 1 mole of H_2O is greater than the enthalpy of 1 mole of CO_2 plus 1 mole of H_2 by 41.16 kJ when all the components are at their standard states. Therefore, the water–gas-shift reaction is an endothermic reaction and requires heat from the surroundings to occur.

4.3.5 Standard Heat of Combustion

Consider the combustion of coal:

$$C(s) + O_2(g) \rightarrow CO_2(g) + \Delta H_r$$

here the pure carbon representing the coal is in solid state, while oxygen and carbon dioxide are both in gas state. The reactants in a combustion reaction possess greater energy than do the products, and this energy is transferred to the surroundings as heat, so the reaction is exothermic. The *heat of combustion* of fuel may be measured by a flow calorimeter. The fuel is mixed with air at an ambient temperature and the mixture flows into a combustion chamber of the calorimeter where reaction occurs. The combustion products enter a water-cooled section and are cooled to the temperature of the reactants. As there is no shaft work produced in the process and potential and kinetic energies are negligible, the change in enthalpy caused by the combustion reaction becomes the heat flowing from the calorimeter

$$q = \Delta H = \text{heat of reaction} \tag{4.29}$$

For example, 2,2,4-trimethylpentane (isooctane), widely used in petrol, has a chemical formula of C_8H_{18} and it reacts with oxygen exothermically and produces 10.86 MJ per mole of isooctane

$$2C_8H_{18}(l) + 25O_2(g) \rightarrow 16CO_2(g) + 18H_2O(g) + 10.86\,\text{MJ/mole} \tag{4.30}$$

Incomplete combustion of petroleum or petrol products results in carbon monoxide. At a constant volume, the heat of combustion of a petroleum product can be approximated by:

$$q_v = 12{,}400 - 2{,}100d^2 \tag{4.31}$$

where q_v is measured in cal/gram and d is the specific gravity at 60°F (16°C). Ethane C_2H_6 is an aliphatic hydrocarbon. The complete combustion of ethane releases 1561 kJ/mole, or 51.9 kJ/g, of heat, and produces carbon dioxide and water according to the chemical equation

$$C_2H_6 + 3/2O_2 \rightarrow 2CO_2 + 3H_2O + 1561\,\text{kJ/mole} \tag{4.32}$$

Gasoline contains about 35 MJ/l (132 MJ/US gal) (higher heating value). This is an average value. Gasoline blends differ, therefore actual energy content varies from season to season and from batch to batch, by up to 4% more or less than the average value [1, 4].

Stable combustion reactions require the right amounts of fuel and oxygen. In theory, there is a theoretical amount of oxygen needed (known as stoichiometric amount) to completely burn a given amount of fuel. In practice, burning conditions are never ideal. Therefore, more air than that of stoichiometric amount is supplied to burn all fuel completely. The amount of air that is more than the theoretical requirement is referred to as *excess air*. Power plant boilers, for example, normally run with about 10–20% excess air. Natural gas-fired boilers may run as low as 5% of the excess air. Pulverized coal-fired boilers may run with 20% excess air. Gas turbines run very lean with up to 300% excess air. If an insufficient amount of air is supplied to the burner, unburned fuel, soot, smoke, and carbon monoxide are exhausted from the boiler. They may lead to surface fouling, pollution, lower combustion efficiency, flame instability and a potential for explosion. To avoid inefficient and unsafe conditions, boilers normally operate at an excess air level.

Example 4.12 **Estimation of standard heats of reaction**

Determine the standard heat of formation for n-pentane: $5C(s) + 6H_2 \rightarrow C_5H_{12}(g)$.

Solution:

Many standard heats of formation come from *standard heats of combustion* measurements in a calorimeter. A combustion reaction occurs between an element or compound and oxygen to form specified combustion products. When organic compounds are made up of carbon, hydrogen, and oxygen only, the combustion products are carbon dioxide and water. The state of the water produced may be either vapor or liquid.

The reaction for formation of n-pentane may result from combination of the following combustion reactions in practice. Stoichiometric coefficients v_i are positive for a product and negative for a reactant.

$5C(s) + 5O_2(g) \rightarrow 5CO_2(g)$	$\Delta H^o_{f298} = (5)(-393.51)$ kJ/mole
$6H_2(g) + 3O_2(g) \rightarrow 6H_2O(l)$	$\Delta H^o_{f298} = (6)(-285.83)$ kJ/mole
$5C(s) + 6H_2 \rightarrow C_5H_{12}$	$\Delta H^o_{f298} = 3535.77$ kJ/mole
$5CO_2(g) + 6H_2O(l) \rightarrow C_5H_{12}(g) + 8O_2(g)$	$\Delta H^o_{f298} = -146.76$ kJ/mole

The net reaction is obtained from the summation of the first three equations by considering that the value of stoichiometric coefficients, v_i, are positive for the products and negative for the reactants. This result is the standard heat of formation of n-pentane listed in Table C1.

Example 4.13 **Estimation of standard heats of combustion from standard heats of formation**

Estimate standard heats of reaction for the combustion of methane:

$$CH_4(g) + 2O_2(g) \rightarrow CO_2(g) + 2H_2O(l)$$

Solution:

The pertinent formation reactions and their heats of formation from Table C1 are:

CH_4: $C(s) + 2H_2(g) \rightarrow CH_4(g)$	$\Delta H^o_{f298} = -74.52$ kJ/mole, $v = -1$
O_2: $O_2(g)$	$\Delta H^o_{f298} = 0$ because oxygen is a stable element
CO_2: $C(s) + O_2(g) \rightarrow CO_2(g)$	$\Delta H^o_{f298} = -393.51$ kJ/mole, $v = 1$
H_2O: $H_2(g) + \frac{1}{2}O_2(g) \rightarrow H_2O(l)$	$\Delta H^o_{f298} = -285.83$ kJ/mole, $v = 2$

Stoichiometric coefficients v_i are positive for a product and negative for a reactant.

Standard heat of reaction:

$$\Delta H_r^o = \sum_i \nu_i \Delta H_{fi}^o : \Delta H_{298}^o = -393.51 + 2(-285.83) - (-74.52)$$
$$= -890.65\ \text{kJ/mole}.$$

This result shows that the enthalpy of 1 mole of CH_4 plus 2 mole of O_2 is less than the enthalpy of 1 mole of CO_2 plus 2 mole of H_2O by 890.65 kJ when all the components are at their standard states. Therefore, the reaction is an exothermic reaction and emits heat to the surroundings.

4.4 Effect of Temperature on the Heat of Reaction

Standard state of formation enthalpies is function of temperature only

$$\mathrm{d}H_{fi}^o = C_{pi}^o \mathrm{d}T \tag{4.33}$$

Multiplying by ν_i and summing over all products and reactants gives

$$d\Delta H^o = \sum_i \mathrm{d}\nu_i H_{fi}^o = \sum_i \nu_i C_{pi}^o \mathrm{d}T \tag{4.34}$$

The standard heat capacity change of reaction is

$$\Delta C_p^o = \sum_i \nu_i C_{pi}^o \tag{4.35}$$

or

$$\Delta C_p^o = \Delta A + \Delta B \times 10^{-3} T + \Delta C \times 10^{-6} T^2$$

where

$$\Delta A = \sum_i \nu_i A_i;\ \Delta B = \sum_i \nu_i B_i;\ \Delta C = \sum_i \nu_i C_i$$

As a result, the effect of temperature on heats of reaction is obtained from

$$\Delta H_{r2}^o = \Delta H_{r1}^o + \int_{T_1}^{T_2} \nu_i C_{pi}^o \mathrm{d}T$$
$$= \Delta H_1^o + \int_{T_1}^{T_2} (\Delta A + \Delta B \times 10^{-3} T + \Delta C \times 10^{-6} T^2)\mathrm{d}T \tag{4.36}$$

where ΔH_1^o is the known value, such as the standard heat of reaction $T_1 = 25°C$ and ΔH_2^o is the standard heat of reaction at another temperature T_2.

Industrial reactions usually are not carried out at standard-state conditions. In real applications, several reactions may occur simultaneously, actual reactions may also not go to completion, and the final temperature may differ from the initial temperature. Also, some inert components may be present in the reactions. Still, calculations of the heat effects will be based on the principles outlined above [4].

Example 4.14 **Estimation of standard heat of reaction at a temperature other than 298 K**

Estimate the standard heat of reaction for the methanol synthesis at 800°C.

$$CO(g) + 2H_2(g) \rightarrow CH_3OH(g)$$

Solution:

By using standard heat of formation data in Eq. (4.26), we have

Stoichiometric coefficients v_i are positive for a product and negative for a reactant.

$$\Delta H^o_{r,298} = -200.66 - (-110.52) = -90.14\,\text{kJ/mole}$$

The heat of formation for hydrogen is zero as it is naturally occurring stable element.

The ideal gas heat capacity parameters for all the components from Table C1 are:

Components	v_i	A	B	C
CO	− 1	3.376	0.557×10^{-3}	0.0
H_2	− 2	3.249	0.422×10^{-3}	0.0
CH_3OH	1	2.211	12.216×10^{-3}	-3.45×10^{-6}

$$\Delta A = (1)(2.211) + (-1)(3.376) + (-2)(3.249) = -7.663$$

$$\Delta B = [(1)(12.216) + (-1)(0.557) + (-2)(0.422)] \times 10^{-3} = 10.815 \times 10^{-3}$$

$$\Delta C = [(1)(-3.45) + (-1)(0.0) + (-2)(0.0)] \times 10^{-6} = -3.45 \times 10^{-6}$$

$$T_1 = 298.15\text{ K}, T_2 = (800 + 273.15)°\text{C} = 1073.15\text{ K}, R = 8.314\,\text{J/mole K}$$

$$\Delta H^o_{r2} = \Delta H^o_{r1} + R\int_{T_1}^{T_2} (\Delta A + \Delta B \times 10^{-3}T + \Delta C \times 10^{-6}T^2)dT$$

$$= -90.14 + R \int_{298.15}^{1073.15} \left(-7.663 + 10.815 \times 10^{-3}T - 3.45 \times 10^{-6}T^2\right)dT$$

$$= -13,249.2 \text{ kJ}$$

The reaction is exothermic and releases energy. The heat of reaction increased with increased temperature.

4.5 Standard Enthalpy Changes

Standard enthalpy changes describe the change in enthalpy observed in a system when going between different states under standard conditions. Some of the standard enthalpy changes are [3]:

- *Enthalpy offormation* is the enthalpy observed when one mole of a compound is formed from its elementary antecedents under standard conditions. The enthalpy change of any reaction under any conditions can be computed using the standard enthalpy of formation of all of the reactants and products.
- *Enthalpy of reaction* defined as the enthalpy observed when one mole of substance reacts completely under standard conditions.
- *Enthalpy of combustion* is the enthalpy observed when one mole of a substance combusts completely with oxygen under standard conditions.
- *Enthalpy of neutralization* is the enthalpy observed when one mole of water is produced when an acid and a base react under standard conditions.
- *Enthalpy offusion* is the enthalpy required to completely change the state of one mole of substance between solid and liquid states under standard conditions.
- *Enthalpy of vaporization* is the enthalpy required to completely change the state of one mole of substance between liquid and gaseous states under standard conditions.
- *Enthalpy of sublimation* is the heat required to completely change the state of one mole of substance between solid and gaseous states under standard conditions.
- *Enthalpy of hydration* is the enthalpy observed when one mole of gaseous ions is completely dissolved in water forming one mole of aqueous ions.

4.6 Adiabatic Flame Temperature

There are two types of *adiabatic flame temperatures*: at constant volume and at constant pressure, describing the temperature of a combustion reaction products theoretically reaching if no energy is lost to the outside environment. Constant-pressure adiabatic flame temperature results from a complete combustion process

Table 4.3 Adiabatic constant-pressure flame temperature T_{ad} of common gases/materials [4]

Fuel	T_{ad} °C	T_{ad} °F
Acetylene (C_2H_2)	2500	4532
Butane (C_4H_{10})	1970	3578
Ethane (C_2H_6)	1955	3551
Hydrogen (H_2)	2210	4010
Methane (CH_4)	1950	3542
Natural gas	1960	3562
Propane (C_3H_8)	1980	3596
Wood	1980	3596
Kerosene	2093	3801
Light fuel oil	2104	3820
Medium fuel oil	2101	3815
Heavy fuel oil	2102	3817
Bituminous coal	2172	3943
Anthracite	2180	3957

that occurs without any heat transfer or changes in kinetic or potential energy. The adiabatic flame temperature of the constant-pressure process is lower than that of the constant volume process. This is because some of the energy released during combustion goes into changing the volume of the control system. The constant-pressure adiabatic flame temperature for many fuels, such as wood, propane, and gasoline in air is around 1950°C [4]. Assuming initial atmospheric conditions (1 atm and 25°C), Table 4.3 lists the adiabatic flame temperatures for various gases for a stoichiometric fuel–air mixture.

Example 4.15 **Maximum flame temperature**

Estimate the maximum temperature that can be reached by the combustion of methane with air at constant pressure. Both the methane and air enter the combustion chamber at 25°C and 1 atm. The combustion reaction is

$$CH_4(g) + 2O_2(g) \rightarrow CO_2(g) + 2H_2O(g)$$

Solution:

Stoichiometric coefficients v_i are positive for a product and negative for a reactant.

The standard heats of formation from Table C1 are:

$$\Delta H^{o}_{f298}(CH_4) = -74.52\,\text{kJ/mole}, n = -1$$

$$\Delta H^{o}_{f298}(O_2) = 0 \text{ because oxygen is an element}, n = -2$$

$$\Delta H^{\text{o}}_{\text{f298}}(\text{CO}_2) = -393.51\ \text{kJ/mole},\ n = 1$$

$$\Delta H^{\text{o}}_{\text{f298}}(\text{H}_2\text{O}) = -241.82\ \text{kJ/mole (vapor)},\ n = 2,\ R = 8.314\ \text{J/mole K}$$

Standard heat of reaction:

$$\Delta H^{\text{o}}_{\text{r}} = \sum_i \nu_i \Delta H^{\text{o}}_{\text{fi}};$$
$$= (-393.51) + 2(-241.82) - (-74.52) = 802.63\ \text{kJ/mole}$$

Burning one mole of methane releases 802.62 kJ. We have 3.76 mole of N_2 for one mole of O_2 in the air. The heat released by the combustion reaction will only be used to heat the products of 1 mole of carbon dioxide, 2 mole of water vapor, and 7.52 mole of nitrogen:

$$\Delta H^{\text{o}}_{\text{r298}} = \int_{298.15}^{T} C_{\text{p,mix}} \text{d}T$$

$$C_{\text{p,mix}} = \sum n_i C^{\text{o}}_{\text{pi}},$$

by $\frac{C^{\text{o}}_{\text{pi}}}{R} = A_i + B_i \times 10^{-3} T$ and

$$\frac{C^{\text{o}}_{\text{p,mix}}}{R} = A' + B' \times 10^{-3} T$$

where n_i is the moles of component i, T the maximum flame temperature in K.

Heat capacity parameters are from Table B1:

	CO_2	H_2O g	N_2
A	5.457	3.47	3.280
B	1.045×10^{-3}	1.450×10^{-3}	0.593×10^{-3}
n_i	1	2	7.52

$$A' = \sum_i n_i A_i = 5.457 + 2(3.47) + 7.52(3.280) = 37.06$$

$$B' = \sum_i n_i B_i [1.045 + 2(1.45) + 7.52(0.593)] 10^{-3} = 8.40 \times 10^{-3}$$

$$802{,}630\ \text{J/mole} = R \int_{298.15}^{T} C^{\text{o}}_{\text{p,mix}} \text{d}T$$

$$= 8.314[37.06(T - 298.15) + (1/2)8.40 \times 10^{-3}(T^2 - 298.15^2)]$$

$T = 2309$ K.

4.7 Air Pollution from Combustion Processes

Emissions from combustion of carbon fuels processes lead to air pollution. The main derivatives of the process are carbon dioxide CO_2, water, particulate matter, nitrogen oxides, sulfur oxides, and some partially combusted hydrocarbons, depending on the operating conditions and the fuel–air ratio. Not all the fuel will be completely consumed by the combustion process; a small amount of fuel will be present after combustion, some of which can react to form oxygenates or hydrocarbons not initially present in the fuel mixture. Increasing the amount of air in the combustion process may lead to complete combustion but tends to increase nitrogen oxides (NO_x) that have been demonstrated to be hazardous to both plant and animal health. Further chemicals released are benzene and 1,3-butadiene that are also particularly harmful. When incomplete burning occurs carbon monoxide (CO) may also be produced [4].

4.8 Heat of Mixing

Upon mixing of pure components, the thermodynamic properties change as shown by the symbol Δ. Figure 4.5 shows the heat of mixing and vapor–liquid equilibrium in ethanol–water mixture. The enthalpy of the mixture is composition dependent. As Fig. 4.5 shows that, at 197.2°F (isotherm), $H_{\mathrm{sat\ liq}} \cong 155$ Btu/lb, while $H_{\mathrm{sat\ vap}} \cong 775$ Btu/lb. When the ethanol mass fraction is $w_e = 0.4$ at 197.2°F, the enthalpy of mixture is around 600 Btu/lb.

The property change of mixing is

$$\Delta H = H - \sum_i x_i H_i \tag{4.37}$$

The partial properties are obtained from the following relationships (see Example 4.16)

$$\Delta H_1 = \Delta H + (1 - x_1)\frac{\mathrm{d}\Delta H}{\mathrm{d}x_1} \tag{4.38}$$

$$\Delta H_2 = \Delta H - x_1\frac{\mathrm{d}\Delta H}{\mathrm{d}x_1} \tag{4.39}$$

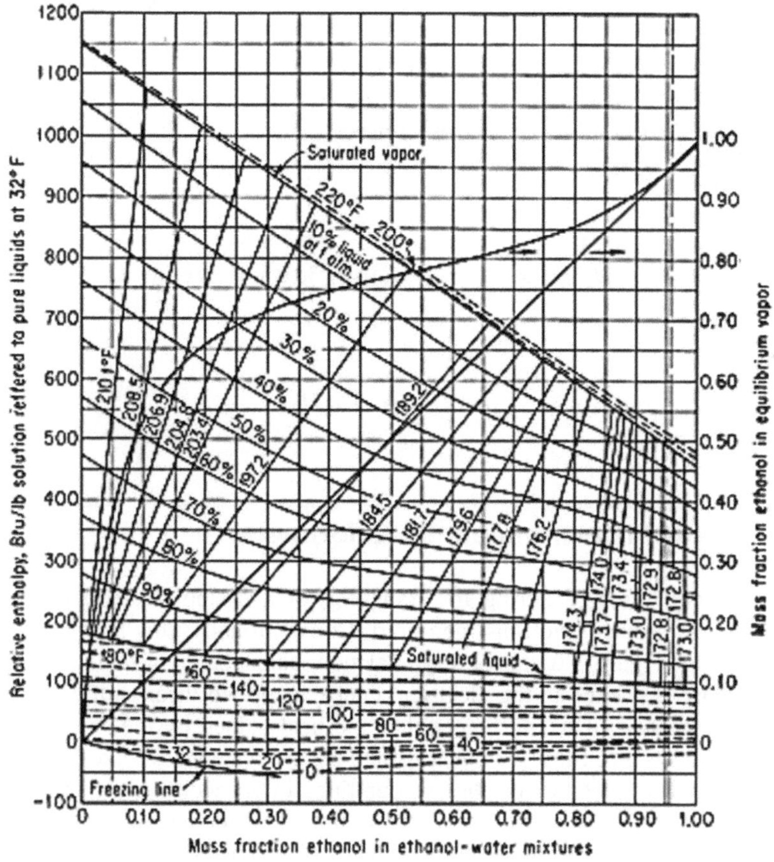

Fig. 4.5 Heat of mixing [1, 4]

Example 4.16 Estimation of partial enthalpies

The heat of mixing (excess enthalpy) for a binary mixture is $H = x_1x_2(2ax_1 + ax_2)$, where a is the parameter in J/mole, x_1 and x_2 are the mole fractions of components 1 and 2, respectively. Derive equations for the partial enthalpies of H_1 and H_2 in terms of x_1.

Solution:

The partial properties from Eqs. 4.39 and 4.40:

$$H_1 = H + (1 - x_1)\frac{\mathrm{d}H}{\mathrm{d}x_1}$$

$$H_2 = H - x_1\frac{\mathrm{d}H}{\mathrm{d}x_1}$$

With $\sum x_i = 1$: $x_2 = 1 - x_1$, and the equation for the heat of mixing H becomes: $H = ax_1 - ax_1^3 H = ax_1 - ax_1^3$.

The differentiation: $\frac{dH}{dx_1} = a - 3ax_1^2$.

Therefore, the partial molar excess enthalpies:

$$H_1 = ax_1 - ax_1^3 + (1 - x_1)\left(a - 3ax_1^2\right) = a - 3ax_1^2 + 2ax_1^3$$

$$H_2 = ax_1 - ax_1^3 - x_1\left(a - 3ax_1^2\right) = 2ax_1^3$$

and

$$H = x_1 H_1 + x_2 H_2$$

4.9 Heat Measurements by Calorimeter

Only the energy change during the transition of a system from one state into another can be defined and thus measured. Conventionally, this heat is measured by calorimeter. Figure 4.6 shows a typical flow calorimeter in which the change of temperature is related to the electrical energy supplied through the heater. The heat lost by the system would be equal to the heat gained by the surroundings in an exothermic system. Similarly, heat gained by the system from the surroundings would be equal to heat lost by the surroundings in an endothermic process. In a typical calorimeter, a hot substance (the system) and its surroundings can reach equilibrium at some final temperature. The heat flows from the system (hot) to the surroundings (cold). The heat lost by the system must be equal to the heat gained by the surroundings if there is no heat loss from the surroundings. No calorimeter is perfectly insulating to heat loss. Once the calorimeter has been calibrated to determine the calorimeter constant, we can use it to determine the specific heat effect [4]

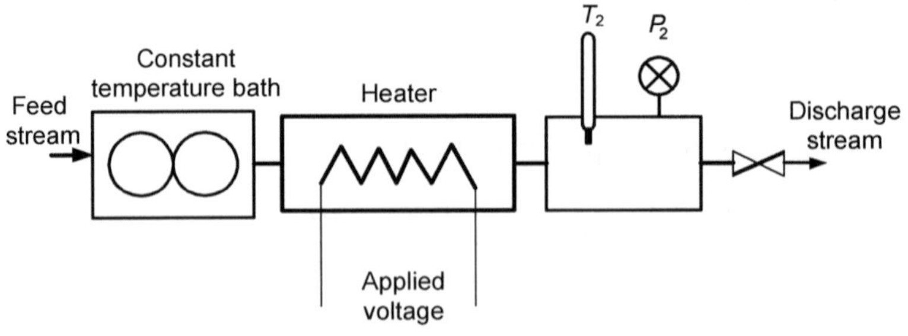

Fig. 4.6 Typical flow calorimeter

Example 4.17 **Measurement of heat capacity of a metal in a calorimeter**

28.2 g of an unknown metal is heated to 99.8°C and placed into a calorimeter containing 150.0 g of water at a temperature of 23.5°C. The temperature can equilibrate, and the final temperature is measured as 25°C. The calorimeter constant has been determined to be 19.2 J/°C. Estimate the specific heat capacity of the metal.

Solution:

Heat transfer is represented by:

Heat lost by the hot metal = heat gained by the water + heat gained by the calorimeter.
$(28.2\ \text{g})(C_{\text{p,m}})(99.8 - 25.0)°\text{C}$

$= (150.0\,\text{g})(4.184\,\text{J/g°C})(25.0 - 23.5)°\text{C} + 19.2\,\text{J/°C}(25.0 - 23.5)°\text{C}.$

Solving specific heat capacity:

$$C_{\text{p,m}} = 969\ \text{J}/\left(2.11 \times 10^3 g°\text{C}\right) = 0.459\ \text{J/g°C}$$

Metals tend to have low specific heat capacities, while water has a relatively high heat capacity.

A bomb calorimeter is a well-insulated tank containing a small combustion chamber in a water bath to measure the heat of combustion of a reaction. Bomb calorimeters must withstand the large pressure within the calorimeter as the heat of reaction is being measured. Electrical energy is used to ignite the fuel. The energy released by combustion raises the temperature of the steel bomb, its contents, and the surrounding water jacket. The temperature change in the water is then accurately measured. This temperature rise, along with a bomb calorimeter constant (which is dependent on the heat capacity of the metal bomb parts) is used to calculate the energy given out by the sample.

In a *differential scanning calorimeter*, conduction heat flows into a sample contained in a small aluminum capsule or a "pan" and is measured by comparing it to the flow into an empty reference pan. The flow of heat into the sample is larger because of its heat capacity C_p of that sample. The difference in the heat flows induces a small temperature difference ΔT across the slab, which is measured using a thermocouple. Similarly, the heat of melting of a sample substance can also be measured in such a calorimeter [4].

4.10 Psychrometric Diagram

Figure 4.7 shows a psychrometric diagram representing physical and thermal properties of moist air in a graphical form. Psychrometric diagrams help determine environmental control concepts such as humidity of air and change of enthalpy resulting in condensation of the moisture.

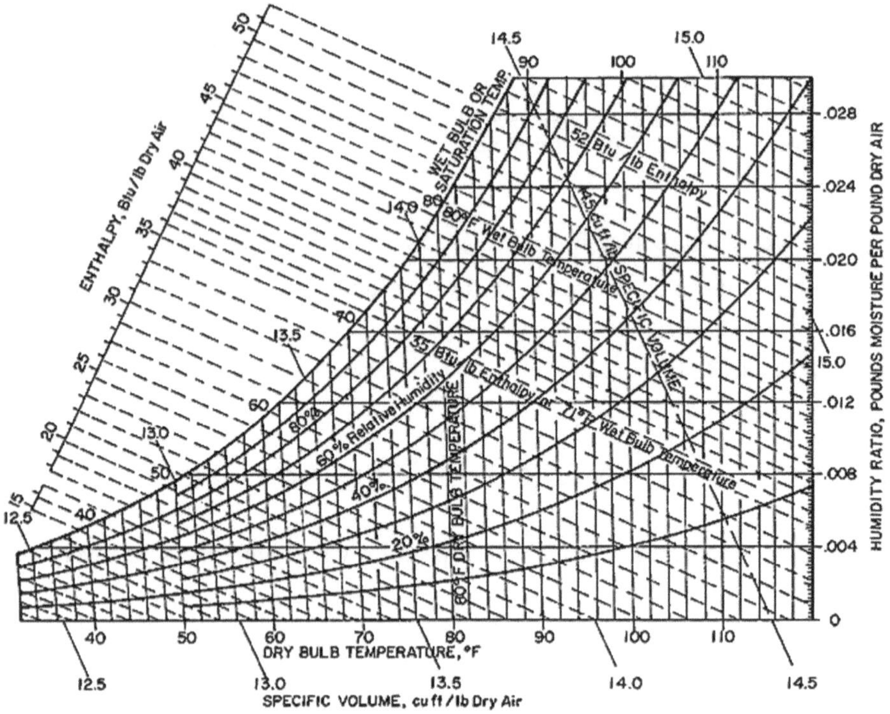

Fig. 4.7 Psychrometric diagram with permission (http://www.engineeringtoolbox.com)

- The *dry-bulb temperature* T_{db} is an ambient air temperature, which can be measured using a normal thermometer freely exposed to the air but shielded from radiation and moisture. The dry-bulb temperature appears as vertical lines in the psychrometric chart.
- The *wet-bulb temperature* T_{wb} is indicated by a moistened thermometer bulb exposed to the airflow. Wet-bulb temperature can be measured by using a thermometer with the bulb of thermometer wrapped in wet cloth. The adiabatic evaporation of water from the thermometer and the cooling effect is indicated by a "wet-bulb temperature" which is lower than the "dry-bulb temperature" in the air.

The rate of evaporation from a wet bandage on the bulb depends on the humidity of the air. The evaporation is reduced when the air contains more water vapor. The wet-bulb temperature and the dry-bulb temperature are identical at 100% relative humidity of the air (the air humidity is at the saturation line). Combining the dry bulb and wet-bulb temperature in a psychrometric diagram determines the state of the humid air. Lines of constant wet-bulb temperatures run diagonally from the upper left to the lower right on the psychrometric diagram.

The *dew point* is the temperature at which water vapor starts to condense out of the air (the temperature at which air becomes completely saturated). Above this temperature, the moisture will stay in the air. For example, if moisture condensates on a cold bottle taken from the refrigerator, the dew-point temperature of the air is above the temperature in the refrigerator. The dew point is given by the saturation line in the psychrometric diagram [4].

The Mollier diagram is used as a basic design tool for engineers for heating, ventilation, and air-conditioning (HVAC) calculations as well as the values of enthalpy of air at various conditions. Figure 4.8 shows the Mollier diagram for water and steam.

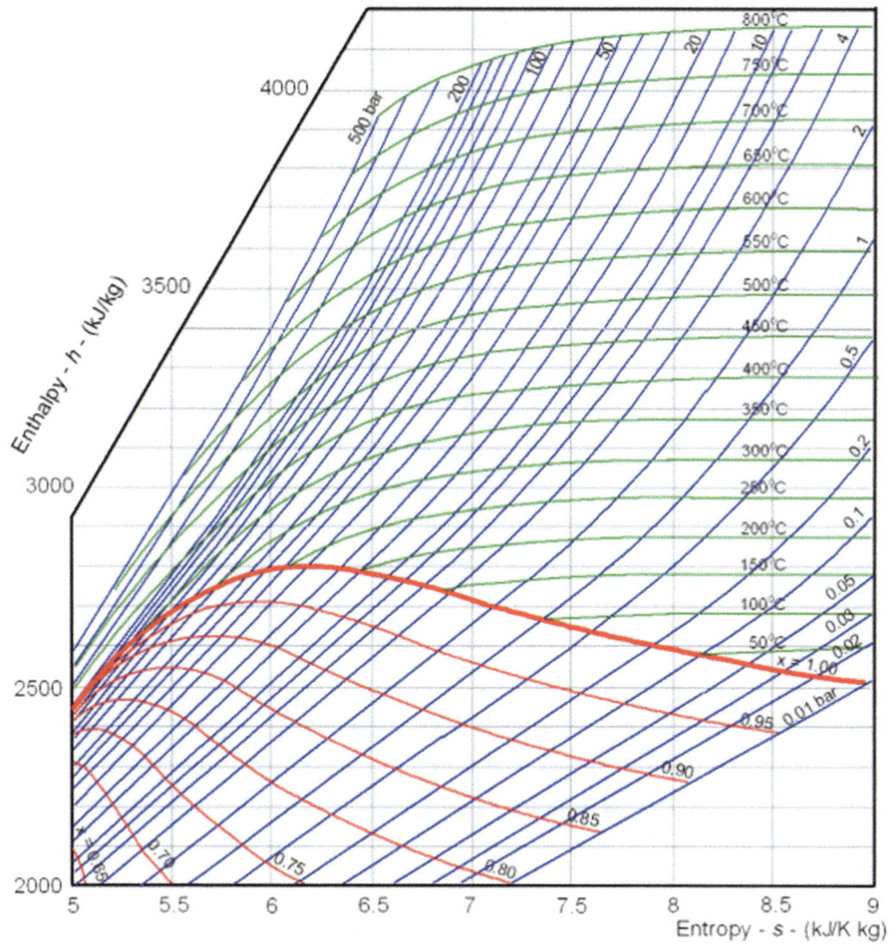

Fig. 4.8 Mollier diagram for water-steam (http://www.engineeringtoolbox.com) (with permission). The Mollier diagram is useful in determining the values of enthalpy and entropy in various adiabatic flow processes, such as nozzles and turbines

Example 4.18 **Determination of air properties on a psychrometric chart**

For a dry-bulb temperature of 74°F and a wet-bulb temperature of 67°F:

(a) Estimate the air properties and its enthalpy.
(b) Estimate the heat to be added if the air is heated to 84°F dry-bulb temperature at constant humidity ratio.
(c) Estimate the dry-bulb temperature for an evaporative cooling starting from the state point reached in part (b).

Solution:

(a) Find the intersection (A) of the dry-bulb temperature of 74°F and wet-bulb temperature of 67°F on the psychrometric chart as shown on the figure below. This intersection (A) is a "state point" for the air.

We read relative humidity as 70% and dew-point temperature as approximately 63.2°F (follow horizontal line, moving left, toward the curved upper boundary of saturation temperatures). The enthalpy is 31.7 Btu/lb dry air.

(b) The heating process moves horizontally to the right along a line of constant humidity ratio (B), and we read the enthalpy as approximately 34.2 Btu/lb dry air and the relative humidity as 50%.

The heat necessary: $q = (34.2 - 31.7)$ Btu/lb dry air = 2.5 Btu/lb dry air.

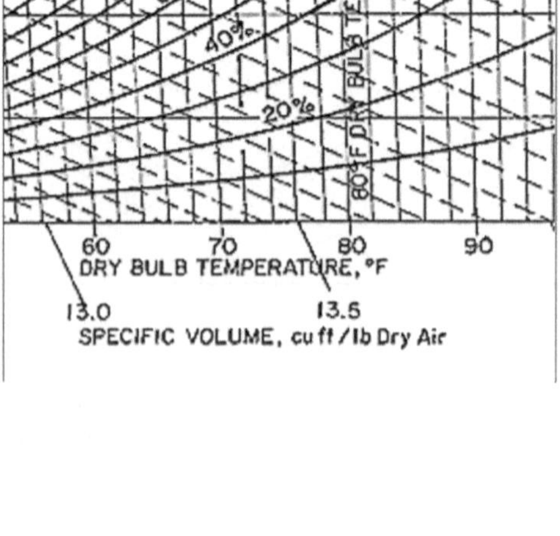

(c) During the evaporative cooling heat contained in the air evaporates water. The process moves upward along the line of constant enthalpy or constant web-bulb temperature. The air temperature (dry bulb) drops to 70°F while water content (humidity) rises to the saturation point (C) with a relative humidity of 100%. Evaporation is often used in hot weather to cool ventilation air.

Example 4.19 **Determination of moisture from the Mollier chart**

Determine the quality and the pressure when the enthalpy and entropy of a saturated steam approximately are 2300 kJ/kg and 6.65 kJ/kg K, respectively.

Solution:

Consider the Mollier diagram for water-steam in Fig. 4.8.

Under the saturation line, we have a two-phase region. In this region we can find the intersection point between the approximate values of enthalpy as 2300 kJ/kg and the entropy as 6.65 kJ/kg K.
At this intersection point, the constant-pressure line of 0.5 bar and constant quality of 0.85 can be obtained. The moisture is 15% corresponding to the quality of the steam $x = 0.85$.

The readings are only approximate values.

Example 4.20 **Determination of enthalpy and entropy from the Mollier chart**

Determine the enthalpy and entropy of a saturated steam using the Mollier chart when the pressure is 4 bar and temperature is 200°C.

Solution:

Consider The Mollier diagram for water-steam in Fig. 4.8.

Find the intersection point which is well above the saturation line on the Mollier chart when the pressure is 4 bar and temperature is 200°C.
At this intersection point, we can read the approximate value of enthalpy as 2850 kJ/kg and approximate value of entropy as 7.2 kJ/kg K.

4.11 Heat Transfer

During the transfer of energy in the form of work, some energy involved may be dissipated as heat. Heat and work are the two mechanisms by which energy can be transferred. Work performed on a body is, by definition, an energy transfer to the body that is due to a change to external parameters of the body, such as the volume, magnetization, and center of mass in a gravitational field [1]. Heat is the energy transferred to the body in any other way.

4.11.1 Conduction Heat Transfer

Heat is transferred by conduction, convection, and radiation. The heat transfer by conduction is based on direct molecular contact, most commonly in solids, while the convection is based on moving particles of liquids and gases. The heat transfer by radiation is based on the emission of electromagnetic waves from warm bodies to their surroundings. Rate of heat transfer by conduction is estimated by

$$\dot{q} = -kA\frac{(T_2 - T_1)}{\Delta x} \tag{4.40}$$

where $\dot{q}$ is the rate of heat transfer, J/s (Watt, W) or Btu/h, k is the thermal conductivity in J/m s °C or Btu/ft s °F, A is the area for heat transfer, $(T_2 - T_1)$ is the temperature difference between state 2 and state 1, and Δx is the distance. As the heat flows from high to low temperature, the value of $(T_2 - T_1)$ is negative. The values of thermal conductivity for various substances are tabulated in Table A1. Table 4.4 shows the thermal conductivities and heat capacities of some gases at 1 atm and some liquids and solids [1].

Table 4.4 Thermal conductivities and heat capacities of some gases(at 1 atm), liquids and solids [1]

Substance	T K	k W/m K
Gases		
Hydrogen, H_2	300	0.1779
Oxygen, O_2	300	0.0265
Carbon dioxide, CO_2	300	0.0166
Methane, CH_4	300	0.0342
Nitrogen monoxide, NO	300	0.0259
Liquids		
Water, H_2O	300	0.6089
	350	0.6622
Ethanol, C_2H_5OH	300	0.1676
Carbon tetrachloride, CCl_4	350	0.0893
Acetone, $(CH_3)_2CO$	300	0.1274
Mercury, Hg	372.2	10.50
Solids		
Lead, Pb	977.2	15.1
Aluminum, Al	373.2	205.9
Copper, Cu	291.2	384.1
Cast iron, Fe	373.2	51.9
Steel	291.2	46.9
Silver	373.2	411.9

4.11.2 Convection Heat Transfer

Rate of heat transfer by convection is estimated by

$$\dot{q} = hA(T_2 - T_1) \tag{4.41}$$

where h is the heat transfer coefficient in J/m^2 s °C or Btu/ft^2 s °F, and. Heat flux, $\dot{q}/A$, is the rate of heat transfer per unit cross-sectional area.

4.11.3 Radiation Heat Transfer

All bodies radiate energy in the form of photons moving in a random direction and frequency. When the photons reach another surface, they may either be absorbed, reflected, or transmitted [1]:

- α = absorptance – fraction of incident radiation absorbed
- τ = transmittance – fraction of incident radiation transmitted
- ρ = reflectance – fraction of incident radiation reflected

From energy considerations, the three coefficients must sum to unity: $\alpha + \tau + \rho = 1$. Real bodies radiate less effectively than black bodies. The measurement of this is the emittance ε defined by

$$\varepsilon = \frac{E}{E_b} \tag{4.42}$$

where E is radiation from the real body, and E_b is radiation from a black body at temperature T. Values of emittance are near unity for rough surfaces such as ceramics or oxidized metals, and roughly 0.02 for polished metals or silvered reflectors. The energy radiated per unit area of a black body is estimated by: $E_b = \sigma T^4$ where σ is the Stefan-Boltzmann constant, $\sigma = 5.67 \times 10^{-8}$ W/m^2 K^4. The level of emittance can be related to absorptance by

$$\frac{E}{E_b} = \varepsilon = \alpha \tag{4.43}$$

The relation $\varepsilon = \alpha$ is known as Kirchhoff's Law. It implies that good radiators are good absorbers. Heat transfer between gray and planar surfaces is [1]

$$\dot{q} = \frac{\sigma\left(T_1^4 - T_2^4\right)}{\frac{1}{\varepsilon_1} + \frac{1}{\varepsilon_2} - 1} \tag{4.44}$$

Example 4.21 **Estimation of radiation heat transfer**

Use a thermos bottle to estimate the heat transfer between thermos silver walls. The temperature of silver walls is $T_1 = 100°C = 373$ K and the surroundings temperature is $T_2 = 20°C = 293$ K. The values of the emittances are $\varepsilon_1 = \varepsilon_2 = 0.02$.

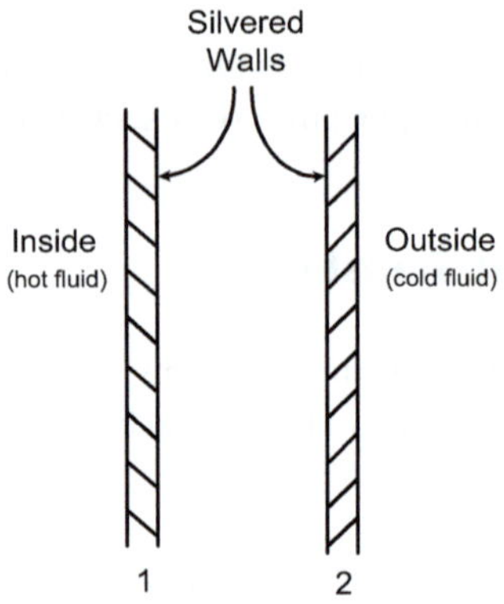

Solution:

The Stefan-Boltzmann constant, $\sigma = 5.67 \times 10^{-8}$ W/m^2 K^4,

$T_1 = 100°C = 373$ K, $T_2 = 20°C = 293.15$ K, and $\varepsilon_1 = \varepsilon_2 = 0.02$.

From Eq. (4.44): $\dot{q} = \frac{\sigma(T_1^4 - T_2^4)}{\frac{1}{\varepsilon_1} + \frac{1}{\varepsilon_2} - 1} = 6.9\,\text{W/m}^2$

The thermos is a good insulator as the heat transfer is very small from the hot wall to cold wall.

4.12 Exergy

An energy source comes with its amount of energy, and its work potential that is the useful energy or available energy. As heat flows out from a finite energy source with a high temperature to a low energy source, its temperature decreases, the quantity of available energy decreases, and so the quality of energy degrades. The amount and quality of energy in a source can be distinguished by the concept of exergy, *Ex*. Exergy is the maximum amount of work theoretically available by bringing a resource into equilibrium with its surroundings through a reversible process. Exergy is also known as availability. Quality of energy is the ability to produce work under the conditions determined by the *natural environment* called *the reference state*. The temperature and pressure of the natural environment are usually the standard-state values, such as 298.15 K and 101.31 kPa. High temperature fluid has a higher potential of heat-to-work conversion than the fluid at a low temperature. That is how exergy provides a means of defining energy quality [2].

The work done during a process depends on the initial state, final state, and the process path chosen between these initial and final states. The initial state is not a variable in an exergy analysis. In the final state, a process reaches thermodynamic equilibrium with the environment and the system's temperature and pressure become the same with the values of temperature and pressure of the environment. So, the system has no kinetic and potential energy related to the environment. The work output is maximized when the process takes place between the initial and final states in a reversible manner. In practical processes, this means that work output is maximized when the process proceeds through a path with the lowest possible *entropy production* between the initial and final states.

Kinetic and potential energies can be converted to work completely. Therefore, exergy of the kinetic energy and potential energy of a system are equal to the kinetic energy and potential energy

$$\Delta Ex_{\mathrm{KE}} = m\frac{\Delta v^2}{2} \tag{4.45}$$

$$\Delta Ex_{\mathrm{PE}} = mg\Delta z \tag{4.46}$$

However, the enthalpy H and internal energy U of a system are not completely available. The exergy change of a closed system during a process is

$$\Delta Ex = \Delta H - T_{\mathrm{o}}\Delta S \tag{4.47}$$

Here, the Δ denotes the difference between the initial and final state (reference state), and T_{o} is the reference temperature. The fluid flow exergy in metric units is

$$\dot{E}x = \dot{m}\left((H - H_{\mathrm{o}}) - T_{\mathrm{o}}(S - S_{\mathrm{o}}) + \frac{\Delta v^2}{2} + g\Delta z\right) \tag{4.48}$$

Exergy change of a fluid flow in an open system is given by

$$\Delta\dot{E}x = \dot{m}\left(\Delta H - T_{\mathrm{o}}\Delta S + \frac{\Delta v^2}{2} + g\Delta z\right) \tag{4.49}$$

Here, as before, the dot over a symbol indicates a quantity per unit time. If the changes is kinetic and potential energy are neglected, then Eq. (4.49) becomes

$$\Delta\dot{E}x = \dot{m}(\Delta H - T_{\mathrm{o}}\Delta S) \tag{4.50}$$

Heat (especially at low temperature) is much lower in exergy due to its low efficiency in its conversion to other forms of energy. The useful (available) work content of the heat is limited by the Carnot factor, η_{Carnot}, therefore, the available heat $q_{\mathrm{available}}$ becomes

$$Ex_{\mathrm{heat}} = q_{\mathrm{available}} = q\eta_{\mathrm{Carnot}} = q\left(1 - \frac{T_{\mathrm{C}}}{T_{\mathrm{H}}}\right) \tag{4.51}$$

where T_{H} and T_{C} are the temperatures for the hot and cold regions, respectively. Equation (4.51) means that a power generator based on heat cannot have efficiency greater than that of a Carnot cycle working between the hot source with T_{H} and cold source with T_{C}. It is possible to increase or lower exergy as exergy is not a conserved property [2].

Problems

4.1 Determine the states of the system containing water under the following conditions by using the steam tables.

T °C	P kPa	H_{vap} kJ/kg	U_{vap} kJ/kg	State
175	30	2830.0	2623.6	
45.83	10	2584.8	2438.0	
450	2400	3352.6	3027.1	
100	98	2676.0	2506.5	

4.2 Determine the states of the system containing water under the following conditions by using the steam tables.

T °F	P psia	H_{vap} Btu/lb	U_{vap} Btu/lb	State
164	5.99	1134.2	1065.4	
500	20	1286.9	1181.6	
600	1000	1517.4	1360.9	
331.37	105	1188.0	1105.8	

4.3 Determine the states of the system containing water under the following conditions by using the steam tables.

T °C	P kPa	V m^3/kg	$H_{\mathrm{sat\ vap}}$ kJ/kg	State
60	19.92	5.50	2609.7	
110	143.27	1.21	2691.3	
340	8200	0.0280	–	
40	47.36	0.001	–	

4.4 Determine the states of the system containing water under the following conditions by using the steam tables.

T °C	*P* kPa	*x*	*H* kJ/kg	State
120.2	200	0.7	2045.8	
80	47.36	1	2643.8	
150	476	0	632.1	
325	500	–	3116.4	

4.5 Determine the states of the system containing refrigerant-134a under the following conditions.

T °F	*P* psia	*H* Btu/lb	State
– 10	16.67	95.0	
5	23.8	102.47	
30	40.78	20.87	
60	70	110.23	

4.6 Determine the states of the system containing water under the following conditions by using the steam tables.

T °C	*P* kPa	*V* m^3/kg	*U* kJ/kg	State
133.5	300	0.50	2196.4	
300	725	0.358	2799.3	
400	6500	0.043	2888.9	
211.1	1950	0.001	900,46	

4.7 Determine the missing property values in the table below for a system that contains water.

P kPa	*V* m^3/kg	*U* kJ/kg	*H* kJ/kg	*x* kg vap/kg total mass	State
9.1		2250			
200				0.85	

4.8 Determine the missing property values in the table below for a system that contains pure refrigerant R-134a.

T °C	*P* kPa	*V* m^3/kg	*U* kJ/kg	*H* kJ/kg	State
30	836				
– 45			204.51		
	300	0.054			
	1124			244.0	
65	1550				

4.9 A 50 kg block of copper (Cu) at 0°C and a 100 kg block of iron (Fe) at 200°C are brought into contact in an insulated space. Predict the final equilibrium temperature.

4.10 Estimate the internal energy and enthalpy of 100 lb of water at 220 psia and 500°F.

4.11 A rigid tank contains 500 lb of saturated steam at 230°F. Estimate internal energy and enthalpy.

4.12 Determine the specific internal energy of 100 L of hydrogen gas at 600 K and 1 atm. Hydrogen gas has a total enthalpy of 19,416 J. Assume that the hydrogen gas is an ideal gas.

4.13 Calculate the change in internal energy when 0.75 m^3 of air at 200 kPa is heated from 20 to 250°C in a constant volume process.

4.14 Calculate the change in internal energy when 15 m^3 of air at 110 kPa is heated from 30 to 400°C in a constant volume process.

4.15 Calculate the change in internal energy when 1.5 m^3 of air at 275 kPa is cooled from 150 to 20°C in a constant volume process.

4.16 Calculate the change in internal energy when 10 m^3 of air at 150 kPa is cooled from 300 to 50°C in a constant volume process.

4.17 Calculate the change in internal energy when 25.1 m^3 of carbon dioxide at 250 kPa is heated from 30 to 350°C in a constant volume process.

4.18 Calculate the change in internal energy when 100 m^3 of carbon dioxide at 150 kPa is heated from 0 to 500°C in a constant volume process.

4.19 Calculate the change in internal energy when 5 m^3 of carbon dioxide at 250 kPa is cooled from 270 to 20°C in a constant volume process.

4.20 Calculate the change in internal energy when 52.8 m^3 of nitrogen at 200 kPa is heated from 15 to 700°C in a constant volume process.

4.21 Calculate the change in internal energy when 0.92 m^3 of nitrogen at 200 kPa is cooled from 300 to 30°C in a constant volume process.

4.22 Calculate the change in internal energy when 32.5 ft^3 of nitrogen at 20 psia is heated from 75 to 300°F in a constant volume process.

4.23 Superheated water vapor at 400°C expands isothermally in a piston-and-cylinder device from 1350 to 770 kPa. Calculate the change in the molar enthalpy and molar internal energy in units of kJ/mole.

4.24 One mole of gas is undergoing an expansion process, in which it expands from 0.03 to 0.06 m^3 at constant pressure of 2 atm. Estimate the change in internal energy if $C_{v,av} = 12.5$ J/mole K.

4.25 A one mole of gas at 300 K expands from 0.03 to 0.06 m^3 at constant temperature. Estimate the change in internal energy of the gas.

4.26 A large rigid tank contains 70 kg of water at 80°C. If 14 kg of the water is in the liquid state, estimate the pressure and enthalpy of the water in the tank.

4.27 A large rigid tank contains 40 lb of water at 200°F. If 2.8 lb of water is in the liquid state, estimate the internal energy and enthalpy of the water in the tank.

4.28 A rigid tank is filled with 0.5 lb of refrigerant-134a at $T = 79.2°F$. If 0.06 lb of the refrigerant is in the liquid state, estimate the internal energy and enthalpy of the refrigerant.

4.29 Heat capacity of toluene in Btu/lbmole °F is given by $C_p = 20.869 + 5.293 \times 10^{-2}T$. Estimate the heat capacity in cal/mole K and estimate the heat necessary if 400 g toluene is heated from 17 to 48°C.

4.30 Determine the enthalpy change and internal energy change of ammonia in J/mole, as it is heated from 30 to 150°C, using the ideal gas heat capacity given by the following equation:$C_p^{ig} = R\left(3.578 + 3.02 \times 10^{-3}T\right)$ T in K (Table B1).

4.31 Determine the enthalpy change and internal energy change of air in kJ/kg, as it is heated from 300 to 750 K, using the following equation for C_p.

$C_p^{ig} = R\left(3.355 + 0.575 \times 10^{-3}T\right)$ T in K (Table B1).

4.32 An ideal gases mixture of carbon dioxide (CO_2), nitrogen (N_2), and Oxygen (O_2) at 200 kPa and 300°C contains 10 mole% CO_2, 60 mole% N_2, and 30 mole% O_2. The mixture is heated to 600°C. Calculate the change in the molar internal energy of the mixture.

4.33 12 kg of benzene is heated from a subcooled liquid at $P_1 = 100$ kPa and $T_1 = 300$ K to a superheated vapor at $T_2 = 450$ K and $P_2 = 100$ kPa. The heating process is at constant pressure. Calculate the total enthalpy, ΔH, in kJ/kg. Assume the n-heptane vapor behaves as an ideal gas and has a constant heat capacity of 85.29 J/mole K. Use the Antoine and Clausius-Clapeyron equations to estimate the heat of vaporization of n-heptane at T_1.

4.34 Water is heated from a subcooled liquid at $P_1 = 200$ kPa and $T_1 = 300$ K to a superheated vapor at $P_2 = 2400$ kPa and $T_2 = 573$ K. Calculate the molar ΔH for each step in the path in kJ/kg.

4.35 Water is cooled from a superheated vapor at $P_1 = 250$ psia and $T_1 = 460°F$ to a subcooled liquid at $P_2 = 15$ psia and $T_2 = 140$ F. Calculate the molar ΔH for each step in the path in Btu/lb.

4.36 In a steady-state flow heating process, 50 kmole of methanol is heated from subcooled liquid at 300 K and 3 bar to superheated state at 500 kPa. Estimate the heat flow rate for the process. The saturation temperature for methanol at 3 bar is 368.0 K.

4.37 A one mole of gas at 300 K expands from 0.03 to 0.06 m^3 at adiabatic conditions. Estimate the change in internal energy of the gas. $C_{v.av} = 12.5$ J/mole K

4.38 A piston-and-cylinder device contains 0.3 kg refrigerant R-134a vapor at 12°C. This vapor is expanded in an internally reversible, adiabatic process until the temperature is − 3°C. Determine the work for this process. The constant heat capacity of refrigerant-134a is $C_{v,av} = 1.7$ kJ/kg °C.

4.39 A piston-and-cylinder device contains 0.5 kg nitrogen at 30°C. The nitrogen is compressed in an internally reversible, adiabatic process until the temperature is 42°C. Determine the work for this process. The ratio of constant heat capacities $\gamma = C_p/C_v = 1.4$ for the nitrogen.

4.40 Twenty kg of hydrogen are heated from 25 to 275°C. Estimate the change in enthalpy in kJ.

4.41 Fifty kg of hydrogen is heated from − 10 to 100°C. Estimate the change in enthalpy in kJ.

4.42 Twenty-five kg of hydrogen are cooled from 20°C to − 50°C. Estimate the change in enthalpy ΔH in Btu.

4.43 Ten kg of carbon dioxide is heated from 30 to 300°C in a steady-state flow process at a pressure sufficiently low that the carbon dioxide may be considered as an ideal gas. (a) Estimate the value of the heat required, (b) Estimate the average heat capacity in J/mole K for the process.

4.44 The molar heat capacity of propane gas in ideal gas state is given as a function of temperature in Kelvin:

$$C_p^{ig}/R = 1.213 + 28.785 \times 10^{-3}T - 8.824T \times 10^{-6} \text{ (Table B1)}$$

Convert this equation to an equation for C_p for temperature in °C.

4.45 The molar heat capacity of methane gas in ideal gas state is given as s function of temperature in Kelvin:

$$C_p^{ig}/R = 1.702 + 9.082 \times 10^{-3}T - 2.164T \times 10^{-6} \text{ (from Table B1)}$$

Convert this equation to an equation for C_p in J/g °C.

4.46 The molar heat capacity of methane gas in ideal gas state is given as s function of temperature in Kelvin:

$$C_p^{ig}/R = 1.702 + 9.082 \times 10^{-3}T - 2.164T \times 10^{-6} \text{ (from Table B1)}$$

Convert this equation to an equation for C_p in J/g °C.

4.47 Ten kg of water at 25°C in a closed vessel of 0.5 m^3 in volume is heated to 120°C and 1000 kPa. Estimate the change of enthalpy and the final quality of the steam in the vessel.

4.48 Twenty-two kg of water at 25°C in a closed vessel of 2.6 m^3 in volume is heated to 120°C and 1000 kPa. Estimate the change of enthalpy and the quality of the steam in the vessel.

4.49 Estimate the enthalpy of a 10 lb steam at 120 psia with a quality of 27%.

4.50 Calculate the change in enthalpy when 150 lb of nitrogen at 25 psia is heated from 50 to 700°F in a constant-pressure process. The heat capacity in Btu/lbmole °F is

$$C_p^{ig} = 6.895 + 0.7624 \times 10^{-3}T - 0.7009 \times 10^{-7}T^2$$

4.51 Calculate the change in enthalpy when 200 lb of nitrogen at 20 psia is cooled from 500 to 100°F in a constant-pressure process. The heat capacity in Btu/lbmole °F is

$$C_p^{ig} = 6.895 + 0.7624 \times 10^{-3}T - 0.7009 \times 10^{-7}T^2$$

4.52 Calculate the change in enthalpy when 500 lb of oxygen at 20 psia is heated from 60 to 450°F in a constant-pressure process. The heat capacity in Btu/lbmole °F is

$$C_p^{ig} = 6.895 + 0.7624 \times 10^{-3}T - 0.7009 \times 10^{-7}T^2$$

4.53 Calculate the change in enthalpy when 265 ft^3 of carbon dioxide at 20 psia is heated from 80 to 360°F in a constant-pressure process. The heat capacity in Btu/lbmole °F is

$$C_p^{ig} = 8.448 + 5.757 \times 10^{-3}T - 21.59 \times 10^{-7}T^2$$

4.54 Calculate the change in enthalpy when 1150 ft^3 of carbon monoxide at 20 psia is cooled from 250 to 75°F in a constant-pressure process. The heat capacity in Btu/lbmole °F is

$$C_p^{ig} = 6.865 + 0.8024 \times 10^{-3}T - 0.7367 \times 10^{-7}T^2$$

4.55 Calculate the change in enthalpy when 500 lb of carbon monoxide at 20 psia is heated from 50 to 290°F in a constant-pressure process. The heat capacity in Btu/lbmole °F is

$$C_p^{ig} = 6.865 + 0.8024 \times 10^{-3}T - 0.7367 \times 10^{-7}T^2$$

4.56 Ten kg of carbon dioxide at 50°C and 1 atm is heated to 200°C at constant pressure. Carbon dioxide is assumed to be an ideal gas. Estimate the changes in internal energy and enthalpy.

4.57 A 50 ft^3 rigid tank is filled with saturated vapor at 150 psia. The steam is cooled as the heat is transferred from the tank and the final pressure of the steam drops to 20 psia. Estimate: (a) the mass of the steam that has condensed, (b) the final temperature of the steam, and (c) the amount of heat transferred from the steam.

4.58 A piston-and-cylinder device contains R-134a which is initially at − 10°C and 200 kPa. The R-134a undergoes an isochoric heating to 80°C. Estimate the work in kJ/kg.

4.59 The primary energy source for cells is the aerobic oxidation of sugar called glucose ($C_6H_{12}O_6$). Estimate standard heat of reaction for the oxidation of glucose:

$$C_6H_{12}O_6(s) + 6O_2(g) \rightarrow 6CO_2(g) + 6H_2O(g)$$

4.60 Synthesis gas is a mixture of H_2 and CO. One of the uses of synthesis gas is to produce methanol from the following reaction: $CO(g) + 2H_2(g) \rightarrow CH_3OH(g)$

Estimate the standard heat of reaction.

4.61 Determine the standard heats of a reaction at 298.15 K for the reactions below:
(a) $N_2(g) + 3H_2(g) \rightarrow 2NH_3(g)$
(b) $CaCO_3(s) \rightarrow CaO(s) + CO_2(g)$

4.62 Estimate the heat of reaction for the combustion of propane:

$$C_3H_8(g) + 5O_2(g) \rightarrow 3CO_2(g) + 4H_2O(g)$$

4.63 Estimate the heat of reaction for the combustion of ethane:

$$C_2H_6+\frac{7}{2}O_2 \rightarrow 2CO_2(g) + 3H_2O(g)$$

4.64 Estimate the heat of reaction for the combustion of methane:

$$CH_4 + 2O_2 \rightarrow CO_2(g) + 2H_2O(g)$$

4.65 Nitric acid (HNO_3) is produced by the oxidation of ammonia. Estimate the standard heat of reaction: $NH_3 + 2O_2 \rightarrow HNO_3 + H_2O$

4.66 Estimate the maximum temperature that can be reached by the combustion of methane with air at constant pressure. Both the methane and air enter the combustion chamber at 25°C and 1 atm. The combustion reaction is $CH_4(g) + 2O_2(g) \rightarrow CO_2(g) + 2H_2O(g)$

4.67 Find the heat of reaction and determine whether the process is exothermic or endothermic for the following chemical reaction: $16H_2S(g) + 8SO_2(g) \rightarrow 16H_2O(l) + 3S_8(s)$.

4.68 The heating value of a fuel varies if the water produced is in a vapor state or liquid state. The higher heating value (HHV) of a fuel is its standard heat of combustion when the water product is liquid; the lower heating value when the water product is vapor. Estimate the HHV and the LHV for natural gas represented by pure methane: $CH_4(g) + 2O_2(g) \rightarrow CO_2(g) + 2H_2O(g)$.

4.69 Estimate the HHV and the LHV for heating oil represented by pure n-decane. As liquid n-decane heat of formation is − 249.7 kJ/mole: $C_{10}H_{22}(l) + 15.5O_2(g) \rightarrow 10CO_2(g) + 11H_2O(g)$.

4.70 What is the standard heat of combustion at 25°C of $C_6H_{12}(g)$ when the reaction products are $CO_2(g)$ and $H_2O(g)$?

$$C_6H_{12}(g) + 9O_2 \rightarrow 6CO_2(g) + 6H_2O(g)$$

4.71 What is the standard heat of combustion at 25°C of $6CH_3OH$ when the reaction products are $CO_2(g)$ and $H_2O(g)$? The reaction is: $6CH_3OH\ (g) + 9O_2 \rightarrow 6CO_2(g) + 12H_2O(g)$.

4.72 Estimate the standard heat of combustion of n-pentane gas at 25°C when the combustion products are $CO_2(g)$ and $H_2O(l)$:
$C_5H_{12}\ (g) + 8O_2\ (g) \rightarrow 5CO_2(g) + 6H_2O(l)$.

4.73 Estimate the standard heat of reaction of the following reactions at 25°C.

(a) $4NH_3(g) + 5O_2(g) = 4NO(g) + 6H_2O(g)$
(b) $C_2H_4(g) + 1/2O_2(g) = 2(CH_2)O(g)$
(c) $CH_3OH(g) + 1/2O_2(g) = HCHO(g) + H_2O(g)$
(d) $C_2H_5OH(l) + O_2(g) = CH_3COOH(l) + H_2(g)$

4.74 A heating oil with an average chemical composition of $C_{10}H_{18}$ is burned with oxygen completely in a bomb calorimeter. The heat released is measured as 43,900 J/g at 25°C. The combustion on the calorimeter takes place at constant volume and produces liquid water. Estimate the standard heat of reaction of the oil at 25°C when the products are H_2O (g) and CO_2 (g).

4.75 Derive the following isentropic relation for ideal gases with constant specific heats.

$$\frac{T_2}{T_1} = \left(\frac{P_2}{P_1}\right)^{(\gamma-1)/\gamma}$$

4.76 A piston-and-cylinder device with a free-floating piston has an initial volume of 0.1 m^3 and contains 0.6 kg of steam at 400 kPa. Heat is transferred into the steam until the temperature reaches 300°C while the pressure remains constant. Determine the heat and work transfer.

4.77 1 kg Superheated steam in a piston cylinder system is initially at 200°C and 1300 kPa. The steam is cooled at constant pressure until half of the mass condenses. Estimate the final temperature and the volume change.

4.78 1.21 lb Superheated refrigerant-134a is initially at 150°F and 140 psia. The refrigerant is cooled to 5°F at constant pressure until it exists as compressed liquid. Estimate the changes in internal energy and total volume.

4.79 Refrigerant-134a with a flow rate of 0.12 kg/s is cooled to 40°C at constant pressure in a condenser by using cooling water. Initially, the refrigerant is 1200 kPa and 50°C. Estimate the heat removed by the cooling water.

4.80 Air with a flow rate of 500 ft^3/s is heated from 100 to 910°F at approximately atmospheric pressure for a combustion process. What is the rate of heat transfer required?

4.81 An oven wall is made of 6-inch-thick forebrick. If the interior of the wall is at 1550°F and the exterior of the wall is at 350°F, estimate the heat loss per hour. The thermal conductivity of the firebrick is 0.81 Btu/h °F ft and assumed remain constant.

4.82 Estimate the thickness (Δx) needed for a cork insulation with $k = 0.04$ J/m s K when the heat transfer is 13.8 W/m^2 between the walls with temperatures $T_1 = 100$°C and $T_2 = 20$°C.

4.83 A resistance heater within a well-insulated tank is used to heat 0.8 kg of water. The heater passes a current of 0.2 ampere (A) for 30 min from a 120-V source. The water in the tank was originally at 25°C. Estimate the temperature of the water after 30 min of heating.

4.84 A resistance heater within a well-insulated tank is used to heat 4 kg of water. The heater passes a current of 0.2 ampere (A) from a 120-V source. The temperature of water in the tank was originally 20°C. Estimate the time necessary for the temperature of the water to reach 45°C.

4.85 A resistance heater within a tank is used to heat water. The heater passes a current of 0.4 ampere (A) for 2 h from a 120-V source. A heat loss of 8.7 kW occurs from the tank to surroundings. Estimate the amount of heat transferred to the water after 2 h of heating.

4.86 A resistance heater within a tank is used to heat 150 kg of water. The heater passes a current of 0.6 ampere (A) for 8 h from a 120-V source. A heat loss of 38.2 kW occurs from the tank to the surroundings. The water in the tank is originally at 5°C. Estimate the amount of heat transferred to the water after 8 h of heating.

4.87 A resistance heater within a well-insulated room is used to heat the inside air. The heater passes a current of 2 ampere (A) from a 120-V source. Estimate the amount of heat transfer to the air after 1 h of heating.

4.88 A resistance heater within a well-insulated room is used to heat the inside air. The heater passes a current of 2 ampere (A) for 3 h from a 120-V source. Estimate the amount of heat transferred to the air after 3 h of heating.

4.89 A resistance heater within a well-insulated room is used to heat the inside air. The heater passes a current of 0.8 ampere (A) for 2 h from a 120-V source. The heat loss from the room is 29.0 kJ. Estimate the amount of heat transferred to the air after 2 h of heating.

4.90 The wall of an oven is made of 10-inch-thick brick. If the interior wall of the oven is at 2000°F and the exterior of the wall is at 70°F, estimate the heat loss per hour. The average value of the thermal conductivity of the brick is 0.81 Btu/h ft °F.

4.91 A vertical plate receives solar radiation of 1000 W/m^2. The plate is insulated at the back and loses heat by convection heat transfer through the airflow. The heat transfer coefficient is 35 W/m^2 K and the air temperature is 22°C. The absorptivity of the surface α is 0.6. Estimate the surface temperature of the plate.

4.92 An insulated frame wall of a house has an average thermal conductivity of 0.0318 Btu/f ft R. The wall is 6 in thick, and it has an area of 160 ft^2. The inside air temperature is 70°F and the heat transfer coefficient for convection between the inside air and the wall is 1.5 Btu/h ft^2°R. On the outside of the wall, the convection heat transfer coefficient is 6 Btu/h ft^2°R and the air temperature is − 10°F. Ignoring radiation, determine the rate of heat transfer through the wall at steady state in Btu/h.

4.93 A small rigid cylindrical vessel contains 1 kg of humid air at 80°C under a total pressure of 1 bar. The humid air inside the vessel contains 0.150 kg of water vapor. Determine the relative humidity of the air in the container.

4.94 A small rigid cylindrical vessel contains 1 kg of humid air at 75°C under a total pressure of 1 bar. The humid air inside the vessel contains 0.10 kg of water vapor. Determine the relative humidity of the air in the container.

4.95 A water heating system delivers 230 kW with a temperature difference of 20°C into circulating water. Estimate the volumetric flow rate mass flow rate of the water.

4.96 Heat is added at a rate of 100 J/sec to a material whose initial temperature is 20°C. At this temperature, the material is a solid with a specific heat of 2.4 J/g °C. The melting point of this material is 36°C, the latent heat of fusion is 340 J/g, and the specific heat at liquid phase is 4.1 J/g °C. If the mass of material is 0.5 kg, how long will it take to reach a final temperature of 40°C?

4.97 A vessel contains ethanol and water in equilibrium. If the composition of the liquid is 35% ethanol, what is the water concentration in the vapor phase and what percent of liquid is present at these conditions?

4.98 A vessel contains ethanol and water in equilibrium. What is the change in mixing enthalpy if the system starts at 90% ethanol vapor and ends at 1% ethanol liquid?

References

1. Bird RB, Stewart WE, Lightfoot EN (2002) Transport phenomena, 2nd edn. Wiley, New York
2. Demirel Y (2013) Thermodynamics analysis. Arab J Sci Eng 38:221–249
3. Demirel Y, Gerbaud V (2025) Nonequilibrium thermodynamics: transport and rate processes in physical, chemical and biological systems, 5th edn. Elsevier, Amsterdam
4. Himmelblau DM, Riggs JB (2012) Basic principles and calculations in chemical engineering, 8th edn. Prentice Hall, Upper Saddle River

Energy Balances

5

Introduction and Learning Objectives: Global conservation law of energy states that the total energy of a system always remains constant. Mass and energy are always conserved in a process, while exergy is not conserved. In energy production systems, energy and mass conservations are solved together. By accounting the balance equations, energy interactions have been analyzed in turbines, compressors, nozzles, pumps, valves, mixers, and heat exchangers. Besides, the Carnot cyclic process is discussed briefly.

The learning objectives of this chapter are:

- Mass, energy, and exergy balance equations
- Balance equations for analyzing the fluid flow processes
- Heat exchangers

5.1 Balance Equations

According to energy conservation law, the total inflow of energy into a system must be equal to the total outflow of energy from the system. The flow of energy can be in various forms, such as work, heat, electricity, light, and other forms. The total energy of a system can be subdivided and classified in various ways. For example, it is sometimes convenient to distinguish potential energy (which is a function of coordinates) from kinetic energy (which is a function of time derivatives of coordinates). It may also be convenient to distinguish gravitational energy, electric energy, and thermal energy from each other.

Figure 5.1 shows an open system with mass and energy interactions with its surroundings. Especially for open systems involving fluid flow processes, energy

Y. Demirel, *Energy*, Green Energy and Technology,
https://doi.org/10.1007/978-3-031-97111-2_5

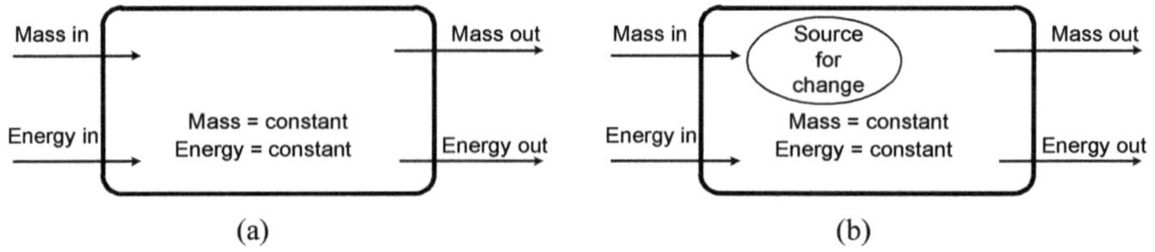

Fig. 5.1 **a** Open steady-state flow process with no internal source of changing mass and energy, **b** open steady-state flow process with an internal source for change in mass and energy, entropy, and exergy. For example, in an internal chemical reaction in a system, a substance may be consumed, and heat of reaction may be released. So, the chemical reaction acts as source for internal change apart from the boundary interactions of the system with its surroundings

balance alone may not fully describe the processes. Therefore, other balance equations of mass, entropy, and exergy may also be required. These balance equations are the main tools of analyses of energy interactions of various processes in closed and open systems. Mass and energy are always conserved in a process. However, entropy and exergy are not conserved and the balance equations of them account for a *source term* (see Fig. 5.1b) [1]. The source term mainly refers to internal change of mass, energy, entropy, and exergy in a system. For example, in an internal chemical reaction in a system, a substance may be consumed, and heat of reaction may be released. So, the chemical reaction acts as a source for internal change apart from the boundary interactions of the system with its surroundings. The conservation equations do not have the source terms and hence are different from the balance equations.

5.2 Mass Balance

In a closed system, mass is constant as the boundary does not permit the transfer of matter between the system and its surroundings. However, mass and energy interactions take place in an open system as seen in Fig. 5.1. The amount of mass flowing through a cross section per unit time is the mass flow rate $\dot{m}$. Here, the dot over the symbol for mass m indicates the amount of mass per unit time. The mass flow rate of a fluid flowing in a pipe is proportional to the cross-sectional area A of the pipe, the density ρ and the average velocity v of the fluid

$$\dot{m} = vA\rho \tag{5.1}$$

In Eq. (5.1), the fluid flow is assumed to be in a one-dimensional flow and the velocity is assumed to be constant across the cross section. When there is no chemical reaction within a system, *mass balance* for an unsteady state flow process is

$$\frac{\mathrm{d}m}{\mathrm{d}t} + \sum_i (\dot{m}_i)_{\text{out}} - \sum_i (\dot{m}_i)_{\text{in}} = 0 \tag{5.2}$$

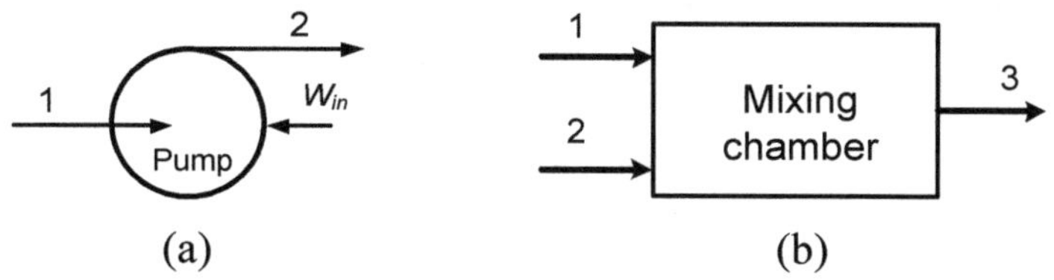

Fig. 5.2 Input and output flows in a pump and mixing chamber

As Eq. (5.2) shows that the total amount of incoming masses is equal to the total amount of outgoing masses in the same units. This is known as the *conservation of mass*. This form of mass conservation equation is also called the *continuity equation* [2]. The conservation of mass principles is a fundamental principle in nature. At steady-state flow processes, the accumulation term (dm/dt) becomes zero

$$\sum_i (\dot{m}_i)_{\text{out}} - \sum_i (\dot{m}_i)_{\text{in}} = 0 \tag{5.3}$$

For example, in a pump, the amount of water entering the pump is equal to the amount of water leaving it at a steady-state pumping process.

Many engineering devices, such as pumps, nozzles, compressors, and turbines, involve a single fluid flow. Usually, we denote the flow at the inlet state by subscript 1 and the flow at the exit state by subscript 2, as seen in for a pump in Fig. 5.2a. Then, for a single fluid flow system mass balance is

$$\dot{m}_1 = \dot{m}_2 \text{ or } \rho_1 v_1 A_1 = \rho_2 v_2 A_2 \tag{5.4}$$

For a mixer chamber shown in Fig. 5.2b, with two inlet flow and one outlet flow, mass balance becomes

$$\dot{m}_1 + \dot{m}_2 = \dot{m}_3 \tag{5.5}$$

Using the mole flow rate $\dot{n}$, the steady-state balance equation is

$$\sum_i (\dot{n}_i)_{\text{out}} - \sum_i (\dot{n}_i)_{\text{in}} = 0 \tag{5.6}$$

The volume of the fluid flowing in a pipe or a duct per unit time is known as the *volumetric flow rate* $\dot{Q}$ in m^3/s or ft^3/h

$$\dot{Q} = vA \tag{5.7}$$

where v is the average velocity of the flow and A is the cross-sectional area of a pipe or duct. The mass and volume flow rates are related by

$$\dot{m} = \rho \dot{Q} \tag{5.8}$$

For *incompressible fluids* such as liquids, the density remains constant. The conservation of mass for incompressible fluid flows is

$$\sum_i (\dot{Q}_i)_{\text{out}} - \sum_i (\dot{Q}_i)_{\text{in}} = 0 \tag{5.9}$$

The conservation of mass relations can be simplified for a single incompressible fluid flow by

$$\dot{Q}_{\text{in}} = \dot{Q}_{\text{out}} \text{ or } v_{\text{in}} A_{\text{in}} = v_{\text{out}} A_{\text{out}} \tag{5.10}$$

Liquids are mostly incompressible fluids. Sometimes, gases also may be assumed as incompressible under suitable conditions.

5.3 Energy Balance

In open and closed systems, energy can cross the boundary in the form of heat q and work W. If the energy exchange is because of temperature difference between a system and its surroundings, this energy exchange appears as heat otherwise it appears as work. When a force acts on a system over a distance, then energy is transferred as work. The heat and work interactions change the internal energy of a closed system ΔU

$$\Delta U = q + W \tag{5.11}$$

According to the modern energy sign convention recommended by the International Union of Pure and Applied Chemistry, heat and work are considered positive for their transfers into the system from their surroundings. Heat and work are negative when they are transferred out of the system to the surroundings. Total internal energy is an extensive property as its value depends on the quantity of material in a system.

Finite exchanges of energy between a system and its surroundings obey the energy balance

$$\Delta(\text{Energy of the system}) + \Delta(\text{Energy of the surroundings}) = 0 \tag{5.12}$$

Figure 5.1 shows the mass and energy interactions between a system and its surroundings. As the energy is conserved, the net rate of change of energy within the system is equal to the net rate of energy exchanged between the system and its surroundings at steady state. Streams in various numbers may flow into and out of the system. These streams may have energy associated with them in the forms of internal, kinetic, and potential energies. All these forms of energy contribute to the energy change of the system [1].

5.3.1 Unsteady-State Flow Systems

When there is no chemical reaction taking place in a system, the general energy balance for an unsteady state open system is

$$\frac{d(mU)}{dt} + \dot{m}\left(\Delta H + \frac{1}{2}\Delta v^2 + \Delta zg\right) = \dot{q} + \dot{W}_s \tag{5.13}$$

where U is the internal energy, H is the enthalpy, v is the average velocity, z is the elevation above a datum level, and g is the acceleration of gravity. The work $\dot{W}_s$ denotes the work other than the pressure work (PV) and is called the shaft work.

5.3.2 Steady-State Flow Systems

For *steady-state flow processes,* the accumulation term is zero ($d(mU)/dt = 0$) and the general energy balance at steady-state fluid flow process in metric units becomes

$$\dot{m}\left(\Delta H + \frac{\Delta v^2}{2} + g\Delta z\right) = \dot{q} + \dot{W}_s \tag{5.14}$$

For the English engineering systems of units, we have

$$\dot{m}\left(\Delta H + \frac{\Delta v^2}{2g_c} + \frac{g}{g_c}\Delta z\right) = \dot{q} + \dot{W}_s \tag{5.15}$$

If the kinetic and potential energy changes are small enough, then Eq. (5.15) reduces to

$$\sum_i (\dot{m}_i H_i)_{\text{out}} - \sum_i (\dot{m}_i H_i)_{\text{in}} = \dot{q} + \dot{W}_s \tag{5.16}$$

Equation (5.15) is widely applicable to many thermal engineering systems. For a single fluid flow process, energy balance becomes

$$\dot{m}\Delta H = \dot{q} + \dot{W}_s \tag{5.17}$$

When the energy is expressed per number of moles n, such as kJ/kmole or Btu/lbmole, the energy balance given Eq. (5.16) becomes

$$\sum_i (\dot{n}_i H_i)_{\text{out}} - \sum_i (\dot{n}_i H_i)_{\text{in}} = \dot{q} + \dot{W}_s \tag{5.18}$$

Consider an adiabatic, steady-state, and one-dimensional flow of a compressible fluid in a duct or in a pipe. If there is no shaft work and the changes in potential and kinetic energies are negligible, then Eq. (5.17) reduces to

$$\Delta H = 0 \tag{5.19}$$

All energy types in Eqs. (5.11) to (5.19) require the same energy units. In the SI system the unit for energy is the joule. Other units are the Btu, ft lbf, and calorie.

Mass and energy are the conserved properties, and they cannot be created or destroyed. However, mass and energy can be converted to each other according to the well-known equation proposed by Einstein

$$E = mc^2 \tag{5.20}$$

where c is the speed of light. Equation (5.20) suggests that the mass of a system is related to its energy. However, for all energy interactions taking place in practice, except the nuclear reactions, the change of mass is extremely small hence negligible.

Example 5.1 **Energy balance in a closed system**

A 60 m^3 room is heated by an electrical heater. The room air originally is 12°C and 100 kPa. The room loses heat at a rate of 0.2 kJ/s. The air has a constant heat capacity of $C_{v,av} = 0.72$ kJ/kg°C. If the electrical heater supplies energy at a rate of 3 kW, estimate the time necessary for the room temperature to reach 22°C.

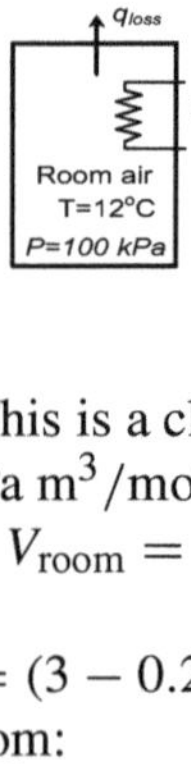

Solution:

Assumptions: The air is ideal gas. This is a closed system of air in the room.

$C_{v,av} = 0.72$ kJ/kg°C, $R = 8.314$ Pa m^3/mole K, $MW_{air} = 29$ kg/kmole.

$P_1 = 100$ kPa, $T_1 = 12$°C (285 K), $V_{room} = 60$ m^3, $T_2 = 22$°C (295 K),

Electrical energy supplied:

$W_e = 3$ kW, $q_{loss} = 0.2$ kW, $q_{net} = (3 - 0.2)$ kW $= 2.8$ kW.

The amount of mass of air in the room:

$m_{air} = MW(\text{moles of air}) = MW\frac{P_1 V_{room}}{RT_1} = 73.4$ kg.

Energy balance: $q_{net}\Delta t = (m\Delta U)_{air} = m_{air}C_{v,av}(T_2 - T_1)$

$$(2.8\text{ kW})(\Delta t) = (73.4\text{ kg})\left(0.72\text{ kJ/kg°C}\right)(22 - 12)\text{°C}$$

$$\Delta t = 189\text{ s} = 3.1\text{ min}$$

5.4 Exergy Balance

The physical exergy change of a closed system is [1, 5]

$$\Delta Ex = \Delta H - T_o \Delta S \tag{5.21}$$

Here the Δ denotes the difference between the initial and final state (final state − initial state), and T_o is the surroundings (dead state) temperature. Exergy change rate of fluid flow in an open system is

$$\Delta \dot{E}x = \dot{m}\left(\Delta H - T_o \Delta S + \frac{\Delta v^2}{2} + g\Delta z\right) \tag{5.22}$$

Exergy is an extensive property and a thermodynamic potential. Exergy is not conserved and always lost during a process. The exergy balance states that the exergy change during a process in a system is equal to the difference between the net exergy transfer and the exergy lost during a process within the system because of irreversibility. The exergy balance is

$$\Delta \dot{E}x = \dot{E}x_{\text{in}} - \dot{E}x_{\text{out}} - \dot{E}x_{\text{loss}} \tag{5.23}$$

$$\text{Change in exergy} = \text{exergy in} - \text{exergy out} - \text{exergy loss} \tag{5.24}$$

Exergy loss is positive in irreversible processes and vanishes in reversible processes. The rate of loss exergy $\dot{E}x_{\text{loss}}$ represents the overall imperfections and is directly proportional to the rate of entropy production due to irreversibility in a process. The rate of exergy loss (exergy destroyed) is

$$\dot{E}x_{\text{loss}} = T_o \dot{S}_{\text{prod}} \tag{5.25}$$

This equation shows that the rate of exergy transferred into the control volume must exceed the rate of exergy transferred out, and the difference is the exergy lost due to irreversibility [5].

Example 5.2 **Exergy loss estimations**

In a mixer, we mix saturated steam (stream 1) at 110°C with superheated steam (stream 2) at 1000 kPa and 300°C. The saturated steam enters the mixer at a flow rate of 1.5 kg/s. The product mixture (stream 3) from the mixer is at 350 kPa and 240°C. The mixer operates at adiabatic conditions. Determine the rate of energy dissipation if the surroundings are at 290 K.

Solution:

Assume that the kinetic and potential energy effects are negligible, this is a steady adaibatic process, and there are no work interactions.

The properties of steam from the steam tables:
Stream 1: Saturated steam: $T_1 = 110°C$, $H_1 = 2691.3$ kJ/kg, $S_1 = 7.2388$ kJ/kg K

Stream 2: Superheated steam: $P_2 = 1000$ kPa, $T_2 = 300°C$, $H_2 = 3052.1$ kJ/kg $S_2 = 7.1251$ kJ/kg K

Stream 3: Superheated steam: $P_3 = 350$ kPa, $T_3 = 240°C$, $H_3 = 2945.7$ kJ/kg, $S_3 = 7.4045$ kJ/kg K

The mass, energy, and entropy balances for the mixer at steady state are:

Mass balance: $\dot{m}_{\text{out}} = \dot{m}_{\text{in}} \rightarrow \dot{m}_1 + \dot{m}_2 = \dot{m}_3$

Energy balance: $\dot{E}_{\text{out}} = \dot{E}_{\text{in}} \rightarrow \dot{m}_1 H_1 + \dot{m}_2 H_2 = \dot{m}_3 H_3$

Entropy balance: $\dot{S}_{\text{in}} - \dot{S}_{\text{out}} + \dot{S}_{\text{prod}} = 0 \rightarrow \dot{m}_1 S_1 + \dot{m}_2 S_2 - \dot{m}_3 S_3 + \dot{S}_{\text{prod}} = 0$

Combining the mass and energy balances we estimate the flow rate of the superheated steam $\dot{m}_2$:

$$\dot{m}_1 H_1 + \dot{m}_2 H_2 = (\dot{m}_1 + \dot{m}_2) H_3$$
$$\dot{m}_2 = \frac{\dot{m}_1 (H_1 - H_3)}{H_3 - H_2} = \frac{1.5\ \text{kg/s}(2691.3 - 2945.7)\ \text{kJ/kg}}{(2945.7 - 3052.1)\ \text{kJ/kg}} = 3.58\ \text{kg/s}$$

The mass flow rate of the warm water

$\dot{m}_3 = \dot{m}_1 + \dot{m}_2 = 1.5 + 3.58 = 5.08$ kg/s.

The rate of entropy production for this adiabatic mixing process:

$$\dot{S}_{\text{prod}} = \dot{m}_3 S_3 - \dot{m}_1 S_1 - \dot{m}_2 S_2$$
$$\dot{S}_{\text{prod}} = (5.08\ \text{kg/s})(7.4045) - (1.5\ \text{kg/s})(7.2388) - (3.58\ \text{kg/s})(7.1251) = 1.25\ \text{kJ/s K}$$

The energy dissipated because of mixing:

$$\dot{E}_{\text{loss}} = T_o \dot{S}_{\text{prod}} = 290\ \text{K}(1.25\ \text{kJ/s K}) = 362.5\ \text{kW}$$

5.5 Fluid Flow Processes

Many thermal engineering devices and processes operate under the steady-state conditions for a long period of time. For example, the processes using pumps, boilers, turbines, and compressors operate at steady state until the system is shut down for maintenance, and they are analyzed as steady-state flow processes. Steady-flow energy equation can be applied to [2, 3]:

- Steam turbine
- Compressor
- Pump
- Nozzle
- Boiler
- Condenser
- Throttling
- Adiabatic mixing

Some examples of unsteady flow processes are filling of a liquefied petroleum gas (LPG) cylinder in a bottling plant and discharging LPG to the supply line to the gas oven.

5.5.1 Turbines, Compressors and Pumps

Figure 5.3 shows a single steady-fluid flow turbine, compressor, and pump. In a steam power plant, the hot and pressurized fluid passes through a turbine and the work is done against the blades, which are attached to the shaft. As a result, the turbine drives the electric generator and produces work. The work is transferred to the surroundings and its sign is negative. Heat transfer effects in a turbine are usually negligible [1–3].

Compressors, pumps, and fans increase the pressure of fluids; therefore, work from the surroundings must be supplied. Pumps use incompressible fluids that are liquids, while compressors and fans use compressible fluids of gases. Fans only slightly increase the pressure of gases. Potential energy changes are usually negligible in all these devices. Kinetic energy and heat effects are usually negligible in turbines.

For a pump, we may neglect the kinetic and potential energies and assume that there is no considerable heat exchange with the surroundings. For turbines, compressors, and pumps, we have the following similar relations for energy balance equation

$$\dot{m}\Delta H = \dot{m}(H_2 - H_1) = \dot{q} + \dot{W} \tag{5.26}$$

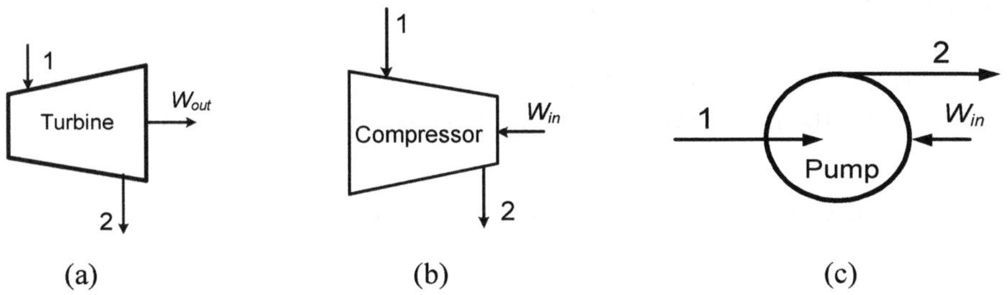

Fig. 5.3 Schematics of **a** turbine, **b** compressor, and **c** pump

For a turbine, the work is transferred to the surroundings and will have a negative sign. However, the work is transferred from the surroundings for compressors and pumps, and the sign of work for them is positive.

Example 5.3 **Turbine calculations**

A superheated steam (stream 1) expands in a turbine from 5000 kPa and 325°C to 150 kPa and 200°C. The steam flow rate is 10.5 kg/s. If the turbine generates 1.1 MW of power, determine the heat loss to the surroundings.

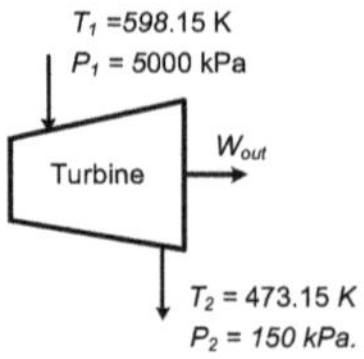

Solution:

Assume that the kinetic and potential energy effects are negligible; this is a steady process.

The properties of steam from the steam tables:

Stream 1: Superheated steam:

$P_1 = 5000$ kPa, $T_1 = 325$°C (598.15 K), $H_1 = 3001.8$ kJ/kg.

Stream 2: Superheated steam:

$P_2 = 150$ kPa, $T_2 = 200$°C (473.15 K), $H_2 = 2872.9$ kJ/kg.

$$\dot{W}_{\text{out}} = -1100\ \text{kW} = -1100\ \text{kJ/s}$$

The mass and energy balances for the turbine are:

Mass balance: $\dot{m}_{\text{out}} = \dot{m}_{\text{in}}$.

Energy balance: $\dot{E}_{\text{out}} = \dot{E}_{\text{in}}\ \dot{m}_1(H_2 - H_1) = \dot{q}_{\text{out}} + \dot{W}_{\text{out}}$.

Heat loss from the energy balance

$$\begin{aligned}\dot{q}_{\text{out}} &= -\dot{W}_{\text{out}} + \dot{m}_1(H_2 - H_1)\\ &= 1100\ \text{kJ/s} + (10.5\ \text{kg/s})(2872.9 - 3001.8)\text{kJ/kg} = -253.45\ \text{kJ/s}\end{aligned}$$

The sign is negative as the heat is lost from the system.

Example 5.4 **Compressor calculations**

Air enters a compressor at 100 kPa, 300 K, and a velocity of 2 m/s through a feed line with a cross-sectional area of 1.0 m^2. The effluent is at 500 kPa and 400 K. Heat is lost from the compressor at a rate of 5.2 kW. If the air behaves as an ideal gas, what is the power requirement of the compressor in kW.

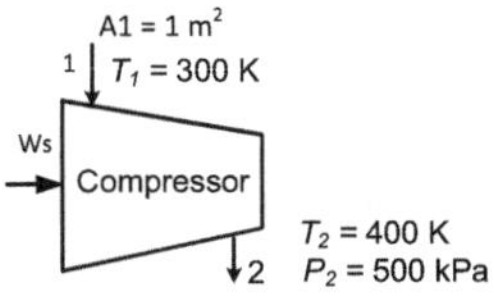

Solution:

Assume: The compressor is at steady state, air is ideal gas, and the change in the potential and kinetic energy of the fluid from the inlet to the outlet is negligible.

$R = 8.314\text{kPa m}^3/\text{kmole K}$, $MW = 29$ kg/kmole, $v_1 = 2$ m/s, $A_1 = 1.0\ \text{m}^2$

$P_1 = 100$ kPa, $T_1 = 300$ K, $P_2 = 500$ kPa, $T_2 = 400$ K, $\dot{q}_{\text{loss}} = -5.2$ kW
For ideal gas: $V_1 = \frac{RT_1}{P_1} = 0.025\ \text{m}^3/\text{mole}$, the specific volume $(1/\rho)_{\text{air}} = \frac{V_1}{MW_{\text{air}}} = 0.86\ \text{m}^3/\text{kg}$

The mass flow rate of air: $\dot{m} = \frac{v_1 A_1}{(1/\rho)} = 1.72\ \text{kg/s}$

From Table D1 for ideal gas of air: $H_2 = 400.9\ \text{kJ/kg}$, $H_1 = 300.2\ \text{kJ/kg}$

Energy balance: $\dot{m}(H_2 - H_1) - \dot{q} = \dot{W}$

$$(1.72\ \text{kg/s})((400.9 - 300.2)\text{kJ/kg}) + 5.2\ \text{kW} = 178.4\ \text{kW}$$

This is the work needed by the compressor.

Example 5.5 **Calculation of shaft work during air compression**

An air compressor compresses air from 100 to 600 kPa whereupon the internal energy of air has increased by 90 kJ/kg. The airflow rate is 9000 kg/h, and specific volumes of air at the inlet and outlet are 0.95 m^3/kg and 0.19 m^3/kg, respectively. The jacket cooling water removes 100 kW of heat generated due to the compression. Estimate the rate of the shaft work input to the compressor.

Solution:

Assume that the air is ideal gas and kinetic and potential energy changes are negligible.

From the first law of thermodynamics: $U_1 + P_1V_1 + q = U_2 + P_2V_2 + W_c$

$$\dot{m} = 9000\ \text{kg/h} = (9000\ \text{kg/h}) \times (\text{h}/3600\text{s}) = 2.5\ \text{kg/s}$$

$$2.5\ \text{kg/s}\big[(-90\ \text{kJ/kg}) + (100 \times 0.95 - 600 \times 0.19)\ \text{kPa m}^3\text{/kg}\big] - 100\ \text{kJ/s} = \dot{W}_c$$

$$2.5\ \text{kg/s}\big[(-90\ \text{kJ/kg}) + (-19)\ \text{kPa m}^3\text{/kg}\big] - 100\ \text{kJ/s} = \dot{W}_c$$

The shaft work to the compressor is – 372.5 kW

***Example 5.6* Pump power calculation**

A pump increases the pressure in liquid water from 100 kPa at 25°C to 4000 kPa. Estimate the minimum horsepower motor required to drive the pump for a flow rate of 0.01 m^3/s.

Solution:

Assume that the pump operates adiabatically and nearly isothermally, changes in potential and kinetic energy are negligible, and water is incompressible.

$$P_1 = 100\ \text{kPa}, T_1 = 298.15\ \text{K}, P_2 = 4000\ \text{kPa},\ Q = 0.01\text{m}^3/\text{s}$$

Energy Balance: $\dot{q} + \dot{W}_{\text{pump}} = \dot{m}\Delta H$

For an adiabatic operation ($\dot{q} = 0$): $\dot{W}_{\text{pump}} = \dot{m}\Delta H$

Since the water is incompressible, the internal energy is only the function of temperature.

$\Delta U = 0$ as the system is isothermal.

$$\Delta H = \Delta U + \Delta(PV) = V\Delta P$$

The specific volume of saturated liquid water at 25°C: $V_1 = 0.001003\ \text{m}^3/\text{kg}$ (steam tables)

The mass flow rate: $\dot{m} = \frac{Q}{V_1} = 9.97\ \text{kg/s}$

$$\begin{aligned}\dot{W}_{\text{pump}} &= \dot{m}V_1(P_2 - P_1)\\ &= (9.97\ \text{kg/s})(0.001003\ \text{m}^3/\text{kg})(4000 - 100)\ \text{kPa}\ \left(\text{kJ/kPa m}^3\right)\\ &= 39\ \text{kJ/s} = 39\ \text{kW}\ = 52.3\ \text{hp}\ (1\ \text{hp} = 745.7\ \text{W}).\end{aligned}$$

***Example 5.7* Pump work calculations**

A centrifugal pump delivers 1.8 m^3/h water to an overhead tank 20 m above the eye of the pump impeller. Estimate the work required by the pump, assuming a steady frictionless flow. Take density of water = 1000 kg/m^3.

Solution:

Assume that the pump operates adiabatically and nearly isothermally, changes in potential and kinetic energy are negligible, and water is incompressible.

Data: $Q = 1.8\ \text{m}^3/\text{h}$ and density of water: $= 1000\,\text{kg/m}^3$.

Energy balance: $\dot{q} + \dot{W}_{\text{pump}} = \dot{m}\Delta H$

$$\Delta\left(H + \frac{1}{2}v^2 + zg\right)\dot{m} = \dot{q} + \dot{W}_s$$

For an adiabatic operation ($\dot{q} = 0$) with no kinetic and potential energy effects:

$$\dot{W}_{\text{pump}} = \dot{m}g\Delta z = \left(\frac{1.8\ \text{m}^3/\text{h}}{3600\ \text{s/h}}\right)(1000)\ \text{kg/m}^3\left(9.81\ \text{m/s}^2\right)(30\ \text{m})$$
$$= 147.15\ \text{Nm/s} = 147.15\ \text{W}$$

5.5.2 Nozzles and Diffusers

Nozzles and diffusers are widely used in the manufacturing sector including in jet engines. A converging nozzle increases the velocity of fluid while the pressure of the fluid is reduced. A diffuser (diverging duct) increases the pressure of a fluid, while the velocity of the fluid is reduced. In nozzle and diffusers, the rate of heat interactions between the system and its surroundings are usually very small, and process may be analyzed as adiabatic ($q = 0$). Nozzle and diffusers generally involve no work, and changes in potential energy are negligible; therefore, the analyses of such processes may require kinetic energy changes [2, 3]. Figure 5.4a shows the schematics of nozzles and diffusers.

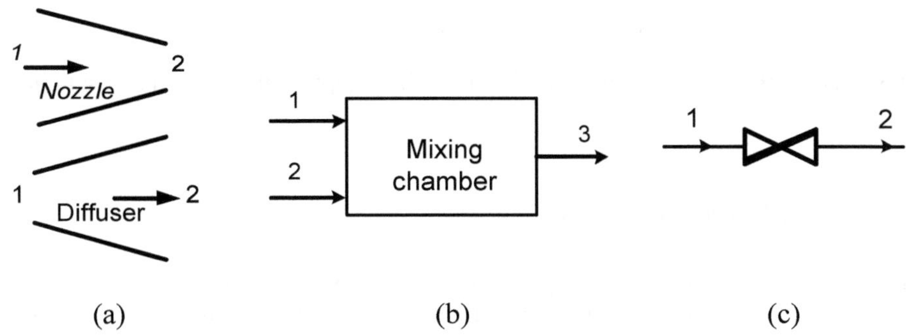

Fig. 5.4 Single steady-flow processes: **a** nozzle and diffuser, **b** mixing chamber, **c** throttle valve

For a converging nozzle the velocity increases, then we consider the kinetic energy, and usually neglect the potential energy. There is no work effect and energy balance becomes

$$\dot{m}\left(\Delta H + \frac{\Delta v^2}{2}\right) = \dot{q} \tag{5.27}$$

Example 5.8 **Nozzle calculations**

In a converging nozzle steam flow is accelerated from 20 m/s to 200 m/s. The steam enters the nozzle at 3000 kPa and 275°C and exits at 1700 kPa. The nozzle inlet cross-sectional area is 0.005m^2. The nozzle loses heat at a rate of 76 kJ/s. Estimate the steam mass flow rate and exit temperature.

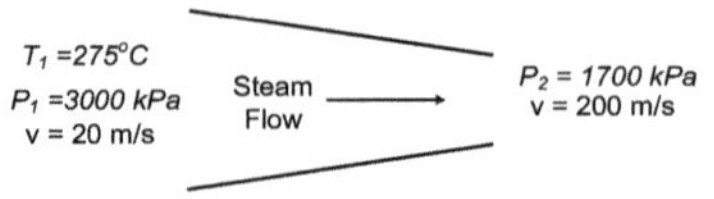

Solution:

Assumptions: This is a steady-state flow.

Inlet : $P_1 = 3000$ kPa, $T_1 = 275$°C, $v_1 = 20$ m/s, $A_1 = 0.005$ m^2, $\dot{q}_{\text{loss}} = -76$ kJ/s

$H_1 = 2928.2$ kJ/kg, $V_1 = 76.078$ cm^3/g $= 0.0761$ m^3/kg (steam tables)

Outlet: $P_2 = 1700$ kPa, $v_2 = 200$ m/s

The steam mass flow rate: $\dot{m}_s = \frac{1}{V_1}A_1v_1 = 1.314$ kg/s

Energy balance: $\dot{m}_s\left[(H_2 - H_1) + \left(\frac{v_2^2}{2} - \frac{v_1^2}{2}\right)\right] = \dot{q}_{\text{loss}}$

Estimate the enthalpy at the exit H_2:

$(1.314\text{kg/s})\left[(H_2 - 2928.2) + \left(\frac{200^2}{2} - \frac{20^2}{2}\right)\right] = -76$ kJ/s.

$H_2 = 2850.56$ kJ/kg, T_2 (at H_2 and P_2) $= 225$°C (approximate from steam tables)

5.5.3 Mixing Chambers

In thermal engineering applications, mixing two streams in a chamber is very common. For example, mixing a hot stream with a cold stream produces a warm stream. Conservation of mass and energy principles require that the sum of incoming mass flow rates and enthalpy rates are equal to the mass flow rates and enthalpy rates of the sum of outgoing streams. For example, for the mixing chamber in

Fig. 5.4b, mass and energy balances become

$$\dot{m}_1 + \dot{m}_2 = \dot{m}_3 \tag{5.28}$$

$$\dot{m}_1 H_1 + \dot{m}_2 H_2 = \dot{m}_3 H_3 \tag{5.29}$$

Mixing chambers are usually well insulated; hence, they are generally adiabatic. The kinetic and potential energy effects are negligible. Mixing chambers do not involve any kind of work production [1].

Example 5.9 **Mixing chamber calculations**

A hot air at 900 K is mixed with cold air at 300 K in a mixing chamber operating under adiabatic conditions. The cold airflow rate is 2.4 kg/s. The outlet stream is warm air at 700 K. Estimate the flow rate of hot air.

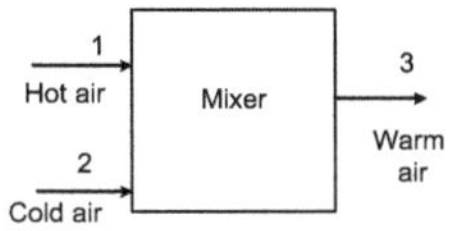

Solution:

Assumptions: This is a steady-state flow process. Kinetic and potential energy changes are negligible. The heat loss from the mixer is negligible. The air streams are ideal gases.

For an ideal gas, enthalpy depends only on the temperature

Hot air in: $\dot{m}_1 =$ unknown, $T_1 = 900$ K, $H_1 = 933.0\ \text{kJ/kg}$ (Table D1)

Cold air in: $\dot{m}_2 = 2.4\ \text{kg/s}$, $T_1 = 300$ K, $H_2 = 300.2\ \text{kJ/kg}$ (Table D1)

Outlet stream: Warm air at 700 K and $H_3 = 713.3\ \text{kJ/kg}$ (Table D1)

Energy and mas balances are: $\dot{m}_1 H_1 + \dot{m}_2 H_2 = \dot{m}_3 H_3$, $\dot{m}_1 + \dot{m}_2 = \dot{m}_3$

Combining and organizing these equations yield the hot air mass flow rate:

$$\dot{m}_1 = \dot{m}_2 \frac{H_3 - H_2}{H_1 - H_3} = (2.4\ \text{kg/s})\left(\frac{713.3 - 300.2}{933.0 - 713.3}\right) = 4.51\ \text{kg/s}$$

5.5.4 Throttling Valve

In a throttling valve (see Fig. 5.4c), a fluid flows through a restriction, such as an orifice, a partly closed valve, capillary tube, or a porous structure with negligible kinetic energy and potential energy changes. Throttling valve causes a considerable drop in pressure of fluid without involving any work. Throttling valves usually operate under adiabatic conditions [3]. When there is no or negligible heat transfer, then the energy balance equation for the single steady flow throttling becomes

$$H_1 = H_2 \tag{5.30}$$

Since the values of enthalpy at the inlet and at the outlet are the same, throttling valve is called as an *isenthalpic process*

$$\text{Internal energy} + \text{Pressure energy} = \text{Constant} \tag{5.31}$$

$$(U + PV)_1 = (U + PV)_2 = \text{Constant} \tag{5.32}$$

A throttling process does not change the temperature of an ideal gas since the enthalpy of an ideal gas depends only on temperature. For most real gases, however, pressure drop at constant enthalpy results in temperature decrease. If a wet steam is throttled to a considerably low pressure, the liquid evaporates, and the steam may become superheated. Throttling of a saturated liquid may cause vaporization (or flashing) of some of the liquid, which produces saturated vapor and liquid mixture. For throttling valves with large exposed surface areas such as capillary tubes, heat transfer may be significant.

Example 5.10 **Throttling process calculations**

(a) A steam at 600 kPa and 200°C is throttled to 101.32 kPa. Estimate the temperature of the steam after throttling.
(b) A wet steam at 1200 kPa and with a quality of 0.96 is throttled to 101.32 kPa. Estimate the enthalpy and the temperature of the throttled steam.

Solution:

Assume: No heat transfer occurs, and kinetic and potential energy changes are negligible in the throttling process.

Data from steam tables:

(a) Inlet: $P_1 = 600$ kPa, $T_1 = 200°\text{C}$, $H_1(600\ \text{kPa}, 200°\text{C}) = 2849.7\ \text{kJ/kg}$

Outlet: $P_2 = 101.32\ \text{kPa}$.

For adiabatic throttling: $H_1 = H_2$.

From steam tables: $T_2(101.32\ \text{kPa}, 2849.7\ \text{kJ/kg}) = 187.1°\text{C}$. (after an interpolation)

The temperature change ($\Delta T = 2.9°C$) is small

(b) Throttling wet steam to a lower pressure may cause the liquid to evaporate and the vapor to become superheated.

Inlet: the wet steam is saturated at: $P_1 = 1200$ kPa, $T_{sat} = 187.9\,°C, x = 0.96$ (steam tables).

$$H_{1\text{sat vap}} = 2782.7 \text{ kJ/kg}, H_{1\text{sat liq}} = 798.4 \text{ kJ/kg}$$
$$H_1 = (1 - x)H_{1\text{sat liq}} + xH_{1\text{sat vap}} = 2683.5 \text{ kJ/kg}$$

Outlet: $P_2 = 101.32$ kPa

$$T_2(101.32 \text{ kPa}, 2683.5 \text{ kJ/kg}) = 103.6^{\circ}\text{C}.$$

Since $T_2 > T_{2\text{ sat}}$ at 101.32 kPa, the throttled steam is superheated

The temperature change ($\Delta T = 84.3°C$) is considerable and results because of evaporation of liquid within the wet steam.

Example 5.11 **Throttling of a refrigerant**

Refrigerant-134a in a refrigerator enters a throttling valve as saturated liquid at 85.8 psia and is throttled to a pressure of 18.8 psia. Estimate the quality of the refrigerant and temperature drop at the exit.

Solution:

Inlet conditions: Saturated liquid (steam tables):

$P_1 = 85.8$ psia and $T_1 = 70°F, H_1 = 33.9$ Btu/lb

Exit conditions: Saturated mixture of liquid and vapor at (steam tables):

$$P_{2\text{sat}} = 18.8 \text{ psia and, } T_{2\text{sat}} = -5°\text{F},$$
$$H_{2\text{sat liq}} = 10.1 \text{ Btu/lb}, H_{2\text{sat vap}} = 101.0 \text{ Btu/lb}$$

Throttling process : $H_1 = H_2 = 33.9 = (1 - x)H_{2\text{sat liq}} + x\,H_{2\text{sat vap}}$

Quality of the saturated mixture: $x = \frac{m_{vap}}{m_{total}} = \frac{33.9-10.1}{101-10.1} = 0.25$

After the throttling process 25.0% of the saturated liquid is evaporated.

Temperature change: $\Delta T = T_2 - T_1 = (-5 - 70)°\text{F} = -75°\text{F}$

5.5.5 Heat Exchangers

In shell and tube heat exchangers, hot and cold fluids exchange heat without mixing. Figure 5.5 shows a typical countercurrent shell and tube heat exchanger where hot and cold streams flow opposite each other; in the tube side, a cold stream 4 enters and leaves the tube as hot stream 3, while hot stream 1 enters the shell side and leaves as cold stream 2. Heat exchangers may have various designs, such as double-pipe and shell and tube, and are widely used in thermal engineering systems [2]. For example, for the heat exchanger shown in Fig. 5.5, mass and energy balances become

$$\dot{m}_1 = \dot{m}_2 \text{ and } \dot{m}_3 = \dot{m}_4 \tag{5.33}$$

$$\dot{m}_1(H_2 - H_1) = \dot{m}_3(H_4 - H_3) \tag{5.34}$$

Heat exchangers involve no work interactions and usually operate with negligible kinetic and potential energy changes for hot and cold fluid flows.

Example 5.12 **Heat exchanger calculations**

Steam is cooled in a condenser using cooling water. The steam enters the condenser at 20 kPa and a quality of 0.9 and exits at 20 kPa as saturated liquid. The steam flow rate is 3 kg/s. The cooling water enters the condenser at 18°C and exits at 40°C. The average heat capacity of the cooling water is $C_{p,\text{av}} = 4.2$ kJ/kg °C. Estimate the mass flow rate of the cooling water.

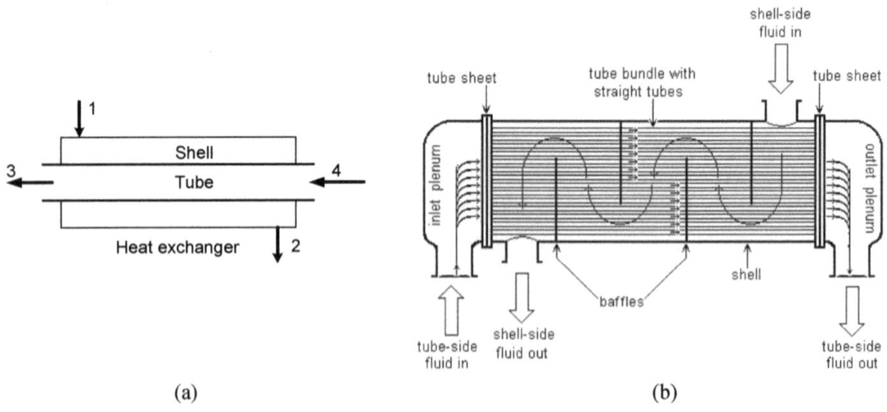

Fig. 5.5 **a** Schematic of a typical heat exchanger where hot and cold fluid flows do not mix with each other, **b** schematic of a countercurrent shell and tube heat exchanger. Baffles help distribute the shell side fluid around the tube bundle

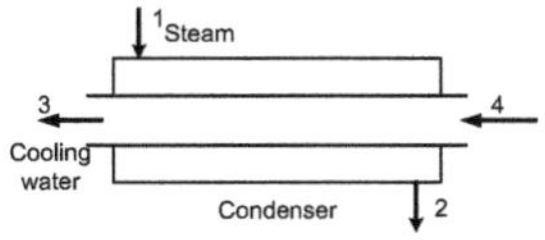

Solution:

Assumptions: This is a steady-state system and kinetic energy, and potential energies are negligible.

Steam flow rate: $\dot{m}_s = 3\ \text{kg/s}$

Cooling water: $T_4 = 18°\text{C}, T_3 = 40°\text{C}, C_{p,\text{av}} = 4.2\ \text{kJ/kg°C}$

Steam is saturated liquid and vapor mixture: $P_1 = 20\ \text{kPa}, x_1 = 0.90$

$$H_{1\text{sat liq}} = 251.4\ \text{kJ/kg}, H_{1\text{sat vap}} = 2609.7\ \text{kJ/kg (steam tables)}$$
$$H_1 = (1 - x_1)H_{1\text{sat liq}} + xH_{1,\text{sat vap}} = 2373.9\ \text{kJ/kg}$$

Exit: saturated liquid at $P_2 = 20\ \text{kPa}$, $H_2 = 251.4\ \text{kJ/kg}$ (steam tables)

Energy balance: heat removed by the cooling water = heat lost by the steam

$$\dot{m}_w C_{p,\text{av}}(T_3 - T_4) = \dot{m}_s(H_1 - H_2)$$
$$\dot{m}_w = \dot{m}_s \frac{(H_1 - H_2)}{C_{p,\text{av}}(T_4 - T_3)} = (3\ \text{kg/s})\left(\frac{2373.9 - 251.4}{(4.2\ \text{kJ/kg°C})(40 - 18)\ °\text{C}}\right) = 69.24\ \text{kg/s}$$

5.5.6 Pipe and Duct Flows

The transport of fluids in pipes and ducts has wide applications in many heating/cooling systems including heating ventilation and air-conditioning (HVAC) systems. Fluids mainly flow through pipes and ducts at steady-state flow conditions. Heat interactions of pipe and duct flows with their surroundings may be considerable, especially for long pipes and ducts. For example, a pipe with a hot fluid flowing inside must be insulated to reduce heat loss to the surroundings.

The velocities of fluid flows in pipes and ducts are usually low and the kinetic energy changes may be negligible. This is especially true when the cross-sectional area remains constant, and the heat effects are small. Kinetic energy changes, however, may be considerable when the cross-sectional area changes for gases flowing in a duct. The potential energy changes may also be considerable when a fluid undergoes a large elevation change as it flows in a pipe or duct [2, 3].

5.6 Energy Balance in a Cyclic Process

Most of the processes described above are part of some important thermal cyclic processes, such as Carnot cycle, heat pump, or refrigeration cycle. These thermal cyclic processes are used for power production as well as for heating ventilation and air-conditioning purposes [3, 4].

Fossil fuel-fired power plants mainly consume coal, natural gas, or oil. Figure 5.6 describes the main processes within the cycle. The main processes are:

- Process 4–1: The water is pumped to a boiler.
- Process 1–2: The high-pressure liquid enters a boiler where it is heated at a constant pressure by an external heat source q_H to become superheated vapor.
- Process 2–3: The superheated vapor expands through a turbine and generates power. This decreases the temperature and pressure of the vapor, and some condensation may occur.
- Process 3–4: The wet vapor then enters a condenser where the condensation of heat is discharged q_C to start a new cycle with the liquid water.

The process analysis yields the following energy balance equations. Pump power needed

$$\dot{W}_{\text{p,in}} = \dot{m}V_4(P_1 - P_4) \tag{5.35}$$

where $\dot{m}$ is the steam unit mass flow rate, V_4 is the specific volume at state 4 that is the saturated liquid water, P_i is the pressures at state i. Enthalpy rate at state 1, H_1, is

$$H_1 = H_4 + W_{p,\text{in}} \tag{5.36}$$

Fig. 5.6 Typical heat engine cycle to produce work

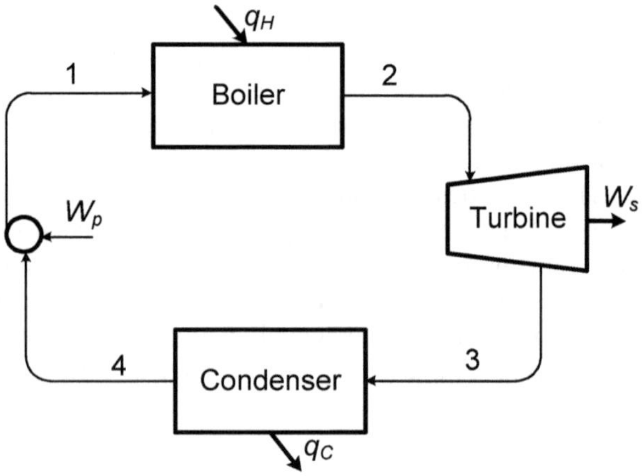

The heat input q_{in} to the pressurized water in the boiler is

$$\dot{q}_{\text{in}} = \dot{m}(H_2 - H_1) \tag{5.37}$$

The discharged heat rate q_{out} at the condenser is

$$\dot{q}_{\text{out}} = \dot{m}(H_4 - H_3) \tag{5.38}$$

The net value of the work $\dot{W}_{\text{net}}$ produced is

$$\dot{W}_{\text{net}} = (\dot{q}_{\text{in}} - \dot{q}_{\text{out}}) \tag{5.39}$$

Problems

5.1. Develop a simple energy balance formulation for a steady-state flow operation: (a) Pump; (b) gas turbine, (c) gas compressor, (d) nozzle, (e) throttle valve.

5.2. A copper ingot of 1 kg is heat treated at 400°C and then quenched in a 50 L oil bath initially at 15°C. Assuming no heat exchange with the surroundings, determine the temperature of the ingot and the oil when they reach thermal equilibrium using the data: $C_{\text{v,copper}} = 0.386$ kJ/kg K, $C_{\text{v,oil}} = 1.8$ kJ/kg K, $\rho_{\text{copper}} = 8900$ kg/m^3, $\rho_{\text{oil}} = 910$ kg/m^3.

5.3. A well-insulated 30-m^3 tank is used to store exhaust steam. The tank contains 0.01 m^3 of liquid water at 30°C in equilibrium with the water vapor. Estimate the internal energy of the tank.

5.4. Superheated steam at a rate of 0.6 lb/s flows through a heater. The steam is at 100 psia and 380°F. If an electrical heater supplies 37.7 Btu into the steam at constant pressure estimate the final temperature of the steam.

5.5. A tank of 4ft^3 contains superheated steam of 1 lb and 120 psia. The tank is heated at constant pressure. The amount of heat added is 200 Btu. Estimate the temperature of the steam after the heating process.

5.6. A 60 m^3 room is heated by an electrical heater. The air room originally was 12°C and 100 kPa. The room loses heat at a rate of 0.2 kJ/s. If the electrical heater supplies 3 kW, estimate the time necessary for the room temperature to reach 22°C.

5.7. A 1-kilowatt heater is used to heat a room with dimensions (3 m)(4.5 m)(4 m). Heat loss from the room is 0.15 kW and the pressure is always atmospheric. The air in the room may be assumed to be an ideal gas. The air has a constant heat capacity of $C_{v,\text{av}} = 0.72$ kJ/kg °C. Initially, the room temperature was 290 K. Determine the temperature in the room after 10 min.

5.8. A tank filled with 10 lb of air is heated. The air originally was at 40 psia and 100°F. After heating, the air pressure becomes 60 psia. Estimate the amount of heat transferred. Heat capacity for the air is 0.171 Btu/lb °F and remains constant during the heating process.

5.9. A 2.0 m^3 tank filled with hydrogen (H_2) at 200 kPa and 500 K. The hydrogen is cooled until its temperature drops to 300 K. Estimate the amount of heat removed from the tank. Heat capacity for the hydrogen is $C_{v,av} =$ 10.4 kJ/kg °C and remains constant during the heating process.

5.10. An electrical resistance heater is used to heat a room from 17°C to 22°C within 30 min. The room has dimensions of $4 \times 4x5$ m^3. The air in the room is at atmospheric pressure. The heat loss from the room is negligible. Estimate the power needed for this temperature change.

5.11. A room with dimensions of (5 m) (5 m) (6 m) is heated with a radiator supplying 8000 kJ/h. Heat loss from the room is 800 kJ/h. The temperature of the air was originally 15°C. Estimate the time necessary to heat the room to 24°C.

5.12. A 2-kilowatt heater is used to heat a room with dimensions (3 m) (4.5 m) (4 m). Heat loss from the room is negligible and the pressure is always atmospheric. The air in the room may be assumed to be an ideal gas, and the heat capacity is 29 J/mol K. Initially, the room temperature was 290 K. Determine the rate of temperature increase in the room.

5.13. A resistance heater within a well-insulated tank is used to heat 20 kg of water. The heater passes a current of 0.4 ampere (A) from a 120-V source. The temperature of water in the tank was originally 10°C. Estimate the time necessary for the temperature of the water to reach 60°C.

5.14. A resistance heater within a tank is used to heat 5 kg of water. The heater passes a current of 0.8 ampere (A) for 2 h from a 120-V source. A heat loss of 15.6 kW occurs from the tank to surroundings. The temperature of water in the tank was 30°C. Estimate the temperature of the water after 2 h of heating.

5.15. A resistance heater within a tank is used to heat 150 kg of water. The heater passes a current of 0.6 ampere (A) for 8 h from a 120-V source. A heat loss of 38.2 kW occurs from the tank to surroundings. The temperature of water in the tank was 5°C. Estimate the temperature of the water after 8 h of heating.

5.16. A resistance heater within a well-insulated 120 m^3 room is used to heat the inside air. The heater passes a current of 2 ampere (A) from a 120-V source. The air originally is at 15°C and 100 kPa. Estimate the time necessary for the inside air temperature to reach 21°C.

5.17. A resistance heater within a well-insulated 85 m^3 room is used to heat the inside air. The heater passes a current of 2 ampere (A) for 3 h from a 120-V source. The temperature of air was 20°C and 100 kPa. Estimate the temperature of the air after 3 h of heating.

5.18. A resistance heater within a well-insulated 105 m^3 room is used to heat the inside air. The heater passes a current of 0.8 ampere (A) for 2 h from a 120-V source. The air originally is at 17°C and 100 kPa. The heat loss from the room is 29.0 kJ. Estimate the temperature of the air after 2 h of heating.

5.19. A condenser in a power plant is used to condense the discharged steam from the turbine by using cooling water. The temperature rise of the cooling water is 12°C. The discharged steam enters the condenser at 10 kPa with a quality of 0.95 and exits as saturated liquid water at constant pressure. If the steam flow rate is 5 kg/s estimate the flow rate of cooling water.

5.20. A condenser in a power plant is used to condense the discharged steam from the turbine by using cooling water at a flow rate of 160 kg/s. Inlet temperature of the cooling water is 15°C. The discharged steam enters the condenser at 10 kPa with a quality of 0.95 and exits as saturated liquid water at constant pressure. If the discharged steam flow rate is 6.5 kg/s, estimate the exit temperature of the cooling water.

5.21. A 20 lb refrigerant-134a flowing in a coil is cooled by an airflow. The refrigerant enters the coil at 140°F and 140 psia and leaves as saturated vapor at 140 psia. The air enters the heat exchanger at 14.7 psia and 480 R. Volumetric flow rate of the air is 150 ft^3/s. Estimate the exit temperature of the air.

5.22. A 5 lb refrigerant-134a flowing in a coil is cooled by airflow. The refrigerant enters the coil at 200°F and 100 psia and leaves as saturated vapor at 100 psia. The air enters the heat exchanger at 14.7 psia and 480 R. Volumetric flow rate of the air is 250 ft^3/s. Estimate the exit temperature of the air.

5.23. Refrigerant-134a is cooled in a condenser using cooling water. The refrigerant enters the condenser at 1000 kPa and 90°C and exits at 1000 kPa and 30°C. The cooling water at a flow rate of 2.0 kg/s enters the condenser at 20°C and exits at 27°C. Estimate the mass flow rate of the refrigerant.

5.24. Steam is cooled in a condenser using cooling water. The steam enters the condenser at 20 kPa and a quality of 0.9 and exits at 20 kPa as saturated liquid. The steam flow rate is 3 kg/s. The cooling water enters the condenser at 18°C and exits at 26°C. Estimate the mass flow rate of the cooling water.

5.25. A compressed liquid water at 20°C and a superheated steam at 150°C and 400 kPa are fed into a mixing chamber operating under adiabatic conditions. The flow rate of compressed liquid water is 12 kg/s. The outlet stream is saturated liquid water at 400 kPa. Estimate the flow rate of superheated steam.

5.26. A superheated steam at a flow rate of 50 kg/s and at 150°C, 200 kPa is mixed with another superheated steam at 350°C and 200 kPa in a mixing chamber operating under adiabatic conditions. The outlet stream is a superheated steam at 225°C and 200 kPa. Estimate the flow rate of superheated steam at 350°C and 200 kPa.

5.27. Hot water at 80°C is mixed with cold water at 20°C in a mixing chamber operating under adiabatic conditions. Flow rate of cold water is 15.0 kg/s. The outlet stream is warm water at 45°C and 100 kPa. Estimate the flow rate of hot water.

5.28. Hot air at 900 R is mixed with cold air at 520 R in a mixing chamber operating under adiabatic conditions. The flow rate of the cold air is 24 lbmole/s. The outlet stream is warm air at 700 R. Estimate the flow rate of hot air.

5.29. Superheated steam at 300°C and 300 kPa is mixed with liquid water with a flow rate of 4 kg/s and at 20°C in a mixing chamber operating under adiabatic conditions. The outlet stream is 60°C and 300 kPa. Estimate the flow rate of superheated steam.

5.30. An open-feed water heater in a steam power plant operates at a steady state with liquid entering at $T_1 = 40°C$ and $P_1 = 7$ bar. Water vapor at $T_2 = 200°C$ and $P_2 = 7$ bars enters in a second feed stream. The effluent is saturated liquid water at $P_3 = 7$ bar. Estimate the ratio of the mass flow rates of the two feed streams.

5.31. Initially, the mixing tank shown below has 100 kg of water at 25°C. Later two other water inlet streams 1 and 2 add water and outlet stream 3 discharges water. The water in the tank is well mixed and the temperature remains uniform and equal to the temperature of outlet stream 3. Stream 1 has a flow rate of 20 kg/h and is at 60°C, while stream 2 has a flow rate of 15 kg/h and is at 40°C. The outlet stream has a flow rate of 35 kg/h. Determine the time-dependent temperature of the water in the mixing tank.

5.32. In a mixer, we mix saturated steam (stream 1) at 110°C with superheated steam (stream 2) at 1000 kPa and 300°C. The saturated steam enters the mixer at a flow rate of 1.5 kg/s. The product mixture (stream 3) from the mixer is at 350 kPa and 240°C. The mixer loses heat at a rate of 2 kW. Determine the rate of energy dissipation if the surroundings are at 300 K.

5.33. In a mixer, we mix liquid water (1) at 1 atm and 25°C with superheated steam (2) at 325 kPa and 200°C. The liquid water enters the mixer at a flow rate of 70 kg/h. The product mixture (3) from the mixer is at 1 atm and 55°C. The mixer loses heat at a rate of 3000 kJ/h. Estimate the flow rate of superheated steam and the rate of entropy production.

5.34. In a steady-state mixing process, 50.25 kg/s of saturated steam (stream 1) at 501.15 K is mixed with 7.363 kg/s of saturated steam (stream 2) at 401.15 K. The mixer is well insulated and adiabatic. Determine the energy dissipation (work loss) if the surroundings are at 298.15 K.

5.35. In a steady-state mixing process, 50.0 kg/s of saturated steam (stream 1) at 501.15 K is mixed with 17.0 kg/s of saturated steam (stream 2) at 423.15 K. The product steam (stream 3) is at 473.15 K. Determine the rate of heat loss.

5.36. In a steady-state mixing process, 15 kmole/s of air (stream 1) at 550 K and 2 atm is mixed with 40 kmole/s of air (stream 2) at 350 K and 1 atm. The product (stream 3) is 300 K and 1 atm. Determine the heat loss.

5.37. A steam turbine produces 3.80 MW. The steam at the inlet is at 1000 psia and 900°F. The steam at the exits is saturated at 12 psia. Estimate the heat loss if the steam flow rate is 12.8 lb/s.

5.38. A steam turbine produces 3.80 MW. The steam at the inlet is at 900 psia and 900°F. The steam at the exits is saturated at 5 psia. Estimate the heat loss if the steam flow rate is 16.2 lb/s.

5.39. A steam turbine consumes 4000 lb/h steam at 540 psia and 800°F. The exhaust steam is at 165 psia. The turbine operation is adiabatic.

(a) Determine the exit temperature of the steam and the work produced by the turbine.
(b) Determine the entropy production if the turbine efficiency is 80%.

5.40. A superheated steam (stream 1) expands in a turbine from 5000 kPa and 325°C to 150 kPa and 200°C. The steam flow rate is 10.5 kg/s. If the turbine generates 1.1 MW of power, determine the heat loss to the surroundings.

5.41. A superheated steam (stream 1) expands in a turbine from 5000 kPa and 325°C to 150 kPa and 200°C. The steam flow rate is 10.5 kg/s. If the turbine generates 1.1 MW of power, determine the heat loss to the surroundings.

5.42. Steam expands in a turbine from 6600 kPa and 300°C to a saturated vapor at 1 atm. The steam flow rate is 9.55 kg/s. If the turbines' power output is 1 MW, estimate the rate of heat loss.

5.43. Steam expands adiabatically in a turbine from 850 psia and 600°F to a wet vapor at 12 psia with a quality of 0.9. Estimate the steam flow rate if the turbines' power output is 2 MW.

5.44. Steam expands adiabatically in a turbine from 800 psia and 650°F to a wet vapor at 10 psia with a quality of 0.95. Estimate the steam flow rate if the turbine's power output is 2 MW.

5.45. Air enters an insulated compressor operating at steady state at 1.0 bar, 300 K with a mass flow rate of 3.6 kg/s and exits at 2.76 bar. Kinetic and potential energy effects are negligible. (a) Determine the minimum theoretical power input required, in kW, and the corresponding exit temperature, in K. (b) If the actual exit temperature is 420 K, determine the power input, in kW, and the isentropic compressor efficiency.

5.46. A compressor increases the pressure of carbon dioxide from 100 to 600 kPa. The inlet temperature is 300 K and the outlet temperature is 400 K. The mass flow rate of carbon dioxide is 0.01 kmole/s. The power required by the compressor is 55 kW. The temperature of the surroundings is 290 K. Determine the minimum amount of work required.

5.47. Hydrogen gas is compressed from an initial state at 100 kPa, 300 K and 5 m^3 to 300 kPa and 370 K. The compression process is polytropic ($PV^\gamma =$ constant). The average heat capacity of hydrogen is $C_{p,\mathrm{av}} = 29.1$ J/mole K. Estimate the total entropy change of the system.

5.48. A compressor receives air at 15 psia and 80°F with a flow rate of 1.2 lb/s. The air exits at 40 psia and 300°F. At the inlet, the air velocity is low but increases to 250 ft/s at the outlet of the compressor. Estimate the power input of the compressor if it is cooled at a rate of 200 Btu/s.

5.49. In an adiabatic compression operation, air is compressed from 20°C and 101.32 kPa to 520 kPa. The airflow rate is 22 mol/s. Estimate the work required under isentropic operation.

5.50. Refrigerant-134a with a flow rate of 0.1 lb/s enters a compressor as saturated vapor at $T_1 = 5$°F ($P_{\mathrm{sat}} = 23.8$ psia) and exits a superheated vapor at 60°F and 40 psia. Estimate the energy transferred during the compression process.

5.51. Air enters a compressor at 100 kPa, 300 K, and a velocity of 6 m/s through a feed line with a cross-sectional area of 3 m^2. The effluent is at 500 kPa and 400 K and has a velocity of 2 m/s. Heat is lost from the compressor at a rate of 160 kW. If the air behaves as an ideal gas, what is the power requirement of the compressor in kW.

5.52. Carbon dioxide at a flow rate of 1.5 kg/s enters a compressor at 300 K and 100 kPa and exits at 480 K. Estimate the volumetric flow rate of CO_2 and the power for compression.

5.53. Propane with a flow rate of 5 lb/s enters a compressor as a saturated vapor at 30°F and 66.5 psia. The propane leaves the compressor as superheated vapor at 140 psia and 100°F. Estimate the heat transfer to propane in the steady-state flow process.

5.54. In a two-stage continuous compression process, methane (stream 1) enters the first compressor at 300 K and 1 bar. The methane (stream 2) leaves the second compressor at 300 K and 60 bar. The flow rate of methane is 0.5 kg/s. The total power input is 400 kW. The intercooler between the compressors uses cooling water. The surroundings are at 295 K. Estimate the minimum work required and the heat removed in the intercooler.

5.55. In a two-stage continuous compression process, methane (stream 1) enters the first compressor at 300 K and 1 bar. The methane (stream 2) leaves the second compressor at 350 K and 80 bar. The flow rate of methane is 0.6 kg/s. The total power input is 450 kW. The intercooler between the compressors uses cooling water. The cooling water enters the cooler of 295 K and leaves at 305 K. The surroundings are at 295 K. Estimate the cooling water rate.

5.56. In an adiabatic compression operation, air is compressed from 20°C and 101.32 kPa to 520 kPa with an efficiency of 0.7. The airflow rate is 20 mol/s. The air is assumed to be an ideal gas. Estimate the ideal work required.

5.57. Refrigerant-134a enters a compressor operating at steady-state as saturated vapor at 2 bars with a volumetric flow rate of 0.019 m^3/s. The refrigerant is compressed to a pressure of 8 bar in an internally reversible process according to $PV^{1.03=}$ constant. Estimate (a) the power required, in kW, and (b) the rate of heat transfer, in kW.

5.58. Air is compressed from 1 bar and 310 K to 8 bars. Estimate the specific work and heat transfer if the air follows a polytropic process path with $\gamma = 1.32$.

5.59. A pump increases the pressure in liquid water from 100 at 25°C to 2500 kPa. What is the minimum horsepower motor required to drive the pump for a flow rate of 0.005 m^3/s?

5.60. A pump increases the pressure in liquid water from 100 kPa at 25°C to 3000 kPa. What is the minimum horsepower motor required to drive the pump for a flow rate of 0.25 m^3/s?

5.61. In a converging nozzle, airflow is accelerated from 20 m/s to 60 m/s. The air at the inlet is at 400 kPa, 127°C. The inlet cross-sectional area of the nozzle is 90 cm^2. At the outlet, the airflow pressure drops to 100 kPa. Estimate the air mass flow rate and exit temperature.

5.62. In a converging nozzle, steam flow is accelerated from 20 m/s to 200 m/s. The steam enters the nozzle at 3000 kPa and 275°C and exits at 900 kPa. The nozzle inlet cross-sectional area is 0.005 m^2. The nozzle loses heat at a rate of 76 kJ/s. Estimate the steam mass flow rate and exit temperature.

5.63. In a converging nozzle, carbon dioxide (CO_2) flow is accelerated to 400 m/s. The CO_2 enters the nozzle at 1000 kPa and 500 K and exits at 900 kPa. Flow rate of the CO_2 is 5000 kg/h. The nozzle inlet cross-sectional area is 0.005m^2. Estimate the inlet velocity and exit temperature of CO_2.

5.64. Steam expands in a nozzle from inlet conditions of 500°F, 250 psia, and a velocity of 260 ft/s to discharge conditions of 95 psia and a velocity 1500 ft/s. If the flow is at 10 lb/s and the process is at steady state and adiabatic, determine the outlet temperature.

5.65. Air with a flow rate of 1 kg/s enters a nozzle at 400 K and 60 m/s and leaves the nozzle at a velocity 346 m/s. The air inlet pressure is 300 kPa, while the pressure at the outlet is 100 kPa. The rate of heat lost in the nozzle is 2.5 kJ/kg. Determine the enthalpy at the outlet.

5.66. Steam enters a nozzle at 30 psia and 300°F and exits as a saturated vapor at 300°F. The steam enters at a velocity of 1467 ft/s, and leaves at 75 ft/s. The nozzle has an exit area of 0.5 ft^2. Determine the rate of energy dissipation when the ambient temperature $T_o = 500$ R.

5.67. A steam enters a nozzle at 500 kPa and 220°C and exits at 400 kPa and 175°C. The steam enters at a velocity of 200 m/s, and leaves at 50 m/s. The nozzle has an exit area of 0.2 m^2. Determine the rate of heat loss to the surroundings.

5.68. A steam enters a nozzle at 4000 kPa and 425°C with a velocity of 50 m/s. It exists at 286.18 m/s. The nozzle is adiabatic and has an inlet area of 0.005 m^2. Estimate the enthalpy of the outlet stream.

5.69. Steam enters a nozzle at 3200 kPa and 300°C with a velocity of 20 m/s. It leaves at 274.95 m/s. The nozzle is adiabatic and has an inlet area of 0.01 m^2. Estimate the enthalpy at the outlet and the amount of loss of work if the surroundings are at 300 K.

5.70. Steam at 8200 kPa and 500°C passes through a throttling process so that the pressure is suddenly reduced to 7400 kPa. What is the expected temperature after the throttle?

5.71. 5 kg/s of superheated steam at 400°C and 1100 kPa is throttled to 125 kPa adiabatically through a valve. Determine the outlet temperature and the work loss if the surroundings are at 298.15 K.

5.72. Refrigerant-134a expands adiabatically in an expansion valve. The refrigerant enters at $P_1 = 140$ psia and $T_1 = 140$°F and exits at 50 psia. The reference conditions are $P_o = 21.2$ psia and $T_o = 0$°F. Determine the loss of exergy and second law efficiency.

5.73. Superheated steam with a flow rate of 2.0 kg/s is throttled to 1700 kPa. The steam is initially at 4100 kPa and 300°C. Estimate the lost work potential of the steam during this throttling process.

5.74. Refrigerant-134a is throttled by a valve. The refrigerant enters the valve as a saturated liquid at 30 psia and exits at 15 psia. Estimate the exit temperature and quality of the refrigerant.

5.75. Steam enters the valve at 1000 psia and 600°F and is throttled by a valve adiabatically. At the exit, the steam is at 850 psia. Estimate the drop in temperature.

5.76. Steam at 900 psia and 700°F is throttled in a valve to 55 psia at a rate of 20 lb/min at steady state. Determine the entropy production due to expansion of steam.

References

1. Çengel YA, Boles MA (2014) Thermodynamics: an engineering approach, 8th edn. McGraw-Hill, New York
2. Çengel YA, Turner R, Cimbala J (2011) Fundamentals of thermal-fluid sciences, 4th edn. McGraw-Hill, New York
3. Chattopadhyay P (2015) Engineering thermodynamics, 2nd edn. Oxford Univ Press, Oxford
4. Demirel Y, Gerbaud V (2025) Nonequilibrium thermodynamics. Transport and rate processes in physical, chemical and biological systems, 5th ed. Elsevier, Amsterdam
5. Demirel Y (2013) Thermodynamics analysis. The Arabian J Sci Eng 38:221–249

6 Energy Production

Introduction and Learning Objectives: After transforming the primary energy resources, the secondary energy in the form of liquid and gas fuels of petroleum products, electricity, heat, as well as biofuels can be produced. For example, chemical energy found in fossil fuels such as coal or natural gas is used in steam power plants to produce electricity. Electricity is also produced by other means such as the kinetic energy of flowing water, wind, solar photovoltaic, and geothermal power. Hydroelectric power is produced when falling water turns an electromagnet that generates electricity. Nuclear power is produced when neutrons bombard heavy atoms, which release energy in a process called fission. The world's power demand is expected to rise considerably. Cogeneration is the production of more than one useful form of energy (such as process heat and electric power) from the same energy source.

The learning objectives of this chapter are:

- Production technologies of electricity and fuels.
- Biofuel production.
- Fuel cell technologies.

6.1 Energy Production

Energy is mainly produced from transforming the primary energy sources into secondary energy needed the most. Primary energy sources are coal, natural gas, crude oil, nuclear, waste, solar, wind, biomass, hydro, and geothermal. For example, the chemical energy stored in fossil fuels and nuclear sources are used to produce approximately 80% of the world's electricity. After refining petroleum oil into its fractions, many useful types of energy including Liquefied Petroleum Gases

Y. Demirel, *Energy*, Green Energy and Technology,
https://doi.org/10.1007/978-3-031-97111-2_6

(LPG), gasoline, kerosene, diesel, and fuel oil are produced. Therefore, secondary energy mainly introduces a transformation process between the primary energy sources and the secondary energy sources. There are various forms of energy production in the world, each with its own risks and benefits [1, 13]. *Cogeneration* refers to the production of more than one useful form of energy, for example power and heat from the same energy source.

6.1.1 Electric Power Production

The fundamental principles of electricity production were discovered during the 1820s and early 1830s by the British scientist Michael Faraday. Based on Faraday's work, electricity is produced by the movement of a loop of wire, or disk of copper between the poles of a magnet [9].

A fossil fuel power plant produces electricity by converting fossil fuel energy into mechanical work. The two major power plant systems are based on the steam turbine cycle and the gas-turbine cycle mostly using fossil fuels. The steam power cycle relies on the Rankine cycle in which high-pressure and high-temperature steam produced in a boiler is expanded through a turbine that drives an electric generator. The discharged steam from the turbine gives up its heat of condensation in a condenser to a heat sink such as water from a river or a lake. The condensate is pumped back into the boiler to restart a new cycle. The heat taken up by the cooling water in the condenser is dissipated mostly through cooling towers into the atmosphere (see Fig. 6.1). Most of the electricity comes from the steam turbines that expand superheated steam produced in the boiler using various heat sources, including fossil fuels, nuclear energy, and renewables. Steam turbines produce a considerable portion of the electric power in the world [2–4].

Assuming the nitrogen remains inert, the complete combustion of a fossil fuel using air as the oxygen source may be represented approximately by

$$C_xH_y + \left(x + \frac{y}{4}\right)O_2 + 3.76\left(x + \frac{y}{4}\right)N_2 \rightarrow xCO_2 + \left(\frac{y}{2}\right)H_2O - 3.76\left(x + \frac{y}{4}\right)N_2 + \text{Heat} \quad (6.1)$$

where C_xH_y represents a fossil fuel consisting of carbon (C) and hydrogen (H) only, with stoichiometric coefficients of x and y, respectively, depending on the fuel type. fossil fuels, such as coal or natural gas, may have other elements such as sulfur and compounds such as minerals beside carbon and hydrogen. A simple example is the combustion of coal (taken here as consisting of pure carbon): $C+O_2 \rightarrow CO_2$. Coal is prepared by grinding it to powder and mixing it with air which preheats the coal to dry excess moisture content and transports it to the furnace.

Example 6.1 illustrates the analysis of an adiabatic steam turbine. Centralized power production became possible when it was recognized that alternating current power lines can transport electricity at low costs across great distances and power transformers can adjust power by raising and lowering the voltage.

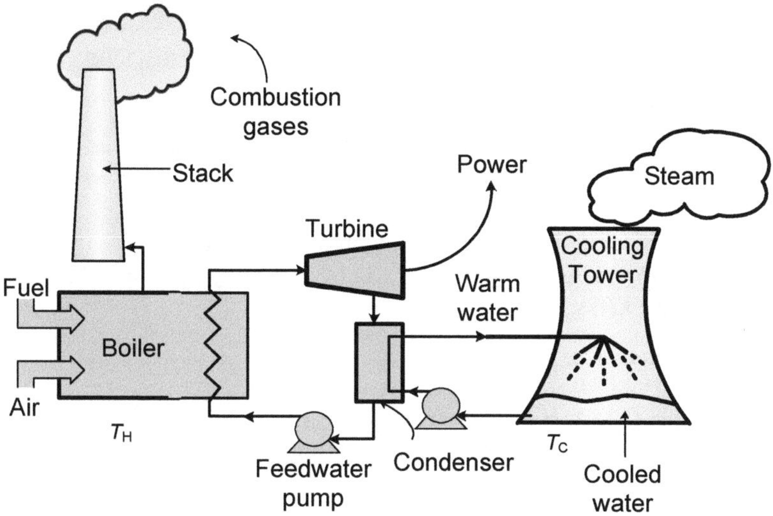

Fig. 6.1 Schematic of a steam power plant

Renewable energy sources such as solar photovoltaics, wind, biomass, hydro, and geothermal can also provide clean and sustainable electricity. However, renewable energy sources are naturally variable, requiring energy storage or a hybrid system to accommodate daily and seasonal changes. One solution is to produce hydrogen through electrolysis by splitting water and to use that hydrogen in a fuel cell to produce electricity [9].

Example 6.1 Power production by an adiabatic steam turbine

Superheated steam at 4100 kPa and 300°C expand adiabatically in a steam turbine and exits at 15 kPa with a quality of $x = 0.9$. The velocity of the steam at the inlet is 50 m/s and at the exit 160 m/s. Elevation at the inlet is 10 m and at the exit 6 m. Estimate the power produced for the steam flow rate of 1kg/s.

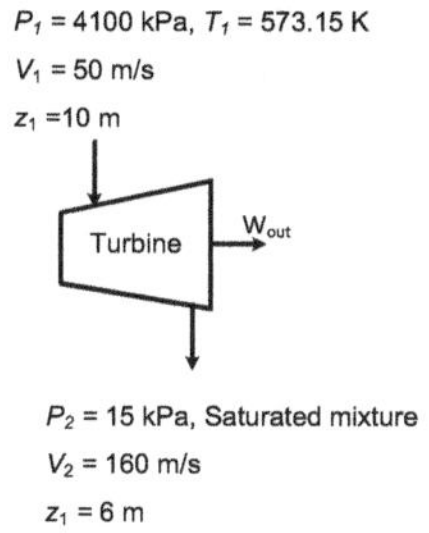

Solution:

Assume: Steady-state adiabatic ($q_{loss} = 0$) process.

Steam flow rate: $\dot{m}_s = 1$ kg/s; using steam tables.

Inlet conditions: $P_1 = 4100$ kPa and $T_1 = 300°C$, $v_1 = 50$ m/s, $z_1 = 10$ m,

$H_1 = 2958.5$ kJ/kg.

Exit conditions: Saturated mixture of liquid and vapor water:
$P_{2,sat} = 15$ kPa, $T_{2,sat} = 54.0°C$, $H_{2sat\ liq} = 226.0$ kJ/kg, $H_{2sat\ vap} = 2599.2$ kJ/kg,

$x = 0.9$, $v_2 = 160$ m/s, $z_2 = 6$ m

$$H_{2mix} = (1 - x)H_{sat\ liq} + xH_{sat\ vap}$$
$$= (1 - 0.9)226\ \text{kJ/kg} + (0.9)2599.2\ \text{kJ/kg} = 2361.9\ \text{kJ/kg}$$

Energy balance:

$$\dot{W}_{out} = \dot{m}\left(\Delta H + \frac{\Delta v^2}{2} + g\Delta z\right)$$

$$\Delta H = H_2 - H_1 = (2361.9 - 2958.5)\text{kJ/kg} = -596.6\ \text{kJ/kg}$$

$$\Delta KE = \frac{v_2^2 - v_1^2}{2} = \left(\frac{160^2 - 50^2}{2}\right)\text{m}^2/\text{s}^2\left(\frac{\text{kJ/kg}}{1000\ \text{m}^2/\text{s}^2}\right) = 11.55\ \text{kJ/kg}$$

$$\Delta PE = g(z_2 - z_1)\left(\frac{\text{kJ/kg}}{1000\ \text{m}^2/\text{s}^2}\right) = 9.81(6 - 10)\text{m}^2/\text{s}^2\left(\frac{\text{kJ/kg}}{1000\ \text{m}^2/\text{s}^2}\right)$$
$$= -0.04\ \text{kJ/kg}$$

$$\dot{W}_{out} = \dot{m}\left(\Delta H + \frac{\Delta v^2}{2} + g\Delta z\right) = (1\ \text{kg/s})(-596.6 + 11.5 - 0.04)\text{kJ/kg}$$
$$= -585.1\ \text{kW}$$

Contribution of kinetic energy is less than 2%, and the contribution of potential energy is negligible. In practice, changes in kinetic and potential energies are assumed to be negligible.

6.1.2 Distributed Energy Productions

Distributed energy production refers to small power plants or storage systems located close to the point of use. Distributed energy resources usually have higher efficiencies through cogeneration and may reduce emissions of carbon dioxide, because of their use of onsite renewable resources and low-greenhouse gas fuels

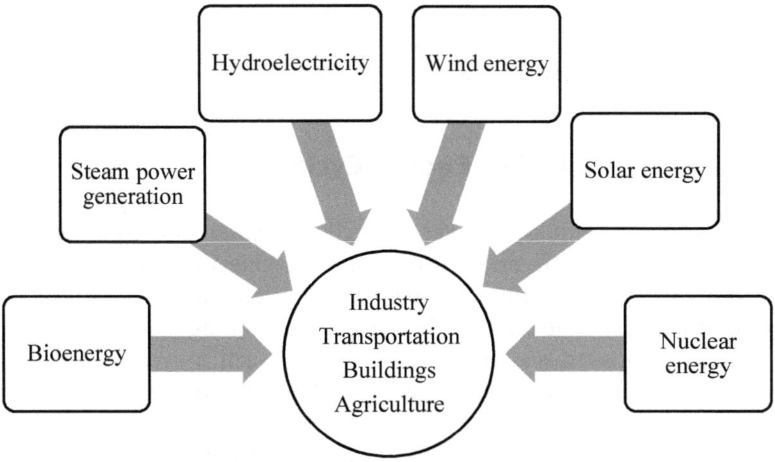

Fig. 6.2 Transition to distributed energy resources from producers into end users

such as natural gas as seen in Fig. 6.2. Some distributed generation technologies, like photovoltaic and fuel cells, can generate electricity with no emissions or less emissions than large-scale *centralized* fossil fuel-fired power plants. Advanced sensors and controls, energy storage and heat exchangers with improved waste heat recovery improve efficiency of distributed energy systems. Other advantages include fuel source flexibility, reduced transmission line losses, enhanced power quality and reliability, and more user control. Distributed generation permits consumers who are producing electricity for their own needs, to send their surplus electrical power back into the power grid [1, 4, 11].

6.2 Transmission of Energy

Electricity grid transmits and distributes power from power plants to end users. Sources include coal or natural gas burning power plants as well as nuclear and solar power plants. A combination of substations, transformers, towers, and cables is used to maintain a constant flow of electricity at the required levels of voltage such as 110 V or 220 V. Grids also have a predefined carrying capacity or load. New small-scale energy sources may be placed closer to the consumers so that less energy is lost during electricity distribution [11].

Crude oil is carried and distributed through long pipelines between the sources and refineries where it is fractioned into many different types of fuels, such as gasoline, kerosene, diesel, and fuel oil. Pipelines can also distribute the refined fractions of the crude oil to end users. Slurry pipelines are sometimes used to transport coal. Oil pipelines are made of steel or plastic tubes with an inner diameter typically from 0.3–4 ft (0.1–1.2 m). For natural gas, pipelines are constructed of carbon steel and vary in size from 0.2–5 ft (0.06–1.5 m) in diameter, depending on the type of pipeline. The gas is pressurized by compressor stations. Most pipelines are buried at a typical depth of about 3–6 ft (0.9–1.8 m) [1].

6.3 Power Producing Engine Cycles

The systems used to produce power are called *engines*. Most power producing engines operate with cyclic processes using a working fluid. Steam power plants use water as the working fluid. Actual engine cycles are complex to analyze. Therefore, it is common to analyze the cycles by assuming that they operate without friction, heat losses, and other complexities. Such a cycle is known as the ideal engine. The analysis of ideal engines yields the major operating parameters controlling the cycle performance.

A *heat engine*, for example, converts heat to mechanical energy by bringing a working fluid from a high temperature T_H to a lower temperature state T_C. Figure 6.3 shows a typical pressure–volume *PV* and temperature-entropy *TS* diagrams of ideal engine cycles. On both the *PV* and *TS* diagrams, the area enclosed with the process curves of a cycle represents the network produced during the cycle, which is equivalent to the net heat transfer for the cycle. As seen in Fig. 6.3, the characteristics of cycles on a *TS* diagram are [5]:

- Heat addition increases pressure and temperature
- Heat rejection process decreases temperature
- The isentropic (internally reversible and adiabatic) process takes place at constant entropy

The area under the heat addition process on a *TS* diagram measures the total heat input, and the area under the heat rejection process measures the total heat output. The difference between these two areas represents the net heat transfer and the network output. Therefore, any modifications that improve the net heat transfer

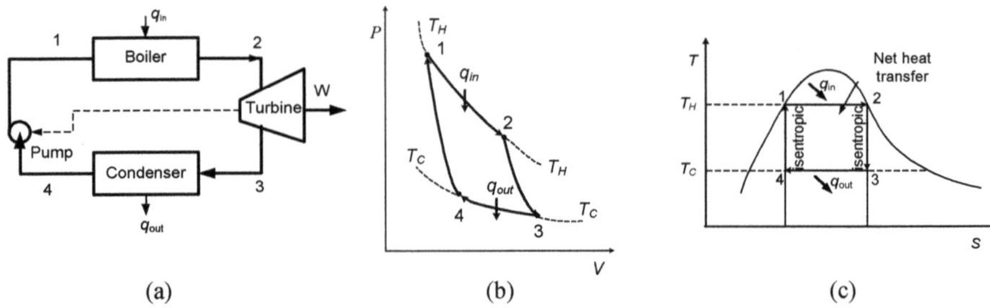

Fig. 6.3 Typical ideal engine cycles: **a** cyclic process, **b** on a pressure–volume (*PV*) diagram, **c** on a temperature-entropy (*TS*) diagram

rate will also improve the network output. Example 6.2 illustrates the analysis of power output from a steam power plant [2, 4, 7, 9].

A heated working fluid at high temperature generates work in the engine while transferring remaining heat to the colder "sink" until it reaches a low temperature state. The working fluid usually is a liquid or gas. During the operation of an engine some of the thermal energy is converted into work and the remaining energy is lost to a heat sink, mainly the general surroundings. Example 6.3 estimates the water mass flow rate for a given power output.

Example 6.2 **Steam power production**

A steam power production plant uses steam at 8200 kPa and 823.15 K. The turbine discharges the steam at 30 kPa. The turbine and pump operate reversibly and adiabatically. Determine the work produced for every kg steam produced in the boiler.

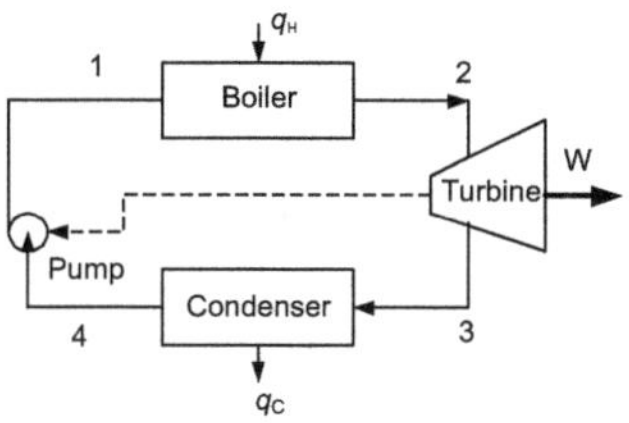

Solution:

Assume that the turbine and pump operate reversibly and adiabatically; no pressure drop across condenser and boiler; only heat transfer occurs at condenser and boiler.

Steam flow rate: $\dot{m}_s = 1$ kg/s.

Data from steam tables:
Turbine inlet: $P_2 = 8200$ kPa, $T_2 = 823\ 15$ K, $H_2 = 3517.8$ kJ/kg, $S_2 = 6.8646$ kJ/kg K.

Turbine outlet: $P_3 = 30$ kPa, $T_3 = 342.27$ K, $V_4 = 0.001022$ m^3/kg.

$H_{3\text{sat liq}} = 289.30$ kJ/kg, $H_{3\text{sat vap}} = 2625.4$ kJ/kg, $S_{3\text{sat liq}} = 0.9441$ kJ/kg K, $S_{3\text{sat vap}} = 7.7695$ kJ/kg K.

For an ideal operation $S_2 = S_3 = 6.8646$. Since S$_3 < S_{3\text{sat vap}}$ the discharged steam is a mixture of liquid and vapor. Solve for fraction of vapor, x, using entropy balance for the exhaust of the turbine:

$$x_3 = \frac{S_2 - S_{3\text{sat liq}}}{S_{3\text{sat vap}} - S_{3\text{sat liq}}} = \frac{6.8646 - 0.9441}{7.7695 - 0.9441} = 0.867$$

$$H_3 = 0.867(2625.4) + (1 - 0.867)(289.3) = 2315.8\,\text{kJ/kg}$$

$$T_2 = 342.27 \text{ K (Saturated)}$$

$$H_1 = H_4 + (P_1 - P_4)V_4 = 289.3 + (8200 - 30)(0.001022) = 297.65\,\text{kJ/kg}$$

$$\begin{aligned} q_{out} &= -(H_4 - H_3) = -(289.3 - 2315.8)\,\text{kJ/kg} \\ &= 2026.5\,\text{kJ/kg(heat received by the cooling medium)} \\ q_{in} &= (H_2 - H_1) = (3517.8 - 297.65)\,\text{kJ/kg} \\ &= 3220.2\text{kJ/kg (heat received by the steam)} \end{aligned}$$

Work output for 1 kg/s steam:

$$\dot{W}_{out} = \dot{m}_s(q_{out} - q_{in}) = -1193.7\,\text{kW(work output by the turbine)}$$

Note that the work produced is negative as it is outgoing work of the system, that is the turbine.

Example 6.3 **Steam flow rate calculation in a power plant**

A steam power plant output is 50 MW. It uses steam (stream 1) at 8200 kPa and 550°C. The discharged stream (stream 2) is saturated at 75 kPa. If the expansion in the turbine is adiabatic, determine the steam flow rate.

Solution:

Assume that kinetic and potential energy are negligible, and the system is in a steady state.

(a) Basis: 1 kg/s steam with the properties from the steam tables:

$\dot{W} = -\,50$ MW $= -\,50{,}000$ kW (turbine is the system).

Turbine inlet:

$H_1 = 3517.8$ kJ/kg, $S_1 = 6.8648$ kJ/kg K at $T_1 = 550°\text{C}$, $P_1 = 8200$ kPa.

Turbine outlet:

$$\begin{aligned} S_{2\text{sat vap}} &= 7.3554\,\text{kJ/kg K}, S_{2\text{sat liq}} = 1.2131\,\text{kJ/kg K at } P_2 = 75\,\text{kPa}. \\ H_{2\text{sat vap}} &= 2663.0\,\text{kJ/kg}, H_{2\text{sat liq}} = 384.45\,\text{kJ/kg}. \end{aligned}$$

At isentropic conditions, then we have $S_2 = S_1 < 7.3554$ kJ/kg K.

Discharged steam is wet steam: x_s (the quality at isentropic operation):

$$x_s = \frac{6.8646 - 1.2131}{7.4570 - 1.2131} = 0.905$$

The discharged steam enthalpy at isentropic conditions H_{2s} is

$$H_{2s} = H_{2\text{sat liq}}(1 - x_s) + H_{2\text{sat vap}}x_s$$

$$= 384.45(1 - 0.905) + 2663(0.905) = 2446.8 \text{ kJ/kg}$$

The steam rate:

$$\dot{m} = \frac{-50000 \text{ kW}}{(H_{2s} - H_1) \text{ kJ/kg}} = 46.7 \text{ kg/s}$$

6.3.1 Carnot Cycle

A Carnot *heat engine* performs the conversion of heat to mechanical energy by bringing a working fluid from a high temperature T_H to a lower temperature state T_C. Figure 6.3 shows a typical pressure–volume *PV* and temperature-entropy *TS* diagrams of ideal engine cycles. On both the *PV* and *TS* diagrams, the area enclosed by the process curves of a cycle represents the net heat transfer to be converted to mechanical energy by the engine.

The Carnot cycle consists of four totally reversible processes shown in Fig. 6.5c:

- Process 1–2: Isothermal heat addition at constant temperature T_H.
- Process 2–3: Isentropic expansion at constant entropy $S_2 = S_3$.
- Process 3–4: isothermal heat rejection at constant temperature T_C.
- Process 4–1: isentropic compression at constant entropy $S_4 = S_1$.

From Fig. 6.3c, we can estimate the amounts of added and rejected heats per unit mass flow rate of a working fluid

$$q_{in} = T_H(S_2 - S_1) \tag{6.2}$$

$$q_{out} = T_C(S_4 - S_3) \tag{6.3}$$

Here q_{in} is the heat received by the steam, while q_{out} is the heat received by the heating medium, for example, cooling water that the system. According to the sign convention considered here, heat received by the system is assumed as positive. The net power output becomes

$$W_{out} = q_{out} - q_{in} \tag{6.4}$$

Here the W_{out} is the work produced by the system that is turbine and hence its sign is negative. Example 6.4 illustrates the analysis of power output from a Carnot cycle [5, 9].

Example 6.4 **Power output from a Carnot cycle**

A Carnot cycle uses water as the working fluid at a steady-flow process. Heat is transferred from a source at 250°C and water changes from saturated liquid to saturated vapor. The saturated steam expands in a turbine with an output at 10 kPa, and a heat of 1045 kJ/kg is discharged from a condenser at 10 kPa. Estimate the net power output of the cycle for a flow rate of 10 kg/s of the working fluid.

Solution:

Assumptions: Kinetic and potential energy changes are negligible.

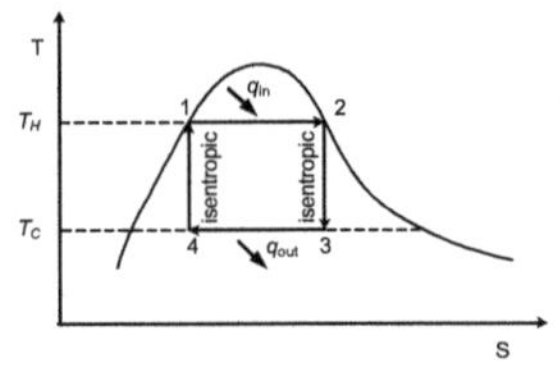

Data from steam tables: $\dot{m} = 10\ \text{kg/s}$.

Turbine inlet:

$P_{sat} = 3977.6$ kPa, $T_H = 250°C = 523.15$ K (from the saturated steam table)

Turbine outlet:

$$T_C = T^{sat}(\text{at } 10\ \text{kPa}) = 45.8°\text{C} = 318.8\,\text{K}.$$

Heat supplied is equivalent to the heat of vaporization at $T_H = 250°C$.

$H_{1\text{sat liq}} = 1085.8\,\text{kJ/kg}, H_{2\text{sat vap}} = 2800.4\ \text{kJ/kg}, \text{ at } P_1 = P_2 = 3977.6\ \text{kPa}.$

$q_{in} = H_{2\text{sat vap}} - H_{2\text{sat liq}} = 1714.6\ \text{kJ/kg}$ (heat received by the steam).

$q_{out} = 1045\ \text{kJ/kg}$ (heat received by the cooling medium).

Total net power output of the cycle: $\dot{W}_{net} = \dot{m}(q_{out} - q_{in}) = -6696.0\,\text{kW}$.

6.3.2 Rankine Cycle

The *Rankine cycle* converts heat into work and generates about 80% of all electric power used throughout the world. Figure 6.4 describes the main processes within the cycle. Common heat sources are the combustion of coal, natural gas, oil, and the fission of nuclear material. The Rankine cycle processes are [3, 5]:

- Process 1–2: The high-pressure liquid enters a boiler where it is heated at constant pressure by an external heat source to become a superheated vapor.

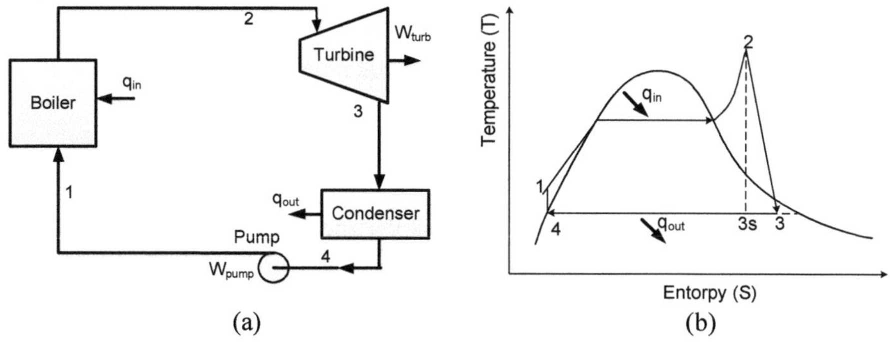

Fig. 6.4 **a** Schematic of Rankine cycle, **b** Rankine cycle processes on a T versus S diagram

- Process 2–3: The vapor expands through a turbine, generating power. This decreases the temperature and pressure of the vapor, and some condensation may occur.
- Process 3–4: The wet vapor discharged from the turbine enters a condenser where it is condensed at a constant pressure.
- Process 4–1: The water is pumped from low to high pressure to start a new cycle.

The Rankine cycle is sometimes referred to as a practical Carnot cycle because in an ideal Rankine cycle the pump and turbine would be *isentropic* and hence maximize the work output shown in Fig. 6.5 as the line 2-3s where the entropy remains constant. The main difference is that the heat addition (in the boiler) and rejection (in the condenser) are isobaric in the Rankine cycle and isothermal in the Carnot cycle.

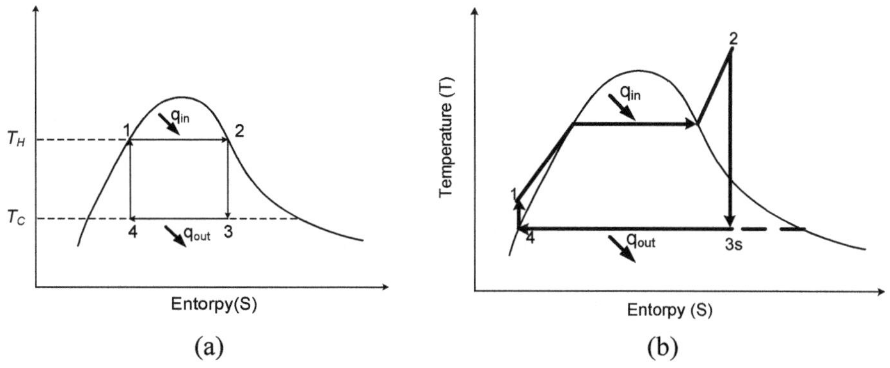

Fig. 6.5 **a** Carnot cycle, **b** Ideal Rankine cycle on a TS diagram, where on both the cycles pumps and turbines operate at constant entropy (isentropic operation)

6.3.3 Analysis of Ideal Rankine Cycle

The analyses of processes yield the following equations for a Rankine cycle: Pump power needed

$$\dot{W}_{p,\text{in}} = \dot{m}V_4(P_1 - P_4) \tag{6.5}$$

where $\dot{m}$ is the steam mass flow rate, V_4 is the specific volume at state 4 that is the saturated liquid water, and P_i is the pressures at state i. Enthalpy rate at state 1 H_1 is

$$\dot{m}H_1 = \dot{m}H_4 + \dot{W}_{p,\text{in}} \tag{6.6}$$

For the isentropic process $S_1 = S_4$ and $S_2 = S_3$. The quality of the discharged wet steam, x_{3s}, shows the molar or mass fraction of vapor ($S_3 < S_{3\text{sat vap}}$)

$$x_{3s} = \frac{\left(S_3 - S_{3\text{sat liq}}\right)}{\left(S_{3\text{sat vap}} - S_{3\text{sat liq}}\right)} \tag{6.7}$$

where S_3 is the entropy at state 3, $S_{3\text{sat vap}}$, and $S_{3\text{sat liq}}$ are the entropy of saturated vapor and saturated liquid at state 3, respectively. Enthalpy rate of the wet steam at state 3 is

$$\dot{m}H_3 = \dot{m}\left[(1 - x_{3s})H_{3\text{sat liq}} + x_{3s}H_{3\text{sat vap}}\right] \tag{6.8}$$

where $H_{3\text{sat vap}}$ and $H_{3\text{sat liq}}$ are the saturated vapor and saturated liquid enthalpies at state 3. The rate of heat input, $\dot{q}_{\text{in}}$, to the pressurized water in the boiler is

$$\dot{q}_{\text{in}} = \dot{m}(H_2 - H_1) \tag{6.9}$$

The discharged heat rate is received by a cooling medium (system) $\dot{q}_{\text{out}}$ at the condenser is

$$\dot{q}_{\text{out}} = -\dot{m}\Delta H = -\dot{m}(H_4 - H_3) \tag{6.10}$$

Net value of the work $\dot{W}_{\text{net}}$ produced by the turbine is

$$\dot{W}_{\text{net}} = (\dot{q}_{\text{out}} - \dot{q}_{\text{in}}) \tag{6.11}$$

The sign of network is negative as the output of the system that is turbine. The performance of a steam turbine will be limited by the quality discharged steam. Discharged steam with a low quality (steam with more moisture) may decrease the life of turbine blades and efficiency of the turbine. The easiest way to overcome this problem is by superheating steam. Example 6.5 illustrates the analysis of an ideal Rankine cycle [5, 9].

Example 6.5 **Analysis of a simple ideal Rankine cycle**

A steam power plant operates on a simple ideal Rankine cycle shown below. The turbine receives steam at 698.15 K and 4100 kPa, while the discharged steam is at 40 kPa. The mass flow rate of steam is 8.5 kg/s. Determine the network output.

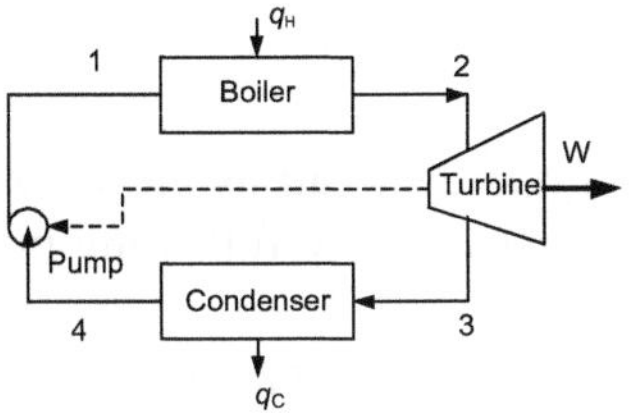

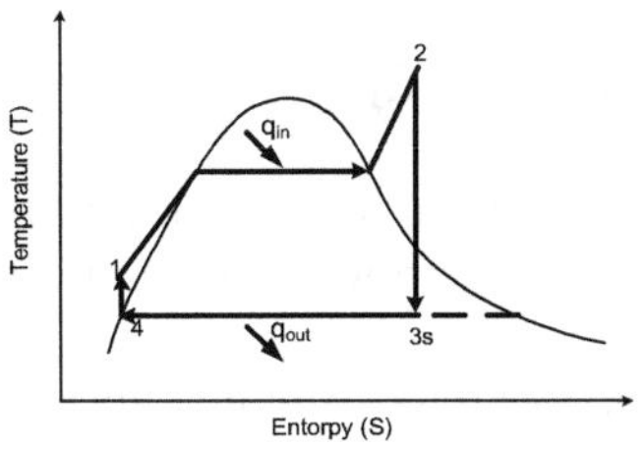

Solution:

Assume that the kinetic and potential energy changes are negligible. From steam tables.

$$\dot{m}_s = 8.5\ \text{kg/s}, P_2 = P_1 = 4100\ \text{kPa}, H_2 = 3272.3\ \text{kJ/kg},$$
$$S_2 = 6.8450\ \text{kJ/kg K (steam tables)}.$$

Saturated steam properties at $P_3 = P_4 = 40$ kPa, $V_4 = 0.001022$ m^3/kg

$$H_{3\text{sat vap}} = 2636.9\ \text{kJ/kg}, H_4 = H_{3\text{sat liq}} = 317.65\ \text{kJ/kg}$$
$$S_{3\text{sat vap}} = 7.6709\ \text{kJ/kg K}; S_{3\text{sat liq}} = 1.0261\ \text{kJ/kg K}$$

Basis: mass flow rate of 1 kg/s:

$$W_{p,\text{in}} = V_1(P_1 - P_4) = (0.001022)(4100 - 40)\left(\frac{1\text{kJ}}{1\ \text{kPa m}^3}\right) = 4.14\ \text{kJ/kg}$$
$$H_1 = H_4 + W_{p,\text{in}} = 321.79\ \text{kJ/kg}$$

Isentropic process $S_1 = S_4$ and $S_3 = S_2$. The quality of the discharged wet steam ($S_3 < S_{3\text{sat vap}}$):

$$x_{3s} = (6.845 - 1.0262)/(7.6709 - 1.0261) = 0.875$$

$$H_3 = 317.65(1 - 0.875) + 2636.9 \times 0.875 = 2356.6\ \text{kJ/kg}$$

$$q_{\text{in}} = H_2 - H_1 = 2950.5\ \text{kJ/kg (positive heat received by the steam)}$$

$$q_{\text{out}} = -\Delta H = -(H_4 - H_3)$$
$$= 2038.9\ \text{kJ/kg (positive heat received by the cooling medium)}$$

With a steam flow rate of 8.5 kg/s, we have

$$\dot{W}_{\mathrm{net}} = \dot{m}(q_{\mathrm{out}} - q_{\mathrm{in}}) = -7748.6\ \mathrm{kW}\ = -7.75\ \mathrm{MW}$$

6.3.4 Brayton Cycle

The gas-turbine cycle relies on the *Brayton cycle* using air as the working fluid as shown in Fig. 6.6. An ideal Brayton cycle consists of the following processes:

- *Isentropic Process*—Ambient air is drawn into the compressor.
- *Isobaric Process*—The compressed air then runs through a combustion chamber where fuel is burned, heating that air at a constant-pressure process.
- *Isentropic Process*—The heated, pressurized air then gives up its energy, expanding through a turbine (or series of turbines).
- *Isobaric Process*—Heat rejection to the surroundings.

Some of the electricity produced from the turbine is used to drive the compressor through a shaft arrangement. The gas turbine requires clean fuels such as natural gas or light fuel oil.

Actual Brayton cycle has the processes:

- Compression (adiabatic)
- Heat addition (isobaric)
- Expansion (adiabatic)
- Heat rejection (isobaric)

Since neither the compression nor the expansion can be truly isentropic, losses through the compressor and the expander may be considerable. Increasing the compression ratio increases the overall power output of the Brayton system. Intercooling the working fluid decreases the amount of work needed for the compression stage overall and increases fuel consumption of the combustion chamber.

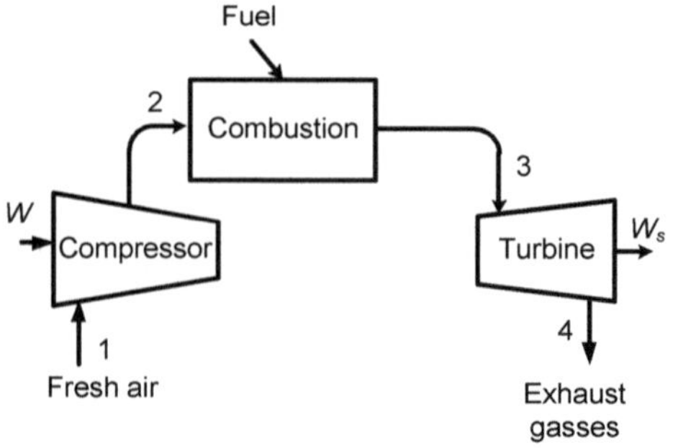

Fig. 6.6 Schematic of ideal Brayton cycle with processes of adiabatic compression, isobaric heat addition, adiabatic expansion, and isobaric heat rejection

To increase the power output for a given compression ratio, air expands through a series of turbines then is passed through a second combustion chamber before expanding to ambient pressure [2, 9].

6.3.5 Stirling Engine

A Stirling engine is a heat engine operating by cyclic compression and expansion of the working fluid at different temperature levels such that there is a net conversion of heat energy to mechanical work (see Fig. 6.7). The working fluid is mainly air although other gases can also be used. The Stirling engine is highly efficient compared to steam engines and can use almost any heat source [1, 9]. The Stirling engine is classified as an external combustion engine, as all heat transfers to and from the working fluid take place through the engine wall. The engine cycle consists of compressing gas, heating the gas, expanding the hot gas, and finally cooling the gas before repeating the cycle. Its practical use is largely confined to low-power domestic applications.

6.3.6 Combined Cycles

A combined cycle power plant uses the Brayton cycle of the gas turbine with the Rankine cycle of a heat recovery steam generator. The combined cycle plants are designed in a variety of configurations composed of gas turbines followed by a steam turbine. They generate power by burning natural gas in a gas turbine and use residual heat to generate additional electricity from steam.

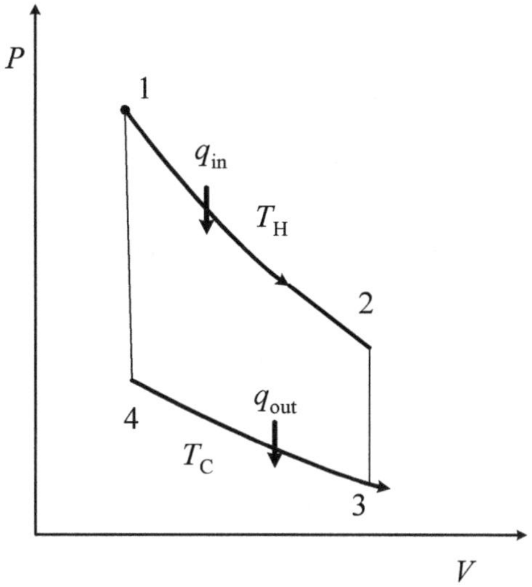

Fig. 6.7 Schematic of Stirling engine

Coal gasification produces a fuel gas that can be used in the gas turbine. By integrating coal gasification with gas turbine and steam cycles, a low pollutant emission can be achieved while using coal. A potential additional advantage of the integrated gasification combined cycle is the capability of capturing carbon dioxide from the fuel gas and making it ready for high-pressure pipeline transportation to a carbon sequestration site [12].

6.4 Improving Power Production in Steam Power Plants

Improvements on power plant operations may increase efficiency and power output while reducing fuel consumption. The modifications for improvements aim at increasing the average temperature at which heat is transferred to the steam in the boiler and decrease the average temperature at which heat is removed in the condenser [2, 5, 13]. Some common modifications in steam power production are summarized in the next sections.

6.4.1 Modification of Operating Conditions of the Condenser and Boiler

Superheating the steam to a high temperature increases the temperature at which the heat flows into the boiler and increases the power output and quality of the discharged steam. Figure 6.8 shows the effects of superheating the steam to higher temperatures and reducing the condenser pressure on the ideal Rankine cycle on a *TS* diagram. The area enclosed by the process curve "1′–2′–3′–4" is larger than that of the area for "1–2–3–4" and represents the network output increase of the cycle under higher boiler temperature and lower condenser pressure. The operating pressures may be as high as 30 MPa (approximately 4500 psia) in modern steam power plants operating at supercritical pressures ($P > 22.09$ MPa). By metallurgical constraints, the temperature of steam is limited. The temperature at the turbine inlet may be as high as 620°C [3, 5].

6.4.2 Reheating the Steam

Superheating the steam to high temperatures enables the expansion of the steam in various stages instead of a single expansion process. Mainly, reheating increases the power output and the steam quality to protect the material. Figure 6.9 shows a typical reheat Rankine cycle. In an ideal reheat Rankine cycle with two-stage expansion, for example, the steam is expanded to an intermediate pressure isentropically in the high-pressure turbine section and sent to the boiler to be reheated. In the low-pressure turbine section, the reheated steam is expanded to the condenser operating pressure. Example 6.6 illustrates a simple analysis of the ideal reheat Rankine cycle.

Fig. 6.8 Effects of superheating the steam to higher temperatures and reducing the condenser pressure on an ideal Rankine cycle

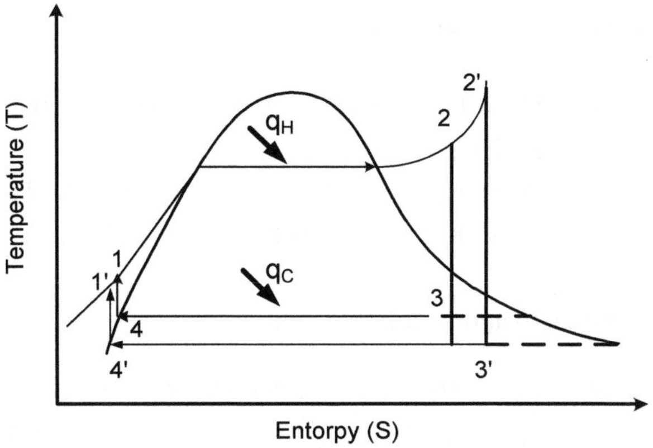

Example 6.6 Simple reheat Rankine cycle in a steam power plant

A simple ideal reheat Rankine cycle is used in a steam power plant shown in Fig. 6.9. Steam enters the turbine at 9000 kPa and 823.15 K and leaves at 4350 kPa and 698.15 K. The steam is reheated at constant pressure to 823.15 K. The discharged steam from the low-pressure turbine is at 10 kPa. The net power output of the turbine is 40 MW. Determine the mass flow rate of steam.

Solution:

Assume that the kinetic and potential energy changes are negligible, and this is a steady process.

Consider Fig. 6.9.

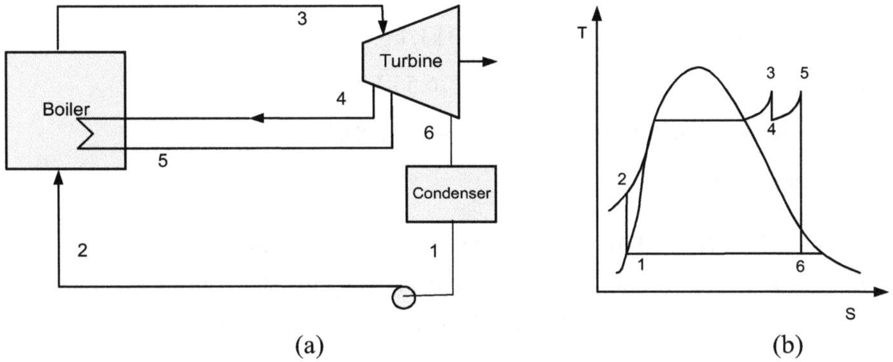

Fig. 6.9 **a** Ideal reheat Rankine cycle, **b** T versus S diagram for the reheat Rankine cycle [5]

Data: From steam tables.

$$V_1 = 0.00101\ \text{m}^3/\text{kg},\ P_3 = 9000\ \text{kPa},\ H_3 = 3509.8\ \text{kJ/kg},\ S_3 = 6.8143\ \text{kJ/kg K}$$
$$P_6 = 10\ \text{kPa}.$$
$$T_4 = 698.15\ \text{K},\ P_4 = 4350\ \text{kPa},\ H_4 = 3268.5\ \text{kJ/kg},\ S_3 = S_4 = 6.8143\ \text{kJ/kg K}$$
$$T_5 = 823.15\ \text{K},\ P_5 = 4350\ \text{kPa},\ H_5 = 3555.2\ \text{kJ/kg},\ S_5 = S_6 = 7.1915\ \text{kJ/kg K}$$

Saturated vapor and liquid data:

$$H_{6\ \text{sat vap}} = H_{1\ \text{sat vap}} = 2584.8\ \text{kJ/kg},\quad H_{1\ \text{sat liq}} = H_{6\ \text{sat liq}} = 191.81\ \text{kJ/kg}$$
$$S_{6\ \text{sat vap}} = 8.1511\ \text{kJ/kg K},\quad S_{6\ \text{sat liq}} = 0.6493\ \text{kJ/kg K}$$

Based on $\dot{m} = 1$ kg/s steam flow rate:

$$W_{p,\text{in}} = V_1(P_2 - P_1) = 0.00101(9000 - 10)\left(\frac{1\text{kJ}}{1\ \text{kPa m}^3}\right) = 9.08\ \text{kJ/kg}$$
$$H_2 = H_1 + W_{p,\text{in}} = 200.9\ \text{kJ/kg}$$

Because this is an isentropic process $S_5 = S_6$ and $S_1 = S_2$.

The quality of the discharged wet steam ($S_6 < S_{6\text{sat vap}}$) after passing through the turbine:

$$x_{6s} = (7.1915 - 0.6493)/(8.1511 - 0.6493) = 0.872$$
$$H_6 = 191.81(1 - 0.872) + 2584.8(0.872) = 2278.7\ \text{kJ/kg}$$

Turbine outputs:

First turbine: $W_{T1} = H_4 - H_3 = -241.3\ \text{kJ/kg}$.

Second turbine: $W_{T2} = H_6 - H_5 = -1276.5\ \text{kJ/kg}$.

The total work by the turbines: $W_{\text{total}} = -1517.8\ \text{kJ/kg}$.

Heat interactions:

$$q_{23,\text{in}} = H_3 - H_2 = 3308.9\ \text{kJ/kg (heat received by steam)}$$
$$q_{45,\text{in}} = H_5 - H_4 = 286.7\ \text{kJ/kg (heat received by the stem during the reheat)}$$
$$q_{\text{out}} = -(H_1 - H_6) = 2086.9\ \text{kJ/kg (heat is received by the cooling medium)}$$
$$q_{\text{in}} = q_{23,\text{in}} + q_{45.\text{in}} = 3595.6\ \text{kJ/kg}$$
$$\dot{W}_{\text{out}} = \dot{m}(q_{\text{out}} - q_{\text{in}}) = \dot{m}(2086.9 - 3595.5)\ \text{kJ/kg}$$
$$= -40{,}000\ \text{kW (work produced by the turbine)}.$$

The flow rate of steam is:

$$\dot{m} = \left|\dot{W}_{\text{out}}\right|/1508.6 = 26.5\ \text{kg/s}$$

In this ideal reheat Rankine cycle, the expanded steam from the first part of the high-pressure section is reheated in the boiler until it reaches the boiler exit temperature. The reheated steam is expanded through the turbine to condenser conditions. The reheating decreases the moisture within the discharged steam.

6.4.3 Regeneration

Increasing the boiler feed water temperature by using the expanding steam is possible in a regenerative cycle. Figure 6.12 shows a typical regenerative Rankine cycle. Steam extracted at intermediate pressures from the turbine is used to heat the boiler water feed. The steam at stage 3 leaves the condenser as a saturated liquid at the condenser operating pressure by adjusting the fraction of steam extracted from the turbine. Regeneration helps deairate the water and control the discharged steam flow rate. Example 6.7 illustrates a simple analysis of the ideal regenerative Rankine cycle.

Example 6.7 **Power output of ideal regenerative Rankine cycle**

A steam power plant is using an ideal regenerative Rankine cycle shown in Fig. 6.10. Steam enters the high-pressure turbine at 8200 kPa and 773.15 K, and the condenser operates at 20 kPa. The steam is extracted from the turbine at 350 kPa to heat the feed water in an open heater. The water is a saturated liquid after passing through the feed water heater. Determine the net power output of the cycle.

Solution:

Assume that the kinetic and potential energy changes are negligible, and this is a steady-state process.

Consider Fig. 6.10.

Basis: steam flow rate is 1 kg/s. From steam tables:

$$\begin{aligned}
V_1 &= 0.001017\ \text{m}^3/\text{kg},\ V_3 = 0.001079\ \text{m}^3/\text{kg},\ P_1 = P_7 = 20\ \text{kPa},\\
H_{7\text{sat vap}} &= H_{1\text{sat vap}} = 2609.9\ \text{kJ/kg},\ H_{7\text{sat liq}} = H_{1\text{sat liq}} = 251.45\ \text{kJ/kg},\\
S_{7\text{sat vap}} &= 7.9094\ \text{kJ/kg K},\ S_{7\text{sat liq}} = 0.8321\ \text{kJ/kg K}\\
P_2 &= P_3 = 350\ \text{kPa},\\
H_{6\text{sat liq}} &= H_{3\text{sat liq}} = 584.27\ \text{kJ/kg},\ H_{6\text{sat vap}} = H_{3\text{sat vap}} = 2731.50\ \text{kJ/kg}.\\
S_{3\text{sat liq}} &= 1.7273\ \text{kJ/kg K},\ S_{3\text{sat vap}} = 6.9392\ \text{kJ/kg K}
\end{aligned}$$

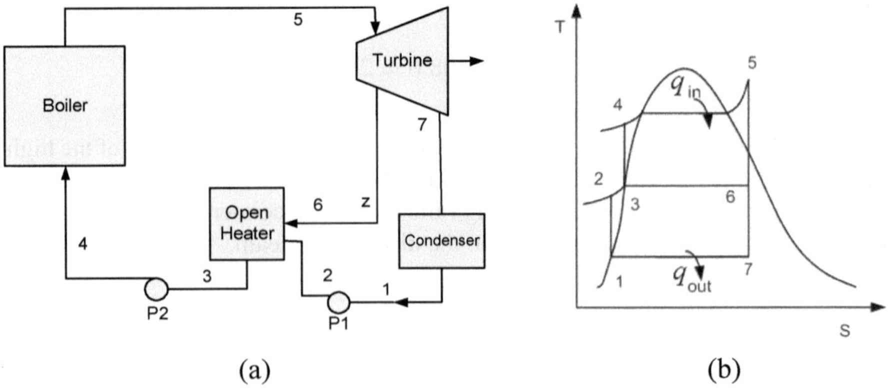

Fig. 6.10 **a** Ideal regenerative Rankine cycle, **b** temperature T versus entropy S diagram for ideal regenerative Rankine cycle. Here P1 and P2 show the circulation pumps

In this ideal regenerative Rankine cycle, the steam extracted from the turbine heats the water from the condenser, and the water is pumped to the boiler. Sometimes, this occurs in several stages. The condensate from the feed water heaters is throttled to the next heater at lower pressure. The condensate of the final heater flashes into the condenser.

$$W_{p1} = V_1(P_2 - P_1) = 0.001017(350 - 20)\left(\frac{1\text{kJ}}{1\ \text{kPa m}^3}\right) = 0.335\ \text{kJ/kg}$$

$$H_2 = H_1 + W_{p1} = 252.78\ \text{kJ/kg}$$

$$W_{p2} = V_3(P_4 - P_3) = 0.001079(8200 - 350)\left(\frac{1\text{kJ}}{1\ \text{kPa m}^3}\right) = 8.47\ \text{kJ/kg}$$

$$H_4 = H_3 + W_{p2} = 592.74\ \text{kJ/kg}$$

Because this is an isentropic process, $S_5 = S_6 = S_7$. We estimate the quality of the discharged wet steam at states 6 ($S_5 < S_{6\text{sat vap}}$) and 7 ($S_5 < S_{7\text{sat vap}}$):

$$x_6 = \frac{6.7124 - 1.7273}{6.9392 - 1.7273} = 0.956$$

$$H_6 = 584.27(1 - 0.956) + 2731.50(0.956) = 2638.06\ \text{kJ/kg}$$

$$x_7 = \frac{6.7124 - 0.8321}{7.9094 - 0.8321} = 0.83$$

$$H_7 = 252.45(1 - 0.83) + 2609.9(0.83) = 2211.18\ \text{kJ/kg}$$

The fraction of steam extracted is estimated from the energy balance: $\dot{m}_6 H_6 + \dot{m}_2 H_2 = \dot{m}_3 H_3$.

In terms of the mass fraction: $z = \dot{m}_6/\dot{m}_3$, the energy balance becomes: $zH_6 + (1 - z)H_2 = H_3$.

The mass fraction: $z = \frac{H_3 - H_2}{H_6 - H_2} = 0.139$

$$q_{\text{in}} = (H_5 - H_4) = 2803.66\ \text{kJ/kg}$$

$q_{\text{out}} = -(1 - z)(H_1 - H_7) = 1687.32\ \text{kJ/kg}$ (heat received by the cooling medium that is the system)

For a steam flow rate of 1 kg/s: $\dot{W}_{\text{net}} = \dot{m}(q_{\text{out}} - q_{\text{in}}) = -1116.34\ \text{kW}$

This work is the output of the turbine that is the system.

6.4.4 Reheat Regenerative Rankine Cycle

In *reheat regenerative Rankine cycle*, shown in Fig. 6.11, the part of steam extracted from the turbine heats the water from the condenser, and the remaining part is reheated in the boiler. Example 6.7 illustrates a simple analysis of reheat regenerative Rankine cycle.

Example 6.8 **Ideal reheat regenerative cycle**
A steam power plant uses an ideal reheat regenerative Rankine cycle. Steam enters the high-pressure turbine at 9000 kPa and 773.15 K and leaves at 850 kPa. The condenser operates at 10 kPa. Part of the steam is extracted from the turbine at 850 kPa to heat the water in an open heater, where the steam and liquid water from the condenser mix and direct contact heat transfer takes place. The rest of the steam is reheated to 723.15 K and expanded in the low-pressure turbine section to the condenser pressure. The water is a saturated liquid after passing through the water heater and is at the heater pressure. The flow rate of steam is 32.5 kg/s. Determine the power produced.

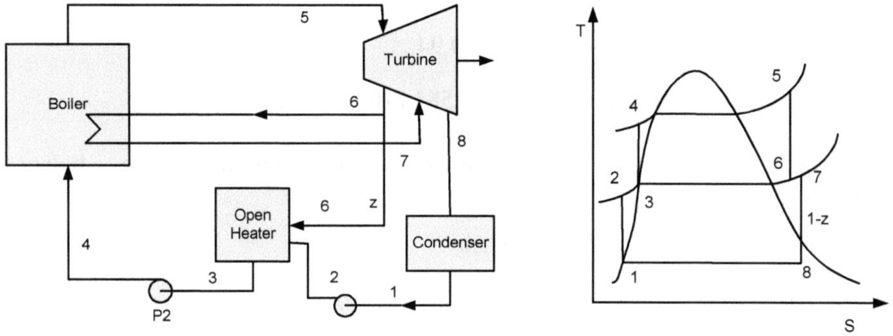

(a) (b)

Fig. 6.11 a Schematic of ideal reheat regenerative Rankine cycle, b *TS* diagram of the cycle

Solution:

Assume negligible kinetic and potential energy changes, and that this is a steady-state process. The surroundings are at 285 K.

Consider Fig. 6.11. From steam tables:

$$
\begin{aligned}
P_5 &= 9000\ \text{kPa}, H_5 = 3386.8\ \text{kJ/kg}, S_5 = 6.6600\ \text{kJ/kg K}, T_5 = 773.15\ \text{K} \\
P_1 &= P_8 = 10\ \text{kPa}, H_{8\text{sat vap}} = 2584.8\ \text{kJ/kg}, \\
H_{8\text{sat liq}} &= 191.83\ \text{kJ/kg}, V_1 = 0.00101\ \text{m}^3/\text{kg}. \\
S_{8\text{sat vap}} &= 8.1511\ \text{kJ/kg K}, S_{8\text{sat liq}} = 0.6493\ \text{kJ/kg K}. \\
P_3 &= 850\ \text{kPa}, H_{3\text{sat liq}} = 732.03\ \text{kJ/kg}, \\
H_{3\text{sat vap}} &= 2769.90\ \text{kJ/kg}, V_3 = 0.001079\ \text{m}^3/\text{kg}. \\
P_7 &= 850\ \text{kPa}, H_7 = 3372.7\ \text{kJ/kg}, S_7 = 7.696\ \text{kJ/kg K}, T_7 = 723.15\ \text{K}. \\
P_6 &= 850\ \text{kPa}, T_6 = 450.0\ \text{K}, \\
H_6 &= 2779.58\ \text{kJ/kg}.
\end{aligned}
$$

Work and enthalpy estimations for a unit mass flow rate of steam yield:

$$
\begin{aligned}
W_{p1} &= V_1(P_2 - P_1) = 0.00101(850 - 10)\left(\frac{1\text{kJ}}{1\ \text{kPa m}^3}\right) = 0.848\ \text{kJ/kg} \\
H_2 &= H_1 + W_{p1} = 192.68\ \text{kJ/kg} \\
W_{p2} &= V_3(P_4 - P_3) = 0.001079(9000 - 850)\left(\frac{1\text{kJ}}{1\ \text{kPa m}^3}\right) = 9.046\ \text{kJ/kg} \\
H_4 &= H_3 + W_{p2} = 741.07\ \text{kJ/kg}
\end{aligned}
$$

Because this is an isentropic process, $S_5 = S_6$ and $S_7 = S_8$, and we estimate the quality of the discharged wet steam at state 8 ($S_7 < S_{8\text{sat vap}}$):

$$
\begin{aligned}
x_8 &= \frac{7.696 - 0.6493}{8.1511 - 0.6493} = 0.94 \\
H_8 &= 191.83(1 - 0.94) + 2584.8(0.94) = 2439.63\ \text{kJ/kg}
\end{aligned}
$$

The fraction of steam extracted is estimated from the energy balance: $\dot{m}_6 H_6 + \dot{m}_2 H_2 = \dot{m}_3 H_3$.

In terms of the mass fraction: $z = \dot{m}_6/\dot{m}_3$.

Energy balance becomes: $zH_6 + (1 - z)H_2 = H_3$.

Mass fraction is: $z = \frac{H_3 - H_2}{H_6 - H_2} = \frac{732.03 - 192.68}{2779.58 - 192.68} = 0.208$.

Turbine work output with $\dot{m} = 1$ kg/s working fluid is:

$$\dot{q}_{\text{in}} = \dot{m}[(H_5 - H_4) + (1 - z)(H_7 - H_6)] = 3115.18 \text{ kW}$$
$$\dot{q}_{\text{reheat}} = \dot{m}z(H_3 - H_6) = 425.89 \text{ kW}$$

$\dot{q}_{\text{out}} = -\dot{m}(1 - z)(H_1 - H_8) = 1779.14$ kW (heat received by the cooling medium that is the system)

Total heat input: $\dot{q}_{\text{total in}} = 3115.18 + 425.89 = 3541.1$ kW

For steam flow rate of 36.5 kg/s, workout:
$\dot{W}_{\text{net}} = \dot{m}_s(q_{\text{out}} - q_{\text{in}}) = -48765.4 \text{ kW} = -48.76 \text{ MW}$.

6.5 Geothermal Power Plants

Figure 6.12 shows the schematic of a geothermal power plant. Geothermal power comes from heat energy buried beneath the surface of the earth. There are three types of geothermal power plants:

- *Dry Steam Power Plants*: The geothermal steam goes directly to a turbine, where it expands and produces power. The expanded steam is injected into the geothermal well.
- *Flash Steam Power Plants*: Geothermal fluids above 360°F (182°C) can be flashed in a tank at low pressure causing some of the fluid to rapidly vaporize. The vapor then expands in a turbine.
- *Binary Cycle Power Plants*: Moderate-temperature geothermal fluids between 85 and 170°C are common. The term "binary" refers to dual-fluid systems, wherein hot geothermal brine is pumped through a heat exchange network to transfer its energy to a working fluid driving a power train. A hot geothermal fluid and a suitable working fluid with a much lower boiling point than geothermal fluid pass through a heat exchanger. The vaporized working fluid drives the turbines; no working fluid is emitted into the atmosphere. The working fluids may be isobutene, isopentane, n-pentane, or ammonia.

At temperatures below about 200°C, binary power systems are favored for relative cost effectiveness. In general, above 200°C, flashing geothermal fluids to produce steam and directly driving turbine/generator is preferred. Example 6.9 illustrates a simple analysis of steam power plants using a geothermal source [1–3].

***Example 6.9* A steam power plant using a geothermal energy source**

A steam power plant uses a geothermal energy source. The geothermal source is available at 220°C and 2320 kPa with a flow rate of 200 kg/s. The hot water goes

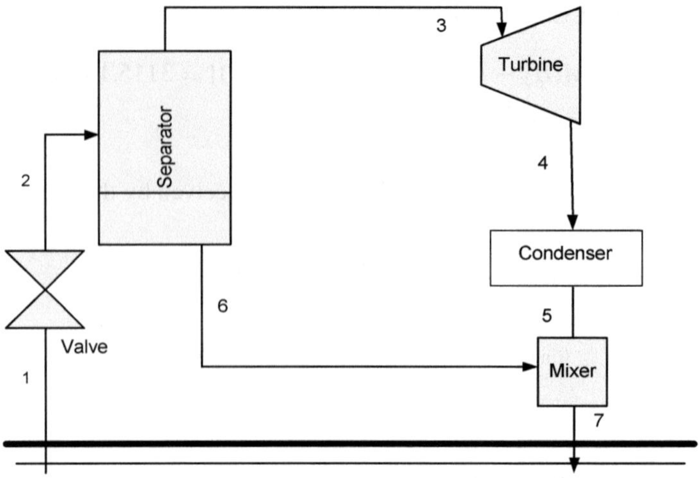

Fig. 6.12 Schematic of geothermal power plant

through a valve and a flash drum. Steam from the flash drum enters the turbine at 550 kPa and 428.62 K. The discharged steam from the turbine has a quality of $x_4 = 0.96$. The condenser operates at 10 kPa. The water is a saturated liquid after passing through the condenser. Determine the work output of turbine.

Solution:

Assume: The kinetic and potential energy changes are negligible, and this is a steady-state process.

$\dot{m}_1 = 200$ kg/s. From steam tables. Consider Fig. 6.13.

$T_1 = 493.15$ K, $P_1 = 2319.8$ kPa, $H_1 = H_2 = 943.7$ kJ/kg, $S_1 = 2.517$ kJ/kg K.

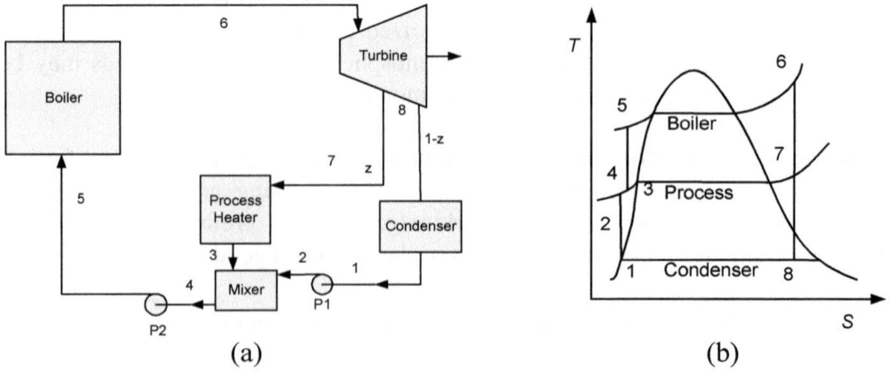

Fig. 6.13 **a** Schematic of ideal cogeneration plant, **b** *TS* diagram

$T_3 = 428.62$ K, $P_3 = 550$ kPa, $H_3 = 2751.7$ kJ/kg, $S_3 = 6.787$ kJ/kg K (saturated).

$H_{3\text{sat vap}} = 2551.7$ kJ/kg, $H_{3\text{sat liq}} = 655.80$ kJ/kg, $S_{3\text{sat vap}} = 6.787$ kJ/kg K, $S_{3\text{sat liq}} = 1.897$ kJ/kg K, $P_4 = 10$ kPa, $H_{4\text{sat vap}} = 2584.8$ kJ/kg, $H_{4\text{sat liq}} = 191.8$ kJ/kg.

In this geothermal power plant, the hot water is flushed, and steam is produced and used in the turbine.

The rate of vapor is estimated from the quality at state 2. The fraction of steam after flashing is:

$$x_2 = \frac{943.7 - 655.8}{2751.7 - 655.8} = 0.159$$

$$S_2 = (1 - 0.159)1.897 + 0.159(6.787) = 2.6756 \text{ kJ/kg K}$$

Steam flow rate is:

$$\dot{m}_3 = x_2(\dot{m}_1) = 0.159(200) = 31.84 \text{ kg/s}$$

From the mass balance around the flash drum, we have $\dot{m}_6 = \dot{m}_1 - \dot{m}_3 = 168.15$ kg/s

Discharged steam has the quality of: $x_4 = 0.96$

$$H_4 = (1 - 0.96)H_{4\text{sat liq}} + (0.96)H_{4\text{sat vap}} = 2489.08 \text{ kJ/kg}$$

From the flash drum at state 6, we have: $\dot{W}_{\text{net}} = \dot{m}_3(H_4 - H_3) = -1993.82$ kW

6.6 Cogeneration

Figure 6.13 shows a typical cogeneration plant. Cogeneration plants produce electric power and process heat from the same heat source. This may lead to the utilization of more available energy and the reduction of waste heat. Cogeneration plants, for example, may use the waste heat from Brayton engines, typically for hot water production or for space heating. The process heat in industrial plants usually needs steam at 500–700 kPa, and 150–200°C. The steam expanded in the turbine to process pressure is used as process heat. Cycles making use of cogeneration may be an integral part of large processes where the energy of the expanded steam from the turbine at intermediate pressure may be fully utilized in producing electricity and heat simultaneously. The *utilization factor* for a cogeneration plant is the ratio of the energy used in producing power and heat to the total energy input. Examples 6.10 and 11 illustrate simple analyses of energy output in cogeneration plants [1, 3, 4].

Example 6.10 **Energy output in a cogeneration plant**

A cogeneration plant uses steam at 8200 kPa and 773.15 K (see Fig. 6.13). One-fourth of the steam is extracted at 700 kPa from the turbine for cogeneration. The extracted steam is condensed and mixed with the water output of the condenser. The rest of the steam expands from 8200 kPa to the condenser pressure of 10 kPa. The steam flow rate produced in the boiler is 60 kg/s. Determine the work output and process heat produced.

Solution:

Assume that the kinetic and potential energy changes are negligible, and this is a steady-state process.

Consider Fig. 6.13. From steam tables.

$$\begin{aligned}
\dot{m} &= 60\ \text{kg/s}, z = 0.25 \\
P_1 &= P_8 = 10\ \text{kPa}, H_{1\text{sat vap}} = 2584.8\ \text{kJ/kg}, H_{1\text{sat liq}} = 191.83\ \text{kJ/kg}, \\
V_1 &= 0.00101\ \text{m}^3/\text{kg}, S_{1\text{sat vap}} = 8.1511\ \text{kJ/kg K}, S_{1\text{sat liq}} = 0.6493\ \text{kJ/kg K}, \\
P_3 &= P_7 = P_2 = P_4 = 700\ \text{kPa}, H_3 = 697.06\ \text{kJ/kg}, \\
S_3 &= 1.9918\ \text{kJ/kg K} \\
P_6 &= 8200\ \text{kPa}, T_6 = 773.15\ \text{K}, H_6 = 3396.4\ \text{kJ/kg}, S_6 = 6.7124\ \text{kJ/kg K}
\end{aligned}$$

In this cogeneration cycle, the steam extracted from the turbine is used as heat. The liquid condensate from the process heat is combined with the output of the condenser.

Basis: mass flow rate = 1 kg/s

$$W_{p1} = V_1(P_2 - P_1) = 0.00101(700 - 10)\left(\frac{1\text{kJ}}{1\ \text{kPa m}^3}\right) = 0.697\ \text{kJ/kg}$$

$$H_2 = H_1 + W_{p1} = 191.83 + 0.697 = 192.53\ \text{kJ/kg}$$

From the energy balance around the mixer, we have $\dot{m}_3/\dot{m}_6 = 0.25$

$$\dot{m}_6 = \dot{m}_4 = 60\,\text{kg/s}, \dot{m}_3 = \dot{m}_7 = 15\,\text{kg/s}, \dot{m}_8 = \dot{m}_1 = 0.75(60) = 45.0\ \text{kg/s}$$

$$\dot{m}_4 H_4 = \dot{m}_2 H_2 + \dot{m}_3 H_3$$

$$H_4 = [45(192.53) + 15(697.06)]/60 = 318.66\ \text{kJ/kg}$$

$$T_4 = 349.15\text{K}, \quad V_4 = 0.001027\,\text{kg/m}^3$$

$$W_{p2} = V_4(P_5 - P_4) = 0.001027(8200 - 700)\left(\frac{1\text{kJ}}{1\ \text{kPa m}^3}\right) = 7.70\ \text{kJ/kg}$$

$$H_5 = H_4 + W_{p2} = 326.36\,\text{kJ/kg}$$

Isentropic processes:

$$S_6 = S_7 = S_8 = 6.7124\,\text{kJ/kgK and } P_7 = 700\,\text{kPa},\ H_7 = 2765.68\,\text{kJ/kg}.$$

We estimate the quality of the discharged wet steam at state 8:

$$x_8 = \frac{6.7124 - 0.6493}{8.1511 - 0.6493} = 0.808$$
$$H_8 = 191.83(1 - 0.808) + 2584.80(0.808) = 2125.87\,\text{kJ/kg}$$

The energy balance yields the fraction of steam extracted

$$\dot{W}_{\text{total}} = \dot{m}_6(H_7 - H_6) + \dot{m}_8(H_8 - H_7) = -66634.44\ \text{kW}$$
$$\sum \dot{W}_{pi} = \dot{m}_1 W_{p1} + \dot{m}_4 W_{p2} = 493.51\ \text{kW}$$

Net work output:

$$\dot{W}_{\text{net}} = \dot{W}_{\text{total}} - \sum \dot{W}_{pi} = -66140.93\ \text{kW (work output of the turbine}$$
$$\dot{q}_{\text{process}} = \dot{m}_7(H_3 - H_7) = -31029.3\ \text{kW(heat discharged from the steam)}$$

Example 6.11 **Estimation of process heat in a cogeneration plant**

A cogeneration plant uses steam at 900 psia and 1000°F to produce power and process heat. The steam flow rate from the boiler is 16 lb/s. The process requires steam at 70 psia at a rate of 3.2 lb/s supplied by the expanding steam in the turbine with a value of $z = 0.2$. The extracted steam is condensed and mixed with the water output of the condenser. The remaining steam expands from 70 psia to the condenser pressure of 3.2 psia. Determine the rate of process heat.

Solution:

Assume that the kinetic and potential energy changes are negligible, and this is a steady-state process.

Consider Fig. 6.13, and steam tables:

$$\dot{m}_6 = 16\ \text{lb/s},\ z = \dot{m}_3/\dot{m}_6 = 0.2,\ \dot{m}_3 = 3.2\text{lb/s}.$$
$$P_1 = P_8 = 3.2\ \text{psia},\ H_{1\text{sat vap}} = 1123.6\ \text{Btu/lb},\ H_{1\text{sat liq}} = 111.95\ \text{Btu/lb}.$$
$$V_1 = 0.01631\ \text{ft}^3/\text{lb},\ S_{1\text{sat vap}} = 0.2051\ \text{Btu/lb R},\ S_{1\text{sat liq}} = 1.8810\ \text{Btu/lb R}.$$
$$P_3 = 70\ \text{psia},\ H_3 = 272.74\ \text{Btu/lb, (Saturated)}.$$

$$P_6 = 900 \text{ psia}, H_6 = 1508.5 \text{ Btu/lb}, S_6 = 1.6662 \text{ Btu/lb R}, T_6 = 1000°\text{F}.$$

In this cogeneration cycle, the steam extracted from the turbine is used in process heat. The liquid condensate from the process heat is combined with the output of the condenser.

$$\dot{m}_2 = \dot{m}_1 = 12.8 \text{lb/s}, \quad \dot{m}_4 = 16 \text{ lb/s}$$

Because these are isentropic processes, $S_6 = S_{7s} = S_{8s} = 1.6662$ Btu/lb R

$$P_7 = 70 \text{psia}, \quad H_{7s} = 1211.75 \text{ Btu/lb}$$

Process heat: $\dot{q}_{\text{process}} = \dot{m}_3(H_3 - H_7) = -3004.8 \text{ Btu/s}$

6.7 Chemical-Looping Combustion Combined Cycle

Chemical-looping combustion with inherent carbon capture may be an emerging power generation technology. Chemical-looping combustion uses an oxygen carrier that transfers oxygen from the air to the fuel to prevent direct contact between them. The oxygen carrier can be alternately oxidized and reduced through the circulation between air and fuel reactor as seen in Fig. 6.14. Well-designed and operated systems offer scalable, diverse, economical, and environmentally sustainable energy production pathways with inherited carbon capture. The fuel may be coal, natural gas, and biomass. The product gas contains mainly CO_2 and water undiluted with nitrogen, and without the production of nitrogen oxides (NO_X) as the high temperatures associated with the use of flame is avoided. The oxidation of the oxygen carrier is strongly exothermic and hence can be used to heat airflow to high temperatures (1000–1200°C) and can drive a gas turbine [6].

6.8 Nuclear Power Plants

Nuclear power is a method in which steam is produced by heating water through a process called *nuclear fission*. Nuclear fission is a nuclear reaction in which the nucleus of an atom splits into smaller parts, often producing free neutrons and photons in the form of *gamma rays* (γ). *Gamma rays* are electromagnetic radiation of frequencies above 10^{19} Hz and therefore have energies above 100 keV and wavelength less than 10 pm. In a nuclear power plant, a reactor contains a core of nuclear fuel, primarily uranium. When atoms of uranium fuel are hit by neutrons, they fission (split) releasing heat and more neutrons. Under controlled conditions, these neutrons can strike and split more uranium atoms [9].

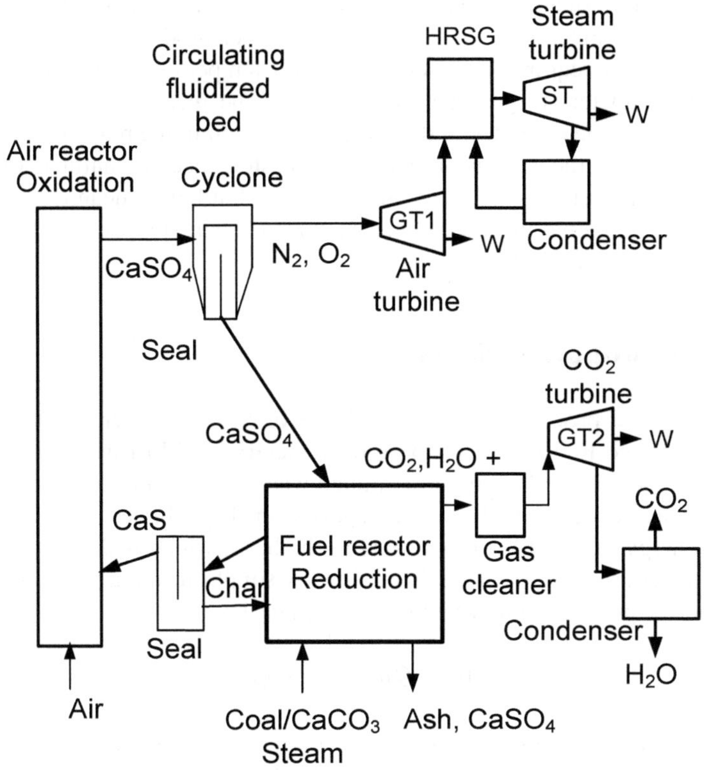

Fig. 6.14 Schematic of chemical-looping combustion combined cycle for coal-fired power production [6]: GT: Gas turbine, HRSH: Heat recovery steam generator, ST: Steam turbine, $CaSO_4$ is the oxygen carrier

Typical fission releases about two hundred million eV of energy (1 eV is 96.485 kJ/mol). By contrast, most chemical oxidation reactions, such as burning coal, release at most a few eV. In a nuclear reactor, the energy is converted to heat as the particles and gamma rays collide with the atoms that make up the reactor and its working fluid, usually water or occasionally heavy water [9].

Fusion power, on the other hand, is generated by nuclear fusion reactions where two light atomic nuclei fuse together to form a heavier nucleus and in doing so, release a large amount of energy. If two light nuclei fuse, they will generally form a single nucleus with a slightly smaller mass than the sum of their original masses. The difference in mass is released as energy according to Albert Einstein's mass-energy equivalence formula $E = mc^2$. The dividing line between "light" and "heavy" is iron-56. Above this atomic mass, energy will generally be released by nuclear fission reactions. Most fusion reactions combine isotopes of hydrogen that are protium, deuterium, or tritium to form isotopes of helium ^{3}He or ^{4}He. Most fusion power plants involve using the fusion reactions to create heat, which is then used to operate a steam turbine, which drives generators to produce electricity. This is similar to most coal, oil, and gas-fired power production plants [9].

Nuclear fusion requires precisely controlled temperature, pressure, and magnetic field parameters to generate net energy. If the reactor was damaged, these parameters would be disrupted and the heat generation in the reactor would rapidly cease. In contrast, the fission products in a fission reactor continue to generate heat for several hours or even days after reactor shutdown, meaning that melting of fuel rods is possible even after the reactor has been stopped due to continued accumulation of heat [1, 14].

6.9 Hydropower Plants

Hydropower is a process in which the force of flowing water is used to spin a turbine connected to a generator to produce electricity. Most hydroelectric power comes from the potential energy of dammed water driving a turbine and generator. The power extracted from the water depends on the volume and the amount of potential energy in water is proportional to the head, which is difference in height between the source and the water's outflow. A simple formula for approximating electric power production at a hydroelectric plant is:

$$\dot{W} = \dot{Q}\rho g \Delta z = \dot{m} g \Delta z \tag{6.12}$$

where $\dot{W}$ is the power, ρ is the density of water (~1000 kg/m^3), Δz is the height, Q is the volumetric flow rate, and g is the acceleration due to gravity. Annual electric energy production depends on the available water supply. In some installations, the water flow rate can vary by a factor of 10:1 over the course of a year. Example 6.12 illustrates the analysis of hydroelectric power output.

Hydropower eliminates the use of fossil fuels and hence carbon dioxide emissions. Hydroelectric plants also tend to have longer economic lives (50 years or longer) than fuel-fired power production. The sale of hydroelectricity may cover the construction costs after 5 to 8 years of full operation [1]. Operating labor cost and maintenance cost are also usually low, as plants are automated. The hydroelectric capacity is either the actual annual energy production or by installed capacity. A hydroelectric plant rarely operates at its full capacity over a full year. The ratio of annual average power output to installed capacity is the *capacity factor* for a hydroelectric power plant. There are large, small, and micro hydropower plant operations, which are summarized below [14]:

- *Large hydroelectric power stations* have outputs from over a few hundred megawatts to more than 10 GW. Many large hydroelectric projects supply public electricity networks. The construction of these large hydroelectric facilities and the changes it makes to the environment are monitored by specialized organizations, such as the International Hydropower Association.

- *Small hydroelectric power plants* have a capacity of up to 10 MW. Some are created to serve specific industrial plants, such as for aluminum electrolytic plants, for which substantial amounts of electricity are needed.
- Micro hydroelectric power installations typically produce up to 100 KW of power. These installations can provide power to an isolated home or small community. Sometimes, micro hydro systems may complement photovoltaic solar energy systems because water flow and available hydropower is highest in the winter when solar energy is at a minimum.

Example 6.12 **Hydroelectric power output**

A hydroelectric plant operates by water falling from a 200 ft height. The turbine in the plant converts potential energy into electrical energy, which is lost by about 5% through the power transmission so the available power is 95%. If the mass flow rate of the water is 396 lb/s, estimate the power output of the hydro plant.

Solution:

$$\dot{W} = \frac{\dot{m} g \Delta z}{g_c}$$

Data: $\Delta z = 200$ ft, water flow rate 396 lb/s; transmission loss = 5%

$$\begin{aligned}\dot{W} &= -\frac{(396 \text{ lb/s})(32.2 \text{ ft}^2\text{/s})(200 \text{ ft})}{32.2 \text{ ft lb /lbf s}^2} \\ &= \left(\frac{1.055 \text{ kW}}{778 \text{ lbf ft/s}}\right) = -107.4 \text{ kW (with the sign convention)}\end{aligned}$$

With the transmission loss of 5% of the available power:

– 107.4 kW (1 – 0.05) = – 102.0 kW.

6.10 Wind Power Plants

The earth is unevenly heated by the sun and the differential heating drives a global atmospheric convection system leading to the wind. *Wind power* is the conversion of wind energy into electricity by using wind turbines. A wind turbine is a device for converting the kinetic energy in the wind into the mechanical energy of a rotating shaft. The generator is usually connected to the turbine shaft through gears which turn the generator at a different speed than the turbine shaft. Power electronic controls and converts the electricity into the correct frequency and voltage to feed into the power grid at 60 Hertz or 50 Hertz [8, 9].

The power produced by a wind turbine is proportional to the kinetic energy of the wind captured by the wind turbine. The kinetic energy of the wind is equal to the product of the kinetic energy of air per unit mass and the mass flow rate of air through the blade span area:

$$\text{Wind power} = (\text{efficiency})(\text{kinetic energy})(\text{mass flow rate of air})$$

$$\dot{W}_{\text{wind}} = \eta_{\text{wind}} \frac{v^2}{2}(\rho A v) = \eta_{\text{wind}} \frac{v^2}{2} \rho \frac{\pi D^2}{4} v \tag{6.13}$$

After rearrangement, we have

$$\dot{W}_{\text{wind}} = \eta_{\text{wind}} \rho \frac{\pi v^3 D^2}{8} = (\text{constant}) v^3 D^2 \tag{6.14}$$

$$\text{constant} = \frac{\eta_{\text{wind}} \rho \pi}{8} \tag{6.15}$$

where ρ is the density of air, v is the velocity of air, D is the diameter of the blades of the wind turbine, and η_{wind} is the efficiency of the wind turbine. Therefore, the power produced by the wind turbine is proportional to the cube of the wind velocity and the square of the blade span diameter. The strength of wind varies, and an average value for a given location does not alone indicate the amount of energy a wind turbine could produce there. To assess the frequency of wind speeds at a location, a probability distribution function is often fit to the observed data.

Although the entire kinetic energy of the wind is available for power generation, Betz's law states that the power output of a wind turbine is maximum when the wind is slowed down to one-third of its initial velocity. The actual efficiency of wind power is

$$\eta_{\text{wind turbine}} = \frac{\text{Shaft power out of turbine into gear box}}{\text{Wind power into turbine blades}} \tag{6.16}$$

This efficiency ranges between 20 and 40% and becomes part of the constant to be used in Eq. (6.15) [1].

Large-scale wind farms are connected to the electric power transmission network, while the smaller facilities are used to provide electricity to isolated locations. Wind energy, as an alternative to fossil fuels, is plentiful, renewable, widely distributed, clean, and produces no greenhouse gas emissions during operation. Wind power is nondispatchable, and for economic operation, all the available output must be taken when it is available. Problems of variability are addressed by grid energy storage, batteries, pumped-storage hydroelectricity, and energy demand management. Wind power has projected a long useful life of the equipment, which may be more than twenty years, but a high cost of capital. The estimated average cost per unit includes the cost of construction of the turbine and transmission facilities [1, 11].

In a wind farm, individual turbines are interconnected with a medium-voltage (often 34.5 kV), power collection system, and communications network. At a substation, this medium-voltage electric current is increased in voltage with a transformer for connection to the high voltage electric power transmission system. Wind-turbine generators need extensive modeling of dynamic electromechanical characteristics to ensure stable behavior. The ratio between annual average power and installed capacity rating of wind power production is the *capacity factor*. Typical capacity factors for wind power change between 20 and 40%. Example 6.13 illustrates the windmill calculations.

Wind energy penetration refers to the fraction of energy produced by wind compared with the total production capacity available. There is not a generally accepted maximum level of wind penetration. An interconnected electricity grid usually includes reserve production and transmission capacity to allow for equipment failures; this reserve capacity can also serve to regulate the varying power production by wind plants. At present, a few grid systems have penetration of wind energy above 5%: Despite the power forecasting methods used, predictability of wind plant output remains low for short-term operation. Pumped-storage hydroelectricity or other forms of grid energy storage can store energy developed by high-wind periods and release it when needed [8, 11].

Aside from the availability of wind itself, other factors include the availability of transmission lines, value of energy, cost of land acquisition, land use considerations, and environmental impact of construction and operations. *Wind power density* is a calculation of the effective power of the wind at a location. A map showing the distribution of wind power density is a first step in identifying possible locations for wind turbines. *Small-scale wind power* has the capacity to produce up to 50 kW of electrical power. Isolated communities may use small-scale wind turbines to displace fossil fuel consumption [8].

Example 6.13 **Windmill power estimations**

A windmill farm supplies a power output of 1 MW for a community. Each windmill has a blade 10 m in diameter. At the location of the windmills, the average velocity of the wind is 11 m/s and the average temperature is 20°C. Estimate the minimum number of windmills to be installed.

Solution:

Air is an ideal gas, and the pressure is atmospheric.

$$\text{Inlet: } v = 11\text{m/s}, \quad R = 8.314\,\text{kPa}\,\text{m}^3/\text{kmole K}, T = 293\,\text{K},$$
$$D = 10\text{m}, MW_{\text{air}} = 29\,\text{kg/kmole}$$

Power output = 1MW = 1000 kW

$$\text{Density of air } \rho = (MW)\frac{P}{RT} = \frac{(29\text{ kg/kmole})101.3\text{ kPa}}{\left(8.314\text{ kPa m}^3\text{/kmole K}\right)(293\text{K})} = 1.2\text{ m}^3/\text{kg}$$

Air mass flow rate:

$$\dot{m} = \rho A v = \rho \pi \frac{D^2}{4} v = 1036.2 \text{ kg/s}$$

Power from each windmill:

$$KE = -\dot{m}\frac{v^2}{2} = 1036.2 \text{ kg/s}\frac{(11 \text{ m/s})^2}{2}\left(\frac{\text{kJ/kg}}{1000 \text{ m}^2/\text{s}^2}\right) = -62.7 \text{ kW}$$

The minimum number of windmills to be installed:

$$1000 \text{ kW}/|62.7|\text{kW} = 16 \text{ windmills}$$

Example 6.14 **Estimation of power available from a wind turbine**

A wind turbine with 10 m rotor diameter is to be installed on a hilltop where the wind blows steadily at an average speed of 15 m/s. Determine the maximum power that can be generated by the turbine operated with a 35% efficiency.

Solution:

Assume that the air is ideal gas at 300 K and 1 atm (101.3 kPa).

$$\text{Inlet}: v = 15 \text{ m/s}, R = 8.314 \text{ kPa m}^3/\text{kmole K}, T = 300\text{K}, D = 10 \text{ m},$$
$$MW_{\text{air}} = 29 \text{ kg/kmole}$$

$$\text{Density of air} \rho = (MW)\frac{P}{RT} = \frac{(29 \text{ kg/kmole})101.3 \text{ kPa}}{(8.314 \text{ kPa m}^3/\text{kmole K})(300 \text{ K})} = 1.18 \text{ m}^3/\text{kg}$$

Air mass flow rate:

$$\dot{m} = \rho A v = \rho \pi \frac{D^2}{4} v = 1389.45 \text{ kg/s}$$

Power from a wind turbine with 35% efficiency η:

$$KE = -\eta \dot{m}\frac{v^2}{2} = (0.35)(1389.45 \text{ kg/s})\frac{(10 \text{ m/s})^2}{2}\left(\frac{\text{kJ/kg}}{1000 \text{ m}^2/\text{s}^2}\right) = -24.31 \text{ kW}$$

6.11 Solar Power Plants

Solar power is derived from the energy of sunlight. Average insolation changes from 150 to 300 W/m^2 or 3.5 to 7.0 kWh/m^2 day. There are two main types of technologies for converting solar energy to electricity: photovoltaic and solar-thermal electric (see Fig. 6.15).

- *Photovoltaic conversion* produces electricity directly from sunlight in a solar cell. There are many types of photovoltaic cells, such as thin film, monocrystalline silicon, polycrystalline silicon, and amorphous cells, as well as multiple types of concentrating solar power. Photovoltaic power initially is used in small and medium-sized applications, from the calculator powered by a single solar cell to off-grid homes powered by photovoltaic modules.
- *Solar-thermal electric production* based on concentrating solar power systems use lenses or mirrors and tracking systems to focus a large area of sunlight into a small beam. The concentrated solar energy heat is used to produce steam to drive turbines and produce electricity. A parabolic trough consists of a linear parabolic reflector that concentrates light onto a receiver positioned along the reflector's focal line. The receiver is a tube positioned right above the middle of the parabolic mirror and is filled with a working fluid. The reflector is made to follow the sun during the daylight hours by tracking along a single axis

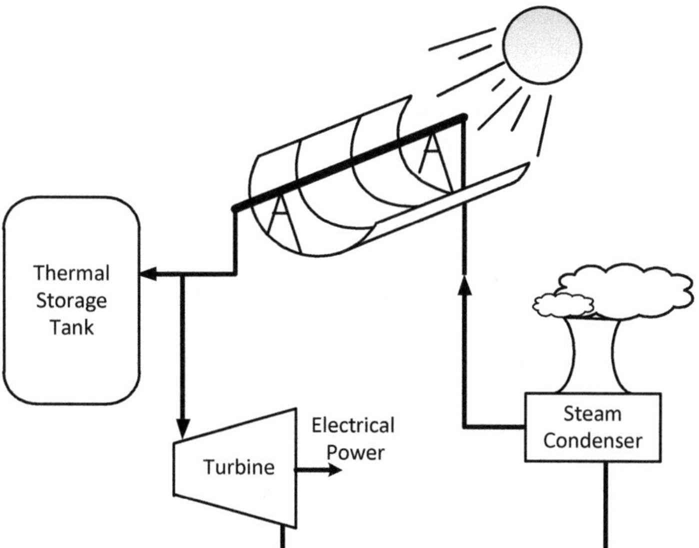

Fig. 6.15 Concentrating solar power technologies offer utility-scale power production [8, 11]

A Stirling solar dish consists of a standalone parabolic reflector that concentrates light onto a receiver positioned at the focal point. The reflector tracks the sun along two axes and captures the sun's energy through a parabolic dish solar concentrator. The concentrated energy drives a Stirling engine, which spins a generator producing electricity. The advantages of Stirling solar over photovoltaic cells are higher efficiency of converting sunlight into electricity and longer lifetime.

Mirrors, called heliostats, track the sun throughout the day and reflect sunlight to the receiver. Early designs used water heated to steam by the sun's energy to drive turbines. New systems use liquid salt because of its thermal characteristics. Salt is usually a mixture of 60% sodium nitrate and 40% potassium nitrate. The mixture melts at 220°C/428°F. Cold salt is pumped from a holding tank through the receiver where the focused sunlight heats it to over 1000°F. The hot salt passes through a heat exchanger that makes steam to drive turbine. Power towers are more cost effective, offer higher efficiency, and better energy storage capability among concentrated solar power technologies [1, 8, 11].

6.12 Fuel Cells

A *fuel cell* oxidizes fuel, such as hydrogen or methane, electrochemically to produce electric power. It consists of two electrodes separated by an electrolyte. The fuel and oxygen are continuously fed into the cell, and the products of reaction are withdrawn continuously. The fuel makes intimate contact with the anode, fuel electrode. Oxygen, usually in air, makes intimate contact with the cathode, oxygen electrode. Half-cell reactions take place at each electrode. The sum of the half-cell reactions is the overall reaction. The type of electrolyte characterizes the type of fuel cell. Schematic fuel cells using hydrogen as fuel are illustrated in Fig. 6.16. When the electrolyte is acidic, the half-cell reactions occurring at the hydrogen electrode (anode) and at the oxygen electrode (cathode) are

$$H_2 \rightarrow 2H^+ + 2e^- \text{(anode)} \tag{6.17}$$

$$\frac{1}{2}O_2 + 2e^- + 2H^+ \rightarrow H_2O(g)\text{(cathode)} \tag{6.18}$$

The electrons with negative charge (e^-) are released at the anode. These electrons produce an electric current which is used by the reaction occurring at the cathode. The electric current is carried out by an external circuit. The cation H^+ migrates from anode to cathode through the electrolyte. The sum of the half-cell reaction is the overall reaction taking place at the fuel cell.

$$H_2 + \frac{1}{2}O_2 \rightarrow H_2O(g) \tag{6.19}$$

A thin solid polymer known as *proton-exchange membrane* serves as an acid electrolyte in the hydrogen/oxygen fuel cell. Each side of the membrane is bonded

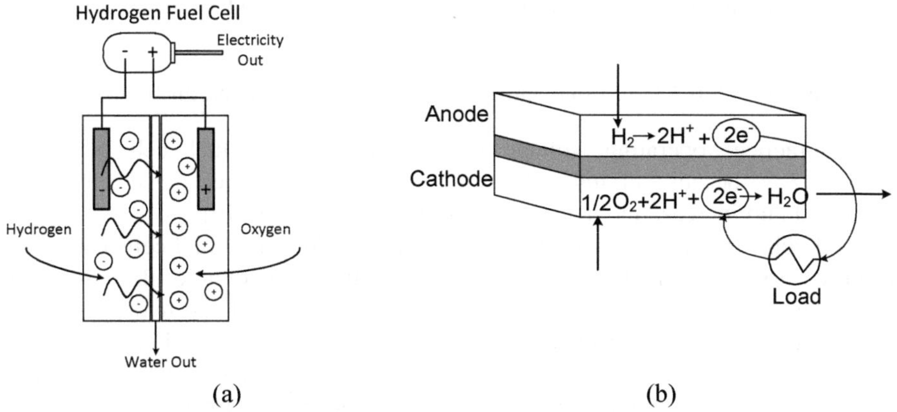

Fig. 6.16 **a** Schematic of a hydrogen fuel cell. The waste product is water. Source, **b** half-cell reactions for a hydrogen/oxygen fuel cell with acidic electrolyte

to a porous carbon electrode impregnated with platinum which serves as a catalyst. The porous electrode provides a very large interface area for the reaction and facilitates the diffusion of hydrogen and oxygen into the cell and the water vapor out of the cell. Cells can be connected in series to make a compact unit with a required level of energy output and operate at a temperature near 60°C [1, 9, 12].

For each mole of hydrogen consumed, 2 mole of electrons pass to the external circuit. Therefore, the electrical energy (work) is the product of the charge transferred and the voltage V of the cell

$$W_e = -2FV = \Delta G \tag{6.20}$$

where F is the Faraday's constant ($F = 96{,}485$ C/mole) and ΔG is the Gibbs energy. The electric work of reversible and isothermal fuel cells is

$$W_e = \Delta H - q = \Delta G \tag{6.21}$$

Consider a hydrogen/oxygen fuel cell operating at 20°C and 1 bar with pure hydrogen and oxygen as reactants and water vapor as product. The overall reaction is the standard formation reaction for water and the from Appendix C Table C1.

$$\Delta H = \Delta H^o_{\text{f, H2O}} = -241{,}818\ \text{J/mole and}\ \Delta G = \Delta G^o_{\text{f, H2O}} = -228{,}572\ \text{J/mole}$$

Therefore, for the hydrogen/oxygen fuel cell the electric work and the voltage are

$$\mathrm{W_e} = -228{,}572\ \text{J/mole and}\ V = \frac{-\Delta G}{2F} = 1.184\ \text{volts}$$

Using the air instead of pure oxygen in a reversible and isothermal fuel cell, we have

$$W_e = -226{,}638\,\text{J/ mole and } V = 1.174 \text{ volts (hydrogen/air fuel cell)}$$

In practice, operating voltage of hydrogen/oxygen fuel cell is around 0.6 – 0.7 V, because of internal irreversibility which reduce the electric work produced and increase the heat transfer to the surroundings [9].

Fuel cells are very efficient, but expensive to build. Small fuel cells can power electric cars. Large fuel cells can provide electricity in remote places with no power lines. Because of the high cost to build fuel cells, large hydrogen power plants may still not be feasible. However, fuel cells are being used in some places as a source of emergency power, from hospitals to wilderness locations. Portable fuel cells are being manufactured to provide longer power for laptop computers, cell phones, and military applications [12].

6.12.1 Direct-Methanol Fuel Cells

Direct-methanol fuel cells are a subcategory of proton-exchange fuel cells in which methanol is used as the fuel. Methanol is toxic, flammable, and a liquid from – 97.0 to 64.7°C at atmospheric pressure. Methanol is fed as a weak solution (usually around 1M, i.e., about 3% in mass). At the anode, methanol and water are adsorbed on a catalyst usually made of platinum and ruthenium particles and lose protons until carbon dioxide is formed. Direct-methanol fuel cells use a methanol solution to carry the reactant into the cell; common operating temperatures are in the range 50–120°C. Water is consumed at the anode and is produced at the cathode. Protons (H^+) are transported across the proton-exchange membrane often made from Nafion to the cathode where they react with oxygen to produce water. Electrons are transported through an external circuit from anode to cathode, providing power to connected devices. The half-reactions taking place at the anode and cathode, and the overall reaction of the cell are

$$CH_2OH + H_2O \rightarrow CO_2 + 6H^+ + 6e^- \quad \text{(Anode)} \tag{6.22}$$

$$3/2O_2 + 6H^+ + 6e^- \rightarrow 3H_2O \quad \text{(Cathode)} \tag{6.23}$$

$$CH_3OH + 3/2O_2 \rightarrow CO_2 + 2H_2O + \text{ electrical energy (Overall reaction)} \tag{6.24}$$

The only waste products with these types of fuel cells are carbon dioxide and water. Currently, platinum is used as the catalyst for both half-reactions. Efficiency is presently quite low for these cells, so they are targeted especially to portable application, where energy and power density are more important than efficiency. These cells need improvements for the loss of methanol and the management of carbon dioxide created at the anode [7, 12].

6.12.2 Microbial Fuel Cell

A *microbial fuel cell* converts chemical energy, available in a bio substrate, directly into electricity. To achieve this, bacteria are used as a catalyst to convert a variety of organic compounds into CO_2, water, and energy. Microorganisms use the produced energy to grow and to maintain their metabolism. A microbial fuel cell can harvest a part of this microbial energy in the form of electricity. A microbial fuel cell consists of an anode, a cathode, a proton-exchange membrane, and an electrical circuit as seen in Fig. 6.17.

The bacteria live in the anode and convert a bio substrate such as glucose and acetate but also wastewater into CO_2, protons, and electrons

$$C_{12}H_{22}O_{11} + 13H_2O \rightarrow 12CO_2 + 48H^+ + 48e^- \qquad (6.25)$$

Due to the ability of bacteria to transfer electrons to an insoluble electron acceptor, microbial fuel cells collect the electrons originating from microbial metabolism. The electrons then flow through an electrical circuit to the cathode. The potential difference (Volt) between the anode and the cathode, together with the flow of electrons (ampere) results in the generation of electrical power. The protons flow through the proton-exchange membrane to the cathode. At the

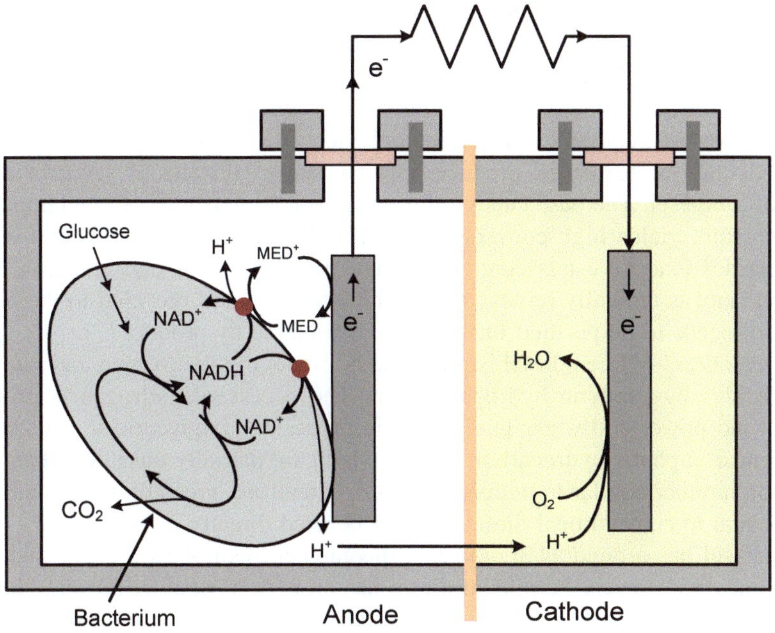

Fig. 6.17 General layout of a microbial fuel cell in which in the anodic compartment the bacteria can bring about oxidative conversions while in the cathodic compartment chemical and microbial reductive processes can occur [1, 12]

cathode the electrons, oxygen, and protons combine to form only water. The two electrodes are at different potentials (about 0.5 V). Microbial fuel cells use inorganic mediators to tap into the electron transport chain of cells and channel electrons produced. Some possible mediators include natural red, methylene blue, or thionine. The mediator exits the cell laden with electrons that it shuttles to an electrode where it deposits them. This electrode becomes the anode negatively charged electrode. After releasing the electrons, the mediator returns to its original oxidized state ready to repeat the process. Therefore, microbial activity is strongly dependent on the redox potential of the anode [12].

6.13 Bioenergy Production

Bioenergy is renewable energy derived from biomass feedstock in a biorefinery platform that can convert various types of biomasses into multi-products including biofuels and bioproducts. Biofuels derived from energy crops are among the most rapidly growing transportation liquid fuels including ethanol and biodiesel [7, 8].

Ethanol Production—Corn and sugarcane are the main energy crops for producing ethanol with the by-products of dried distillers' grain soluble and bagasse. The most common process uses yeast to ferment the glucose into ethanol and carbon dioxide. The resulting solution is distilled to get fuel-grade ethanol to be blended with conventional gasoline. Lignocellulosic biomass can also be used as feedstock for ethanol production by a more complicated process than conventional ethanol made from fermentation. This process can decouple increasing ethanol production from energy crops from food supply chain [7].

Biodiesel and Green Diesel Production—Biodiesel is produced from transesterification of vegetable oils or animal fats with methanol catalytically. One hundred units of feedstock produce approximately 10 units of glycerin and 90 units of biodiesel. The base catalyzed transesterification occurs at low temperature and pressure, yields high conversion (~98%). Animal and plant fats and oils are triglycerides that are esters containing three free fatty acids and glycerol. Residual methanol is typically removed through distillation and recycled to the reactor. Glycerol needs to be purified further to be a valuable by-product [7].

Green Diesel—Green diesel is produced by removing the oxygen and saturating triglycerides by catalytic hydrogeneration. The necessary hydrogen is produced using wind power and water in electrolytic processes. Triglycerides acids contain long, linear aliphatic hydrocarbon chains, which are partially unsaturated and have a carbon number similar to petroleum diesel. Therefore, green diesel has a heating value equal to conventional diesel and can be used directly in vehicles.

Biodiesel has around 11% oxygen, whereas petroleum-based diesel and green diesel have no oxygen. Petroleum diesel has around 10 ppm sulfur and biodiesel and green diesel have less than 1 ppm sulfur. Green diesel yield depends on both feedstock type and the level of hydro isomerization required to achieve product cloud point specification [7].

Energy from Solid Waste—Municipal solid waste contains biomass like paper, cardboard, food scraps, grass clippings, leaves, wood, and others. Municipal solid waste that is land filled or burned in special waste-to-energy plants, which produce heat to make steam to heat buildings or to produce electricity. Such plants help reduce the amount of solid waste. Solid waste incinerators can burn solid waste without electricity production.

6.14 Other Energy Production Opportunities

There are other energy production systems and processes either under development or operation with limited scale [9]. Some of these energy production opportunities are:

Solid-state generation (without moving parts) is of particular interest in portable applications. This area is largely dominated by thermoelectric devices, though thermionic and thermophotovoltaic systems have been developed as well.

- *Piezoelectric devices* are used for power generation from mechanical strain, particularly in power harvesting.
- *Betavoltaics* is another type of solid-state power generator which produces electricity from radioactive decay.
- *Fluid-based magnetohydrodynamic* power generation has been studied as a method for extracting electrical power from nuclear reactors and from more conventional fuel combustion systems.
- *Osmotic power* is another possibility when salt and sweet water merges in rivers and seas.
- Electromagnetic energy harvesting.
- Harvesting circuits.
- Thermoelectrics.
- Microbatteries.

Problems

6.1. In a steam power plant, steam at 200 psia and 600°F enters a turbine and exits at 5 psia and 200°F. The steam enters the turbine through a 2.5-inch-diameter pipe with a velocity of 11 ft/s and exits through a 9-inch-diameter pipe. Estimate the power produced by the turbine.

6.2. A turbine operates at adiabatic and steady-state conditions. At the inlet, steam at 600°C and 1100 kPa enters the turbine. The steam flow rate is 3 kg/s. The inlet tube diameter is 10 cm. After expanding in the turbine, the steam exits through a pipe with a diameter of 25 cm. At the exit the steam is at 300°C and 110 kPa. Estimate the work produced by the turbine.

6.3. A turbine operates at adiabatic and steady-state conditions. At the inlet, steam at 550°C and 1500 kPa enters the turbine. The steam flow rate is 4 kg/s. The inlet tube diameter is 12 cm. After expanding in the turbine, the steam exits through a pipe with a diameter of 28 cm. At the exit, the steam is saturated at 110 kPa (at 102.3°C). Estimate the work produced by the turbine.

6.4. A turbine operates at adiabatic and steady-state conditions. At the inlet, steam at 600°C and 4000 kPa enters the turbine. The steam flow rate is 10 kg/s. The inlet tube diameter is 10 cm. After expanding in the turbine, the steam exits through a pipe with a diameter of 30 cm. At the exit, the steam is at 150 kPa and 120°C. Estimate the work produced by the turbine.

6.5. A turbine operates at adiabatic and steady-state conditions. At the inlet, steam at 600°C and 8000 kPa enters the turbine. The steam flow rate is 15 kg/s. The inlet tube diameter is 9 cm. After expanding in the turbine, the steam exits through a pipe with a diameter of 35 cm. At the exit, the steam is saturated at 100 kPa (99.63°C). Estimate the work produced by the turbine.

6.6. Superheated steam at 4100 kPa and 300°C expand adiabatically in a steam turbine and exits at 15 kPa with a quality of $x = 0.87$. The velocity of the steam at the inlet is 50 m/s and at the exit 160 m/s. Elevation at the inlet is 10 m and at the exit 6 m. Estimate the power produced for the steam flow rate of 1 kg/s.

6.7. A turbine serves as an energy source for a small electrical generator. The turbine operates at adiabatic and steady-state conditions. At the inlet, steam at 600°C and 1100 kPa enters the turbine. The steam flow rate is 3 kg/s. At the exit, the steam is 300°C and 110 kPa. Estimate the work produced by the turbine.

6.8. A steam at 8000 kPa and 400°C expands in a turbine. At the exit, the steam is at 20 kPa with a quality of 0.9. Estimate the power output if the steam flow rate is 11.5 kg/s.

6.9. Steam at 8400 kPa and 400°C expands in a turbine. At the exit, the steam is at 15 kPa with a quality of 0.92. Estimate the power output if the steam flow rate is 11.5 kg/s.

6.10. Steam expands in a turbine. The steam enters the turbine at 9000 kPa and 450°C and exits at 10 kPa with a quality of 0.95. If the turbine produces a power of 6.5 MW, estimate the steam flow rate.

6.11. A hot exhaust gas is heating a boiler to produce superheated steam at 100 psia and 400°F. In the meantime, the exhaust gas is cooled from 2500 to 350°F. Saturated liquid water (stream 1) at 14.7 psia enters the boiler with a flow rate of 200 lb/h. Superheated steam (stream 2) is used in a turbine and discharged as saturated steam (stream 3) at 14.7 psia. Determine the molar flow rate of the exhaust gas needed and the maximum work produced. Assume that the surroundings are at 70°F, and the heat capacity of the flue gas is

$$C_p = 7.606 + 0.0006077T, \text{ where } T \text{ is in Rankine and } C_p \text{ is in Btu/(lbmole R)}.$$

6.12. Steam expands in a turbine. The steam enters the turbine at 1000 psia and 800°F and exits as a saturated vapor at 5 psia. The turbine produces a power of 5 MW. If the steam flow rate is 20 lb/s, estimate the heat loss from the turbine.

6.13. A turbine serves as an energy source for a small electrical generator. The turbine operates at adiabatic and steady-state conditions. At the inlet a steam at 550°C and 1500 kPa enters the turbine. The steam flow rate is 4 kg/s. At the exit, the steam is saturated at 110 kPa (at 102.3°C). Estimate the work produced by the turbine.

6.14. A turbine serves as an energy source for a small electrical generator. The turbine operates at adiabatic and steady-state conditions. At the inlet a steam at 600°C and 4000 kPa enters the turbine. The steam flow rate is 10 kg/s. At the exit, the steam is at 150 kPa and 120°C. Estimate the work produced by the turbine.

6.15. A turbine serves as an energy source for a small electrical generator. The turbine operates at adiabatic and steady-state conditions. At the inlet steam at 600°C and 8000 kPa enters the turbine. The steam flow rate is 15 kg/s. At the exit, the steam is saturated at 100 kPa (99.63°C). Estimate the work produced by the turbine.

6.16. Steam at 8200 kPa and 823.15 K (state 1) is being expanded to 30 kPa in continuous operation. Determine the final temperature (state 2), entropy produced, and work produced per kg of steam for an isothermal expansion through a turbine.

6.17. Steam enters an adiabatic turbine at 5000 kPa and 450°C and leaves as a saturated vapor at 140 kPa. Determine the work output per kg of steam flowing through the turbine if the process is reversible and changes in kinetic and potential energies are negligible.

6.18. Steam at 8200 kPa and 823.15 K (state 1) is being expanded to 30 kPa in continuous operation. Determine the final temperature (state 2), entropy produced, and work produced per kg of steam for an adiabatic expansion through a turbine.

6.19. A steady-flow adiabatic turbine receives steam at 650 K and 8200 kPa and discharges it at 373.15 K and 101.32 kPa. If the flow rate of the steam is 12 kg/s, determine (a) the maximum work and (b) the work loss if the surroundings are at 298.15 K.

6.20. A turbine discharges steam from 6 MPa and 400°C to saturated vapor at 360.15 K while producing 480 kJ/kg of shaft work. The temperature of surroundings is 300 K. Determine maximum possible production of power in kW.

6.21. A Carnot cycle uses water as the working fluid at a steady-flow process. Heat is transferred from a source at 400°C and water changes from saturated liquid to saturated vapor. The saturated steam expands in a turbine at 10 kPa, and a heat of 1150 kJ/kg is transferred in a condenser at 10 kPa. Estimate the net power output of the cycle for a flow rate of 10 kg/s of the working fluid.

6.22. A Carnot cycle uses water as the working fluid at a steady-flow process. Heat is transferred from a source at 400°C and water changes from saturated liquid to saturated vapor. The saturated steam expands in a turbine at 30 kPa, and a heat of 1150 kJ/kg is transferred in a condenser at 30 kPa. Estimate the net power output of the cycle for a flow rate of 14.5 kg/s of the working fluid.

6.23. A steam power production plant uses steam at 8200 kPa and 823.15 K. The turbine discharges the steam at 30 kPa. The turbine and pump operate reversibly and adiabatically. Determine the work produced for every kg steam produced in the boiler.

6.24. A steam power plant operates on a simple ideal Rankine cycle shown below. The turbine receives steam at 698.15 K and 4400 kPa, while the discharged steam is at 15 kPa. The mass flow rate of steam is 12.0 kg/s. Determine the network output.

6.25. A reheat Rankine cycle is used in a steam power plant. Steam enters the high-pressure turbine at 9000 kPa and 823.15 K and leaves at 4350 kPa. The steam is reheated at constant pressure to 823.15 K. The steam enters the low-pressure turbine at 4350 kPa and 823.15 K. The discharged steam from the low-pressure turbine is at 10 kPa. The steam flow rate is 24.6 kg/s. Determine the net power output.

6.26. A reheat Rankine cycle is used in a steam power plant. Steam enters the high-pressure turbine at 10,000 kPa and 823.15 K and leaves at 4350 kPa. The steam is reheated at constant pressure to 823.15 K. The steam enters the low-pressure turbine at 4350 kPa and 823.15 K. The discharged steam from the low-pressure turbine is at 15 kPa. The steam flow rate is 38.2 kg/s. Determine the net power output.

6.27. A simple ideal reheat Rankine cycle is used in a steam power plant shown below. Steam enters the turbine at 9200 kPa and 823.15 K and leaves at 4350 kPa and 698.15 K. The steam is reheated at constant pressure to 823.15 K. The discharged steam from the low-pressure turbine is at 15 kPa. The net power output of the turbine is 75 MW. Determine the mass flow rate of steam.

6.28. A steam power plant is using an actual regenerative Rankine cycle. Steam enters the high-pressure turbine at 11,000 kPa and 773.15 K, and the condenser operates at 10 kPa. Steam is extracted from the turbine at 475 kPa to heat the water in an open heater. The water is a saturated liquid after passing through the water heater. The steam flow rate is 65 kg/s. Determine the work output.

6.29. A steam power plant is using an actual regenerative Rankine cycle. Steam enters the high-pressure turbine at 10,000 kPa and 773.15 K, and the condenser operates at 30 kPa. Steam is extracted from the turbine at 475 kPa to heat the water in an open heater. The water is a saturated liquid after passing through the water heater. The steam flow rate is 45.6 kg/s. Determine the work output.

6.30. A steam power plant is using an ideal regenerative Rankine cycle shown below. Steam enters the high-pressure turbine at 8400 kPa and 773.15 K, and the condenser operates at 10 kPa. Steam is extracted from the turbine at 400 kPa to heat the feed water in an open heater. The water is a saturated liquid after passing through the feed water heater. Determine the net power output of the cycle.

6.31. A steam power plant uses an actual reheat regenerative Rankine cycle. Steam enters the high-pressure turbine at 11,000 kPa and 773.15 K, and the condenser operates at 10 kPa. The steam is extracted from the turbine in 2000 kPa to heat the water in an open heater. The steam is extracted at 475 kPa for process heat. The water is a saturated liquid after passing through the water heater. Determine the work output for a flow rate of steam of 66.0 kg/s.

6.32. A steam power plant uses an ideal reheat regenerative Rankine cycle. Steam enters the high-pressure turbine at 9400 kPa and 773.15 K and leaves at 850 kPa. The condenser operates at 15 kPa. Part of the steam is extracted from the turbine at 850 kPa to heat the water in an open heater, where the steam and liquid water from the condenser mix and direct contact heat transfer takes place. The rest of the steam is reheated to 723.15 K and expanded in the low-pressure turbine section to the condenser pressure. The water is a saturated liquid after passing through the water heater and is at the heater pressure. The flow rate of steam is 20 kg/s. Determine the power produced

6.33. A steam power plant is using an actual reheat regenerative Rankine cycle. Steam enters the high-pressure turbine at 10,800 kPa and 773.15 K, and the condenser operates at 15 kPa. The steam is extracted from the turbine in 2000 kPa to heat the water in an open heater. The steam is extracted at 475 kPa for process heat. The water is a saturated liquid after leaving the water heater. The steam flow rate is 30.8 kg/s. Determine the power produced.

6.34. A steam power plant uses a geothermal energy source. The geothermal source is available at 220°C and 2320 kPa with a flow rate of 180 kg/s. The hot water goes through a valve and a flash drum. Steam from the flash drum enters the turbine at 550 kPa and 428.62 K. The discharged steam from the turbine has a quality of $x_4 = 0.95$. The condenser operates at 40 kPa. The water is a saturated liquid after passing through the condenser. Determine the network output.

6.35. A steam power plant uses a geothermal energy source. The geothermal source is available at 220°C and 2320 kPa with a flow rate of 50 kg/s. The hot water goes through a valve and a flash drum. Steam from the flash drum enters the turbine at 550 kPa and 428.62 K. The discharged steam from the turbine has a quality of $x_4 = 0.90$. The condenser operates at 15 kPa. The water is a saturated liquid after passing through the condenser. Determine the network output.

6.36. A cogeneration plant is using steam at 5500 kPa and 748.15 K to produce power and process heat. The amount of process heat required is 10000 kW.

Twenty percent of the steam produced in the boiler is extracted at 475 kPa from the turbine for cogeneration. The extracted steam is condensed and mixed with the water output of the condenser. The remaining steam expands from 5500 kPa to the condenser conditions. The condenser operates at 10 kPa. Determine the network output.

6.37. A cogeneration plant uses steam at 8400 kPa and 773.15 K. One-fourth of the steam is extracted at 600 kPa from the turbine for cogeneration. After it is used for process heat, the extracted steam is condensed and mixed with the water output of the condenser. The rest of the steam expands from 600 kPa to the condenser pressure of 10 kPa. The steam flow rate produced in the boiler is 60 kg/s. Determine the work output.

6.38. A cogeneration plant uses steam at 900 psia and 1000°F to produce power and process heat. The steam flow rate from the boiler is 40 lb/s. The process requires steam at 70 psia at a rate of 5.5 lb/s supplied by the expanding steam in the turbine. The extracted steam is condensed and mixed with the water output of the condenser. The remaining steam expands from 70 psia to the condenser pressure of 3.2 psia. Determine the work output.

6.39. One kmole of air is initially at 1 atm, −13°C, performs a power cycle consisting of three internally reversible processes in series. Step 1–2: Adiabatic compression to 5 atm. Step 2–3: Isothermal expansion to 1 atm. Step 3–1: Constant-pressure compression. Determine the network output.

6.40. A steam power plant output is 62 MW. It uses steam (stream 1) at 8200 kPa and 550°C. The discharged stream (stream 2) is saturated at 15 kPa. If the expansion in the turbine is adiabatic, and the surroundings are at 298.15 K, determine the steam flow rate.

6.41. A steam power plant output is 55 MW. It uses steam (stream 1) at 8400 kPa and 500°C. The discharged stream (stream 2) is saturated at 30 kPa. If the expansion in the turbine is adiabatic, and the surroundings are at 298.15 K, determine the steam flow rate.

6.42. In a hydropower plant, a hydro turbine operates with a head of 33 m of water. Inlet and outlet conduits are 1.70 m in diameter. If the outlet velocity of the water is 4.6 m/s, estimate the power produced by the turbine.

6.43. In a hydropower plant, a hydro turbine operates with a head of 46 m of water. Inlet and outlet conduits are 1.80 m in diameter. If the outlet velocity of the water is 5.5 m/s, estimate the power produced by the turbine.

6.44. Consider a hydropower plant reservoir with an energy storage capacity of 1.5×10^6 kWh. This energy is to be stored at an average elevation of 40 m relative to the ground level. Estimate the minimum amount of water must be pumped back to the reservoir.

6.45. Consider a hydropower plant reservoir with an energy storage capacity of 2.0×10^6 kWh. This energy is to be stored at an average elevation of 160 m relative to the ground level. Estimate the minimum amount of water must be pumped back to the reservoir.

6.46. A farm of windmills supplies a power output of 2 MW for a community. Each windmill has a blade 10 m in diameter. At the location of the windmills, the average velocity of the wind is 11 m/s and the average temperature is 20°C. Estimate the minimum number of windmills to be installed.

6.47. A farm of windmills supplies a power output of 3 MW for a community. Each windmill has blades 11 m in diameter. At the location of the windmills, the average velocity of the wind is 14 m/s and the average temperature is 20°C. Estimate the minimum number of windmills to be installed.

6.48. A farm of windmills supplies a power output of 4.2 MW for a community. Each windmill has blades 10 m in diameter. At the location of the windmills, the average velocity of the wind is 15 m/s and the average temperature is 20°C. Estimate the minimum number of windmills to be installed.

References

1. Çengel YA, Boles MA (2014) Thermodynamics: an engineering approach, 8th edn. McGraw-Hill, New York
2. Chattopadhyay P (2015) Engineering thermodynamics, 2nd edn. Oxford Univ Press, Oxford
3. Clark JA (2014) Combined Heat and Power (CHP): Integration with industrial processes. In: Anwar S (ed) (2014) Encyclopedia of energy engineering and technology, 2nd edn. CRC Press, Boca Raton
4. Demirel Y, Gerbaud V (2025) Nonequilibrium thermodynamics transport and rate processes in physical, chemical and biological systems, 5th edn. Elsevier, Amsterdam
5. Demirel Y, Matzen M, Winters C, Gao X (2015) Capturing and using CO_2 as feedstock with chemical-looping and hydrothermal technologies. Int J Energy Res 39:1011–1047
6. Demirel Y (2025) Biofuels. In comprehensive energy systems, I. Dincer (Ed.), 2nd edn. Elsevier, Amsterdam, Vol 2, 344–384
7. Dincer I, Ratlamwala T (2013) Development of novel renewable energy-based hydrogen production systems: a comparative study. Energy Convers Management 7:77–87
8. IEA (2024) International energy agency. World Energy Outlook 2024. https://www.iea.org/reports/world-energy-outlook-2024
9. Jaffe RL, Taylor W (2018) The physics of energy. Cambridge Univ. Press, Cambridge
10. Moran MJ, Shapiro HN, Boettner DD, Bailey MB (2014) Fundamentals of engineering thermodynamics, 8th edn. Wiley, New York
11. Rabiee MM (2014) Modular smart power system. In: Anwar S (ed) Encyclopedia of energy engineering and technology, 2nd ed. CRC Press, Boca Raton
12. Vielstich W et al (eds) (2009) Handbook of fuel cells: advances in electrocatalysis, materials, diagnostics and durability, Vol 6. Wiley, New York
13. Wang X, Demirel Y (2018) Feasibility of power and methanol production by an entrained-flow coal gasification system. Energy Fuels 32:7595–7610
14. Wood MB (2014) Nuclear energy: technology. In: Anwar S (ed) Encyclopedia of energy engineering and technology, 2nd ed. CRC Press, Boca Raton

Energy Conversion

7

Introduction and Learning Objectives: Energy can be converted from one form to another in a device or in a system. For example, batteries convert chemical energy to electrical energy and operate mobile electronic equipment. A dam converts gravitational potential energy to kinetic energy of moving water used on the blades of a turbine, which ultimately converts the kinetic energy to electric energy through an electric generator. Heat engines are devices converting heat into work. With each energy conversion, part of the energy becomes dispersed and less useful and hence unavailable for further use. This chapter discusses energy conversion and thermal efficiency of turbines, compressors, and heat engines. Heat engines include Carnot, Rankine, Brayton, Otto, and Diesel cycles.

The learning objectives of this chapter are:

- Energy conversion
- Thermal efficiencies
- Improving the efficiency of energy conversions
- Energy conversion in biological systems

7.1 Energy Conversion

In most processes, energy is constantly changing from one form to another. This is called an *energy conversion.* Examples include the living systems converting the solar energy to chemical energy by synthesizing food from water and carbon dioxide through photosynthesis. The mechanical energy of a waterfall can also be converted to electromagnetic energy in a generator. An internal combustion engine converts the chemical energy in gasoline into heat, which is then transformed into the kinetic energy that moves a vehicle. A solar photovoltaic (PV) cell converts

Y. Demirel, *Energy*, Green Energy and Technology,
https://doi.org/10.1007/978-3-031-97111-2_7

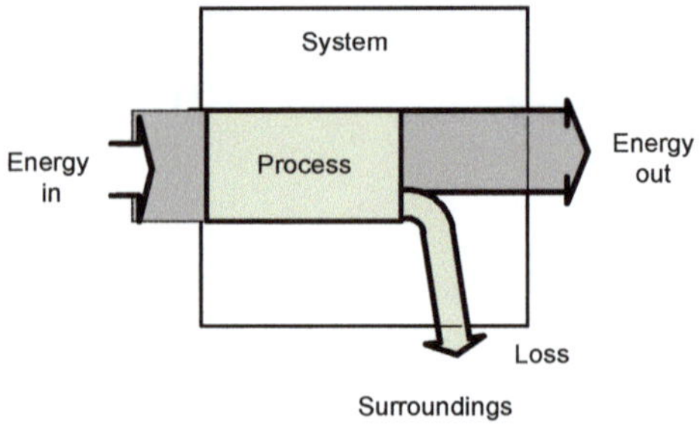

Fig. 7.1 Schematic of the energy usage and conversion in a process. Output energy at a new form is always lower than input energy. The total energy input is recovered in various other forms and hence the energy is conserved

solar radiation into electrical energy. The energy that enters a conversion device or a process is turned into other forms of energy, so an equal quantity of energy before and after the conversion is conserved.

Energy is most usable where it is most concentrated as in highly structured chemical bonds in gasoline and sugar. All other forms of energy may be completely converted to heat, but the conversion of heat to other forms of energy cannot be complete. Due to inefficiencies such as friction, heat loss, and other factors, thermal efficiencies of energy conversion are typically much less than 100%. For example, only 35–40% of the heat can be converted to electricity in a steam power plant and a typical gasoline automobile engine operates at around 25% efficiency. With each energy conversion, a part of the energy is lost usually in less useful and dispersed form, as illustrated in Fig. 7.1. This shows that there are limitations to the efficiency for energy conversion. Only a part of energy may be converted to useful work and the remainder of the energy must be reserved to be transferred to a thermal reservoir at a lower temperature [2]. There are many different processes and devices that convert energy from one form to another. Table 7.1 shows a short list of such processes and devices.

When electric current flows in a circuit, it can transfer energy to do work. Devices convert electrical energy into many useful forms, such as heat (electric heaters), light (light bulbs), motion (electric motors), sound (loudspeaker), and information technological processes (computers). Electric energy is one of the most useful forms of output energy, which can be produced by various mechanical and/or chemical devices. There are several fundamental methods of directly transforming other forms of energy into electrical energy [2, 4, 8]:

- *Static electricity* is produced from the physical separation and transport of charge. Electrons are mechanically separated and transported to increase their electric potential and imbalance of positive and negative charges leads to static electricity. For example, lightning is a natural example of static discharge. Also, low conductivity fluids in pipes can build up static electricity.
- *Electromagnetic induction*, where an electrical generator, dynamo, or alternator transform kinetic energy into electricity. Almost all commercial electrical generations are done using electromagnetic induction, in which mechanical energy forces an electrical generator to rotate. There are many different methods of

Table 7.1 Some of the processes and devices converting energy from one form to another [2, 4]

Process and device	Energy input	Useful energy output
Steam engine	Heat	Mechanical energy
Photosynthesis	Solar energy	Chemical energy
Hydroelectric dams	Gravitational potential energy	Electric energy
Windmills	Mechanical energy	Electric energy
Electric generator	Mechanical energy	Electric energy
Diesel or petrol engine	Chemical energy	Mechanical energy
Electric motor	Electric energy	Mechanical energy
Fuel cells	Chemical energy	Electric energy
Battery	Chemical energy	Electric energy
Electric bulb	Electric energy	Heat and light
Resistance heater	Electric energy	Heat
Ocean thermal power	Heat	Electric energy
Bioluminescence	Chemical energy	Light energy
Nerve impulse	Chemical energy	Electrical energy
Muscular activity	Chemical energy	Mechanical energy
Geothermal power	Heat	Electric energy
Wave power	Mechanical energy	Electric energy
Friction	Kinetic energy	Heat
Thermoelectric	Heat	Electric energy
Piezoelectrics	Strain	Electric energy

developing mechanical energy, including heat engines, hydro, wind, and tidal power.

- *Electrochemistry* is the direct transformation of chemical energy into electricity, as in a battery, fuel cell, or nerve impulse.
- *The photoelectric effect* is the transformation of light into electrical energy, as in solar cells.
- *The thermoelectric effect* is the direct conversion of temperature differences to electricity, as in thermocouples, thermopiles, and thermionic converters.
- *The piezoelectric effect* is the electricity from the mechanical strain of electrically anisotropic molecules or crystals.
- *Nuclear transformation* is the creation and acceleration of charged particles. The direct conversion of nuclear energy to electricity by beta decay is used only on a small scale.

7.2 Series of Energy Conversions

It takes a whole series of energy conversions in various processes before a useful form of energy becomes available. For example, when you use your computer, there are the following energy conversion processes involve:

- Chemical energy stored in coal is released as heat when the coal is burned.
- The heat is used to produce steam which is converted into mechanical energy in a turbine.
- The generator converts mechanical energy into electric energy that travels through the power lines into your home.
- From the power outlet at home, the computer receives that electric energy.

In a conventional internal combustion engine, these energy transformations are involved:

- Potential energy in the fuel is converted to kinetic energy of expanding gas after combustion.
- The kinetic energy of expanding gas is converted to piston movement and hence to rotary crankshaft movement.
- Rotary crankshaft movement is passed into the transmission assembly to drive the wheels of a car.

7.3 Conversion of Chemical Energy of Fuel to Heat

Chemical energy of a fuel is converted to heat during a combustion or oxidation reaction. Combustion reactions are exothermic and release heat. For example, standard heat of combustion of propane (C_3H_8) can be estimated by using the standard heats of formation tabulated in Table C1 in Appendix C. The combustion reaction of 1 mol propane is:

$$C_3H_8(g) + 5O_2(g) \rightarrow 3CO_2(g) + 4H_2O(g) + \text{Heat} \tag{7.1}$$

When the products and reactants are at their standard states (ideal gas state at 1 bar and 25°C) the standard eat of reaction is

$$\Delta H_\text{r}^\text{o} = \sum_i \nu_i \Delta H_\text{fi}^\text{o} = \nu_{CO_2} \Delta H_{\text{f}CO_2}^\text{o} + \nu_{H_2O} \Delta H_{\text{f}H_2O}^\text{o} + \nu_{C_3H_8} \Delta H_{\text{f}C_3H_8}^\text{o} \tag{7.2}$$

where ν_i is the stoichiometric coefficient of substance i, which is positive for a product and negative for a reactant. For example, for the reaction above: $\nu_{CO_2} = 3$, $\nu_{H_2O} = 4$, $\nu_{C_3H_8} = -1$. The standard heat of formation for stable elements such as oxygen, used in the combustion of propane, is zero.

7.3.1 Heating Value of Fuels

The *heating value* of a fuel is the amount of heat released during the combustion in kJ/kg, Btu/m^3, or kcal/kg. The heating values for fuels are expressed as the *higher heating value* (HHV), *lower heating value* (LHV), or *gross heating value* (GHV). Higher heating value is determined by bringing all the products of combustion back to the original precombustion temperature and condensing any water vapor produced. The combustion process of a fuel consisting of carbon and hydrogen can be approximately represented by

$$[\mathrm{C}+\mathrm{H}](\mathrm{fuel})+[\mathrm{O_2}+\mathrm{N_2}](\mathrm{Air}) \rightarrow \mathrm{CO_2}+\mathrm{H_2O}(\mathrm{liquid})+\mathrm{N_2}+\ \mathrm{Heat}(\mathrm{HHV}) \tag{7.3}$$

$$[\mathrm{C}+\mathrm{H}](\mathrm{fuel})+[\mathrm{O_2}+\mathrm{N_2}](\mathrm{Air}) \rightarrow \mathrm{CO_2}+\mathrm{H_2O}(\mathrm{vapor})+\mathrm{N_2}+\mathrm{Heat}(\mathrm{LHV}) \tag{7.4}$$

where C = Carbon, H = Hydrogen, O = Oxygen, and N = Nitrogen. Gross heating value considers the heat used to vaporize the water during the combustion reaction as well as the water existing within the fuel before it has been burned. This value is especially important for fuels like wood or coal, which contain some amount of water prior to burning. Higher and lower heating values of some common fuels, and energy density of some fuels are tabulated in Tables 2.7, 2.8, and 2.9.

Lower heating value (or *net calorific value*) is determined by subtracting the heat of vaporization of the water vapor from the higher heating value. A common method of relating higher heating value to lower heating value is:

$$\mathrm{LHV}=\mathrm{HHV}-\left(\Delta H_{\mathrm{vap}}\right)\left(n_{\mathrm{H_2O,\,out}}/n_{\mathrm{fuel,\,in}}\right)\left(\mathrm{MW}_{\mathrm{H_2O,\,out}}/\mathrm{MW}_{\mathrm{fuel,in}}\right) \tag{7.5}$$

or

$$\mathrm{LHV}=\mathrm{HHV}-\left(\Delta H_{\mathrm{vap}}\right)\left(m_{\mathrm{H_2O,\,out}}/m_{\mathrm{fuel,in}}\right)$$

where ΔH_{vap} is the heat of vaporization of water (in kJ/kg or Btu/lb), $n_{\mathrm{H_2O,out}}$ is the moles of water vaporized and $n_{\mathrm{fuel,in}}$ is the number of moles of fuel combusted, $\mathrm{MW}_{\mathrm{H_2O}}$ is the molecular weight of water, and $\mathrm{MW}_{\mathrm{fuel}}$ is the molecular weight of fuel. Example 7.1 illustrates the estimation of lower heating value from higher heating value, while Example 7.2 shows the estimation of heating values from the standard heat of combustion.

***Example 7.1* Estimation of lower heating value from higher heating value**

Higher heating value of natural gas is measured as 23,875 Btu/lb around room temperature (70°F). Convert this higher heating value to lower heating value. Heat of vaporization of water: ΔH_{vap} = 1055 Btu/lb (70°F).

Solution:

Assume that the natural gas is represented by methane CH_4.

The combustion of methane: $CH_4 + 3/2O_2 \rightarrow CO_2 + 2H_2O$.

Higher heating value: HHV (methane) = 23,875 Btu/lb,

From Table A1:

$$MW_{H_2O} = (H_2) + 1/2(O_2) = 2 + 16 = 18\,\text{lb/lbmole}$$

$$MW_{CH_4} = (C) + 4(H) = 12 + 4 = 16\,\text{lb/lbmole}$$

$$n_{CH_4} = 1\ \text{lbmole and}\ n_{H_2O} = 2\,\text{lbmole and}\ n_{H_2O}/n_{CH_4} = 2/1$$

Heat of vaporization: $\Delta H_{vap} = 1055\,\text{Btu/lb}(70°\text{F})$.

Lower heating value:

$$\text{LHV} = \text{HHV} - (\Delta H_{vap})(n_{H_2O,\,out}/n_{fuel,in})(MW_{H_2O,\,out}/MW_{fuel,in})$$

$$\text{LHV} = 23{,}875\,\text{Btu/lb} - (1055\,\text{Btu/lb})(2/1)(18/16) = 21{,}500\,\text{Btu/lb}$$

***Example 7.2* Estimating the heating values from the standard heat of combustion**

The combustion reaction of 1 mole propane is:

$$C_3H_8(g) + 5O_2(g) \rightarrow 3CO_2(g) + 4H_2O(g)$$

Estimate the higher and lower heating values of 1 kg of propane at room temperature. Heat of vaporization is: $\Delta H_{vap} = 2442\,\text{kJ/kg}(25\,°\text{C})$.

Solution:

When the products and reactants are at their standard states (ideal gas state at 1 bar and 25°C) the standard heat of reaction is expressed by

$$\Delta H_r^o = \sum_i \nu_i \Delta H_{fi}^o = \nu_{CO_2}\Delta H_{fCO_2}^o + \nu_{H_2O}\Delta H_{fH_2O}^o + \nu_{C_3H_8}\Delta H_{fC_3H_8}^o$$

Stoichiometric coefficients: $\nu_{CO_2} = 3,\ \nu_{H_2O} = 4,\ \nu_{C_3H_8} = -1$.

Using the standard heats of formation form Table C1, the heat of combustion of one mole of propane (C_3H_8) is estimated by

$$\Delta H^{o}_{r298} = 3(-393.51)\,\text{kJ/mole} + 4(-241.818)\,\text{kJ/mole} \quad - (-104.7)\,\text{kJ/mole} = -2043.1\,\text{kJ/mole}$$

Here the standard heat of reaction for oxygen is zero, as it is a naturally existing molecule in the environment.

Molecular weight of propane: MW = 44g/mole.

Heat released after the combustion of 1 kg of propane is the lower heating value since the water product is at vapor state:

$$\Delta H^{o}_{r298} = \text{LHV} = (2043.1\,\text{kJ/mole})(\text{mole}/0.044\,\text{kg}) = 46{,}430\,\text{kJ/kg}$$

The higher heating values:

$$\text{MW}_{H_2O} = (H_2) + 1/2(O_2) = 2 + 16 = 18\,\text{g/mole}$$

$$\text{MW}_{C_3H_8} = 3(C) + 8(H) = 36 + 8 = 44\,\text{g/mole}$$

$$n_{C_3H_8} = 1\,\text{mole and}\,n_{H_2O} = 4\,\text{mole}$$

$$\Delta H_{vap} = 2442\,\text{kJ/kg}(25°\text{C})$$

Higher heating value becomes:

$$\text{HHV} = \text{LHV} + (\Delta H_{vap})(n_{H_2O,\,out}/n_{C_3H_8,in})(\text{MW}_{H_2O,\,out}/\text{MW}_{C_3H_8,\,in})$$

$$\text{HHV} = 46{,}430\,\text{kJ/kg} + (2442\,\text{kJ/kg})(4/1)(18/44) = 50{,}426\,\text{kJ/kg}$$

7.4 Thermal Efficiency of Energy Conversions

Thermal efficiency is a measure of the amount of thermal energy that can be converted to another useful form. For an energy conversion device like a boiler or furnace, the thermal efficiency η_{th} is the ratio of the amount of useful energy to the energy that went into the conversion process

$$\eta_{th} = \frac{q_{useful}}{q_{in}} \tag{7.6}$$

No energy conversion process is 100% efficient. For example, most incandescent light bulbs are only 5–10% efficient because most of the electric energy is lost as heat to the surroundings. An electric resistance heater has a thermal conversion efficiency close to 100%. When fuels are used, thermal efficiency usually involves a comparison of the total chemical energy in the fuels, and the useful energy extracted from the fuels. The definition of heating value (HHV or LHV) as input energy affects the value of efficiency.

If a water flow in a boiler produces steam, the heat gained by the water can be expressed as the summation of sensible heat increase plus latent heat of vaporization

$$\dot{q} = \dot{m}\left(C_{p,\mathrm{av}}\Delta T + \Delta H_{\mathrm{vap}}\right) \tag{7.7}$$

where $\dot{q}$ is the total heat gained, $\dot{m}$ is the mass flow rate, $C_{\mathrm{p,av}}$ is the average specific heat capacity, ΔT is the temperature difference between inlet and outlet of the water in the boiler, and ΔH_{vap} is the heat of vaporization. The vaporization process needs heat, and the condensation process releases heat as shown below

$$\mathrm{H_2O\ (liquid)} + \Delta \mathrm{H_{vap}} \rightarrow \mathrm{H_2O\ (vapor)} \quad \text{Vaporization} \tag{7.8}$$

$$\mathrm{H_2O\ (vapor)} \rightarrow \mathrm{H_2O\ (liquid)} + \Delta H_{\mathrm{cond}} \ \text{Condensation} \tag{7.9}$$

The heat of vaporization is equal to the heat of condensation at the same temperature and pressure.

$$\Delta H_{\mathrm{vap}} = -\Delta H_{\mathrm{cond}} \ \text{(At the same temperature and pressure)} \tag{7.10}$$

The flue gas from boilers contains the water in vapor form and the actual amount of heat available to the boiler is the lower heating value. Accurate control of the air supply is essential to boiler efficiency. Too much air cools the furnace and carries away useful heat, while the combustion will be incomplete with little air and unburned fuel will be carried over and produced smoke. *Net calorific value* of a fuel excludes the energy in the water vapor discharged to the stack in the combustion process. For heating systems their peak steady-state thermal efficiency is often stated as, for example, this furnace is 90% efficient. However, a more detailed measure of seasonal energy effectiveness is the Annual Fuel Utilization Efficiency (AFUE), which is discussed in Sect. 9.4.

7.5 Ideal Fluid Flow Energy Conversions

For fluid flow systems, enthalpy rather than internal energy is used since the fluid flow work (PV) considered within the enthalpy. Figure 7.2 shows an open steady-state flow system.

The general relation between heat and work for fluid flow is expressed by:

$$\dot{m}\left(\Delta H + \frac{\Delta v^2}{2} + g\Delta z\right) = \dot{q} + \dot{W}_s \tag{7.11}$$

Heat transfer and work transfer should be distinguished based on the entropy transfer:

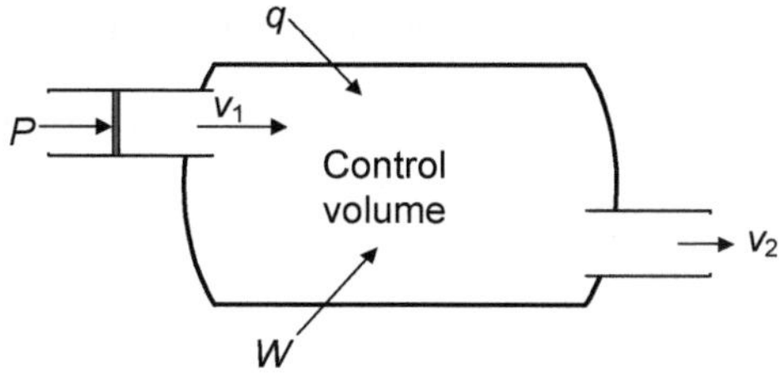

Fig. 7.2 Fluid flow with heat and work interactions between the control volume and surroundings: $\dot{m}(\Delta H + \Delta v^2/2 + g\Delta z) = \dot{q} + \dot{W}_s$ at constant pressure

- Energy interactions that are accompanied by entropy transfer are known as heat transfer.
- Energy interactions that are not accompanied by entropy transfer are known as work.

For a process producing work, such a turbine as shown in Fig. 7.3a, the *ideal work* is the maximum possible work (isentropic) produced. However, for a process requiring work, such as a compressor shown in Fig. 7.3b, the ideal work is the minimum amount of required work (isentropic). These limiting values of work occur when the change of state in the process is accomplished reversibly [3, 4]. For such processes, friction, heat loss, and other losses are negligible. When the changes in kinetic and potential energies are negligible, Eq. (7.11) becomes

$$\dot{m}\Delta H = \dot{q} + \dot{W}_s \tag{7.12}$$

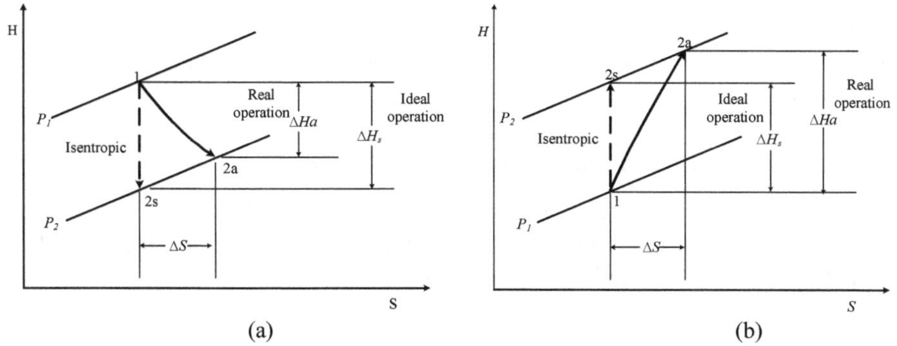

Fig. 7.3 Fluid flow work **a** in a turbine ($P_1 > P_2$), **b** in a compressor ($P_1 < P_2$); ideal operation is isentropic in a turbine and produces maximum work; ideal operation is isentropic in a compressor and requires minimum work

For a uniform surrounding temperature T_o and using the definition of heat flow as

$$\dot{q} = T_o \Delta(\dot{m}S) \tag{7.13}$$

ideal work becomes

$$\dot{W}_{\text{ideal}} = \dot{W}_{\text{rev}} = \Delta(\dot{m}H) - T_o \Delta(\dot{m}S) \tag{7.14}$$

A reversible process, however, is hypothetical and used for determination of the ideal work limit and comparing it with the actual work. When the ideal work is produced (Fig. 7.3a), we have the *adiabatic or isentropic efficiency* defined by

$$\eta_{\text{Ts}} = \frac{W_{\text{prod}}}{W_{\text{ideal}}} = \frac{H_{2a} - H_1}{H_{2s} - H_1} = \frac{\Delta H_a}{\Delta H_s} \quad \text{(work produced)} \tag{7.15}$$

With the ideal work required (Fig. 7.3b), such as in a compressor, the *isentropic efficiency* η_s is

$$\eta_{\text{Cs}} = \frac{W_{\text{ideal}}}{W_{\text{req}}} = \frac{H_{2s} - H_1}{H_{2a} - H_1} = \frac{\Delta H_s}{\Delta H_a} \quad \text{(work required)} \tag{7.16}$$

Reversible work $\dot{W}_{\text{rev}}$ in terms of the values of exergy [5, 10] for a process between the specified initial and final states is

$$\dot{W}_{\text{rev}} = \dot{m}(\text{Ex}_1 - \text{Ex}_2) + \left(1 + \frac{T_o}{T}\right)\dot{q} \tag{7.17}$$

where T_o is the temperature of the surroundings. Example 7.3 illustrates the maximum expansion work calculations, while Example 7.4 calculates the isentropic efficiency.

***Example 7.3* Maximum work (ideal work) calculations**

One mole of air expands from an initial state of 500 K and 10 atm to the ambient conditions of the surroundings at 300 K and 1 atm. The average heat capacity of the air is $C_{p,av} = 29.5$ J/mole K. Estimate the maximum work.

Solution:
Assume: Air is ideal gas system, and the change of state is completely reversible.

$$T_1 = 500\,\text{K}, P_1 = 10\,\text{atm}, \quad T_2 = 300\,\text{K}, P_2 = 1\,\text{atm},$$

$$C_{p,av} = 29.5\,\text{J/mole K}\,, \quad R = 8.314\,\text{J/mole K}$$

For an ideal gas, enthalpy is independent of pressure and its change is:

$$\Delta H = \int_1^2 C_{\mathrm{p,av}} \mathrm{d}T = C_{\mathrm{p,av}}(T_2 - T_1) = 5900\,\mathrm{J/mole}$$

Change in entropy: $\Delta S = C_{p,\mathrm{av}} \ln\left(\frac{T_2}{T_1}\right) - R \ln\left(\frac{P_2}{P_1}\right) = 34.21\,\mathrm{J/mole\,K}$

Maximum work is ideal work:

$W_{\mathrm{ideal}} = \Delta H - T_{\mathrm{o}}\,\Delta S = -16{,}163\,\mathrm{J/mole} = -557.9\,\mathrm{J/g}$

(with $\mathrm{MW_{air}} = 28.97\,\mathrm{g/mole}$)

The negative sign indicates that this expansion work is transferred to the surroundings.

Example 7.4 **Isentropic turbine efficiency**

An adiabatic turbine is used to produce electricity by expanding a superheated steam at 4100 kPa and 350°C. The steam leaves the turbine at 40 kPa and 100°C. The steam mass flow rate is 8 kg/s. Determine the isentropic efficiency of the turbine.

Solution:

Assume: Steady-state adiabatic operation. The changes in kinetic and potential energies are negligible.

Inlet conditions: superheated steam (steam tables)

$$P_1 = 4100\,\mathrm{kPa},\ T_1 = 623.15\,\mathrm{K},\ H_1 = 3092.8\,\mathrm{kJ/kg},\ S_1 = 6.5727\,\mathrm{kJ/kg\,K}$$

Exit conditions: saturated steam (steam tables)

$$P_2 = 40\,\mathrm{kPa},\ T_2 = 373.15\,\mathrm{K},\ H_{2\mathrm{a}} = 2683.8\,\mathrm{kJ/kg},$$
$$S_{2\,\mathrm{sat\,vap}} = 7.6709\,\mathrm{kJ/kg\,K},\ S_{2\,\mathrm{sat\,liq}} = 1.2026\,\mathrm{kJ/kg\,K}$$

$$H_{2\mathrm{sat\,vap}} = 2636.9\,\mathrm{kJ/kg},\ H_{2\mathrm{sat\,liq}} = 317.6\,\mathrm{kJ/kg}\ \text{at 40 kPa (steam tables)}$$

For the isentropic operation $S_1 = S_2 = 6.5727\,\mathrm{kJ/kg\,K}$

Since $S_{2\mathrm{sat\,liq}} < S_2 = S_{2\mathrm{sat\,vap}}$ the steam at the exit is saturated liquid–vapor mixture, and the quality of that mixture $x_{2\mathrm{s}}$ is

$$x_{2\mathrm{s}} = \frac{S_2 - S_{2\mathrm{sat\,liq}}}{S_{2\mathrm{sat\,vap}} - S_{2\mathrm{sat\,liq}}} = \frac{6.5727 - 1.2026}{7.6709 - 1.2026} = 0.83$$

$$H_{2\mathrm{s}} = (1 - x_{2\mathrm{s}})H_{2\mathrm{sat\,liq}} + x_{2\mathrm{s}}H_{2\mathrm{sat\,vap}} = 2243.1\,\mathrm{kJ/kg}\quad \text{(Isentropic)}$$

and

$$H_{2a} = 2683.8\,\text{kJ/kg (Real operation)}$$

$$H_1 = 3092.8\,\text{kJ/kg}$$

Isentropic efficiency becomes: $\eta_{Ts} = \frac{H_{2a}-H_1}{H_{2s}-H_1} = 0.48\,(\text{or}\,48\%)$.

7.6 Lost Work

The difference between the ideal work and the actual work occurs because of irreversibility within the selected path between the initial and final states for a process. The extent of irreversibility is equivalent to exergy lost and is a measure of lost work potential [5] defined as a difference between the actual and ideal works and related to the rate of entropy production. Actual work is

$$\dot{W}_{act} = \left[\dot{m}\left(\Delta H + \frac{\Delta v^2}{2} + \Delta gz\right)\right] - \dot{q} \tag{7.18}$$

For surrounding temperature T_o, ideal work is

$$\dot{W}_{ideal} = \left[\dot{m}\left(\Delta H + \frac{\Delta v^2}{2} + \Delta gz\right)\right] - T_o\Delta(\dot{m}S) \tag{7.19}$$

Then the lost work

$$-\dot{W}_{lost} = \dot{W}_{act} - \dot{W}_{ideal} = \dot{q} - T_o\Delta(\dot{m}S) \tag{7.20}$$

The lost work occurs because of the rate of entropy production $\dot{S}_{prod}$, which is defined as

$$\dot{S}_{prod} = \Delta(\dot{m}S) - \frac{\dot{q}}{T_o} \tag{7.21}$$

Therefore Eq. (7.20) becomes

$$-\dot{W}_{lost} = T_o\dot{S}_{prod} \tag{7.22}$$

Since the rate of entropy production is always positive ($\dot{S}_{prod} > 0$) for irreversible processes, then $\dot{W}_{lost} > 0$. Only when a process is completely reversible, then $\dot{W}_{lost} = 0$ since $\dot{S}_{prod} = 0$. The rate of entropy production and hence the amount of lost work will increase as the irreversibility increases for a process [5, 6]. This leads to an increase in the amount of energy that is unavailable and hence waste. Example 7.5 illustrates the lost work calculations, while 7.6 estimates the minimum power required in a compressor.

Example 7.5 **Estimation of lost work**

A turbine discharges steam from 6 MPa and 400°C to saturated vapor at 360.15 K while producing 500 kW of shaft work. The temperature of surroundings is 290 K. Determine maximum production of power in kW and the amount of work lost.

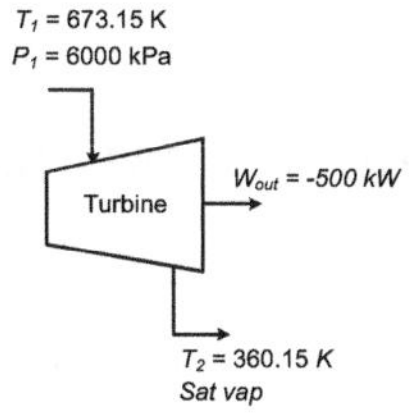

Solution:

Assume: The turbine operates at a steady state. Kinetic and potential energy changes are negligible.

Basis: 1 kg/s steam flow rate

Inlet: superheated steam: $H_1 = 3180.1\,\mathrm{kJ/kg},\ S_1 = 6.5462\,\mathrm{kJ/kg\,K}$ (steam tables)

Outlet: saturated steam: $H_2 = 2655.3\,\mathrm{kJ/kg},\ S_2 = 2655.3\,\mathrm{kJ/kg\,K}$ (steam tables)

$$\dot{W}_{\mathrm{out}} = -500\,\mathrm{kW},\quad T_{\mathrm{o}} = 290\,\mathrm{K}$$

$$\Delta H = H_2 - H_1 = (2655.3 - 3180.1)\,\mathrm{kJ/kg} = -524.8\,\mathrm{kJ/kg}$$

The amount of heat transfer: $q = -W + (H_2 - H_1) = -24.8\,\mathrm{kJ/kg}$.

We can determine the entropy production from an entropy balance on the turbine operating at a steady state that exchanges heat only with the surroundings:

$$S_{\mathrm{prod}} = S_2 - S_1 - \frac{q}{T_{\mathrm{o}}} = 7.5189 - 6.5462 + 24.8\,\mathrm{kJ/kg/(290)\,K} = 1.06\,\mathrm{kJ/kg\,K}$$

Lost work: $\dot{W}_{\mathrm{lost}} = \dot{m}T_{\mathrm{o}}S_{\mathrm{prod}} = 307.4\ \mathrm{kJ/s}$

Maximum work output:

$\dot{W}_{\mathrm{ideal}} = \dot{W}_{\mathrm{max}} = \dot{W}_{\mathrm{out}} - T_{\mathrm{o}}\dot{S}_{\mathrm{prod}} = -500 - 307.4 = -807.4\,\mathrm{kJ/kg}$.

Example 7.6 **Estimation of a minimum power required in a compressor**

A compressor receives air at 15 psia and 80°F with a flow rate of 1.0 lb/s. The air exits at 40 psia and 300°F. Estimate the minimum power input to the compressor. The surroundings are at 520 R.

Solution:

Assume that potential energy effects are negligible and steady process.

Basis: airflow rate = $\dot{m} = 1\ \mathrm{lb/s}$. The surroundings are at 520 R

The properties of air from Table D1 after conversions from SI units:

State 1: $P_1 = 15\,\mathrm{psia},\ T_1 = 540\,\mathrm{R},\ H_1 = 129.0\,\mathrm{Btu/lb},\ S_1 = 0.6008\,\mathrm{Btu/lb\,R}$.

State 2: $P_2 = 40\,\mathrm{psia},\ T_2 = 760\,\mathrm{R},\ H_2 = 182.0\,\mathrm{Btu/lb},\ S_2 = 0.6831\,\mathrm{Btu/lb\,R}$.

Compressor work:

$\dot{W}_s = \dot{m}(H_2 - H_1) = (1\,\mathrm{lb/s})(182.08 - 129.06)\mathrm{Btu/lb} = 53.0\,\mathrm{Btu/s}$.

The entropy production: $\dot{S}_{\mathrm{prod}} = \dot{m}(S_2 - S_1) = 0.0823\,\mathrm{Btu/s\,R}$.

Lost work:

$$\dot{W}_{\mathrm{lost}} = T_o \dot{S}_{\mathrm{prod}} = 42.8\,\mathrm{Btu/s}$$

$$\text{Minimum work required} = (53.0 - 42.8)\,\mathrm{Btu/s} = 10.2\,\mathrm{Btu/s}$$

7.7 Efficiency of Mechanical Energy Conversions

Transfer of mechanical energy is usually accomplished through a rotating shaft. When there is no loss (for example, in the form of friction), mechanical energy can be converted completely from one mechanical form to another. The mechanical energy conversion efficiency is

$$\eta_{\mathrm{mech}} = \frac{\text{energy out}}{\text{energy in}} \tag{7.23}$$

For example, a mechanical energy conversion efficiency of 95% shows that 5% of the mechanical energy is converted to heat because of friction and other losses [4, 6].

In fluid flow systems, a pump receives shaft work usually from an electric motor and transfers this shaft work partly to the fluid as mechanical energy and partly to the frictional losses as heat. As a result, the fluid pressure or velocity, or elevation is increased. A turbine, however, converts the mechanical energy of a fluid to shaft work. The energy efficiency for a pump is

$$\eta_{\mathrm{pump}} = \frac{\text{mechanical energy out}}{\text{mechanical energy in}} = \frac{\dot{E}_{\text{mech out}}}{\dot{W}_{\text{shaft in}}} \tag{7.24}$$

Mechanical energy efficiency of a fan is the ratio of kinetic energy of air at the exit to the mechanical power input. If a fan, for example, is using a 50W motor and producing an airflow velocity of 15 m/s and the air mass flow rate of 0.3 kg/s, then the mechanical energy efficiency of the fan becomes.

$$\eta_{\text{fan}} = \frac{\dot{E}_{\text{mech out}}}{\dot{W}_{\text{shaft in}}} = \frac{(0.3\ \text{kg/s})(15\ \text{m/s})^2/2}{50\ \text{W}} = 0.675$$

Here, the velocity at the inlet is zero, $v_1 = 0$, and the pressure energy and potential energy are zero ($\Delta P = 0$, and $\Delta z = 0$). The energy efficiency for a turbine is

$$\eta_{\text{turb}} = \frac{\text{mechanical energy output}}{\text{mechanical energy (extracted from fluid) in}} = \frac{W_{\text{shaft out}}}{\dot{E}_{\text{mech in}}} \tag{7.25}$$

The motor efficiency and the generator efficiency are

$$\eta_{\text{motor}} = \frac{\text{mechanical power output}}{\text{electrical energy input}} = \frac{\dot{W}_{\text{shaft out}}}{\dot{W}_{\text{elect in}}} \quad \text{(motor)} \tag{7.26}$$

$$\eta_{\text{gen}} = \frac{\text{electrical power output}}{\text{mechanical power input}} = \frac{\dot{W}_{\text{elect out}}}{\dot{W}_{\text{shaft in}}} \quad \text{(generator)} \tag{7.27}$$

A hydraulic turbine is combined with its generator in power production cycles, as shown in Fig. 7.4. A pump is also combined with its motor. Therefore, an overall efficiency for turbine-generator and pump-motor systems is

$$\eta_{\text{turb-gen}} = \eta_{\text{turb}}\eta_{\text{gen}} = \frac{\dot{W}_{\text{elect out}}}{\Delta\dot{E}_{\text{mech in}}} \quad \text{(turbine-generator systems)} \tag{7.28}$$

The overall efficiency of a hydraulic turbine-generator shows the fraction of the mechanical energy of the water converted to electrical energy.

$$\eta_{\text{pump-motor}} = \eta_{\text{pump}}\eta_{\text{motor}} = \frac{\Delta\dot{E}_{\text{mech}}}{\dot{W}_{\text{elect in}}} \quad \text{(pump-motor systems)} \tag{7.29}$$

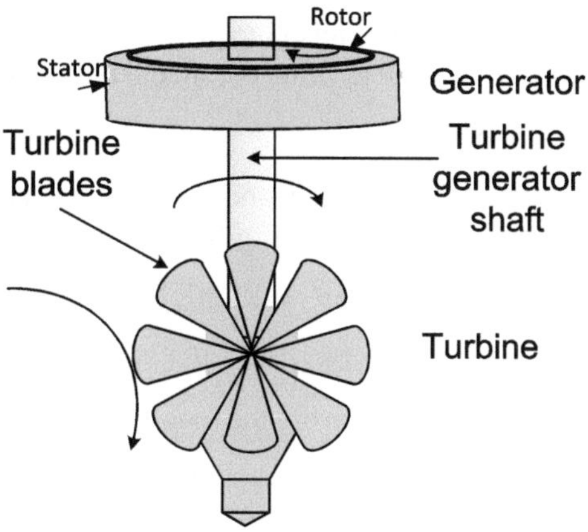

Fig. 7.4 Typical turbine and generator. The overall efficiency of a hydraulic turbine-generator is the ratio of the thermal energy of the water converted to the electrical energy; for a turbine efficiency of 0.8 and a generator efficiency of 0.9, we have: $\eta_{\text{turbgen}} = \eta_{\text{turb}}\eta_{\text{gen}} = (0.8)(0.95) = 0.76$

The overall efficiency of a pump-motor system shows the fraction of the electrical energy converted to mechanical energy of the fluid [2, 4]. Example 7.7 illustrates the estimation of heat loss in an electric motor, while Example 7.8 illustrates the estimation of mechanical efficiency of a pump.

Example 7.7 **Heat loss in an electric motor**

An electric motor attached to a pump draws 10.0 A at 110 V. At steady load, the motor delivers 1.32 hp of mechanical energy. Estimate the heat loss from the motor.

Solution:

Assume that the load to motor is steady.

$$I = 10.0\text{A},\ V = 110\,\text{V},\ \dot{W}_s = 1.32\,\text{hp} = 983\,\text{W}$$

The electric power received by the motor is used to create pump work and heat:

Power received:

$$\dot{W}_{\text{e}} = \text{IV} = 1100\,\text{W}$$

Heat loss:

$$\dot{q}_{\text{loss}} = \dot{W}_{\text{s}} - \dot{W}_{\text{e}} = (983 - 1100) = -117\,\text{W}$$

Only 983 kW of the 1100 kW is converted to the pump work.

Example 7.8 **Mechanical efficiency of a pump**

The pump of a water storage tank is powered with a 16-kW electric motor operating with an efficiency of 90%. The water flow rate is 55 l/s. The diameters of the inlet and exit pipes are the same, and the elevation difference between the inlet and outlet is negligible. The absolute pressures at the inlet and outlet are 100 kPa and 300 kPa, respectively. Determine the mechanical efficiency of the pump.

Solution:

Assume: Steady-state one-dimensional and adiabatic flow. Kinetic and potential energy changes are negligible.

$\rho_{\text{water}} = 1000\ \text{kg/m}^3 = 1\ \text{kg/l}$; Mass flow rate of water:

$\dot{m}_{\text{water}} = \rho\dot{Q} = (1\ \text{kg/l})(55\ \text{l/s}) = (1\ \text{kg}/1)(551/\text{s}) = 55\ \text{kg/s}.$

Electric motor delivers the mechanical shaft work:

$\dot{W}_{\text{pump}} = \eta_{\text{motor}}\dot{W}_{\text{electric}} = (0.9)(16\ \text{kW}) = 14.4\ \text{kW}.$

Increase in the mechanical energy of the water:

$$\Delta \dot{E}_{\text{mech}} = \dot{m}\left(\frac{P_2 - P_1}{\rho} + \frac{v_2^2 - v_1^2}{2} + g(z_2 - z_1)\right).$$

After neglecting kinetic and potential energies:

$$\Delta \dot{E}_{\text{mech}} = \dot{m}\left(\frac{P_2 - P_1}{\rho}\right) = (55\ \text{kg/s})\left(\frac{(300 - 100)\ \text{kPa}}{1000\ \text{kg/m}^3}\right)\left(\frac{\text{kJ}}{\text{kPa m}^3}\right) = 11.0\,\text{kW}$$

The mechanical efficiency of the pump is:

$$\eta_{\text{pump}} = \frac{\Delta \dot{E}_{\text{mech}}}{\dot{W}_{\text{pump}}} = \frac{11\ \text{kW}}{14.4\ \text{kW}} = 0.763\ \text{or}\ 76.3\%\ .$$

Only 11 kW of the 14.4 kW received by the pump is converted to the pump work. The remaining 3.4 kW is lost as heat because of friction.

7.8 Conversion of Thermal Energy by Heat Engines

A heat engine is a device that converts heat into work. It does so by operating between two temperature limits T_H and T_C. That is, a heat engine is a device that can operate continuously to produce work receiving heat from an elevated temperature source T_H and rejecting nonconverted heat to a low temperature sink T_C. Heat engines transform thermal energy q_{in} into mechanical energy or work W_{out}. Examples of heat engines include the steam engine, heat pump, gasoline (petrol) engine in an automobile, diesel engine, and gas power cycles. All these heat engines drive the mechanical motion of the engine by the expansion of heated gases. Some heat engines operate with phase change cycles, such as steam power production by Rankine cycle where liquid water changes to vapor after adding heat. After expansion in the cycle, the vapor condenses into liquid water. Other types of heat engines operate without phase change cycles, such as Brayton cycle where hot gas is cooled after the expansion in the heat engine. Typical gas power cycles consist of compressing cool gas, heating the gas, expanding the hot gas, and finally cooling the gas before repeating the cycle. Table 7.2 lists important engines and their cycle processes. Typical thermal efficiency for power plants in the industry is around 33% for coal and oil-fired plants and up to 50% for combined-cycle gas-fired plants [2, 4, 8].

Each process in a cycle is at one of the following states:

- Isothermal (at constant temperature)
- Isobaric (at constant pressure)
- Isochoric (at constant volume)
- Adiabatic (no heat is added or removed)

Table 7.2 Comparison of the processes for some of the heat engines

Cycle	Compression	Heat addition	Expansion	Heat rejection
Rankine	Adiabatic	Isobaric	Adiabatic	Isobaric
Carnot	Isentropic	Isothermal	Isentropic	Isothermal
Ericsson	Isothermal	Isobaric	Isothermal	Isobaric
Stirling	Isothermal	Isochoric	Isothermal	Isochoric
Diesel	Adiabatic	Isobaric	Adiabatic	Isochoric
Otto	Adiabatic	Isochoric	Adiabatic	Isochoric
Brayton	Adiabatic	Isobaric	Adiabatic	Isobaric

Table 7.3 Various types of cycles for heat engines

External combustion cycles	Without phase change: Brayton, Carnot, Ericsson, Stirling With phase change: Rankine, Two-phased Stirling
Internal combustion cycles	Without phase change: Diesel, Otto
Refrigeration cycles	Vapor-compression
Heat pump cycles	Vapor-compression

Engine types vary as shown in Table 7.3. In the combustion cycles, once the mixture of air and fuel is ignited and burnt, the available energy can be transformed into work by the engine. In a reciprocating engine, the high-pressure gases inside the cylinders drive the engine's pistons. Once the available energy has been converted to mechanical work, the remaining hot gases are vented out and this allows the pistons to return to their previous positions [2, 4, 6].

Part of available heat input is not converted into work but is dissipated as waste heat q_{out} into the environment. The thermal efficiency of a heat engine is the percentage of heat energy that is converted into work

$$\eta_{th} = \frac{|W_{out}|}{q_{in}} = 1 - \frac{|q_{out}|}{q_{in}} \tag{7.30}$$

The absolute signs for work out and heat out are for the sign convention that assumes the output of work or heat from a system is negative. The efficiency of even the best heat engines is low: usually below 50% and often far below. So, the energy lost to the environment by heat engines is a major waste of energy resources, although modern cogeneration, combined cycle, and energy recycling schemes are beginning to use this waste heat for other purposes. This inefficiency can be attributed to three causes:

- Thermal efficiency of any heat engine is limited by Carnot's efficiency.
- Specific types of engines may have lower limits on their efficiency due to the inherent irreversibility of the engine cycle they use.

- The nonideal behavior of real engines, such as mechanical friction and losses in the combustion process, may cause further efficiency losses.

Thermal efficiency of heat engine cycles cannot exceed the limit defined by the Carnot cycle, which states that the overall efficiency is dictated by the difference between the lower and upper operating temperatures of the engine. This limit assumes that the engine is operating in ideal conditions that are reversible with no heat loss. A car engine's real-world fuel economy is usually measured in the units of miles per gallon (or fuel consumption in liters per 100 km). Even when aided with turbochargers and design aids, most engines retain an *average* efficiency of about 18–20% [4, 8]. The efficiency of rocket engines is better, up to 70%, because they operate at very high temperatures and pressures, and can have very high expansion ratios. For stationery and shaft engines, fuel consumption is measured by calculating the *brake specific fuel consumption* which measures the mass flow rate of fuel consumed divided by the power produced.

Engine efficiency alone is only one factor. For a more meaningful comparison, the overall efficiency of the entire energy supply chain from the fuel source to the consumer should be considered. Although heat wasted by heat engines is usually the largest source of inefficiency, factors such as the energy cost of fuel refining and transportation and energy loss in electrical transmission lines may offset the advantage of a more efficient heat engine [2]. Engines must be optimized for other goals besides efficiency such as low pollution. Vehicle engines must also be designed for low emissions, adequate acceleration, fast starting, light weight, and low noise. These requirements may lead to compromises in design that may reduce efficiency. Large stationary electric generating plants have fewer of these competing requirements, so the Rankine cycles are significantly more efficient than vehicle engines. Therefore, replacing internal combustion vehicles with electric vehicles, which run on a battery recharged with electricity generated by burning fuel in a power plant, has the theoretical potential to increase the thermal efficiency of energy use in transportation, thus decreasing the demand for fossil fuels [3, 4].

Real engines have many departures from ideal behavior that waste energy, reducing actual efficiencies far below the theoretical values because of:

- Friction of moving parts
- Inefficient combustion
- Heat loss from the combustion chamber
- Nonideal behavior of the working fluid
- Inefficient compressors and turbines
- Imperfect valve timing

Example 7.9 **Carnot limit in heat engines**

In a cyclic process, a heat engine takes in energy from a hot reservoir q_H and converts this energy into mechanical work W, discharges some energy q_C to the surroundings and returns to its original state.

Solution:

Conservation of energy shows: $q_H = W + q_C$.

The entropy added when heat is absorbed, and entropy decreased when energy is discharged are

$$\Delta S_H = \left(\frac{q}{T}\right)_H + \Delta S'_H$$

$$\Delta S_C = \left(\frac{q}{T}\right)_C - \Delta S'_C$$

where S' is the entropy production due to irreversibility and always positive. The net change in entropy is zero in the cyclic process $\Delta S_H = \Delta S_C$ and in the reversible operation, we have

$$\frac{q_H}{q_C} = \frac{T_H}{T_C}$$

Defining the efficiency of a heat engine as: $\eta = \frac{W}{q_H}$.

And using $q_H = W + q_C$, we have: $\eta = \frac{W}{q_H} = \frac{q_H - q_C}{q_H} \leq \frac{T_H - T_C}{T_H}$.

This relation shows the limit on the thermal efficiency of a heat engine formulated by Carnot and known as Carnot limit, which is applied to cyclic operation of heat engines.

$$\eta_C = \frac{T_H - T_C}{T_H} > \frac{W}{q_H}$$

This relation also brings out an important difference between thermal energies associated with entropy which is not the case for mechanical energy. Table below compares the thermal efficiency of various heat engines [8]

Energy source	T_h K	T_c K	η_C %	η %
Steam turbine nuclear plant	600	300	50	33
Steam turbine coal plant	800	300	60	33
Combined-cycle gas turbine	1500	300	83	45
Auto engine (Otto cycle)	2300	300	87	25
Solar energy (photovoltaic module)	6000	300	95	15–20

Example 7.10 **Thermal efficiency of a heat engine**

Heat is transferred to a heat engine from a furnace at a rate of 90 MW. The waste heat is discharged to the surroundings at a rate of 55 MW. Estimate the net power output and thermal efficiency of the engine if all the other power losses are neglected.

Solution:

Assume: Steady-state process with negligible heat losses.

$$\dot{q}_{\mathrm{in}} = 90\,\mathrm{MW}, \quad \dot{q}_{\mathrm{out}} = 55\,\mathrm{MW}(\text{system is surrounding that receives this heat})$$

Net power output: $\dot{W}_{\mathrm{net}} = \dot{q}_{\mathrm{out}} - \dot{q}_{\mathrm{in}} = -35\,\mathrm{MW}(\text{the system is heat engine})$

Thermal efficiency: $\eta_{\mathrm{th}} = \frac{|\dot{W}_{\mathrm{net}}|}{\dot{q}_{\mathrm{in}}} = \frac{35}{90} = 0.388\ (\text{or } 38.8\%)$

The heat engine can convert 38.8% of the heat transferred from the furnace.

Example 7.11 **Fuel consumption of a car**

The overall efficiencies are about 25–28% for gasoline car engines, 34–38% for diesel engines, and 40–60% for large power plants [3]. A car engine with a power output of 120 hp has a thermal efficiency of 24%. Determine the fuel consumption of the car if the fuel has a higher heating value of 20,400 Btu/lb.

Solution:

Assume: The car has a constant power output.

Net heating value ≅ (20,400) Btu/lb (0.9) = 18,360 Btu/lb (Approximate)

Car engine power output and efficiency: $\dot{W}_{\mathrm{net}} = 120\,\mathrm{hp}$, $\eta_{\mathrm{th}} = 0.24$

$$\eta_{\mathrm{th}} = \frac{\dot{W}_{\mathrm{net}}}{\dot{q}_{\mathrm{in}}} \rightarrow \dot{q}_{\mathrm{in}} = \frac{\dot{W}_{\mathrm{net}}}{\eta_{\mathrm{th}}} = \frac{120\,\mathrm{hp}}{0.24}\left(\frac{2545\,\mathrm{Btu/h}}{\mathrm{hp}}\right)$$

$$\dot{q}_{\mathrm{in}} = 1{,}272{,}500\,\mathrm{Btu/h}$$

Net heating value ≅ higher heating value (0.9) = 18,360 Btu/lb

$$\begin{aligned}\text{Fuel consumption} &= \dot{q}_{\mathrm{in}}/\text{net heating value}\\ &= (1{,}272{,}500\,\mathrm{Btu/h})/(18{,}360\,\mathrm{Btu/lb}) = 69.3\,\mathrm{lb/h}\end{aligned}$$

Assuming an average density of 0.75 kg/l:

$$\rho_{\mathrm{gas}} = (0.75\,\mathrm{kg/l})(2.2\,\mathrm{lb/kg})(1/0.264\ \mathrm{gal}) = 6.25\,\mathrm{lb/gal}$$

Fuel consumption in terms of gallon: $(69.3\,\mathrm{lb/h})/(6.25\,\mathrm{lb/gal}) = 11.1\,\mathrm{gal/h}$.

7.8.1 Air-Standard Assumptions

Internal combustion cycles of Otto and Diesel engines as well as the gas turbines are well-known examples of engines that operate on gas cycles. In the gas power cycles, the working fluid remains gas for the entire cycle. In the analysis of gas power cycles, the following assumptions known as *the air-standard assumptions* are used:

- Working fluid is air and always behaves as an ideal gas.
- All processes in the cycle are internally reversible (isentropic).
- The heat addition process uses an external heat source.
- The heat rejection process restores the working fluid to its original state.

It is also assumed that the air has a constant value for the ratio of specific heats determined at room temperature (25°C or 77°F). These air-standard assumptions are called the *cold-air-standard assumptions,* which simplify the analysis of gas power cycles without significantly deviating from the actual cycles [2–4].

7.8.2 Isentropic Processes of Ideal Gases

The entropy change of ideal gas is

$$\Delta S = S_2 - S_1 = \int_1^2 C_p(T)\frac{dT}{T} - R\ln\left(\frac{P_2}{P_1}\right) \tag{7.31}$$

Under isentropic conditions ($\Delta S = 0$) and constant heat capacity $C_{p,av}$, Eq. (7.31) become

$$0 = C_{p,av}\ln\left(\frac{T_2}{T_1}\right) - R\ln\left(\frac{P_2}{P_1}\right) \tag{7.32}$$

For ideal gas, the heat capacities are related by

$$C_p - C_v = R \tag{7.33}$$

From Eqs. (7.32) and (7.33) the following relations may be derived

$$\left(\frac{T_2}{T_1}\right) = \left(\frac{V_1}{V_2}\right)^{(\gamma-1)} \quad \text{and} \quad \left(\frac{T_2}{T_1}\right) = \left(\frac{P_2}{P_1}\right)^{(\gamma-1)/\gamma} \tag{7.34}$$

where $\gamma = C_p/C_v$.

Under isentropic conditions ($\Delta S = 0$) and variable heat capacities, Eq. (7.31) becomes

$$0 = S_{T2} - S_{T1} - R\ln\left(\frac{P_2}{P_1}\right) \tag{7.35}$$

where S_{T1} and S_{T2} are the values of entropy at temperatures T_1 and T_2, respectively. After rearranging, Eq. (7.35) yields,

$$\frac{P_2}{P_1} = \exp\left(\frac{S_{T2} - S_{T1}}{R}\right) = \frac{\exp(S_{T2}/R)}{\exp(S_{T1}/R)} = \frac{P_{r2}}{P_{r1}} \tag{7.36}$$

where P_{r1} is called the *relative pressure* given by exp. (S_{T1}/R), which is dimensionless quantity. Using the ideal gas relation

$$\frac{P_1V_1}{T_1} = \frac{P_2V_2}{T_2} \tag{7.37}$$

the relative specific volume V_r ($V_r = T/P_r$) is derived by

$$\frac{V_2}{V_1} = \frac{T_2/P_{r2}}{T_1/P_{r1}} = \frac{V_{r2}}{V_{r1}} \tag{7.38}$$

The values of P_r and V_r are tabulated for the air against temperature in Appendix D, Table D1. Equations (7.35) to (7.38) account for the variation of specific heats with temperature and are valid only for isentropic processes of ideal gases. Equations (7.36) and (7.38) are useful in the analysis of gas power cycles operating with isentropic processes [2, 4, 8].

7.8.3 Conversion of Mechanical Energy by Electric Generator

In electricity generation, an *electric generator* converts mechanical energy to electrical energy using *electromagnetic induction.* Electromagnetic induction is the production of electric potential (voltage) across a conductor moving through a magnetic field. A generator forces electrons in the windings to flow through the external electrical circuit. The source of mechanical energy may be water falling through a turbine, waterwheel, an internal combustion engine, and a wind turbine. The reverse is the conversion of electrical energy into mechanical energy, which is done by an electric motor. Figure 7.4 shows a schematic view of a generator.

The efficiency of a generator is determined by the power of the load circuit and the total power produced by the generator. For most commercial electrical generators, this ratio can be as high as 95%. The losses typically arise from the transformer, the copper windings, magnetizing losses in the core, and the rotational friction of the generator. The overall efficiency of a hydraulic turbine-generator is the ratio of the thermal energy of the water converted to the electrical energy and obtained by

$$\eta_{\text{turb-gen}} = \eta_{\text{turb}} \eta_{\text{gen}} \tag{7.39}$$

Michael Faraday discovered the operating principle of electromagnetic generators. The principle, later called Faraday's law, is that an electromotive force is generated in an electrical conductor that encircles a varying magnetic flux. The two main parts of a generator are rotor and stator. The rotor is the rotating part, while stator is the stationary part. The armature of generator windings generates electric current. The armature can be on either the rotor or the stator. Because power transferred into the field circuit is much less than in the armature circuit, alternating current generators mostly have the field winding on the rotor and the stator as the armature winding.

An electric generator or electric motor uses magnetic field coils. If the field coils are not powered, the rotor in a generator can spin without producing any usable electrical energy. When the generator first starts to turn, the small amount of magnetism present in the iron core provides a magnetic field to get it started, generating a small current in the armature. This flows through the field coils, creating a larger magnetic field which generates a larger armature current. This "bootstrap" process continues until the magnetic field in the core levels off due to saturation and the generator reaches a steady-state power output. Large power station generators often utilize a separate smaller generator to excite the field coils of the larger generator [4, 8].

The *dynamo* was the first electrical generator capable of delivering power for industry. The dynamo uses electromagnetic principles to convert mechanical rotation into the direct current. A dynamo consists of a stationary structure, which provides a constant magnetic field, and a set of rotating windings which turn within that field. A *commutator* is a rotary switch in certain types of electric motors or electrical generators that periodically reverse the current direction between the rotor and the external circuit [8].

7.8.4 Carnot Engine Efficiency

A steam power plant is a typical heat engine converting low-grade energy (heat) to high-grade energy (electricity). In the steam power plant, heat is added to boiler to produce steam. The steam is then expanded through a steam turbine where the steam's enthalpy drop is converted to shaft work and electricity eventually as a positive work output. The steam is then exhausted into a condenser that serves as a sink where the steam is condensed.

Carnot *heat engine* converts heat to mechanical energy by bringing a working fluid from an elevated temperature state T_H to a lower temperature state T_C. Figure 7.5 shows a typical pressure–volume *PV* and temperature-entropy *TS* diagrams of ideal engine cycles. On both the *PV* and *TS* diagrams, the area enclosed

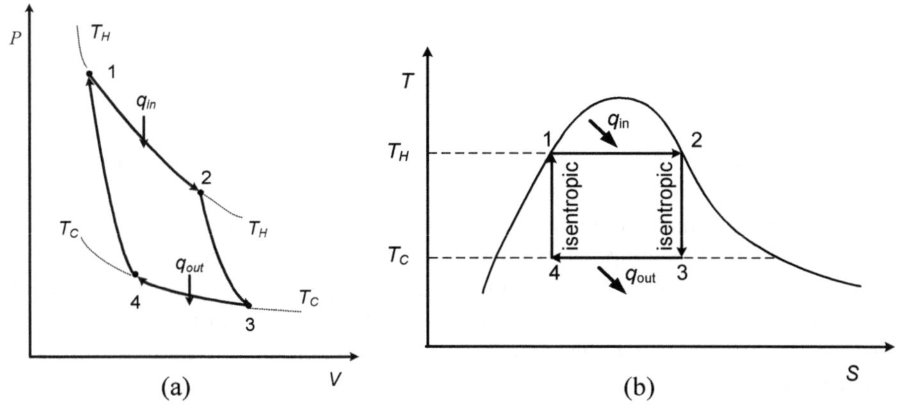

Fig. 7.5 Typical ideal engine cycles **a** on a pressure–volume *PV* diagram, **b** on a temperature-entropy *TS* diagram

by the process curves of a cycle represents the net heat transfer to be converted to mechanical energy by the engine. Therefore, any modifications that improve the net heat transfer rate will also improve the thermal efficiency [4, 5, 8].

A heat source heats the working fluid in an elevated temperature state. The working fluid generates work in the engine while transferring remaining heat to the colder sink until it reaches a low temperature state. The working fluid usually is a gas or liquid. During the operation of an engine some of the thermal energy is converted into work and the remaining energy is lost to a heat sink, the general surroundings.

Carnot cycle is composed of four totally reversible processes shown in Fig. 7.5:

- Process 1-2: Isothermal heat addition at constant temperature T_{H}.
- Process 2-3: Isentropic expansion at constant entropy $S_2 = S_3$.
- Process 3-4: Isothermal heat rejection at constant temperature T_{C}.
- Process 4-1: Isentropic compression at constant entropy $S_4 = S_1$.

Thermal efficiency of the Carnot engine is calculated by

$$\eta_{\mathrm{th}} = 1 - \frac{|q_{\mathrm{out}}|}{q_{\mathrm{in}}} \tag{7.40}$$

From Fig. 7.5b, we can estimate the amounts of added q_{in} and rejected heat q_{out} values as

$$q_{\mathrm{in}} = T_{\mathrm{H}}(S_2 - S_1) \tag{7.41}$$

$$q_{\mathrm{out}} = T_{\mathrm{C}}(S_4 - S_3) \tag{7.42}$$

Using Eqs. (7.41) and (7.42) in Eq. (7.40), we have

$$\eta_{th} = 1 - \frac{T_C(S_4 - S_3)}{T_H(S_2 - S_1)} = 1 - \frac{T_C}{T_H} \tag{7.43}$$

Since the power cycles are isentropic and we have $S_2 = S_3$ and $S_4 = S_1$. Equation (7.43) shows that heat engines efficiency is limited by Carnot's efficiency which is equal to the temperature difference between the hot (T_H) and cold (T_C) divided by the temperature at the hot end, all expressed in absolute temperatures. The Carnot cycle is a standard against which the actual or ideal cycle can be compared. The Carnot cycle has the highest thermal efficiency of all heat engines operating between the same heat source temperature T_H and the same sink temperature T_C.

The cold side of any heat engine is close to the ambient temperature of the environment of around 300 K, such as a lake, a river, or the surrounding air. Therefore, most efforts to improve the thermodynamic efficiencies of various heat engines focus on increasing the temperature of the hot source within material limits. The highest value of T_H is limited by the maximum temperature that the components of the heat engine, such as pistons or turbine blades, can withstand.

The thermal efficiency of Carnot engine is independent of the type of working fluid, or whether the cycle operates in a closed or steady-flow system. For example, if an automobile engine burns gasoline at a temperature of $T_H = 1200$ K and the air at an exhaust temperature is $T_C = 800$ K, then its maximum efficiency is:

$$\eta_{Carnot} = 1 - \frac{800\ \text{K}}{1200\ \text{K}} = 0.33\ \text{or}\ 33\%$$

Reversible engine cycles are difficult to achieve as it requires very large heat exchangers and needs a very long time. Practical engines have efficiencies far below the Carnot limit. For example, the average automobile engine is less than 35% efficient. As Carnot's theorem only applies to heat engines, devices that convert the fuel's energy directly into work without burning it, such as fuel cells, can exceed Carnot efficiency. Actual cycles involve friction, pressure drops as well as heat losses in various processes in the cycle; hence they operate at lower thermal efficiencies [11].

7.8.5 Endoreversible Heat Engine Efficiency

A Carnot engine must operate at an infinitesimal temperature gradient, and therefore the Carnot efficiency assumes the infinitesimal limit. Endoreversible engine gives an upper bound on the energy that can be derived from a real process that is lower than that predicted by the Carnot cycle and accommodates the heat loss occurring as heat is transferred. This model predicts how well real-world heat

Table 7.4 Efficiencies of engines operating on the Carnot cycle, endoreversible cycle, and actual cycle [2, 4]

Power plant	T_C °C	T_H °C	η (Carnot)	η (endoreversible)	η (observed)
Coal-fired power plant	25	565	0.64	0.40	0.36
Nuclear power plant	25	300	0.48	0.28	0.30
Geothermal power plant	80	250	0.33	0.18	0.16

engines can perform [2] by

$$\eta_{\max} = 1 - \sqrt{\frac{T_C}{T_H}} \tag{7.44}$$

Table 7.4 compares the efficiencies of engines operating on the Carnot cycle, endoreversible cycle, and actual cycle. As can be seen, the values of observed efficiencies of actual operations are close to those obtained from the endoreversible cycles.

7.8.6 Rankine Engine Efficiency

The Rankine cycle is used in steam turbine power plants. A turbine converts the kinetic energy of a moving fluid to mechanical energy. The steam is forced against a series of blades mounted on a shaft connected to the generator. The generator, in turn, converts its mechanical energy to electrical energy based on the relationship between magnetism and electricity. The most of the world's electric power is produced with this cycle. Since the cycle's working fluid changes from liquid to vapor and back to liquid water during the cycle, their efficiency depends on the properties of water. The thermal efficiency of modern steam turbine plants with reheat cycles can reach 47%, and in combined-cycle plants it can approach 60%.

Figure 7.6 describes the main processes within the Rankine cycle. The main processes of the Rankine cycle are:

- Process 4-1: The water is pumped from low to high pressure.
- Process 1-2: The high-pressure liquid enters a boiler where it is heated at constant pressure by an external heat source to become a vapor.
- Process 2-3: The vapor expands through a turbine, generating power. This decreases the temperature and pressure of the vapor, and some condensation may occur.
- Process 3-4: The wet vapor then enters a condenser where it is condensed at a constant pressure to become a saturated liquid.

The Rankine cycle is sometimes referred to as a practical Carnot cycle (see Fig. 7.5) because in an ideal Rankine cycle the pump and turbine would be *isentropic* and produce no entropy and hence maximize the work output. The main

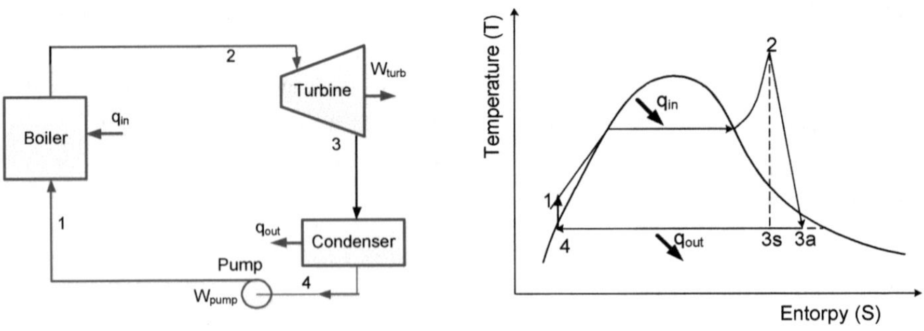

(a) (b)

Fig. 7.6 **a** Schematic of Rankine cycle, **b** Rankine cycle processes on a T versus S diagram; Process 2-3s shows isentropic expansion, while process 2-3a shows a real expansion process where entropy is not constant and the amount of waste heat increases

difference is that heat addition (in the boiler) and rejection (in the condenser) are isobaric in the Rankine cycle and isothermal in the Carnot cycle [4]. Example 7.12 analyzes the power output of steam power plants.

Example 7.12 Steam turbine efficiency and power output

An adiabatic turbine is used to produce electricity by expanding a superheated steam to 4100 kPa and 350°C. The steam leaves the turbine at 40 kPa. The steam mass flow rate is 8 kg/s. If the isentropic efficiency is 0.75, determine the actual power output of the turbine.

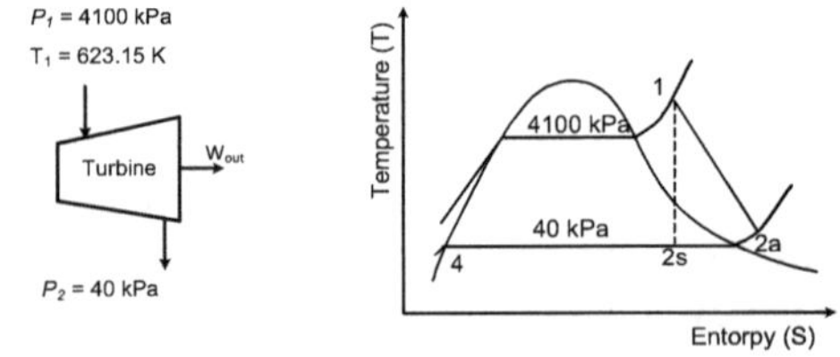

Solution:

Assume: Steady-state adiabatic operation. The changes in kinetic and potential energies are negligible.

Data from steam tables:

Inlet conditions:

$P_1 = 4100\,\text{kPa},\ T_1 = 623.15\,\text{K},\ H_1 = 3092.8\,\text{kJ/kg},\ S_1 = 6.5727\,\text{kJ/kg K}.$

Exit conditions:

$$P_2 = 40\,\text{kPa},\ S_{2\text{sat vap}} = 7.6709\,\text{kJ/kg K},\ S_{2\text{sat liq}} = 1.0261\,\text{kJ/kg K},$$

$$H_{2\text{sat vap}} = 2636.9\,\text{kJ/kg},\ H_{2\text{sat liq}} = 317.6\,\text{kJ/kg at } 40\,\text{kPa}$$

For the isentropic operation $S_1 = S_2 = 6.5727$ kJ/kg K

Since $S_1 < S_{2\text{sat vap}}$ the steam at the exit is saturated liquid–vapor mixture

Quality of the saturated mixture: $x_{2s} = \frac{S_2 - S_{2\text{sat liq}}}{S_{2\text{sat vap}} - S_{2\text{sat liq}}} = \frac{6.5727 - 1.0261}{7.6709 - 1.0261} = 0.83$

$$H_{2s} = (1 - x_{2s})H_{2\text{sat liq}} + x_{2s}H_{2\text{sat vap}} = 2243.1\,\text{kJ/kg}$$

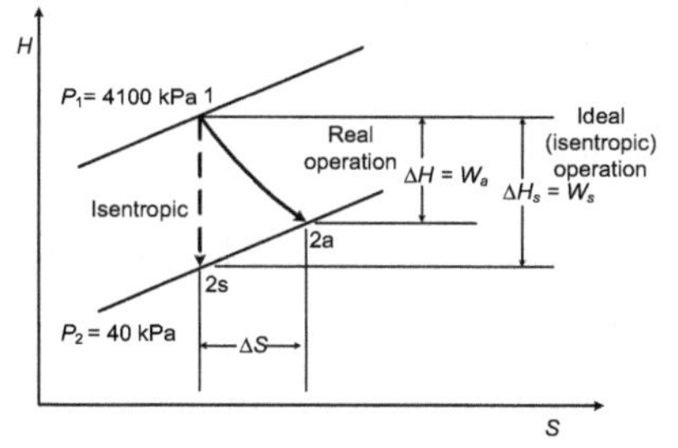

Using isentropic efficiency of 0.75, we can estimate the actual enthalpy H_{2a},

$$\eta_t = \frac{H_1 - H_{2a}}{H_1 - H_{2s}} = 0.75, \quad H_{2a} = 2455.5\,\text{kJ/kg}$$

Actual power output which is not isentropic $\dot{W}_a$ (see the figure above)

$$\begin{aligned}\dot{m}\Delta H &= \dot{W}_a = \dot{m}(H_{2a} - H_1) = (8\text{kg/s})(2455.5 - 3092.8) \\ &= -5098.4\,\text{kW} = -5.098\,\text{MW}\end{aligned}$$

Maximum power output: $\dot{W}_s$ (see figure above) is only achievable by isentropic expansion.

$$\begin{aligned}\dot{m}\Delta H &= \dot{W}_s = \dot{m}(H_{2s} - H_1) = (8\,\text{kg/s})(2243.1 - 3092.8) \\ &= -6797.6\ \text{kW} = -6.797\,\text{MW}\end{aligned}$$

The enthalpy-entropy *HS* diagram above explains the difference between real and ideal turbine operations. The signs are negative for work output.

Example 7.13 **Estimation of thermal efficiency of a Rankine cycle**

A Rankine cycle shown below uses natural gas to produce 0.12 MW power. The combustion heat supplied to a boiler produces steam at 10,000 kPa and 798.15 K, which is sent to a turbine. The turbine efficiency is 0.7. The discharged steam from the turbine is at 30 kPa. The pump efficiency is 0.75. Determine:

(a) The thermal efficiency of an ideal Rankine cycle.
(b) The thermal efficiency of an actual Rankine cycle.

Solution:

Assume that kinetic and potential energy changes are negligible, and the system is at steady state. Consider Fig. 7.6.

(a) The basis is 1 kg/s steam.

$$\text{Turbine in}: H_2 = 3437.7\,\text{kJ/kg},\ S_2 = 6.6797\,\text{kJ/kg K, at } T_2 = 798.15\,\text{K},$$
$$P_2 = 10{,}000\,\text{kPa (steam tables)}$$

$$\text{Turbine out}: H_{4\text{sat liq}} = H_{3\text{sat liq}} = 289.3\,\text{kJ/kg},$$
$$H_{4\text{sat vap}} = H_{3\text{sat vap}} = 2625.4\,\text{kJ/kg; at } P_3 = P_4 = 30\,\text{kPa},$$
$$S_{3\text{satliq}} = 0.9441\,\text{kJ/kg K},\ S_{3\text{satvap}} = 7.7695\,\text{kJ/kg K; (steam tables)}$$
$$\eta_{\text{turb}} = 0.70;\ \eta_{\text{pump}} = 0.75;\ V = 1020\,\text{cm}^3/\text{kg at } T = 342.15\,\text{K}.$$

For the ideal Rankine cycle, the operation is isentropic, and $S_2 = S_3$. However, $S_2 < S_{3\text{sat vap}}$, and the discharged steam from the turbine is wet steam.

The quality of the wet steam x_{3s}: $x_{3s} = \frac{S_2 - S_{3\text{sat liq}}}{S_{3\text{sat vap}} - S_{3\text{sat liq}}} = 0.84$.

The enthalpy of the wet steam H_{3s}:

$$H_{3s} = (1 - x_{3s})H_{3\text{sat liq}} + x_{3s}H_{3\text{sat liq}} = 2252.4\,\text{kJ/kg}$$

Ideal operation:

$$W_s = \Delta H_s = H_{3s} - H_2 = -1185.3\,\text{kJ/kg} \quad \text{(isentropic expansion)}.$$

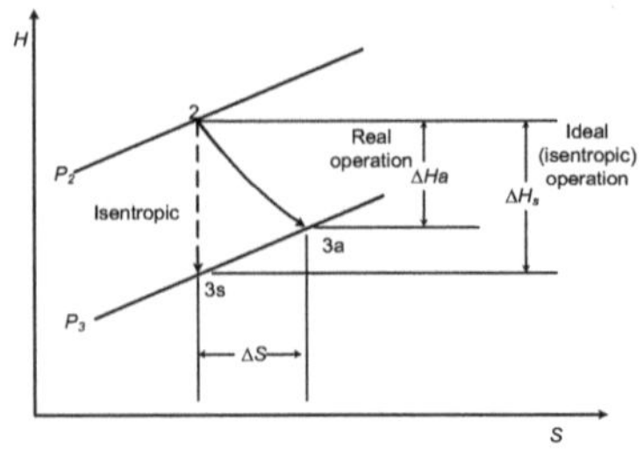

$$P_3 = P_4,\ P_1 = P_2$$

From the isentropic pump operation, we have: $W_{ps} = V(P_2 - P_3) = 10.2\,\mathrm{kJ/kg}$.

So the enthalpy at point 1: $H_{1s} = \Delta W_{ps} + H_4 = 10.2 + 289.3 = 299.5\,\mathrm{kJ/kg}$.

Heat required in the boiler becomes: $q_{in} = H_2 - H_{1s} = 3138.2\,\mathrm{kJ/kg}$.

Net work for the ideal Rankine cycle:

$$W_{net,s} = W_p + \Delta H_s = (10.2 - 1185.3)\,\mathrm{kJ/kg} = -1175.1\,\mathrm{kJ/kg}$$

Efficiency of the ideal Rankine cycle: $\eta_{\text{ideal cycle}} = \frac{|W_{net,s}|}{q_{in}} = 0.374$ or 37.4%

(b) Actual cycle efficiency: $\frac{\Delta H_a}{\Delta H_s} = \eta_{t,actual}$

With the turbine efficiency of $\eta_t = 0.7$, we have: $\Delta H_a = \eta_t \Delta H_s = 829.7\,\mathrm{kJ/kg}$.

Pump efficiency: $\frac{\Delta W_{ps}}{\Delta W_{pa}} = \eta_{pump}$: we have: $W_{pa} = \frac{W_{ps}}{\eta_{pump}} = 13.6\,\mathrm{kJ/kg}$.

The net work for the actual cycle: $W_{net,act} = W_{pa} + \Delta H_a = 816.1\,\mathrm{kJ/kg}$

$$H_{1a} = \Delta W_{pa} + H_4 = (13.6 + 289.3)\mathrm{kJ/kg} = 302.9\mathrm{kJ/kg}$$

$$q_{in,act} = H_2 - H_{1a} = (3437.7 - 302.9)\,\mathrm{kJ/kg} = 3134.81\,\mathrm{kJ/kg}$$

Therefore, the efficiency of the actual cycle: $\eta_{actual} = \frac{|W_{net,act}|}{q_{in,act}} = 0.260$ or 26%

Comparison of the two efficiencies shows that both operations have low efficiencies, although the actual cycle is less efficient than the ideal Rankine cycle (37.4%). The actual cycles involve friction, pressure drops, as well as heat losses in various processes; therefore, have lower thermal efficiency than those of ideal cycles.

7.8.7 Brayton Engine Efficiency

The Brayton cycle was first proposed by George Brayton in around 1870. The Brayton cycle is used for gas turbines operating on an open cycle as shown in Fig. 7.7. The Brayton engine consists of three main components: a compressor, a combustion chamber, and a turbine. The air after being compressed in the compressor is heated by burning fuel in it. The heated air expands in a turbine and produces power. The two main applications of gas-turbine engines are jet engines and electric power production. Jet engines take a large volume of hot gas from a combustion process and feed it through a nozzle which accelerates the plane to high speed (see Fig. 7.8). Gas-turbine cycle engines employ a continuous combustion system where compression, combustion, and expansion occur simultaneously at various places in the engine. Combustion takes place at constant pressure. The fuel must release sufficient heat of combustion to produce necessary power.

Part of the power produced in the turbine of a gas-turbine power plant is used to drive the compressor. The ratio of the compressor work to the turbine work is called the *back-work ratio*. Sometimes, more than one-half of the turbine work may be used by the compressor [2, 4].

The Brayton cycle is analyzed as an open system. By neglecting the changes in kinetic and potential energies, the energy balance on a unit-mass basis is

$$\Delta H = (q_{\text{in}} - q_{\text{out}}) + (W_{\text{in}} - W_{\text{out}}) \tag{7.45}$$

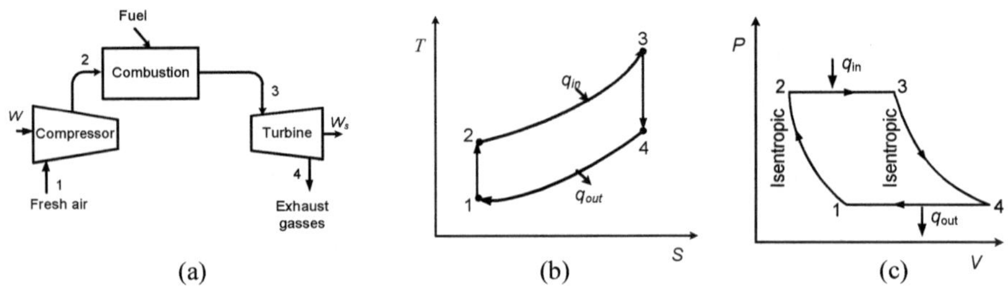

Fig. 7.7 **a** Schematic of open Brayton cycle, **b** ideal Brayton cycle on a *TS* diagram with processes of isentropic adiabatic compression, isobaric heat addition, isentropic adiabatic expansion, and isobaric heat rejection, **c** ideal Brayton cycle on a *PV* diagram

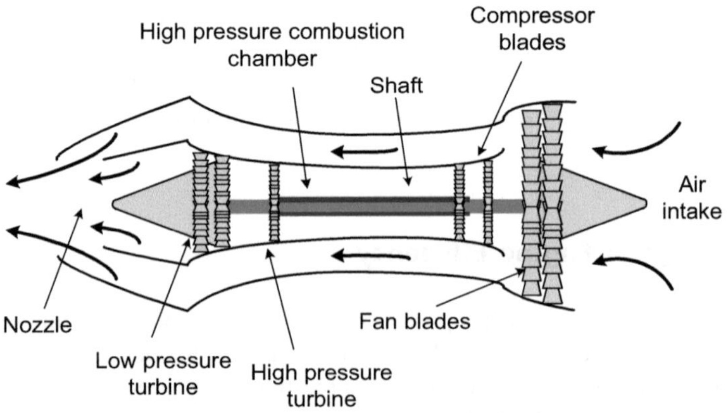

Fig. 7.8 Schematic of a jet engine using gas power cycle

Assuming constant heat capacity, thermal efficiency can be derived using the heat added q_{in} and heat rejected q_{out}

$$\eta_{Brayton} = 1 - \left(\frac{|q_{out}|}{q_{in}}\right) = 1 - \left(\frac{C_{p,av}(T_4 - T_1)}{C_{p,av}(T_3 - T_2)}\right) \quad (7.46)$$

Upon rearrangement, Eq. (7.46) reduces to

$$\eta_{Brayton} = 1 - \left(\frac{T_1}{T_2}\right)\left(\frac{T_4/T_1 - 1}{T_3/T_2 - 1}\right) \quad (7.47)$$

Figure 7.7 shows that processes 1-2 and 3-4 are isentropic, and $P_2 = P_3$ and $P_4 = P_1$, thus the previously derived equations for isentropic process are expressed by

$$\left(\frac{T_2}{T_1}\right) = \left(\frac{P_2}{P_1}\right)^{(\gamma-1)/\gamma} \quad \text{and} \quad \left(\frac{T_3}{T_4}\right) = \left(\frac{P_3}{P_4}\right)^{(\gamma-1)/\gamma}$$

Therefore, the thermal efficiency is

$$\eta_{Brayton} = 1 - \left(\frac{1}{r_p^{(\gamma-1)/\gamma}}\right) \quad \text{(for constant heat capacity)} \quad (7.48)$$

where r_p is the compression ratio (P_2/P_1) and $\gamma = C_p/C_v$. Equation (7.48) shows that the Brayton cycle depends on the pressure ratio of the gas turbine and the ratio of specific heats of the working fluid. The typical values of pressure ratio change between 5 and 20 [2]. As seen from Fig. 7.7, part of the work produced is used to drive the compressor. The back-work ratio r_{bw} shows the part of the produced energy is diverted to the compressor

$$r_{bw} = \frac{W_{comp.in}}{W_{turb.out}} \quad (7.49)$$

Example 7.14 illustrates the analysis of an ideal Brayton cycle when the heat capacity is temperature dependent, while Example 7.15 illustrates the analysis of a real Brayton cycle. Example 7.16 illustrates the analysis of an ideal Brayton cycle with constant specific heats.

***Example 7.14* Simple ideal Brayton cycle calculations with variable specific heats**

A power plant is operating on an ideal Brayton cycle with a pressure ratio of $r_p = 9$. The fresh air temperature is 295 K at the compressor inlet and 1300 K at the end of the compressor (inlet of the turbine). Using the standard-air assumptions determine the thermal efficiency of the cycle.

Solution:

Assume that the cycle is at a steady-state flow and the changes in kinetic and potential energy are negligible. Heat capacity of air is temperature dependent, and the air is an ideal gas.

Consider Fig. 7.7.

Basis: 1 kg air. Using the data from the Appendix: Table D1.

Process 1-2 isentropic compression.

At $T_1 = 295$ K, $H_1 = 295.17$ kJ/kg; $P_{r1} = 1.3068$ (P_r is the relative pressure defined in Eq. 7.36).

$$\frac{P_{r2}}{P_{r1}} = \frac{P_2}{P_1} = P_r \rightarrow P_{r2} = (9)(1.3068) = 11.76$$

Approximate values from Table D1 for the compressor exit at $P_{r2} = 11.76$, $T_2 = 550$ K, $H_2 = 555.74$ kJ/kg.

Process 3-4 isentropic expansion in the turbine as seen on the *TS* diagram in Fig. 7.7.

$T_3 = 1300$ K, $H_3 = 1395.97$ kJ/kg, $P_{r3} = 330.9$ (From Table D1)

$$\frac{P_{r4}}{P_{r3}} = \frac{P_4}{P_3} \rightarrow P_{r4} = \left(\frac{1}{9}\right)(330.9) = 36.76$$

Approximate values from Table D1 at the exit of turbine $P_{r4} = 36.76$:

$T_4 = 745$ K and $H_4 = 761.87$ kJ/kg.

The work input to the compressor for 1 kg/s mass flow rate:

$$\begin{aligned}\dot{W}_{\text{comp.in}} &= \dot{m}(H_2 - H_1) = (555.74 - 295.17)\text{kW} \\ &= 260.6\text{kW (work received from the compressor)}\end{aligned}$$

The work output of the turbine:

$$\begin{aligned}\dot{W}_{\text{turb. out}} &= \dot{m}(H_4 - H_3) = (761.87 - 1395.97)\text{kW} \\ &= -634.1\text{kW(work produced from the turbine)}\end{aligned}$$

The net work out: $W_{net} = W_{out} - W_{in} = (63.41 - 260.6) = -373.53\ \text{kJ/kg}$.

The back-work ratio r_{bw} becomes: $r_{bw} = \frac{W_{comp.in}}{W_{turb.out}} = \frac{260.6}{634.1} = 0.41$ or 41%

This shows that 41% of the turbine output has been used in the compressor.

The amount of heat added: $q_{in} = H_3 - H_2 = 1395.97 - 555.74 = 840.23\ \text{kJ/kg}$.

The amount of heat rejected by the hot gas:

$q_{\text{out}} = H_1 - H_4 = 295.17 - 761.87 = -466.7\,\text{kJ/kg}.$

The thermal efficiency: $\eta_{\text{th}} = \frac{|W_{\text{net}}|}{q_{\text{in}}} = 1 - \frac{|q_{\text{out}}|}{q_{\text{in}}} = 0.444$ or 44%

Example 7.15 **Thermal efficiency of an actual Brayton cycle with variable specific heats**

A power plant is operating on an ideal Brayton cycle with a pressure ratio of $r_{\text{p}} = 9$. The fresh air temperature is 295 K at the compressor inlet and 1300 K at the end of the compressor and at the inlet of the turbine. Assume the gas-turbine cycle operates with a compressor efficiency of 85% and a turbine efficiency of 85%. Determine the thermal efficiency of the cycle.

Solution:

Assume that the cycle is at a steady-state flow and the changes in kinetic and potential energy are negligible. Heat capacity of air is temperature dependent, and the air is an ideal gas. The standard-air assumptions are applicable.

Basis: 1 kg air.

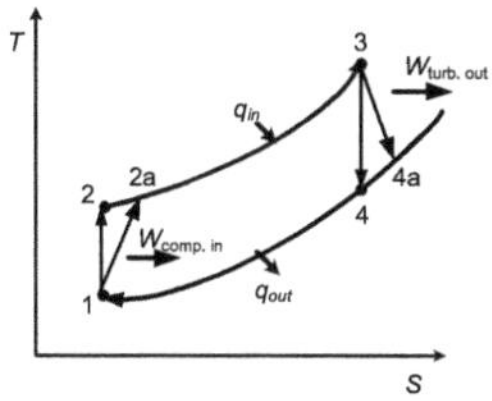

At $T_1 = 295\,\text{K}$, $H_1 = 295.17\,\text{kJ/kg}$; $P_{\text{r1}} = 1.3068$; (Table D1)

P_{r} is the relative pressure (Eq. 7.36).

$$\frac{P_{\text{r2}}}{P_{\text{r1}}} = \frac{P_2}{P_1} = P_{\text{r}} \rightarrow P_{\text{r2}} = (9)(1.3068) = 11.76$$

Approximate values from Table D1 for the compressor exit at $P_{\text{r2}} = 11.76$:

$$T_2 = 550\,\text{K} \quad \text{and} \quad H_2 = 555.74\,\text{kJ/kg}$$

Process 3-4 isentropic expansion in the turbine as seen on the *TS* diagram above

$$T_3 = 1300\,\text{K}, \quad H_3 = 1395.97\,\text{kJ/kg}, \quad P_{\text{r3}} = 330.9$$

(from Table D1)

$$\frac{P_{r4}}{P_{r3}} = \frac{P_4}{P_3} \rightarrow P_{r4} = \left(\frac{1}{9}\right)(330.9) = 36.76$$

Approximate values from Table D1 at the exit of turbine at 36.76:

$T_4 = 745$ K and $H_4 = 761.87$ kJ/kg.

The work input to the compressor for a mass flow rate of 1 kg/s:

$$\dot{W}_{\text{comp.in}} = \dot{m}(H_2 - H_1) = (555.74 - 295.17)\,\text{kW} = 260.57\,\text{kW}$$

The work output of the turbine:

$$\dot{W}_{\text{turb·out}} = \dot{m}(H_4 - H_3) = (761.87 - 1395.97)\,\text{kW} = -634.10\,\text{kW}$$

From the efficiency definitions for compressor and turbine, we have:

$$\eta_C = \frac{W_{Cs}}{W_{Ca}} \rightarrow W_{Ca} = \frac{W_{Cs}}{\eta_C} = \frac{260.6\,\text{kJ/kg}}{0.85}$$
$$= 306.6\,\text{kW(actual compression work)}$$

(where $\dot{W}_{\text{comp.in}} = \dot{W}_{Cs} = 260.6\,\text{kW}$)

The work output of the turbine with 85% efficiency

$$\eta_T = \frac{W_{Ta}}{W_{Ts}} \rightarrow W_{Ta} = \eta_T(W_{Ts}) = 0.85(-634.1\,\text{kW}) = -539.0\,\text{kW}$$

(where $\dot{W}_{\text{turb·out}} = \dot{W}_s - 634.1$ kW in ideal operation in Example 7.13)

The net work out: $\dot{W}_{\text{net}} = \dot{W}_{\text{out}} - \dot{W}_{\text{in}} = (-539.0 + 306.6) = -232.4\,\text{kW}$.

The back-work ratio r_{bw} becomes: $r_{bw} = \left|\frac{\dot{W}_{\text{comp.in}}}{\dot{W}_{\text{turb.out}}}\right| = \left|\frac{306.6}{539.0}\right| = 0.568$.

This shows that the compressor now consumes 56.8% of the turbine output. The value of back-work ratio increased from 41 to 56.8% because of friction, heat losses, and other nonideal conditions in the cycle.

Enthalpy at the exit of compressor:

$$W_{Ca} = H_{2a} - H_1 \rightarrow H_{2a} = W_{Ca} + H_1$$
$$= (306.6 + 295.2)\,\text{kJ/kg} = 601.8\,\text{kJ/kg}$$

Heat added: $q_{\text{in}} = H_3 - H_{2a} = 1395.97 - 601.8 = 794.2\,\text{kJ/kg}$.

The thermal efficiency: $\eta_{\text{th}} = \frac{W_{\text{net}}}{q_{\text{in}}} = \frac{|232.4\,\text{kJ/kg}|}{794.2\,\text{kJ/kg}} = 0.292(\text{or } 29.2\%)$.

The actual Brayton-gas cycle thermal efficiency drops to 0.292 from 0.444. Efficiencies of the compressor and turbine affect the performance of the cycle. Therefore, for better cycle thermal efficiency, significant improvements are necessary for the compressor and turbine operations [6].

	Ideal	Actual
W_{net} kJ/kg	– 373.5	– 232.4
W_C kJ/kg	260.6	306.6
W_T kJ/kg	– 634.1	– 539.0
q_{in} kJ/kg	840.2	794.2
q_{out} kJ/kg	– 466.7	– 561.8
η_{th} %	44.4	29.2

Example 7.16 **Ideal Brayton cycle with constant specific heats**

A power plant is operating on an ideal Brayton cycle with a pressure ratio of $r_p = 9$. The air temperature is 300 K at the compressor inlet and 1200 K at the end of the compressor. Using the standard-air assumptions and $\gamma = 1.4$, determine the thermal efficiency of the cycle.

Solution:

Assume that the cycle is at a steady-state flow and the changes in kinetic and potential energy are negligible. The specific heat capacities are constant, and the air is an ideal gas.

Consider Fig. 7.7:

Basis: 1 kg air and $\gamma = 1.4$. Using the data from the Appendix: Table D1

Process 1-2 isentropic compression: $T_1 = 300$ K, $H_1 = 300.2$ kJ/kg

From Eq. 7.34: $\frac{T_2}{T_1} = \left(\frac{P_2}{P_1}\right)^{(\gamma-1)/\gamma} \rightarrow T_2 = 300(9)^{0.4/1.4} = 562\,\text{K}.$

Process 3-4 isentropic expansion in the turbine as seen on the *TS* diagram above

$T_3 = 1200$ K and $\frac{T_4}{T_3} = \left(\frac{P_4}{P_3}\right)^{(\gamma-1)/\gamma} \rightarrow T_4 = 1200\left(\frac{1}{9}\right)^{0.4/1.4} = 640.5\,\text{K}.$

From Eq. 7.46: $\eta_{th} = 1 - \frac{q_{out}}{q_{in}} = 1 - \frac{|C_p(T_1 - T_4)|}{C_p(T_3 - T_2)} = 1 - \frac{640.5 - 300}{1200 - 562} = 0.466\,\text{or}\,46.6\%$

From Eq. (7.48): $\eta_{Brayton} = 1 - \left(\frac{1}{r_p^{(\gamma-1)/\gamma}}\right) = 0.463\,\text{or}\,46.3\%$

The results of efficiency calculations are close to each other; for a constant heat capacity it is easy to use Eq. (7.48) directly to estimate the efficiency using the compression ratio r_p.

7.8.8 Otto Engine Efficiency

Otto cycle is named after Nikolaus A. Otto who manufactured a four-stroke engine in 1876 in Germany. Otto cycle describes the functioning of a typical ideal cycle for spark-ignition reciprocating piston engine. This cycle is common in automobile internal combustion engines using a chemical fuel with oxygen from the air. The combustion process results in heat, steam, carbon dioxide, and other chemicals at high temperatures. Petroleum fractions, such as diesel fuel, gasoline, and petroleum gases, are the most common fuels. Bioethanol, biodiesel, and hydrogen can also be used as fuels with modified engines.

Internal combustion engines require either spark ignition or compression ignition of the compressed air–fuel mixture [2, 4, 8]. Gasoline engines take in a mixture of air and gasoline and compress it to around 12.8 bar (1.28 MPa), then use a high-voltage electric spark to ignite the mixture. The compression level in diesel engines is usually twice or more than a gasoline engine. Diesel engines will take in the air only, and shortly before peak compression, a small quantity of diesel fuel is sprayed into the cylinder via a fuel injector that allows the fuel to instantly ignite. Most diesels also have a battery and charging system. Diesel engines are heavier, noisier, and more powerful than gasoline engines. They are also more fuel-efficient in most circumstances and are used in heavy road vehicles, some automobiles, ships, railway locomotives, and light aircraft.

Figure 7.9 shows the Otto engine cycles on *PV* and *TS* diagrams. The ideal cycle paths are:

- Process 1-2 is an isentropic compression of the air as the piston moves from bottom dead center to top dead center (*Intake stroke*).
- Process 2-3 is a constant volume heat transfer to the air from an external source while the piston is at top dead center. This process leads to the ignition of the fuel–air mixture (*Compression stroke*).
- Process 3-4 is an isentropic expansion (*Power stroke*).
- Process 4-1 completes the cycle by a constant volume process in which heat is rejected from the air while the piston is at bottom dead center (*Exhaust stroke*).

In the case of a four-stroke Otto cycle, technically there are two additional processes as shown in Fig. 7.9a: one for the exhaust of waste heat and combustion products (by isobaric compression), and one for the intake of cool oxygen-rich air (by isobaric expansion). However, these are often omitted in simplified analysis. Processes 1-2 and 3-4 do work on the system but no heat transfer occurs during adiabatic expansion and compression. Processes 2-3 and 4-1 are isochoric (constant volume) therefore heat transfer occurs but no work is done as the piston volume does not change.

Idealized *PV* diagram in Fig. 7.9a of the Otto cycle shows the combustion heat input q_{in} and the waste exhaust output q_{out}. The power stroke is the top curved line, and the bottom is the compression stroke.

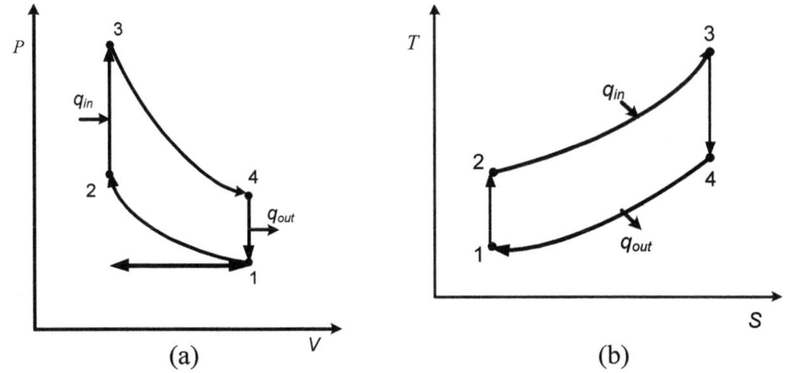

Fig. 7.9 Thermodynamic diagrams of Otto cycle; **a** Otto cycle on a *PV* diagram, **b** Otto cycle on a *TS* diagram

The Otto cycle is analyzed as a close system. The energy balance on a unit-mass basis is

$$\Delta U = (q_{in} - q_{out}) + (W_{in} - W_{out}) \tag{7.50}$$

Assuming that the heat capacity is constant, thermal efficiency can be derived by the heat added q_{in} and heat rejected q_{out}

$$\eta_{Otto} = 1 - \left(\frac{|q_{out}|}{q_{in}}\right) = 1 - \left(\frac{|C_{v,av}(T_1 - T_4)|}{C_{v,av}(T_3 - T_2)}\right) \tag{7.51}$$

In an ideal Otto cycle, there is no heat transfer during the process 1-2 and 3-4 as they are reversible adiabatic processes. Heat is supplied only during the constant volume processes 2-3 and heat is rejected only during the constant volume processes 4-1. Upon rearrangement:

$$\eta_{Otto} = 1 - \left(\frac{T_1}{T_2}\right)\left(\frac{T_4/T_1 - 1}{T_3/T_2 - 1}\right) \tag{7.52}$$

Since $T_4/T_1 = T_3/T_2$ (see Fig. 7.8), Eq. (7.52) reduces to

$$\eta_{Otto} = 1 - \left(\frac{T_1}{T_2}\right) \tag{7.53}$$

From the isentropic equations of ideal gases, we have

$$C_v \ln\left(\frac{T_2}{T_1}\right) - R \ln\left(\frac{V_2}{V_1}\right) = 0 \tag{7.54}$$

$$\left(\frac{T_2}{T_1}\right) = \left(\frac{V_1}{V_2}\right)^{(\gamma-1)} = r^{(\gamma-1)} \tag{7.55}$$

where r is the compression ratio (V_1/V_2) and $\gamma = C_p/C_v$, and $R = C_p - C_v$ for an ideal gas. Then the thermal efficiency becomes

$$\eta_{Otto} = 1 - \left(\frac{1}{r^{(\gamma-1)}}\right) \quad \text{(constant specific heats)} \tag{7.56}$$

The specific heat ratio of the air–fuel mixture γ varies with the fuel but is generally close to the air value of 1.4, and when this approximation is used the cycle is called an *air-standard cycle*. However, the real value of γ for the combustion products of the fuel/air mixture is approximately 1.3.

Equation (7.56) shows that the Otto cycle depends upon the compression ratio $r = V_1/V_2$. At the higher compression ratio, the efficiency is higher and the temperature in the cylinder is higher. However, the maximum compression ratio is limited to 10:1 for typical automobiles. Usually this does not increase much because of the possibility of auto ignition, which occurs when the temperature of the compressed fuel/air mixture becomes too high before it is ignited by the flame front leading to *engine knocking*.

Knocking-under elevated temperature and pressure hydrocarbon fuels break down or combust before ignition. This is called "*knock*" reducing the efficiency of engines and possibly causing damage. To prevent the knock a compression ratio between 9:1 and 10.5:1 is used in spark-ignition engines. For example, with a compression ratio of 10:1 the ideal Otto cycle efficiency of spark-ignition engine is around 50%. In real operations, the efficiency may be reduced to 35% or even lower. Due to various inefficiencies, typical power delivered to the drive train is 25% of the energy from fuel [8].

***Example 7.17* Estimation of the efficiency of Otto engine**

Compute the efficiency of an air-standard Otto cycle if the compression ratio is 9 and $\gamma = 1.4$. Compare this with a Carnot cycle operating between the same temperature limits if the lowest temperature of the cycle is 80°F and the highest temperature is 1000°F.

Solution:

The efficiency of the Otto cycle is a function only of the compression ratio.

$$\eta_{Otto} = 1 - \left(\frac{1}{r^{(\gamma-1)}}\right) \quad \text{(constant specific heats)}$$

$$\eta_{Otto} = 1 - \left(\frac{1}{9^{(1.4-1)}}\right) = 0.5847(58.47\%)$$

For the Carnot cycle: $\eta_{Carnot} = 1 - \frac{T_C}{T_H} = 1 - \frac{(460+80)\,\mathrm{R}}{(460+1000)\,\mathrm{R}} = 0.63.01(63.01\%)$

Example 7.18 **Estimation of the work output of Otto engine**

Assuming that the value of $C_v = 0.17$ Btu/lb$_m$°R and the compression ratio of 9 estimate: (a) the heat added and the heat rejected per pound of fluid for the case in which the peak temperature of the cycle is 1200°F and the lowest temperature of the cycle is 80°F, (b) the work output, and (c) the efficiency of the Otto cycle based on these values.

Solution:

Assume that $\gamma = 1.4$.

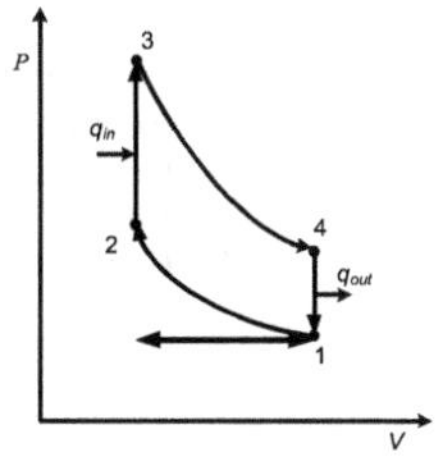

$$\frac{T_3}{T_2} = \left(\frac{V_2}{V_3}\right)^{\gamma-1} = (9)^{0.4} = 2.408\, T_3 = 2.408(540) = 1300\,\text{R}$$

$$q_{\text{in}} = C_v(T_3 - T_2) = 0.17(1660 - 1300) = 61.20\,\text{Btu/lb}$$

To determine T_4, therefore:

$$\frac{T_4}{T_3} = \left(\frac{V_3}{V_2}\right)^{\gamma-1} = \left(\frac{1}{9}\right)^{0.4} = \frac{1}{2.408} \text{ and } T_4 = (1660/2.408) = 689.36\text{R}$$

$$\begin{aligned} q_{\text{out}} &= -C_v(T_1 - T_4) = -0.17(540 - 689.36) \\ &= 25.39\,\text{Btu/lb(heat received by the surrounding)} \end{aligned}$$

The net work out from the engine that is the system is:

$$W_{\text{out}} = q_{\text{out}} - q_{\text{in}} = (25.39 - 61.20)\,\text{Btu/lb} = -35.81\,\text{Btu/lb}$$

Therefore, the efficiency is: $\eta = \frac{|W_{\text{out}}|}{q_{\text{in}}} = \frac{|-38.51|}{61.20} = 0.585(58.5\%)$.

Example 7.19 **Efficiency calculations of ideal Otto engine with variable specific heats**

An ideal Otto cycle operates with a compression ratio ($r = V_{\text{max}}/V_{\text{min}}$) of 8.8. Air is at 101.3 kPa and 280 K at the start of compression (state 1). During the constant volume heat addition process, 1000 kJ/kg of heat is transferred into the air from a source at 1900 K. Heat is discharged to the surroundings at 280 K. Determine the thermal efficiency of energy conversion.

Solution:

Assume that the surroundings are at 280 K and the kinetic and potential energy changes are negligible. The specific heats are temperature dependent.

Heat capacity of air is temperature dependent, and the air is an ideal gas. The *PV* diagram below shows the cycle and the four processes.

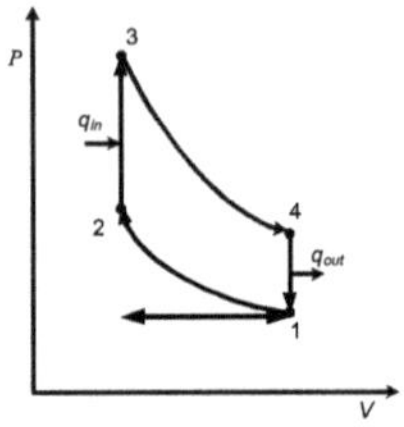

Processes in the cycle:
1-2 Isentropic compression
2-3 Constant volume heat transfer
3-4 Isentropic expansion
4-1 Constant volume heat discharge

Basis: 1 kg air. Using the data from the Appendix:

$q_{in} = 1000$ kJ/kg, $U_1 = 199.75$ kJ/kg, $V_{r1} = 783.0$ at $T = 280$ K from Table D1

$$\frac{V_{r2}}{V_{r1}} = \frac{V_2}{V_1} = \frac{1}{8.8} \rightarrow V_{r2} = \frac{V_{r1}}{r} = \frac{783}{8.8} = 88.97$$

At the value of $V_{r2} = 88.97$, the air properties from Table D1:

$$U_2 = 465.5\,\text{kJ/kg and } T_2 = 640\,\text{K}$$

From isentropic compression of air, we estimate

$$P_2 = P_1\left(\frac{T_2}{T_1}\right)\left(\frac{V_1}{V_2}\right) = 101.3\left(\frac{640}{280}\right)8.8$$
$$= 2037\,\text{kPa (ideal gas equation)}$$

The heat transferred in the path 2-3:

$q_{in} = 1000\,\text{kJ/kg} = U_3 - U_2 \rightarrow U_3 = 1465.5\,\text{kJ/kg}$.

At $U_3 = 1465.5$ kJ kg, we estimate T_3 and V_{r3} by interpolation using the data below:

T K	U kJ/kg	V_r
1750	1439.8	4.328
1800	1487.2	3.994
1777	1465.5	4.147

We estimate the pressure at state 3:

$P_3 = P_2\left(\frac{T_3}{T_2}\right)\left(\frac{V_2}{V_3}\right) = 2037\left(\frac{1777}{640}\right)(1) = 5656\,\text{kPa}.$

Internal energy of air at state 4: $\left(\frac{V_{r4}}{V_{r3}}\right) = \left(\frac{V_4}{V_3}\right) = 8.8 \rightarrow V_{r4} = (8.8)(4.147) = 36.5$

At $V_{r4} = 36.5$, approximate values from Table D1 for:

$U_4 = 658$ kJ/kg and $T_4 = 880$ K.

Process 4-1 is a constant heat discharge. We estimate the discharged heat q_{out}

$$q_{out} = -(U_1 - U_4) \rightarrow q_{out} = 458.2\ \text{kJ/kg (heat received by the surroundings)}$$

$$P_4 = P_3\left(\frac{T_4}{T_3}\right)\left(\frac{V_3}{V_4}\right) = 5656\left(\frac{880}{1777}\right)\left(\frac{1}{8.8}\right) = 318.3\,\text{kPa}$$

The net work output: $W_{net} = q_{out} - q_{in} = (458.2 - 1000)\text{kJ/kg} = -541.8\,\text{kJ/kg}.$

So the thermal efficiency: $\eta_{th} = \frac{|W_{net}|}{q_{in}} = 0.542$ or 54.2%

Although all the processes are internally reversible, the heat transfer and discharge take place at finite temperature difference and are irreversible.

Example 7.20 **Efficiency calculations of an ideal Otto cycle with constant specific heats**

An ideal Otto cycle operates with a compression ratio (V_{max}/V_{min}) of 8. The air is at 101.3 kPa and 300 K at the start of compression (state 1). During the constant volume heat addition process, 730 kJ/kg of heat is transferred into the air from a source at 1900 K. Heat is discharged to the surroundings at 300 K. Determine the thermal efficiency of energy conversion. The average values of specific heats are $C_{p,av} = 1.00$ kJ/kg K and $C_{v,av} = 0.717$ kJ/kg K.

Solution:

Assume: The air-standard assumptions are applicable. The changes in kinetic and potential energy s are negligible. The specific heats are constant.

The *PV* diagram below shows the cycle and the four states.

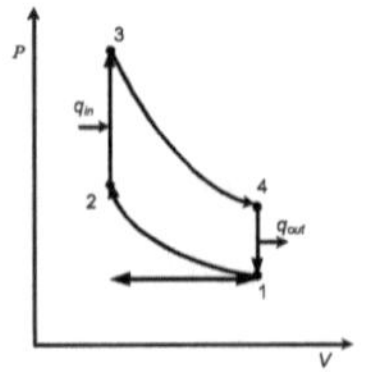

Processes in the cycle:
1-2 Isentropic compression
2-3 Constant volume heat transfer
3-4 Isentropic expansion
4-1 Constant volume heat discharge

Basis: 1 kg air.

$q_{\text{in}} = 730$ kJ/kg; Using the data from Table D1 at $T_o = 300$ K, $U_1 = 214.1$ kJ/kg.

The average specific heats:

$C_{\text{p,av}} = 1.0$ kJ/kg K and $C_{\text{v,av}} = 0.717$ kJ/kg K and $\gamma = 1/0.717 = 1.4$.

From Eq. 7.34: $\frac{T_2}{T_1} = \left(\frac{V_1}{V_2}\right)^{\gamma-1} \rightarrow T_2 = \frac{V_1}{V_2} T_1 = 8(300\text{ K})^{0.4} = 689\,\text{K}.$

Estimate T_3 from the heat transferred at constant volume in process 2-3:

$$q_{\text{in}} = U_3 - U_2 = C_{\text{v,av}}(T_3 - T_2) \rightarrow T_3 = 1703\,\text{K}$$

Process 3-4 isentropic expansion: $T_4 = T_3\left(\frac{V_3}{V_4}\right)^{\gamma-1} = 1703.0\left(\frac{1}{8}\right)^{0.4} = 741.3\,\text{K}.$

Process 4-1 : $q_{\text{out}} = -(U_1 - U_4) = -C_{\text{v,av}}(T_1 - T_4) = 316.4\,\text{kJ/kg}$
(heat discharged) $q_{\text{in}} = 730\text{kJ/kg}$

Net heat transfer is equal to the net work output:

$W_{\text{net}} = (q_{\text{out}} - q_{\text{in}}) = (316.4 - 730.0) = -413.6\text{kJ/kg}.$

So the thermal efficiency is $\eta_{\text{th}} = \frac{|W_{\text{net}}|}{q_{\text{in}}} = 0.566\,\text{or}\,56.6\%$

7.8.9 Diesel Engine Efficiency

Figure 7.10 shows an ideal Diesel cycle on the *PV* and *TS* diagrams. The compression ignited engine is first proposed by Rudolph Diesel in 1890. Most diesel engines use a similar cycle to spark-ignited gasoline engine, but with a compression heating ignition system, rather than needing a separate ignition system. In combustion-ignited engine, the air is compressed to a temperature that is above the autoignition temperature of the fuel. Therefore, the combustion starts as the fuel is injected into the hot air at constant pressure. This variation is called the Diesel cycle.

The Diesel cycle is analyzed as a piston cylinder of a close system. The amount of heat transferred to the working fluid air at constant pressure on a unit-mass basis

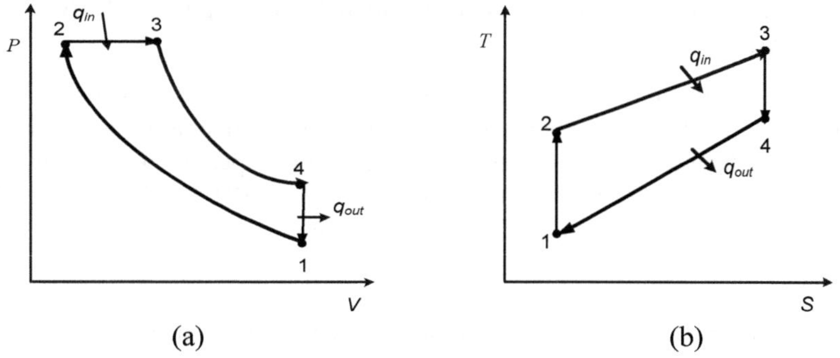

Fig. 7.10 Ideal Diesel cycle **a** on *PV* diagram, **b** on *TS* diagram; the cycle follows the numbers 1-4 in clockwise direction

is

$$q_{\text{in}} = P_2(V_3 - V_2) + (U_3 - U_2) = H_3 - H_2 = C_{\text{p,av}}(T_3 - T_2) \tag{7.57}$$

The amount of heat rejected at constant volume is

$$q_{\text{out}} = U_1 - U_4 = C_{\text{v,av}}(T_1 - T_4) \tag{7.58}$$

Alternatively, thermal efficiency can be derived by the heat added q_{in} and heat rejected q_{out}

$$\eta_{\text{Diesel}} = 1 - \left(\frac{|q_{\text{out}}|}{q_{\text{in}}}\right) = 1 - \left(\frac{(T_4 - T_1)}{\gamma(T_3 - T_2)}\right) = 1 - \left(\frac{T_1(T_4/T_1 - 1)}{\gamma T_2(T_3/T_2 - 1)}\right) \tag{7.59}$$

where $\gamma = C_{\text{p}}/C_{\text{v}}$. In an ideal Diesel cycle, the cutoff ratio r_{c} is defined as the ratio of the cylinder volume after and before the combustion

$$r_{\text{c}} = \frac{V_3}{V_2} \tag{7.60}$$

After using the cutoff ratio r_{c} and compression ratio r with the isentropic ideal gas relations, the thermal efficiency of energy conversion becomes

$$\eta_{\text{Diesel}} = 1 - \frac{1}{r^{(\gamma-1)}}\left(\frac{r_{\text{c}}^{\gamma} - 1}{\gamma(r_{\text{c}} - 1)}\right) \quad \text{(with constant specific heats)} \tag{7.61}$$

Equation (7.61) shows that as the cutoff ratio r_{c} decreases, the efficiency of the Diesel cycle increases. For the limiting ratio of $r_{\text{c}} = 1$, the value of the bracket in Eq. (7.61) becomes unity, and the efficiencies of the Otto and Diesel cycles become identical. Thermal efficiencies of large diesel engines vary from 35 to 40% [2, 4].

Example 7.21 **Thermal efficiency of an ideal Diesel engine with the constant specific heats**

An ideal Diesel cycle has an air-compression ratio of 20 and a cutoff ratio of 2. At the beginning of the compression, the fluid pressure, temperature, and volume are 14.7 psia, 70°F, and 120 in^3, respectively. The average specific heats of air at room temperature are $C_{p,av} = 0.24$ Btu/lb R and $C_{v,av} = 0.171$ Btu/lb R. Estimate the thermal efficiency with the cold-air-standard assumptions.

Solution:

Assume: The cold-air-standard assumptions are applicable. Air is an ideal gas. The changes in kinetic and potential energies are negligible. Consider Fig. 7.10.
$P_1 = 14.7$ psia, $T_1 = 70$°F, and $V_1 = 120$ in^3, $C_{p,av} = 0.24$ Btu/lb R, and $C_{v,av} = 0.171$ Btu/lb R.

$R = 10.73$ psia ft^3/lbmole R, $MW = 29$ lb/lbmole, $\gamma = C_{p,av}/C_{v,av} = 1.4$, $r = 20$, and $r_c = 2$.

Air mass: $m_{air} = MW\frac{PV}{RT} = 0.0052$ 1b.

The volumes for each process:

$V_1 = V_4 = 120$ in^3, $V_2 = V_1/r = 6$ in^3, $V_3 = V_2\, r_c = 12$ in^3

Process 1-2: Isentropic compression: $\left(\frac{T_2}{T_1}\right) = \left(\frac{V_1}{V_2}\right)^{(\gamma-1)} \rightarrow T_2 = 1756.6\,\text{R}$ (Eq. 7.34).

Process 2-3: Heat addition at constant pressure for ideal gas:

$\left(\frac{T_3}{T_2}\right) = \left(\frac{V_3}{V_2}\right) \rightarrow T_3 = 3513\,\text{R}$

$$q_{in} = m(H_3 - H_2) = mC_{p,av}(T_3 - T_2) = 2.19\,\text{Btu}$$

Process 3-4: Isentropic expansion: $\left(\frac{T_4}{T_3}\right) = \left(\frac{V_3}{V_4}\right)^{(\gamma-1)} \rightarrow \quad T_4 = 1398\,\text{R}.$

Process 4-1 Heat rejection at constant volume for ideal gas

$$q_{out} = -m(H_1 - H_4) = mC_{v,av}(T_4 - T_1) = 0.772\,\text{Btu (heat received by the surroundings)}$$

Network output is equal to the difference between heat input (q_{in}) and waste heat (q_{out})

$$W_{out} = q_{out} - q_{in} = (0.772 - 2.19) = -1.418\,\text{Btu}$$

Thermal efficiency: $\eta_{th} = \frac{|W_{net}|}{q_{in}} = 0.647(\text{or } 64.7\%)$.

Thermal efficiency can also be estimated from Eq. (7.61):

$$\eta_{\text{Diesel}} = 1 - \frac{1}{r^{(\gamma-1)}}\left(\frac{r_c^{\gamma} - 1}{\gamma(r_c - 1)}\right) = 0.647(\text{or } 64.7\%).$$

Example 7.22 **Thermal efficiency of an ideal diesel engine with variable specific heats**

An ideal Diesel cycle has an air-compression ratio of 18 and a cutoff ratio of 2. At the beginning of the compression, the fluid pressure and temperature are 100 kPa, 300 K, respectively. Utilizing the cold-air-standard assumptions, determine the thermal efficiency.

Solution:

Assume: The cold-air-standard assumptions are applicable. Air is the ideal gas. The changes in kinetic and potential energies are negligible. The specific heats depend on temperature.

Consider Fig. 7.10: $P_1 = 100$ kPa, $T_1 = 300$ K, $r = 18$, and $r_c = V_3/V_2 = 2$.

At 300 K, $U_1 = 214.1$ kJ/kg, $V_{r1} = 621.2$ (V_r is the relative specific volume) (Table D1).

From Eq. 7.38: $\frac{V_{r2}}{V_{r1}} = \frac{V_2}{V_1} = \frac{1}{18} \rightarrow V_{r2} = \frac{V_{r1}}{r} = \frac{621.2}{18} = 34.5$

At this value of $V_{r2} = 34.5$, approximate values for: $H_2 = 932.9$ kJ/kg, $T_2 = 900$ K (Table D1).

The heat transferred at constant pressure in process 2-3:

$\frac{V_3}{T_3} = \frac{V_2}{T_2} \Longrightarrow T_3 = \frac{V_3}{V_2} T_2 = 2T_2 = 1800\,\text{K}$

$$H_3 = 2003.3\,\text{kJ/kg}, \quad V_{r3} = 4.0 \quad (\text{Table D1})$$

$$q_{\text{in}} = H_3 - H_2 = 1070.4\,\text{kJ/kg}$$

Processes 3-4: Next, we need to estimate the internal energy of air at state 4:

$$\left(\frac{V_{r4}}{V_{r3}}\right) = \left(\frac{V_4}{V_3}\right) = \left(\frac{V_4}{2V_2}\right) = \left(\frac{18}{2}\right) \rightarrow V_{r4} = (9)(4.0) = 36$$

At $V_{r4} = 36$, $U_4 = 659.7\,\text{kJ/kg}$ (Table D1).

The process 4-1 is heat discharge at constant volume. We estimate the discharged heat q_{out}

$$\begin{aligned} q_{\text{out}} &= -(U_1 - U_4) = 659.7 - 214.1 \\ &= 445.6\,\text{kJ/kg (heat received by the surroundings)} \end{aligned}$$

The net heat transfer is equal to the net work output:

$W_{net} = q_{out} - q_{in} = 445.6 - 1070.4 = -624.8\,\mathrm{kJ/kg}$.

So the thermal efficiency is $\eta_{th} = \frac{|W_{net}|}{q_{in}} = 0.583$ or 58.3%.

7.8.10 Ericsson and Stirling Engine Efficiency

The Ericsson engine is an "*external combustion engine*" because it is externally heated. The four processes that occur between the four stages of the ideal Ericsson cycle are:

- Process 1-2: Isothermal compression. The compressed airflows into a storage tank at constant pressure.
- Process 2-3: Isobaric heat addition. The compressed airflows through the regenerator and picks up heat.
- Process 3-4: Isothermal expansion. The cylinder expansion space is heated externally, and the gas undergoes isothermal expansion.
- Process 4-1: Isobaric heat removal. Before the air is released as exhaust, it is passed back through the regenerator, thus cooling the gas at a low constant pressure, and heating the regenerator for the next cycle.

A *Stirling engine* operates by cyclic compression and expansion of air or other gas at different temperature levels and converts the heat of hot gas into mechanical work. The Stirling engine has high efficiency compared to steam engines and can use almost any heat source [2, 4]. Stirling cycle has four totally reversible processes:

- Process 1-2 Isothermal heat addition from external source.
- Process 2-3 Internal heat transfer from working fluid to regenerator at constant volume.
- Process 3-4 Isothermal heat rejection.
- Process 4-1 Internal heat transfer from regenerator back to working fluid at constant volume.

Heat addition during process 1-2 at T_H and heat rejection process 3-4 at T_C are both isothermal. For an isothermal process heat transfer is

$$q = T\Delta S \tag{7.62}$$

The change in entropy of an ideal gas at isothermal conditions is

$$\Delta S_{1-2} = -R\ln\frac{P_2}{P_1} \tag{7.63}$$

Using Eqs. (7.62) and (7.63) the heat input and output are estimated by

$$q_{in} = T_H(S_2 - S_1) = T_H\left(-R\ln\frac{P_2}{P_1}\right) = RT_H\ln\frac{P_1}{P_2} \quad (7.64)$$

$$q_{out} = T_C(S_4 - S_3) = T_C\left(-R\ln\frac{P_4}{P_3}\right) = RT_C\ln\frac{P_3}{P_4} \quad (7.65)$$

The thermal efficiency of the Ericsson cycle becomes

$$\eta_{Ericsson} = 1 - \frac{|q_{out}|}{q_{in}} = 1 - \frac{RT_C\ln(P_4/P_3)}{RT_H\ln(P_1/P_2)} = 1 - \frac{T_C}{T_H} \quad (7.66)$$

Since $P_1 = P_4$ and $P_3 = P_2$.

Both the ideal Ericsson and Stirling cycles are external combustion engines. Both the cycles are reversible, as is the Carnot cycle and have the same thermal efficiency between the same temperature limits when using an ideal gas as the working fluid

$$\eta_{Ericsson} = \eta_{Stirling} = \eta_{Carnot} = 1 - \frac{T_C}{T_H} \quad (7.67)$$

Both the cycles utilize regeneration to improve efficiency. Between the compressor and the expander heat is transferred to a thermal energy storage device called the regenerator during one part of the cycle and is transferred back to the working fluid during another part of the cycle. Equation (7.67) is valid for both the closed and steady-state flow cycles of the engines.

7.8.11 Atkinson Engine Efficiency

The *Atkinson engine* is an internal combustion engine and is used in some modern hybrid electric applications. Figure 7.11 shows the cyclic processes in the Atkinson engine. In the engine, the expansion ratio can differ from the compression ratio and the engine can achieve greater thermal efficiency than a traditional piston engine [2, 4]. Expansion ratios are obtained from the ratio of the combustion chamber volumes when the piston is at the bottom dead center and top dead center.

In Atkinson cycle, the intake valve allows a reverse flow of intake air. The goal of the modern Atkinson cycle is to allow the pressure in the combustion chamber at the end of the power stroke to be equal to atmospheric pressure; when this occurs, all the available energy has been utilized from the combustion process. For any given portion of air, the greater expansion ratio allows more energy to be converted from heat to useful mechanical energy; hence the engine is more efficient. The disadvantage of the four-stroke Atkinson cycle is the reduced power output as a smaller portion of the compression stroke is used to compress the intake of air. Atkinson cycles with a supercharger to make up for the loss of power

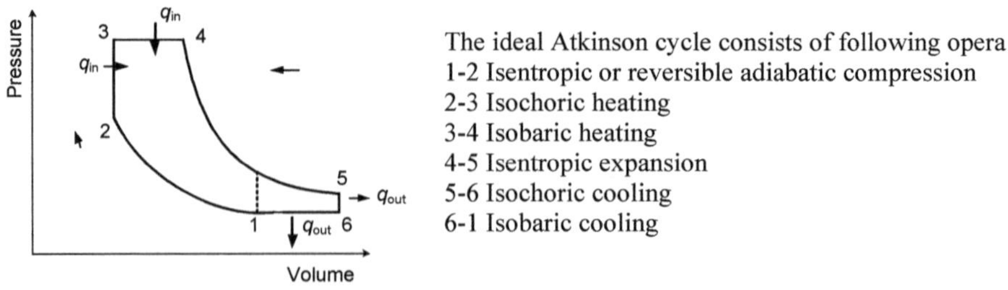

Fig. 7.11 Ideal Atkinson gas cycle

density are known as *Miller engines* [2, 4]. The power of the Atkinson engine can be supplemented by an electric motor. This forms the basis of an Atkinson cycle-based hybrid electric drivetrain.

7.9 Improving Efficiency of Heat Engines

The following may be some options toward increasing the efficiency of heat engines [2, 6]:

- Increased hot side temperature is the approach used in modern combined-cycle gas turbines. The melting point of the construction of materials and environmental concerns regarding NO_x production limit the maximum temperature on the heat engines.
- Lowering the output temperature by using mixed chemical working fluids such as using a 70/30 mix of ammonia and water as its working fluid may also increase efficiency. This mixture allows the cycle to generate useful power at lower temperatures.
- Use of supercritical fluids such as CO_2 as working fluids may increase efficiency.
- Exploit the chemical properties of the working fluid, such as nitrogen dioxide (NO_2), which has a natural dimer as di-nitrogen tetra oxide (N_2O_4). At low temperature, the N_2O_4 is compressed and then heated. The increasing temperature causes each N_2O_4 to break apart into two NO_2 molecules. This lowers the molecular weight of the working fluid, which drastically increases the efficiency of the cycle. Once the NO_2 has expanded through the turbine, it is cooled and recombined into N_2O_4. This is then fed back to the compressor for another cycle.
- New fuels, such as hydrogen, may have a positive impact on the performance of the engines since the energy density of hydrogen is higher than that of electric batteries.

7.10 Hydroelectricity

Hydroelectricity is the production of electrical power using the gravitational force of falling or flowing water. Figure 7.12 shows a schematic of a conventional hydroelectric dam, in which hydroelectric power comes from the potential energy of dammed water driving a water turbine and generator. The power extracted from the water depends on the volume and on the difference in height between the source and the water's outflow. This height difference is called the head. The amount of potential energy in water is proportional to the head. A large pipe called the penstock delivers water to the turbine. Example 7.23 illustrates the efficiency calculations for a hydraulic turbine.

The pumped-storage method produces electricity to supply high peak demands by moving water between reservoirs at different elevations. At times of low electrical demand, excess generation capacity is used to pump water into the higher reservoir. When there is higher demand, water is released back into the lower reservoir through a turbine. Compared to wind farms, hydroelectricity power plants have a more predictable load factor. If the project has a storage reservoir, it can generate power when needed. Hydroelectric plants can be easily regulated to follow variations in power demand [4, 7]. Example 7.24 illustrates the pumped energy in a hydropower plant.

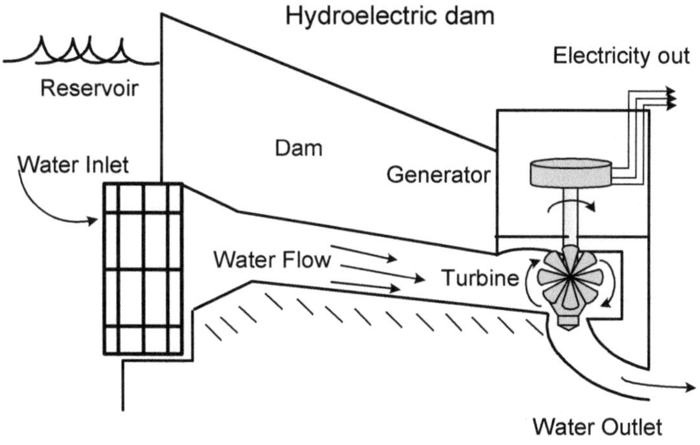

Fig. 7.12 Cross section of a conventional hydroelectric dam

Example 7.23 **Efficiency of a hydraulic turbine**

Electricity is produced by a hydraulic turbine installed near a large lake. The average depth of the water in the lake is 45 m. The water mass flow rate is 600 kg/s. The produced electric power is 220 kW. The generator efficiency is 95%. Determine the overall mechanical efficiency of the turbine-generator and the shaft work transferred from the turbine to the generator.

Solution:

Assume: The mechanical energy of water at the turbine exit is small and negligible. The density of the water is 1000 kg/m^3.

$$\dot{m} = 600\,\text{kg/s}, \quad \eta_{\text{gen}} = 0.95, \quad \dot{W}_{\text{out}} = -220\,\text{kW}$$

$$\dot{W}_{\text{fluid}} = \dot{W}_{\text{in}} = PE = \dot{m}gz = (600\text{kg/s})\left(9.81\,\text{m/s}^2\right)(45\,\text{m})\left(\text{kJ/kg} = 1000\,\text{m}^2/\text{s}^2\right)$$
$$= 264.9\,\text{kW}$$

The overall mechanical efficiency of the turbine generator:

$$\eta_{\text{overall}} = \eta_{\text{turb}}\eta_{\text{gen}} = \frac{|\text{energy out}|}{\text{energy in}} = \frac{|220\text{ kW}|}{264.9\text{ kW}}$$
$$= 0.83\left(\eta_{\text{gen}} = 0.95\right) \text{ and } \eta_{\text{turb}} = \frac{\eta_{\text{overall}}}{\eta_{\text{gen}}} = 0.87$$

The shaft power transferred from the turbine:

$W_{\text{turb}} = \eta_{\text{turb}} W_{\text{fluid}} = 0.87(264.9\text{ kW}) = 230.5\,\text{kW}.$

The lake supplies 264.9 kW of mechanical energy to the turbine. Only 87% of the supplied energy is converted to shaft work. This shaft work drives the generator, and 220 kW is produced.

Example 7.24 **Pumped energy in a hydropower plant**

Consider a hydropower plant reservoir with an energy storage capacity of 1×10^6 kWh. This energy is to be stored at an average elevation of 60 m relative to the ground level. Estimate the minimum amount of water has to be pumped back to the reservoir.

Solution:

Assume that the evaporation of the water is negligible.

$$PE = 1 \times 10^6\,\text{kWh}, \ \Delta z = 60\,\text{m}, \quad g = 9.8\,\text{m/s}^2$$

Energy of the work potential of the water: $PE = mg\Delta z$.

Amount of water: $m = \frac{PE}{g\Delta z} = \frac{1\times10^6\text{kWh}}{9.8\text{ m/s}^2(60\text{ m})}\left(\frac{3600\text{ s}}{\text{hr}}\right)\left(\frac{1000\text{ m}^2/\text{s}^2}{1\text{ kJ/kg}}\right) = 6.122 \times 10^9$ kg.

7.11 Wind Electricity

The power produced by a wind turbine is proportional to the kinetic energy of the wind captured by the wind turbines

$$\dot{W}_{\text{wind}} = \eta_{\text{wind}} \rho \frac{\pi v^3 D^2}{8} \tag{7.68}$$

where ρ is the density of air, v is the velocity of air, D is the diameter of the blades of the wind turbine, and η_{wind} is the efficiency of the wind turbine. Therefore, the power produced by the wind turbine is proportional to the cube of the wind velocity and the square of the blade span diameter. The strength of wind varies, and an average value for a given location does not alone indicate the amount of energy a wind turbine could produce [2, 7]. Example 7.25 illustrates the efficiency calculations for wind turbines.

Example 7.25 **Efficiency of a wind turbine**

A wind-turbine generator with a 25-foot-diameter blade produces 0.5 kW of electric power. In the location of the wind turbine, the wind speed is 11 miles per hour. Determine the efficiency of the wind-turbine generator.

Solution:

Assume: The wind flow is steady. The wind flow is one-dimensional and incompressible. The frictional effects are negligible.

$$\rho_{\text{air}} = 0.076\,\text{lb/ft}^3,\ D = 25\,\text{ft, actual power production} = 0.5\,\text{kW}$$

Kinetic energy can be converted to work completely. The power potential of the wind is its kinetic energy.

Average speed of the wind: $v_1 = (11\,\text{mph})\left(\frac{1.4667\,\text{ft/s}}{1\,\text{mph}}\right) = 16.13\,\text{ft/s}.$

The mass flow rate of air: $\dot{m} = \rho v_1 A = \rho v_1 \frac{\pi D^2}{4} = 601.4\,\text{lb/s}.$

$$\dot{W}_{\text{max}} = \dot{m}\frac{v_1^2}{2g_c} = (601.4\,\text{lb/s})\frac{(16.13\,\text{ft/s})^2}{2}\left(\frac{1\,\text{lbf}}{32.2\,\text{lb}_\text{m}\,\text{fts}^2}\right)\left(\frac{1\,\text{kW}}{737.56\,\text{lbf ft/s}}\right) = 3.29\,\text{kW}$$

This is the available energy to the wind turbine. The turbine-generator efficiency is

$$\eta_{\text{wind turb}} = \frac{\dot{W}_{\text{act}}}{\dot{W}_{\text{max}}}\frac{0.5\,\text{kW}}{3.29\,\text{kW}} = \mathbf{0.152\ or\ 15}.2\%$$

Only 15.2% of the incoming kinetic energy is converted to electric power. The remaining part leaves the wind turbine as outgoing kinetic energy.

7.12 Geothermal Electricity

Geothermal electricity refers to the energy conversion of geothermal energy to electric energy. Technologies in use include dry steam power plants, flash steam power plants, and binary cycle power plants. Geothermal power is sustainable because the heat extraction is negligible compared with the earth's heat content.

Geothermal electric plants have until recently been built exclusively where elevated temperature geothermal resources are available near the surface. The development of binary cycle power plants and improvements in drilling and extraction technology may enable enhanced geothermal systems over a much greater geographical range [4].

The thermal efficiency of geothermal electric plants is low, around 10–23%, because geothermal fluids are at a low temperature compared with steam from boilers. This low temperature limits the efficiency of heat engines in extracting useful energy during the generation of electricity. Exhaust heat is wasted, unless it can be used directly, for example, in greenhouses, timber mills, and district heating. To produce more energy than the pumps consume, electricity generation requires elevated temperature geothermal fields and specialized heat cycles [2, 4, 7].

7.13 Ocean Thermal Energy Conversion

Ocean thermal energy conversion uses the difference between cooler deep and warmer shallow or surface ocean waters to run a heat engine and produce useful work, usually in the form of electricity (see Fig. 7.13). Warm surface seawater is pumped through a heat exchanger to vaporize the fluid. The expanding vapor turns the turbogenerator. Chilly water, pumped through a second heat exchanger, condenses the vapor into a liquid, which is then recycled through the system. In the tropics, the temperature difference between surface and deep water is a modest 20–25°C. Ocean thermal energy conversion systems is still considered an emerging technology with a thermal efficiency of 1–3%, which is well below the theoretical maximum for this temperature difference of between 6 and 7% [7]. The most used heat cycle for ocean thermal energy conversion systems is the Rankine cycle using a low-pressure turbine system. Closed-cycle engines use working fluids such as ammonia or R-134a. Open-cycle engines use vapor from the seawater itself as the working fluid.

7.14 Thermoelectric Effect

The thermoelectric effect involves energy conversions between heat and electricity as a temperature difference creates an electric potential or electric potential creates a heat flow leading to a temperature difference. There are two thermoelectric effects:

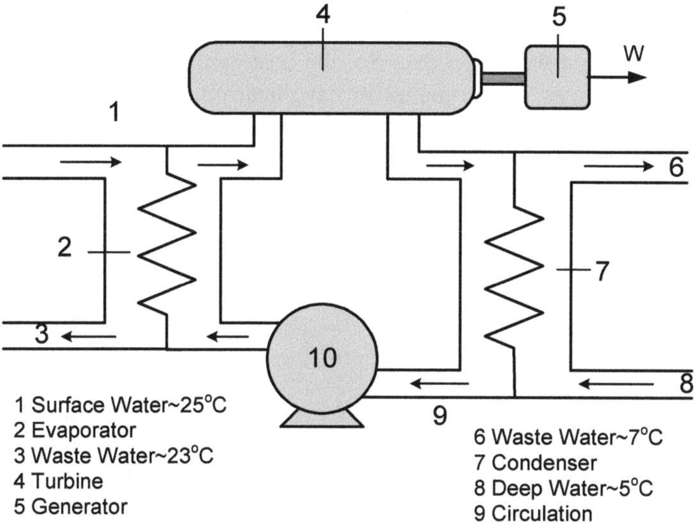

Fig. 7.13 Diagram of a closed-cycle ocean thermal energy conversion (OTEC) plant

- The *Seebeck effect* refers to conversion of heat to electricity.
- The *Peltier effect* refers to conversion of electricity to heat.

Measuring temperature by *thermocouples* operating between a hot and a cold junction is based on the Seebeck effect. A commonly used thermoelectric material in such applications is Bismuth telluride (Bi_2Te_3). The thermoelectric efficiency approaches the Carnot limit.

Thermoelectric materials can be used as refrigerators, called "thermoelectric coolers," or "Peltier coolers," although it is far less common than vapor-compression refrigeration. Compared to a vapor-compression refrigerator, the main advantages of a Peltier cooler are its lack of moving parts or circulating liquid, and its small size and flexible shape. Another advantage is that Peltier coolers do not require refrigerant liquids, which can have harmful environmental effects. The main *disadvantage* of Peltier coolers is that they cannot simultaneously have low cost and high-power efficiency. Advances in thermoelectric materials may lead to the creation of Peltier coolers that are both cheap and efficient [6].

7.15 Efficiency of Heat Pumps and Refrigerators

Heat pumps, refrigerators, and air conditioners need outside work to move heat from a colder to a warmer region, so their function is the opposite of a heat engine. Since they are heat engines, these devices are also limited by Carnot efficiency. A heat pump can be used for heating or cooling as part of *heating, ventilation, and air-conditioning* applications. The heat pump can heat and when necessary, it uses the basic refrigeration cycle to cool. To do that, a heat pump can change which coil is the condenser and which the evaporator by controlling the flow direction of the refrigerant by reversing valve. So, the heat can be pumped in either direction. In cooler climates it is common to have heat pumps that are designed only to heat. In heating mode, the outside heat exchanger is the evaporator, and the indoor exchanger is the condenser to discharge heat to the inside air. In cooling mode, however, the outside heat exchanger becomes the condenser, and the indoor exchanger is the evaporator to absorb heat from the inside air. The next sections discuss the heat pumps and refrigerators [1].

7.15.1 Heat Pumps

A heat pump is a reversed heat engine that receives heat from a low temperature reservoir and rejects it to a high temperature reservoir by consuming an external work [1]. The efficiency of a heat pump cycle is usually called the *coefficient of performance.* The external work supplied for obtaining that desired effect is to supply heat to the hot body. A heat pump uses a fluid which absorbs heat as it vaporizes and releases the heat when it condenses. Figure 7.14 shows a typical heat pump drawing heat from the ambient air. A heat pump requires external work to extract heat q_C (q_{in}) from the outside air (cold region) and deliver heat q_H (q_{out}) to the inside air (hot region). The most common heat pump is a phase change heat pump. During the cycle of such a heat pump, the compressor pumps gas through the condenser where it gets cooled down and finally condenses into liquid phase after releasing heat of condensation q_{out}. Then the liquid flows through an expansion valve where its pressure and temperature both drop considerably. Further it flows through the evaporator where it warms up and evaporates to gaseous phase again by extracting heat of vaporization q_{in} from the surroundings.

In heat pumps, the work energy W_{in} provided mainly in the form of electricity is converted into heat, and the sum of this energy and the heat energy that is moved from the cold reservoir (q_C) is equal to the total heat energy added to the hot reservoir q_H.

$$\dot{q}_H = \dot{q}_C + \dot{W}_{in} \tag{7.69}$$

Thermal efficiency of heat pumps is measured by the coefficient of performance COP_{HP}.

$$COP_{HP} = \frac{\dot{q}_H}{\dot{W}_{in}} \tag{7.70}$$

The amount of heat they move can be greater than the input work. Therefore, heat pumps can be a more efficient way of heating than simply converting the input work into heat, as in an electric heater or furnace. The limiting value of Carnot

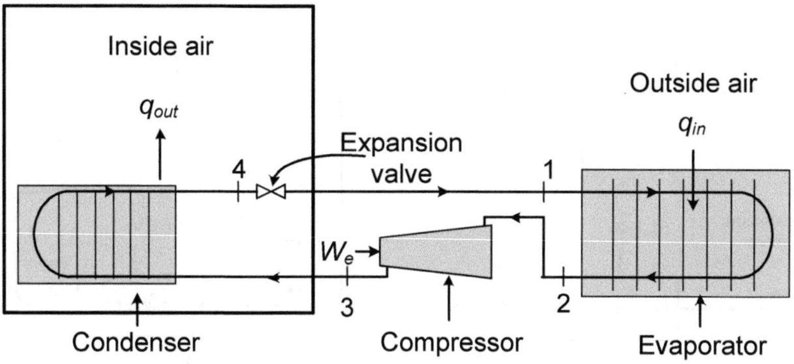

Fig. 7.14 Typical heat pump for heating a room; compressor pumps gas from a low temperature region to a high temperature region. Within the evaporator, the heat of vaporization is absorbed from the outside surroundings. Compressor increases the pressure and temperature of the vapor and pumps it through the cycle. In the condenser, the vapor condenses and releases heat of condensation to the inside surroundings. The expansion valve causes a flash of the liquid to boil because of sudden drop in pressure; complete vaporization of the liquid takes place in the evaporator

efficiency for the heat pump is

$$\mathrm{COP_{HP}} \leq \frac{T_{\mathrm{H}}}{T_{\mathrm{H}} - T_{\mathrm{C}}} \text{(heating)} \tag{7.71}$$

Equation (7.71) shows that the COP decreases with increasing temperature difference between the hot and cold regions.

Figure 7.15 compares the flow directions of the working fluid when the heat pump is used for heating and for cooling. A reversing valve reverses the direction of the fluid flow so the inside coil in the summer is used as the outside coil in the winter. This means that the working fluid is evaporated in the inside coil extracting heat from the warm air inside and is condensed outside discharging heat to warm outside surroundings in the summer [1, 3]. Example 7.26 illustrates the analysis of a heat pump.

Example 7.26 Effect of the COP on heat pump performance

A heat pump provides 60 MJ/h to a house. If the compressor requires an electrical energy input of 5 kW, calculate the COP. If electricity costs \$0.08 per kWh and the heat pump operates 100 h per month, how much money does the homeowner save by using the heat pump instead of an electrical resistance heater?

Solution:

The heat pump operates at a steady state.

COP for a heat pump with a heat supply of 60 MJ/h = 16.66 kW for $W_{\mathrm{HP}} = 5$ kW

$$\mathrm{COP_{HP}} = \frac{q_{\mathrm{out}}}{W_{\mathrm{HP}}} = 3.33$$

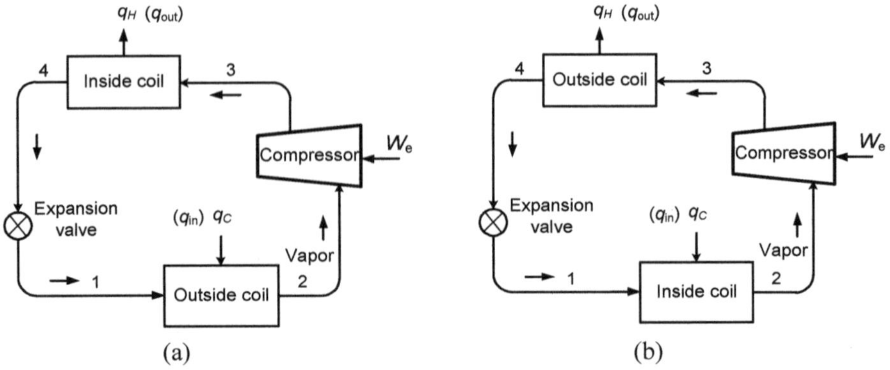

Fig. 7.15 Heat pump can be used: **a** to heat a house in the winter, and **b** to cool in the summer. A reversing valve reverses the direction of the fluid flow so the inside coil in the summer is used as outside coil in the winter

An electrical resistance heater converts all the electrical work supplied W_e into heat q_H. Therefore, in order to get 16.66 kW into your home, you must buy 16.66 kW of electrical power.

Cost of resistance heater:

$$\text{Power} = (16.66\,\text{kW})$$
$$\begin{aligned}&\text{Cost}(\$/\text{month})\\&= \text{electricity}(\$0.08/\text{kWh})\,\text{power}(16.66\,\text{kW})\,\text{time}(100\,\text{h/month})\\&= \$133.3/\text{month}\end{aligned}$$

Cost of heat pump with a power of 5 kW:

$$\begin{aligned}&\text{Cost}(\$/\text{month})\\&= \text{electricity}(\$0.08/\text{kWh})\,\text{power}(5.0\ \text{kW})\,\text{time}(100\,\text{h/month})\\&= \$40.0/\text{month}\end{aligned}$$

Therefore, monthly saving is $\$(133.3 - 40.0)/\text{month} = \$93.3/\text{month}$.

Electrical resistance heaters are not popular, especially in cold climates. The thermal efficiency of a heat pump drops significantly as the outside temperature falls. When the outside temperature drops far enough that the $COP_{HP} \sim 1$, it becomes more practical to use the resistance heater.

7.15.2 Refrigerators

A refrigerator is similar to a heat pump that extracts heat from the cold body/space and delivers it to high temperature body/surroundings using an external work. Figure 7.16 shows a typical refrigeration cycle. A refrigerator is a heat engine in which work is required to extract energy from the freezing compartment and discharge that heat to the room through a condenser at the back of the refrigerator. This leads to further cooling of the cold region. One of the common refrigerants is 1,1,1,2-tetrafluoroethane (CF_3CH_2F) known as R-134a. R-134a has a boiling point temperature of − 26.2°C (− 15°F) and a latent heat of 216.8 kJ/kg at 1.013 bar. It is compatible with most existing refrigeration equipment. R-134a has no harmful influence on the ozone layer of the earth's atmosphere. It is noncorrosive and nonflammable. R-134a is used for medium-temperature applications, such as air conditioning and commercial refrigeration. Another refrigerant is pentafluoroethane (C_2HF_5) known as R-125, which is used in low- and medium-temperature applications. With a boiling point of − 55.3°F at atmospheric pressure, R-125 is nontoxic, nonflammable, and noncorrosive [2, 4, 6].

Thermal efficiency of refrigerators and air conditioners is called the coefficient of performance COP_R:

$$COP_R = \frac{\dot{q}_C}{\dot{W}_{in}} \tag{7.72}$$

The limiting value of the Carnot efficiency for the refrigeration processes is

$$COP_R \leq \frac{T_C}{T_H - T_C} \quad \text{(cooling)} \tag{7.73}$$

Equation (7.73) shows that the COP_R decreases with increasing temperature difference between the hot and cold regions. Example 7.25 illustrates the analysis of a refrigerator. When the desired effect is cooling, the heat resulting from the input work is just an unwanted by-product. In everyday usage the efficiency of air conditioners is often rated by the *Seasonal Energy Efficiency Ratio* (SEER) [1, 4].

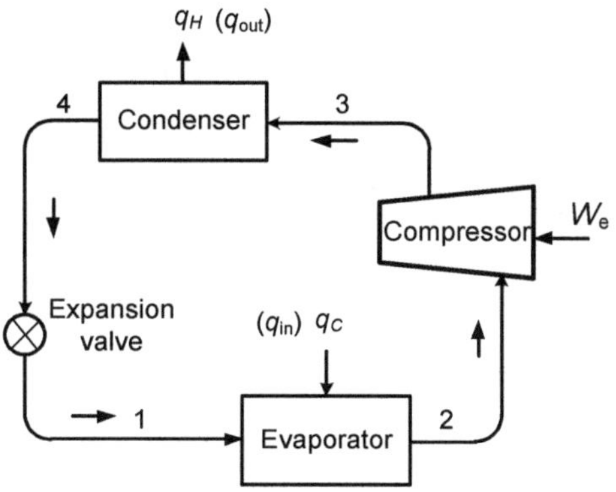

Fig. 7.16 Typical refrigeration cycle; the evaporator is the freezing compartment of the fridge, where the refrigerant absorb heat to be vaporized. The condenser is usually at the back of the fridge, where the heat of condensation is discharged to outside hot surroundings

Example 7.27 **Analysis of a refrigeration cycle**

In a refrigeration cycle, the superheated R-134a (state 2) enters a compressor at 263.15 K and 0.18 MPa. The R-134a (state 3) leaves the compressor at 313.15 K and 0.6 MPa, and enters a condenser, where it is cooled by cooling water. The R-134a (state 4) leaves the condenser at 293.15 K and 0.57 MPa as saturated liquid and enters a throttling valve. The partially vaporized R-134a (state 1) leaves the valve at 0.293 MPa. The cycle of R-134a is completed when it passes through an evaporator to absorb heat from the matter to be refrigerated. The flow rate of R-134a is 0.2 kg/s. The total power input is 60 kW. The surroundings are at 290 K. Determine the coefficient of performance and the exergy loss of the cycle.

Solution:

Assume that kinetic and potential energy changes are negligible, and the system is at a steady state.

Consider Fig. 7.16.

From Tables E1 to E2 the data for R-134a:

$$H_2 = 242.06\,\text{kJ/kg}, \quad S_2 = 0.9362\,\text{kJ/(kg K)}, \text{ at } T_2 = 263.15\,\text{K},$$
$$P_2 = 0.18\,\text{MPa(superheated vapor)}$$

$$H_3 = 278.09\,\text{kJ/kg}, \quad S_3 = 0.9719\,\text{kJ/(kg K)}, \text{ at } T_3 = 313.15\,\text{K},$$
$$P_3 = 0.6\,\text{MPa(superheated vapor)}$$

$$H_4 = 77.26\,\text{kJ/kg}, \text{ at } T_4 = 293.15\,\text{K}, \; P_4 = 0.571\,\text{MPa (saturated liquid)}$$

$$P_1 = 0.293\,\text{MPa}, \; T_1 = 273.15\,\text{K}, \; W_{\text{in}} = 60\,\text{kW}, \quad \dot{m}_r = 0.2\,\text{kg/s}$$

$$T_o = 290\,\text{K}, \; T_{\text{evaporator}} = 273\,\text{K}, \; T_{\text{condenser}} = 290\,\text{K}$$

The throttling process where $H_4 = H_1 = 77.26$ kJ/kg (Stage 1) causes partial vaporization of the saturated liquid coming from the condenser.

The vapor part of the mixture ('quality,') can be obtained using the enthalpy values at 0.293 MPa

$$H_{1\text{sat liq}} = 50.02\,\text{kJ/kg}, \quad H_{1\text{sat vap}} = 247.23\,\text{kJ/kg},$$
$$S_{1\text{sat liq}} = 0.1970\,\text{kJ/kg K}, \; S_{1\text{sat vap}} = 0.9190\,\text{kJ/kg K}$$

$$x_1 = \frac{77.26 - 50.02}{247.23 - 50.02} = 0.138$$

Then the value of entropy:

$S_1 = (1 - 0.138)S_{1\text{sat iq}} + 0.138S_{1\text{sat vap}} = 0.2967\,\text{kJ/kg K}.$

For the cycle, the total enthalpy change is zero.

At the compressor, outside energy W_{in} is needed.

At the evaporator, heat transfer, q_{in}, is used to evaporate the refrigerant R-134a.

The heat absorbed within the evaporator is: $\dot{q}_{\text{in}} = \dot{m}_r(H_2 - H_1) = 32.96\,\text{kW}.$

The energy balance indicates that the total energy $(W_{\text{in}} + q_{\text{in}})$ is removed:

$\dot{W}_{\text{in}} + \dot{q}_{\text{in}} = \dot{q}_{\text{out}} = 92.96\,\text{kW}$

$$\dot{W}_{\text{ideal,in}} = \dot{m}_{\text{r}}(H_3 - H_2) = 7.20\,\text{kW}$$

Coefficient of performance (COP) of the refrigerator:

$\text{COP} = \frac{\dot{q}_{\text{in}}}{\dot{W}_{\text{ideal,in}}} = \frac{H_2 - H_1}{H_3 - H_2} = 4.57.$

The total work (energy) loss:

$\dot{E}x_{\text{total}} = \dot{W}_{\text{in}} + \left(1 - \frac{T_0}{T_{\text{evaporater}}}\right)\dot{q}_{\text{in}} - \left(1 - \frac{T_0}{T_{\text{condenser}}}\right)(-\dot{q}_{\text{out}}) = 58.94\,\text{kW}.$

Exergy analysis identifies the performance of individual processes. Finding ways to improve the thermodynamic performance of individual steps is equally important.

Example 7.28 **Heat rejection by a refrigerator**

The food compartment of a refrigerator is maintained at 4°C by removing heat from it at a rate of 350 kJ/min. If the required power input of the refrigerator is 1.8 kW, determine (a) the coefficient of performance (COP) of the refrigerator, (b) the rate of heat discharged to the surroundings.

Solution:

Assume: Steady-state operation.

$$\dot{q}_c = 350\,\text{kJ/min}, \quad \dot{W}_{\text{net,in}} = 1.8\,\text{kW},$$

$$\text{COP}\frac{\dot{q}_{\text{C}}}{\dot{W}_{\text{net,in}}}\frac{350\,\text{kJ/min}}{1.8\,\text{kW}}\left(\frac{\text{kW}}{60\,\text{kJ/min}}\right)_{\text{R}} = 3.2$$

This means that 3.2 kJ of heat is removed from the refrigerator for each kJ of energy supplied.

(b) Energy balance

$$\dot{q}_{\text{H}} = \dot{q}_{\text{C}} + \dot{W}_{\text{net,in}} = 350\,\text{kJ/min} + 1.8\,\text{kW}\left(\frac{60\,\text{kJ/min}}{\text{kW}}\right) = 458\,\text{kJ/ min}$$

Both the heat removed from the refrigerator space and the unused energy supplied to the refrigerator as electrical work are discharged to surrounding air.

Example 7.29 **Coefficient of performance of a vapor-compression refrigeration cycle**

An ideal vapor-compression refrigeration cycle uses refrigerant-134a. The compressor inlet and outlet pressures are 120 and 900 kPa. The mass flow rate of refrigerant is 0.04 kg/s. Determine the coefficient of performance.

Solution:

Assume: Steady-state operation. The changes in kinetic and potential energies are negligible.

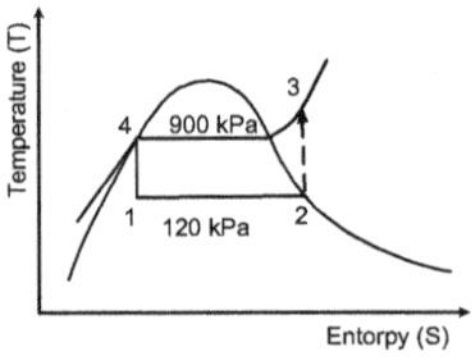

Solution:

Assume: Steady-state adiabatic operation. The changes in kinetic and potential energies are negligible.

The refrigerant mass flow rate = 0.04 kg/s

$$P_2 = 120\,\text{kPa},\ H_2 = 233.86\,\text{kJ/kg},\ S_2 = 0.9354\,\text{kJ/kg K}$$

$P_3 = 900\,\text{kPa}$, for the isentropic compression:

$S_3 = S_2 = 0.9354\,\text{kJ/kg K} \rightarrow H_3 = 276.7\,\text{kJ/kg}$,

$$P_4 = 900\,\text{kPa},\ H_{4\text{sat liq}} = 99.56\,\text{kJ/kg},\ H_{1\text{sat iq}} = H_{4\text{sat liq}}\ \text{(throttling)}$$

Power input to the compressor:

$\dot{W}_{\text{in}} = \dot{m}(H_3 - H_2) = (0.04\,\text{kg/s})(276.7 - 233.86) = 1.71\,\text{kW}$.

Heat removed from the refrigerator:

$\dot{q}_C = \dot{m}(H_2 - H_1) = (0.04\,\text{kg/s})(233.86 - 99.56)\,\text{kJ/kg} = 5.37\,\text{kW}.$

Heat discharged from the refrigerator:

$\dot{q}_H = \dot{m}(H_3 - H_4) = (0.04\,\text{kg/s})(276.7 - 99.56)\,\text{kJ/kg} = 7.08\,\text{kW}$

$$\text{COP}_R = \frac{\dot{q}_C}{\dot{W}_{in}} = \frac{5.37\,\text{kW}}{1.71\,\text{kW}} = 3.14$$

This refrigerator is capable of removing 3.14 units of energy from the refrigerated space for each unit of electric energy it consumes.

7.16 Efficiency of Fuel Cells

A *fuel cell* is an electrochemical cell that converts chemical energy of a fuel into electric energy the reaction between a fuel supply and an oxidizing agent. A hydrogen fuel cell uses hydrogen as fuel and oxygen (usually from air) as an oxidant. Other fuels include hydrocarbons and alcohol. A typical fuel cell produces a voltage from 0.6 to 0.7 V at full rated load. To deliver the desired amount of energy, the fuel cells can be combined in series and parallel circuits, where series circuits yield higher voltage, and parallel circuits allow a higher current to be supplied. Such a design is called a *fuel cell stack*. The cell surface area can be increased to allow stronger current from each cell [10].

The efficiency of a fuel cell depends on the amount of power drawn from it. Drawing more power means drawing more current and hence increasing the losses in the fuel cell. Most losses appear as a voltage drop in the cell, so the efficiency of a cell is almost proportional to its voltage. For this reason, it is common to show graphs of voltage versus current (polarization curves) for fuel cells. A typical cell running at 0.7 V has an efficiency of about 50%, meaning that 50% of the energy content of the hydrogen is converted into electrical energy and the remaining 50% will be converted into heat. Fuel cells are not heat engines and so their efficiency is not limited by Carnot cycle efficiency. Consequently, they can have high efficiencies in converting chemical energy to electrical energy, especially when they are operated at low power density, and using pure hydrogen and oxygen as reactants. Fuel cell vehicles running on compressed hydrogen may have a power-plant-to-wheel efficiency of 22% if the hydrogen is stored as high-pressure gas [10]. Table 7.5 compares the efficiency of several fuel cells.

The overall efficiency (electricity to hydrogen and back to electricity) of such plants (known as *round-trip efficiency*) is between 30 and 50%, depending on conditions. While a much cheaper lead-acid battery might return about 90%, the electrolyzer/fuel cell system can store indefinite quantities of hydrogen and is therefore better suited for long-term storage [10].

Table 7.5 Types of fuel cell using polymer membrane as electrolyte [4.10]

Fuel cell	Power output	T °C	Cell efficiency %
Proton-exchange membrane fuel cell	100 W–500 kW	150–120 (Nafion)	50–70
Direct-methanol fuel cell	100 mW–1 kW	90–120	20–30
Microbial fuel cell	Low	< 40	Low

7.17 Energy Conversions in Biological Systems

Energy production by way of biochemical reactions, its conversion, and storage is the basis of *bioenergetics* of living systems [9]. Outside energy is supplied with the intake of food or with solar radiation. Living systems convert part of this energy to produce electrons and protons. Flow of protons leads to the production of adenosine triphosphate (ATP). The hydrolysis of ATP is coupled with synthesizing protein molecules, transporting ions and substrates, producing mechanical work, and other metabolic activity.

All living systems must convert energy to chemical energy in the form of energy-rich chemical compounds such as ATP. The two biochemical cyclic processes for such conversions are the oxidative phosphorylation in animals and the photosynthesis in plants. These cycles are discussed briefly in the next sections.

7.17.1 Energy Conversion by Oxidative Phosphorylation

In oxidative phosphorylation, the electrons are removed from food molecules in electron transport chain. A series of proteins in the mitochondria use the energy released from passing electrons from reduced molecules like NADH onto oxygen to pump protons across the membrane in mitochondria. The flow of protons causes the rotation of stalk subunit of a large protein called the ATPase. The rotation of ATPase changes the shape of the active site and synthesizes adenosine triphosphate (ATP) from adenosine diphosphate (ADP) and inorganic phosphorus (Pi)

$$\mathrm{ADP} + \mathrm{Pi} = \mathrm{ATP} \tag{7.74}$$

ATP is an energy-rich compound having three phosphate groups attached to a nucleoside of adenine called adenosine. Of the three phosphate groups, the terminal one has a weak linkage. This phosphate group can break spontaneously whenever ATP forms a complex with an enzyme. The breaking up of this bond releases chemical energy causing an immediate shift in the bond energy giving rise to ADP. The energy of ATP is used for all the activity of living cells, such as transport of ions and molecules, synthesis of new proteins and other substances, and growth and development. ATP therefore acts as "energy currency of the cell" and is used to transfer chemical energy between different biochemical reaction cycles [6, 9].

7.17.2 Energy from Photosynthesis

Photosynthesis is the synthesis of carbohydrates from sunlight and carbon dioxide (CO_2). The capture of solar energy is similar in principle to oxidative phosphorylation, as the proton motive force then drives ATP synthesis. The electrons needed to drive this electron transport chain come from light-gathering proteins called photosynthetic reaction centers. In plants, cyanobacteria and algae, oxygenic photosynthesis splits water, with oxygen produced as a waste product. This process uses the ATP and NADPH produced by the photosynthetic reaction centers. This carbon-fixation reaction is carried out by the enzyme RUBisCO [6]. http://en.wikipedia.org/wiki/Metabolism-cite_note-50#cite_note-50.

7.17.3 Metabolism

Metabolism is the set of biochemical reactions that occur in living organisms to maintain life. Metabolism is usually divided into two categories catabolism and anabolism. Catabolism breaks down organic matter, for example to harvest energy in cellular respiration. Anabolism uses energy to construct components of cells such as proteins and nucleic acids. Adenosine triphosphate (ATP) is used to transfer chemical energy between different biochemical reaction cycles. ATP in cells is continuously regenerated and acts as a bridge between catabolism and anabolism, with catabolic reactions generating ATP and anabolic reactions consuming it [6, 9].

7.17.4 Biological Fuels

Biological fuels can be categorized in three groups: carbohydrates (CH), representing a mixture of mono-, di-, and polysaccharides, fats (F), and proteins (Pr) [14]. Carbohydrates are straight-chain aldehydes or ketones with many hydroxyl groups that can exist as straight chains or rings. Carbohydrates are the most abundant biological molecules, and play numerous roles, such as the storage and transport of energy (starch, glycogen) and structural components such as cellulose in plants [6].

The fuel value is equal to the heat of reaction of combustion (oxidation). Carbohydrates and fats can be completely oxidized while proteins can only be partially oxidized and hence a lower fuel value. The energy expenditure may be calculated from the energy balance. Assume that (i) carbohydrate (CH), fat (F), and protein (Pr) are the only compounds involved in the oxidation process, (ii) the other compounds are stationary, and (iii) the uptake and elimination of oxygen, carbon dioxide, and nitrogen is instantaneous.

7.17.5 Converting Biomass to Biofuels

Technologies for converting biomass to biofuels include *thermochemical* and *biochemical conversions:*

- *Thermochemical conversion* involves applying heat to break down biomass into chemical intermediates that can be used to make fuel substitutes. These thermal technologies are known for well over a century and are used primarily in transforming coal into fuels. Gasification combined with catalytic conversion of syngas (carbon monoxide and hydrogen) to fuels, for example, hydrogen, liquid fuels, mixed alcohols, or dimethyl ether. Pyrolysis of biomass produces biooils that could serve as intermediates in a petroleum refinery.
- *Biochemical conversion* focuses on conversion of organic materials including animal waste, plant residues into fuels and bioproducts by enzymes, bacteria, and microorganisms. For example, anaerobic digestion or fermentation converts organic waste. Biogas, biofuel, and bioproducts

Optimal combinations of both biochemical and thermochemical fuel productions may lead to greater energy efficiency in the transformation of the energy of biomass into usable fuels.

Accessing energy in lignocellulosic biomass increases the potential supply of biofuels. Figure 7.17 provides a schematic representation of the conversion process of cellulosic feedstock to biofuels. Cellulosic biomass needs pretreatment and enzymatic processes before it can be converted to biofuel by fermentation. Critical technology advancements need for biological catalysts that can cost-effectively break down the carbohydrate polymers in biomass to sugars and microbes that can ferment all of the sugars in biomass to ethanol [6–9].

Government incentives for biodiesel have spurred significant growth in the vegetable oil-based biodiesel fuel substitute. Converting vegetable oil to biodiesel is a simple thermochemical process in which the natural oil is synthesized with methanol to form fatty acid methyl ester [7, 9].

Problems

7.1 A pump delivers water from the bottom of a storage tank open to the atmosphere containing water at 75°F. The water in the storage tank is 10 ft deep and the bottom of the tank is 50 ft above the ground. The pump delivers the water at 80°F and 100 psia to another tank, which is 10 ft above the ground. If the water flow rate is 5000 lb/h and pump efficiency is 80%, estimate the power in hp required.

7.2 The pump of a water storage tank is powered with an 18-kW electric motor operating with an efficiency of 90%. The water flow rate is 40 L/s. The diameters of the inlet and exit pipes are the same, and the elevation difference between the inlet and outlet is negligible. The absolute pressures at

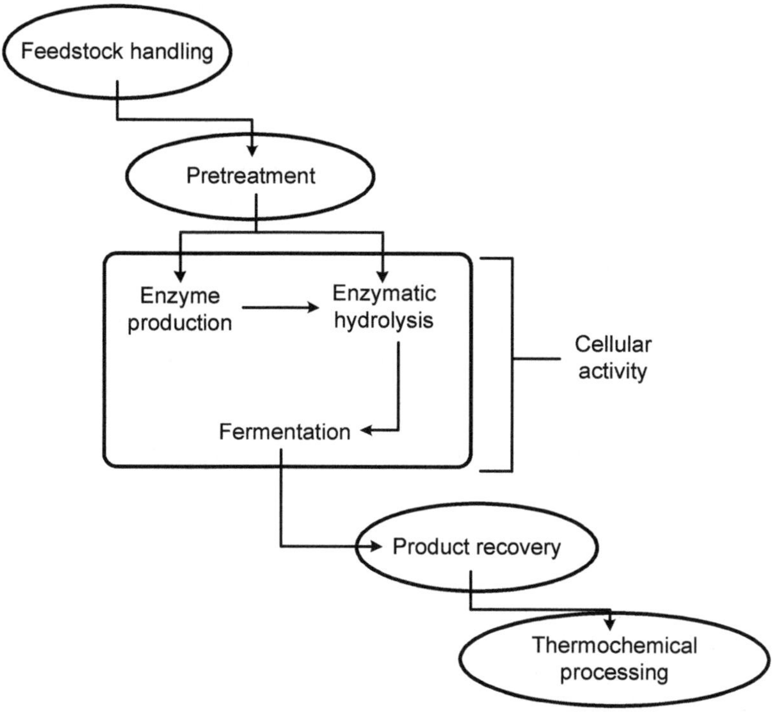

Fig. 7.17 Process schematic of biological conversion of lignocellulosic biomass to ethanol

the inlet and outlet are 100 kPa and 400 kPa, respectively. Determine the mechanical efficiency of the pump.

7.3 The pump of a water storage tank is powered with an 18-kW electric motor operating with an efficiency of 85%. The water flow rate is 30 L/s. The diameters of the inlet and exit pipes are the same, and the elevation difference between the inlet and outlet is negligible. The absolute pressures at the inlet and outlet are 100 kPa and 500 kPa, respectively. Determine the mechanical efficiency of the pump.

7.4 A wind-turbine generator with a 24-foot-diameter blade produces 0.35 kW of electric power. In the location of the wind turbine, the wind speed is 11 miles per hour. Determine the efficiency of the wind-turbine generator.

7.5 A wind-turbine generator with a 27-foot-diameter blade produces 0.4 kW of electric power. In the location of the wind turbine, the wind speed is 10 miles per hour. Determine the efficiency of the wind-turbine generator.

7.6 A hydroelectric plant operates by water falling from a 25 m height. The turbine in the plant converts potential energy into electrical energy with an assumed efficiency of 85%. Electricity is lost by about 8% through power transmission. Estimate the mass flow rate of the water necessary to power a 500 W light bulb.

7.7 A hydroelectric plant is generating 125 kW of power. If their turbines only convert the potential energy form the water with a 75% efficiency, find the height change for the water flow rate of 480 kg/s must fall in order to continue to make 125 kW of power.

7.8 A hydroelectric plant operates by water falling from a 20 m height. The turbine in the plant converts potential energy into electrical energy with an assumed efficiency of 80%. Electricity is lost by about 8% through power transmission. Estimate the mass flow rate of the water necessary to power a 1500 W light bulb.

7.9 A hydroelectric plant operates by water falling from a 100 ft height. The turbine in the plant converts potential energy into electrical energy with an assumed efficiency of 82%. Electricity is lost by about 6% through power transmission. Estimate the mass flow rate of the water necessary to power a 3500 W light bulb.

7.10 A hydroelectric plant operates by water falling from a 50 ft height. The turbine in the plant converts potential energy into electrical energy with an assumed efficiency of 85%. The electricity is lost by about 5% through power transmission so the available power is 95%. Estimate the mass flow rate of the water necessary to power a 3500 W light bulb.

7.11 A hydroelectric plant operates by water falling from a 75 ft height. The turbine in the plant converts potential energy into electrical energy with an assumed efficiency of 85%. Power is lost by about 5% through power transmission so the available power is 95%. If the mass flow rate of the water is 396 lb/s, estimate the power output of the hydro plant.

7.12 An electric motor attached to a pump draws 10.2 amperes at 110 volts. At steady load the motor delivers 1.32 hp of mechanical energy. Estimate the rate of heat transfer from the motor.

7.13 An electric motor attached to a pump draws 12 amperes at 110 volts. At steady load, the motor delivers 1.5 hp of mechanical energy. Estimate the rate of heat transfer from the motor.

7.14 Air enters an insulated compressor operating at steady state at 1.0 bar, 300K with a mass flow rate of 3.6 kg/s and exits at 2.76 bar. Kinetic and potential energy effects are negligible. (a) Determine the minimum theoretical power input required, in kW, and the corresponding exit temperature, in K. (b) If the actual exit temperature is 420 K, determine the power input, in kW, and the isentropic compressor efficiency.

7.15 Air enters an insulated compressor operating at steady state at 1.0 bar, 300K with a mass flow rate of 2.5 kg/s and exits at 2.6 bar. Kinetic and potential energy effects are negligible. (a) Determine the minimum theoretical power input required, in kW, and the corresponding exit temperature, in K. (b) If the actual exit temperature is 420 K, determine the power input, in kW, and the isentropic compressor efficiency.

7.16 A compressor increases the pressure of carbon dioxide from 100 kPa to 600 kPa. The inlet temperature is 300 K and the outlet temperature is 400 K. The mass flow rate of carbon dioxide is 0.01 kmole/s (440 g/s). The power required by the compressor is 55 kW. The temperature of the surroundings is 290 K. Determine the minimum amount of work required and the coefficient of performance.

7.17 A compressor increases the pressure of carbon dioxide from 100 kPa to 500 kPa. The inlet temperature is 300 K and the outlet temperature is 400 K. The mass flow rate of carbon dioxide is 0.015 kmole/s (660 g/s). The power required by the compressor is 60 kW. The temperature of the surroundings is 290 K. Determine the minimum amount of work required and the coefficient of performance.

7.18 A compressor receives air at 15 psia and 80°F with a flow rate of 1.2 lb/s. The air exits at 40 psia and 300°F. At the inlet, the air velocity is low but increases to 250 ft/s at the outlet of the compressor. Estimate the power input to the compressor if it is cooled at a rate of 200 Btu/s.

7.19 A compressor receives air at 15 psia and 80°F with a flow rate of 1.5 lb/s. The air exits at 50 psia and 300°F. At the inlet, the air velocity is low but increases to 250 ft/s at the outlet of the compressor. Estimate the power input to the compressor if it is cooled at a rate of 150 Btu/s.

7.20 In an adiabatic compression operation, air is compressed from 20°C and 101.32 kPa to 520 kPa with an efficiency of 0.7. The airflow rate is 22 moles/s (637.3 g/s). Assume that the air remains the ideal gas during compression. The surroundings are at 298.15 K. Determine the thermodynamic efficiency η_{th} and the rate of energy dissipated $\dot{E}_{loss}$.

7.21 In an adiabatic compression operation, air is compressed from 25°C and 101.32 kPa to 560 kPa with an efficiency of 0.7. The airflow rate is 20 moles/s (579.4 g/s). Assume that the air remains the ideal gas during compression. The surroundings are at 298.15 K. Determine the thermodynamic efficiency η_{th} and the rate of energy dissipated $\dot{E}_{loss}$.

7.22 The power required to compress 0.05 kg/s of steam from a saturated vapor state at 50°C to a pressure of 800 kPa at 200°C is 15 kW. Find the conversion rate of power input to heat loss from the compressor.

7.23 The power required to compress 0.1 kg/s of steam from a saturated vapor state at 80°C to a pressure of 1000 kPa at 200°C is 28 kW. Find the conversion rate of power input to heat loss from the compressor.

7.24 A steam turbine consumes 4000 lb/h steam at 540 psia and 800°F. The exhaust steam is at 165 psia. The turbine operation is adiabatic.
(a) Determine the exit temperature of the steam and the work produced by the turbine.
(b) Determine the thermal efficiency.

7.25 A steam turbine consumes 3800 lb/h steam at 540 psia and 800°F. The exhaust steam is at 180 psia. The turbine operation is adiabatic.
(a) Determine the exit temperature of the steam and the work produced by the turbine.
(b) Determine the thermal efficiency.

7.26 A superheated steam (stream 1) expands in a turbine from 5000 kPa and 325°C to 150 kPa and 200°C. The steam flow rate is 15.5 kg/s. If the turbine generates 1.1 MW of power, determine the heat loss to the surroundings and thermal efficiency.

7.27 Superheated steam (stream 1) expands in a turbine from 6000 kPa and 325°C to 150 kPa and 200°C. The steam flow rate is 16.2 kg/s. If the turbine generates 1.2 MW of power, determine the heat loss to the surroundings and thermal efficiency.

7.28 Steam expands in a turbine from 6600 kPa and 300°C to a saturated vapor at 1 atm. The steam flow rate is 9.55 kg/s. If the turbine generates a power of 1 MW, determine the thermal efficiency.

7.29 Steam expands in a turbine from 7000 kPa and 300°C to a saturated vapor at 1 atm. The steam flow rate is 14.55 kg/s. If the turbine generates a power of 1.5 MW, estimate the thermal efficiency.

7.30 Steam expands adiabatically in a turbine from 850 psia and 600°F to a wet vapor at 12 psia with a quality of 0.9. The turbine produces a power output of 1500 Btu/s. Estimate the thermal efficiency for a steam flow rate of 12.8 lb/s.

7.31 Steam expands adiabatically in a turbine from 900 psia and 600°F to a wet vapor at 10 psia with a quality of 0.9. The turbine produces a power output of 1550 Btu/s. Estimate the thermal efficiency for a steam flow rate of 14.2 lb/s.

7.32 A turbine produces 65000 kW electricity with an efficiency of 70%. It uses a superheated steam at 8200 kPa and 550°C. The discharged stream is a saturated mixture at 75 kPa. If the expansion in the turbine is adiabatic, and the surroundings are at 298.15 K, determine the thermodynamic efficiency and the work loss.

7.33 A turbine produces 70 MW electricity with an efficiency of 70%. It uses a superheated steam at 8800 kPa and 550°C. The discharged stream is a saturated mixture at 15 kPa. If the expansion in the turbine is adiabatic, and the surroundings are at 298.15 K, determine the thermodynamic efficiency and the work loss.

7.34 A Carnot cycle uses water as the working fluid at a steady-flow process. Heat is transferred from a source at 250°C and water changes from saturated liquid to saturated vapor. The saturated steam expands in a turbine at 10 kPa, and heat is transferred in a condenser at 10 kPa. Estimate the thermal efficiency and net power output of the cycle.

7.35 A Carnot cycle uses water as the working fluid at a steady-flow process. Heat is transferred from a source at 200°C and water changes from saturated liquid to saturated vapor. The saturated steam expands in a turbine at 20 kPa, and heat is transferred in a condenser at 20 kPa. Estimate the thermal efficiency of the cycle and the amount of heat transferred in the condenser for a flow rate of 5.5 kg/s of the working fluid.

7.36 A Carnot cycle uses water as the working fluid at a steady-flow process. Heat is transferred from a source at 200°C and water changes from saturated liquid to saturated vapor. The saturated steam expands in a turbine at 10 kPa, and heat is transferred in a condenser at 10 kPa. Estimate the thermal efficiency of the cycle and the amount of heat transferred in the condenser for a flow rate of 7.5 kg/s of the working fluid.

7.37 A Carnot cycle uses water as the working fluid at a steady-flow process. Heat is transferred from a source at 400°F and water changes from saturated liquid to saturated vapor. The saturated steam expands in a turbine at 5 psia, and heat is transferred in a condenser at 5 psia. Estimate the thermal efficiency of the cycle and the amount of heat transferred in the condenser for a flow rate of 10 lb/s of the working fluid.

7.38 Consider a simple ideal Rankine cycle. If the turbine inlet temperature and the condenser pressure are kept the same, discuss the effects of increasing the boiler pressure on: (a) turbine power output, (b) heat supplied, (c) thermal efficiency, (d) heat rejected.

7.39 Consider a simple ideal Rankine cycle. If the boiler and the condenser pressures are kept the same, discuss the effects of increasing the temperature of the superheated steam on: (a) turbine power output, (b) heat supplied, (c) thermal efficiency, (d) heat rejected.

7.40 A steam power production plant burns fuel at 1273.15 K (T_H), and cooling water is available at 290 K (T_C). The steam produced by the boiler is at 8200 kPa and 823.15 K. The condenser produces a saturated liquid at 30 kPa. The turbine and pump operate reversibly and adiabatically. Determine the thermal efficiency of the cycle for the steam flow rate of 1k/s.

7.41 A steam power plant operates on a simple ideal Rankine cycle. The boiler operates at 3000 kPa and 350°C. The condenser operates at 30 kPa. The mass flow rate of steam is 22 kg/s. Estimate the thermal efficiency of the cycle and the net power output.

7.42. A steam power plant operates on a simple ideal Rankine cycle shown below. The turbine receives steam at 698.15 K and 4100 kPa, while the discharged steam is at 40 kPa. The mass flow rate of steam is 3.0 kg/s. In the boiler, heat is transferred into the steam from a source at 1500 K (T_H). In the condenser, heat is discharged to the surroundings at 298 K (T_C). The condenser operates at 298 K. Determine the thermal efficiency of the cycle.

7.43 A steam power plant operates on a simple ideal Rankine cycle. The boiler operates at 5000 kPa and 300°C. The condenser operates at 20 kPa. The mass flow rate of steam is 25 kg/s. Estimate the thermal efficiency of the cycle and the net power output.

7.44 A steam power plant operates on a simple ideal Rankine cycle. The boiler operates at 10000 kPa and 400°C. The condenser operates at 30 kPa. The power output of the cycle is 140 MW. Estimate the thermal efficiency of the cycle and the mass flow rate of the steam.

7.45 A steam power plant is operating on the simple ideal Rankine cycle. The steam mass flow rate is 20 kg/s. The steam enters the turbine at 3500 kPa and 400°C. Discharge pressure of the steam from the turbine is 15 kPa. Determine the thermal efficiency of the cycle.

7.46 A simple ideal Rankine cycle is used in a steam power plant. Steam enters the turbine at 6600 kPa and 798.15 K. The net power output of the turbine is 35 kW. The discharged steam is at 10 kPa. Determine the thermal efficiency.

7.47 A steam power plant operates on a simple ideal Rankine cycle. The boiler operates at 10000 kPa and 500°C. The condenser operates at 10 kPa. The power output of the cycle is 175 MW. Turbine operates with an isentropic efficiency of 0.80, while the pump operates with an isentropic efficiency of 0.90. Estimate the thermal efficiency of the cycle and the heat transferred in the condenser.

7.48 A steam power plant operates on a simple ideal Rankine cycle. The boiler operates at 10000 kPa and 400°C. The condenser operates at 10 kPa. The power output of the cycle is 145 MW. Turbine operates with an isentropic efficiency of 0.85, while the pump operates with an isentropic efficiency of 0.95. Estimate the thermal efficiency of the cycle and the heat transferred in the condenser.

7.49 A steam power plant operates on a simple ideal Rankine cycle. The boiler operates at 10000 kPa and 400°C. The condenser operates at 10 kPa. The mass flow rate of steam is 110 kg/s. Turbine operates with an isentropic efficiency of 0.85, while the pump operates with an isentropic efficiency of 0.95. Estimate the thermal efficiency and the power output of the cycle.

7.50 A steam power plant operates on a simple ideal Rankine cycle. The boiler operates at 8000 kPa and 400°C. The condenser operates at 20 kPa. The mass flow rate of steam is 80 kg/s. Turbine operates with an isentropic efficiency of 0.85, while the pump operates with an isentropic efficiency of 0.90. Estimate the thermal efficiency and the power output of the cycle.

7.51 A steam power plant shown below uses natural gas to produce 0.1 MW power. The combustion heat supplied to a boiler produces steam at 10000 kPa and 798.15 K. The turbine efficiency is 0.7. The discharged steam from the turbine is at 30 kPa and is sent to a condenser. The condensed water is pumped to the boiler. The pump efficiency is 0.90. Determine the thermal efficiency of the cycle.

7.52 A steam power plant shown below uses natural gas to produce 0.12 MW power. The combustion heat supplied to a boiler produces steam at 10000 kPa and 798.15 K. The turbine efficiency is 0.75. The discharged steam from the turbine is at 30 kPa and is sent to a condenser. The condensed water is pumped to the boiler. The pump efficiency is 0.85. Determine the thermal efficiency of an ideal and actual Rankine cycles.

7.53 A steam power plant shown below uses natural gas to produce 0.12 MW power. The combustion heat supplied to a boiler produces steam at 9000 kPa and 798.15 K. The turbine efficiency is 0.8. The discharged steam from the turbine is at 10 kPa and is sent to a condenser. The condensed water is pumped to the boiler. The pump efficiency is 0.9. Determine the thermal efficiency of an ideal and actual Rankine cycles.

7.54 A simple ideal reheat Rankine cycle is used in a steam power plant shown below. Steam enters the turbine at 9000 kPa and 823.15 K and leaves at

4350 kPa and 698.15 K. The steam is reheated at constant pressure to 823.15 K. The discharged steam from the low-pressure turbine is at 10 kPa. The net power output of the turbine is 40 MW. In the boiler, heat is transferred into the steam from a source at 1600 K (T_H). In the condenser, heat is discharged to the surroundings at 298 K (T_C). The condenser operates at 298 K. The turbine efficiency is 0.8 and the pump efficiency is 0.9. Determine the mass flow rate of steam and the thermal efficiency of the cycle.

7.55 A steam power plant is using an ideal regenerative Rankine cycle shown below. Steam enters the high-pressure turbine at 8200 kPa and 773.15 K, and the condenser operates at 20 kPa. The steam is extracted from the turbine at 350 kPa to heat the feed water in an open heater. The water is a saturated liquid after passing through the feed water heater. The work output of the turbine is 50 MW. In the boiler, heat is transferred into the steam from a source at 1600 K (T_H). In the condenser, heat is discharged to the surroundings at 285 K (T_C). Determine the thermal efficiency of the cycle.

7.56 A steam power plant is using an ideal reheat regenerative Rankine cycle shown below. Steam enters the high-pressure turbine at 9000 kPa and 773.15 K and leaves at 850 kPa. The condenser operates at 10 kPa. Part of the steam is extracted from the turbine at 850 kPa to heat the water in an open heater, where the steam and liquid water from the condenser mix and direct contact heat transfer takes place. The rest of the steam is reheated to 723.15 K and expanded in the low-pressure turbine section to the condenser pressure. The water is a saturated liquid after passing through the water heater and is at the heater pressure. The work output of the turbine is 75 MW. In the boiler, heat is transferred into the steam from a source at 1600 K (T_H). In the condenser, heat is discharged to the surroundings at 285 K (T_C). Determine the thermal efficiency of the cycle.

7.57 A steam power plant uses a geothermal energy source. The geothermal source is available at 220°C and 2320 kPa with a flow rate of 200 kg/s. The hot water goes through a valve and a flash drum. Steam from the flash drum enters the turbine at 550 kPa and 428.62 K. The discharged steam from the turbine has a quality of $x_4 = 0.96$. The condenser operates at 10 kPa. The water is a saturated liquid after passing through the condenser. Determine the thermal efficiency of the cycle.

7.58 A geothermal power production plant produces 7 MW power. Inlet temperature of the hot geothermal liquid source is 150°C. The flow rate of hot liquid water is 220 kg/s. The reference state is at 25°C. Estimate the second law efficiency of the plant.

7.59 A reheat Rankine cycle is used in a steam power plant. Steam enters the high-pressure turbine at 9000 kPa and 823.15 K and leaves at 4350 kPa. The steam is reheated at constant pressure to 823.15 K. The steam enters the low-pressure turbine at 4350 kPa and 823.15 K. The discharged steam from the low-pressure turbine is at 10 kPa. The net power output of the turbine is 65 MW. The isentropic turbine efficiency is 80%. The pump efficiency is 95%. In the boiler, heat is transferred into the steam from a source at 1600 K. In the condenser, heat is discharged to the surroundings at 298 K. The condenser operates at 298 K. Determine the thermal efficiency.

7.60 A reheat Rankine cycle is used in a steam power plant. Steam enters the high-pressure turbine at 10000 kPa and 823.15 K and leaves at 4350 kPa. The steam is reheated at constant pressure to 823.15 K. The steam enters the low-pressure turbine at 4350 kPa and 823.15 K. The discharged steam from the low-pressure turbine is at 15 kPa. The net power output of the turbine is 65 MW. The isentropic turbine efficiency is 80%. The pump efficiency is 95%. In the boiler, heat is transferred into the steam from a source at 1600 K. The condenser operates at 298 K. Determine the thermal efficiency.

7.61 A steam power plant is using an actual regenerative Rankine cycle. Steam enters the high-pressure turbine at 11000 kPa and 773.15 K, and the condenser operates at 10 kPa. The steam is extracted from the turbine at 475 kPa to heat the water in an open heater. The water is a saturated liquid after passing through the water heater. The work output of the turbine is 90 MW. The pump efficiency is 95% and the turbine efficiency is 75%. In the boiler, heat is transferred into the steam from a source at 1700 K. In the condenser, heat is discharged to the surroundings at 285 K. Determine the thermal efficiency.

7.62 A steam power plant uses an actual regenerative Rankine cycle. Steam enters the high-pressure turbine at 10000 kPa and 773.15 K, and the condenser operates at 15 kPa. The steam is extracted from the turbine at 475 kPa to heat the water in an open heater. The water is a saturated liquid after passing through the water heater. The work output of the turbine is 90 MW. The pump efficiency is 90% and the turbine efficiency is 80%. In the boiler, heat is transferred into the steam from a source at 1700 K. In the condenser, heat is discharged to the surroundings at 285 K. Determine the thermal efficiency.

7.63 A steam power plant uses an actual reheat regenerative Rankine cycle. Steam enters the high-pressure turbine at 11000 kPa and 773.15 K, and the condenser operates at 10 kPa. The steam is extracted from the turbine at 2000 kPa to heat the water in an open heater. The steam is extracted at 475 kPa for process heat. The water is a saturated liquid after passing through the water heater. The work output of the turbine is 90 MW. The turbine efficiency is 80%. The pumps operate isentropically. In the boiler, heat is transferred into the steam from a source at 1700 K. In the condenser, heat is discharged to the surroundings at 290 K. Determine the thermal efficiency of the plant.

7.64 A steam power plant uses an actual reheat regenerative Rankine cycle. Steam enters the high-pressure turbine at 10000 kPa and 773.15 K, and the condenser operates at 15 kPa. The steam is extracted from the turbine at 2000 kPa to heat the water in an open heater. The steam is extracted at 475 kPa for process heat. The water is a saturated liquid after passing through the water heater. The work output of the turbine is 85 MW. The turbine efficiency is 80%. The pumps operate isentropically. In the boiler, heat is transferred into the steam from a source at 1700 K. In the condenser, heat is discharged to the surroundings at 290 K. Determine the thermal efficiency of the plant.

7.65 A steam power plant operates on a regenerative cycle. Steam enters the turbine at 700 psia and 800°F and expands to 1 psia in the condenser. Part of the steam is extracted at 60 psia. The efficiencies of the turbine and pump are 0.80 and 0.95, respectively. If the mas flow rate of steam is 9.75 lb/s, estimate the thermal efficiency of the turbine.

7.66 A steam power plant operates on a regenerative cycle. Steam enters the turbine at 750 psia and 800°F and expands to 5 psia in the condenser. Part of the steam is extracted at 60 psia. The efficiencies of the turbine and pump are 0.80 and 0.90, respectively. If the mas flow rate of steam is 10.5 lb/s, estimate the thermal efficiency of the turbine.

7.67 A cogeneration plant shown below uses steam at 900 psia and 1000°F to produce power and process heat. The steam flow rate from the boiler is 16 lb/s. The process requires steam at 70 psia at a rate of 3.2 lb/s supplied by the expanding steam in the turbine. The extracted steam is condensed and mixed with the water output of the condenser. The remaining steam expands from 70 psia to the condenser pressure of 3.2 psia. In the boiler, heat is transferred into the steam from a source at 3000 R. In the condenser, heat is discharged to the surroundings at 540 R. If the turbine operates with an efficiency of 80% and the pumps with an efficiency of 85%, determine the thermal efficiency of the cycle.

7.68 A power plant is operating on an ideal Brayton cycle with a pressure ratio of $r_p = 9$. The fresh air temperature is 300 K at the compressor inlet and 1200 K at the end of the compressor and at the inlet of the turbine. Using the standard-air assumptions, determine the thermal efficiency of the cycle.

7.69 A power plant is operating on an ideal Brayton cycle with a pressure ratio of $r_p = 9$. The fresh air temperature is 300 K at the compressor inlet and 1200 K at the end of the compressor and at the inlet of the turbine. Assume the gas-turbine cycle operates with a compressor efficiency of 80% and a turbine efficiency of 80%. Determine the thermal efficiency of the cycle.

7.70 The net work of a power cycle is 8×10^6 Btu/s and the heat transfer to the cold reservoir, q_C, is 12×10^6 Btu/s. The hot source operates at 1400 R and the cold source temperature is 560 R. What is the ratio of the achieved thermal efficiency to the maximum thermal efficiency of the cycle?

7.71 The net work of a power cycle is 25 MW and the heat transfer to the cold reservoir, q_C, is 67 MW. The hot source operates at 573 K and the cold source temperature is 285. What is the ratio of the achieved thermal efficiency to the maximum thermal efficiency of the cycle?

7.72 The net work of a power cycle is 52000 Btu/s and the heat transfer to the cold reservoir, q_C, is 69000 Btu/s. The hot source operates at 1450 R and the cold source temperature is 550R. What is the ratio of the achieved thermal efficiency to the maximum thermal efficiency of the cycle?

7.73 The net work of a power cycle is 75000 Btu/s and the heat transfer to the cold reservoir, q_C, is 85000 Btu/s. The hot source operates at 1500 R and the cold source temperature is 530 R. What is the ratio of the achieved thermal efficiency to the maximum thermal efficiency of the cycle?

7.74 An ideal Otto cycle operates with a compression ratio (V_{max}/V_{min}) of 8.9. Air is at 101.3 kPa and 300 K at the start of compression (state 1). The maximum and minimum temperatures in the cycle are 1360 K and 300 K. Specific heats depend on the temperature. Determine the thermal efficiency of the cycle and the thermal efficiency of a Carnot cycle working between the same temperature limits.

7.75 An ideal Otto cycle operates with a compression ratio (V_{max}/V_{min}) of 9.2. Air is at 101.3 kPa and 300K at the start of compression (state 1). During the constant volume heat addition process, 730 kJ/kg of heat is transferred into the air from a source at 1900 K. Heat is discharged to the surroundings at 280K. Determine the thermal efficiency of energy conversion.

7.76 An ideal Otto cycle operates with a compression ratio (V_{max}/V_{min}) of 9. Air is at 101.3 kPa and 295 K at the start of compression (state 1). During the constant volume heat addition process, 900 kJ/kg of heat is transferred into the air from a source at 1800 K. Heat is discharged to the surroundings at 295 K. Determine the thermal efficiency of the energy conversion.

7.77 An ideal Otto cycle operates with a compression ratio ($= V_{max}/V_{min}$) of 8.5. Air is at 101.3 kPa and 285 K at the start of compression (state 1). During the constant volume heat addition process, 1000 kJ/kg of heat is transferred into the air from a source at 1800 K. Heat is discharged to the surroundings at 280 K. Determine the thermal efficiency of energy conversion.

7.78 An ideal Otto cycle operates with a compression ratio ($= V_{max}/V_{min}$) of 10. Air is at 101.3 kPa and 295 K at the start of compression (state 1). During the constant volume heat addition process, 1000 kJ/kg of heat is transferred into the air from a source at 1800 K. Heat is discharged to the surroundings at 295 K. Determine the thermal efficiency of the energy conversion.

7.79 An ideal Otto cycle operates with a compression ratio ($r = V_{max}/V_{min}$) of 8. Air is at 101.3 kPa and 300 K at the start of compression (state 1). The maximum and minimum temperatures in the cycle are 1300 K and 300 K, respectively. Determine the thermal efficiency of energy conversion and the thermal efficiency of the Carnot engine. The average specific heats are $C_{p,av} = 1.00$ kJ/kg K and $C_{v,av} = 0.717$ kJ/kg K.

7.80 An ideal Otto cycle operates with a compression ratio ($r = V_{max}/V_{min}$) of 8.8. Air is at 101.3 kPa and 300 K at the start of compression (state 1). The maximum and minimum temperatures in the cycle are 1400 K and 300 K, respectively. Determine the thermal efficiency of energy conversion and the thermal efficiency of the Carnot engine. The average specific heats are $C_{p,av} = 1.00$ kJ/kg K and $C_{v,av} = 0.717$ kJ/kg K.

7.81 One kmole of carbon dioxide is initially at 1 atm and −13°C, performs a power cycle consisting of three internally reversible processes in series. Step 1-2: Adiabatic compression to 5 atm. Step 2-3: Isothermal expansion to 1 atm. Step 3-1: Constant-pressure compression. Determine the net work, in Btu per lb_m and the thermal efficiency.

7.82 An ideal Diesel cycle has an air-compression ratio of 18 and operating with maximum temperature of 2660 R. At the beginning of the compression, the fluid pressure and temperature are 14.7 psia, 540, respectively. The average specific heats of air at room temperature are $C_{p,av} = 0.24$ Btu/lb R and $C_{v,av} = 0.171$ Btu/lb R. Utilizing the cold-air-standard assumptions, estimate the thermal efficiency.

7.83 An ideal Diesel cycle has an air-compression ratio of 18 and operating with maximum temperature of 2200 K. At the beginning of the compression, the fluid pressure and temperature are 100 kPa, 290 K, respectively. The average specific heats of air at room temperature are $C_{p,av} = 1.0$ kJ/kg K and $C_{v,av} = 0.\ 718$ kJ/kg K. Utilizing the cold-air-standard assumptions, estimate the thermal efficiency of the cycle.

7.84 An ideal Diesel cycle has an air-compression ratio of 16 and a cutoff ratio of 2. At the beginning of the compression, the fluid pressure, temperature, and volume are 100 kPa, 300 K, respectively. The average specific heats of air at room temperature are $C_{p,av} = 1.005$ kJ/kg K, $C_{v,av} = 0.7181$ kJ/kg K. Utilizing the cold-air-standard assumptions, determine the thermal efficiency of the cycle.

7.85 An ideal Diesel cycle has an air-compression ratio of 16 and a cutoff ratio of 2. At the beginning of the compression, the fluid pressure, temperature, and volume are 100 kPa, 290 K, respectively. The average specific heats of air at room temperature are $C_{p,av} = 1.005$ kJ/kg K, $C_{v,av} = 0.7181$ kJ/kg K. Utilizing the cold-air-standard assumptions, determine the thermal efficiency of the cycle.

7.86 A refrigerator using tetrafluoroethane (R-134a) as refrigerant operates with a capacity of 10000 Btu/h. The refrigerated space is at 15°F. The evaporator and condenser operate with a 10°F temperature difference in their heat transfer. Cooling water enters the condenser at 70°F. Therefore, the evaporator is at 5°F, and the condenser is at 80°F. Determine the ideal and actual power necessary if the compressor efficiency is 85%.

7.87 In a refrigeration cycle, the superheated R-134a (state 2) enters a compressor at 200°F and 90.0 psia. The R-134a (state 3) leaves the compressor at 360°F and 140 psia, and enters a condenser, where it is cooled by cooling water. The R-134a (state 4) leaves the condenser at 90.5°F and 120 psia as saturated liquid and enters a throttling valve. The partially vaporized R-134a (state 1) leaves the valve at 100 psia. The cycle of R-134a is completed when it passes through an evaporator to absorb heat from the matter to be refrigerated. The flow rate of R-134a is 0.2 lb/s. The total power input is 85 Btu/s. The cooling water enters the condenser at 80°F and leaves at 115°F. The surroundings are at 210°F. Determine the overall exergy loss.

7.88 In a tetrafluoroethane (R-134a) refrigeration cycle, the superheated R-134a (state 1) enters a compressor at 253.15 K and 0.14 MPa. The R-134a (state 2) leaves the compressor at 303.15 K and 0.5 MPa, and enters a condenser, where it is cooled by cooling water. The R-134a (state 3) leaves the condenser at 299.87 K and 0.75 MPa and enters a throttling valve. The partially vaporized R-134a (state 4) leaves the valve at 0.205 MPa. The cycle is completed by passing the R-134 through an evaporator to absorb heat from the matter to be refrigerated. The R-134a (state 1) leaves the evaporator as superheated vapor. The flow rate of R-134a is 0.16 kg/s. The total power input is 750 kW. Estimate the coefficient of performance.

7.89 A refrigerator using tetrafluoroethane (R-134a) as a refrigerant operates with a capacity of 250 Btu/s. Cooling water enters the condenser at 70°F. The evaporator is at 10°F, and the condenser is at 80°F. The refrigerated space is at 20°F. Determine the ideal and actual power necessary if the compressor efficiency is 75%. Assume that the kinetic and potential energy changes are negligible.

7.90 A refrigerator using tetrafluoroethane (R-134a) as a refrigerant operates with a capacity of 2500 kW. Cooling water enters the condenser at 280 K. Evaporator is at 271.92 K, and the condenser is at 299.87 K. The refrigerated space is at 280 K. Determine the ideal and actual power necessary if the compressor efficiency is 80%. Assume that the kinetic and potential energy changes are negligible.

7.91 A refrigeration cycle has a COP = 3.0. For the cycle, q_H = 2000 kJ. Determine q_C and W_{net}, each in kJ.

7.92 A refrigeration cycle has a COP = 2.5. For the cycle, q_H = 1500 kJ. Determine q_C and W_{net}, each in kJ.

7.93 In a pentafluoroethane (R-125) refrigeration cycle, the saturated R-125 (state 1) enters a compressor at 250 K and 3 bar. The R-125 (state 2) leaves the compressor at 320 K and 23.63 bar, and enters a condenser, where it is cooled by cooling water. The R-125 (state 3) leaves the condenser as saturated liquid at 310 K and 18.62 bar and enters a throttling valve. The partially vaporized R-125 (state 4) leaves the valve at 255 K and 3.668 bar. The cycle is completed by passing the R-125 through an evaporator to absorb heat from the matter to be refrigerated. The R-125 leaves the evaporator as saturated vapor. The evaporator temperature is 275.15 K. The flow rate of R-125 is 0.75 kg/s. The total power input is 60 kW. The cooling water enters the condenser at 293.15 K and leaves at 295.15K. The surroundings are at 298.15 K. Determine the coefficient of performance.

7.94 A refrigerator using tetrafluoroethane (R-134a) as refrigerant operates with a capacity of 10000 Btu/h. The refrigerated space is at 15°F. The evaporator and condenser operate with a 10°F temperature difference in their heat transfer. Cooling water enters the condenser at 70°F. Therefore, the evaporator is at 5°F, and the condenser is at 80°F. Determine the coefficient of performance if the compressor efficiency is 85%.

7.95 In a refrigeration cycle, the superheated R-134a (state 2) enters a compressor at 200°F and 90.0 psia. The R-134a (state 3) leaves the compressor at 360°F and 140 psia, and enters a condenser, where it is cooled by cooling water. The R-134a (state 4) leaves the condenser at 90.5°F and 120 psia as saturated liquid and enters a throttling valve. The partially vaporized R-134a (state 1) leaves the valve at 100 psia. The cycle of R-134a is completed when it passes through an evaporator to absorb heat from the matter to be refrigerated. The flow rate of R-134a is 0.2 lb/s. The total power input is 85 Btu/s. The cooling water enters the condenser at 80°F and leaves at 115°F. The surroundings are at 210°F. Determine the coefficient of performance.

7.96 In a tetrafluoroethane (R-134a) refrigeration cycle, the superheated R-134a (state 1) enters a compressor at 253.15 K and 0.14 MPa. The R-134a (state 2) leaves the compressor at 303.15K and 0.5 MPa, and enters a condenser, where it is cooled by cooling water. The R-134a (state 3) leaves the condenser at 299.87 K and 0.75 MPa and enters a throttling valve. The partially vaporized R-134a (state 4) leaves the valve at 0.205 MPa. The cycle is completed by passing the R-134 through an evaporator to absorb heat from the matter to be refrigerated. The R-134a (state 1) leaves the evaporator as superheated vapor. The flow rate of R-134a is 0.16 kg/s. The total power input is 750 kW. Estimate the coefficient of performance.

7.97 A refrigerator using tetrafluoroethane (R-134a) as a refrigerant operates with a capacity of 250 Btu/s. Cooling water enters the condenser at 70°F. The evaporator is at 10°F, and the condenser is at 80°F. The refrigerated space is at 20°F. Determine the coefficient of performance if the compressor efficiency is 75%. Assume that the kinetic and potential energy changes are negligible.

7.98 A refrigerator using tetrafluoroethane (R-134a) as refrigerant operates with a capacity of 2500 kW. Cooling water enters the condenser at 280 K. Evaporator is at 271.92K, and the condenser is at 299.87 K. The refrigerated space is at 280 K. Determine the ideal and actual power necessary if the compressor efficiency is 80%. Assume that the kinetic and potential energy changes are negligible.

References

1. Aye L (2014) Heat pumps. In: Anwar S (ed) Encyclopedia of energy engineering and technology, 2nd edn. CRC Press, Boca Raton
2. Çengel YA, Boles MA (2014) Thermodynamics: an engineering approach, 8th edn. McGraw-Hill, New York
3. Çengel YA, Turner R, Cimbala J (2011) Fundamentals of thermal-fluid sciences, 4th edn. McGraw-Hill, New York
4. Chattopadhyay P (2015) Engineering thermodynamics, 2nd edn. Oxford University Press, Oxford

5. Demirel Y (2013) Thermodynamics analysis. Arab J Sci Eng 38:221–249
6. Demirel Y, Gerbaud V (2025) Nonequilibrium thermodynamics. In: Transport and rate processes in physical, chemical and biological systems, 5th edn. Elsevier, Amsterdam
7. IRENA (2020) Global renewables outlook: energy transformation 2050 (edition: 2020). International Renewable Energy Agency, Abu Dhabi
8. Jaffe RL, Taylor W (2018) The physics of energy. Cambridge University Press, Cambridge
9. Küçük K, Tevatia R, Sorgüven E, Demirel Y, Özilgen M (2015) Bioenergetics of growth and lipid production in *Chlamydomonas reinhardtii*. Energy 83:503–505
10. Li X, Gholamreza Karimi G (2014) Fuel cells: intermediate and high temperature. In: Anwar S (ed) Encyclopedia of energy engineering and technology, 2nd edn. CRC Press, Boca Raton
11. Wang X, Demirel Y (2018) Feasibility of power and methanol production by an entrained-flow coal gasification system. Energy Fuels 32:7595–7610. https://doi.org/10.1021/acs.energyfuels.7b03958

Energy Storage 8

Introduction and Learning Objectives: Devices or physical media can store some form of energy to perform a useful operation later or at a different location. Energy storage reduces the mismatches between the energy production and demand. For example, if it is stored the solar energy would still be available during the night. Also, the stored energy may be a supplement during the peak demand for energy. Besides, a stored energy can be transported. A battery, for example, makes it possible to use many electronic devices. This chapter begins by underlying the importance of energy storage and discusses thermal, electric, chemical, and mechanical energy storage systems. Solar energy storage by sensible and/or latent heat and for short- and long-term applications is discussed briefly. Some common phase changing materials and usage of them for the latent heat storage technique are described. Underground thermal energy systems are discussed briefly. Capacitor, hydroelectric, and battery are discussed in storing electricity. Chemical energy storage by biosynthesis is briefly discussed. Later, mechanical energy storage by compressed air, flywheel, hydraulic, and springs is discussed.

The learning objectives of this chapter are:

- Thermal energy storage
- Phase changing materials
- Electrical energy storage
- Chemical energy storage
- Mechanical energy storage

Y. Demirel, *Energy*, Green Energy and Technology,
https://doi.org/10.1007/978-3-031-97111-2_8

8.1 Energy Storage

Devices or physical media can store some form of energy to perform a useful operation later time. A battery stores readily convertible chemical energy to operate a mobile phone, a hydroelectric dam stores energy in a reservoir as gravitational potential energy, and ice storage tanks store thermal energy to meet peak demand for cooling. Battery and fossil fuels such as coal and gasoline store chemical energy. Synthesis of some energy-rich biomolecules such as adenosine triphosphate stores and transforms chemical energy in living systems. Sunlight is also captured by plants as *chemical potential energy* when carbon dioxide and water are converted into of carbohydrates.

Two important physical media for storage and regulation of thermal energy are water and hydrogen. Table 8.1 lists some physical properties of water and hydrogen, including the heats of formation of water in liquid and vapor states. Their material and physical properties make them unique in not only storing and transforming energy but also regulation of the temperature of the earth. Hydrogen represents a store of potential energy which can be released by nuclear fusion in the sun in the form of light. Solar energy may be stored as chemical energy by the photosynthesis after it strikes the earth. Also, when water evaporates from oceans and is deposited high above sea level, it drives turbine/generators to produce electricity, and hence sunlight may again be stored as gravitational potential energy [2, 3].

8.1.1 Ecological Regulation by Water

Approximately 80% of the earth's surface is water. Water has a high specific heat capacity of 4.185 kJ/kg °C, which is higher than that of soil and rock (see Table 8.1). This enables the water to store and retain large amount of heat making the oceans one of the Earth's temperature regulators. Therefore, the temperature of water changes more slowly than that of earth surface. Beside that water has large heat of vaporization (2257 kJ/kg at 20°C) than can be transferred over long distances in vapor form and released after condensing as rain (see Fig. 8.1). Accumulation of the rain at reservoirs stores large amount of potential energy, which drives turbines to produce electricity. The oceans control the earths' energy balance as it absorbs nearly four times more insolation than the earth.

Table 8.1 Physical properties of water and hydrogen [1]

Compound	MW	ρ kg/m^3	$C_{p,av}$ kJ/kg K	T_m °C	T_b °C	ΔH_v kJ/kg	ΔH_m kJ/kg
Water (H_2O)	18.02	1000	4.18	0.0	100.0	2257	333.7
Hydrogen (H_2)	2.02	70.7	10.0	– 259.2	– 252.8	445.7	59.5

MW Molecular weight, T_m Temperature of melting, $C_{p,av}$ Average heat capacity, ρ Density, ΔH_m Heat of fusion, ΔH_{vap} Heat of vaporization

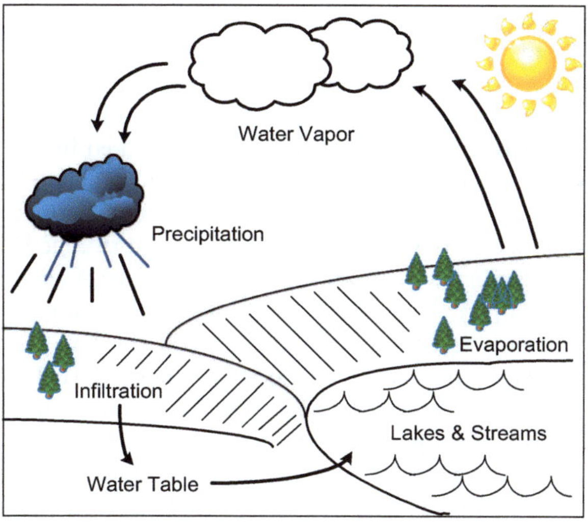

Fig. 8.1 Water circulation on the earth

The liquid water also helps to control the amount of carbon dioxide in the atmosphere by dissolving it into "carbonic acid" (soda water). Some of the dissolved carbon dioxide would combine with minerals in the water and settle to the ocean floor to form limestone. Water is also essential for all life forms as most animals and plants contain more than 60% water by volume, which helps organisms to regulate their body temperature. Therefore, water regulates the temperature of the earth as well as the temperature of living systems [1, 7].

8.1.2 Hydrogen

Hydrogen is a versatile storage medium for its conversion to useful work in a fuel cell to produce electricity and can be an electrical power storage medium. Hydrogen can also be used in an internal combustion engine as a clean fuel emitting water only. Hydrogen must first be manufactured by other energy sources and may become a significant energy storage medium for renewable energies including solar and wind power. Hydrogen can be manufactured by steam reforming of fossil fuels or by electrolysis of water using renewable power. Large quantities of gaseous hydrogen can be stored in underground caverns and depleted oil or gas fields for many years and can function as multi-purpose energy storage [10].

8.2 Types of Energy Storage

A method of energy storage may be chosen based on stability, ease of transport, energy capture and release. Main energy storage technologies involve the store of energy in thermal, electrical, chemical, and mechanical forms:

- *Thermal*—Thermal energy can be stored by sensible and latent heats at temperatures above or below the ambient temperatures. Heat is also stored for a short or long-term applications. Thermal energy from the sun, for example, can be captured by solar collectors and stored in a reservoir for daily or seasonal use. Thermal energy for cooling can be stored in ice.
- *Electrical*—Dams can be used to store hydroelectricity by accumulating large amounts of water at high altitude. The water then turns a turbine and generates electricity. Also by using pumped-storage hydroelectricity stores energy by pumping water back into the reservoir. A capacitor can store electric charge as a part of electrical energy production system. Battery can also store electric energy to be used in portable electrical devices.
- *Chemical*—Stable chemical compounds such as fossil fuels store chemical energy. Renewable energy can be stored in chemical and fuel productions [10]. Biological systems can store energy in chemical bonds of energy-rich molecules, such as glucose and adenosine triphosphate (ATP). Other forms of chemical energy storage include hydrogen, synthetic hydrocarbon fuel, and batteries.
- *Mechanical*—Energy can be stored in pressurized gases such as in compressed air, and used to operate vehicles and power tools. Hydraulic accumulator, flywheel, and springs can also store mechanical energy.

8.3 Thermal Energy Storage

The thermal storage medium may be maintained at a temperature above (hotter) or below (colder) than that of the ambient temperature of 25°C. In hot climates, the primary applications of thermal energy storage are cold storage because of large electricity demand for air conditioning. Thermal energy storage may be planned for short term or seasonal. Some advantages of utilizing thermal energy storage are [2, 3, 7]:

- Reduced energy consumptions and carbon footprint
- Reduced initial equipment and maintenance costs
- Reduced pollutant emissions such as CO_2
- Increased flexibility of operation, efficiency, and effectiveness of equipment utilization
- Process application in portable and rechargeable way at the required temperature
- Isothermal and higher storage capacity per unit weight
- Energy from any source (thermal or electrical) when needed

Thermal energy is generally stored in the form of sensible heat and latent heat. Figure 8.2 compares the characteristics of heat storage by sensible and latent heats. Temperature keeps increasing as a material absorbs and store sensible heat until it reaches a temperature where a phase changing occurs. Under their melting points, the solid–liquid phase changing materials store sensible heat and their temperature rises as they absorb the heat (see Fig. 8.2a). When phase changing materials reach their melting temperature they absorb large amounts of latent heat at an almost constant temperature until melting is completed. If the outside heat supply continues, then the melted material starts to store sensible heat. A material may freeze at a temperature lower than its actual freezing temperature. This is called the *subcooling*. Also, a material may melt at a temperature higher than its actual melting temperature. This is called the *superheating*. Subcooling and superheating of a phase changing material create an interval for temperature control (see Fig. 8.2b). This temperature interval between the subcooling and superheating may be reduced by adding nucleating agents to phase changing material.

Stored heat is usually transferred to a *heat transfer fluid* such as water or air in a heat exchanger system as seen in Fig. 8.3. When air is used as heat transfer fluid, the cold inside room air extracts the heat as it flows through a heat storage system and the warm air out of the storage is fed directly into the room.

Nighttime low-cost electricity may be used to store energy as sensible heat or/and latent heat. The stored energy is then used for the cooling/heating needs during the peak demand hours. Thermal energy storage systems can be designed using chilled water, ice, and encapsulated phase change material. Phase change material

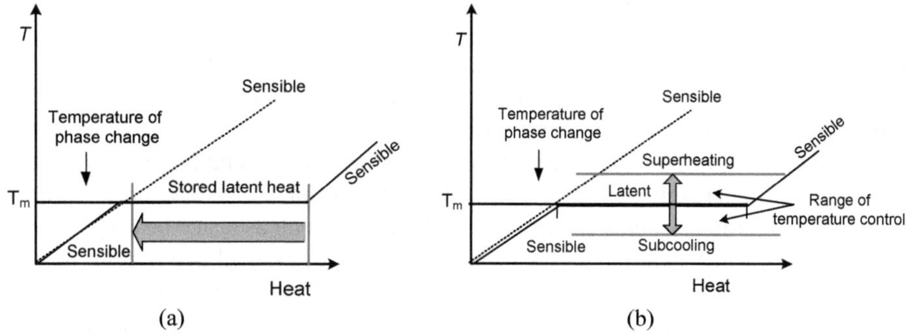

Fig. 8.2 **a** Comparison of sensible heat with latent heat storage, **b** temperature control during phase change energy storage; subcooling and super heating of a phase changing material creates a range for temperature control; this range may be reduced by adding nucleating agents to phase changing material

Fig. 8.3 Direct space heating using air as the heat transfer fluid. Cold inside room airflows through a heat storage system and the warm air is fed directly into the room for heating

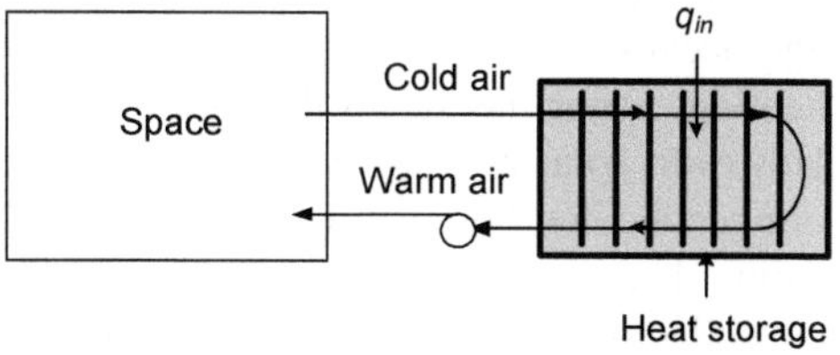

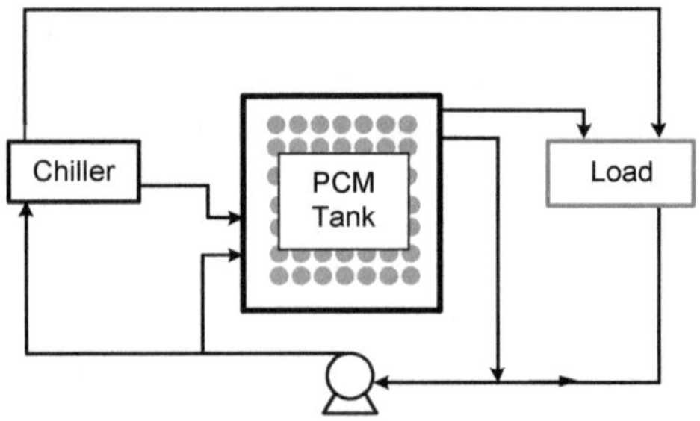

Fig. 8.4 Configuration of a thermal energy storage system with phase change material (PCM)

modules installed in a chilled water tank has proven to increase the energy density and improve the performance of water and ice based thermal energy storage systems. Figure 8.4 shows a typical arrangement of cooling using phase change material. A phase change material in the supplementary tank is charged by the chiller during off-peak hours. During peak demand hours the cooling is achieved by the stored energy in the phase change material energy storage tank as the heat exchange fluid passes through phase change material tank. This stabilizes the load during peak demand hours and shifts the load to the off-peak period [13].

8.3.1 Solar Energy Storage

Due to time-dependence of sun exposure, the efficiency of solar-thermal systems relies on the well-integrated thermal energy storage technology. For example, in utilizing solar water collectors with thermal energy storage, the higher efficiency is achieved by bridging the gap between solar heat availability and hot water demand. Solar-thermal energy is usually captured by active solar collectors and transferred to insulated storage systems for various applications, such as space heating, domestic or process water heating, as seen in Fig. 8.5. Most practical active solar heating systems have storage for a few hours to a day's worth of energy collected. There are also a growing number of seasonal thermal storage systems being used to store summer energy for space heating during winter. Water-based technology for thermal energy storage is practical because of the large heat of fusion of water (see Table 8.1). The original definition of a "ton" of cooling capacity was the heat to melt one ton of ice every 24 h. This definition has since been replaced by one ton heating ventilation and air-conditioning (HVAC) capacity of 12,000 Btu/h [9].

Many of the early applications utilized solar water collectors with water storage because of the advantages of increased efficiency and reduced size. For home heating, water has to go through a heat exchanger to release its heat to a room. However, solar heating systems that use air as the transfer medium are being recommended for home usage because of direct heating, less potential for damage from a leak or frozen water. In *direct heating*, the warm air coming from the solar air collector can be fed to a room for heating. Also air collectors and ducting are usually cheaper and require less maintenance [6, 13].

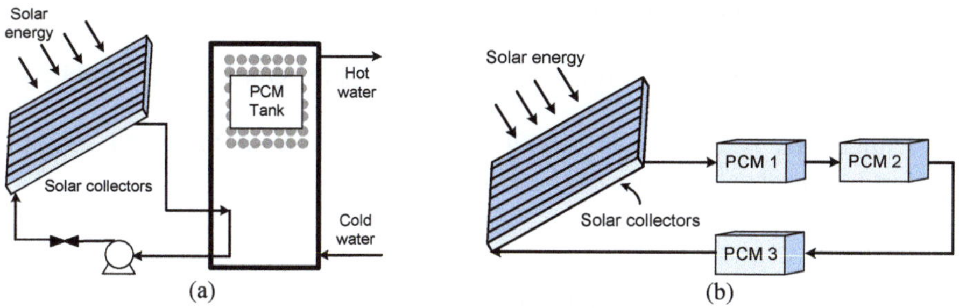

Fig. 8.5 Configuration of solar energy storage in the forms of sensible and latent heats: **a** solar water heater with heat storage with phase changing material (PCM), **b** hot and cold energy storage of solar energy using various phase changing materials working at high and low temperatures; PCM 1 to PCM 3 refer to phase changing materials with different heat storage characteristics

8.3.2 Sensible Heat Storage

Sensible heat is stored by raising its temperature of a material at constat pressure is

$$q = mC_{\mathrm{p,av}}\Delta T = \rho V C_{\mathrm{p,av}}\Delta T \tag{8.1}$$

where q is the sensible heat stored, V is the volume, ρ is the density, m is the mass, $C_{\mathrm{p,av}}$ is the average specific heat capacity of material, and ΔT is the temperature change. Energy density or volumetric heat capacity is defined by the product of density and heat capacity: $\rho C_{\mathrm{p,av}}$. Table 8.2 shows some materials used for heat storage in the form of sensible heating. Stored heat is usually transferred by a heat transfer fluid such as water or air to the location where to be used as seen in Fig. 8.5. Example 8.1 illustrates the sensible heat storage calculations.

Table 8.2 Some materials used for heat storage in the form of sensible heating, volumetric heat capacity, or energy density are defined by $\rho C_{\mathrm{p}} = \mathrm{J/m^3\,K}$ [6, 13]

Material	Temperature range °C	Density ρ kg/m^3	Heat capacity $C_{\mathrm{p,av}}$ J/kg °C	Energy density $\rho C_{\mathrm{p,av}}$ kJ/m^3°C
50% Ethylene glycol 50% water[a]	0–100	1075	3480	3741
Dowtherm A	12–260	867	2200	1907
Therminol 66	– 9–343	750	2100	1575
Water	0–100	1000	4190	4190
Granite	–	2400	790	1896
Draw salt: 50% $NaNO_3$–50%KNO_3	220–540	1733	1550	2686
Molten salt: 50% KNO_3–40% $NaNO_2$–7% $NaNO_3$[a]	142–540	1680	1560	2620

[a] Weight percentages

***Example 8.1* Sensible heat storage calculations**

(a) Estimate sensible heat stored in 2 m^3 water and 2 m^3 granite heated from 20 to 40°C.
(b) Estimate the heat stored in 1000 lb of water heated from 20 to 30°F. Assume that average heat capacity is: $C_{p,av} = 1.0\,Btu/lb°F$.

Solution:

Assume that the values of heat capacity remain constant.

(a) Sensible heat storage by water and granite:

Use Eq. (8.1) with the data for heat capacity $C_{p,av}$ and density ρ from Table 8.2:

$$\text{Water: } q = \rho V C_{p,av} \Delta T$$
$$= (1000\,kg/m^3)(2\,m^3)(4190 J/kg°C)(40 - 20)°C = 167{,}600\,kJ.$$

$$\text{Granite : } q = \rho V C_{p,av} \Delta T$$
$$= (2400\,kg/m^3)(2\,m^3)(790\,J/kg°C)(40 - 20)°C = 75840\,kJ.$$

(b) Heat stored in 1000 lb of water when the average specific heat of water is 1.0 Btu/lb °F:

$$q = m C_{p,av} \Delta T = (1000\,lb)(1.0\,Btu/lb°F)(30 - 20)°F = 10000\,Btu = 10{,}550\,kJ.$$

8.3.3 Latent Heat Storage by Phase Changing Material

Latent heat storage can be achieved through solid–solid, solid–liquid, solid–gas, and liquid–gas phase changes. Liquid–gas transitions have a higher heat of transformation than solid–liquid transitions. However, liquid–gas phase changes are not practical because of the large volumes or high pressures required to store the materials when they are in their gas phase states. Solid–solid phase changes are typically very slow and have a rather low heat of transformation. Therefore, the main phase change used for heat storage is the solid–liquid change. Figure 8.6 shows the change of enthalpy of water from solid to vapor state on enthalpy-temperature diagram. Between − 30 and 0°C the sensible heat of water increases. At 0°C the water starts to melt by absorbing heat of melting. After the melting is completed, the liquid water is heated until 100°C [3, 11].

Phase changing materials used for storing heat are chemical substances that undergo a solid–liquid transition at temperatures within the desired range for heating and cooling purposes. During the transition process, the material absorbs energy as it goes from a solid to a liquid state and releases energy as it goes back to a solid from liquid state. A phase changing material should possess:

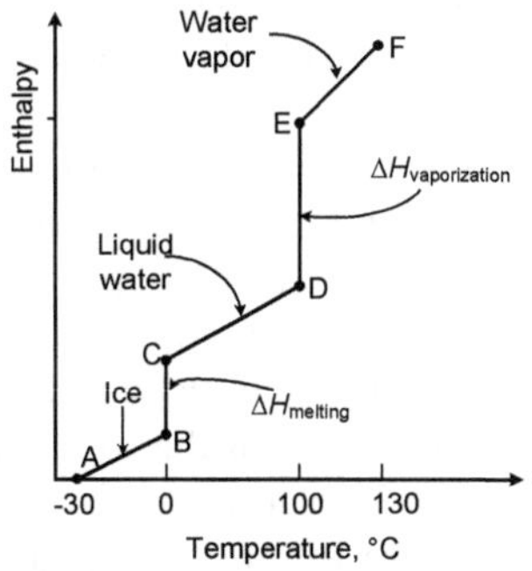

- Process A-B: Increase in sensible heat of ice between – 30°C and 0°C
- Process B-C: Phase change form solid to liquid: melting of ice at 0°C; heat of melting ($\Delta H_{\text{melting}}$) is required.
- Process C-D: Increase in sensible heat of liquid water between 0°C –100°C.
- Process D-E: Phase change from liquid to vapor at 100°C; heat of vaporization ($\Delta H_{\text{vaporization}}$) is required.
- Process E-F: Increase in sensible heat of water vapor between 100°C and 130°C.

Fig. 8.6 Changes of enthalpy values of water from solid state (ice) to vapor state between – 30 and 130°C. Between – 30 and 0°C the sensible heat of water increases. At 0°C the water starts to melt by absorbing heat of melting. After the melting process is completed, the liquid water is heated until 100°C at which the transformation from liquid to vapor starts. After absorbing the heat of vaporization all the liquid water becomes vapor. Between 100 and 130°C, the water in vapor state increases its sensible heat

- Melting temperature in the desired operating temperature range
- High latent heat of fusion per unit volume
- High specific heat, density, and thermal conductivity
- Small changes of volume and vapor pressure on phase change at operating temperatures
- Congruent melting (solid compositions is the same with the composition of the liquid melt)
- Chemical stability and complete reversible freeze/melt cycles
- Noncorrosive, nontoxic, nonflammable, and nonexplosive
- Low cost and availability

Latent heat storage can reduce temperature fluctuations and offers a higher heat storage capacity per volume/mass. The temperature and the amount of energy stored can be adjusted by selecting a specific phase changing material. Total heat stored by a solid-to-liquid phase changing material between initial and final temperatures is

$$q_{\text{stored}} = \text{Solid sensible heat} + \text{Latent heat} + \text{Liquid sensible heat}$$

$$q_{\text{stored}} = mC_{\text{ps,av}}(T_{\text{m}} - T_{\text{i}}) + m\Delta H_{\text{m}} + mC_{\text{pl,av}}(T_{\text{f}} - T_{\text{m}}) \quad (T_{\text{f}} > T_{\text{m}} > T_{\text{i}}) \quad (8.2)$$

where m is the mass of phase changing material, $C_{\text{ps,av}}$ and $C_{\text{pl,av}}$ are the average heat capacities for solid and liquid phases, respectively, T_{m} is the temperature of melting, T_{i} and T_{f} are the initial and final temperatures, respectively, and ΔH_{m} is the heat of melting. Stored heat is transferred by a heat transfer fluid such as water or air in a heat exchanger. Consider a phase changing material, which is melted; when the heat transfer fluid temperature T_{f} is lower than the melting temperature ($T_{\text{f}} < T_{\text{m}}$), the phase changing material solidifies and releases its heat of melting to the heat transfer fluid. When the temperature of heat transfer fluid is higher than the melting temperature ($T_f > T_m$), phase changing material starts melting and stores heat.

Various phase changing materials are available in any required temperature range from – 5°C up to 190°C storing 5 to 14 times more heat per unit volume than conventional storage materials such as water [3, 11]. Tables 8.3 and 8.4 present short lists of some common materials for heat storage. The most commonly used phase change materials are salt hydrates, fatty acids, esters, and various paraffins (such as octadecane). The chemical composition of salts is varied in the mixture to achieve required phase change temperature and stability in phase transformations. Special nucleating agents may be added to the mixture to minimize phase change salt separation and eliminating sub cooling. Salt hydrates have high volumetric latent heat storage capacity, sharp melting point, high heat of fusion, and thermal conductivity. Salt hydrates are nontoxic, nonflammable, and economical. They however can be corrosive and have a finite life of around 10,000 cycles. All salt-based phase changing material solutions must be encapsulated to prevent water evaporation or uptake. The packaging material should conduct heat well. It should be durable enough to withstand frequent changes in the storage material's volume as phase changes occur. Packaging must also resist leakage and corrosion. Materials of encapsulation include stainless steel, polypropylene, and polyolefin.

Organic phase changing materials, such as paraffin (C_nH_{2n+2}) and fatty acids ($CH_3(CH_2)_{2n}COOH$), freeze without much subcooling with self-nucleating properties [11]. They are chemically stable with relatively high heat of fusion, safe, nonreactive, and recyclable. On the other hand, they have low volumetric latent heat storage capacity and low thermal conductivity in their solid state so requiring high heat transfer rates during the freezing cycle. They can be expensive and

Table 8.3 Comparison of typical storage densities of various materials for energy storage in the forms of sensible and latent heats [3, 6, 11, 13]

Method/material	kJ/l	kJ/kg	Temperature °C
Sensible heat			
Granite	50	17	$\Delta T = 20$
Water	84	84	$\Delta T = 20$
Latent heat of melting			
Water	330	330	0
Lauric acid		178	42–44
Capric acid		153	32
Butyl stearate		140	19
Paraffin	180	200	5–130
Paraffin C18	196	244	28
Salthydrate	300	200	5–130
Salt	600–1500	300–700	300–800
Latent heat of evaporation			
Water	2452	2450	1 atm and 25°C

Table 8.4 Properties of some phase change materials (PCMs) [3, 6, 11, 13]

PCM	T_m °C	ΔH_m kJ/kg	$C_{pl,av}$ kJ/kg K	$C_{ps,av}$ kJ/kg K	k_l W/m K	k_s W/m K	ρ_l kg/m³	ρ_s kg/m³
Water	0	333	4.19	2.0	0.595	2.2	1000	920
n-Hexadecane	18	235	2.1	1.95	0.156	0.43	765	835
Octadecane $CH_3(CH_2)_{16}CH_3$	28	243	2.2	1.8	0.15	0.42	775	~900
n-Eicosane	36.4	248	2.4	1.92	0.146	0.426	769	910
n-Docosane $CH_3(CH_2)_{20}CH_3$	44.5	196–252	~2.5	~1.9	0.15	0.4	778	920
Paraffins	10–70	125–240	1.7	1.7	0.15	0.25	~700	~900

T_m Temperature of melting, H_m Heat of fusion, $C_{pl,av}$ Average heat capacity at liquid state, $C_{ps,av}$ Average heat capacity at solid state, ρ_l Density at liquid state; ρ_s Density at solid state

flammable at high temperatures. Sometimes mixtures of phase change materials may be more beneficial. *Thermal composite* is a term given to combinations of phase change materials and other (usually solid) structures. A simple example is a copper-mesh immersed in paraffin. Table 8.4 shows a short list of some phase change materials and their thermo-physical properties. Example 8.2 illustrates the heat storage estimations for home heating.

***Example 8.2* Solar energy storage by phase changing material**

A typical square two-story home with a roof surface area of 1260 ft^2 and a wall surface area of 2400 ft^2 is to be heated with solar energy storage using a salt hydrate as phase change material. It presently has as an insulation of 6 inches in the roof and 1 inch in the walls. Inside temperature will be held at 70°F and expected outside low temperature is 10°F. Average solar radiation is 650 Btu/ft^2. Each solar collector has the total absorber area of 20 ft^2. Estimate the total energy needed from the storage amount of salt hydrate and number of solar air collectors.

Solution:
Assume that the approximate thermal conductivity of the walls and roof is: $k = 0.025$ Btu/h °F ft^2.
Average solar radiation is 650 Btu/ft^2 and the cost of solar air collector is \$1.1/ft^2. The salt hydrate costs around \$0.15/lb and its heat of melting is 145 Btu/lb.

Heating requirement of building with present insulation:

Heat loss from roof with ($\Delta x = 6/12 = 0.5$ ft); $\Delta T = (10 - 70) = -60$°C

$$\dot{q}_{\text{loss, roof}} = -kA\frac{\Delta T}{\Delta x}$$
$$= -\left(0.025\,\text{Btu/h°F ft}^2\right)\left(1260\,\text{ft}^2\right)\left(-60\text{°F}/0.5\,\text{ft}\right) = 3780\,\text{Btu/h}$$

Heat loss from walls ($\Delta x = 1$ in $= 0.083$ft)

$$\dot{q}_{\text{loss,wall}} = -kA\frac{\Delta T}{\Delta x}$$
$$= -\left(0.025\,\text{Btu/h°F ft}^2\right)\left(2400\,\text{ft}^2\right)\left(-60°\text{F}/0.083\,\text{ft}\right) = 43{,}373\,\text{Btu/h}$$

Total heat loss = Total heating requirement

$$\dot{q}_{\text{total loss}} = 3780\,\text{Btu/h} + 43{,}373\,\text{Btu/h} = 47{,}153\,\text{Btu/h}$$

Total heat storage material needed:

$m_s = (47{,}153\,\text{Btu/h})(24\,\text{h})/145\,\text{Btu/lb} = 7804.0\,\text{lb}$

Size and cost of collector to meet present heating requirements:

Area of collectors $= (47{,}153\,\text{Btu/h})(24\,\text{h})/\left(650\,\text{Btu/ft}^2\right) = 1740\,\text{ft}^2$.

Number of collectors: $\left(1740\,\text{ft}^2\right)/\left(12\,\text{ft}^2\right) = 87$.

Annual cost of heating may be optimized by compromising between the two opposing effects of capital costs of thermal energy storage system and fuel as shown below.

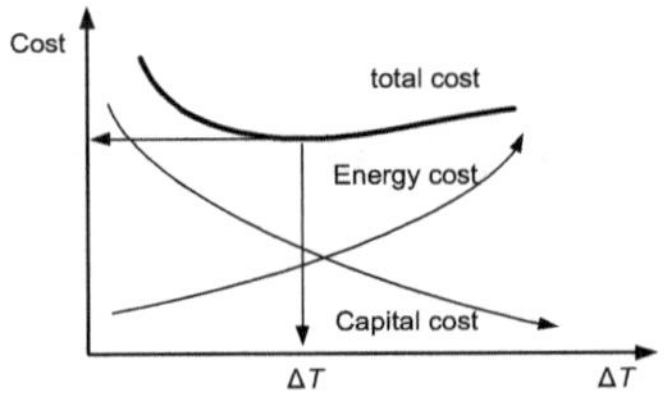

As the temperature difference between hot and cold space increases, the capital cost may decrease while the energy (fuel) cost increases. The total cost should be optimized for a required value for ΔT.

8.3.4 Cold Storage

Ice storage is the thermal energy storage using ice. It is practical because of the large heat of fusion of water. One metric ton of water (one cubic meter) can store 334 MJ or 317,000 Btu, 93 kWh, or 26.4 ton-h. The original definition of a "ton" of cooling capacity was the heat to melt one ton of ice every 24 h. One ton heating ventilation and air conditioning (HVAC) capacity is known as 12,000 Btu/h. A small storage unit can hold enough ice to cool a large building for a day or a week, by using, for example, off-peak power and other such intermittent energy source.

Cold thermal energy storage may be an economically viable technology. It has become a key component in many successful thermal systems. Cold is valuable commodity. Generally a cold energy storage system comprises three basic operations in a full cycle: charging, storing and discharging. A typical cold storage system consists of a tank containing a fixed quantity of storage fluid and heat transfer coil through which a heat transfer fluid is circulated. Usually, kinetic and potential energies and pump work are negligible compare to cold input or cold accumulation and hence ignored. Therefore, an energy balance over an entire cycle of the cold storage becomes:

$$\text{Cold input} - (\text{Cold loss} + \text{Cold recovered}) = \text{Cold accumulation}$$

Here cold input is the heat removed from the cold thermal energy storage source by the heat transfer fluid during charging, the cold recovered is the heat removed from the heat transfer fluid by the cold storage medium, the cold loss is the heat gained from the environment to the storage medium during charging, storing, and discharging, and the cold accumulation is the decrease in internal energy of the storage medium during one complete cycle [6, 13].

8.3.5 Molten Salt Technology

Molten salt can be employed as a heat store to retain heat collected by a solar tower so that it can be used to generate electricity. The molten salt is a mixture of 60% sodium nitrate and 40 percent potassium nitrate, called *saltpeter*. It is nonflammable, nontoxic and is used in the chemical and metals industries as a heat transfer fluid. The salt is kept as liquid at 288°C (550°F) in an insulated storage tank and is pumped through panels in a solar collector where the focused sun heats it to 566°C (1051°F). It is then sent to a well-insulated storage tank. When electricity is needed, the hot salt is pumped to a conventional steam-generator to produce superheated steam for a turbine/generator of a power plant [7, 13].

8.3.6 Seasonal Thermal Energy Storage

A *seasonal thermal storage* retains heat deposited during the hot summer months for use during colder winter weather for heating. For cooling, the winter is used to store cold heat to be used in the summer. The heat is typically captured using solar collectors, although other energy sources are sometime used separately or in parallel [2, 3]. Seasonal (or "annualized") thermal storage can be divided into three broad categories:

- Low-temperature systems use the soil adjoining the building as a heat store medium. At depths of about 20 feet (6 m) temperature is naturally "annualized" at a stable year-round temperature. Two basic techniques can be employed:

- In the *passive seasonal heat storage*, solar heat is directly captured by the structure's spaces through windows and other surfaces in summer and then passively transferred by conduction through its floors, walls, and sometimes, roof into adjoining thermally buffered soil. It is then *passively* returned by conduction and radiation as those spaces cool in winter.
- The *active seasonal heat storage* concept involves the capture of heat by solar collectors or geothermal sources and deposited in the earth or other storage masses or mediums. A heat transfer fluid in a coil embedded in the storage medium is used to deposit and recover heat.
- Warm-temperature systems also use soil or other heat storage mediums to store heat, but employ active mechanisms of solar energy collection in summer to store heat and extract in winter. Water circulating in solar collectors transfers heat to the storage units beneath the insulated foundation of buildings. A ground source heat pump may be used in winter to extract the heat from the storage system. As the heat pump starts with a relatively warm temperature, the coefficient of performance may be high.
- High-temperature systems are essentially an extension of the building's HVAC and water heating systems. Water or a phase change material is normally the storage medium.

Advantages of seasonal storage systems:

- Seasonal storage is a renewable energy utilization system.
- Stored thermal energy would be available in desired amount, time, and location.
- Nontoxic, noncorrosive, and inexpensive storage medium can be selected for a required application.
- Seasonal heat storage needs relatively moderate initial and maintenance costs.
- Size and capacity of heat storage system can be adjusted based on the size of space to be conditioned.
- It can be used as preheater and precooler of existing heating and cooling systems.
- It is possible to use water as heat transfer fluid during charging the storage with water solar heaters, and airflow for discharging mode.

8.3.7 Seasonal Solar-Thermal Energy Storage for Greenhouse Heating

The three common processes in a seasonal thermal energy storage for greenhouse heating are [2, 3]:

- Charging: The charging is for capturing solar energy by solar air heaters and feeding the warm air to the storage unit through the coils embedded in the heat storage tank.

- Storing: The storage process is for storing the captured solar energy by sensible and latent heat using a phase change material in a well-insulated heat storage tank.
- Discharging: The discharge process, on the other hand, is for recovering the stored heat by the cold air flowing through the heat storage tank and delivering the warm air to the greenhouse directly when necessary.

Figure 8.7 shows the schematics of the seasonal solar energy storage systems by paraffin for greenhouse heating with a data acquisition and control system to activate the required process, such as discharging process [2, 3, 12].

Figure 8.8 shows the charging and discharging operations within the three units of:

- Unit 1: Solar air heaters
- Unit 2: Heat storage
- Unit 3: Greenhouse

Ambient air is heated to around 60–75°C by the solar radiation using solar air collectors. The warm air from the solar collectors flows in spiral coils embedded inside the storage tank and heats the paraffin. When the temperature of greenhouse drops below a set point of around 14°C, the discharge process is activated so that cold air from the greenhouse flows through the hot and melted paraffin, recovers a part of the latent heat, and delivers it to the greenhouse until raising the temperature to a required level. The data acquisition system controls the activation and

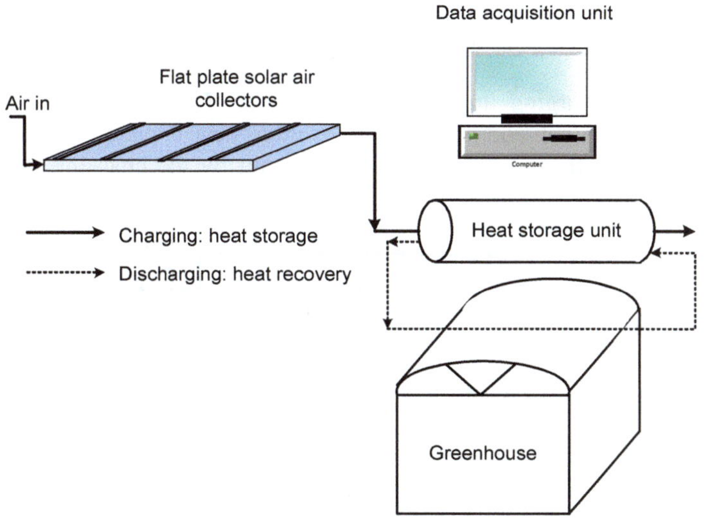

Fig. 8.7 Seasonal heat storage system using paraffin as a phase change material for heating a greenhouse [2, 12]

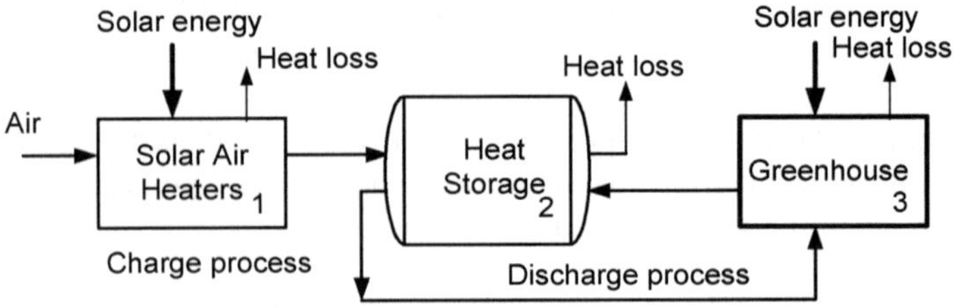

Fig. 8.8 Schematic structure with three components of the seasonal heat storage system: Unit 1 is for capturing solar energy to heat the air and send it to heat storage, Unit 2 is heat storage by sensible and latent heats of paraffin, and Unit 3 is the greenhouse. The discharge process takes place between the green house and heat storage when the temperature in the greenhouse drops below a set value

deactivation of the discharge process. Energy losses occur in various levels in all the three units [2].

Figure 8.9 shows typical temperature profiles of paraffin and airflow in charging and discharging modes. Here, the paraffin starts with the cold-solid temperature of T_{sc} and ends with the storage-hot temperature of T_{sh}, representing the cold and hot level of temperatures within the storage unit. An average temperature of T_s may be assumed for thermal energy storage calculations. The m_c is the airflow rate for charging and m_d is the airflow rate for discharging, while T_{ci} and T_{co} are the inlet and outlet temperatures for charging (loading) and T_{di} and T_{do} are inlet and outlet temperatures for discharging (recovery) process, respectively.

The latent heat storage system undergoes a temperature difference of $(T_{sh} - T_{sc})$. The solar heat captured through the series of solar collectors is

$$\dot{q}_c = \dot{m}_c C_{p,av}(T_{ci} - T_{co}) \tag{8.3}$$

where $\dot{m}_c$ is the charging fluid flow rate, T_{ci} and T_{co} are the inlet and outlet temperatures of the air, and $C_{p,av}$ is the average heat capacity of air. The heat captured will be delivered to the heat storage unit by the charging process. The

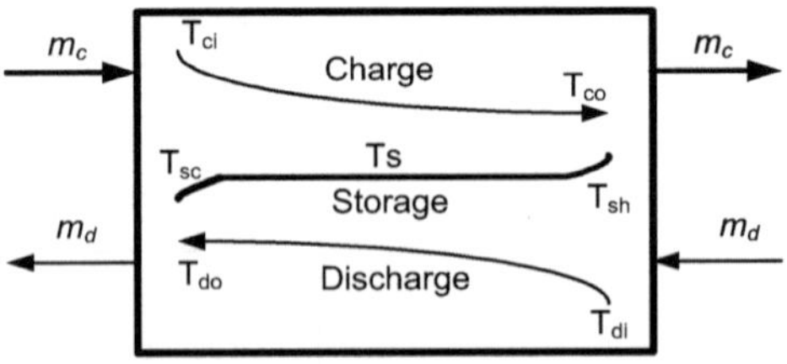

Fig. 8.9 Approximate temperature profiles of airflow and heat storage material; m_c is the airflow rate for charging and m_d is the airflow rate for discharging, T_{ci} and T_{co}: inlet and outlet temperatures for charging (loading); T_{di} and T_{do}: inlet and outlet temperatures for discharging (recovery); T_{sc} and T_{sh}: cold and hot temperatures of paraffin temperatures, respectively

heat stored by the paraffin q_s will be (i) sensible heat until the temperature of the solid paraffin reaches its lower temperature of melting, (ii) latent heat, and (iii) sensible heat until the temperature of the melted paraffin reaches the temperature T_{sh}. The amount of stored heat q_s is

$$q_s = m_s\left[C_{ps,av}(T_s - T_{sc}) + \Delta H_m + C_{pl,av}(T_{sh} - T_s)\right] \tag{8.4}$$

where ΔH_m is the heat of melting, T_{sc} and T_{sh} are the lowest and highest melting points of the paraffin, and $C_{ps,av}$ and $C_{pl,av}$ denote the average heat capacities of solid and liquid states of the paraffin, respectively. T_s is an average temperature of the paraffin, which may be approximated as $T_s = (T_{sc} + T_{sh})/2$.

The thermal efficiency is the ratio of amount of heat stored to the amount of heat captured at a certain time interval

$$\eta_s = \frac{\text{Actual heat stored}}{\text{Maximum energy gain}} = \frac{\dot{q}_s}{\dot{q}_c} \tag{8.5}$$

The recovered heat through the discharge process is

$$\dot{q}_d = \dot{m}_d C_{p,av}(T_{di} - T_{do}) \tag{8.6}$$

where T_{di} is the temperature of air from the greenhouse, T_{do} is the temperature of the air leaving the heat storage system, and $C_{p,av}$ is the average heat capacity of air in the discharge process. The approximate thermal efficiency of the discharged process is

$$\eta_d = \frac{\dot{q}_d}{\dot{q}_s} \tag{8.7}$$

All the temperatures are time dependent, and the charging and discharging cycles need to be monitored over the time of operation. Example 8.3 illustrates the latent heat storage calculations [13].

Example 8.3 Latent heat storage calculations

As a phase change material of 60 kg octadecane is heated from 20 to 30°C by a solar energy system, which supplies heat at a rate of 2 kW. Assume that the octadecane is fully melted. Estimate the minimum size of the storage unit and the time necessary for the charging process.

Solution:

Assume that there is no heat loss from the thermal energy storage system.

$$T_{sc} = 20°\text{C}, T_{sh} = 30°\text{C}, T_s = 28°\text{C}, m_s = 60.0\,\text{kg},$$
$$\dot{q}_{net} = 2\,\text{kW}, \dot{q}_{loss} = 0\,\text{kW}$$

Use data from Table 8.4

PCM	T_m °C	ΔH_m kJ/kg	$C_{pl,av}$ kJ/kg K	$C_{ps,av}$ kJ/kg K	k_l W/m K	k_s W/m K	ρ_l kg/m^3	ρ_s kg/m^3
Octadecane $CH_3(CH_2)_{16}CH_3$	28	243	2.2	1.8	0.15	0.42	775	~900

Total heat stored:

$$q_s = m_s\left[C_{ps,av}(T_s - T_{sc}) + \Delta H_m + C_{pl,av}(T_{sh} - T_s)\right]$$

$$q_s = 60\,\text{kg}[(1.8\,\text{kJ/kg K})(28 - 20)°\text{C} + 243\,\text{kJ/kg} + (2.2\,\text{kJ/kg K})(30 - 28)°\text{C}] = 15{,}708\,\text{kJ}$$

$$V_{tank} = m/\rho_s = 60\,\text{kg}/(775\,\text{kg/m}^3) = 0.08\,\text{m}^3$$

Energy balance: energy supplied = energy stored + energy lost

$$\text{Energy supplied} = (\dot{q}_{net})(\Delta t) = q_s = 15{,}708\,\text{kJ}$$

$$\Delta t = 7854\,\text{s} = 2.2\,\text{h}$$

Contributions of sensible heats:

$$q_s = 60\,\text{kg}[(1.8\,\text{kJ/kg K})(28 - 20)°\text{C} + (2.2\,\text{kJ/kg K})(30 - 28)°\text{C}]/15{,}708\,\text{kJ}$$
$$q_s = (864 + 264)\,\text{kJ}/15708\,\text{kJ} = 0.055 + 0.017$$

Contributions of sensible heats are 5.5% for solid state and 1.7% for liquid state; therefore, the main contribution toward heat storage comes from latent heat of octadecane. The charging process needs 2.2 h to supply the heat of 15,708 kJ required.

8.3.8 Underground Thermal Energy Storage Systems

Underground thermal energy storage (UTES) can store large amount of low-temperature heat for space heating and cooling as well as for preheating and precooling. Common energy sources include winter ambient air, heat pump reject water, solar energy, and process heat. Underground thermal energy storage may supply all or part of heating and/or cooling requirements of the buildings or processes. A heat pump may be used to decrease or increase the storage temperature for cooling or heating. Underground thermal energy may store energy, which is actively gathered, or store waste, or by-product energy, which is called the double effect storage and more likely to be more economical [9, 13]. Underground thermal energy storage encompasses both aquifer thermal energy storage (ATES) and borehole thermal energy storage (BTES) systems. These systems are discussed within the next sections.

8.3.9 Aquifer Thermal Energy Storage

Aquifers are underground, geological formations and can have gravel, sand, or rocks. Aquifer thermal energy storage may be used on a short-term or long-term basis for:

- The sole source of energy for partial storage
- A temperature useful for direct application
- Combination with a dehumidification system, such as desiccant cooling

Cold storage water is usually supplied at 2–5°C, with a cooling power typically ranging from 200 kW to 20 MW, and with the stored cooling energy of 29 GWh. Cold storage underground is now a standard design option in several countries such as Sweden. The duration of storage depends on the local climate and the type of building and/or process.

Aquifer thermal energy storage is used extensively for various applications such as heating of greenhouses. In summer, the greenhouse is cooled with ground water, pumped from an aquifer, which is the cold source. This heats the water, which is then stored by the aquifer system. In winter, the warm water is pumped up to supply heat to the greenhouse. The now cooled water is returned to the cold source. The combination of cold and heat storage with heat pumps has an additional benefit for greenhouses, as it may be combined with humidification. In the closed circuit system, the hot water is stored in one aquifer, while the cold water is stored in another. The water is used to heat or cool the air, which is moved by fans. Such a system can be completely automated.

Chemical changes in ground water due to temperature and pressure variations with aquifer thermal energy storage may cause operational and maintenance problem. These problems are avoidable and manageable. Flushing is a recommended practice to maintain the efficiency of well. Potential environmental concerns over the use of earth energy heat pump and UTES are:

- The possible leakage of the heat exchanger fluid into the natural environment
- Thermally induce biochemical effects on ground water quality
- Ecological distress due to chemical and thermal pollution
- External contaminants entering the ground water

All these and other possible problems are addressed within the guidelines and standards for planning, construction, and operation of underground thermal energy storage.

Ground source heat pump systems can significantly lower the heating and cooling operating costs and can qualify for renewable energy credits under sustainable building rating programs. Approximately, 65% less energy consumption is attainable with existing technologies at reasonable costs, and a heat pump supported

Table 8.5 Four types of application of UTES and their typical performances [9, 13]

Application	Energy source %	Payback period year[a]	COP[b]
Direct heating and cooling	90–95	0–2	20–40
Heating and cooling with heat pump	80–87	1–3	5–7
Heat pump supported heating only	60–75	4–8	3–4
Direct cooling only	90–97	0–2	20–60

[a] Payback period is the time in years to recover the cost of investment fully
[b] COP is the coefficient of performance

underground thermal energy storage can play useful role in achieving these reductions [5, 9]. Table 8.5 displays the performances of various types of underground thermal energy storage applications.

8.3.10 Borehole Thermal Energy Systems

Borehole thermal energy storage (BTES) applications involve the use of boreholes, typically 5–200 m deep, and operate in closed loop, in which there is no contact between the natural ground water and heat exchanger fluid. Typically a borehole thermal energy storage includes one or more boreholes containing borehole heat exchangers, such as U-tubes, through which waste energy (hot or cold) is circulated and transferred to the ground for storage. Borehole thermal energy storage is typically applied for combined heating and cooling often supported with heat pumps for a better usage of the low-temperature heat from the storage. For example, there are more than 300,000 boreholes in Sweden, delivering around 27% of all space heating requirements [5].

In a solar energy integrated application, solar heated water is pumped into a borehole thermal energy storage system consisting of 144 boreholes, each 37 m (121 ft) deep, and operates with the ground temperature of around 90°C (194°F). During the winter, the hot water flows from the borehole thermal energy storage system to the houses through a distribution network. Once inside the house, it flows through a heat exchanger with coil units, over which air is blown. The hot air then heats the house. Each house also has an independent solar-thermal system installed on its sloped roof to provide domestic hot water. This system has a 90% solar fraction, meaning 90% of the energy required to heat the air and water within the community is provided by the sun. This results in an emission reduction of over five tons of CO_2 per house per year [5].

8.4 Electric Energy Storage

A *capacitor* is a device for storing electric charge and used as parts of electrical systems. Figure 8.10 shows a flat plate model for a capacitor. The forms of practical capacitors vary widely, but all contain at least two conductors separated by a nonconductor. For example, a capacitor consists of metal foils separated by a layer of insulating film. When there is a potential difference (voltage) across the conductors, a static electric field develops across the dielectric, causing positive charge to collect on one plate and negative charge on the other plate. Energy is stored in the electrostatic field. An ideal capacitor is characterized by a single constant value called the capacitance, measured in Farads. Capacitance is the ratio of the electric charge on each conductor to the potential difference between them. Capacitors may be used to produce high intensity releases of energy, such as a camera's flash.

The dielectric is just e die electrical insulator. Examples of dielectric mediums are glass, air, paper, vacuum, and even a semiconductor depletion region chemically identical to the conductors. A capacitor is assumed to be self-contained and isolated, with no net electric charge and no influence from any external electric field. The conductors thus hold equal and opposite charges on their facing surfaces, and the dielectric develops an electric field. In SI units, a capacitance of one farad means that one coulomb of charge on each conductor causes a voltage of one volt across the device. In practice the dielectric between the plates passes a small amount of leakage current and has an electric field strength limit, resulting in a breakdown voltage [7, 8].

The capacitor is a reasonably general model for electric fields within electric circuits as shown in Fig. 8.11. An ideal capacitor is fully characterized by a constant capacitance C, defined as the ratio of charge $\pm Q$ on each conductor to the voltage V between them:

$$C = \frac{Q}{V} \tag{8.8}$$

Work must be done by an external influence to transport charge between the conductors. When the external influence is removed the charge separation persists

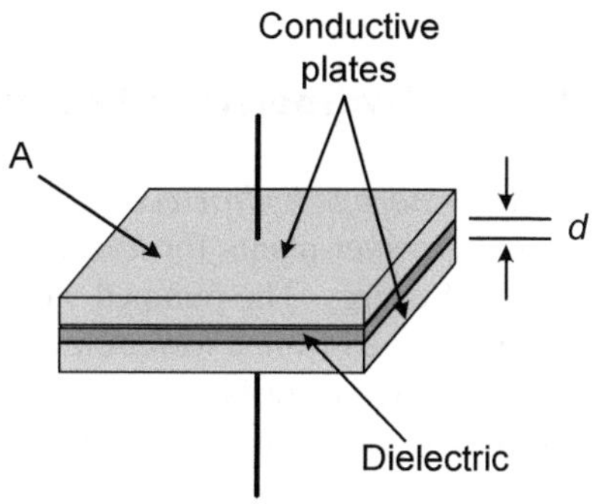

Fig. 8.10 Parallel plate model for a capacitor. The capacitance increases with area and decreases with separation; A is the surface area of the capacitor, and d is the thickness of dielectric

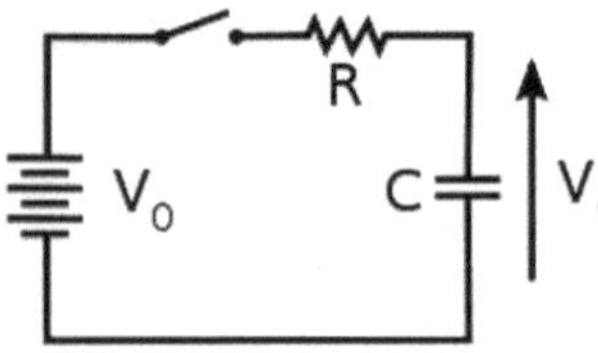

Fig. 8.11 Simple resistor–capacitor circuit demonstrates charging of a capacitor

in the electric field and energy is stored. The energy is released when the charge is allowed to return to its equilibrium position. The work done in establishing the electric field, and hence the amount of energy stored is

$$W = \int_0^Q V\mathrm{d}Q = \int_0^Q \frac{Q}{C}\mathrm{d}Q = \frac{1}{2}\frac{Q^2}{C} = \frac{1}{2}VQ \tag{8.9}$$

As the capacitor reaches equilibrium with the source voltage, the voltage across the resistor and the current through the entire circuit decay exponentially. *Discharging* of a charged capacitor demonstrates exponential decay [7, 8].

The capacitance increases with area and decreases with separation. The capacitance is therefore greatest in devices made from materials with a high permittivity, large plate area, and small distance between plates. The maximum energy is a function of dielectric volume, permittivity, and dielectric strength per distance. So, increasing the plate area and decreasing the separation between the plates while maintaining the same volume does not change the amount of energy the capacitor can store [7].

The dual of the capacitor is the inductor, which stores energy in the magnetic field rather than the electric field. Inductors consist of coils of wire for concentrating the magnetic field and collecting the induced voltage. Generated voltage is proportional to the rate of change in current in a circuit and the proportionality coefficient is the inductance L in henry in SI units

$$V = L\frac{\mathrm{d}I}{\mathrm{d}t} \tag{8.10}$$

8.4.1 Hydroelectric Energy Storage

Pumped-storage hydroelectricity is a type of hydroelectric power generation used by some power plants for *load balancing*. 1 kg mass elevated to 1000 m can store 9.8 kJ of energy. The pumped-storage hydroelectricity stores energy in the form of water, pumped from a lower elevation reservoir to a higher elevation. Low-cost off-peak electric power is used to run the pumps. The stored water is released through turbines to produce electric power. Although the losses of the pumping process makes the plant a net consumer of energy overall, the system increases revenue by selling more electricity during periods of *peak demand*, when electricity prices are highest. Example 8.4 illustrates the electric energy storage by the pumped energy in a hydropower plant [8].

Example 8.4 **Pumped energy in a hydropower plant**

One method of meeting the additional electric power demand at peak usage is to pump some water from a source such as a lake back to the reservoir of a hydropower plant at a higher elevation when the demand or the cost of electricity is low. Consider a hydropower plant reservoir with an energy storage capacity of 0.5×10^6 kWh. This energy is to be stored at an average elevation of 40 m relative to the ground level. Estimate the minimum amount of water has to be pumped back to the reservoir.

Solution:

Assume that the evaporation of the water is negligible.

$$PE = 0.5 \times 10^6 \,\mathrm{kWh}, \quad \Delta z = 40\,\mathrm{m}, \quad g = 9.8\,\mathrm{m/s^2}$$

Energy of the work potential of the water: $PE = mg\Delta z$.

Amount of water: $m = \frac{PE}{g\Delta z} = \frac{0.5\times 10^6\,\mathrm{kWh}}{9.8\,\mathrm{m/s^2}(40\,\mathrm{m})}\left(\frac{3600\,\mathrm{s}}{\mathrm{h}}\right)\left(\frac{1000\,\mathrm{m^2/s^2}}{1\,\mathrm{kJ/kg}}\right) = 4.59 \times 10^9\,\mathrm{kg}.$

8.4.2 Electric Energy Storage in Battery

A electrical *battery* is one or more electrochemical cells that can store electric energy in the form chemical energy as well as convert stored chemical energy into electrical energy. Alessandro Volta invented the first battery in 1800. Batteries come in various sizes, such as miniature cells and car batteries. There are two types of batteries: *primary batteries*, which are disposable and designed to be used once and discarded, and *secondary batteries*, which are rechargeable and designed to be recharged and used multiple times. Rechargeable batteries can have their chemical reactions reversed by supplying electrical energy to the cell.

Zinc-carbon batteries and alkaline batteries are the common types of disposable batteries. Generally, these batteries have higher energy densities than rechargeable batteries. Rechargeable batteries, therefore, can store electric energy in the form of chemical energy. For example, they can be used to store electrical energy in battery to be used in electric vehicles. Battery of electric vehicles can be charged from the grid. Because the stored energy is derived from electricity, it is possible to use other forms of alternative energy such as wind, solar, geothermal, nuclear, or hydroelectric in electric vehicles.

The oldest form of rechargeable battery is the lead-acid battery used in cars. Low manufacturing cost and its high surge current levels (450 A) make lead-acid battery use common where a large capacity (over approximately 10 Ah) is required. The anode of the battery is lead metal while the cathode is made up of lead 4 oxide. During the discharge, the chemical reaction that takes place at the anode, negative electrode, is

$$Pb(s) + SO_4^{-2}(aq) \rightarrow PbSO_4(s) + 2e^- \quad (8.11)$$

The chemical reaction that takes place at the cathode, positive electrode, is

$$PbO_2(s) + SO_4^{-2}(aq) + 4H^+(aq) + 2e^- \rightarrow PbSO_4(s) + 2H_2O(l) \quad (8.12)$$

During discharge, sulfuric acid is used up and lead sulfate is deposited on both electrodes. The battery becomes "dead" when the electrodes are fully covered with lead sulfate. During recharging the chemical reaction that takes place at the anode is

$$PbSO_4(s) + 2H_2O(l) \rightarrow PbO_2(s) + SO_4^{-2}(aq) + 4H^+(aq) + 2e^- \quad (8.13)$$

The chemical reaction that takes place at the cathode is

$$PbSO_4(s) + 2e^- \rightarrow Pb(s) + SO_4^{-2}(aq) \quad (8.14)$$

By passing a current the battery can be recharged. Recharging reverses both reactions and leads to sulfuric acid regeneration and the removal of the lead sulfate from the electrodes. Lead-acid batteries should not be discharged to below 20% of their full capacity, because internal resistance will cause heat and damage when they are recharged [8].

8.4.3 Rechargeable Battery for Electric Car

Electric car uses a bank of onboard batteries, generally lithium-ion batteries, to generate power for the motor. The choice of material used for the anode, cathode, and electrolyte determines the performance, safety, and lifespan of lithium-ion batteries. Lithium-ion batteries (see Fig. 8.13) are also lighter. Many electric cars utilize lithium-ion batteries, which contains layers of lithium isolated between plates which are made from a combination of cobalt and oxygen atoms [7, 8].

The batteries in electric cars must be charged from an external power source. Charging time depends on the type and capacity of the batteries and the type of outlet they are charging from. A standard 120 V outlet can provide about 1.5 kW/h of charge. If you plug it into a 240 V circuit then about 6.6 kW/h can be charged, drastically cutting charge time. When calculating efficiency this must be considered.

Battery life in cars is measured in cycles rather than time or distance traveled. A battery cycle is one charge and one discharge and the number of cycles is usually expressed as a range because how the battery is used and charged can affect its life. There are various kinds of batteries commonly considered and used in electric vehicles as well as for general purposes [8]:

- Nickel–cadmium (NiCd) batteries are the rechargeable battery that are used by power tools, portable electronics and electric toys all use NiCd batteries because of their long life and low cost. With a 40–60 Wh/kg (watt-hour per kilogram) energy-to-weight ratio and a life cycle of around 1000–2000 cycles, they are inexpensive source of power.
- Nickel-metal hydride (NiMH) batteries have better energy to weight ratio than NiCd at 30–80 Wh/kg and are free of the toxic element of cadmium. Nickel-metal hydride batteries have a life of about 800–1000 cycles. Low-capacity nickel metal hydride batteries (1700–2000 mAh) can be charged for about 1000 cycles, whereas high capacity NiMH batteries (above 2500 mAh) can be charged for about 500 cycles.
- Lithium-ion battery powers the laptop or other battery powered consumer electronics. An efficient 100–160 Wh/kg energy to weight ratio makes these the battery of choice in most high power/low weight applications. Lithium-ion batteries have a life of approximately 1200 cycles [7].

The standard hybrid vehicle increases fuel efficiency by adding additional battery capacity and charging capability. Plug-in vehicles thereby operate on energy derived from the electrical grid rather than from gasoline. Electric motors are more efficient than internal combustion engines; however, much of the electricity is produced from coal, which has a low thermal efficiency and higher carbon-to-energy ratio than petroleum. A computer inside the vehicle determines when it is most efficient for the vehicle to be fueled by electricity or by liquid fuel (either petroleum or biofuels).

Hybrid vehicles have improved efficiency in three ways. First, the gasoline engine can be sized slightly smaller and therefore operate in a more efficient range because the electric motor can supplement the engine when needed. Second, the gasoline engine can be shut down during periods when it would be operating inefficiently, particularly idling at stops, but also at slow speeds. Finally, it enables regenerative braking whereby some of the vehicle's kinetic energy is converted into electricity by the electric motor acting as a generator and charging the battery.

In plug-in hybrid vehicles, the small battery is replaced with a larger battery pack. Plug-in hybrid vehicles use both electric and gasoline drive and can be plugged into a standard home voltage. The vehicles have onboard controls to use the appropriate power source for maximum efficiency. Plug-in hybrid vehicles with modest battery packs could run primarily on the electric motor with stored energy for around 20 miles/day, thereby reducing the gasoline use [8].

8.5 Chemical Energy Storage

The free energy available from the electron-transport cycle is used in producing the proton flow through the inner membrane in mitochondria to synthesize adenosine triphosphate (ATP) in living cells. ATP is an energy-rich compound having three phosphate groups attached to a nucleoside of adenine, called adenosine (adenine

+ pentose sugar). The terminal one of the three phosphate groups has a weak bond linkage and can break spontaneously whenever ATP forms a complex with an enzyme. The breaking up of this bond releases chemical energy causing an immediate shift in the bond energy giving rise to adenosine diphosphate. ATP formation therefore acts as chemical energy storage of the living cell [4].

In plants, ATP is used in biochemical cycles to synthesize carbohydrates such as glucose from carbon dioxide and water. The glucose is stored mainly in the form of starch granules consisting of a large number of glucose units. This polysaccharide is produced by all green plants as chemical energy store. Starch is processed to produce many of the sugars in processed foods.

A *biological battery* generates electricity from sugar in a way that is similar to the processes observed in living organisms. The battery generates electricity through the use of enzymes that break down carbohydrates, which are essentially sugar. A similarly designed sugar drink can power a phone using enzymes to generate electricity from carbohydrates that covers the phone's electrical needs.

8.5.1 Bioenergy Sources

Carbohydrates, fats, and proteins are the major metabolic fuels (sources of energy) of living systems. Chemical energy released by oxidation of these fuels is stored as energy-rich molecules such as adenosine triphosphate in living cells. Energy contents of the major metabolic fuels are:

- Carbohydrates release about 4 kcal/g (17 kJ/g).
- Fats release about 9 kcal/g (37 kJ/g).
- Proteins release about 4 kcal/g (17 kJ/g).

In a human body excess energy is stored mainly by:

- Fat-Triacylglycerol (Triglyceride): Triacylglycerol is the major energy store of the body.
- Glycogen: Energy stored by glycogen is relatively small but it is critical. For example muscle glycogen is oxidized for muscle contraction.

Metabolism is the set of biochemical reactions that occur in living organisms to maintain life and consists of two major pathways of catabolism and anabolism. In catabolism, organic matter is broken down, for example, to harvest energy. In anabolism, the energy is used to construct components of cells such as proteins and nucleic acids. Daily energy expenditure includes the energy required for the basal metabolic rate and for the physical activity in every day. Basal metabolic rate for a person is approximately 24 kcal/kg (101 kJ/kg) body weight per day [4, 7].

8.5.2 Energy Storage in Biofuels

Various biofuels or biomass can be helpful in reducing the use of hydrocarbon fuels. Some chemical processes, such as Fischer–Tropsch synthesis, can convert the carbon and hydrogen in coal, natural gas, biomass, and organic waste into short hydrocarbons suitable as replacements for existing hydrocarbon fuels. Many hydrocarbon fuels have the advantage of being immediately usable in existing engine technology and existing fuel distribution infrastructures. A long-term high oil price may make such synthetic liquid fuels economical on a large-scale production despite some of the energy in the original source is lost in the conversion processes.

Carbon dioxide and hydrogen can be converted into methane and other hydrocarbon fuels [10] with the help of energy from another source preferably a renewable energy source. As hydrogen and oxygen are produced in the electrolysis of water,

$$2H_2O \rightarrow 2H_2 + O_2 \tag{8.15}$$

hydrogen would then be reacted with carbon dioxide to produce methane and water in the Sabatier reaction

$$CO_2 + 4H_2 \rightarrow CH_4 + 2H_2O \tag{8.16}$$

Produced water would be recycled back to the electrolysis stage, reducing the need for new pure water. In the electrolysis stage oxygen would also be stored for methane combustion in a pure oxygen environment (eliminating nitrogen oxides). In the combustion of methane, carbon dioxide and water are produced.

$$CH_4 + 2O_2 \rightarrow CO_2 + 2H_2O \tag{8.17}$$

Produced carbon dioxide would be recycled back to the Sabatier process. Methane production, storage, and adjacent combustion would recycle all the reaction products, creating a cycle.

Methane is the simplest hydrocarbon with the molecular formula CH_4. Methane can be stored more easily than hydrogen, and the technologies for transportation, storage, and combustion infrastructure are highly developed. Methane would be stored and used to produce electricity later. Besides that, methanol can also be synthesized from carbon dioxide and hydrogen. If the hydrogen is produced by the electrolysis using the electricity from the wind power, then the electricity is stored by the production of methane or methanol.

8.5.3 Energy Storage in Voltaic Cell

Each voltaic cell consists of two half-cells connected in series by a conductive electrolyte containing anions and cations. One half-cell includes electrolyte and the positive electrode called the cathode. Negatively charged ions called the anions migrate to the cathode. The other half-cell includes electrolyte and the negative electrode called anode. The positively charged ions called the cations migrate to anode. In the redox reaction that powers the battery, cations are reduced (electrons are added) at the cathode, while anions are oxidized (electrons are removed) at the anode. The electrodes do not touch each other but are electrically connected by the electrolyte. Figure 8.12 shows the half-cells of a copper-zinc battery [7].

In a copper-zinc battery, the tendency for Zn to lose electron is stronger than that for copper. When the two cells are connected by a *salt bridge* and an *electric conductor* as shown in Fig. 8.13 to form a closed circuit for electrons and ions to flow, copper ions (Cu^{2+}) actually gains electron to become copper metal. Electrons flow through the electric conductors connecting the electrodes and ions flow

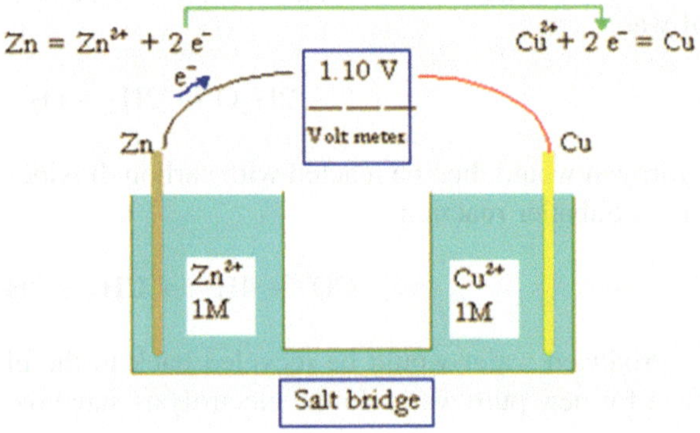

Fig. 8.12 Half-cells of a copper-zinc voltaic cells; Zn electrode to loose electron according to the reaction so it is cathode: $Zn = Zn^{2+} + 2e^{-}$

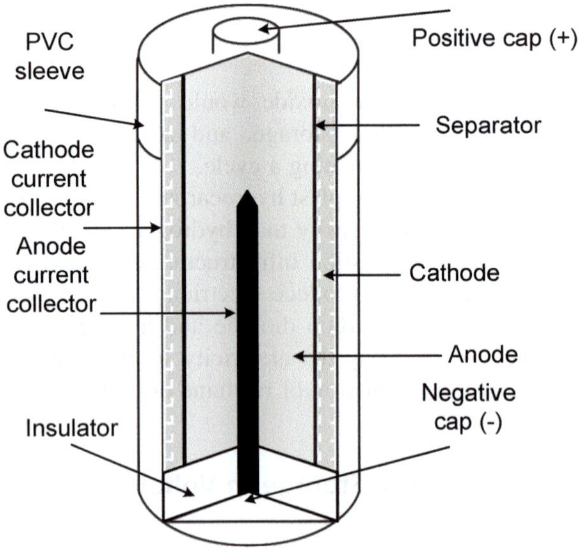

Fig. 8.13 Schematic of a dry cell

through the salt bridge. The overall reaction of the cell is

$$Zn + Cu^{2+} = Zn^{2+} + Cu \tag{8.18}$$

The net electromotive force of the cell is the difference between the reduction potentials of the half-reactions. The electrical driving force of a cell is known as the *terminal voltage* (ΔV) and is measured in volts. An ideal cell has negligible internal resistance, so it would maintain a constant terminal voltage until exhausted. If such a cell maintained 1.5 V and stored a charge of one coulomb then on complete discharge it would perform 1.5 J of work. A battery is a collection of multiple electrochemical cells.

Other portable rechargeable batteries include several "dry cell" types using a paste in place of liquid electrolyte and are therefore useful in appliances such as mobile phones and laptop computers. Cells of this type include nickel–cadmium (NiCd), nickel metal hydride (NiMH) and lithium-ion (Li-ion) cells. Lithium-ion has the highest share of the dry cell rechargeable market. Meanwhile, NiMH has replaced NiCd in most applications due to its higher capacity, but NiCd remains in use in power tools, two-way radios, and medical equipment. Figure 8.13 shows the main layers of a dry cell [7, 8].

The battery capacity printed on a battery is usually the product of 20 h multiplied by the maximum constant current that a new battery can supply for 20 h at 68°F (20°C), down to a predetermined terminal voltage per cell. A battery rated at 100 Ah will deliver 5 A over a 20-h period at room temperature. The relationship between current, discharge time, and capacity for a lead-acid battery is approximated by Peukert's law:

$$t = \frac{Q_B}{I^k} \tag{8.19}$$

where Q_B is the capacity when discharged at a rate of 1 amp, I is the current drawn from battery, t is the amount of time (in hours) that a battery can sustain, and k is a constant around 1.3. When discharging at low rate, the battery's energy is delivered more efficiently than at higher discharge rates, but if the rate is too low, it will self-discharge during the long time of operation, again lowering its efficiency. Sodium-sulfur batteries are being used to store wind power [7].

8.6 Mechanical Energy Storage

Energy may also be stored in pressurized gases or alternatively in a vacuum. Compressed air, for example, may be used to operate vehicles and power tools. Large-scale compressed air energy storage facilities may be used to supplement demands on electricity generation by providing energy during peak hours and storing energy during off-peak hours. Such systems save on expensive generating capacity since it only needs to meet average consumption [9].

8.6.1 Compressed Air Energy Storage

Compressed air energy storage (CAES) technology stores low-cost off-peak energy, in the form of compressed air in an underground reservoir. The air is then released during peak load hours and heated with the exhaust heat of a standard combustion turbine. This heated air is converted to energy through expansion turbines to produce electricity. Energy storage systems often use large underground caverns due to the very large volume, and thus the large quantity of energy that can be stored with only a small pressure change. The cavern space can be compressed adiabatically with little temperature change and heat loss. Compressed air energy storage can also be employed on a smaller scale such as exploited by air cars and air-driven locomotives. Compressed air energy storage application using sequestered carbon dioxide is also possible for energy storage [9].

Compression of air generates a lot of heat and the air gets warmer. If no extra heat is added, the air will be much colder after decompression, which requires heat. If the heat generated during compression can be stored and used again during decompression, the efficiency of the storage improves considerably. Compressed air energy storage can be achieved in adiabatic, diabatic, or isothermic processes:

- *Adiabatic storage* retains the heat produced by compression and returns it to the air when the air is expanded to generate power. Heat can be stored in a fluid such as hot oil (up to 300°C) or molten salt solutions (600°C).
- *Diabatic storage* dissipates the extra heat with intercoolers (thus approaching isothermal compression) into the atmosphere as waste. Upon removal from storage, the air must be reheated prior to expansion in the turbine to power a generator which can be accomplished with a natural gas fired burner.
- *Isothermal compression* and expansion approaches attempt to maintain operating temperature by constant heat exchange to the environment. They are only practical for low power levels and some heat losses are unavoidable. Compression process is not exactly isothermal and some heat losses will occur.

The ideal gas law for an isothermal process is

$$PV = nRT = \text{constant} \tag{8.20}$$

$$W_{AB} = nRT\ln(V_B/V_A) = nRT\ln(P_A/P_B) \tag{8.21}$$

where A and B are the initial and final states of the system, and W represents the maximum energy that can be stored or released. In practice, no process is perfectly isothermal and the compressors and motors will have heat losses. Compressed air can transfer power at very high flux rates, which meets the principal acceleration and deceleration objectives of transportation systems, particularly for hybrid vehicles. Examples 8.5 and 8.6 illustrate the simple analysis of mechanical energy storage by compressed air.

Example 8.5 **Maximum air compressed energy storage**

Determine the maximum available energy if we compress 1450 kg air from 100 to 1200 kPa at 300 K at isothermal conditions with a heat loss of 24,000 kJ.

Solution:

Assume that air is the ideal gas.

$$\mathrm{MW} = 29\,\text{kg/kmole}, \quad R = 8.314\,\text{kJ/kmole K},$$

$$P_1 = 100\,\text{kPa}, \quad P_2 = 1200\,\text{kPa}, \quad m = 1450\,\text{kg}$$

Number of moles of air $= 1450/28.97 = 50\,\text{kmole}$

Mechanical energy stored:

$$W = nRT\ln\left(\frac{P_1}{P_2}\right)_{\max} = -(50\,\text{kmole})(8.314\,\text{kJ/kmole K})(300\,\text{K})\ln(100/1200)$$

$$= 309{,}892.7\,\text{kJ}$$

$$q_{\text{loss}} = 24{,}000\,\text{kJ}$$

Net energy to be stored:

$$W_{\max} - q_{\text{loss}} = (309892.7 - 24{,}000)\,\text{kJ} = 285{,}892.7\,\text{kJ} = 285.9\,\text{MJ}$$

Example 8.6 **Maximum air compressed energy storage in a large cavern**

An underground cavern will be used to store energy of a compressed air. If the cavern has a volume of 29,000 m^3. Determine the value of stored energy by the compression of air from 100 to 1500 kPa at 300 K at isothermal conditions with a heat loss of 55,000 kJ.

Solution:

Assume that air is ideal gas. Cavern space is under ideal gas conditions and at standard temperature and pressure (25°C and 1 atm).

$$\mathrm{MW} = 28.97\,\text{kg/kmole}, \quad R = 8.314\,\text{kJ/kmole K}, \quad P_1 = 100\,\text{kPa}, \quad P_2 = 1500\,\text{kPa}$$

Under ideal gas conditions and at standard temperature and pressure (25°C and 1 atm).

1 Mole of air occupies 22.4 l and 1 kmol occupies 2.24×10^4 l

$$\text{Volume} = 29{,}000\,\text{m}^3 = 2.9 \times 10^7\,\text{l}$$

Approximate number of moles of air = $2.9 \times 10^7\,1/2.24 \times 10^4\,1 = 1294.6\,\text{kmole}$

Energy stored:

$$W = -nRT \ln\left(\frac{P_1}{P_2}\right)$$
$$= -(1294.6\,\text{kmole})(8.314\,\text{kJ/kmole K})(300\,\text{K})$$
$$\ln(100/1500) = 8{,}744{,}270.6\,\text{kJ}$$
$$q_{\text{loss}} = 55{,}000\,\text{kJ}$$

Net energy to be stored $= W_{\max} - q_{\text{loss}} = 8{,}689{,}270.6\,\text{kJ}$ or $8689.3\,\text{MJ}$

8.6.2 Flywheel Energy Storage

Flywheel energy storage works by accelerating a rotor (flywheel) to a very high speed and maintaining the energy in the system as rotational energy. When energy is extracted from the system, the flywheel's rotational speed is reduced as a consequence of the principle of conservation of energy. Most flywheel energy storage systems use electricity to accelerate and decelerate the flywheel, but devices that directly use mechanical energy are also being developed. Advanced flywheel energy storage systems have rotors made of high strength carbon filaments, suspended by magnetic bearings, and spinning at speeds from 20,000 to over 50,000 rpm in a vacuum enclosure [9].

Example 8.7 **Energy storage in a flywheel**

Estimate the energy stored in a flywheel of steel disk with a mass of 5 kg and radius of 0.2 m rotating at 40,000 rpm.

Solution:

The moment of inertia of the disk of density ρ radius r and height z is

$$I = \frac{\pi \rho z r^4}{2} = \frac{mr^2}{2} = \frac{5\,\text{kg}(0.2\,\text{m})^2}{2} = 0.1\,\text{kg}\,\text{m}^2$$
$$\omega = 40{,}000(2\pi/60) = 4200/\text{s}$$

Approximate stored energy is

$$E = \frac{I\omega^2}{2} = \frac{0.1\,\text{kg}\,\text{m}^2(4200/\text{s})^2}{2} = 0.88\,\text{MJ}$$

8.6.3 Hydraulic Accumulator

A *hydraulic accumulator* is an energy storage device in which a noncompressible hydraulic fluid is held under pressure by an external source. That external source can be a spring, a raised weight, or a compressed gas. An *accumulator* is used in a hydraulic system so that the pump doesn't need to be so large to cope with extremes of demand. Also, the supply circuit can respond more quickly to any temporary demand and to smooth pulsations. Compressed gas accumulators are the most common type. These are also called hydropneumatic accumulators [9].

8.6.4 Springs

A *spring* is an elastic material used to store mechanical energy. When a spring is compressed or stretched, the force it exerts is proportional to its change in length. The *spring constant* of a spring is the change in the force it exerts divided by the change in deflection of the spring. An extension or compression spring has units of force divided by distance, for example lbf/in or N/m. The stiffness (or rate) of springs in parallel is additive, as is the compliance of springs in series. Depending on the design and required operation, any material can be used to construct a spring, as long as the material has the required combination of rigidity and elasticity [9].

Problems

8.1 (a) Estimate sensible heat stored in 6 m^3 water and 6 m^3 granite heated from 20 to 40°C.
(b) Estimate the heat stored in 500 lb of water heated from 20 to 30°F.

8.2 (a) Estimate sensible heat stored in 1.1 m^3 Dowtherm A and 1.1 m^3 therminol 66 heated from 20 to 240°C.
(b) Estimate the heat stored in 500 lb of molten salt heated from 150 to 300°C.

8.3 (a) Estimate sensible heat stored in 2.5 m^3 mixture of 50% ethylene glycol and 50% water and 2.5 m^3 water heated from 20 to 65°C.
(b) Estimate the heat stored in 1500 lb of draw salt heated from 250 to 400°C.

8.4 A typical square two-story home with a roof surface area of 1260 ft^2 and a wall surface area of 2400 ft^2 is to be heated with solar energy storage using a salt hydrate as phase change material. It presently has as an insulation of 6 inches in the roof and 1 inch in the walls. Inside temperature will be held at 70°F and expected outside low temperature is 10°F. Average solar radiation is 300 Btu/ft^2 and the cost of solar air collector is \$4.5/ft^2. The salt hydrate costs around \$0.5/lb. Estimate the costs of salt hydrate and solar air collectors.

8.5 A large building with a roof surface area of 3200 ft^2 and a wall surface area of 4200 ft^2 is to be heated with solar energy storage using a salt hydrate as phase change material. It presently has as an insulation of 6 inches in the roof and 1 inch in the walls. Inside temperature will be held at 70°F and expected outside low temperature is 10°F. Average solar radiation is 350 Btu/ft^2, and the cost of solar air collector is \$3.4/ft^2. The salt hydrate costs around \$0.4/lb. Estimate the costs of salt hydrate and solar air collectors.

8.6 Energy from a thermal energy storage system is used to heat a room from 15 to 25°C. The room has the dimensions of $(4 \times 4 \times 5)$ m^3. The air in the room is at atmospheric pressure. The heat loss from the room is negligible. Estimate the amount of n-hexadecane needed if the heat of melting is the stored energy for this temperature change.

8.7 Energy from a thermal energy storage system is used to heat a room from 12 to 27°C. The room has the dimensions of $(4 \times 5 \times 5)$ m^3. The air in the room is at atmospheric pressure. The heat loss from the room is negligible. Estimate the amount of n-octadecane needed if the heat of melting is the stored energy for this temperature change.

8.8 Energy from a thermal energy storage system is used to heat a room from 14 to 24°C. The room has the dimensions of $(5 \times 5 \times 5)$ m^3. The air in the room is at atmospheric pressure. The heat loss from the room is negligible. Estimate the amount of paraffin needed if the heat of melting is the stored energy for this temperature change.

8.9 A room with dimensions of $(5 \times 5 \times 6)$ m^3 is heated with a thermal energy storage supplying a heating rate of 80 kJ/h. Heat loss from the room is 2 kJ/h. The air is originally at 15°C. Estimate the time necessary to heat the room to 24°C.

8.10 A 0.50-kW solar-thermal energy storage system is used to heat a room with dimensions 3.0 m $\times$ 4.5 m $\times$ 4.0 m. Heat loss from the room is 0.15 kW and the pressure is always atmospheric. The air in the room may be assumed to be an ideal gas. The air has a constant heat capacity of $C_{v,av} = 0.72$ kJ/kg °C. Initially the room temperature is 15°C. Determine the temperature in the room after 30 min.

8.11 A 60 m^3 room is heated by a thermal energy storage system. The room air originally is at 12°C and 100 kPa. The room loses heat at a rate of 0.2 kJ/s. If the thermal energy storage system supplies 0.8 kW, estimate the time necessary for the room temperature to reach 22°C.

8.12 A superheated steam at a rate of 0.6 lb/s flows through a heater. The steam is at 100 psia and 380°F. If a heat storage system supplies 37.7 Btu/s into the steam at constant pressure, estimate the final temperature of the steam.

8.13 A superheated steam at a rate of 0.9 lb/s flows through a heater. The steam is at 100 psia and 380°F. If a heat storage system supplies 25.4 Btu/s into the steam at constant pressure, estimate the final temperature of the steam.

8.14 A thermal energy storage system is used to heat 20 kg of water in a well-insulated tank. The storage system supplies a heat at a rate of 1.5 kW. The water in the tank is originally at 10°C. Estimate the time necessary for the temperature of the water to reach 50°C and the amount of salt hydrate with a heat of melting of 200 kJ/kg.

8.15 A thermal energy storage system is used to heat 120 kg of water in a well-insulated tank. The storage system supplies a heat at a rate of 1.5 kW. The water in the tank is originally at 15°C. Estimate the time necessary for the temperature of the water to reach 50°C and the amount of n-hexadecane with a heat of melting of 235 kJ/kg (Table 8.4).

8.16 A thermal energy storage system is used to heat 150 kg of water. The heat is stored by latent heat of paraffin. The storage supplies heat at a rate of 2.3 kW. A heat loss of 0.35 kW occurs from the tank to surroundings. The water in the tank is originally at 20°C. Estimate the temperature of the water after 2 h of heating and amount of paraffin required.

8.17 A thermal energy storage system is used to heat 50 kg of water. The heat is stored by latent heat of paraffin. The storage supplies heat at a rate of 2.0 kW. A heat loss of 0.4 kW occurs from the tank to surroundings. The water in the tank is originally at 5°C. Estimate the temperature of the water after 3 h of heating and amount of paraffin required.

8.18 As a phase change material 20 kg octadecane is heated from 20 to 30°C by a solar energy system, which supplies heat at a rate of 2 kW. Assume that the octadecane is fully melted. Estimate the time necessary for the process and the minimum size of the storage unit.

8.19 As a phase change material 50 kg n-hexadecane is heated from 12 to 26°C by a solar energy system, which supplies heat at a rate of 1.5 kW. Assume that the n-hexadecane is fully melted. Estimate the time necessary for the process and the minimum size of the storage unit.

PCM	T_m °C	ΔH_m kJ/kg	$C_{pl, av}$ kJ/kg K	$C_{ps,av}$ kJ/kg K	k_l W/m K	k_s W/m K	ρ_l kg/m^3	ρ_s kg/m^3
Water	0	333	4.19	2.0	0.595	2.2	1000	920
n-Hexadecane	18	235	2.1	1.95	0.156	0.43	765	835

8.20 As a phase change material 120 kg octadecane is heated from 20 to 30°C by a solar energy system, which supplies heat at a rate of 2 kW. Assume that the octadecane is fully melted. Estimate the time necessary for the process and the minimum size of the storage unit.

8.21 As a phase change material 60 kg n-eicosane is heated from 20 to 42°C by a solar energy system, which supplies heat at a rate of 1.8 kW. Assume that the n-eicosane is fully melted. Estimate the time necessary for the process and the minimum size of the storage unit.

8.22 As a phase change material 100 kg docosane is heated from 20 to 55°C by a solar energy system, which supplies heat at a rate of 1.2 kW. Assume that the docosane is fully melted. Estimate the time necessary for the process and the minimum size of the storage unit.

8.23 As a phase change material 30 kg docosane is heated from 20 to 55°C by a solar energy system, which supplies heat at a rate of 1.1 kW. Assume that the docosane is fully melted. Estimate the time necessary for the process and the minimum size of the storage unit.

8.24 One method of meeting the additional electric power demand at peak usage is to pump some water from a source such as a lake back to the reservoir of a hydropower plant at a higher elevation when the demand or the cost of electricity is low. Consider a hydropower plant reservoir with an energy storage capacity of 2×10^6 kWh. This energy is to be stored at an average elevation of 60 m relative to the ground level. Estimate the minimum amount of water has to be pumped back to the reservoir.

8.25 One method of meeting the additional electric power demand at peak usage is to pump some water from a source such as a lake back to the reservoir of a hydropower plant at a higher elevation when the demand or the cost of electricity is low. Consider a hydropower plant reservoir with an energy storage capacity of 1.2×10^6 kWh. This energy is to be stored at an average elevation of 52 m relative to the ground level. Estimate the minimum amount of water has to be pumped back to the reservoir.

8.26 One method of meeting the additional electric power demand at peak usage is to pump some water from a source such as a lake back to the reservoir of a hydropower plant at a higher elevation when the demand or the cost of electricity is low. Consider a hydropower plant reservoir with an energy storage capacity of 3.45×10^6 kWh. This energy is to be stored at an average elevation of 44 m relative to the ground level. Estimate the minimum amount of water has to be pumped back to the reservoir.

8.27 Determine the maximum available energy from compression of 725 kg air from 100 to 1200 kPa at 300 K at isothermal conditions with a loss of 24,000 kJ.

8.28 Determine the maximum available energy if we compress of 1100 kg air from 100 to 1100 kPa at 290 K at isothermal conditions with a loss of 21,200 kJ.

8.29 Determine the maximum available energy if we compress of 540 kg air from 100 to 1100 kPa at 290 K at isothermal conditions with a loss of 12000 kJ.

8.30 Determine the maximum available energy if we compress of 1450 kg air compression from 100 to 1200 kPa at 300 K at isothermal conditions with a loss of 24,000 kJ.

8.31 An underground cavern will be used to store energy of a compressed air. If the cavern has a volume of 58,000 m^3 determine the maximum available energy from compression of air from 100 to 1500 kPa at 300 K at isothermal conditions with a loss of 55,000 kJ.

8.32 An underground cavern will be used to store energy of a compressed air. If the cavern has a volume of 90,000 m^3 determine the maximum available energy from compression of air from 100 to 1400 kPa at 300 K at isothermal conditions with a loss of 50,000 kJ.

8.33 An underground cavern will be used to store energy of a compressed air. If the cavern has a volume of 58,000 ft^3 determine the maximum available energy from compression of air from 15 psia to 600 psia at 540 R at isothermal conditions with a loss of 50,000 Btu.

8.34 An underground cavern will be used to store energy of a compressed air. If the cavern has a volume of 150,000 ft^3 determine the maximum available energy from compression of air from 15 to 500 psia at 540 R at isothermal conditions with a loss of 100,000 Btu.

8.35 A well-insulated 30-m^3 tank is used to store exhaust steam. The tank contains 0.01 m^3 of liquid water at 30°C in equilibrium with the water vapor. Determine the amount of wet-exhaust steam, in kg, from a turbine at 1 atm at the end of an adiabatic filling process. The wet steam has the quality of 90%, and the final pressure within the tank is 1 atm. Assume that heat transfer between the liquid water and the steam is negligible.

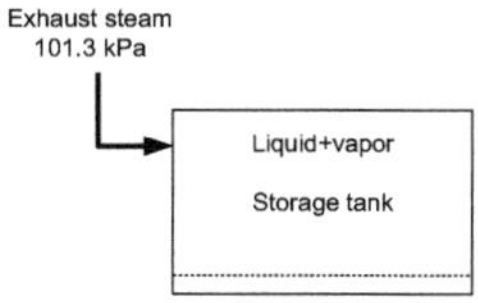

References

1. Çengel YA, Turner R, Cimbala J (2011) Fundamentals of thermal-fluid sciences, 4th edn. McGraw-Hill, Ney York
2. Demirel Y, Ozturk HH (2006) Thermoeconomics of seasonal heat storage system. Int J Energy Res 30:1001–1012
3. Demirel Y (2007) Heat storage by phase changing materials and thermoeconomics. In: Paksoy HO (ed) Thermal energy storage for sustainable energy consumption NATO science series. Springer, Dordrecht, vol 234, pp 133–151
4. Demirel Y, Gerbaud V (2025) Nonequilibrium thermodynamics: transport and rate processes in physical, chemical and biological systems, 5th edn. Elsevier, Amsterdam
5. Dincer I, Rosen MA (2007) A unique borehole thermal storage system at University of Ontario Institute of Technology. In: Paksoy HO (ed) Thermal energy storage for sustainable energy consumption. NATO Science Series, Springer, Dordrecht
6. Kenisarin M, Mahkamov K (2007) Solar energy storage using phase change materials. Renewable Sustain Energy Rev 11:1913–1965
7. Jaffe RL, Taylor W (2018) The physics of energy. Cambridge University Press, Cambridge
8. Loisel R, Mercier A, Gatzen C, Elms N, Petric H (2010) Valuation framework for large scale electricity storage in a case with wind curtailment. Energy Policy 38:7323–7337
9. Marloth R (2014) Energy storage. In: Anwar S (ed) Encyclopedia of energy engineering and technology, 2nd edn. CRC Press, Boca Raton
10. Matzen M, Alhajji M, Demirel Y (2015) Chemical storage of wind energy by renewable methanol production: feasibility analysis using a multi-criteria decision matrix. Energy 93:343–353

11. Mehling E, Cabeza LF (2007) Phase changing materials and their basic properties. In: Paksoy HO (ed) Thermal energy storage for sustainable energy consumption. NATO Science Series, Springer, Dordrecht
12. Ozturk HH, Demirel Y (2004) Exergy-based performance analysis of packed bed solar air heaters. Int J Energy Res 28:423–432
13. Sharma A, Tyagi VV, Chen CR, Buddhi D (2009) Review on thermal energy storage with phase change materials and applications. Renewable Sustain Energy Rev 13:318–345

9 Energy Conservation

Introduction and Learning Objectives: Energy conservation can be achieved through increased efficiency in energy and deceased use in total energy. Energy recovery also may be a part of energy conservation through captured and hence reduced waste energy. Energy conservation is an important part of controlling climate change by reducing greenhouse gas emissions. This chapter discusses the possibilities in conserving energy in power production, in compressor work, and in furnaces and boilers. This chapter also explores the energy conservation options in heating and cooling of houses as well as in lighting and driving. Some worked examples show it is always possible to conserve energy by process improvements and by increasing efficiency of processes.

The learning objectives of this chapter are energy conservation in:

- Power production
- Compressor work
- Furnaces and boilers
- Heating and cooling of houses as well as in lighting and driving

9.1 Energy Conservation and Recovery

Energy conservation mainly refers to reducing energy consumption and increasing efficiency in energy usage. Energy conservation may lead to increased security, financial gain, and environmental protection. For example, electric motors consume considerable amounts of electrical energy and operate at efficiencies between 70 and 90%. Therefore, using an electric motor operating with better efficiency will conserve energy throughout its useful life [2, 5, 7].

Y. Demirel, *Energy*, Green Energy and Technology,
https://doi.org/10.1007/978-3-031-97111-2_9

Energy recovery leads to reducing the energy input by reducing the overall waste energy from a system. For example, a waste energy, mainly in the form sensible or latent heat, from a subsystem may be usable in another part of the same system. Therefore, energy recovery may be a part of energy conservation. There is a large potential for energy recovery in industries and utilities leading to reduced use of fossil fuels and hence greenhouse gas emission. Some examples of energy recovery are:

- Hot water from processes such as power plants and steel mills may be used for heating homes and offices in the nearby area. Energy conservation through insulation or improved buildings may also help. Low temperature heat recovery would be more effective for a short distance from producer to consumer.
- Regenerative brake is used in electric cars and trains, where the part of kinetic energy is recovered and stored as chemical energy in a battery.
- Active pressure reduction systems where the differential pressure in a pressurized fluid flow is recovered rather than converted to heat in a pressure reduction valve.
- Energy recycling.
- Water heat recycling.
- Heat recovery steam generator (HRSG).
- Heat regenerative cyclone engine.
- Thermal diode.
- Thermoelectric modules.

9.2 Conservation of Energy in Industrial Processes

Most of the industrial processes depend on stable and affordable energy supply to be competitive. Some of these industrial processes produce energy, while others use energy. Energy conservation in the industrial sector will reduce wasted energy and GHG emissions.

9.2.1 Energy Conservation in Power Production

Section 6.7 discusses some possible modifications in improving the efficiency of power plants. Here, some examples underline the importance of energy conservation of Rankine cycle and Brayton cycle. In gas-turbine engines, the temperature of the exhaust gas T_4 is often higher than the temperature of the gas leaving the compressor T_2 as seen in Fig. 9.1. Therefore, the gas leaving the compressor can be heated in a regenerator by the hot exhaust gases as shown in Fig. 9.1b. A regenerator is a counter-current heat exchanger, which is known as *recuperator*, and recovers waste heat. The thermal efficiency of the Brayton cycle increases because of regeneration because the portion of energy of the exhaust gases is used to preheat the gas entering the combustion chamber. Thus, in turn, regeneration

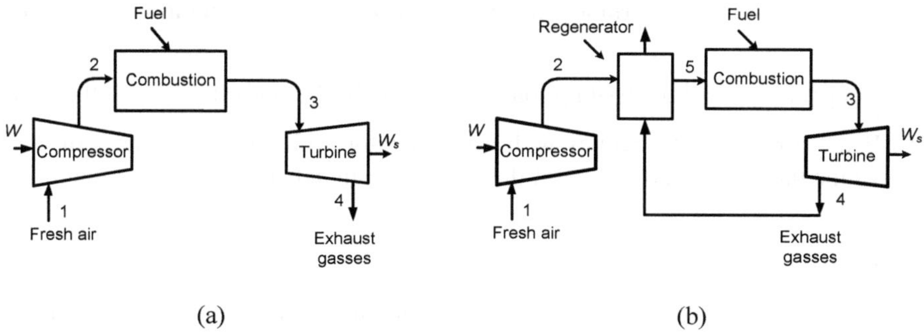

Fig. 9.1 **a** Simple Brayton cycle, **b** Brayton cycle with regeneration; the condition for regeneration is $T_4 > T_2$

can reduce the fuel input required for the same work output from the cycle. The addition of a regenerator (operating without thermal losses) does not affect the network output of the cycle. A regenerator with higher effectiveness will conserve more fuel. The effectiveness ε of the regenerator operating under adiabatic conditions is

$$\varepsilon = \left(\frac{H_5 - H_2}{H_4 - H_2}\right) \tag{9.1}$$

The enthalpies in Eq. (9.1) are shown in Fig. 9.1b. Under the cold-air standard temperature assumptions, the effectiveness ε of the regenerator in Fig. 9.1b is

$$\varepsilon \simeq \left(\frac{T_5 - T_2}{T_4 - T_2}\right) \tag{9.2}$$

The regeneration is possible only when $T_4 >> T_2$. The effectiveness of most regenerators used in practical engine operations is below 0.85 [2, 3].

Under the cold-air standard temperature assumptions, thermal efficiency of an ideal Brayton cycle with regeneration depends on the ratios of the temperatures and the pressures, and estimated by

$$\eta_{\text{th. regen.}} = 1 - \left(\frac{T_1}{T_3}\right)(r_p)^{(\gamma-1)/\gamma} \tag{9.3}$$

where r_p is the compression ratio (P_2/P_1) and $\gamma = C_p/C_v$.

Example 9.1 **Energy conservation by regeneration in a Brayton cycle**

A power plant is operating on an ideal Brayton cycle with a pressure ratio of $r_p = 9$. The fresh air temperature at the compressor inlet is 295 K. The air temperature at the inlet of the turbine is 1300 K. The cycle operates with a compressor efficiency of 80% and a turbine efficiency of 80%. The cycle operates 360 days per year.

(a) Using the standard-air assumptions, determine the thermal efficiency of the cycle.

(b) If the power plant operates with a regenerator with an effectiveness of 0.78, determine the thermal efficiency of the cycle and annual conservation of fuel.

Solution:

Assume that the cycle is at steady-state flow and the changes in kinetic and potential energy are negligible. Heat capacity of the air is temperature dependent, and the air is an ideal gas.

(a) Basis: 1 kg/s air. Data given: $\eta_{\text{turb}} = 0.8$, $\eta_{\text{comp}} = 0.8$, $r_p = P_2/P_1 = 9$.

Process 1–2: isentropic compression.

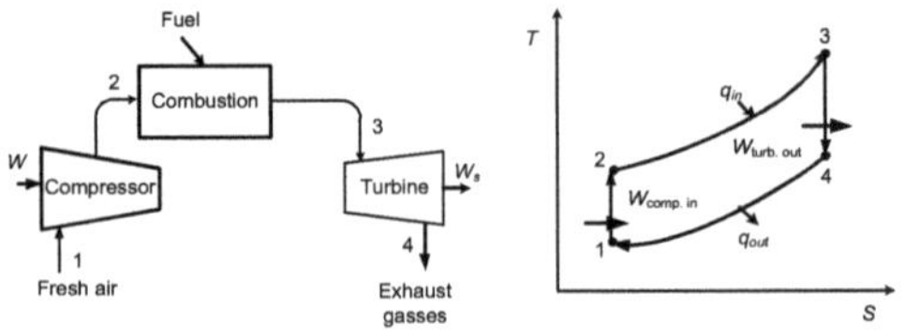

Data from (Table D1):

$$T_1 = 295 \text{ K}, \ H_1 = 295.17 \text{ kJ/kg}, \ P_{r1} = 1.3068$$

P_r shows the relative pressure defined in Eq. (7.36):

$\frac{P_{r2}}{P_{r1}} = \frac{P_2}{P_1} = r_p \rightarrow P_{r2} = (9)(1.3068) = 11.76.$

Approximations from Table D1 for the compressor exit:

$P_{r2} = 11.76$: $T_2 = 550$ K and $H_2 = 555.74$ kJ/kg.

Process 3–4: isentropic expansion in the turbine as seen on the *TS* diagram above

$$T_3 = 1300 \text{ K}, H_3 = 1395.97 \text{ kJ/kg}, P_4/P_3 = 1/r_p = 1/9, P_{r3} = 330.9$$

$$\frac{P_{r4}}{P_{r3}} = \frac{P_4}{P_3} \rightarrow P_{r4} = \left(\frac{1}{9}\right)(330.9) = 36.76$$

Approximate values from Table D1 at the exit of turbine:

at $P_{r4} = 36.76$, $T_4 = 745$ K, $H_4 = 761.87$ kJ/kg.

The work input to the compressor:

$$W_{\text{comp.in}} = \frac{H_2 - H_1}{\eta_{\text{comp}}} = \frac{(335.74 - 295.17)\ \text{kJ/kg}}{0.8} = 325.7\ \text{kJ/kg}$$

The work output of the turbine:

$$W_{\text{turb.out}} = \eta_{\text{turb}}(H_4 - H_3) = 0.8(761.87 - 1395.97) = -507.3\ \text{kJ/kg}$$

The back-work ratio r_{bw}:

$$r_{bw} = \frac{W_{\text{comp.in}}}{|W_{\text{turb.out}}|} = \frac{325.7}{507.3} = 0.64$$

This shows that 64% of the turbine output has been used in the compressor.

Work output:

$$W_{\text{net}} = W_{\text{out}} - W_{\text{in}} = -181.6\ \text{kJ/kg}$$

Enthalpy at the compressor outlet:

$$H_{2\text{a}} = H_1 + W_{\text{comp.in}} = (295.17 + 325.70)\ \text{kJ/kg} = 620.87\ \text{kJ/kg}$$

The amount of heat added:

$$q_{\text{in}} = H_3 - H_{2a} = 1395.97 – 620.87 = 775.1\ \text{kJ/kg}$$

The thermal efficiency:

$$\eta_{\text{th}} = \frac{|W_{\text{net}}|}{q_{\text{in}}} = \frac{181.6}{775.1} = 0.234 \text{ or } 23.4\%$$

The temperature of the exhaust air, T_4, and the actual enthalpy are estimated from the energy balance:

$$W_{\text{turb out}} = H_{4a} - H_3$$
$$H_{4\text{a}} = 1395.97 - 507.3 = 888.67\ \text{kJ/kg} \rightarrow T_4 = 860\ \text{K (From Table D1)}$$

Advances in the compressor and turbine designs with minimal losses increase the efficiency of these components. In turn, a significant increase in the thermal efficiency of the cycle is possible.

(b) $T_4 = 860$ K and $T_2 = 550$ K, since $T_4 > T_2$ regeneration is possible.

Regeneration with effectiveness of $\varepsilon = 0.78$:

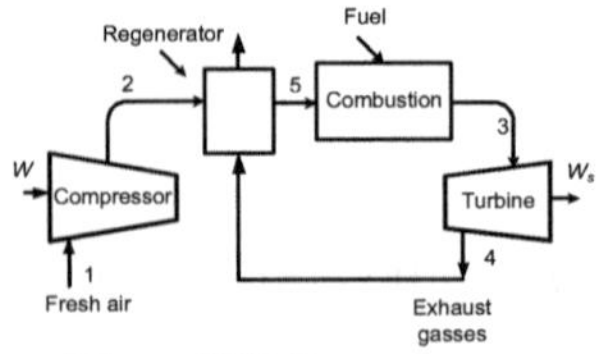

$$\varepsilon = \left(\frac{H_5 - H_2}{H_4 - H_2}\right) = \frac{H_5 - 620.9}{888.67 - 620.9} = 0.78 \rightarrow H_5 = 829.80\ \text{kJ/kg}$$
$$q_{in} = H_3 - H_5 = 1395.97\text{–}829.8 = 566.2\ \text{kJ/kg}$$

This represents a conservation of 840.2 – 566.2 = 274.0 kJ/kg from the fuel required.

The thermal efficiency:

$$\eta_{th} = \frac{|W_{net}|}{q_{in}} = \frac{181.6}{566.2} = 0.32 \text{ or } 32\%$$

Days of operation 360, hours of operation per year = 360(24) = 8640 h/year.

Conserved fuel: (274 kJ/kg) (1 kg/s) (8640 h/year) = 2,367,360 kWh/year.

After the regeneration, the thermal efficiency has increased from 23 to 32% in the actual Brayton cycle operation. The addition of a regenerator (operating without thermal losses) does not affect the net work output of the cycle

9.2.2 Conservation of Energy by Process Improvements

Some of the possible modifications in operation of steam power plants to increase the efficiency are [1, 2, 5]:

- *Increasing the efficiency of a Rankine cycle by reducing the condenser pressure.* Example 9.2 illustrates the estimation of efficiency of a Rankine cycle operating at two different discharge pressures. The thermal efficiency increased from 27.6 to 33.4% by reducing the condenser pressure from 78.5 to 15.0 kPa. However, the quality of the discharged steam decreased from 0.9 to 0.84, which is not desirable for the blades of the turbine. Savings are considerable since the power output increased. Example 9.3 compares the actual thermal efficiency with the maximum thermal efficiency obtained from fully reversible Carnot cycle.
- *Increasing the efficiency of a Rankine cycle by increasing the boiler pressure.* Example 9.4 illustrates the estimation of thermal efficiency of a Rankine cycle operating at a higher pressure. The thermal efficiency increased from 0.276

to 0.326 by increasing the boiler pressure from 3500 to 9800 kPa. However, the quality of the discharged steam decreased from 0.9 to 0.80, which is not desirable for the blades of the turbine.

- *Increasing the efficiency of a Rankine cycle by increasing the boiler temperature.* Example 9.5 illustrates the estimation of thermal efficiency of a Rankine cycle operating at a higher temperature. The thermal efficiency increased from 0.276 to 0.298 by increasing the boiler temperature from 400 to 525°C. The quality of the discharged steam increased from 0.9 to 0.96, which is desirable for the protection of the turbine blades. Example 9.6 compares the estimated thermal efficiency with the maximum thermal efficiency obtained from the full reversible Carnot cycle.

***Example 9.2* Increasing the efficiency of a Rankine cycle by reducing the condenser pressure**

A steam power plant is operating on the simple ideal Rankine cycle. The steam mass flow rate is 20 kg/s. The steam enters the turbine at 3500 kPa and 400°C. Discharge pressure of the steam from the turbine is 78.5 kPa.

(a) Determine the thermal efficiency of the cycle.

(b) If the pressure of the discharge steam is reduced to 15 kPa determine the thermal efficiency.

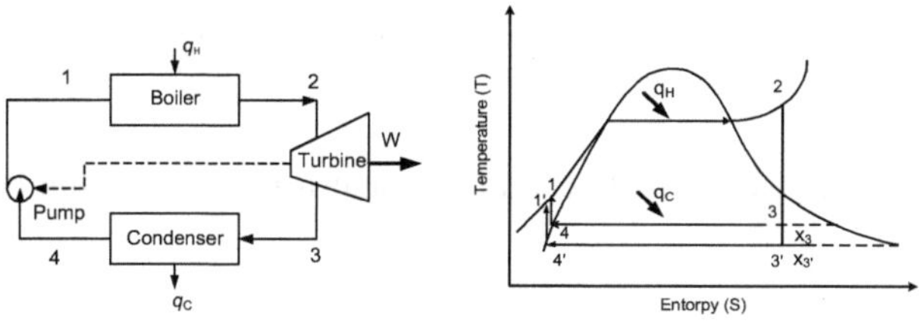

Solution:

Assume that the cycle is at steady-state flow and the changes in kinetic and potential energy are negligible. Efficiency of pump and turbine is 100%.

(a) $\dot{m}_s = 20.0\,\text{kg/s}$. Using the data from the steam tables.

Superheated steam:

$$P_2 = P_1 = 3500\,\text{kPa}, H_2 = 3224.2\,\text{kJ/kg}, S_2 = 6.8443\,\text{kJ/kg K}, T_2 = 400°\text{C}.$$

Saturated steam:

$$P_3 = P_4 = 78.5\,\text{kPa}(T_{\text{sat}} = 366.15\,\text{K}), V_4 = 0.001038\,\text{m}^3/\text{kg}.$$
$$H_{3\text{sat vap}} = 2665.0\,\text{kJ/kg}, H_4 = H_{3\text{sat liq}} = 389.6\,\text{kJ/kg},$$
$$S_{3\text{sat vap}} = 7.4416\,\text{kJ/kg K}, S_{3\text{sat liq}} = 1.2271\,\text{kJ/kgK}.$$

Basis: 1 kg/s steam. With a pump efficiency of $\eta_{\text{pump}} = 100\%$

$$W_{\text{p,in}} = V_1(P_1 - P_4) = (0.001038)(3500 - 78.5)\left(\frac{1\text{kJ}}{1\ \text{kPa m}^3}\right) = 3.55\ \text{kJ/kg}$$
$$H_1 = H_4 + W_{p,\text{in}} = 393.1\ \text{kJ/kg}$$

Isentropic process $S_1 = S_4$ and $S_3 = S_2$.

The quality of the discharged wet steam ($S_2 < S_{3\text{sat vap}}$): 6.8443 < 7.4416

$$x_3 = (6.8463 - 1.2271)/(7.4416 - 1.2271) = 0.90$$
$$H_3 = 389.6(1 - 0.9) + 2665.0 \times 0.9 = 2437.5\ \text{kJ/kg}$$

Heat interactions:

$$q_{\text{in}} = H_2 - H_1 = 3224.2{-}393.1 = 2831.1\ \text{kJ/kg}$$

$$q_{\text{out}} = -(H_4 - H_3) = -(389.6{-}2437.5) = 2048.0\ \text{kJ/kg}$$

The thermodynamic efficiency of the cycle:

$$\eta_{\text{th}} = 1 - \frac{|q_{\text{out}}|}{q_{\text{in}}} = 0.276 \text{ or } 27.6\%.$$

Therefore, the plant uses only 27.6% of the heat it received in the boiler.

Turbine work output:

$$W_{\text{out}} = H_3 - H_2 = 3224.2{-}2437.5 = -786.7\ \text{kJ/kg}$$

(b) Steam properties:

$$P_3 = P_4 = 15\ \text{kPa},\ T_{\text{sat}} = 327.15\ \text{K},\ V_4 = 0.001014\ \text{m}^3/\text{kg}$$
$$H_{3\text{sat vap}} = 2599.2\ \text{kJ/kg},\ H_4 = H_{3\text{sat liq}} = 226.0\ \text{kJ/kg},$$
$$S_{3\text{sat vap}} = 8.0093\ \text{kJ/kg K},\ S_{3\text{sat liq}} = 0.7550\ \text{kJ/kg K}$$

With a pump efficiency of $\eta_{\text{pump}} = 100\%$

$$W_{p,\text{in}} = V_1(P_1 - P_4) = (0.001014)(3500 - 15)\left(\frac{1\text{kJ}}{1\ \text{kPa m}^3}\right) = 3.53\ \text{kJ/kg}$$
$$H_1 = H_4 + W_{p,in} = 226.0 + 3.53 = 229.5\ \text{kJ/kg}$$

Isentropic process $S_1 = S_4$ and $S_3 = S_2$.

The quality of the discharged wet steam ($S_3 < S_{3satvap}$): 6.8443 < 8.0093

$$x_{3'} = (6.8443 - 0.7550)/(8.0093 - 0.7550) = 0.84$$
$$H_{3'} = 226.0(1 - 0.84) + 2599.2 \times 0.84 = 2219.5 \text{ kJ/kg}$$

Heat interactions : $q_{in} = H_2 - H_1 = 3224.2 - 229.5 = 2994.7 \text{ kJ/kg}$
$q_{in} = H_2 - H_1 = 3224.2 - 229.5 = 2994. \text{ kJ/kg}$

$$q_{out} = -(H_4 - H_{3'}) = -(226.0 - 2219.5.5) = 1993.5 \text{ kJ/kg}$$
(heat discharged by the condenser)

The thermal efficiency of the cycle:

$$\eta_{th} = 1 - \frac{|q_{out}|}{q_{in}} = 0.334 \text{ or } 33.4\%$$

Therefore, the plant uses only 33.4% of the heat it received in the boiler:

Turbine work output:

$$W_{out} = H_{3'} - H_2 = 2219.5 - 3224.2 = -1004.7 \text{ kJ/kg (heat produced by the turbine)}$$

Cycle work out:

$$W_{net} = (q_{out} - q_{in}) = (1993.5 - 2994.7) = -1001.2 \text{ kJ/kg}$$

For a 360 days of operation = (360)(24) = 8640 h/year

The annual increase in power output:

(20 kg/s)(1001.2 – 783.1) kJ/kg (8640) h/year = 37,687,680 kWh/year.

The thermal efficiency increased from 0.276 to 0.334 by reducing the condenser pressure from 78.5 kPa to 15.0 kPa. However, the quality of the discharged steam decreased from 0.9 to 0.84, which is not desirable for the blades of the turbine. Savings are considerable.

Example 9.3 **Maximum possible efficiency calculation in Example** 9.2

Estimate the maximum possible efficiency for parts (a) and (b) in Example 9.2 and compare them with those obtained in parts (a) and (b) in Example 9.2.

Solution:

(a) The thermal efficiency of a Carnot cycle operating between the same temperature limits:

$$T_{\min} = T_3 = 366.15\,\mathrm{K} \text{ and } T_{\max} = T_2 = 673.15\,\mathrm{K}$$

$$\eta_{\mathrm{th,Carnot}} = 1 - \frac{T_{\min}}{T_{\max}} = 1 - \frac{366.15\,\mathrm{K}}{673.15\,\mathrm{K}} = 0.456 \text{ or } 45.6\%$$

The difference between the two efficiencies occurs because of the large temperature differences. The ratio of the efficiencies, $\eta_{\mathrm{th}}/\eta_{\mathrm{th,Carnot}} = 0.276/0.456 = 0.60$ shows that only 60% of the possible efficiency is achieved in the cycle.

(b) The thermal efficiency of a Carnot cycle operating between the same temperature limits.

$T_{\min} = T_{3'} = 327.15$ K and $T_{\max} = T_2 = 673.15$ K

$$\eta_{th,\mathrm{Carnot}} = 1 - \frac{T_{\min}}{T_{\max}} = 1 - \frac{327.15}{673.15} = 0.514 \text{ of } 51.4\%$$

The ratio of the efficiencies, $\eta_{\mathrm{th}}/\eta_{\mathrm{th,Carnot}} = 0.334/0.51 = 0.65$ shows that only 65% of possible efficiency is achieved in the cycle.

Example 9.4 **Increasing the efficiency of a Rankine cycle by increasing the boiler pressure**

A steam power plant is operating on the simple ideal Rankine cycle. The steam mass flow rate is 20 kg/s. The steam enters the turbine at 3500 kPa and 400°C. Discharge pressure of the steam from the turbine is 78.5 kPa. If the pressure of the boiler is increased to 9800 kPa while maintaining the turbine inlet temperature at 400°C, determine the change in the thermal efficiency.

Solution:

Assume that the cycle is at steady-state flow and the changes in kinetic and potential energy are negligible. Pump efficiency of $\eta_{\mathrm{pump}} = 1$.

Basis: 1 kg/s steam. Using the data from the steam tables.

From Example 9.2 part a:

$$\eta_{\mathrm{th}} = 0.276 \text{ and } W_{\mathrm{net}} = 783.1\,\mathrm{kJ/kg} \text{ at } 3500\,\mathrm{kPa}.$$

(a) Superheated steam:
$P_2 = P_1 = 9800$ kPa; $H_2 = 3104.2$ kJ/kg : $S_2 = 6.2325$ kJ/kg K,
$T_2 = 400°\mathrm{C}$

Saturated steam:

$$P_3 = P_4 = 78.5\ \text{kPa},\ T_{sat} = 366.15\ \text{K},\ V_4 = 0.001038\ \text{m}^3/\text{kg}.$$
$$H_{3\text{sat vap}} = 2665.0\ \text{kJ/kg};\ H_4 = H_{3\text{sat liq}} = 389.6\ \text{kJ/kg};$$
$$S_{3\text{sat vap}} = 7.4416\ \text{kJ/kg K};\ S_{3\text{sat liq}} = 1.2271\ \text{kJ/kg K}$$
$$W_{p,\text{in}} = V_1(P_1 - P_4) = (0.001038)(9800 - 78.5)\left(\frac{1\text{kJ}}{1\text{kPa m}^3}\right) = 10.1\ \text{kJ/kg}$$
$$H_1 = H_4 + W_{\text{p,in}} = 389.6 + 10.1 = 399.7\ \text{kJ/kg}$$

Isentropic process $S_1 = S_4$ and $S_3 = S_2$.

The quality of the discharged wet steam x_3: ($S_3 < S_{3\text{sat vap}}$): 6.2325 < 7.4416

$$x_3 = (6.2325 - 1.2271)/(7.4416 - 1.2271) = 0.80$$
$$H_3 = 389.6(1 - 0.8) + 2665.0 \times 0.8 = 2210.0\ \text{kJ/kg}$$

Heat interactions:

$$q_{\text{in}} = H_2 - H_1 = 3104.2 - 399.7 = 2704.5\ \text{kJ/kg}$$

$$q_{\text{out}} = -(H_4 - H_3) = -(389.6 - 2210.0)$$
$$= 1820.4\ \text{kJ/kg (heat discharged by the condenser)}$$

The thermodynamic efficiency of the cycle:

$$\eta_{\text{th}} = 1 - \frac{q_{\text{out}}}{q_{\text{in}}} = 0.326\ \text{or}\ 32.6\%$$

Therefore, the plant uses only 32.6% of the heat it received in the boiler.

The thermal efficiency increased from 0.276 to 0.326 by increasing the boiler pressure from 3500 to 9800 kPa. However, the quality of the discharged steam decreased from 0.9 to 0.80, which is not desirable for the blades of the turbine.

Turbine work output:

$$W_{\text{out}} = H_3 - H_2 = (2210.0 - 3104.2)\ \text{kJ/kg} = -894.2\ \text{kJ/kg}$$

Cycle work out:

$$W_{\text{net}} = (q_{\text{out}} - q_{\text{in}}) = (1820.4 - 2704.5)\ \text{kJ/kg} = -884.1\ \text{kJ/kg}$$

(b) $\dot{m}_s = 20.0$ kg/s, and for a 360 days of operation

$= (360)(24) = 8640$ h/year.

Annual increase in power output:

(20 kg/s)(884.1 – 783.1)kJ/kg (8640) h/year = 17,452,800 kWh/year.

Example 9.5 **Increasing the efficiency of a Rankine cycle by increasing the boiler temperature**

A steam power plant is operating on the simple ideal Rankine cycle. The steam mass flow rate is 20 kg/s. The steam enters the turbine at 3500 kPa and 400°C. Discharge pressure of the steam from the turbine is 78.5 kPa. If the temperature of the boiler is increased to 525°C while maintaining the pressure at 3500 kPa, determine the change in thermal efficiency.

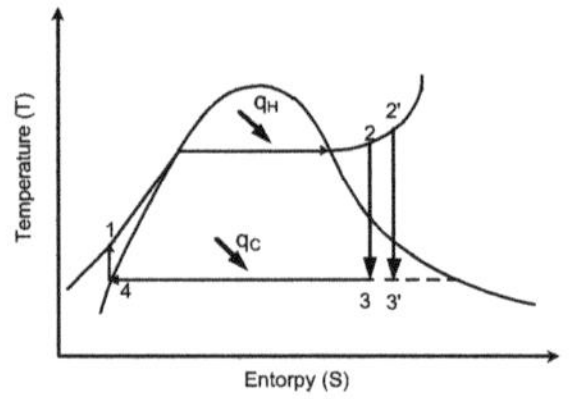

Solution:

Assume that the cycle is at steady-state flow and the changes in kinetic and potential energy are negligible.

(a) Basis: 1 kg/s steam. Using the data from the steam tables:

From Example 9.2 part a:

$$\eta_{\text{th}} = 0.276 \text{ and} W_{\text{net}} = 783.1 \text{ kJ/kg at } 400°\text{C}.$$

Superheated steam:

$$P_2 = P_1 = 3500 \text{ kPa}, H_2 = 3506.9 \text{ kJ/kg}, S_2 = 7.2297 \text{ kJ/kg K}, T_2 = 525°\text{C}.$$

Saturated steam:

$$P_3 = P_4 = 78.5 \text{ kPa}, T_{\text{sat}} = 366.15 \text{ K}, V_4 = 0.001038 \text{ m}^3/\text{kg}.$$
$$H_{3\text{sat vap}} = 2665.0 \text{ kJ/kg}, H_{3\text{sat liq}} = 389.6 \text{ kJ/kg},$$
$$S_{3\text{sat vap}} = 7.4416 \text{ kJ/kg K}, S_{3\text{sat liq}} = 1.2271 \text{ kJ/kg K}$$
$$W_{\text{p,in}} = V_1(P_1 - P_4) = (0.001038)(3500 - 78.5)\left(\frac{1\text{kJ}}{1\text{kPa m}^3}\right) = 3.55 \text{ kJ/kg}$$

$$H_1 = H_4 + W_{p,in} = 389.6 + 3.55 = 393.15 \text{ kJ/kg}$$

Isentropic process $S_1 = S_4$ and $S_3 = S_2$.

The quality of the discharged wet steam ($S_2 < S_{3\text{sat vap}}$): 7.2297 < 7.4416

$$x_{3'} = (7.2297 - 1.2271)/(7.4416 - 1.2271) = 0.96$$
$$H_{3'} = 389.6(1 - 0.96) + 2665.0 \times 0.96 = 2574.0 \text{ kJ/kg}$$

Heat interactions:

$$q_{\text{in}} = H_2 - H_1 = 3506.9\text{–}393.1 = 3113.8 \text{ kJ/kg}$$

$$q_{\text{out}} = -(H_4 - H_{3'}) = -(389.6\text{–}2574.0)$$
$$= 2184.4 \text{ kJ/kg (heat received by the cooling medium)}$$

The thermodynamic efficiency of the cycle is

$\eta_{\text{th}} = 1 - \frac{q_{\text{out}}}{q_{\text{in}}} = 0.298$ or $29.8\% \eta_{\text{th}} = 1 - \frac{q_{\text{out}}}{q_{\text{in}}} = 0.298$ or 29.8%

Therefore, the plant uses only 29.8% of the heat it received in the boiler.

The thermal efficiency increased from 0.276 to 0.298 by increasing the boiler temperature from 400°C to 525°C. The quality of the discharged steam increased from 0.9 to 0.96, which is desirable for the protection of the turbine blades.

Turbine work out:

$$W_{\text{out}} = H_{3'} - H_2 = (2574.0\text{–}3506.9) = -932.9 \text{ kJ/kg}$$

Cycle work out:

$$W_{\text{net}} = (q_{\text{out}} - q_{\text{in}}) = (2184.4 - 3113.8) = -929.4 \text{ kJ/kg}$$

(b) $\dot{m}_s = 20.0$ kg/s and for a 360 days of operation

$= (360)(24) = 8640$ h/year

Annual increase in power output:

(20 kg/s)(929.4 – 783.1) kJ/kg (8640) h/year = 25.280,640 kWh/year.

Example 9.6 **Estimation of maximum possible efficiencies in Example** 9.5

Estimate the maximum possible efficiency for parts (a) and (b) in Example 9.5 and compare them with those obtained in parts (a) and (b) in Example 9.5.

Solution:

(a) The thermal efficiency of a Carnot cycle operating between the temperature limits of $T_{min} = 366.15$ K and $T_{max} = 673.15$ K:

$$\eta_{th,Carnot} = 1 - \frac{T_{min}}{T_{max}} = 1 - \frac{366.15 \text{ K}}{673.15 \text{ K}} = 0.45 \text{ or } 45\%$$

The ratio of the efficiencies, $\eta_{th}/\eta_{th,Carnot} = 0.276/0.45 = 0.61$ shows that only 61% of the possible efficiency is achieved in the cycle with the boiler temperature of 400°C.

(b) The thermal efficiency of a Carnot cycle operating between the temperature limits of $T_{min} = 366.15$ K and $T_{max} = 798.15$ K:

$$\eta_{th,Carnot} = 1 - \frac{T_{min}}{T_{max}} = 1 - \frac{366.15 \text{ K}}{798.15 \text{ K}} = 0.54 \text{ or } 54\%$$

The ratio of the efficiencies, $\eta_{th}/\eta_{th,Carnot} = 0.298/0.54 = 0.55$ shows that only 55% of the possible efficiency is achieved in the cycle with the boiler temperature of 525°C.

9.2.3 Energy Conservation in the Compression and Expansion Work

It is possible to save energy in the compression work by minimizing the friction, turbulence, heat transfer and other losses. A practical way of energy conservation is to keep the specific volume of the gas small during the compression work. This is possible by maintaining the temperature of the gas low as the specific volume is proportional to temperature. Therefore, cooling the gas as it is compressed may reduce the cost of compression work in a multistage compression with intercooling as seen in Fig. 9.2. The gas is cooled to the initial temperature between the compression stages by passing the gas through a heat exchanger called the intercooler. Energy recovery by intercooling may be significant especially when a gas is to be compressed to very high pressure. Example 9.7 illustrates the energy conservation in a two-stage compression work by intercooling.

As Fig. 9.2 shows, the work saved varies with the value of intermediate pressure P_x, which needs to be optimized. The total work input for a two-stage compression process is the sum of the work inputs for each stage

$$W_{comp} = W_{comp1} + W_{comp2} \tag{9.4}$$

$$W_{comp} = \frac{\gamma R T_1}{MW(\gamma - 1)}\left[\left(\frac{P_x}{P_1}\right)^{(\gamma-1)/\gamma}\right] + \frac{\gamma R T_1}{MW(\gamma - 1)}\left[\left(\frac{P_2}{P_x}\right)^{(\gamma-1)/\gamma}\right]$$

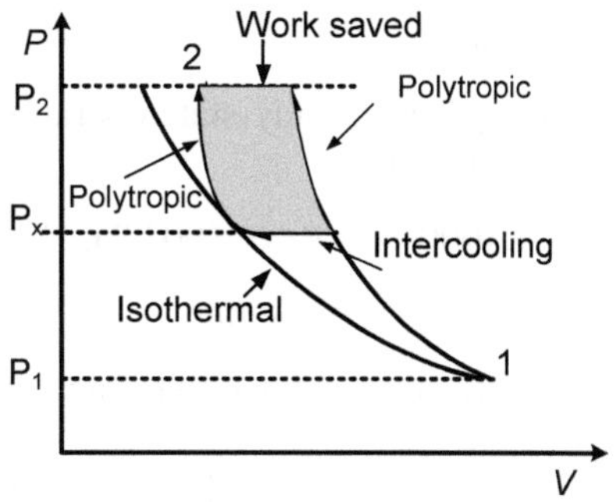

Fig. 9.2 Energy conservation in the compression work by intercooling; the work saved appears between two polytropic compressions starting at the second stage with the pressure P_x

In Eq. (9.4), P_x is the only variable. The optimum value of P_x is obtained by differentiation of Eq. (9.4) with respect to P_x and setting the resulting expression equal to zero. Then, the optimum value of P_x becomes

$$P_x = (P_1 P_2)^{1/2} \text{ or } \frac{P_x}{P_1} = \frac{P_2}{P_x} \quad (9.5)$$

Therefore, energy conservation will be maximum, when the pressure ratio across each stage of the compressor is the same and compression work at each stage becomes identical

$$W_{\text{comp1}} = W_{\text{comp2}} \quad (9.6)$$

$$W_{\text{comp}} = \frac{2\gamma R T_1}{MW(\gamma - 1)} \left[\left(\frac{P_x}{P_1} \right)^{(\gamma-1)/\gamma} \right] \quad (9.7)$$

Example 9.8 illustrates the estimation of minimum and actual power required by a compressor. Example 9.9 illustrates how to produce power out of a cryogenic expansion process. Reduction of pressure by using throttling valve wastes energy. Replacing the throttling valve with a turbine produces power and hence conserves electricity [1].

Example 9.7 **Energy conservation in a two-stage compression work by intercooling**

Air with a flow rate of 2 kg/s is compressed in a steady state and reversible process from an inlet state of 100 kPa and 300 K to an exit pressure of 1000 kPa. Estimate the work for (a) polytropic compression with $\gamma = 1.3$, and (b) ideal two-stage polytropic compression with intercooling using the same polytropic exponent of $\gamma = 1.3$, (c) estimate conserved compression work by intercooling and electricity per year if the compressor is operated 360 days per year.

Solution:

Assumptions: steady-state operation; air is ideal gas; kinetic and potential energies are negligible.

$P_2 = 1000\,\text{kPa}, P_1 = 100\,\text{kPa}, T_1 = 300\,\text{K}, \gamma = 1.3, MW_{\text{air}} = 29\,\text{kg/kmole}.$

Basis 1 kg/s airflow rate.

(a) Work needed for polytropic compression with $\gamma = 1.3$

$$W_{\text{comp}} = \frac{\gamma R T_1}{MW(\gamma - 1)}\left[\left(\frac{P_2}{P_1}\right)^{(\gamma-1)/\gamma}\right]$$

$$= \frac{1.3(8.314\ \text{kJ/kmol K})(300\ \text{K})}{29\ \text{kg/kmol}\ (1.3 - 1)}\left[\left(\frac{1000}{300}\right)^{(1.3-1)/1.3} - 1\right] = 261.3\ \text{kJ/kg}$$

(b) Ideal two-stage polytropic compression with intercooling ($\gamma = 1.3$)

$$P_x = (P_1 P_2)^{1/2} = (100 \times 1000)^{1/2} = 316.2\ \text{kPa}$$

$$W_{\text{comp}} = \frac{2\gamma R T_1}{MW(\gamma - 1)}\left[\left(\frac{P_x}{P_1}\right)^{(\gamma-1)/\gamma}\right] = \frac{2(1.3)(8.314\ \text{kJ/kmole K})(300\ \text{K})}{29\ \text{kg/kmole}\ (1.3 - 1)}$$

$$\left[\left(\frac{316.2}{100}\right)^{(1.3-1)/1.3} - 1\right] = 226.8\ \text{kJ/kg}$$

Recovered energy = (261.3 – 226.8) kJ/kg = 34.5 kJ/kg.

Reduction in energy use: $\frac{261.3-226.8}{261.3} = 0.13$ or 13%

(c) Conservation of compression work = (2 kg/s)(261.3 – 226.8) kJ/kg = 69 kW.

Yearly conserved work: (69 kW) (8640 h/year) = 596,160 kWh/year.

The compression work has been reduced by 13% when two stages of polytropic compression are used instead of single polytropic compression and conserved 596,160 kWh/year.

Example 9.8 **Compressor efficiency and power input**

An adiabatic compressor is used to compress air from 100 kPa and 290 K to a pressure of 900 kPa at a steady-state operation. The isentropic efficiency of the compressor is 80%. The airflow rate is 0.4 kg/s. Determine the minimum and actual power needed by the compressor.

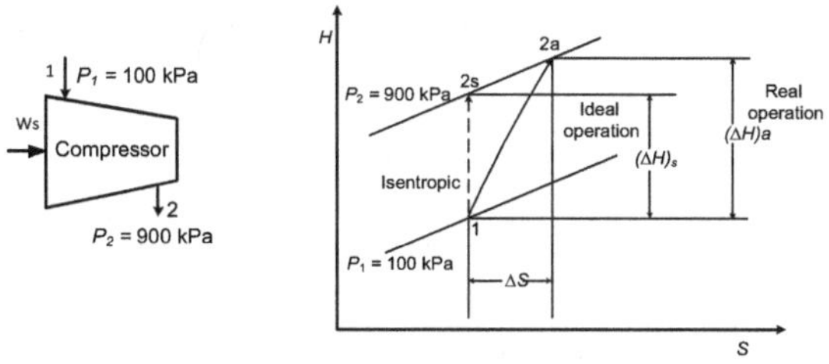

Solution:

Assume: Steady-state adiabatic operation. Air is the ideal gas. The changes in kinetic and potential energies are negligible.

Enthalpies of ideal gas depend on temperature only.

The air mass flow rate = 0.4 kg/s, $\eta_C = 0.8$

Inlet conditions (Table D1):

$P_1 = 100\,\text{kPa}, T_1 = 290\,\text{K}, H_1 = 290.16\,\text{kJ/kg}, Pr_1 = 1.2311.$

Exit conditions (Table D1):

$$P_2 = 900\,\text{kPa}$$

$$Pr_2 = Pr_1\frac{P_2}{P_1} = 1.2311\left(\frac{900\text{ kPa}}{100\text{ kPa}}\right) = 11.1$$

From Table D1: for $\text{Pr}_2 = 11.1$, we find $T_2 = 540$ K and $H_{2\text{s}} = 544.35$ kJ/kg

$$\eta_C = \frac{H_{2s} - H_1}{H_{2a} - H_1} = 0.8 \rightarrow H_{2\text{a}}$$

$H_{2\text{a}} = 607.9$ kJ/kg and $T_2 = 600$ K (Table D1) (Approximate).

As seen from the *TS* diagram above $H_{2\text{a}}$ is the actual enthalpy at the exit.

Actual power required:

$$\dot{m}\Delta H_a = \dot{W}_{\text{net,in}} = \dot{m}(H_{2a} - H_1) = (0.4\text{ kg/s})(607.9 - 290.16)\text{ kJ/kg} = 127.1\text{ kW}$$

(work received by the compressor).

Minimum power required:

$$\dot{m}\Delta H_s = \dot{W}_{\text{net,in}} = \dot{m}(H_{2s} - H_1) = (0.4\text{ kg/s})(544.3 - 290.1)\text{ kJ/kg} = 101.7\text{ kW}$$

If the operation was ideal the rate of conserved energy would be 25.4 kW or 20% savings.

Example 9.9 **Energy conservation in expansion by replacing a throttle valve with a turbine**

A cryogenic manufacturing plant handles liquid methane at 115 K and 5000 kPa at a rate of 0.3 m^3/s. In the plant a throttling valve reduces the pressure of liquid methane to 1000 kPa. A new process considered replaces the throttling valve with a turbine to produce power while reducing the pressure to 1000 kPa at 110 K. Using the data for the properties of liquid methane below, estimate the power that can be produced by the turbine.

T K	P kPa	H kJ/kg	S kJ/kg K	C_p kJ/kg K	ρ kg/m^3
110	1000	209.0	4.875	3.471	425.8
110	2000	210.5	4.867	3.460	426.6
110	5000	215.0	4.844	3.432	429.1
120	1000	244.1	5.180	3.543	411.0
120	2000	245.4	5.171	3.528	412.0
120	5000	249.6	5.145	3.486	415.2

Source Çengel and Boles [1]

Solution:

Assumptions: steady state and reversible operation; adiabatic turbine, methane is the ideal gas; kinetic and potential energies are negligible.

(a) $P_1 = 5000$ kPa, $T_1 = 115$ K, $Q_1 = 0.30$ m^3/s.

$$H_1 = 232.3\,\text{kJ/kg} = (215.0 + 249.6)/2,$$
$$\rho_1 = 422.15\ \text{kg/m}^3 = (429.1 + 415.2)/2.$$
$$P_2 = 1000\ \text{kPa},\ H_2 = 209.0\,\text{kJ/kg}.$$

Unit cost of electricity = \$0.09/kWh.

Mass flow rate:

$$\dot{m} = \rho\dot{Q}_1 = \left(422.15\ \text{kg/m}^3\right)\left(0.3\ \text{m}^3/\text{s}\right) = 126.6\ \text{kg/s}$$

Power produced:

$$\dot{W}_{\text{out}} = \dot{m}(H_1 - H_2) = (126.6\ \text{kg/s})(232.5 - 209.0)\ \text{kJ/kg} = 2949.8\ \text{kW}$$

Annual power production:

$$\dot{W}_{\text{out}}\Delta t = (2949.8\ \text{kW})(360)(24)\text{h/year} = 25{,}486{,}099\ \text{kWh/year}$$

Reduction of pressure by using throttling valve wastes the potential of power production. Replacing the valve with a turbine will produce power and hence conserve electricity.

9.2.4 Conservation of Energy by High-Efficiency Electric Motors

Practically all compressors are powered by electric motor. Electric motors cannot convert the electrical energy they consume into mechanical energy completely. Electric motor efficiency is defined by

$$\eta_{\text{Motor}} = \frac{\text{Mechanical power}}{\text{Electrical power}} = \frac{\dot{W}_{\text{comp}}}{\dot{W}_{\text{elect}}} \tag{9.8}$$

Motor efficiency range: $0.7 < \eta_{\text{motor}} < 96$. The portion of electric energy that is not converted to mechanical power is converted to heat, which is mostly unusable. For example, if no transmission losses occur [7, 8]:

- A motor with an efficiency of 80% will draw an electrical power of 1/0.8 = 1.25 kW for each kW of shaft power it delivers.
- If the motor is 95% efficient, then it will draw 1/0.95 = 1.05 kW only to deliver 1 kW of shaft work.
- Therefore, between these two motors, electric power conservation is

$$1.25(\text{h}_{\text{motor}} = 95\%)\ \text{kW} - 1.05(\text{h}_{\text{motor}} = 95\%) = 0.20\ \text{kW}.$$

High-efficiency motors are more expensive, but their operation saves energy. Saved energy is estimated by

$$\dot{W}_{\text{elect.saved}} = \dot{W}_{\text{elect.std}} - \dot{W}_{\text{elect.efficient}} = \dot{W}_{\text{comp}}\left(\frac{1}{\eta_{\text{std}}} - \frac{1}{\eta_{\text{efficient}}}\right) \tag{9.9}$$

$$\dot{W}_{\text{elect.saved}} = (\text{Rated power})(\text{Load factor})\left(\frac{1}{\eta_{\text{std}}} - \frac{1}{\eta_{\text{efficient}}}\right) \tag{9.10}$$

where the *rated power* is the nominal power delivered at full load of the motor and listed on its label. *Load factor* is the fraction of the rated power at which the motor normally operates. Annual saving is

$$\text{Annual energy saving} = \left(\dot{W}_{\text{elect.saved}}\right)(\text{Annual operation hours}) \tag{9.11}$$

- A compressor that operates at partial load causes the motor to operate less efficiently. The efficiency of the motor will increase with the load.
- Using the cold outside air for compressor intake lowers the compressor work and conserves energy.

9.3 Energy Conservation in Home Heating and Cooling

Space heating and cooling as well as water heating at home account for a considerable part of the utility bill. To conserve energy, one should identify from where the home is losing energy, assign priorities, and form an efficiency plan that improves efficiency and reduces costs. For example, the potential energy savings from reducing drafts in a home may range from 5 to 30% per year. Heat loss through the ceiling and walls in your home could be very large if the insulation levels are less than the recommended minimum. Inspect heating and cooling equipment annually, or as recommended by the manufacturer. If the unit is more than 15 years old, the newer and energy-efficient units may reduce the cost [1, 2]. Use of natural gas has the largest source of fuel for heating houses in the USA. In colder climates, windows that are gas filled with low emissivity coatings on the glass reduce heat loss. In warmer climates, windows with selective coatings may reduce heat gain [7]. Setting the thermostat low in the winter and high in the summer as comfortably as possible may reduce the cost of heating and cooling. Also, the followings can reduce the cost:

- Clean or replace filters on furnaces once a month or as needed.
- Clean baseboard heaters and radiators and make sure that they are not blocked.
- During summer, keep the window coverings closed during the day to lower solar gain.
- Select energy-efficient products when you buy new heating and cooling equipment.

9.3.1 Home Heating by Fossil Fuels

Residential furnaces have a heat input rate of less than 225,000 Btu/h (66,000 watts) and residential boilers have a heat input rate of less than 300,000 Btu/h (88,000 watts). The residential furnace is an appliance that provides heated air with a blower to circulate air through the duct distribution system. The residential boiler is a cast-iron, steel, aluminum, or copper pressure vessel heat exchanger designed to burn fossil fuels to heat to a suitable medium such as water. Hot water can be distributed via baseboard radiators or radiant floor systems, while steam is distributed through pipes to steam radiators. Most furnaces and boilers operate on natural gas, oil, or propane. Steam boilers operate at a higher temperature than hot water boilers. Oil-fired furnaces and boilers can use heating oil blended with biodiesel, which produces less pollution than pure heating oil.

A condensing furnace or boiler condenses the water vapor produced in the combustion process and uses the heat from this condensation. Although condensing units cost more than noncondensing units, the condensing unit can reduce the consumption of fuel and the cost over the 15-to-20-year life of the unit. Old furnaces and boilers can be retrofitted to increase their efficiency. Some retrofitting options include installing programmable thermostats, upgrading ductwork in forced-air systems, and adding zone control for hot water systems. Still the costs of retrofits should be compared with the cost of a new boiler or furnace [6, 7].

9.3.2 Home Heating by Electric Resistance

Electric resistance heating converts nearly 100% of the energy in the electricity to heat without emissions. However, most electricity is produced from oil, gas, or coal by converting only about 30% of the fuel's energy into electricity. Because of production and transmission losses, electric heat is often more expensive than heat produced using combustion appliances, such as natural gas, propane, and oil furnaces. Heat pumps are preferable in most climates, as they easily cut electricity use by 50% when compared with electric resistance heating. It is also possible to use heat storage systems to avoid heating during times of peak power demand [6, 7, 9].

Blowers (large fans) in electric furnaces move air over a group of three to seven electric resistance coils, called elements, each of which are typically rated at five kilowatts. A built-in thermostat prevents overheating and may shut the furnace off if the blower fails or if a dirty filter blocks the airflow.

When operated in heating mode, a heat pump is more efficient than operating resistance heaters. Because an electric heater can convert only the input electrical energy directly to output heat energy with none of the efficiency or conversion advantages of a heat pump. Likewise, when a heat pump operates near its most inefficient outside temperature, typically 0°F, the heat pump will perform close to the same as a resistance heater. Example 9.10 illustrates a simple analysis of heating a house by heat pump. Example 9.11 discusses the energy conservation in house heating by Carnot heat pump.

9.3.3 Home Heating by Solar Systems

Active or passive solar systems can be used for residential heating. There are two basic types of active solar heating systems using either liquid or air heated in the solar collectors. The solar units are environmentally friendly and can now be installed on your roof to blend with the architecture of a house. *Liquid-based systems* use heat water or an antifreeze solution, whereas *air-based systems* heat air in an air collector. A circulating pump transports the fluid through the collector, so its temperature increases 10–20°F (5.6–11°C). The flow rate through a solar water collector should be between 0.02 and 0.03 gallons per minute per square foot of collector. Both systems absorb solar radiation and transfer it directly to the interior space or to a storage system. Liquid systems are more often used when storage is included and are well suited for boilers with hot water radiators and heat pumps. These collectors easily last decades.

Active solar heating systems may reduce the cost more when they are used for most of the year. The economics of an active space heating system improves if it also heats domestic water. Heating homes with active solar energy systems can significantly reduce fossil fuel consumptions, air pollution, and emission of greenhouse gases [2, 5, 6, 9].

Passive solar heaters do not have fans or blowers. In passive solar building design, windows, walls, and floors are made to collect, store, and distribute solar energy in the form of heat in the winter and reject solar heat in the summer. A passive solar building takes advantage of the local climate. Elements to be considered include window placement, thermal insulation, thermal mass, and shading.

Example 9.10 Heating a house with heat pump

A heat pump is used to heat a house and maintain it at 18°C. On a day where the outside temperature is −2°C, the house is losing heat at a rate of 75,000 kJ/h. The heat pump operates with a coefficient of performance (COP) of 2.8. Determine:

(a) Power needed by the heat pump,

(b) The rate of heat absorbed from the surrounding cold air.

Solution:

Assume: Steady-state operation.

(a) House is maintained at 18°C.

Heat pump must supply the same amount of lost heat from the cold source:

$q_H = 75{,}000$ kJ/h $= 20.83$ kW.

Power required by the heat pump:

$$\mathrm{COP_{HP}} = \frac{\dot{q}_H}{\dot{W}_{net,in}} \rightarrow \dot{W}_{net,in} = \frac{\dot{q}_H}{\mathrm{COP}} = \frac{20.83\ \mathrm{kW}}{2.8} = 7.44\ \mathrm{kW}$$

(b) Energy balance:

$$\dot{q}_H - \dot{q}_C = \dot{W}_{net,in} = 7.44\ \mathrm{kW}$$
$$\dot{q}_C = \dot{q}_H - \dot{W}_{net,in} = (20.83 - 7.44)\ \mathrm{kW} = 13.39\ \mathrm{kW}$$

13.39 KW is Extracted from the Outside. The House is Paying Only for the Energy of 7.44 kW that is Supplied as Electrical Work to the Heat Pump.
If we have to use electricity in a resistance heater we have to supply 20.83 kW, so the energy conserved by using heat pump instead of electric heater is 13.39/20.83 = 0.64 (or 64%).

***Example 9.11* Energy conservation in house heating by Carnot heat pump**

A Carnot heat pump is used to heat a house during the winter. The house is maintained 20°C. The house is estimated to be losing heat at a rate of 120,000 kJ/h when the outside temperature is – 4°C. Determine the minimum power needed by the heat pump and the rate of heat absorbed from the surrounding cold air.

Solution:

Assume: Steady-state operation.
Temperatures of hot and cold sources: $T_{\mathrm{H}} = 273.15 + 20°\mathrm{C} = 293.15$ K, $T_{\mathrm{C}} = 273.15 - 4°\mathrm{C} = 269.15$ K.

The amount of heat to be supplied to warm inside room:

$q_H = 120{,}000$ kJ/h $= 33.33$ kW.

A minimum amount of power is possible only for a fully reversible heat pump.

This heat pump is Carnot heat pump.

Coefficient of performance:

$$\mathrm{COP_{HP}} = \frac{1}{1-(T_{\mathrm{C}}/T_{\mathrm{H}})} = \frac{1}{1-(269.15\ \mathrm{K}/293.15\ \mathrm{K})} = 12.2$$

Since the house is losing heat at a rate of 120,000 kJ/h to maintain the house 20°C, the heat pump must supply the same amount of heat from the cold source:

Energy balance:

$$\mathrm{COP_{HP}} = \frac{\dot{q}_{\mathrm{H}}}{\dot{W}_{\mathrm{net,in}}} \rightarrow \dot{W}_{\mathrm{net,in}} = \frac{\dot{q}_{\mathrm{H}}}{\mathrm{COP_{HP}}} = 2.73\ \mathrm{kW}$$

$$\dot{q}_{\mathrm{C}} = \dot{q}_{\mathrm{H}} - \dot{W}_{\mathrm{net,in}} = (33.3\text{–}2.73)\ \mathrm{kW} = 30.57\ \mathrm{kW}$$

The house pays for only 2.73 kW. If we use electricity for heating by an electric resistance heater, the rate of heat necessary is 33.3 kW. 30.6 kW of the necessary heat is supplied from the surrounding cold air. Therefore, the energy conserved is 30.6/33.3 = 0.92 (or 92%).

9.4 Energy Efficiency Standards

Thermal efficiency of residential furnaces and boilers is measured by annual fuel utilization efficiency (AFUE). Annual fuel utilization efficiency is the ratio of heat output of the furnace or boiler compared to the total energy consumed by them over a typical year. Annual fuel utilization efficiency does not account for the circulating air and combustion fan power consumptions and the heat losses of the

Table 9.1 Some typical values of annual fuel utilization efficiency (AFUE) for furnace and boilers [7, 9]

Fuel	Furnace/boiler	AFUE %
Heating oil	Retention head burner	70– 78
	Mid-efficiency	83–89
Electric heating	Central or baseboard	100
Natural gas	Conventional	55–65
	Mid-efficiency	78–84
	Condensing	90–97
Propane	Conventional	55–65
	Mid-efficiency	79–85
	Condensing	88–95
Firewood	Conventional	45–55
	Advanced	55–65
	State of the art	75–90

distributing systems of duct or piping. An AFUE of 90% means that 90% of the energy in the fuel becomes heat for the home and the other 10% escapes up the chimney and elsewhere. Heat losses of the duct system or piping can be as much as 35% of the energy for output of the furnace. Table 9.1 shows some typical values of AFUE for furnace and boiler using various fossil fuels and electricity [7, 9]. Some of the minimum allowed AFUE ratings in the USA are:

- Noncondensing fossil-fueled, warm-air furnace is 78%
- Fossil-fueled boiler is 80%
- Gas-fueled steam boiler is 75%

The annual savings from the replacement of a heating system with a more efficient one may be estimated by using Table 9.2 by assuming that both systems have the same heat output. For older units, actual savings in upgrading to a new system could be much higher than indicated in Table 9.2. AFUE is calculated using ASHRAE Standard 103 (ASHRAE 2007). A furnace with a thermal efficiency of 78% may yield an AFUE of only 64% or so, for example, under the standards' test conditions.

9.4.1 Efficiency of Air Conditioner

The energy efficiency ratio (EER) of a particular cooling device is the ratio of *output* cooling (in Btu/hr) to *input* electrical power (in Watts) at a given operating point. The efficiency of air conditioners is often rated by the *seasonal energy efficiency ratio* (SEER). The SEER rating of a unit is the cooling output in Btu during a typical cooling season divided by the total electric energy input in watt-hours during the same period. The coefficient of performance (COP) is an instantaneous

Table 9.2 Assuming the same heat output, estimated savings for every \$100 of fuel costs by increasing an existing heating equipment efficiency [9]

Existing AFUE	New and upgraded system AFUE							
	60%	65%	70%	75%	80%	85%	90%	95%
55%	\$8.3	\$15.4	\$21.4	\$26.7	\$31.2	\$35.3	\$38.9	\$42.1
60%	–	\$7.7	\$14.3	\$20.0	\$25.0	\$29.4	\$33.3	\$37.8
65%	–	–	\$7.1	\$13.3	\$18.8	\$23.5	\$27.8	\$31.6
70%	–	–	–	\$6.7	\$12.5	\$17.6	\$22.2	\$26.3
75%	–	–	–	–	\$6.5	\$11.8	\$16.7	\$21.1
80%	–	–	–	–	–	\$5.9	\$11.1	\$15.8
85%	–	–	–	–	–	–	\$5.6	\$10.5

measure (i.e., a measure of power divided by power), whereas both EER and SEER are averaged over a duration of time. The time duration considered is several hours of constant conditions for EER, and a full year of typical meteorological and indoor conditions for SEER. Typical EER for residential central cooling units = 0.875 × SEER. A SEER of 13 is approximately equivalent to a COP of 3.43, which means that 3.43 units of heat energy are removed from indoors per unit of work energy used to run the heat pump. SEER rating more accurately reflects overall system efficiency on a seasonal basis and EER reflects the system's energy efficiency at peak day operations.

Air conditioner sizes are often given as "tons" of cooling where 1 ton of cooling is equivalent to 12,000 Btu/hr (3,500 W). This is approximately the power required to melt one ton of ice in 24 h. Example 9.12 illustrates the estimation of electric cost of an air conditioner [6, 7].

Example 9.12 Electricity cost of air conditioner

Estimate the cost of electricity for a 5,000 Btu/h (1,500 W) air-conditioning unit operating, with a SEER of 10 Btu/Wh. The unit is used for a total of 1500 h during an annual cooling season and the unit cost of electricity is \$0.14/kWh.

Solution:

Air conditioner sizes are often given as "tons" of cooling where 1 ton of cooling is equivalent to 12,000 Btu/h (3,500 W).

The unit considered is a small unit and the annual total cooling output would be:

(5000 Btu/h) × (1500 h/year) = 7,500,000 Btu/year.
With a seasonal energy efficiency ratio (SEER) of 10, the annual electrical energy usage:

(7,500,000 Btu/year)/(10 Btu/Wh) = 750,000 Wh/year = 750 kWh/year.

With a unit cost of electricity of \$0.14/kWh, the annual cost: (750 kWh/year)(\$0.14/kWh) = \$105/year.

The average power usage may also be calculated more simply by:

Average power = (Btu/h)/(SEER, Btu/Wh) = 5000/10 = 500 W = 0.5 kW.

With the electricity cost of \$0.14/kWh, the cost per hour:

(0.5 kW) (\$0.14)/kWh = \$0.07/h.

For 1500 h/year, the total cost: (\$0.07/h) (1500 h/year) = \$105/year.

9.4.2 High Efficiency for Cooling

The refrigeration process with the maximum possible efficiency is the Carnot cycle. The coefficient of performance (COP) of an air conditioner using the Carnot cycle is:

$$\mathrm{COP_{Carnot}} = \frac{T_C}{T_H - T_C} \tag{9.12}$$

where T_C is the indoor temperature and T_H is the outdoor temperature in K or R. The EER is calculated by multiplying the COP by 3.413 which is the conversion factor from Btu/h to Watts:

$$\mathrm{EER_{Carnot}} = 3.413(\mathrm{COP_{Carnot}}) \tag{9.13}$$

For an outdoor temperature of 100°F (311 K) and an indoor temperature of 95°F (308 K), the above equation gives a COP of 103, or an EER of 350. This is about 10 times as efficient as a typical home air conditioner available today. The maximum EER decreases as the difference between the inside and outside air temperature increases.

For example:

For T_H = 120 F (49 C) = 322.15 K, and T_C = 80°F (27°C) = 300.15 K, we have

$\mathrm{COP_{Carnot}}$ = 300.15K/(322.15–300.15)K = 13.6 or

EER = (3.413) (13.6) = 46.4

The maximum SEER can be calculated by averaging the maximum values of EER over the range of expected temperatures for the season.

Central air conditioners should have a SEER of at least 14. Substantial energy savings can be obtained from more efficient systems. For example:

By upgrading from SEER 9 to SEER 13
Reduction in power consumption = (1 − 9/13) = 0.30

This means that the power consumption is reduced by 30%. Residential air condition units may be available with SEER ratings up to 26. Example 9.13 illustrates the calculation of the annual cost of power used by an air conditioner. Example 9.14 discusses possible saving in cooling by using a unit operating at a higher SEER rating [9, 10].

Example 9.13 **Calculating the annual cost of power for an air conditioner**
Estimate the annual cost of electric power consumed by a 6-ton air-conditioning unit operating for 2000 h per year with a SEER rating of 10 and a power cost of \$0.16/kWh.

Solution:

Air conditioner sizes are often given as "tons" of cooling:

1. Ton of Cooling = 12,000 Btu/h (3500 W).

This is approximately the power required to melt one ton of ice in 24 h.

The annual cost of electric power consumption:

(6)(12,000 Btu/h) = 72,000 Btu/h

$$\text{Cost} = \frac{(\text{Size, Btu/h})(\text{time, h/year})\left(\text{Cost of energy, \$/kWh}\right)}{(\text{SEER, Btu/Wh})(1000\ \text{W/kW})}$$

= (72,000 Btu/h)(2000 h/year)(\$0.16/kWh)/[(10 Btu/Wh)(1000 W/kW)]

= \$2304/year.

For the temperatures of hot and cold sources:

$T_\text{H} = 273.15 + 20°\text{C} = 293.15\text{K},$
$T_\text{C} = 273.15 - 4°\text{C} = 269.15\text{K}, \text{ and } T_\text{H} - T_\text{C} = 24\text{K}$

$$\text{EER}_\text{Carnot} = 3.41\left(\frac{T_\text{C}}{T_\text{H} - T_\text{C}}\right) = 3.41(269.13/24) = 38.24$$

The maximum EER decreases as the difference between the inside and outside air temperature increases.

Example 9.14 **Saving the cost of cooling with a unit operating at a higher SEER rating**

A 4 tons of current residential air conditioner operates with a seasonal energy efficiency ratio (SEER) rating of 10. This unit will be replaced with a newer unit operating with a SEER rating of 22. The unit operates 130 days with an average of 10 h per day. Average inside and outside temperatures are 21°C and 3°C, respectively. The unit cost of energy is \$0.16/kWh. Estimate the saving in the cost of electricity and the maximum energy efficiency ratio.

Solution:

Cooling load of 4 tons: (4)(12,000 Btu/h) = 48,000 Btu/h (14,000 W).
(130 days/year)(10 h/day) = 1300 h/year.
The estimated cost of electrical power:
SEER = 10, and an energy cost of \$0.16/kWh, using 130 days of 10 h/day operation:
Cost = (48,000 Btu/h)(1300 h/year)(\$0.16/kWh)/[(10 Btu/Wh)(1000 W/kW)] = \$998.4/year.
SEER = 22, and an energy cost of \$0.16/kWh, using 130 days of 10 h/day operation:
Cost = (48,000 Btu/h)(1300 h/year)(\$0.16/kWh)/[(22 Btu/Wh)(1000 W/kW)] = \$453.8/year.
Annual saving = \$998.4/year − \$453.8/year = \$44.6/year.
The ratio of typical EER to maximum EER/EER_{max} = 19.25/52.31 = 0.368.
Typical EER for the current cooling unit:
EER = 0.875 (SEER) = 0.875 (10) = 8.75.
Typical EER for the new cooling unit:
EER = 0.875 (SEER) = 0.875 (22) = 19.25.
Maximum value of EER:

The temperatures of hot and cold sources: $T_H = 273.15 + 21°C = 294.15$ K, $T_C = 273.15 + 3°C = 276.15$ K.

$T_H - T_C = 18$ K

$$EER_{Carnot} = 3.41\left(\frac{T_C}{T_H - T_C}\right) = 3.41(276.13/18) = 52.31$$

The maximum EER decreases as the difference between the inside and outside air temperature increases. The annual savings are significant after using a more efficient unit. The value of EER for the new unit is only 36.6% of the maximum value of EER.

9.4.3 Fuel Efficiency

Fuel efficiency means the efficiency of a process that converts chemical energy contained in a fuel into kinetic energy or work. The increased fuel efficiency is especially beneficial for fossil fuel power plants or industries dealing with combustion of fuels. In the transportation sector, fuel efficiency is expressed in miles per gallon (mpg) or kilometers per liter (km/l). Fuel efficiency is dependent on many parameters of a vehicle, including its engine parameters, aerodynamic drag, and weight. Hybrid vehicles offer higher fuel efficiency use two or more power sources for propulsion, such as a small combustion engine combined with and electric motors [10, 11]. Example 9.15 compares the heating by electricity and by natural gas. Examples 9.16 and 9.17 discuss the amounts of coal necessary in a coal-fired steam power plant in two different efficiency values for combustion and generator efficiency. The examples show that with the increased combustion efficiency and generator efficiency, the required amount of coal is reduced to 23.9 ton/h from 36.6 ton/h. This leads to around 35% savings of coal.

Example 9.15 **Comparison of energy sources of electricity with natural gas for heating**

The efficiency of electric heater is around 73%, while it is 38% for natural gas heaters. A house heating requires 4 kW. The unit cost of electricity is \$0.1/kWh, while the natural gas costs \$0.60/therm. Estimate the rate of energy consumption for both the electric and gas heating systems.

Solution:

Energy supplied by the electric heater: $\dot{q}_{\text{used}} = \dot{q}_{\text{in}}\eta_{\text{th}} = (4\ \text{kW})(0.73) = 2.92\ \text{kW}$.
The unit cost of energy is inversely proportional to the efficiency:
Unit cost of electric energy = (unit cost of energy)/(efficiency)
= \$0.1/0.73 = \$0.137/kWh.
Therm = 29.3 kWh (Table 1.9).
Energy input to the gas heater (at the same rate of used energy that is 2.92 kW):

$$\dot{q}_{\text{in, gas}} = \dot{q}_{\text{used}}/\eta_{\text{th}} = 2.92/0.38 = 3.84\ \text{kW}(= 13100\ \text{Btu/h})$$

Therefore, a gas burner should have a rating of at least 13,100 Btu/h to have the same performance as the electric unit.
Unit cost of energy from the gas = cost of energy/efficiency:

$$[(\$0.60/\text{therm})/(29.3\ \text{kWh/therm})]/0.38 = \$0.054/\text{kWh}.$$

Ratio of unit cost of gas to electric energy; \$0.054/\$0.137 = 0.39.
The cost of utilized natural gas is around 39% of the electricity cost; therefore, heating with an electric heater will cost more for heating.

Example 9.16 **Overall efficiency and required amount of coal in a coal-fired steam power plant**

An adiabatic turbine is used to produce electricity by expanding superheated steam at 4100 kPa and 350°C. The power output is 50 MW. The steam leaves the turbine at 40 kPa and 100°C. If the combustion efficiency is 0.75 and the generator efficiency is 0.9, determine the overall plant efficiency and the amount of coal supplied per hour.

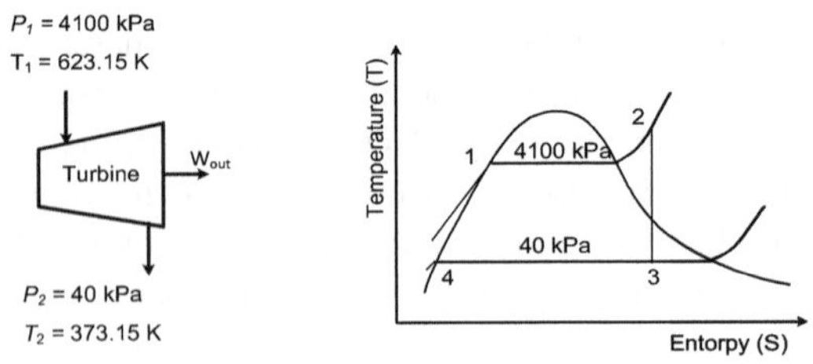

Solution:

Assume: Steady-state adiabatic operation. The changes in kinetic and potential energies are negligible.

Basis: stems flow rate = 1 kg/s.

Combustion efficiency $\eta_{\text{comb}} = 0.75$, Generator efficiency $\eta_{\text{gen}} = 0.9$

Power output = 50 MW = 50,000 kW.

Data from steam tables. Turbine inlet conditions:

$P_2 = 4100$ kPa, $T_2 = 623.15$ K, $H_2 = 3092.8$ kJ/kg, $S_2 = 6.5727$ kJ/kg K.

Turbine outlet conditions:

$$P_4 = P_3 = 40\,\text{kPa},\ T_3 = 373.15\,\text{K},\ V_4 = 1.027\ \text{cm}^3/\text{g} = 0.001027\ \text{m}^3/\text{kg}.$$
$$S_{4\text{sat vap}} = 7.6709\,\text{kJ/kg K},\ S_{4\text{sat liq}} = 1.2026\,\text{kJ/kg K at 40 kPa}.$$
$$H_{4\text{sat vap}} = 2636.9\,\text{kJ/kg},\ H_{4\text{sat liq}} = 317.6\,\text{kJ/kg at 40 kPa}$$

$$W_{p,\text{in}} = V_4(P_1 - P_4) = \left(0.001027\ \text{m}^3/\text{kg}\right)(4100{-}40)\ \text{kPa}\left(1\ \text{kJ}/1\text{kPa m}^3\right) = 4.2\ \text{kJ/kg}$$
$$H_1 = H_4 + W_{p,\text{in}} = 317.6 + 4.2 = 321.8\ \text{kJ/kg}$$

For the isentropic operation $S_3 = S_2 = 6.5727$ kJ/kg K

Since $S_2 < S_{3\text{sat vap}}$: $6.5727 < 7.6709$ the steam at the exit is saturated liquid–vapor mixture.

Quality of that mixture: $x_3 = \frac{S_2 - S_{2\text{sat liq}}}{S_{2\text{sat vap}} - S_{2\text{sat liq}}} = \frac{6.5727 - 1.2026}{7.6709 - 1.2026} = 0.83$

$$H_3 = (1 - x_3)H_{3\text{sat liq}} + x_3 H_{3\text{sat vap}} = 2243.1\ \text{kJ/kg}$$

Heat interactions: $\dot{q}_{\text{in}} = \dot{m}(H_2 - H_1) = (3092.8 - 321.8)\ \text{kJ/kg} = 2771.0\ \text{kJ/s}$

$$\dot{q}_{\text{out}} = -\dot{m}(H_4 - H_3) = -(1\text{kg/s})(317.6 - 2243.1)\ \text{kJ/kg}$$
$$= 1925.5\ \text{kJ/s (heat received by the cooling medium that is the system in the condenser)}$$

Thermal efficiency: $\eta_{\text{th}} = 1 - \frac{q_{\text{out}}}{q_{\text{in}}} = 0.305$ or 30.5%

Overall plant efficiency:

$\eta_{\text{overall}} = \eta_{\text{th}}\eta_{\text{comb}}\eta_{\text{gen}} = (0.305)(0.75)(0.9)$

$= 0.205$ or 20.5%

The rate of coal energy required: $\dot{E}_{\text{coal}} = \frac{|\dot{W}_{\text{net}}|}{\eta_{\text{overall}}} = \frac{50{,}000 \text{ kW}}{0.205} = 243{,}900 \text{ kJ/s}.$

Energy of coal (bituminous) = 24,000 kJ/kg (Table 2.6)

$$\dot{m}_{\text{coal}} = \frac{\dot{E}_{\text{coal}}}{\eta_{\text{overall}}} = \frac{243{,}900}{24{,}000}\left(\frac{1 \text{ tonne}}{1000 \text{ kg}}\right) = 0.0101 \text{ tonne/s } = 36.6 \text{ tonne/h}$$

Example 9.17 **Required amount of coal in a coal-fired steam power plant**

An adiabatic turbine is used to produce electricity by expanding superheated steam at 4100 kPa and 350°C. The steam flow rate is 42 kg/s. The steam leaves the turbine at 40 kPa and 100°C. If the combustion efficiency is 0.77 and the generator efficiency is 0.95, determine the overall plant efficiency and the amount of coal supplied per hour.

Solution:

Assume: Steady-state adiabatic operation. The changes in kinetic and potential energies are negligible.
Basis: stems flow rate = 1 kg/s.
Combustion efficiency $\eta_{\text{comb}} = 0.77$.

Generator efficiency $\eta_{\text{gen}} = 0.95$.

Turbine inlet:

$$P_2 = 4100 \text{ kPa},\ T_2 = 623.15 \text{ K},\ H_2 = 3092.8 \text{ kJ/kg},\ S_2 = 6.5727 \text{ kJ/kg K (steam tables)}$$

Turbine exit:

$$\begin{aligned} P_4 &= P_3 = 40\text{kPa},\ T_3 = 373.15\text{K},\ V_4 = 1.027 \text{ cm}^3/\text{g} = 0.001027 \text{ m}^3/\text{kg}. \\ S_{4\text{sat vap}} &= 7.6709 \text{ kJ/kg K},\ S_{4\text{sat liq}} = 1.2026 \text{ kJ/kg K at 40 kPa}. \\ H_{4\text{sat vap}} &= 2636.9 \text{ kJ/kg},\ H_{4\text{sat liq}} = 317.6 \text{ kJ/kg at 40 kPa} \end{aligned}$$

$$\begin{aligned} W_{p,\text{in}} &= V_4(P_1 - P_4) = (0.001027 \text{ m3/kg})(4100\text{–}40) \text{ kPa} \\ &\quad \left(1 \text{ kJ/1kPa m}^3\right) = 4.2 \text{ kJ/kg} \\ H_1 &= H_4 + W_{p,in} = 317.6 + 4.2 = 321.8 \text{ kJ/kg} \end{aligned}$$

For the isentropic operation $S_3 = S_2 = 6.5727$ kJ/kg K

Since $S_2 < S_{3\text{sat vap}}$: 6.5727 < 7.6709 the steam at the exit is saturated liquid–vapor mixture.

Quality of the mixture:

$$x_3 = \frac{S_2 - S_{2\text{sat liq}}}{S_{2\text{sat vap}} - S_{2\text{sat liq}}} = \frac{6.5727 - 1.2026}{7.6709 - 1.2026} = 0.83$$

$$H_3 = (1 - x_3)H_{3\text{sat liq}} + x_3 H_{3\text{sat vap}} = 2243.1 \text{ kJ/kg}$$

Heat interactions:

$$q_{\text{in}} = (H_2 - H_1) = (3092.8 - 321.8) \text{ kJ/kg} = 2771.0 \text{ kJ/kg}$$

$$q_{\text{out}} = -(H_4 - H_3) = -(317.6 - 2243.1) \text{ kJ/kg} = 1925.5 \text{ kJ/kg}$$
(heat received by the cooling medium)

Cycle work out:

$$\dot{W}_{\text{net}} = \dot{m}(q_{\text{out}} - q_{\text{in}}) = (42 \text{ kg/s})(1925.5\text{–}2771.0) \text{ kJ/kg} = -35{,}511 \text{ kW}$$
(work produced)

The thermal efficiency:

$$\eta_{\text{th}} = 1 - \frac{q_{\text{out}}}{q_{\text{in}}} = 0.305 \text{ or } 30.5\%$$

The overall plant efficiency:

$$\eta_{\text{overall}} = \eta_{\text{th}}\eta_{\text{comb}}\eta_{\text{gen}} = (0.305)(0.77)(0.95) = 0.223$$

The rate of coal energy supply:

$$\dot{E}_{\text{coal}} = \frac{|\dot{W}_{\text{net}}|}{\eta_{\text{overall}}} = \frac{35{,}511}{0.223} = 159{,}165 \text{ kJ/s}$$

Energy of coal (bituminous) = 24,000 kJ/kg (Table 2.6)

$$\dot{m}_{\text{coal}} = \frac{\dot{E}_{\text{coal}}}{\eta_{\text{overall}}} = \frac{159{,}165}{24{,}000}\left(\frac{1 \text{ mton}}{1000 \text{ kg}}\right) = 0.0066 \text{ ton/s} = 23.8 \text{ mton/h}$$

With the increased combustion efficiency and generator efficiency, the required amount of coal is reduced to 23.9 ton/h from 36.6 ton/h. This leads to around 35% savings of coal.

9.4.4 Fuel Efficiency of Vehicles

The fuel efficiency of vehicles can be expressed in miles per gallon (mpg) or liters (l) per km. Diesel engines generally achieve greater fuel efficiency than gasoline engines. Passenger car diesel engines have energy efficiency of up to 41% but more typically 30%, and petrol engines of up to 37.3%, but more typically 20%. The higher compression ratio is helpful in raising the energy efficiency, but diesel fuel also contains approximately 10% more energy per unit volume than gasoline which contributes to the reduced fuel consumption for a given power output [8, 11].

Fuel efficiency directly affects emissions causing pollution. Cars can run on a variety of fuel sources, such as gasoline, natural gas, liquefied petroleum gases, biofuel, or electricity. All these create various quantities of atmospheric pollution. A kilogram of carbon produces approximately 3.63 kg of CO_2 emissions. Typical average emissions from combustion of gasoline and diesel are [11]:

- Gasoline combustion emits 19.6 lb CO_2/US gal gasoline or (2.32 kg CO_2/l gasoline)
- Diesel combustion emits 22.5 lb CO_2/US gal diesel or (2.66 kg CO_2/l diesel)

These values are only the CO_2 emissions of the final forms of fuel products and do not include additional CO_2 emissions created during the drilling, pumping, transportation, and refining steps of the fuel production. Examples 9.18 and 9.19 illustrate the estimation of fuel consumption of a car and show that the conservation in fuel and reduction in emission of CO_2 are significant when the fuel efficiency of car increases.

Example 9.18 **Fuel consumption of a car**

The overall efficiencies are about 25–28% for gasoline car engines, 34–38% for diesel engines, and 40–60% for large power plants [2]. A car engine with a power output of 120 hp has a thermal efficiency of 24%. Determine the fuel consumption of the car if the fuel has a higher heating value of 20,400 Btu/lb.

Solution:

Assume: the car has a constant power output.

Gasoline car engine:

$$\eta_{th} = \frac{|\dot{W}_{net}|}{\dot{q}_{in}} \rightarrow \dot{q}_{in} = \frac{|\dot{W}_{net}|}{\eta_{th}} = \frac{120\ \text{hp}}{0.24}\left(\frac{2545\ \text{Btu/h}}{\text{hp}}\right)$$

$$\dot{q}_{in} = 1{,}272{,}500\ \text{Btu/h}$$

Net heating value = higher heating value (1 – 0.1) = 18,360 Btu/lb (approximately)

$$\text{Fuel consumption} = \dot{q}_{\text{in}}/\text{net heating value} = 1{,}272{,}500\ \text{Btu/h}/18360\ \text{Btu/lb} = 69.3\ \text{lb/h}$$

Assuming an average gasoline density of 0.75 kg/l:

$\rho_{\text{gas}} = (0.75\ \text{kg/l})(2.2\ \text{lb/kg})(\text{l}/0.264\ \text{gal}) = 6.25\ \text{lb/gal}$.

Fuel consumption in terms of gallon:

(69.3 lb/h)/(6.25 lb/gal) = 11.1 gal/h.

Diesel engine with an efficiency of 36%:

$$\eta_{\text{th}} = \frac{|\dot{W}_{\text{net}}|}{\dot{q}_{\text{in}}} \rightarrow \dot{q}_{\text{in}} = \frac{|\dot{W}_{\text{net}}|}{\eta_{\text{th}}} = \frac{120\ \text{hp}}{0.36}\left(\frac{2545\ \text{Btu/h}}{\text{hp}}\right) = 848{,}333\ \text{Btu/h}$$

$$\dot{q}_{in} = 848{,}333\ \text{Btu/h}$$

9.4.5 Energy Conservation in Driving

Some possible energy conservation steps are:

- Speeding, rapid acceleration, and braking waste gas. It can lower your gas mileage by 33 percent at highway speeds and by 5% around town. Gas mileage usually decreases rapidly at speeds above 60 mph and observing the speed limit may lead to fuel saving of 7–23% [11].
- An extra 100 pounds in your vehicle could reduce fuel efficiency by up to 2 percent.
- Using cruise control helps save gas.
- Fixing a serious maintenance problem, such as a faulty oxygen sensor, can improve your mileage by as much as 40 percent.
- Gas mileage may be improved by up to 3.3 percent by keeping your tires inflated to the proper pressure.

The most efficient machines for converting energy to rotary motion are electric motors, as used in electric vehicles. However, electricity is not a primary energy source, so the efficiency of electricity production has also to be considered. In the future, hydrogen cars powered either through chemical reactions in a fuel cell that create electricity to drive electrical motors or by directly burning hydrogen in a combustion engine may be commercially available. These vehicles have near zero pollution from the exhaust pipe. Potentially the atmospheric pollution could be minimal, provided the hydrogen is made by electrolysis using electricity from nonpolluting sources such as solar, wind, or hydroelectricity. In addition to the energy cost of the electricity or hydrogen production, transmission and/or storage losses to support large-scale use of such vehicles should also be accounted [11].

Example 9.19 **Fuel conservation with a more fuel-efficient car**

Assume two cars one with 11 l/100 km city and 9 l/100 km highway, and the other 6.5 l/100 km in city traffic and at 5 l/100 km motorway are used. Estimate the annual fuel saving and emission reduction achieved with the more fuel-efficient car traveling an average of 15,000 km per year.

Solution:

Fuel saving for every 100 km city driving in liters = (11 – 6.5) l = 4.5 l.

Fuel saving for every 100 km highway driving in liters = (9 – 5) l = 4.0 l.

Average saving for 100 km is (4.0 + 4.5)l/2 = 4.2 l.

A fuel-efficient car can conserve an average of about 4.2 l gasoline for every 100 km.

For a car driving 15,000 km per year, the amount of fuel conserved would be:

(15,000 km/year) (4.2 l/100 km) = 630 l/year (167 gal/year).

Gasoline combustion emits 19.4 lb CO_2/US gal or (2.32 kg CO_2/l).

Annual emission reduction = (2.32 kg CO_2/l) (630 l/year)

= 1462 kg/year (or 3294 lb/year).

Both the conservation in fuel and reduction in emission of CO_2 are significant with increased fuel efficiency of cars globally.

9.4.6 Regenerative Braking

In conventional braking systems, the excess kinetic energy is converted to heat by friction in the brake pads and therefore wasted. Regenerative braking is the conversion of the vehicle's kinetic energy into chemical energy, which is stored in the battery and used in driving again in battery-powered and hybrid gas/electric vehicles. The recovery is around 60% [8, 10]. The advanced algorithms in the motor control the motor torque for both driving and regenerative braking. A torque command is derived from the position of the throttle pedal. The motor controller converts this torque command into the appropriate 3-phase voltage and current waveforms to produce the commanded torque in the motor in the most efficient way. When the torque serves to slow the vehicle then energy is returned to the battery.

Negative torque applied to the rear wheels can cause a car to become unstable since regenerative braking is a source of negative torque. The traction control system limits regenerative break if the rear wheels start to slip. The control system provides the driver instant positive and negative torque commands. Regenerative braking is limited when the batteries are fully charged, and motor controller will limit regenerative torque in this case. It is possible that with the increased use of battery-powered vehicles, safe regenerative breaking will be an efficient way of converting and recovering the kinetic energy.

9.4.7 Energy Conservation in Lighting

Energy for lighting accounts for about 10% of the household electric bill. Compact fluorescent bulbs use about 75% less energy than standard lighting, produce 75% less heat, and last up to 10 times longer. Compact fluorescent bulbs contain a very small amount of mercury sealed within the glass tubing. Although linear fluorescent and compact fluorescent bulbs cost a bit more than incandescent bulbs initially, over their lifetime they are cheaper because of the less electricity they use [9, 10]. Example 9.20 illustrates the energy conservation by using the compact fluorescent bulbs.

Example 9.20 **Conservation of energy with compact fluorescent bulbs**

Assume that an average residential rate of electricity is $0.14/kWh and a household consumes about 10,000 kWh per year. If the lighting is provided by compact fluorescent bulbs only, estimate the conservation of energy and saving per year.

Solution:

Assume that 11% of the energy budget of household is for lighting, and compact fluorescent bulbs consume about 75% less energy than standard lighting.

Annual energy use and cost with standard lighting:

(10,000 kWh/year) (0.11) = 1100 kW/year.

(1100 kW/year)($0.14/kWh) = $154/year.

Annual energy use and cost with compact fluorescent bulbs:

(1100 kW/year) (1– 0.75) = 275 kWh/year.

(275 kWh/year) ($0.14/kWh) = $38.5/year.

Annual energy and cost savings:

(1100 – 275) kWh = 825 kWh.

(154 – 38.5) $/year = $115 5/year.

Conservation of energy and cost with compact fluorescent bulbs are considerable.

9.5 Energy Conservation in Electricity Distribution and Smart Grid

A *smart grid* is a form of electricity network using digital technology. Smart grid delivers electricity to consumers to control appliances at homes, optimize power flows, reduce waste, and maximize the use of lowest-cost production resources. The smart grid is envisioned to overlay the ordinary electrical grid with an information and net metering system that includes smart meters. The increased data transmission capacity has made it possible to apply sensing, measurement and control devices with two-way communications to electricity production, transmission, distribution, and consumption parts of the power grid. These devices could communicate information about grid condition to system users, operators, and automated devices. Therefore, the average consumer can respond dynamically to changes in grid condition [7, 10].

A smart grid includes an intelligent monitoring system that keeps track of all electricity flowing in the system. It also has the capability of integrating renewable electricity such as solar and wind. When power is least expensive the user can allow the smart grid to turn on selected home appliances such as washing machines or some processes in a factory. At peak times, it could turn off selected appliances to reduce demand [2, 5]. Some smart grid functions are:

- Motivate consumers to actively participate in operations of the grid.
- Provide higher quality power that will save energy wasted from outages.
- Accommodates all generation and storage options.
- Enable electricity markets to flourish by running more efficiently.
- Enable higher penetration of intermittent power generation sources.

9.5.1 Standby Power

Standby power refers to the electricity consumed by many appliances when they are switched off or in standby mode. The typical power loss per appliance is low (from 1 to 25 W) but when multiplied by the billions of appliances in houses and in commercial buildings, standby losses represent a significant fraction of total world electricity use. Standby power may account for consumption between 7 and 13% of household power consumption. Technical solutions to the problem of standby power exist in the form of a new generation of power transformers that use only 100 milliwatts in standby mode and thus can reduce standby consumption by up to 90%. Another solution is the "smart" electronic switch that cuts power when there is no load and restores it immediately when required [8, 10].

9.6 Energy Harvesting

Despite the high initial costs, energy harvesting is often more effective in cost and resource utilization when done on a group of houses, co-housing, local district, or village rather than on an individual basis. This leads to the reduction of electricity transmission and distribution losses, which may be about 7% of the energy transferred. The net zero fossil energy consumption requires locations of geothermal, micro-hydro, solar, and wind resources to sustain the concept. One of the key areas of debate in zero energy building design is the balance between energy conservation and the distributed point-of-use harvesting of renewable energy such as solar and wind energy. Wide acceptance of zero or close to zero energy building technology may require more government incentives, building code regulations, or significant increases in the cost of conventional energy [2, 5, 6]. Some factors of energy harvesting are:

- Isolation for building owners from future energy price increases.
- Increased comfort due to more-uniform interior temperatures.
- Reduced total cost of ownership due to improved energy efficiency.

9.7 Energy Conservation and Exergy

Energy conservation through exergy concepts for some steady-state flow processes are [3, 5]:

- Exergy is lost by irreversibility associated with pressure drops, fluid friction, and stream-to-stream heat transfer due to temperature differences.
- In a steam power plant, exergy transfers are due to work and heat, and exergy is lost within the control volume.
- In a waste heat recovery system, we might reduce the heat transfer irreversibility by designing a heat recovery steam generator with a smaller stream-to-stream temperature difference, and/or reduce friction by designing a turbine with a higher efficiency.
- A cost-effective design may result from consideration of the trade-offs between possible reduction of exergy loss and potential increase in operating cost.

9.8 Energy Recovery on Utilities Using Pinch Analysis

A process may have available hot and cold streams. When the hot and cold streams exchange heat between them, heat is recovered and the hot and cold utility requirements may be minimized. Pinch analysis yields optimum energy integration of process heat and utilities by matching the hot and cold streams with a network of heat exchangers. In the pinch analysis, hot and cold streams can only exchange

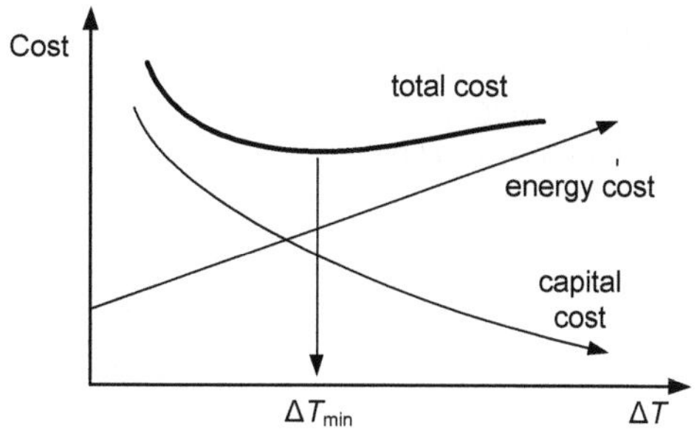

Fig. 9.3 Optimum ΔT_{min} from energy cost and capital cost changes

energy up to a minimum allowable temperature difference ΔT_{min}, which is called the pinch point leading to the minimum driving force for heat transfer [2, 3, 5].

An increase in ΔT_{min} leads to higher energy and lower capital costs (a smaller heat exchange area as seen in Fig. 9.3) toward more sustainable engineering [4]. For example, an increase of 5°C from a value of $\Delta T_{min} = 10$°C decreases the heat exchanger area by 11% and increases the required minimum energy by about 9%. To find the optimum value of ΔT_{min}, the total annual cost is plotted against. An optimum ΔT_{min} appears at the minimum total annual cost of energy and capital. The optimum value for ΔT_{min} is generally in the range of 3–40°C and needs to be established for a process.

9.8.1 Composite Curves

Temperature-enthalpy diagrams called *composite curves* represent the thermal characteristics of hot and cold streams and the amount of heat transferred between them (see Fig. 9.4). The enthalpy change rate for each stream is

$$q = \Delta H = mC_p\Delta T = MC\Delta T \tag{9.15}$$

where ΔH is the enthalpy change rate, $\dot{m}$ is the mass flow rate, C_p is the heat capacity, ΔT is the temperature change in a stream, and MC the heat capacity rate m C_p. The enthalpy change rates are added over each temperature interval that includes one or more of the streams. This leads to hot and cold composite curves of the streams shown in Fig. 9.4. If $\dot{m}C_p$ is constant, q versus T would be a straight line

$$\Delta T = \frac{1}{\dot{m}C_p}q \tag{9.16}$$

Enthalpy changes rather than absolute enthalpies are estimated, and the horizontal location of a composite line on the diagram is arbitrarily fixed. One of the two curves is moved horizontally until the distance of the closest vertical approach matches the selected value of ΔT_{min}. The pinch point is the location of ΔT_{min} on the adjusted composite diagram where the hot and cold curves most

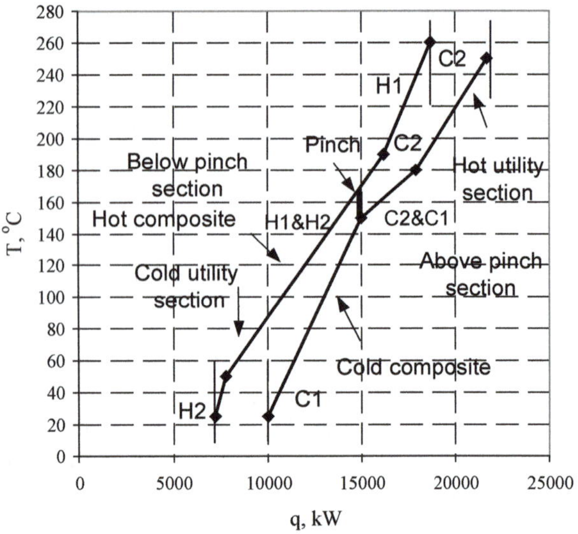

Fig. 9.4 Hot and cold composite curves for $\Delta T_{min\cdot} = 20\,^{\circ}C$, using two hot (H) streams and two cold (C) streams

closely approach to each other in temperature in a vertical direction. The overshoot of the hot composite curve represents the minimum cold utility ($q_{c,\min}$) required, and the overshoot of the cold composite curve represents the minimum hot utility ($q_{h,\min}$) required. Example 9.21 illustrates the use of pinch analysis in process heat integration and minimizing the hot and cold utilities.

Above the pinch, only the hot utility is required, while only the cold utility is required below the pinch. These diagrams enable engineers to minimize the expensive utilities. Pinch analysis may also lead to optimum integration of evaporators, condensers, furnaces, and heat pumps by reducing the utility requirements. Pinch analysis is utilized widely in industry leading to considerable energy savings. Figure 9.5 displays the importance of minimum temperature approach between available hot and cold utilities. The values of $\Delta T_{\min}$ increase the amount of hot and cold utilities increase together with the operating cost. On the other hand, as the values of $\Delta T_{\min}$ decreases, the investment cost increases. So, these two opposing costs should be optimized as seen in Fig. 9.3.

Example 9.21 **Energy conservation by the pinch analysis**

In a process available, hot and cold process streams and their heat capacities are shown below.

Hot and cold process stream conditions.

Streams		T_{in} °C	T_{out} °C	$C = \dot{m}C_p$ kW/°C	$q = \dot{m}C_p\Delta T$ kW
C1	Cold 1	25	180	40	6200
C2	Cold 2	150	250	55	5500
H1	Hot 1	260	50	35	– 7350

(continued)

(continued)

Streams		T_{in} °C	T_{out} °C	$C = \dot{m}C_p$ kW/°C	$q = \dot{m}C_p\Delta T$ kW
H2	Hot 2	190	25	25	– 4125

Construct the balanced composite curves for the process with $\Delta T_{min} = 20°C$ and $\Delta T_{min} = 10°C$, and compare the amounts of hot and cold utilities needed.

Solution:

Assume that heat capacities of hot and cold streams are constant.

A starting enthalpy change rate is chosen as 10,000 kW at 25°C for the streams to be heated, while for the streams to be cooled, the base value chosen is 15,000 kW at 260°C. Initial temperature interval with overlaps is shown below:

Initial temperature interval					Initial enthalpy selection	
Streams	Temperature interval °C		C kW/°C	q kW	T °C	q kW
					25	10,000 (arbitrary)
C1	25	150	40	5000	150	15,000
C1&C2	150	180	95	2850	180	17,850
C2	180	250	55	3850	250	21,700
				11,700		
					260	15,000 (Arbitrary)
H1	260	190	35	– 2450	190	12,550
H2&H1	190	50	60	– 8400	50	4150
H2	50	25	25	– 625	25	3525
				– 11,475		

This table may be converted to the hot and cold composite curves below.

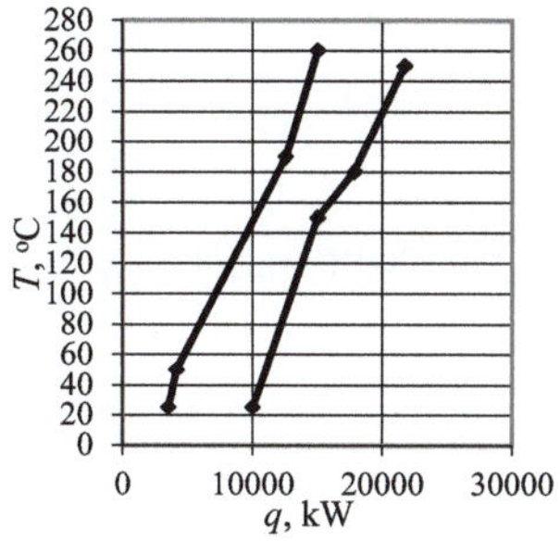

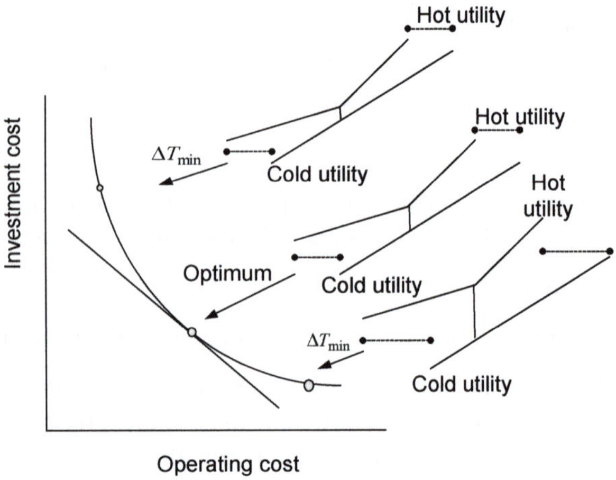

Fig. 9.5 Principle of pinch technology [3]

Here, the composite curves move to each other so that at the pinch point minimum temperature difference would $\Delta T_{\text{min}} = 20\,^{\circ}\text{C}$, vthe diagram is called a *balanced composite diagram.*

Revised temperature interval					Revised enthalpy selection	
					q kW	*T* °C
Stream	Required temperature interval °C		*C* kW/°C	*Q* kW		
C1	25	150	40	5000	15,000	150
C1&C2	150	180	95	2850	17,850	180
C2	180	250	55	3850	21,700	250
				11,700		
					18,650	260
H1	260	190	35	– 2450	16,200	190
H2&1	190	50	60	– 8400	7800	50
H2	50	25	25	– 625	7175	25
				– 11,475		

Composite diagram with $\Delta T_{\text{min}} = 20\,^{\circ}\text{C}$ and $\Delta T_{\text{min}} = 10\,^{\circ}\text{C}$ approach temperatures are shown below.

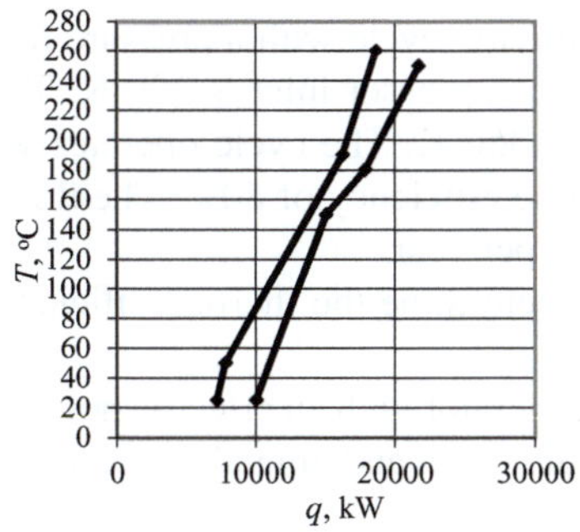

$\Delta T_{\min} = 20°C$

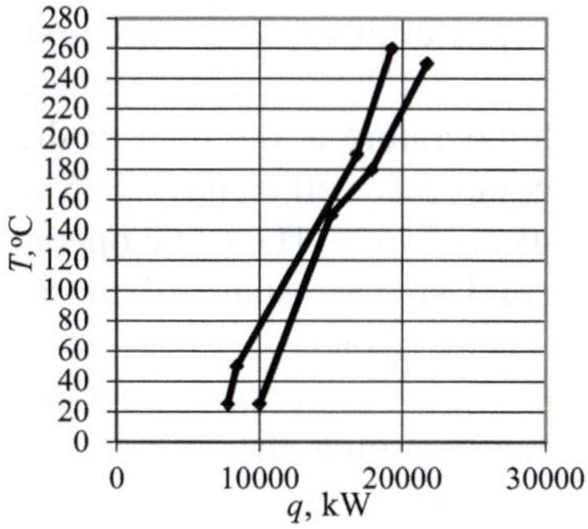

$\Delta T_{\min} = 10°C$

For approach temperatures of $\Delta T_{\text{min}} = 10°C$ and 20°C estimated minimum hot and cold utilities are:

Approach temperatures: ΔT_{min}, °C	10	20
Hot utility (i.e., steam): $q_{\text{hot,min}}$, kW	2450	3050
Cold utility (i.e., cooling water): $q_{\text{cold,min}}$, kW	2225	2825

This simple analysis shows that the smaller approach temperature reduces the utilities needed considerably.

Problems

9.1. A power plant is operating on an ideal Brayton cycle with a pressure ratio of $r_p = 9$. The fresh air temperature at the compressor inlet is 295 K. The air temperature at the inlet of the turbine is 1300 K. The cycle operates with a compressor efficiency of 80% and a turbine efficiency of 80%. The flow rate gas is 3 kg/s. The cycle operates 360 days per year.

a. Using the standard-air assumptions, determine the thermal efficiency of the cycle.
b. If the power plant operates with a regenerator with an effectiveness of 0.80, determine the thermal efficiency of the cycle.

9.2. A power plant is operating on an ideal Brayton cycle with a pressure ratio of $r_p = 9$. The fresh air temperature at the compressor inlet is 300 K. The air temperature at the inlet of the turbine is 1300 K. The cycle operates with a compressor efficiency of 85% and a turbine efficiency of 85%. The flow rate gas is 5 kg/s. The cycle operates 360 days per year.

(a) Using the standard-air assumptions, determine the thermal efficiency of the cycle.
(b) If the power plant operates with a regenerator with an effectiveness of 0.80, determine the thermal efficiency of the cycle and annual conservation of fuel.

9.3. A power plant is operating on an ideal Brayton cycle with a pressure ratio of $r_p = 8$. The fresh air temperature at the compressor inlet is 300 K. The air temperature at the inlet of the turbine is 1350 K. The cycle operates with a compressor efficiency of 85% and a turbine efficiency of 80%. The flow rate gas is 5 kg/s. The cycle operates 360 days per year.
  (a) Using the standard-air assumptions, determine the thermal efficiency of the cycle.
  (b) If the power plant operates with a regenerator with an effectiveness of 0.7, determine the thermal efficiency of the cycle and annual conservation of fuel.

9.4. A power plant is operating on an ideal Brayton cycle with a pressure ratio of $r_p = 15$. The fresh air temperature at the compressor inlet is 290 K. The air temperature at the inlet of the turbine is 1400 K. The cycle operates with a compressor efficiency of 90% and a turbine efficiency of 80%. The flow rate gas is 4.5 kg/s. The cycle operates 360 days per year.
  (a) Using the standard-air assumptions, determine the thermal efficiency of the cycle.
  (b) If the power plant operates with a regenerator with an effectiveness of 0.75, determine the thermal efficiency of the cycle and annual conservation of fuel.

9.5. A power plant is operating on an ideal Brayton cycle with a pressure ratio of $r_p = 14$. The fresh air temperature at the compressor inlet is 290 K. The air temperature at the inlet of the turbine is 1400 K. The cycle operates with a compressor efficiency of 90% and a turbine efficiency of 80%. The flow rate gas is 4.5 kg/s. The cycle operates 360 days per year.
  (a) Using the standard-air assumptions, determine the thermal efficiency of the cycle.
  (b) If the power plant operates with a regenerator with an effectiveness of 0.75, determine the thermal efficiency of the cycle and annual conservation of fuel.

9.6. A power plant is operating on an ideal Brayton cycle with a pressure ratio of $r_p = 7.5$. The fresh air temperature at the compressor inlet is 290 K. The air temperature at the inlet of the turbine is 1250 K. The cycle operates with a compressor efficiency of 90% and a turbine efficiency of 85%. The flow rate gas is 6.5 kg/s. The cycle operates 360 days per year.
  (a) Using the standard-air assumptions, determine the thermal efficiency of the cycle.
  (b) If the power plant operates with a regenerator with an effectiveness of 0.75, determine the thermal efficiency of the cycle and annual conservation of fuel.

9.7. A power plant is operating on an ideal Brayton cycle with a pressure ratio of $r_p = 11.0$. The fresh air temperature at the compressor inlet is 290 K. The air temperature at the inlet of the turbine is 1350 K. The cycle operates with a compressor efficiency of 85% and a turbine efficiency of 80%. The flow rate gas is 6.0 kg/s. The cycle operates 360 days per year.

(a) Using the standard-air assumptions, determine the thermal efficiency of the cycle.
(b) If the power plant operates with a regenerator with an effectiveness of 0.76, determine the thermal efficiency of the cycle and annual conservation of fuel.

9.8. A power plant is operating on an ideal Brayton cycle with a pressure ratio of $r_p = 10.0$. The fresh air temperature at the compressor inlet is 290 K. The air temperature at the inlet of the turbine is 1350 K. The cycle operates with a compressor efficiency of 85% and a turbine efficiency of 80%. The flow rate gas is 6.0 kg/s. The cycle operates 360 days per year.
(a) Using the standard-air assumptions, determine the thermal efficiency of the cycle.
(b) If the power plant operates with a regenerator with an effectiveness of 0.76, determine the thermal efficiency of the cycle and annual conservation of fuel.

9.9. A steam power plant is operating on the simple ideal Rankine cycle. The steam mass flow rate is 20 kg/s. The steam enters the turbine at 3500 kPa and 400°C. Discharge pressure of the steam from the turbine is 78.5 kPa.
(a) Determine the thermal efficiency of the cycle.
(b) If the pressure of the discharge steam is reduced to 15 kPa determine the thermal efficiency.

9.10. A steam power plant is operating on the simple ideal Rankine cycle. The steam mass flow rate is 35 kg/s. The steam enters the turbine at 4000 kPa and 350°C. Discharge pressure of the steam from the turbine is 101.3 kPa.
(a) Determine the thermal efficiency of the cycle.
(b) If the pressure of the discharge steam is reduced to 15 kPa determine the thermal efficiency.

9.11. A steam power plant is operating on the simple ideal Rankine cycle. The steam mass flow rate is 30 kg/s. Steam enters the turbine at 3000 kPa and 400°C. Discharge pressure of the steam from the turbine is 67.5 kPa.
(a) Determine the thermal efficiency of the cycle.
(b) If the pressure of the discharge steam is reduced to 15 kPa determine the thermal efficiency.

9.12. A steam power plant is operating on the simple ideal Rankine cycle. The steam mass flow rate is 30 kg/s. The steam enters the turbine at 2500 kPa and 300°C. The discharge pressure of the steam from the turbine is 62.5 kPa.
(a) Determine the thermal efficiency of the cycle.
(b) If the pressure of the discharge steam is reduced to 15 kPa determine the thermal efficiency.

9.13. A steam power plant is operating on the simple ideal Rankine cycle. The steam mass flow rate is 35 kg/s. The steam enters the turbine at 3000 kPa and 350°C. Discharge pressure of the steam from the turbine is 78.5 kPa.
(a) Determine the thermal efficiency of the cycle.
(b) If the pressure of the boiler is increased to 10,000 kPa while maintaining the turbine inlet temperature at 350°C, determine thermal efficiency.

9.14. A steam power plant is operating on the simple ideal Rankine cycle. The steam mass flow rate is 42 kg/s. The steam enters the turbine at 2500 kPa and 300°C. Discharge pressure of the steam from the turbine is 15.0 kPa.
(a) Determine the thermal efficiency of the cycle.
(b) If the pressure of the boiler is increased to 9000 kPa while maintaining the turbine inlet temperature at 300°C, determine the thermal efficiency.

9.15. A steam power plant is operating on the simple ideal Rankine cycle. The steam mass flow rate is 50 kg/s. The steam enters the turbine at 4000 kPa and 400°C. Discharge pressure of the steam from the turbine is 15.0 kPa.
(a) Determine the thermal efficiency of the cycle.
(b) If the pressure of the boiler is increased to 11,000 kPa while maintaining the turbine inlet temperature at 400°C, determine the thermal efficiency.

9.16. A steam power plant is operating on the simple ideal Rankine cycle. The steam mass flow rate is 35 kg/s. Steam enters the turbine at 3500 kPa and 300°C. Discharge pressure of the steam from the turbine is 78.5 kPa.
(a) Determine the thermal efficiency of the cycle.
(b) If the temperature of the boiler is increased to 500°C while maintaining the pressure at 3500 kPa, determine thermal efficiency.

9.17. A steam power plant is operating on the simple ideal Rankine cycle. The steam mass flow rate is 40 kg/s. The steam enters the turbine at 3000 kPa and 300°C. Discharge pressure of the steam from the turbine is 15 kPa.
(a) Determine the thermal efficiency of the cycle.
(b) If the temperature of the boiler is increased to 500°C while maintaining the pressure at 3000 kPa, determine the thermal efficiency.

9.18. A steam power plant is operating on the simple ideal Rankine cycle. The steam mass flow rate is 40 kg/s. The steam enters the turbine at 6000 kPa and 350°C. Discharge pressure of the steam from the turbine is 15 kPa.
(a) Determine the thermal efficiency of the cycle.
(b) If the temperature of the boiler is increased to 500°C while maintaining the pressure at 6000 kPa, determine thermal efficiency.

9.19. A steam power plant is operating on the simple ideal Rankine cycle. The steam mass flow rate is 40 kg/s. The steam enters the turbine at 9000 kPa and 400°C. Discharge pressure of the steam from the turbine is 15 kPa.
(a) Determine the thermal efficiency of the cycle.
(b) If the temperature of the boiler is increased to 550°C while maintaining the pressure at 9000 kPa, determine thermal efficiency.

9.20. Estimate the maximum possible efficiency for parts (a) and (b) in Problem 9.18 and compare them with those obtained in parts (a) and (b) in Problem 9.18.

9.21. Estimate the maximum possible efficiency for parts (a) and (b) in Problem 9.19 and compare them with those obtained in parts (a) and (b) in Problem 9.19.

9.22. A steam power plant is operating on the simple ideal Rankine cycle. The steam mass flow rate is 20 kg/s. Steam enters the turbine at 3000 kPa and 400°C. Discharge pressure of the steam from the turbine is 78.5 kPa. If the pressure of the boiler is increased to 9000 kPa while maintaining the turbine inlet temperature at 400°C, determine the thermal efficiency.

9.23. A steam power plant is operating on the simple ideal Rankine cycle. The steam mass flow rate is 20 kg/s. The steam enters the turbine at 4000 kPa and 400°C. Discharge pressure of the steam from the turbine is 15 kPa. If the pressure of the boiler is increased to 9500 kPa while maintaining the turbine inlet temperature at 400°C, determine the thermal efficiency.

9.24. A steam power plant is operating on the simple ideal Rankine cycle. The steam mass flow rate is 20 kg/s. The steam enters the turbine at 5000 kPa and 350°C. Discharge pressure of the steam from the turbine is 15 kPa. If the pressure of the boiler is increased to 10,000 kPa while maintaining the turbine inlet temperature at 350°C, determine the thermal efficiency.

9.25. A steam power plant is operating on the simple ideal Rankine cycle. The steam mass flow rate is 20 kg/s. The steam enters the turbine at 9800 kPa and 350°C. Discharge pressure of the steam from the turbine is 78.5 kPa. If the temperature of the boiler is increased to 600°C while maintaining the pressure at 9800 kPa, determine the thermal efficiency.

9.26. A steam power plant is operating on the simple ideal Rankine cycle. The steam mass flow rate is 20 kg/s. Steam enters the turbine at 8000 kPa and 325°C. Discharge pressure of the steam from the turbine is 78.5 kPa. If the temperature of the boiler is increased to 550°C while maintaining the pressure at 8000 kPa, determine thermal efficiency.

9.27. Estimate the maximum possible efficiency for parts (a) and (b) in Problem 9.25 and compare them with those obtained in parts (a) and (b) in Problem 9.25.

9.28. Estimate the maximum possible efficiency for parts (a) and (b) in Problem 9.26 and compare them with those obtained in parts (a) and (b) in Problem 9.26.

9.29. Air with a flow rate of 4 kg/s is compressed in a steady state and reversible process from an inlet state of 100 kPa and 300 K to an exit pressure of 1000 kPa. Estimate the work for (a) polytropic compression with $\gamma = 1.3$, and (b) ideal two-stage polytropic compression with intercooling using the same polytropic exponent of $\gamma = 1.3$, (c) estimate conserved compression work by intercooling if the compressor is operated 360 days per year.

9.30. Air is compressed in a steady state and reversible process from an inlet state of 100 kPa and 285 K to an exit pressure of 800 kPa. The mass flow rate of air is 8 kg/s. Estimate the work for (a) polytropic compression with $\gamma = 1.35$, and (b) ideal two-stage polytropic compression with intercooling using the same polytropic exponent of $\gamma = 1.35$, (c) estimate conserved compression work by intercooling and electricity per year if the compressor is operated 360 days per year.

9.31. Air is compressed in a steady state and reversible process from an inlet state of 110 kPa and 290 K to an exit pressure of 900 kPa. The mass flow rate of air is 10 kg/s. Estimate the work for (a) polytropic compression with $\gamma = 1.3$, and (b) ideal two-stage polytropic compression with intercooling using the same polytropic exponent of $\gamma = 1.3$, (c) estimate conserved compression work by intercooling and electricity per year if the compressor is operated 360 days per year.

9.32. Air is compressed in a steady state and reversible process from an inlet state of 110 kPa and 290 K to an exit pressure of 900 kPa. The mass flow rate of air is 10 kg/s. Estimate the work for (a) polytropic compression with $\gamma = 1.2$, and (b) ideal two-stage polytropic compression with intercooling using the same polytropic exponent of $\gamma = 1.2$, (c) estimate conserved compression work by intercooling and electricity per year if the compressor is operated 360 days per year.

9.33. Air is compressed in a steady state and reversible process from an inlet state of 100 kPa and 290 K to an exit pressure of 900 kPa. The mass flow rate of air is 5 kg/s. Estimate the work for (a) polytropic compression with $\gamma = 1.25$, and (b) ideal two-stage polytropic compression with intercooling using the same polytropic exponent of $\gamma = 1.25$, (c) estimate conserved compression work by intercooling and electricity per year if the compressor is operated 360 days per year.

9.34. Air is compressed in a steady state and reversible process from an inlet state of 100 kPa and 290 K to an exit pressure of 900 kPa. The mass flow rate of air is 3.5 kg/s. Estimate the work for (a) polytropic compression with $\gamma = 1.3$, and (b) ideal two-stage polytropic compression with intercooling using the same polytropic exponent of $\gamma = 1.3$, (c) estimate conserved compression work by intercooling and electricity per year if the compressor is operated 360 days per year.

9.35. Natural gas contains mostly methane gas. In a steady state and reversible process, natural gas is compressed from an inlet state of 100 kPa and 290 K to an exit pressure of 1000 kPa. The mass flow rate of natural gas is 8 kg/s. Estimate the work for (a) polytropic compression with $\gamma = 1.3$, and (b) ideal two-stage polytropic compression with intercooling using the same polytropic exponent of $\gamma = 1.3$, (c) estimate conserved compression work by intercooling and electricity per year if the compressor is operated 350 days per year.

9.36. Natural gas contains mostly the methane gas. In a steady state and reversible process, natural gas is compressed from an inlet state of 100 kPa and 290 K to an exit pressure of 900 kPa. The mass flow rate of natural gas is 5 kg/s. Estimate the work for (a) polytropic compression with $\gamma = 1.2$, and (b) ideal two-stage polytropic compression with intercooling using the same polytropic exponent of $\gamma = 1.2$, (c) estimate conserved compression work by intercooling and electricity per year if the compressor is operated 360 days per year.

9.37. In a steady state and reversible process, propane gas is compressed from an inlet state of 100 kPa and 300 K to an exit pressure of 900 kPa. The mass flow rate of propane is 3 kg/s. Estimate the work for (a) polytropic compression with $\gamma = 1.3$, and (b) ideal two-stage polytropic compression with intercooling using the same polytropic exponent of $\gamma = 1.3$, (c) estimate conserved work by intercooling and electricity per year if the compressor is operated 340 days per year.

9.38. In a steady state and reversible process, hydrogen gas is compressed from an inlet state of 100 kPa and 300 K to an exit pressure of 1100 kPa. The mass flow rate of hydrogen is 3 kg/s. Estimate the work for (a) polytropic compression with $\gamma = 1.3$, and (b) ideal two-stage polytropic compression with intercooling using the same polytropic exponent of $\gamma = 1.3$, (c) estimate conserved work by intercooling and electricity per year if the compressor is operated 350 days per year.

9.39. In a steady state and reversible process, carbon dioxide gas is compressed from an inlet state of 100 kPa and 290 K to an exit pressure of 1000 kPa. The mass flow rate of carbon dioxide is 4 kg/s. Estimate the work for (a) polytropic compression with $\gamma = 1.3$, and (b) ideal two-stage polytropic compression with intercooling using the same polytropic exponent of $\gamma = 1.3$, (c) estimate conserved compression work by intercooling and electricity per year if the compressor is operated 360 days per year.

9.40. An adiabatic compressor is used to compress air from 100 kPa and 290 K to 900 kPa at a steady-state operation. The isentropic efficiency of the compressor is 80%. The airflow rate is 0.55 kg/s. Determine the minimum and actual power needed by the compressor.

9.41. An adiabatic compressor is used to compress air from 100 kPa and 290 K to 1100 kPa at a steady-state operation. The isentropic efficiency of the compressor is 80%. The airflow rate is 0.35 kg/s. Determine the minimum and actual power needed by the compressor.

9.42. An adiabatic compressor is used to compress air from 100 kPa and 290 K to 1400 kPa at a steady-state operation. The isentropic efficiency of the compressor is 85%. The airflow rate is 0.5 kg/s. Determine the minimum and actual power needed by the compressor.

9.43. An adiabatic compressor is used to compress air from 100 kPa and 290 K to 1600 kPa at a steady-state operation. The isentropic efficiency of the compressor is 83%. The airflow rate is 0.4 kg/s. Determine the minimum and actual power needed by the compressor.

9.44. Estimate the power conservation when an electric motor with an efficiency of 78% is replaced with another motor operating at 88% efficiency. Both the motors drive compressors and must deliver a power of 24 kW for an average 2500 h/year.

9.45. Estimate the power conservation when an electric motor with an efficiency of 74% is replaced with another motor operating at 89% efficiency. Both the motors drive compressors and must deliver a power of 36 kW for an average 8000 h/year.

9.46. Estimate the power conservation when an electric motor with an efficiency of 74% is replaced with another motor operating at 89% efficiency. Both the motors drive compressors and must deliver a power of 18 kW for an average 80 h per day.

9.47. A cryogenic manufacturing plant handles liquid methane at 115 K and 5000 kPa at a rate of 0.15 m^3/s. In the plant, a throttling valve reduces the pressure of liquid methane to 2000 kPa. A new process considered replaces the throttling valve with a turbine to produce power while reducing the pressure. Using the data for the properties of liquid methane below, estimate (a) the power that can be produced by the turbine, and (b) the savings in electricity usage per year if the turbine operates 360 day per year.

T K	*P* kPa	*H* kJ/kg	*S* kJ/kg K	C_p kJ/kg K	*P* kg/m^3
110	1000	209.0	4.875	3.471	425.8
110	2000	210.5	4.867	3.460	426.6
110	5000	215.0	4.844	3.432	429.1
120	1000	244.1	5.180	3.543	411.0
120	2000	245.4	5.171	3.528	412.0
120	5000	249.6	5.145	3.486	415.2

Source [1]

9.48. A cryogenic manufacturing plant handles liquid methane at 115 K and 5000 kPa at a rate of 0.2 m^3/s. In the plant, a throttling valve reduces the pressure of liquid methane to 2000 kPa. A new process considered replaces the throttling valve with a turbine to produce power while reducing the pressure. Using the data for the properties of liquid methane given in Problem 9.47, estimate (a) the power that can be produced by the turbine, (b) the savings in electricity usage per year if the turbine operates 300 days per year.

9.49. A heat pump is used to heat a house and maintain it at 18°C. On a day where the outside temperature is −2°C, the house is losing heat at a rate of 79,200 kJ/h. The heat pump operates with a coefficient of performance (COP) of 3.5. Determine (a) power needed by the heat pump, (b) the rate of heat absorbed from to the surrounding cold air.

9.50. A heat pump is used to heat a house and maintain it at 20°C. On a day where the outside temperature is 0°C, the house is losing heat at a rate of 34,500 kJ/h. The heat pump operates with a coefficient of performance (COP) of 3.0. Determine (a) power needed by the heat pump, (b) the rate of heat absorbed from to the surrounding cold air.

9.51. A heat pump is used to heat a house and maintain it at 20°C. On a day where the outside temperature is 4°C, the house is losing heat at a rate of 65,500 kJ/h. The heat pump operates with a coefficient of performance (COP) of 3.9. Determine (a) power needed by the heat pump, (b) the rate of heat absorbed from to the surrounding cold air.

9.52. A Carnot heat pump is used to heat a house during the winter. The house is maintained 20°C. The house is estimated to be losing heat at a rate of 108,000 kJ/h when the outside temperature is – 4°C. Determine the minimum power needed by the heat pump and the rate of heat absorbed from the surrounding cold air.

9.53. A Carnot heat pump is used to heat a house during the winter. The house is maintained 20°C. The house is estimated to be losing heat at a rate of 78,000 kJ/h when the outside temperature is 2°C. Determine the minimum power needed by the heat pump and the rate of heat absorbed from the surrounding cold air.

9.54. Estimate the cost of electricity for a 10,000 Btu/h (3000 W) air-conditioning unit operating, with a SEER of 10 Btu/Wh. The unit is used for a total of 1500 h during an annual cooling season.

9.55. Estimate the cost of electricity for a 12,000 Btu/h air-conditioning unit operating, with a SEER of 14 Btu/Wh. The unit is used for a total of 2500 h during an annual cooling season.

9.56. Estimate the cost of electricity for a 9000 Btu/h air-conditioning unit operating, with a SEER of 12 Btu/Wh. The unit is used for a total of 2000 h during an annual cooling season.

9.57. (a) Estimate the annual cost of electric power consumed by a 6-ton air-conditioning unit operating for 1000 h/year with a SEER rating of 10. (b) Estimate the value of EER for hot and cold temperatures of 20 and −4°C, respectively.

9.58. (a) Estimate the annual cost of electric power consumed by a 4-ton air-conditioning unit operating for 2500 h/year with a SEER rating of 12 (b) Estimate the value of EER for hot and cold temperatures of 21°C and −10°C, respectively.

9.59. (a) Estimate the annual cost of electric power consumed by a 9-ton air-conditioning unit operating for 3000 h/ year with a SEER rating of 14. (b) Estimate the value of EER for hot and cold temperatures of 21°C and −5°C, respectively.

9.60. A 4 tons of current residential air conditioner operates with a SEER rating of 10. This unit will be replaced with a newer unit operating with a SEER rating of 22. The unit operates 130 days with an average of 10 h per day. Average inside and outside temperatures are 20°C and − 4°C, respectively. Estimate the saving in the electricity and the maximum energy efficiency ratio.

9.61. A 4 tons of current residential air conditioner operates with a SEER rating of 10. This unit will be replaced with a newer unit operating with a SEER rating of 22. The unit operates 120 days with an average of 9 h per day. Average inside and outside temperatures are 20 and −0°C, respectively. Estimate the saving in the electricity and the maximum energy efficiency ratio.

9.62. A 4 tons of current residential air conditioner operates with a SEER rating of 10. This unit will be replaced with a newer unit operating with a SEER rating of 20. The unit operates 120 days with an average of 7 h per day. Average inside and outside temperatures are 22°C and − 10°C, respectively. Estimate the saving in the electricity and the maximum energy efficiency ratio.

9.63. The efficiency of an open burner is around 70% for electric heater units and 40% for natural gas units. We operate a 4-kW electric burner. Estimate the rate of energy consumption by the burner and the utilized energy for both electric and gas burners.

9.64. The efficiency of an open burner is around 72% for electric heater units and 39% for natural gas units. We operate a 6-kW electric burner at one location. Estimate the rate of energy consumption by the burner and the utilized energy for both electric and gas burners.

9.65. The efficiency of an open burner is around 69% for electric heater units and 42% for natural gas units. We operate a 10-kW electric burner at one location. Estimate the rate of energy consumption by the burner and unit costs of the utilized energy for both electric and gas burners.

9.66. An adiabatic turbine is used to produce electricity by expanding a superheated steam at 4100 kPa and 350°C. The power output is 60 MW. The steam leaves the turbine at 40 kPa and 100°C. If the combustion efficiency is 0.70 and the generator efficiency is 0.9, determine the overall plant efficiency and the amount of coal supplied per hour.

9.67. An adiabatic turbine is used to produce electricity by expanding a superheated steam at 4100 kPa and 350°C. The power output is 60 MW. The steam leaves the turbine at 40 kPa and 100°C. If the combustion efficiency is 0.70 and the generator efficiency is 0.9, determine the overall plant efficiency and the amount of coal supplied per hour.

9.68. An adiabatic turbine is used to produce electricity by expanding a superheated steam at 5800 kPa and 400°C. The power output is 55 MW. The steam leaves the turbine at 40 kPa and 100°C. If the combustion efficiency is 0.72 and the generator efficiency is 0.9, determine the overall plant efficiency and the amount of coal supplied per hour.

9.69. An adiabatic turbine is used to produce electricity by expanding a superheated steam at 4100 kPa and 350°C. The steam flow rate is 42 kg/s. The steam leaves the turbine at 40 kPa and 100°C. If the combustion efficiency is 0.75 and the generator efficiency is 0.90, determine the overall plant efficiency and the amount of coal supplied per hour.

9.70. An adiabatic turbine is used to produce electricity by expanding a superheated steam at 4100 kPa and 350°C. The steam flow rate is 42 kg/s. The steam leaves the turbine at 40 kPa and 100°C. If the combustion efficiency is 0.75 and the generator efficiency is 0.90, determine the overall plant efficiency and the amount of coal supplied per hour.

9.71. The overall efficiencies are about 25–28% for gasoline car engines, 34–38% for diesel engines, and 40–60% for large power plants. Compare the energy necessary for gasoline and diesel engines. The efficiency for diesel is 36%. A car engine with a power output of 240 hp has a thermal efficiency of 24%. Determine the fuel consumption of the car if the fuel has a higher heating value of 20,400 Btu/lb.

9.72. The overall efficiencies are about 25–28% for gasoline car engines, 34–38% for diesel engines, and 40–60% for large power plants [2]. Compare the energy necessary for gasoline and diesel engines. The efficiency for diesel is 35%. A car engine with a power output of 180 hp has a thermal efficiency of 26%. Determine the fuel consumption of the car if the fuel has a higher heating value of 20,400 Btu/lb.

9.73. Fuel consumption of the two cars is one with 11 l/100 km city and 9 l/100 km highway, and the other 6.5 l/100 km in city traffic and at 5 l/100 km highway. Estimate the annual fuel saving and emission reduction achieved with the more fuel-efficient car traveling an average7500 km per year.

9.74. Fuel consumption of the two cars is one with 10 l/100 km city and 8 l/100 km highway, and the other 6.0 l/100 km in city traffic and at 5 l/100 km highway. Estimate the annual fuel saving and emission reduction achieved with the more fuel-efficient car traveling an average10000 km per year.

9.75. Fuel consumption of the two cars is one with 12 l/100 km city and 9 l/100 km highway, and the other 7.0 l/100 km in city traffic and at 6 l/100 km highway. Estimate the annual fuel saving and emission reduction achieved with the more fuel-efficient car traveling an average12000 km per year.

9.76. Assume that a household consumes about 5000 kWh per year. If the lighting is provided by compact fluorescent bulbs only, estimate the conservation of energy and saving per year

9.77. Assume that a household consumes about 14,000 kWh per year. If the lighting is provided by compact fluorescent bulbs only, estimate the conservation of energy and saving per year

9.78. In a process available hot and cold process streams and their heat capacities are shown below.

Hot and cold process stream conditions.

Streams		T_{in} °C	T_{out} °C	$C = \dot{m}C_p$ kW/°C
C1	Cold 1	20	180	40
C2	Cold 2	160	250	55
H1	Hot 1	280	60	35
H2	Hot 2	190	20	25

Construct the balanced composite curves for the process with $\Delta T_{min} = 20\,°C$ and $\Delta T_{min} = 10\,°C$, and compare the amounts of hot and cold utilities needed.

References

1. Çengel YA, Boles MA (2014) Thermodynamics: an engineering approach, 8th edn. McGraw-Hill, New York
2. Demirel Y (2025) Energy conservation. In: Dincer I (ed) Comprehensive energy systems, 2nd edn, Elsevier, Amsterdam, vol 9, pp 1–51
3. Demirel Y, Gerbaud V (2025) Nonequilibrium thermodynamics: transport and rate processes in physical, chemical and biological systems, 5th edn. Elsevier, Amsterdam
4. Demirel Y, Rosen MA (2023) Sustainable engineering. Process intensification, energy analysis, artificial intelligence, CRC Press, Taylor & Francis
5. Dincer I, Midilli A (2014) Energy conservation. In: Anwar S (ed) Encyclopedia of energy engineering and technology, 2nd edn. CRC Press, Boca Raton
6. Guan L, Chen G (2014) Solar energy: building operations use. In: Anwar S (ed) Encyclopedia of energy engineering and technology, 2nd edn. CRC Press, Boca Raton
7. Herold KE (2014) Waste heat recovery applications: absorption heat pumps. In: Anwar S (ed) Encyclopedia of energy engineering and technology, 2nd edn. CRC Press, Boca Raton

8. Jaffe RL, Taylor W (2018) The physics of energy. Cambridge Univ. Press, Cambridge
9. Krigger J, Dorsi C (2008) The homeowner's handbook to energy efficiency: a Guide to big and small improvements. Saturn Resource Management, Helena
10. McCardell SB (2014) Effective energy use: rewards and excitement. In: Anwar S (ed) Encyclopedia of energy engineering and technology, 2nd edn. CRC Press, Boca Raton
11. Uhrig RE (2014) Greenhouse gas emissions: gasoline, hybrid-electric, and hydrogen-fueled vehicles. In: Anwar S (ed) Encyclopedia of energy engineering and technology, 2nd edn. CRC Press, Boca Raton

Energy Coupling

10

Introduction and Learning Objectives: Various mechanisms and devices can facilitate coupling between energy-producing process and energy requiring processes. For example, in living systems, synthesis of adenosine triphosphate requires energy and therefore must couple with an energy providing process of electron transport chain. On the other hand, the hydrolysis of adenosine triphosphate supplies the energy needed in living systems. This chapter discusses bioenergetics briefly as a representative analysis of energy coupling in biochemical cycles, and energy expenditure.

The learning objectives of this chapter are to discuss:

- Bioenergetics.
- Energy coupling in biochemical cycles.
- Control of energy coupling.
- Energy metabolism.
- Energy expenditure.

10.1 Energy Coupling and Gibbs Free Energy

The Gibbs free energy of a system is related to enthalpy and entropy

$$G = H - TS \tag{10.1}$$

The change in the Gibbs free energy is

$$\Delta G = \Delta H - T\Delta S \tag{10.2}$$

Y. Demirel, *Energy*, Green Energy and Technology,
https://doi.org/10.1007/978-3-031-97111-2_10

The Gibbs free energy is a state function hence its change between two states does not depend on how this change takes place between the states, Any reaction for which ΔG is negative would be favorable, or spontaneous, while any reaction for which ΔG is positive would be unfavorable

$$\Delta G < 0. \quad \text{Favorable, or spontaneous reactions} \tag{10.3}$$

$$\Delta G > 0 \quad \text{Unfavorable, or nonspontaneous reactions} \tag{10.4}$$

Reactions are classified as either *exothermic* ($\Delta H < 0$) releasing energy or *endothermic* ($\Delta H > 0$) requiring energy. Reactions can also be classified as *exergonic* ($\Delta G < 0$) decreasing free energy or *endergonic* ($\Delta G > 0$) increasing free energy during the reaction. Consider the following two reactions:

$$1.\,\mathrm{A} \rightarrow \mathrm{B} \quad \Delta G_{\mathrm{AB}} = (G_{\mathrm{B}} - G_{\mathrm{A}}) \ll 0, \ \text{exergonic-releases energy} \tag{10.5}$$

$$2.\,\mathrm{C} \rightarrow \mathrm{D} \quad \Delta G_{\mathrm{AB}} = (G_{\mathrm{B}} - G_{\mathrm{A}}) > 0, \ \text{endergonic-requires energy} \tag{10.6}$$

Then the first reaction can drive the second reaction if they are coupled and satisfy the following inequality

$$|\Delta G_{\mathrm{AB}}| > |\Delta G_{\mathrm{CD}}|. \tag{10.7}$$

Then the coupled reactions would have total Gibbs energy with a negative value hence become exergonic and favorable reaction

$$(\Delta G_{\mathrm{AB}} + \Delta G_{\mathrm{CD}}) < 0 \tag{10.8}$$

Energy coupling may require some special mechanisms, such as enzymes and configurations in living systems [3, 4], which are briefly discussed in the next section.

10.2 Energy Coupling in Living Systems

Coupling can occur when a flow proceeds without its primary driving force, or opposite to the direction imposed by its primary driving force. For example, the living cells can pump ions flow from low to high concentration levels. Since the favorable transport of materials take place always from high to low concentrations, this opposite transportation of substance is possible only if it is coupled with an energy providing process. In living systems, this energy is provided by the hydrolysis of the adenosine triphosphate (ATP). Energy coupling in living systems means that the metabolic pathways intersect in such a way that energy released from the favorable reactions of catabolism can be used to drive the energy requiring reactions of the anabolic pathways. This transfer of energy from catabolism to anabolism would be possible through coupling.

The chloroplasts in plants use the free energy of the sun to initiate electron transfer cycles and proton gradients to produce ATP in photosynthesis. In animals, respiration cycle initiates electron transfer cycle and proton gradient to produce the ATP in the oxidative phosphorylation. ATP is an energy-rich biomolecule, which stores and provides energy necessary in living systems. Photosynthesis and oxidative phosphorylation represent some of the major energy coupling systems in living systems [5].

10.3 Bioenergetics

Bioenergetics is concerned with energy production, conservation, and conversion processes in living systems [4, 6]. The outside energy is used in the synthesis of an energy-rich molecule called adenosine triphosphate (ATP) from adenosine diphosphate (ADP) and molecular phosphorous (Pi). The chemical formula of ATP is shown in Fig. 10.1. ATP has three phosphate groups attached to adenosine, in which the terminal phosphate has a weak linkage and can break spontaneously whenever ATP forms a complex with an enzyme and releases chemical energy. ATP therefore acts as an energy source of the living cell.

ATP synthesis is an endothermic reaction as it needs energy from outside. So, the APT synthesis is coupled with electron transport chain. Therefore, the electron transport chain drives the ATP synthesis. Synthesizing of ATP is matched and synchronized to cellular ATP utilization through the hydrolysis (combining with water) of ATP

$$\mathrm{ADP} + P_i + n\mathrm{H}^+_{\mathrm{in}} = \mathrm{ATP} + \mathrm{H_2O} + n\mathrm{H}^+_{\mathrm{out}}, \tag{10.9}$$

where "in" and "out" denote two phases separated by a membrane, and n is the ratio H^+/ATP, showing the level of transmembrane proton transport for each ATP to be synthesized. ATP utilization is coupled to synthesizing proteins, enzymes, transporting ions, substrates, and producing mechanical work. ATP molecules, therefore, store and transfer energy in the living systems. For example, some of the internal mechanical work involves pumping blood by the heart. Continuous

$$\mathrm{ADP} + \mathrm{Pi} \rightarrow \mathrm{ATP}$$

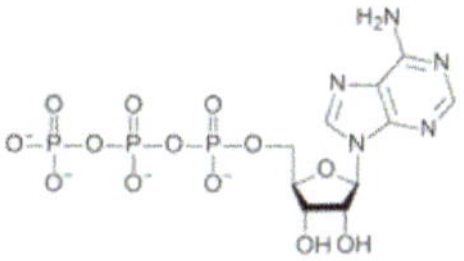

Fig. 10.1 Chemical formula of adenosine triphosphate (ATP). ATP is an energy-rich compound having three phosphate groups attached to adenosine. Of the three phosphate groups, the terminal one has a weak linkage. This phosphate group can break spontaneously whenever ATP forms a complex with an enzyme. The breaking up of this bond releases chemical energy needed in all living systems

biochemical cycles and transport processes can maintain a stationary state by the regulated production and transfer of energy of the ATP.

Most biochemical reactions in bioenergetics occur in pathways, in which other reactions continuously add substrates and remove products. The rate of these reactions depends on the enzymes that catalyze the reactions [4, 5].

10.3.1 Mitochondria

Mitochondria are organelles typically ranging in size from 0.5 to 1 micrometer in length, found in the cytoplasm of eukaryotic cells. Mitochondria contain inner and outer membranes constructed with phospholipids into which specific proteins are embedded. Mitochondrial membranes produce two compartments called the intermembrane space and matrix space enclosed by the inner membrane (Fig. 10.2a). The inner membrane has numerous folds called *cristae* where the synthesis of ATP takes place [4].

10.3.2 Electron Transport Chain and Adenosine Triphosphate (ATP) Synthesis 3

The inner membrane of the mitochondria houses the electron transport chain and ATP synthesis shown in Fig. 10.2c. The tricarboxylic acid cycle, also called the citric acid cycle, produces electron transport chain, which is the major energy-producing pathway. The electrons transport chain causes the proton flow across the inner membrane of the mitochondria and ATP is synthesized by the enzyme

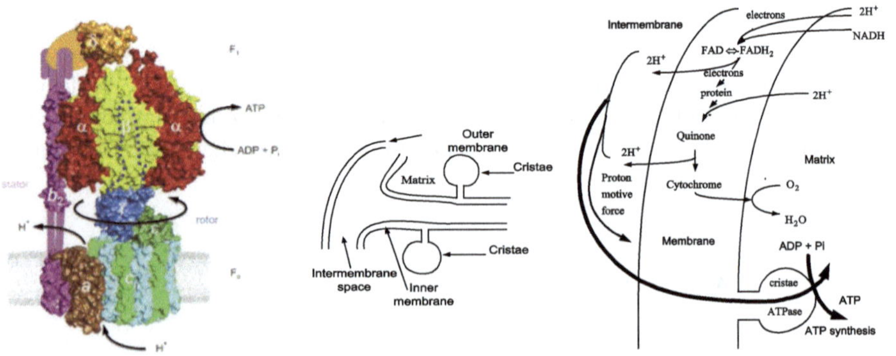

(a) (b) (c)

Fig. 10.2 **a** ATPases are a class of enzymes with F_0 and F_1 subunits that catalyze the synthesis and hydrolysis of adenosine triphosphate (ATP). The cristae have the major coupling factors F_1 (a hydrophilic protein) and F_0 (a hydrophobic lipoprotein complex). F_1 and F_0 together comprise the ATPase (also called ATP synthase), **b** inner membrane structure of the mitochondria, **c** electron transport chain and oxidative phosphorylation of adenosine triphosphate [4, 5] called the ATPases shown in Fig. 10.2a. ATPases have F_o and F_1 subunits that catalyze the synthesis and hydrolysis of ATP. The cristae have the major coupling factors F_1 (a hydrophilic protein) and F_o (a hydrophobic lipoprotein complex) [4].

In animals, the respiration chain generates energy by the oxidation of reducing equivalents of nutrients that are nicotinamide adenine nucleotides NADH and the flavin nucleotides $FADH_2$. When ADP levels are higher than ATP, the cell needs energy, and hence, NADH is oxidized rapidly, and the tricarboxylic acid cycle is accelerated. When the ATP level is higher than ADP, the electron transport chain slows down.

Photosynthesis, driven by light energy, leads to the production of ATP through electron transfer and photosynthetic phosphorylation. Photosynthetic energy conservation takes place in the thylakoid membrane of plant chloroplasts. These membranes facilitate the interactions between the redox system and the synthesis of ATP and are referred to as coupling membranes [4, 5].

10.3.3 Active Transport

Active transport is the transport of a substance from low to high concentration levels and hence against its concentration gradient. This process is also known as uphill transport and needs to be coupled with an energy providing process. The hydrolysis of ATP provides energy for all cellular activity including active transport of ions and molecules. If the process uses chemical energy, such as the energy released from the hydrolysis of adenosine triphosphate, it is termed *primary active transport*. *Secondary active transport* involves the use of an electrochemical gradient produced within the cell.

Hydrolysis of one mole of ATP is an exergonic reaction releasing 31 kJ/mol at pH $= 7$. This energy drives various energy-dependent metabolic reactions and the transport of various ions such as K^+ and Na^+ as seen in Fig. 10.3a. These concentration gradients of K^+ and Na^+ are established by the active transport of both ions. The same enzyme, called the Na^+/K^+ ATPase uses the energy released from the hydrolysis of ATP to transport 3 Na^+ ions out of the cell for each 2 K^+ ions pumped into the cell against their concentration gradients (see Fig. 10.3a). Almost one-third of all the energy generated by the mitochondria in animal cells is used in active transport. In *indirect active transport,* other transporter molecules use the energy already stored in the gradient of a directly pumped ions [4, 5].

10.4 Simple Analysis of Energy Coupling

Figure 10.3b illustrates electron transportation cycle driving the ATP production through the energy coupling. The approximate representative linear phenomenological relations of the oxidative phosphorylation are: [4]

$$J_o = L_o X_o + L_{op} X_p \tag{10.10}$$

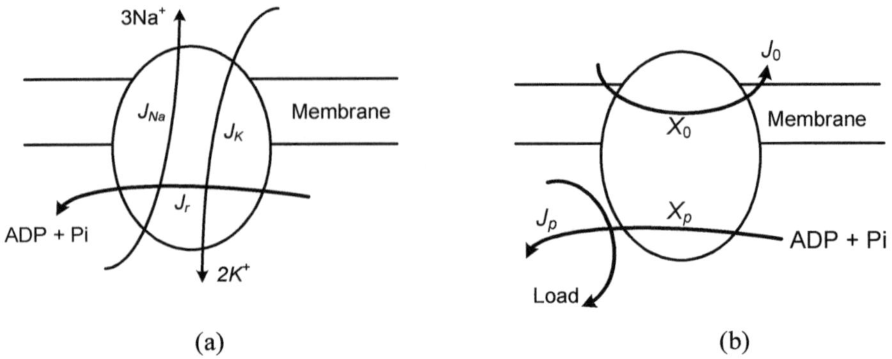

Fig. 10.3 **a** Active transport of Na^+ and K^+ ions coupled to hydrolysis of ATP [4]; **b** energy coupling between respiration cycle and oxidative phosphorylation taking place in the inner membrane of mitochondria. Here J_o and J_p are the oxygen flow rate and the ATP production rate, respectively, while X_o and X_p show the redox potential for oxidizable substances and the phosphate potential, respectively. The flow J represents a load (i.e., osmotic work) coupled to ATP production

$$J_p = L_{po}X_o + L_pX_p \tag{10.11}$$

Here J_o is the oxygen flow rate, J_p is the ATP production rate, X_o is the redox potential for oxidizable substances and X_p is the phosphate potential [4]. The flows (fluxes) are shown by J_o and J_p, while X_o and X_p are called the forces. The L_{op} and L_{po} represent the cross coefficients, while L_o and L_p are called the primary coefficients. The cross coefficients, L_{op} and L_{po}, obey Onsager's reciprocal rules, which state that $L_{op} = L_{po}$.

The *degree of coupling* q is

$$q = \frac{L_{op}}{(L_pL_o)^{1/2}} \tag{10.12}$$

and indicates the extent of overall coupling of the ATP production driven by the respiration cycle in the inner membrane of mitochondria. By defining the *phenomenological stoichiometry* Z by

$$Z = \left(\frac{L_o}{L_p}\right)^{1/2} \tag{10.13}$$

and by dividing Eq. (10.10) by Eq. (10.11), we obtain the flow ratio j and the force ratio x in terms of Z

$$j = \frac{J_o}{J_pZ} \tag{10.14}$$

$$x = \frac{X_oZ}{X_p} \tag{10.15}$$

The flow ratio j and force ratio x can be related to each other by the degree of coupling q as follows

$$j = \frac{x + q}{qZ + 1} \tag{10.16}$$

The energy dissipation Ψ is expressed by [5, 6]

$$\Psi = J_o X_o + J_p X_p = \text{input power} + \text{output power} > 0 \tag{10.17}$$

The efficiency of the linear energy coupling of oxidative phosphorylation may be defined as the ratio of output power to input power and may be expressed in terms of the degree of coupling

$$\eta = -\frac{\text{output power}}{\text{input power}} = \frac{\text{driven process}}{\text{driving process}} = -\frac{J_p X_p}{J_o X_o} = -\frac{x + q}{q + (1/x)} \tag{10.18}$$

The value of optimum efficiency is a function of the degree of coupling only and is expressed by

$$\eta_{opt} = \frac{q^2}{\left(1 + \sqrt{1 - q^2}\right)^2} = \tan^2(\alpha/2) \tag{10.19}$$

where $\alpha = \arcsin(q)$ The value of x at η_{opt} becomes

$$x_{opt} = -\frac{q}{1 + \sqrt{1 + q^2}} \tag{10.20}$$

Therefore, the optimum value of ratio of forces x depends only on the value of degree of coupling [4.5]. Example 10.1 illustrates a simple representative analysis of energy coupling in photosynthesis.

Example 10.1 **Efficiency of energy conversion of photosynthesis**

Consider a model process with an energy exchange between a photon and a composite particle. In this over-simplified model, energy is exchanged through an excited state of the chloroplast by which energy-rich electron/proton pairs from the water react with carbon dioxide. This produces carbohydrates and oxygen molecules, and heat is dissipated away. The flows are related by an approximation [4]

$$\left(\frac{J_D}{J_q}\right)^2 = \frac{k_B n D}{k} \tag{10.21}$$

where $n \sim 3.3 \times 10^{28}$ molecules/m^3, which is a typical value for water and condensed matter in general, J_D mass flow of carbon dioxide (rate of diffusion), J_q is the heat flow, D is the diffusion coefficient, k is the thermal conductivity, and k_B is the Boltzmann constant.

Only a small part of the free energy of photons is available for photosynthesis, and the rest is dissipated. The efficiency of energy conversion in photosynthesis is low and varies in the range 2.4–7.5%. The efficiency of energy conversion is

$$\eta = \frac{J_D}{J_q}\frac{\Delta G}{h\nu} \tag{10.22}$$

where ΔG is the Gibbs energy per molecule and $h\nu$ is the energy per photon. Some approximate values for these driving forces and the coefficients are:

- $\Delta G \sim 7.95 \times 10^{-19}$ J/unit carbohydrate
- $(h\nu) \sim 2.92 \times 10^{-19}$ J/a solar photon (pertaining to red light with a wavelength of 680 nm, which is best absorbed by chlorophyll-α).
- Based on the thermal conductivity of water: $k = 0.607$ W/m K.
- Assuming that the intercellular diffusion of carbon dioxide could be the limiting process in photosynthesis, we have $D \sim 1.95 \times 10^{-9}$ m^2/s based on carbon dioxide in water or $D \sim 0.67 \times 10^{-9}$ m^2/s based on glucose in water.
- $k_B = 1.3806503 \times 10^{-23}$ m^2 kg/s^2 K.

Using these approximate values in the following equation

$$\eta = \frac{J_D}{J_q}\left(\frac{7.95 \times 10^{-19}}{2.92 \times 10^{-19}}\right) = 2.72\left(\frac{k_B n D}{k}\right)^{1/2} \tag{10.23}$$

the approximate predicted values of efficiency of energy conversion become [4]

η (predicted) %	η (measured) %	D m^2/s
2.4	4.9	0.10 10^{-9}
6.1	6.2	0.67 10^{-9}
7.5	~ 7	1.0 10^{-9}

Any plant growing under ideal conditions, the efficiency is expected to be close to 7%.

10.5 Variation of Energy Coupling

The value of degree of coupling q can be calculated from the measurements of oxygen flows at static head (sh) $(J_o)_{sh}$ in which the net rate of ATP vanishes, and at uncoupled state (unc) $(J_o)_{unc}$ where the proton gradient vanishes, and the respiration cycles is uncoupled from the oxidative phosphorylation [4–6]

$$q = \sqrt{1 - (J_{\mathrm{o}})_{\mathrm{sh}}/(J_{\mathrm{o}})_{\mathrm{unc}}} \tag{10.24}$$

For example, the overall degree of energy coupling is higher for rat liver mitochondrion (0.955 ± 0.021) than for brain (0.937 ± 0.026) or heart (0.917 ± 0.037).

Required degrees of coupling of oxidative phosphorylation vary when the ATP production is coupled to a load such as hydrolysis of the ATP (see Fig. 10.3b) [4]. The optimum production functions f for the ATP and output power production are given in terms of the degree of coupling

$$f = \tan^m(\alpha/2)\cos(\alpha) \tag{10.25}$$

The ATP production and power output occur at certain values of degrees of coupling:

- Maximum ATP production at optimal efficiency occurs at $q_{\mathrm{f}} = 0.786$, $f = (J_{\mathrm{p}})_{\mathrm{opt}}$
- Maximum power output at optimal efficiency occurs at $q_{\mathrm{p}} = 0.91$, $f = (J_{\mathrm{p}}X_{\mathrm{p}})_{\mathrm{opt}}$.
- Efficient ATP production at minimal energy cost occurs at $q_{\mathrm{f}}^{\mathrm{ec}} = 0.953$, $f = \eta(J_{\mathrm{p}})_{\mathrm{opt}}$.
- Economic power output at minimal energy cost occurs at $q_{\mathrm{p}}^{\mathrm{ec}} = 0.972$, $f = \eta(J_{\mathrm{p}}X_{\mathrm{p}})_{\mathrm{opt}}$

With the consideration of conductance matching, [5] determined four production functions, which are given in Table 10.1. Fig. 10.4 shows the effect of degree of coupling on optimum efficiency. The sensitivity of oxidative phosphorylation to a fluctuating ATP utilization is minimal at a degree of coupling $q = 0.95$. This means that the phosphate potential is highly buffered with respect to fluctuating energy demands at the degree of coupling, which is very close to the value of $q_{\mathrm{f}}^{\mathrm{ec}}$, at which net ATP production of oxidative phosphorylation occurs at the minimal energy cost [4].

The degree of coupling may be regulated based on the nature of the output required from the energy coupling system. In the heart and brain, the experimental value of q for the cellular respiration pathway is close to the value of $q_{\mathrm{f}}^{ec} = 0.953$, which suggests that the pathway is optimized to economical ATP production. In the brain, the value of $q_{\mathrm{p}} = 0.91$ suggests maximized cellular energy state. However, in the heart, the degree of coupling is 0.786, which is between q_{p} and q_{f}, and consistent with the maximum ATP production necessary for preserving the cellular energy state [4–6].

The optimum output power $(J_1X_1)_{\mathrm{opt}}$ and the efficiency $(J_1X_1\eta)_{\mathrm{opt}}$ are calculated from the plots of J_1X_1 versus x and $J_1X_1\eta$ versus x, respectively. A transition from q_{p} to $q_{\mathrm{p}}^{\mathrm{ec}}$ causes a 12% drop in output power (J_1X_1) and a 51% increase in efficiency.

Table 10.1 Production functions with the consideration of conductance matching [4]

Production function	Loci of the optimal efficiency states	q	Energy cost
Optimum rate of ATP production: $J_p = (q+x)ZL_oX_o$	From the plot of J_p versus x: $(J_p)_{opt} = \tan(\alpha/2)\cos\alpha ZL_oX_o$	$q_f = 0.786$ α = 51.83°	No η = cons.
Optimum output power of OP: $J_pX_p = x(q+x)L_oX_o^2$	From the plot of J_pX_o versus x: $(J_pX_p)_{opt} = \tan^2(\alpha/2)\cos\alpha L_oX_o^2$	$q_p = 0.910$ α = 65.53°	No η = cons.
Optimum rate of ATP production at minimal energy cost: $J_p\eta = -\frac{x(q+x)^2}{xq+1}ZL_oX_o$	From the plot of $J_1\ \eta$ versus x: $(J_p\eta)_{opt} = \tan^3(\alpha/2)\cos\alpha ZL_oX_o$	$q_f^{ec} = 0.953$ α = 72.38°	Yes
Optimum output power of OP at minimal energy cost: $J_pX_p\eta = -\frac{x^2(q+x)^2}{xq+1}L_oX_o^2$	From the plot of $J_1\ X_1\ \eta$ versus x: $(J_pX_p\eta)_{opt} = \tan^4(\alpha/2)\cos\alpha L_oX_o^2$	$q_p^{ec} = 0.972$ α = 76.34°	Yes

OP Oxidative phosphorylation, q_f, q_p are the degree of couplings at optimized ATP production and power output, respectively, and q_f^{ec}, q_p^{ec} are the degree of couplings at optimized ATP production and power output with minimal energy cost, respectively

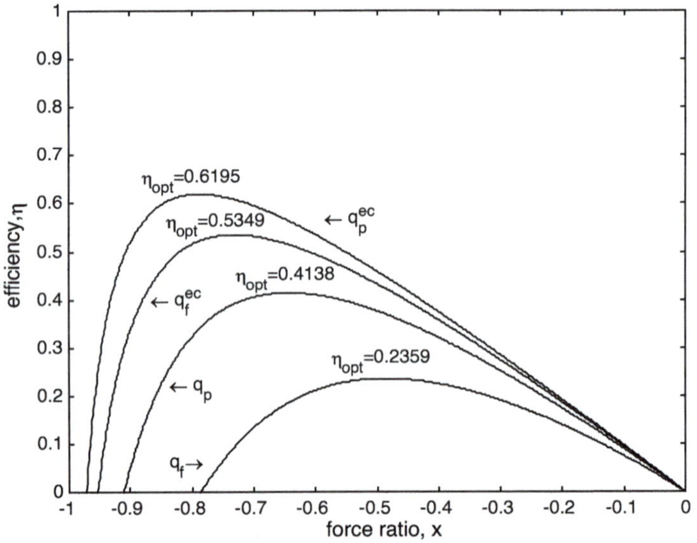

Fig. 10.4 Efficiency versus force ratio at various level of energy couplings; the change of efficiencies η, given in Eq. (10.18), in terms of flow ratio $x(X_oZ/X_p)$ and for the degrees of couplings q_f, q_p, q_f^{ec}, and q_p^{ec} [4]

10.5.1 Regulation of Energy Coupling

Regulation implies a physiological outcome because of manipulating a function. The physiological reasons for these regulations are mainly to synchronize to ATP supply efficiently to ATP demand in workload in bioenergetics, as well as to respond to the external stimuli [4, 5]. The electrochemical proton gradient across the membrane is one of the important mechanisms for regulating the rate of respiration and ATP synthesis. Various substrates regulate the metabolism of energy; fatty acids may regulate and tune the degree of coupling by inducing uncoupling, leading to optimum efficiency of oxidative phosphorylation. A fluctuating ATP/ADP ratio and deviations from the optimal efficiency of oxidative phosphorylation are largely overcome by some buffering enzymes. For example, the mitochondrial creatine kinase is a key enzyme of aerobic energy metabolism, and is involved in buffering, transporting, and reducing the transient nature of the system [5].

The sensitivity of the force (the phosphate potential) to the cellular ATP utilization is minimal at $q = 0.95$. This indicates that the phosphate potential is highly buffered with respect to the changing energy demand to maximize the kinetic stability and efficiency at the same degree of coupling. The standard Gibbs energy of ATP production is obtained as 31.3 kJ/mole at $T = 20°C$, pH = 8.0, pMg = 2.5, and 0.08 M ionic strength, and the standard enthalpy of the reaction is 28.1 kJ/mole [4].

10.5.2 Uncoupling

ATP synthesis is matched to cellular ATP utilization for osmotic work of (downhill and uphill) transport, or mechanical work such as muscle contraction and rotation of bacterial flagellum. The uncoupling of the mitochondrial electron transport chain from the phosphorylation of ADP is physiological and optimizes the efficiency and fine tunes the degree of coupling of oxidative phosphorylation. Fatty acids facilitate the net transfer of protons from intermembrane space into the mitochondrial matrix, hence lowering the proton electrochemical potential gradient and mediating weak uncoupling. Uncoupling proteins generally facilitate the dissipation of the transmembrane electrochemical potentials of H^+ or Na^+ produced by the respiratory chain and result in an increase in the H^+ and Na^+ permeability of the coupling membranes. Some uncoupling is favorable for the energy-conserving function of cellular respiration [4, 8].

10.5.3 Slippages and Leaks

Slippage results when one of two coupled reactions in a cyclic process proceeds without its counterpart, which is also called intrinsic uncoupling. On the microscopic level, individual enzymes cause slippage by either passing a proton without contributing to ATP synthesis, or hydrolyzing ATP without contributing to proton pumping. On the macroscopic level, the measured degree of coupling may be different from the expected coupling.

In terms of energy conversion, a slip may decrease efficiency; it may, however, allow dynamic control and regulation of the enzyme over the varying ranges of driving and driven forces. Mitochondrial energy metabolism may be regulated by the slippage of proton pumping. Slips and leaks occur in parallel. The rate of leakage of the coupling ions depends on the magnitude of the thermodynamic force operating within the system in the absence of the other force. In oxidative phosphorylation, leaks cause a certain uncoupling of two consecutive pumps, such as electron transport and ATP synthase, and may be described as the membrane potential-driven backflow of protons across the bilayer [4].

10.6 Metabolism

Metabolism is the set of biochemical reaction networks that occur in living organisms to maintain life. Metabolism consists of two pathways that are catabolism and anabolism. In catabolism, organic matter is broken down, for example, to harvest energy in cellular respiration. In anabolism, energy is used to construct components of cells such as proteins and nucleic acids. In metabolic pathways, one chemical is transformed through a series of steps into another chemical by a sequence of enzymes. Enzymes allow some processes to drive desirable reactions that require energy by coupling them to spontaneous processes that release energy. One central coenzyme is ATP and is used to transfer chemical energy between different chemical reactions. ATP in cells is continuously regenerated and acts as a bridge between catabolism and anabolism, with catabolic reactions generating ATP and anabolic reactions consuming it through energy coupling [1, 6, 7].

10.6.1 Catabolism

Catabolism is the set of metabolic cyclic processes that break down and oxidize large molecules, including food molecules to simpler molecules, such as carbon dioxide and water. Catabolic reactions provide the energy and components needed by anabolic reactions.

The most common set of catabolic reactions in animals can be separated into three main stages (see Fig. 10.5). In the first, large organic molecules such as proteins, polysaccharides, or lipids are digested into their smaller components outside cells. Next, these smaller molecules are taken up by cells and converted to yet smaller molecules, usually acetyl coenzyme A (acetyl-CoA), which releases some energy. Finally, the acetyl group on the CoA is oxidized to water and carbon dioxide in the citric acid cycle and electron transport chain, releasing the energy that is stored by reducing the coenzyme nicotinamide adenine dinucleotide (NAD^+) into NADH.

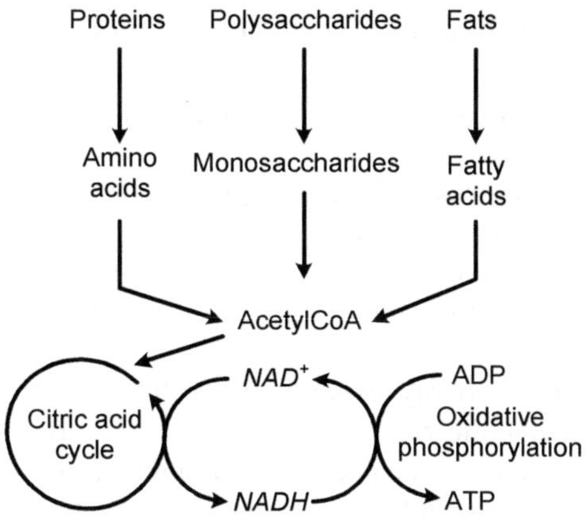

Fig. 10.5 Simplified outline of the catabolism of biofuels proteins, carbohydrates, and fats leading to energy coupling systems [1, 4, 5]

These digestive enzymes include proteases that digest proteins into amino acids, as well as glycoside hydrolases that digest polysaccharides into monosaccharides as shown in Fig. 10.5.

The major pathway of breakdown is glycolysis, where sugars such as glucose and fructose are converted into pyruvate and some ATP molecules are produced. This oxidation releases carbon dioxide as a waste product. Fatty acids release more energy upon oxidation than carbohydrates because carbohydrates contain more oxygen in their structures. Amino acids are either used to synthesize proteins and other biomolecules or are oxidized to urea and carbon dioxide as a source of energy [5, 6].

10.6.2 Anabolism

Anabolism is the set of constructive metabolic processes where the energy released by catabolism is used to synthesize complex molecules in three basic stages. Firstly, the production of precursors such as amino acids, monosaccharides, and nucleotides, secondly, their activation into reactive forms using energy from ATP, and thirdly, the assembly of these precursors into complex folded molecules such as proteins, polysaccharides, lipids and nucleic acids [5, 6].

10.7 Bioenergy Sources

Carbohydrates, fats, and proteins are the major sources of chemical energy of the body and are obtained from the daily diet. Heat released by the oxidation of these fuels is used to maintain body's temperature:

- Carbohydrates release about 4 kcal/g (17 kJ/g).
- Fats release about and 9 kcal/g (39 kJ/g).

- Proteins release about 4 kcal/g (17 kJ/g).

Excess of the body's immediate energy needs is stored mainly by:

- Fat-triacylglycerol (triglyceride): adipose triacylglycerol is the major energy store of the body
- Glycogen: Energy stored by glycogen is relatively small, but it is critical. For example, muscle glycogen is oxidized for muscle contraction.

Daily energy expenditure is the amount of energy required daily for the basal metabolic rate and physical activity. Basal metabolic rate for a person is approximately 24 kcal/kg body weight per day and the basic physical activity requires approximately 30% of basal metabolic rate [4, 7].

Carbohydrates are straight-chain aldehydes or ketones with many hydroxyl groups that can exist as straight chains or rings. Carbohydrates are the most abundant biological molecules, and play numerous roles, such as the storage and transport of energy (starch, glycogen), and structural components such as cellulose in plants, and chitin in animals.

The fuel value is equal to the heat of reaction of combustion (oxidation). Carbohydrates and fats can be completely oxidized while proteins can only be partially oxidized and hence have lower fuel values. The energy expenditure may be calculated from the energy balance. Assume that (i) carbohydrate (CH), fat (F), and protein (Pr) are the only compounds involved in the oxidation process, (ii) the other compounds are stationary, and (iii) the uptake and elimination of oxygen, carbon dioxide, and nitrogen is instantaneous. Then, from the energy balance, we have the energy expenditure $\dot{E}$ as

$$\dot{E} = \sum_i (\dot{m}\Delta H_\mathrm{r})_i = (\dot{m}\Delta H_\mathrm{r})_\mathrm{CH} + (\dot{m}\Delta H_\mathrm{r})_\mathrm{F} + (\dot{m}\Delta H_\mathrm{r})_\mathrm{Pr} = \dot{q} + \dot{W} \tag{10.26}$$

where the specific enthalpy values are:

$\Delta H_{\mathrm{r,CH}} = -17\,\mathrm{kJ/g}$, $\Delta H_{\mathrm{r,Fat}} = -39\,\mathrm{kJ/g}$, $\Delta H_{\mathrm{r,CH}} = -17\,\mathrm{kJ/g}$. Example 10.2 illustrates the oxidation of glucose, while Examples 10.3–10.5 discusses the energy expenditure estimations.

Example 10.2 **Oxidation of glucose**

(a) Estimate the reaction enthalpy for the isothermal and isobaric oxidation of glucose at 310 K and 1 atm.
(b) Estimate the energy expenditure for oxidation of 390 g/day glucose at steady conditions.

Solution:

Assume: The energy for transferring gaseous components to aqueous solutions is small. The temperature correction for the heat of reaction is negligible.

$$C_6H_{12}O_6(aq) + 6O_2(g) \rightarrow 6CO_2(g) + 6H_2O(l)$$

where the symbol (aq) denotes aqueous, (g) is the gas, and (l) is the liquid. If the control volume is a piece of tissue, the reaction above may take place in an aqueous solution (aq).

From Table C1, we obtain the enthalpy of formations for the components of the reaction above at the standard state (298 K and 1 atm)

$$\begin{aligned} \Delta H_r^o(298\,\text{K}) &= 6(-393) + 6(-286) - (-1264) - 6(0) \\ &= -2810 \text{ kJ/mole} \sim \Delta H_r(310\,\text{K}) \\ MW(\text{glucose}) &= 180 \text{ g/gmole} \end{aligned}$$

The energy expenditure $\dot{E}$ at a glucose consumption of 390 g/day is

$$\begin{aligned} \dot{n}_G &= \frac{(390\,\text{g/day})/(180\,\text{g/mole})}{(24\,\text{h/day})(3600\,\text{s/h})} = 0.025 \times 10^{-3}\,\text{mole/s} \\ \dot{E} &= \dot{n}_G(-\Delta H_r) = \left(0.025 \times 10^{-3}\,\text{mole/s}\right)(2810\,\text{kJ/mole}) \\ &= 70.2 \times 10^{-3}\,W = 70.2\,\text{mW} \end{aligned}$$

Example 10.3 **Daily energy expenditure**

A dietary history of an adult of 146 lb shows that he consumes approximately 150 g of carbohydrate, 50 g of protein, and 60 g of fat daily. The adult exercises regularly and consumes an average 480 kcal daily. Estimate the energy expenditure of this adult and comment if the energy intake is excessive.

Solution:

Daily energy expenditure is the amount of energy required daily and includes the energy required for the basal metabolic rate and for physical activity. Basal metabolic rate for a person is approximately 24 kcal/kg (101 kJ/kg) body weight per day [9].

Energy intake:

Carbohydrates: 4 kcal/g $\rightarrow$ (150 g/day)(4 kcal/g) $=$ 600 kcal/day

Fats: 9 kcal/g $\rightarrow$ (60 g/day) (9 kcal/g) $=$ 540 kcal/day.

Proteins: 4 kcal/g $\rightarrow$ (50 g/day) (4 kcal/g) $=$ 200 kcal/day.

Total energy intake $=$ (600 $+$ 540 $+$ 200) kcal/day $=$ 1340 kcal/day

Total energy intake of 1340 kcal/day is less than basal metabolic rate of 2074 kcal.

Example 10.4 **Energy expenditure in small organisms**

Consider a small organism with a body weight of 10 μg consuming 4.0×10^{-9} mole oxygen every hour at steady state and eliminating 3.6×10^{-9} mole carbon dioxide,

0.4×10^{-9} mole N (as ammonia). The external work of the organism is 50×10^{-9} W. Estimate the energy expenditure and the heat loss using the heat of reactions given below.

Hydrocarbon (glucose)(G) :

$C_6H_{12}O_6 + 6O_2(g) \rightarrow 6CO_2 + 6H_2O$ -2870 kJ/mole

Fat(F):

$C_{55}H_{104}O_6 + 78O_2 \rightarrow 55CO_2 + 52H_2O$ -34300 kJ/mole

Protein(Pr):

$C_{32}H_{48}O_{10}N_8 + 33O_2 \rightarrow 32CO_2 + 8NH_3 + 12H_2O$ -14744 kJ/mole

Solution:

The energy expenditure may be calculated with the given reaction enthalpies:

$$\dot{E} = 2870(\dot{n})_G + 34300(\dot{n})_F + 14744(\dot{n})_{Pr} \text{ (kJ/h)}$$

The number of mole flow rates is obtained from the reaction stoichiometric coefficients:

For example, for oxygen:

- one mole/h of glucose consumes 6 mole/h of O_2
- one mole/h of fat consumes 78 mole/h of O_2
- one mole/h of protein consumes 33 mole/h of O_2

With the similar procedure for carbon dioxide and nitrogen, we have

$$\dot{n}_{O_2} = 6\dot{n}_G + 78\dot{n}_F + 33\dot{n}_{Pr} = 4 \times 10^{-9} \text{ mole/h}$$
$$\dot{n}_{CO_2} = 6\dot{n}_G + 55\dot{n}_F + 32\dot{n}_{Pr} = 3.61 \times 10^{-9} \text{ mole/h}$$
$$\dot{n}_N = 8\dot{n}_{Pr} = 0.4 \times 10^{-9} \text{ mole/h}$$

After solving the equations above, we have the energy expenditure:

$$\begin{aligned}\dot{E} &= [2870(0.194)_G + 34{,}300(0.0152)_F + 14{,}744(0.05)_{Pr}10^{-9} \\ &= 1820 \times 10^{-9} \text{ kJ/h} = 0.505\ \mu\text{W}\end{aligned}$$

With $\dot{W} = 50 \times 10^{-9}\text{W} = 0.05\ \mu\text{W}$

Heat loss: $\dot{q} = \dot{E} - \dot{W} = (0.505 - 0.050)\ \mu\text{W} = 0.46\ \mu\text{W}$

Example 10.5 **Energy expenditure in an adult organism**

An adult organism has an oxygen uptake of about 21.16 mol over 24 h, and the associated elimination of carbon dioxide and nitrogen is 16.95 mol and 5.76 g, respectively. If the adult has performed 1.2 MJ of external work over the same period and his energy expenditure at rest is $\dot{E}_o = 65$ W, estimate his energy expenditure, heat loss, and net efficiency for the external work using the biological fuel parameters given below [1, 4].

Fuel	Specific reaction enthalpy kJ/g	Specific turnover		
		O_2	CO_2	N
		mmole/g	mmole/g	g/g
CH	− 17	33.3	33.3	–
Fat	− 39	90.6	63.8	–
Pr	− 17	43.3	34.43	0.16

CH carbohydrate, *Pr* Protein

Solution:

The energy expenditure may be calculated from the energy balance. Assume that (i) carbohydrate (CH), fat (F), and protein (Pr) are the only compounds involved in the oxidation process, (ii) the other compounds are stationary, and (iii) the uptake and elimination of oxygen, carbon dioxide, and nitrogen is instantaneous. From the first law of thermodynamics, we have (kJ/day)

Using the specific enthalpy values of the fuels: (a)

The conservation of mass with the parameters from above table:

$$\dot{n}_{O_2} = 33.3\dot{m}_{CH} + 90.6\dot{m}_F = 21{,}160\,\text{mmole/day}$$
$$\dot{n}_{CO_2} = 33.3\dot{m}_{CH} + 63.8\dot{m}_F = 16{,}950\,\text{mmole/day}$$

From these equations, we obtain

$$\dot{m}_{Pr} = 5.76/0.16 = 36\,\text{g/day}$$
$$\dot{m}_{CH} = 208\,\text{g/day}$$
$$\dot{m}_F = 157\,\text{g/day}$$

From Eq. (10.27), we have the total energy expenditure:

basal metabolic rate + physical exercise

$$17(208) + 39(157) + 17(36) = -\dot{q} + \left(-\dot{W}\right) = \dot{E} = 10271\,\text{kJ/day}$$

Since the work (part of physical exercise) is 1.2 MJ/day = 1200 kJ/day

Heat loss:

$$\dot{q} = (10271 - 1200)\,\text{kJ/day} = 9071\,\text{kJ/day}$$

Basal metabolic rate (energy expenditure at rest):

$$\dot{E}_o = 65\,W = 5616\,\text{kJ/day}$$

A net available energy for physical exercise: (10271 – 5616) kJ/day

Then the bioenergetics efficiency is defined by

$$\eta_B = \frac{\dot{W}}{\left(\dot{E} - \dot{E}_o\right)} = \frac{1200}{(10271 - 5616)} = 0.257 \text{ or } 25.7\,\%$$

$\dot{E}_o$ is the energy expenditure during resting and $\dot{E}$ is the total energy.

Problems

10.1. Estimate the approximate daily energy expenditure of an adult of 55 kg.
10.2. Estimate the approximate daily energy expenditure of a child of 20 kg.
10.3. Estimate the approximate daily energy expenditure of an adult of 51 kg.
10.4. The dietary history of an adult of 65 kg shows that he eats approximately 250 g of carbohydrate, 120 g of protein, and 160 g of fat daily. The adult exercises regularly and consumes an average of 600 kcal daily. Estimate the energy expenditure of this adult and comment if the energy intake is excessive.
10.5. The dietary history of a child of 25 kg shows that he eats approximately 150 g of carbohydrate, 80 g of protein, and 100 g of fat daily. The child exercises regularly and consumes an average of 450 kcal daily. Estimate the energy expenditure of this adult and comment if the energy intake is excessive.
10.6. The dietary history of an adult of 75 kg shows that he eats approximately 275 g of carbohydrate, 160 g of protein, and 190 g of fat daily. The child exercises regularly and consumes an average of 700 kcal daily. Estimate the energy expenditure of this adult and comment if the energy intake is excessive.
10.7. Discuss the consequences of hydrolyzing ATP rapidly.
10.8. Discuss the consequences of intake of uncouple to the human body.
10.9. Discuss the consequences of a lack of oxygen flow to the human body.
10.10. An organism loses heat at a rate of 1.64 W and performs an external work of 1.4 Nm/min. (a) Estimate the energy expenditure of the organism, (b) if the organism can pump 100 ml/min of fluid against a pressure drop of 25 mm Hg (3.34 kPa) with an efficiency of 10%, estimate the efficiency of pumping.

10.11. An organism loses heat at a rate of 2.1 W and performs an external work of 5 Nm/min. (a) Estimate the energy expenditure of the organism, (b) if the organism can pump 52 ml/min of fluid against a pressure drop of 2.5mm Hg (0.334 kPa) with an efficiency of 10%, estimate the efficiency of pumping.

10.12. (a) Estimate the reaction enthalpy for the isothermal and isobaric oxidation of glucose at 310 K and 1 atm.

(b) Estimate the energy expenditure for oxidation of 390 g/day glucose at steady conditions.

10.13. Consider a small organism with a body weight of 10 μg consuming 8.0×10^{-9} mol oxygen every hour at steady state and eliminating 7.1×10^{-9} mol carbon dioxide, 0.8×10^{-9} mol N (as ammonia). The external work of the organism is 80×10^{-9} W. Estimate the energy expenditure and heat loss using the heat of reactions given below.

Hydrocarbon (Glucose) (G):

$C_6H_{12}O_6 + 6O_2(g) \rightarrow 6CO_2 + 6H_2O$ – 2870 kJ/mole

Fat (F):

$C_{55}H_{104}O_6 + 78O_2 \rightarrow 55CO_2 + 52H_2O$ – 34300 kJ/mole

Protein (Pr) :

$C_{32}H_{48}O_{10}N_8 + 33O_2 \rightarrow 32CO_2 + 8NH_3 + 12H_2O$ – 14744 kJ/mole

10.14. An adult organism has an oxygen uptake of about 21.16 mole over 24 hours, and the associated elimination of carbon dioxide and nitrogen is 16.95 mol and 5.76 g, respectively [9]. If the adult has performed 1.25 MJ of external work over the same period and his energy expenditure at rest is $\dot{E}_o = 72$ W, estimate his energy expenditure, heat loss, and net efficiency for the external work using the biological fuel parameters given below.

Fuel	Specific reaction enthalpy kJ/g	Specific turnover		
		O_2 mmole/g	CO_2 mmole/g	N g/g
CH	– 17	33.3	33.3	–
Fat	– 39	90.6	63.8	–
Pr	– 17	43.3	34.43	0.16

CH carbohydrate, Pr Protein

Source (Garby L, Larsen PS (1995) Bioenergetics, Cambridge Univ Press, Cambridge)

10.15. An adult organism has an oxygen uptake of about 22.5 mole over 24 hours, and the associated elimination of carbon dioxide and nitrogen is 17.4 mole and 6.6 g, respectively. If the adult has performed 1.02 MJ of external work over the same period and his energy expenditure at rest is $\dot{E}_o = 60$ W, estimate his energy expenditure, heat loss, and net efficiency for the external work using the biological fuel parameters given below.

References

1. Alberts B, Johnson A, Lewis J, Morgan D, Raff M, Roberts K, Walter P (2014) Molecular biology of the cell, 6th edn. Garland, New York
2. Allen JW, Tevatia R, Demirel Y, DiRusso CC, Black PN (2018) Induction of oil accumulation by heat stress is metabolically distinct from N stress in the green microalgae *Coccomyxa subellipsoidea* C169. PLoS ONE 13(e0204505):1–20
3. Demirel Y (2012) Energy coupling. In: Terzis G, Arp R (eds) Information and living systems in philosophical and scientific perspectives. MIT Press, Cambridge, pp 25–53
4. Demirel Y, Gerbaud V (2025) Nonequilibrium thermodynamics: transport and rate processes in physical, chemical and biological systems, 5th edn. Elsevier, Amsterdam
5. Demirel Y (2010) Nonequilibrium thermodynamics modeling of coupled biochemical cycles in living cells. J Non-Newtonian Fluid Mechanics 165:953–972
6. Küçük K, Tevatia R, Sorgüven E, Demirel Y, Özilgen M (2015) Bioenergetics of growth and lipid production in *Chlamydomonas reinhardtii.* Energy 83:503–510
7. Marks DB (1999) Biochemistry. Kluwer, New York
8. Singh R, White D, Demirel Y, Kelly R, Noll K, Blum P (2018) Uncoupling fermentative synthesis of molecular hydrogen from biomass formation in *Thermotoga maritime.* Appl Environ Microbiol 84:e00998–e1018

11 Sustainability in Energy Technologies

Introduction and Learning Objectives: "Sustainability" is maintaining or improving the material and social conditions for human health and the environment over time without exceeding the ecological capabilities that support them. The dimensions of sustainability are economic, environmental, and societal. Energy demand continues to grow worldwide while extraction of fossil fuels becomes more difficult and expensive. The ways we produce, convert, store, and use energy are changing earth's climate. Nearly two-thirds of global GHG emissions are related to production and consumption of energy. This makes the sustainability in energy sector critical to mitigating global warming and climate change. Increasingly resource-intensive consumption across the world requires novel and sustainable energy technologies and management. Therefore, this chapter focuses on sustainability and life cycle analysis in energy sector addresses the implications of depletion of natural energy and material sources and environmental change.

The learning objectives of this chapter are:

- Sustainability in energy systems
- Life cycle analysis of energy systems
- Exergy and sustainability

11.1 Sustainability

Sustainability is maintaining or improving the material and social conditions for human health and the environment over time without exceeding the ecological capabilities that support them. Energy is one of the main drivers of technology and development and hence the demand for energy continues to grow worldwide even as extraction of fossil fuels becomes more difficult and expensive. The ways

Y. Demirel, *Energy*, Green Energy and Technology,
https://doi.org/10.1007/978-3-031-97111-2_11

we produce, convert, store, and use energy are changing earth's climate and environment hence the ways of human's life as well as the next generation's future. Therefore, energy management is a global challenge. Nearly two-thirds of global GHG emissions are associated with the production and consumption of energy. This makes the sustainability of energy technology critical to mitigating adverse effects of global warming and climate change [17, 18]. Some major actions needed in reducing the growth of atmospheric GHG emissions are:

- Increase energy efficiency
- Increase the usage of renewable energy
- Develop technologies for carbon capture and storage (CCS)
- Develop effective energy storage technologies
- Recognize leadership role in transforming the global energy sector

Zero energy building and green building efforts use resources more efficiently and reduce a building's negative impact on the environment. Zero energy buildings may have a much lower ecological impact compared with other "green" buildings that require imported energy and/or fossil fuel. "Green building certification" programs require reducing the use of nonrenewable energy considerably [18].

11.1.1 Natural Earth Cycles

The Carbon Cycle—Plants use carbon dioxide and energy from the sun to form carbohydrates in photosynthesis

$$6CO_2 + 6H_2O \leftrightarrow C_6H_{12}O_6 + O_2 \tag{11.1}$$

Carbohydrates, in turn, are oxidized to produce water and carbon dioxide. Atmospheric carbon dioxide levels have increased considerably since the industrial revolution.

Nitrogen Cycle—Atmospheric nitrogen is converted to ammonia or ammonium ion by nitrogen-fixing bacteria that live in legume root nodules or in soil. Atmospheric nitrogen is converted to nitrogen oxides during combustion at high temperatures

$$N_2 \rightarrow NH_3 \text{ or } NH_4^+ \tag{11.2}$$

$$NO_2^- + H_2O \rightarrow H_2NO_3 \tag{11.3}$$

Ammonia and ammonium are oxidized by soil bacteria first to nitrite ions and then to nitrate ions. After plants take up nitrogen from the soil in the form of nitrate ions, the nitrogen is passed along the food chain. The remaining nitrogen is released as ammonium ions or ammonia gas.

NO_x *compounds*—Molecules of nitrogen and atmospheric oxygen combine at very high temperatures to form nitric oxides. Lightening or the combustion chambers of engines are effective in causing this conversion of nitrogen to NO_x compounds. Once in the atmosphere, NO_x compounds may react with additional oxygen to form nitrogen dioxide, a red-brown toxic gas that causes irritation to the eyes and respiratory systems [18, 20].

11.1.2 Ozone Formation and Destruction

High-energy ultraviolet (UV) photons react with oxygen, splitting them into highly reactive oxygen atoms. These are free radicals in the stratosphere and can combine with oxygen molecules to form ozone. Each ozone molecule can absorb a UV photon with a wavelength of less than 320 nm and prevents potentially harmful UV rays from reaching the earth's surface. High-energy UV photons in the stratosphere split chlorine radicals from chlorofluorocarbons CFC's by breaking their C–Cl bond. The freed chlorine radicals are very reactive and can destroy ozone by converting it to diatomic oxygen. Some *ozone depleting substances* are hydrochlorofluorocarbons, chlorofluorocarbons, and halons (https://www.epa.gov/ozone-layer-protection/ozone-depleting-substances).

11.1.3 Greenhouse Gases

Greenhouse gases (GHGs) can trap heat in the atmosphere causing global warming. Some important GHGs are carbon dioxide (CO_2), methane (CH_4), nitrous oxide (N_2O), and fluorinated gases. Carbon dioxide is emitted to the atmosphere after burning fossil fuels, through solid waste, biological systems, and certain chemical reactions. Production of coal, natural gas, and oil emit methane. Livestock, agricultural processes, and decompose of organic waste also emit methane. Many agricultural activities, industrial processes, burning of fossil fuels and solid waste, as well as wastewater treatment emit nitrous oxides. Some industrial processes emit fluorinated gases of hydrofluorocarbons, perfluorocarbons, sulfur hexafluoride, and nitrogen trifluoride. These gases are powerful GHGs, although emitted in small quantities [18].

Carbon Tracking—It allows the calculation of CO_2 equivalent emissions after specifying "CO_2 emission factor data source" and "ultimate fuel source." The CO_2 emission factor data source can be from European Commission decision of "2007/589/EC" or United States Environmental Protection Agency (EPA) Rule of E9-5711. Table 11.1 shows the emission rates from some of the fuel sources estimated by the emission factor data source [8–10].

Global CO_2 emissions due to fossil fuel use can be calculated as

$$CO_{2e} = (C_f/E_f)\left(MW_{CO_2}/MW_C\right) \tag{11.4}$$

Table 11.1 Emission rates (lb/MMBtu[a]) for various CO_2 emission factor data sources and fuel sources [9]

Fuel source	US-EPA-Rule-E9-5711 lb/MMBtu[a]	EU-2007/589/EC lb/MMBtu[a]
Natural gas	130.00	130.49
Coal bituminous	229.02	219.81
Coal anthracite	253.88	228.41
Crude oil	182.66	170.49
Biogas	127.67	0

[a] MMBtu = one million Btu

where e_{CO_2} is the CO_2 emission in kgCO_2/kWh, C_f is the carbon content in the fuel (kg_C/kg_{fuel}) and E_f is the energy content of the fuel (kWh/kg_{fuel}). One mton of carbon is equivalent to: $MW_{CO_2}/MW_C = 44/12 = 3.7$ tonne of carbon dioxide. Table 11.2 shows typical emission of carbon dioxide from the combustion of various fuels.

Table 11.2 Emission of carbon dioxide from the combustion of various fuels [18]

Fuel	Specific carbon kg_c/kg_{fuel}	Specific energy kWh/kg_{fuel}	Specific CO_2 emission kg_{CO_2}/kg_{fuel}	Specific CO_2 emission kg_{CO_2}/kWh
Coal (bituminous/ anthracite)	0.75	7.50	2.3	0.37
Gasoline	0.90	12.5	3.3	0.27
Light oil	0.70	11.7	2.6	0.26
Diesel	0.86	11.8	3.2	0.24
LPG—liquid petroleum gas	0.82	12.3	3.0	0.24
Natural gas, methane	0.75	12.0	2.8	0.23
Crude oil				0.26
Kerosene				0.26
Wood[a]				0.39
Peat[a]				0.38
Lignite				0.36

[a] Commonly viewed as a biofuel

11.1.4 Greenhouse Effects

Gases containing carbon in the upper atmosphere and small particles in the form of smoke in the lower atmosphere can trap heat escaping from earth's surface, while they allow sunlight to pass. GHGs, therefore, cause global warming and affect environmental and health adversely, including climate change, respiratory disease from smog and air pollution. Climate change causes extreme weather behavior, including flooding, and wildfires [20, 21].

Climate Change—Global warming and climate change are estimated based on how long GHGs remain in the atmosphere and how effectively they trap heat. Much of the world is covered with ocean water, which heats up by the trapped heat and evaporates more into the clouds. Accumulated moisture in the atmosphere may form into heavy rains and energy-intensive storms. In addition, warmer atmosphere melts glaciers and polar ice caps and cause the rising sea levels [18, 21].

Global Warming Potential—Global warming potential (GWP) measures of how much energy the emissions of 1unit mass of GHG emissions would trap over a given period, relative to the emissions of 1 ton of carbon dioxide. The period is usually 100 years. GWP compares emission levels of various GHGs. Carbon dioxide remains for a long time and has a GWP of 1 as a reference substance. Methane has a GWP of 28–36 over 100 years that shows methane can trap more energy than that of carbon dioxide. Nitrous oxides have a GWP of 265–298 for a 100 year. Chlorofluorocarbons, hydrochlorofluorocarbons, hydrofluorocarbons, perfluorocarbons, and sulfur hexafluoride gases have the very high values of GWP. Chlorofluorocarbon-12 has a global warming potential of 8500, while chlorofluorocarbon-11 has a GWP of 5000. Various hydrochlorofluorocarbons and hydrofluorocarbons have GWPs ranging from 93 to 12,100. These values are calculated over a 100-year period [17, 18].

Global warming potential can be estimated based on data from the three popular standards: (1) the IPCC's 2nd (SAR), (2) 4th (AR4) Assessment Reports, and (3) the US EPA's (CO2E-US) proposed rules from 2009 [8–10].

Since the early 1800s, it is known that various atmospheric gases, acting like the glass in a greenhouse, transmit incoming sunlight but absorb outgoing infrared radiation, thus raising the average air temperature at the earth's surface. Carbon dioxide is clearly the most influential greenhouse gas because of its high level of emissions. The most compelling evidence we have for climate change lies in the so-called paleoclimatic data obtained from ancient ice core samples in Greenland and Antarctica. By analyzing air bubbles that were trapped in the ice when it formed, scientists can determine the content of greenhouse gases and even the average temperature at each point in time. Figure 11.1 shows that over the past 420,000 years, the CO_2 content in the atmosphere has varied cyclically between about 180 ppm and 290 ppm by volume with a period of about 100,000 years in conjunction with variations in the earth's orbit. The earth's temperature has closely followed the greenhouse gas concentration.

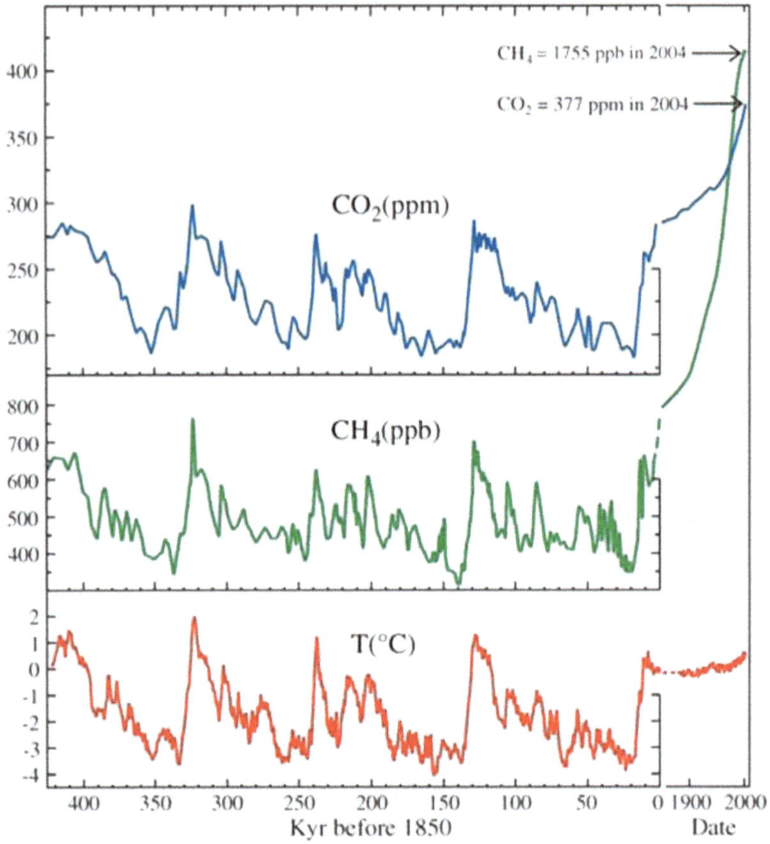

Fig. 11.1 Paleoclimatic data from ice cores shows recent increases in carbon dioxide and methane. The temperature, though increasing, has not yet reached record levels but will likely do so by midcentury [23]

Around 1850, when the CO_2 level was about 280 ppm, the level began to increase and now reached the value of 380 ppm, which indicates a 36% increase over the preindustrial value. Increase in temperature can release CO_2 from the ground and seawater so the two effects reinforce each other. The possible consequences of these increases include ice melts, sea level rises, and severe storms because of the additional energy in the atmosphere. As the ice melts, the resulting water and ground absorb more sunlight, thus exacerbating the global warming [23].

11.1.5 Tackling Global Warming

Electricity production and transportation fuels account a large part of total GHG emissions. For example, the use of renewable energy for electricity generation does not cause additional carbon dioxide emissions and is sustainable. The major challenges with greatly expanded use of renewables are the cost, intermittency of supply, and distance between the resources and the end use. Some possible renewable energy sources are concentrating solar power, geothermal electric plants, wind power, distributed rooftop photovoltaic, and solar hot water heaters. Electric vehicles represent an important improvement as they can be plugged into the grid to be recharged and reduce the amount of gasoline use. In addition, using E85 (85%–15% blend of ethanol and gasoline) may help reduce carbon dioxide emission [17, 18].

11.1.6 Why the Sustainability Matters?

Use of renewable energy technologies (including solar, wind, biomass) and alternative ways of using traditional fossil and nuclear fuels are growing but constrained by various factors including cost, infrastructure, and public acceptance. For example, biomass as a renewable resource can be used to produce biofuels and hence can reduce net GHG emissions by recycling most of the carbon [14]. The following can have part of energy policies toward sustainability:

- Increase energy efficiency with energy integration.
- Increase the usage of renewable energy.
- Include technologies for carbon capture, storage, and utilization.
- Auto thermal processes with heat recovery steam generation (HRSG).

Within the food-energy-water nexus, the world will need 80% more energy, 55% more water, and 60% more food in 2050 [12]. These projections emphasize the role of energy in sustainability for global welfare and protection of ecology. To mitigate the adverse effects of climate change due to GHG emissions, there may be a need for global efforts, including:

- Combine the first and second-generation biomass to produce ethanol with less water and fertilizers and increase the yield.
- Use of waste lignin for producing valuable chemicals and second-generation-based biofuels to be more feasible.
- Municipal wastewater-based algal technology will use less water and nutrients to reduce the cost of algal-based biodiesel production.
- Consequently, biofuels will reduce the water intensity in power production sector and help decouple food production from fossil fuel usage for a more secure water and food supplies.

11.1.7 United Nation's Sustainable Development Goals

In 2015, UN General Assembly formally adopted the universal, integrated and transformative 2030 Agenda for Sustainable Development Goals (SDGs) consisting of a set of 17 SDGs and 169 associated targets (https://sustainabledevelopment.un.org/sdgs). The SDGs are: (1) no poverty, (2) zero hunger, (3) good health and well-being, (4) quality education, (5) gender equality, (6) clean water and sanitation, (7) affordable and clean energy, (8) decent work and economic growth, (9) industry, innovation, and infrastructure, (10) reduced inequalities, (11) sustainable cities and communities, (12) responsible consumption and production, (13) climate action, (14) life below water, (15) life on land, (16) peace justice and strong institutions, and (17) partnerships for the goals. The SDGs were adopted by 193 countries as the world's shared plan to protect the planet, end extreme poverty, and reduce inequality. Achieving the SDGs requires coherent and global efforts that need to be backed by practical commitments and actions [27].

11.1.8 Sustainability and Energy

The green energy transition can achieve environmental goals, improve economic development, job creation, and lead to social equity and welfare for society. Many renewable energy technologies are distributed and allow citizens to be more involved and empowered in the energy transition. The energy transition is at the heart of these objectives.

The IRENA's Transition Energy Scenario (TES) would result in a 70% reduction in carbon dioxide emissions by 2050 achieved by mobilizing resources and initiating policies. Additionally, deep decarbonization perspective emphasizes the important role of emerging technologies in reducing emissions stemming from energy production, transport, industrial processes, and wasteful behaviors. The energy transition is providing the interactions of the energy system with the wider economy, socioeconomics, and the natural environment [20, 21].

Energy transition avoids the use of polluting fuels and creates a climate-resilient economy benefiting all including strong employment and welfare gains. However, the energy transition will generate diverse outcomes for regions and countries stemming from the depth, strength, and diversity of their national supply chains, as well as dependency on fossil fuels and other commodities, technologies, and trade patterns [21].

11.1.9 Sustainability and Ecology

Climate change refers to changes in temperatures, emissions, and climate. *Conservation ecology* addresses changes in populations related to the small population sizes of various species. *Conservation biology* applies science to the conservation of ecosystems, species, and populations. *Source depletion* refers to the exhaustion of raw materials. The collective environmental regulations and technical advances, such as pollution control and prevention and waste minimization have greatly diminished adverse environmental impacts of energy production processes. From a sustainability viewpoint, the most important factors that determine the suitability of energy technology are [18]:

- Energy use per unit of economic value-added product.
- Type of energy used (renewable or nonrenewable).
- Materials use (or resource depletion).
- Fresh water use.
- Waste and pollutants production.
- Environmental impacts of product/process/service.
- Assessment of overall risk to human health and the environment.

Example 11.1 **Carbon dioxide emission from natural gas combustion**

When a hydrocarbon fuel is burned, almost all of the carbon in the fuel burns completely to form carbon dioxide CO_2, which causes the greenhouse effect. On average, 0.59 kg of CO_2 is produced for each kWh of electricity generated from a power plant that burns natural gas. A typical new household uses about 7000 kWh of electricity per year. Determine the amount of CO_2 production in a city with 200,000 households.

Solution:

A kWh is kW times hours. It is the amount of energy consumed in an hour by a device that uses 1 kW of power.
Data: $m_{CO_2} = 0.59$ kg CO_2/kWh, 200,000 houses, power = 7000 kWh/house per year.

The total mass of CO_2 produced per year is the product of the rate of CO_2 production per kWh and the number of houses:

$$m_{\text{total } CO_2} = \left(\frac{7000 \text{ kWh}}{\text{house year}}\right)\left(\frac{0.59 \text{ kg } CO_2}{\text{kWh}}\right)(200{,}000 \text{ Houses})$$

$m_{\text{total } CO_2} = 8.26 \times 10^8$ kg CO_2/year = 826.0 Mton CO_2/year.

11.2 Sustainability Metrics

The dimensions of sustainability are economic, environmental, and societal as seen in Fig. 11.2. Thus, eco-efficiency metrics are indicative of changes in economic and environmental aspects, and are indicated by the intersection of the economic and environmental dimensions. The sustainability metrics, indicated by the intersection of all three dimensions, are truly representative of progress toward

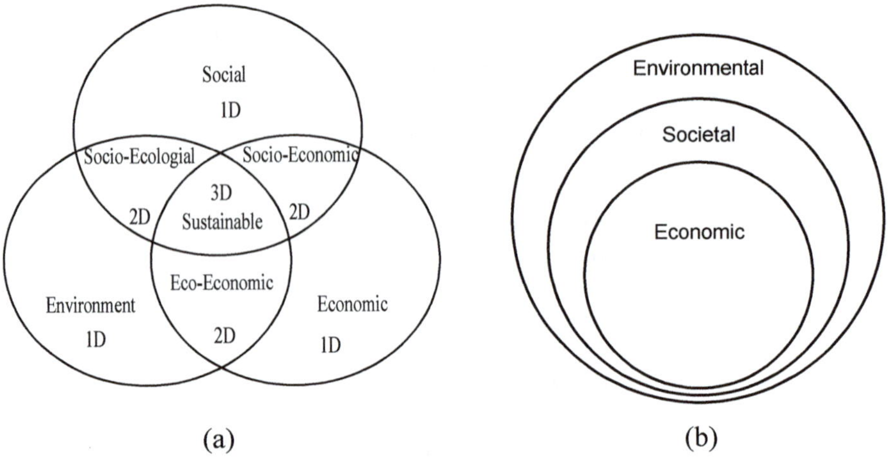

Fig. 11.2 **a** Three dimensions of sustainability; **b** economic and societal dimensions are constrained with the environment

sustainability. One- and two-dimensional metrics, while useful, cannot alone certify progress toward sustainability. Economic and societal indicators may also be constrained with the environment, as seen in Fig. 11.2b. Suitable assessment tools are needed for the development and design of sustainable energy systems for environment. A set of sustainability metrics that are applicable to a specific energy process [9]:

- Material intensity (nonrenewable resources of raw materials, solvents/unit mass of product or value).
- Energy intensity (nonrenewable energy/unit mass of product or value).
- Potential environmental impact (pollutants and emissions/unit mass of product or value).
- Potential chemical risk (toxic emissions/unit mass of product or value).

The first two metrics are associated with the process operation focusing on the inputs to the process. The remaining two metrics represent the outputs from the process as chemical risk to human health in the process environment and the potential environmental impact of the process on the surrounding environment.

The sustainability metrics reflect the three elements of sustainability: (i) Environmental, (ii) economic, and (iii) social. Most of the sustainability metrics are calculated as ratios of a measure of impact independent of the scale of operation that is the ratio of impact per unit mass or economic value of the primary product [17].

Example 11.2 **Consumption of coal and emission of carbon dioxide from coal**

A large public computer lab runs six days per week from Monday through Saturday. Each computer uses a power of around 240 W. If the computer lab contains 45 computers and each is on for 12 h a day, during the course of the year how much CO_2 will the local coal power plant have to release to the atmosphere in kg to keep these computers running?

Solution:

Data: Coal (bituminous/anthracite): 0.37 kg of CO_2/kWh. Data from Table 11.2.

Electricity required:

$$\left(\frac{240\,\text{W}}{\text{computer}}\right)\left(\frac{\text{kW}}{1000\,\text{W}}\right)\left(\frac{52\,\text{weeks}}{\text{year}}\right)\left(\frac{6\,\text{days}}{\text{week}}\right)\left(\frac{12\,\text{h}}{\text{day}}\right)\left(\frac{45\,\text{computers}}{\text{lab}}\right) = 40{,}435.2\,\text{kWh}$$

We know that 40,435.2 kWh is needed from the lab per year, and since the power plant is using coal, we can use its emission value of 0.37 kg of CO_2/kWh (Table 11.2).

(40,435.2 kWh) (0.37 kg CO_2/kWh) = 14,961 kg CO_2 released.

Example 11.3 **Reducing air pollution by geothermal heating**

A district uses natural gas for heating. Assume that average NO_x and CO_2 emissions from a gas furnace are 0.0045 kg/therm and 6.4 kg/therm, respectively. It is considered to replace the gas heating system with a geothermal heating system. The projected saving by the geothermal heating system would be 20×10^6 therms of natural gas per year. Determine the amount of NO_x and CO_2 emissions the geothermal heating system would save every year.

Solution:

therm = 29.3 kWh (Table 1.9).

Reduction in NO_x emission = (0.0045 kg/therm) (20×10^6 therm/year)

$= 9.0 \times 10^4$ kg/year.

Reduction in CO_2 emission = (6.4 kg/therm) (20×10^6 therm/year)

$= 12.8.0 \times 10^7$ kg/year.

A typical car produces about 8.5 kg NO_x and 6000 kg of CO_2 per year. Replacing the gas heating system by the geothermal heating system is equivalent to taking 10,600 cars off the road for NO_x emission and taking 21,000 cars of the road for CO_2 emission. Therefore, the proposed geothermal heating would have a positive impact on the air pollution.

11.3 Sustainability Index

The AIChE's Sustainability Index (https://www.aiche.org/ifs/resources/sustainability-index) has seven key elements that help determine how a company's sustainability efforts are perceived by the community, shareholders, customers, and peers.

The indices benchmark well-defined performance metrics and indicators, including environment, health and safety performance, innovation, and societal measures to:

1. Benchmark a company's performance among peers.
2. Assess a company's performance against well-defined metrics on an ongoing basis.
3. Measure progress toward best practices at regular intervals.
4. Access unbiased, expert interpretation of publicly available technical data.
5. Better understand public perception of your company's sustainability efforts.

1. Sustainability innovation considers:
   - General Research and Development (R&D) commitment as evident in the amount of R&D expenditure per net sales,
   - Sustainable products and processes with superior environmental, social and economic performance,
   - Sustainability approaches in R&D and innovation processes,
   - R&D effectiveness is reflected in the number of patents issued and commercialization of new products that enhance environmental and social sustainability.
2. Strategic commitment considers:
   - Stated commitment to excellence in environmental and social performance throughout a company's value chain.
   - Commitment to voluntary codes and standards, including responsible care, global compact, and others.
   - Sustainability reporting of sustainability performance.
   - Sustainability goals and programs that are specific and challenging.

- Third-party ratings with respected agencies' ratings on company-wide sustainability management and reporting.

3. Environment performance considers:
   - Resource use with intensity of energy, material and water consumption, and use of renewable sources of energy and materials for GHG emissions.
   - Other emissions with air emissions, wastewater, and hazardous waste releases.
   - Compliance management with environmental liability, fines and penalties, and environmental capital investment.
4. Safety performance considers:
   - Employee safety with recordable and days-away-from work injury rates.
   - Process safety records incidents, normalized by number of employees, and occurrence of major safety incidents.
   - Plant security with plant security management system representing completion of a responsible care plant security audit.
5. Product stewardship considers:
   - Assurance system incorporates of a responsible care product safety process, and engagement of value chain partners to assure product safety
   - Risk communication incorporates of a responsible care risk communication process
   - Legal proceedings related to product safety, risk, and toxicity.
6. Social responsibility considers:
   - Stakeholder partnerships and engagement at the project, facility, and corporate levels.
   - Social investment through employment, philanthropy and community development projects.
   - Image in the community indicated by reputable awards and recognition programs, including "most admired" and "best employer" ratings.
7. Value chain management considers:
   - Environmental management systems.
   - Supply chain management.

11.4 Sustainability Impact Indicators

There are two sustainability impacts and impact indicators that are related to the depletion of resources and environmental change shown in Table 11.3. Defining and using of a sustainability framework and the associated potential impact indicators are essential for selecting and evaluating energy systems for the future [17]. A set of indicators provides technical, sustainability, and societal perspectives on energy system choices. Such indicators would allow comparison of diverse energy systems, would provide a common basis for evaluation, and would inform public and stakeholders on technology development and energy policy. A focus on a specific energy system such as biomass to energy may provide technical, sustainability, and societal perspectives. A simplified and specific representation of the

Table 11.3 Sustainability impacts and impact indicators

Impact	Potential impact indicators
Resource depletion	
Depletion of nonrenewable natural resources	Abiotic resource depletion potential
Depletion of nonrenewable primary energy resources	Nonrenewable primary energy consumption potential
Water usage	Water resource depletion potential
Environmental	
Greenhouse gas effect-CO_2e	Global warming potential
Air pollution	
Air acidification	Acidification potential
Photochemical oxidation	Photochemical ozone creation potential
Exhaustion of the ozone layer	Ozone layer depletion potential
Water pollution	
Eutrophication	Eutrophication potential (nitrification)
Soil pollution	
Soil contamination	Soil contamination potential
Toxic hazards	
For humans	Human toxicity potential
For aquatic ecosystems	Aquatic toxicity potential
For sedimentary ecosystems	Sedimentary toxicity potential
For terrestrial ecosystem	Terrestrial toxicity potential
Economics	
Feasibility	Net present value Payback period Rate of return
Earnings	Cash flow
Inflation	Interest rates
Cost of production	Consumer index
Development	Gross Domestic Product (GDP)
Societal	
Workplace	Employment
Society	Welfare
Health and quality of life	Health insurance, education, well-being

biomass to energy may be biomass to ethanol energy system to address the aspects and impacts that need to be considered in developing metrics such as value chain of supply, production, use and fate, social and environmental dimensions of sustainability, context of resources, time, and location. For example, the sustainability indicators would help measure sustainability and risks in biofuel production [22, 24]. Sustainable energy includes:

- Renewable energy sources, such as biomass, solar power, wind power, wave power, geothermal power, and tidal power.
- The strategies and technologies that improve energy efficiency or otherwise reduce the environmental impact.

11.4.1 Resource Depletion

These indicators display a balanced view of the environmental impact of resource usage, emissions, effluents, and waste produced.

Energy—(i) total net primary energy usage rate = imports–exports GJ/year, (ii) total net primary energy sourced from renewables %, (iii) total net primary energy usage per kg product kJ/kg.

Material (excluding fuel and water)—(i) total raw materials used, including packaging ton/year, (ii) raw material recycled from other company operations ton/year, (iii) raw material recycled from consumer ton/year, (iv) raw material used which poses health, safety or environmental hazard ton/year (describe hazard), (v) total raw materials used per kg product kg/kg, (vi) fraction of raw materials recycled within company kg/kg, (vii) fraction of raw materials recycled from consumers kg/kg, (viii) hazardous raw material per kg product kg/kg.

Water—(i) water used in cooling. ton/year, (ii) water used in process ton/year, (iii) other water used, ton/year and. (iv) water recycled internally ton/year, (v) net water consumed = total used – recycled ton/year, (vi) net water consumed per unit mass of product kg/kg.

Land—(i) land occupied by operating unit m^2 (include land needed for all activities), (ii) other land affected by unit's activities m^2 (land used in mining raw material or in dumping waste product) total land m^2, (iii) land restored to original condition m^2/year, (iv) total land occupied + affected for value added (a) m^2/($/year), (v) rate of land restoration (restored per year /total) (b) (m^2/ year)/m^2. The areas of land occupied and affected are those at the start of the reporting period, and the land restored is that area restored during the reporting period.

11.4.2 Environmental Burden

The environmental burden (EB) caused by the emission of a range of substances is calculated by adding up the weighted factor known as the "potency factor." A single substance contributes differently to different burdens, each substance will have several different potency factors. Environmental burden is estimated by [17]:

$$\text{Environmental Burden (EB)}_i = \sum m_n \text{ (tonnes)} \times \text{Potency Factor (pf)}_{i,n} \quad (11.5)$$

where the EB_i is the ith environmental burden and n is the substance and m is the mass amount of that substance.

Atmospheric impact—(i) Global warming burden; (ii) atmospheric acidification burden, (iii) human health burden, (iv) ozone depletion burden, (v) photochemical ozone burden. The environmental burden of global warming is measured by ton/year carbon dioxide equivalent and atmospheric acidification is measured by ton/year sulfur dioxide equivalent (Table 11.4). The potential of certain gasses released to form acid rain and acids is the potency factor for the atmospheric acidification. The potential factors of global warming are based on 100-year integrated time interval.

Human health (carcinogenic)—There are no commonly accepted potency factors for human health. Carcinogenic effects are offered as a default set, but other sets can also be used if they are more appropriate [17]. The potency factor for this category in Table 11.5 is from the reciprocal of the Occupational Exposure Limits (OEL). The OEL for benzene has been chosen as the normalizing factor for this category: Potency factor of substance = (OEL benzene/OEL substance). Chemicals with an OEL greater than 500 mg/m^3 have a minimal impact on the total weighted impact. The unit of EB is ton/year benzene equivalent [17].

Stratospheric ozone depletion—The potency factor is based on the ozone depletion potential in the upper atmosphere relative to chlorofluorocarbon—CFC-11. The unit of environmental burden is ton/year CFC-11 equivalent (CFC-11 is trichlorofluoromethane) (see Table 11.6).

Photochemical ozone (smog) Formation—Potency factors for photochemical ozone formation are obtained from the potential of substances to create ozone photochemically. The unit of environmental burden is ton/year ethylene equivalent (see Table 11.7).

Aquatic Impact

Aquatic acidification—Environmental burdens for emissions to water addresses aquatic acidification. The potency factor is the mass of hydrogen ion released by unit mass of acid that is the number of hydrogen ions released divided by the molecular weight. The unit of environmental burden is ton/year of H^+ ions released (see Table 11.8).

Aquatic oxygen demand—The Stoichiometric Oxygen Demand (StOD) may be the potency factor representing the potential of emissions to remove dissolved oxygen that would otherwise support aquatic life. An alternative potency factor is Chemical Oxygen Demand (COD). StOD is expressed as tons of oxygen required per ton of substance. The unit of environmental burden is ton/year oxygen (see Table 11.9).

Table 11.4 Potency factors for global warming and atmospheric acidification [17]

Substance	Potency factor, pf
Global warming	
Carbon dioxide	1
Carbon monoxide	3
Carbon tetrachloride	1400
Chlorodifluoromethane, R22	1700
Chloroform	4
Chloropentafluoroethane, R115	9300
Dichlorodifluoromethane, R12	8500
Dichlorotetrafluoroethane, R114	9300
Difluoroethane	140
Hexafluoroethane	9200
Methane	21
Methylene chloride	9
Nitrous oxide	310
Nitrogen oxides (NO_x)	40
Tetrafluoroethane, R125	2800
Perfluoro methane	6500
Tetrafluoroethene	1300
Trichloroethane (1,1,1)	140
Trichlorofluoromethane, R11	4000
Trichloro trifluoroethane, R113	5000
Trifluoroethane, R143a	3800
Trifluoroethane, R23	11,700
Volatile organic compounds, VOC	11
Atmospheric acidification	
SO_2	1
Ammonia	1.88
HCL	0.88
HF	1.66
NO_2	0.7
H_2SO_4 mist	0.65

Table 11.5 Human health potency factors [17]

Compound	CAS number	Potency factor, pf
Acrylamide	79-06-1	53.3
Acrylonitrile	107-13-1	3.6
Arsenic and compounds except arsine	7440-38-0	160
Azodicarbonate	123-77-3	16
Benzene	71-43-2	1
Bis (chloromethyl) ether	542-88-1	3,200
Buta-1,3-diene	106-99-0	0.73
Cadmium oxide fume	1306-19-0	640
Carbon disulfide	136-23-6	0.5
1 Chloro-2,3-epoxypropane	106-89-8	8.4
1,2-dibromoethane	106-93-4	4.1
1,2-dichloroethane	107-06-2	0.76
Dichloromethane	75-09-2	0.05
2-2′-Dichloro-4,4′-methylene dianiline	101-14-4	3,200
Diethyl sulfate	64-67-5	50
Dimethyl sulfate	77-78-1	3.8
2-Ethoxyethanol	110-80-5	0.43
2-Ethoxyethyl acetate	111-15-9	0.3
Ethylene oxide	75-21-8	1.7
Formaldehyde	50-00-0	6.4
Hydrazine	30-07-2	533.3
Iodomethane	74-88-4	1.3
Maleic anhydride	108-31-6	16
2-Methoxyethanol	109-86-4	1
2-Methoxyethyl acetate	110-49-6	0.64
4-4′-methylenedianiline	101-77-9	200
2-Nitropropane	79-46-9	0.8
Phthalic anhydride	85-44-9	4
Polychlorinated biphenyls (PCB)	1336-36-3	160
Propylene oxide	75-56-9	1.33
Styrene	100-42-5	0.04
o-Toluidine	95-53-4	18
Triglycidyl isocyanurate (TGIC)	2451-62-9	160
Trimellite anhydride 400	552-30-7	400
Vinylidene chloride	75-35-4	0.4

Table 11.6 Potency factors for ozone depletion [17]

Substance	Potency factor, pf
CFC-11; CFC-12; CFC-13	1.0
CFC 113	0.8
CFC 114	1.0
CFC 115	0.6
CFC 111, CFC 112	1.0
CFC 112, CFC 213; CFC 214	1.0
CFC 215; CFC 216; CFC 217	1.0
Halon-1211	3.0
Halon-1301	10.0
Halon-2402	6.0
Carbon tetrachloride	1.1
1,1,1-trichloroethane	0.1
Methyl bromide	0.7
HCFC-21	0.04
HCFC-22	0.055
HCFC-31	0.02
HCFC-121	0.04
HCFC-122	0.08
HCFC-123	0.02
HCFC-124	0.022
HCFC-131	0.05
HCFC-132	0.05
HCFC-133	0.06
HCFC-141	0.07
HCFC-141b	0.11
HCFC-142	0.07
HCFC-142b	0.065
HCFC-151	0.005
HCFC-221	0.07
HCFC-222	0.09
HCFC-223	0.08
HCFC-224	0.09
HCFC-225	0.07
HCFC-226	0.1
HCFC-231	0.09
HCFC-232	0.1

(continued)

Table 11.6 (continued)

Substance	Potency factor, pf
HCFC-233	0.23
HCFC-234	0.28
HCFC-235	0.52
HCFC-241	0.09
HCFC-242	0.13
HCFC-243	0.12
HCFC-244	0.14
HCFC-252	0.04
HCFC-253	0.03

Example 11.4 **Calculation of the stoichiometric oxygen demand**

Estimate the values of StOD for acetic acid, phenol, and ammonium ion.

Solution:

Acetic acid CH_3COOH with a molecular weight of 60, we have

StOD $= 16(2 \times 2 + 0.5 \times 4 - 2)/60 = 1.07$-tonne equivalent O_2 per ton equivalent acetic acid.

Another example, Phenol C_6H_5OH with a molecular weight of 94

StOD $= 16(2 \times 6 + 0.5 \times 6 - 1)/94 = 2.38$ te O_2 per ton equivalent (te) of phenol.

For ionic species, the charge of the ionic unit is considered. For the ammonium ion (NH^{4+}), for example, we remove an H^+ ion and calculate on the NH_3, so that the ionic balance is not disturbed.

StOD $= 16(0.5 \times 3 + 2.5 \times 1)/17 = 3.76$ te O_2 per tonne equivalent (te) of ammonia $= 3.56$ te O_2 per te of ammonium ion.

Ecotoxicity to aquatic life (values for sea water conditions)—The potency factor of metals is equal to the reciprocal of the Environmental Quality Standard (EQS) divided by the reciprocal of the EQS of copper. The unit of environmental burden is ton/year copper equivalent (see Table 11.10).

Potency factors of other substances (see Table 11.11) are equal to the reciprocal of the environment quality standard (EQS) divided by the reciprocal of the EQS of formaldehyde. The unit of environmental burden is ton/year formaldehyde equivalent.

Eutrophication—Eutrophication is defined as the potential for over-fertilization of water and soil, which can result in increased growth of biomass. The species considered are to be responsible for eutrophication. The unit of environmental burden is ton/year phosphate equivalent (see Table 11.12).

Table 11.7 Potency factors for petrochemical ozone depletion [17]

Substance	Potency factors, pf
Methane	0.034
Ethane	0.14
Propane	0.411
n-Butane	0.6
i-Butane	0.426
n-Pentane	0.624
i-Pentane	0.598
n-Hexane	0.648
2-Methylpentane	0.778
3-Methylpentane	0.661
2,2-Dimethylbutane	0.321
2,3-Dimethylbutane	0.943
n-heptane	0.77
2-Methylhexane	0.719
3-Methylhexane	0.73
n-Octane	0.682
2-Methylheptane	0.694
n-Nonane	0.693
2-Methyloctane	0.706
n-Decane	0.680
2-Methylnonane	0.657
n-Undecane	0.616
n-Dodecane	0.577
Cyclohexane	0.595
Methyl cyclohexane	0.732
Ethylene	1.0
Propylene	1.08
1-Butene	1.13
2-Butene	0.99
2-Pentene	0.95
1-Pentene	1.04
2-Methylbut-1-ene	0.83
3-Methylbut-1-ene	1.18
2-Methylbut-2-ene	0.77
Butylene	0.703
Isoprene	1.18

(continued)

Table 11.7 (continued)

Substance	Potency factors, pf
Styrene	0.077
Acetylene	0.28
Benzene	0.334
Toluene	0.771
o-Xylene	0.831
m-Xylene	0.08
p-Xylene	0.948
Ethylbenzene	0.808
n-Propyl benzene	0.713
i-Propyl benzene	0.744
1,2,3-Trimethylbenzene	1.245
o-Ethyl toluene	0.846
m-Ethyl toluene	0.985
p-Ethyl toluene	0.935
3,5-Dimethylethylbenzene	1.242
3,5-Diethyltoluene	1.195
Formaldehyde	0.554
Acetaldehyde	0.65
Propionaldehyde	0.755
Butyraldehyde	0.77
i-Butyraldehyde	0.855
Valeraldehyde	0.887
Methylethylketone	0.511
Methyl-i-butyl ketone	0.843
Cyclohexanone	0.529
Methyl alcohol	0.205
Ethyl alcohol	0.446
i-Propanol	0.216
n-Butanol	0.628
i-Butanol	0.591
Diacetone alcohol	0.617
Cyclohexanol	0.622
Methyl acetate	0.046
Ethyl acetate	0.328
n-Propyl acetate	0.481
i-Propyl acetate	0.291

(continued)

Table 11.7 (continued)

Substance	Potency factors, pf
n-Butyl acetate	0.291
Formic acid	0.003
Acetic acid	0.156
Propionic acid	0.035
Butyl glycol	0.629
Propylene glycol methyl ether	0.518
Dimethyl ether	0.263
Methyl-t-butyl ether	0.268
Methyl chloride	0.035
Methylene chloride	0.031
Methyl chloroform	0.002
Tetrachloroethylene	0.035
Trichloroethylene	0.075
Vinyl chloride	0.272
1,1-Dichloroethylene	0.232
cis 1,2-Dichloroethylene	0.172
trans 1,2-Dichloroethylene	0.101
Nitrogen dioxide	0.028
Sulfur dioxide	0.048
Carbon monoxide	0.027

Table 11.8 Potency factor for aquatic acidification [17]

Substance	Potency factor, pf
Sulfuric acid	0.02
Hydrochloric acid	0.027
Hydrogen fluoride	0.05
Acetic acid	0.02

Impact to land—Two major impacts are hazardous solid waste and nonhazardous solid waste.

11.4.3 Economics

Economic indicators determine if a process (plant) generates profit and is compatible with other existing plants and operations producing similar products. Three profitability criteria with discounted approach with time value of the money are:

Table 11.9 Potency factor for aquatic oxygen demand [17]

Substance	Potency factor, pf
Acetic acid	1.07
Acetone	2.09
Ammonium nitrate in solution	0.8
Chlorotrifluoroethane	0.54
1,2-Dichloroethane (EDC)	0.81
Ethylene	1
Ethylene glycol	1.29
Ferrous ion	0.14
Methanol	1.5
Methyl methacrylate	1.5
Methylene chloride	0.47
Phenol	0.238
Vinyl chloride	1.28

Table 11.10 Potency factor for ecotoxicity to aquatic life (values for sea water conditions) [17]

Substance	Potency factor, pf
Arsenic	0.2
Cadmium	2.0
Chromium	0.33
Copper	1
Iron	0.005
Lead	0.2
Manganese	0.1
Mercury	16.67
Nickel	0.17
Vanadium	0.05

(1) *Time*-Discounted payback period (DPBP)

$$\text{Simple Payback Period} = \frac{\text{Total investment}}{\text{Average annual cash flows}} \tag{11.6}$$

For profitable projects, the shortest payback period is the most desirable: DPBP $\leq n$. Here n number of years of competitive operation.

(2) *Cash*-Net present value (NPV) refers to cumulative cash position at the end of the project

Table 11.11 Potency factors for other substances [17]

Substance	Potency factor, pf
Ammonia	0.24
Benzene	0.17
Carbon tetrachloride	0.42
Chloride	0.5
Chlorobenzene	1
Chloroform	0.42
Cyanide	1.0
1,2-Dichloroethane	0.5
Formaldehyde	1
Hexachlorobenzene 1	166.67
Hexachlorobutadiene	50
Methylene chloride	0.5
Nitrobenzene	0.25
Nitrophenol	0.5
Toluene	0.125
Tetrachloroethylene	0.5
Trichloroethylene (TRI)	0.5
Xylenes	0.17

Table 11.12 Potency factor for eutrophication [17]

Substance	Potency factors, pf
NO2	0.2
NO	0.13
NOx	0.13
Ammonia	0.33
Nitrogen 0	42
PO4 (III-)	1
Phosphorus	3.06

$$\mathrm{NPV} = \sum_{j=1}^{n} \frac{\mathrm{CF}_j}{(1+i)^j} \tag{11.7}$$

where CF_i is the discounted cash flow for year j, i is the interest rate for the bank loan, n is the total competitive operational time. For a profitable project the highest positive value of NPV is the most desirable: $\mathrm{NPV} \geq 0$.

(3) *Discount rate*-Discounted cash flow rate of return (DCFROR) is the interest rate for which the value of NPV is zero

$$0 = \sum_{j=1}^{n} \frac{CF_j}{(1+i^*)^j} \tag{11.8}$$

where i^* is the discounted cash flow rate of return (percent). For a profitable project it is required $i^* > i$. The internal interest rate is determined by the management and represents the minimum acceptable rate of return for the new project:

$$\text{Rate of return : DCFROR} = \frac{\text{Average annual net profit}}{\text{FCI}} \tag{11.9}$$

Interest rate: DCFROR > internal interest rate for profitable project. Here, FCI is the fixed capital investment and the net profit is the profit after tax.

11.4.4 Society

Indicators of social performance reflect the industry's attitude to treatment of public in the form of employees, suppliers, contractors, customers, and its impact on society at large [16, 17].

Workplace—Employment situation: (i) number of employees who have resigned or been made redundant per year, (ii) number of direct employees promoted per year, (iii) working hours lost through absence per year (all unplanned causes—strikes, sickness, and absenteeism but not holiday or training), (iv) indicative wage and benefit package, (v) indicative wage and benefit package for lowest paid 10% of employees.

Society—Some indicators are: (i) coordination with external stakeholders concerning company operations per year, (ii) indirect benefit to the community resulting from presence of operating unit per year, (iii) complications registered from members of the public concerning the process or products per year, (iv) number of successful legal actions taken against company or employees for work-related incidents or practices. External stakeholders include customers, residents and other community groups, local government, nongovernmental organizations. This metric represents company efforts in communicating with external stakeholders.

Health and quality of life—A major social benefit arising from a successful process industry unit includes the dissemination of skills and know-how which are used in the community to create wealth and enhance quality of life. It is difficult to quantify these benefits, but the net value to community may be estimated by (i) freely published information and know-how, (ii) training given to contractors and suppliers, (iii) training given to employee [17].

11.5 Sustainability in Energy Systems

World energy requirements are expected to rise by roughly 17% by 2020. This level of energy requirement requires a sustainable energy production method. While renewable technologies like solar and wind power are likely the most environmentally friendly methods of energy production, current high costs and makes the timing of renewable energy sources overtaking fossil fuels uncertain [18, 21].

Energy is a global challenge since the ways we produce, convert, store, and use energy are changing earth's climate and environment hence the ways of human's life as well as the next generation's future. Sustainable energy systems aim the largest triangle for sustainable energy production with the inclusion of all the three dimensions of environment, social, and economics (see Fig. 11.3). Use of emerging renewable energy technologies (including solar, wind) and alternative ways of using traditional fossil and nuclear fuels are growing but constrained by various factors including cost, infrastructure, public acceptance, and others [21].

Capturing and storing CO_2 from power plants and industrial processes add significant capital and operating costs without much economics return [14, 18, 20]. Operating costs of existing technologies for capturing the CO_2 from dilute flue gases are high; the flue gases of existing coal-fired power plants contain only 12–14 percent CO_2 and those from existing natural-gas-fired power plants contain only 3–4 percent CO_2. One of the primary challenges is to make carbon capture and storage (CCS) viable for fossil fuel power plants; CO_2 must be isolated from emissions sources and compressed to a supercritical state in order for CO_2 to be transported and stored. A few industrial processes, such as ethanol and ammonia productions, yield emissions that are nearly pure CO_2, mitigating the technical challenge and energy intensity of CO_2 capture. The US EPA is expected to further regulate emissions, solid waste, and cooling water intake that will affect the US electric power sector, particularly the fleet of coal-fired power plants. To comply with those new regulations, existing coal-fired plants may need extensive emission and environmental control retrofits [4, 28].

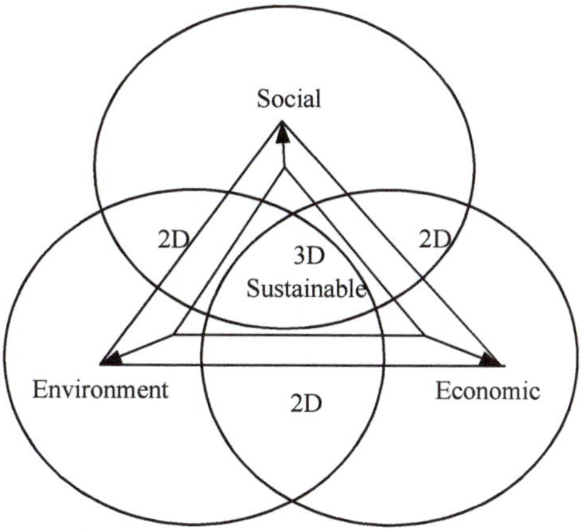

Fig. 11.3 Targeting larger triangle for sustainable energy production with the inclusion of all the three dimensions

11.5.1 Sustainable Energy Systems Design

Sustainable design takes account of environmental considerations, still can be commercially successful, functional, and economical. Some important factors in the sustainability of energy system design are: (i) energy use per unit of economic value-added product, (ii) type of energy used (renewable or nonrenewable). (iii) resource depletion), (iv) fresh water use, (v) waste and pollutants production, and (vi) assessment of overall risk to human health and the environment. Some design concerns to address the environmental issues [1, 2, 28]:

- *The greenhouse effect*: We have an increase of the amount of CO_2 produced by the burning of fossil fuels and need to reduce it by developing alternative energy sources to fossil fuels, improving energy efficiency,
- *Ozone layer*: The ozone layer acts as a filter in the stratosphere, protecting the earth from harmful ultraviolet radiation from the sun. Chlorofluorocarbons (CFCs) destroy stratospheric ozone and cause greenhouse effect.
- *Tropical deforestation*: Forest acts as absorbers of carbon dioxide. Deforestation harms species, wide variety of animal and plant life, disrupt local climates, causes loss of habitat, and contributes to the emissions.
- *Waste*: Waste ends up in landfill sites. It is incinerated and may be simply dumped at sea. Most of this waste does not simply biodegrade into harmless substances. Incineration can generate energy but release toxic gases such as dioxins. Reuse, recycling, and remanufacturing waste are essential.
- *Water pollution*: The growth of population and the increasing use of water for industrial purposes cause insufficient supply of clean water.

11.5.2 Sustainable Engineering Principles in Energy Systems

Sustainable engineering transforms existing engineering disciplines and practices to those that promote sustainability. To fully implement sustainable engineering solutions, engineers use the following principles in energy systems:

- Integrate environmental impact assessment tools.
- Conserve and improve natural ecosystems while protecting human health and well-being.
- Ensure that all material and energy inputs and outputs are as inherently safe and benign as possible.
- Minimize depletion of natural resources and waste.
- Develop and apply engineering solutions in line with local geography, aspirations, and cultures.
- Actively engage with communities and stakeholders.
- Use material and energy inputs that are renewable.

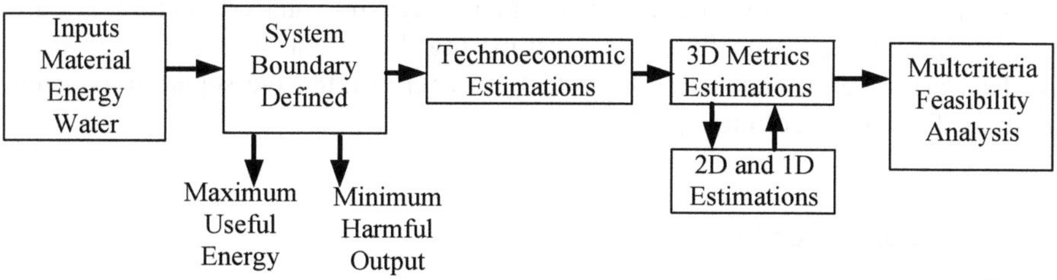

Fig. 11.4 Feasibility analysis with technoeconomic analysis and sustainability metrics [22, 28]

Figure 11.4 displays the major steps toward a comprehensive feasibility analysis incorporating the technoeconomic analysis and sustainability metrics. This may make it possible to count the damage caused because of harmful emissions from the fossil fuels used in energy systems [23, 24].

The plan-do-check-act management system provides a methodical structure for solving problems and implementing solutions. During the plan stage, the problem is identified and analyzed, and a plan is developed. Then, during the do stage, steps are taken to implement the planned actions. To ensure that the best possible solution has been developed, a series of verifications and analyses are conducted during the check stage, and potential improvements are identified. In the final step, the act stage, necessary changes are applied.

11.5.3 Thermodynamic Analysis in Design of Energy Systems

A typical distillation column resembles a heat engine delivering separation work by using heat at a high temperature in the reboiler and discharging most of it to the environment at a lower temperature in the condenser [8, 9]. The "column targeting tool" of Aspen Plus simulation package is based on the Practical Near-Minimum Thermodynamic Condition approximation representing a practical and close to reversible operation [9, 11]. The column targeting tool performs (i) thermal, (ii) exergy, and (iii) hydraulic analyses capabilities that can help identify the targets for appropriate column modifications in order to: (i) reduce utilities cost, (ii) improve energy efficiency, (iii) reduce emissions of GHG.

The column targeting tool exploits the capabilities for thermal and hydraulic analyses of distillation columns to identify the targets for possible column retrofits in: (1) feed stage location, (2) reflux ratio, (3) feed conditioning, and (4) side condensing and/or reboiling to reduce the cost and improve energy efficiency. The "carbon tracking" (CT) and global warming potential options can help quantify the reduction in CO_2 emission in a simulation environment. CTT produces "column grand composite curves" and "exergy loss profiles," which help the engineer suggest retrofits for column configurations and operating conditions. The exergy loss profiles represent inefficient use of available energy due to irreversibility and should be reduced by suitable modifications. Consequently, smaller exergy loss means less waste energy. A comparative thermodynamic analysis before and after the retrofits may be a useful tool for estimating the extent of possible reductions in the waste energy and emission of CO_2 in energy-intensive separation systems such as distillation columns [1, 2].

11.5.4 Case Studies

Wind power-based electrolytic hydrogen production—The generation of renewable electricity suffers from intermittent and fluctuating character and necessitates the storage. Wind power-based electrolytic hydrogen may serve as chemical storage for renewable electricity [4, 25]. Figure 11.5 shows the schematic of wind power-based hydrogen production. Alkaline electrolysis technologies are the most mature commercial systems. These electrolyzers have the energy efficiencies of 57–75%. The typical current density is 100–300 mA/cm^2 [15, 24].

For producing one kg H_2, approximately 26.7 kg water is necessary. The total greenhouse gas emission is around 0.97 kg CO_2e/kg H_2. The hydrogen production cost is highly dependent on the electricity price, which may be around 75% of the final cost [24].

Sustainable Methanol Production—Methanol synthesis needs carbon-rich feedstock, hydrogen, and a catalyst, mainly $Cu/ZnO/Al_2O_3$. Commercial process of methanol production from natural gas is the most efficient process with a typical energy efficiency of 75%. The coal-based syngas process has the highest emission of greenhouse gasses (GHG), which is around 2.8–3.8 kg CO_2/kg methanol. Methanol from biomass or flue gas CO_2 is at least 2–3 times more expensive than the fossil fuel-based methanol (Table 11.13) [25, 26].

Converting CO_2 into chemicals is challenging, and inherently carries costs for the energy and hydrogen supply. Figure 11.6 shows a schematic of wind

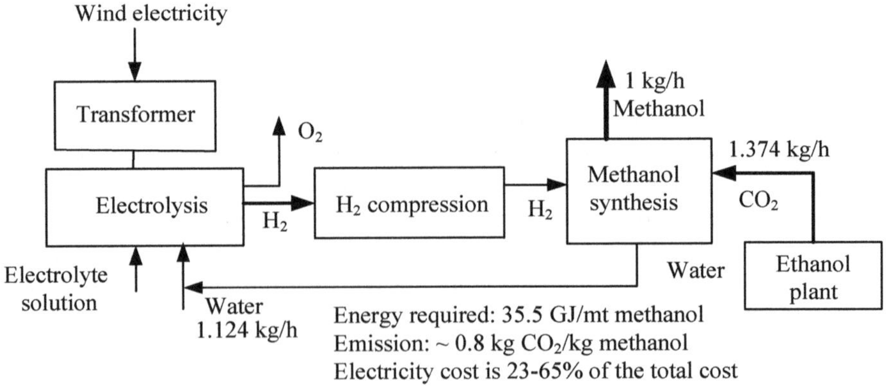

Fig. 11.5 Schematic for alkaline electrolysis of water for hydrogen production with compression, storage and delivery [15, 25]

Table 11.13 Sustainability metrics for the integral methanol plant [25]

Metrics	Integral methanol
Material intensity	
CO_2 used/Unit product. kg/kg	1.39
H_2 used/Unit product, kg/kg	0.19
Energy intensity	
Net duty/unit product, MWh/kg	9.55
Net cost/Unit product, $/kg	828.67
Environmental impact[a]	
Total CO_2e/Unit product, kg/kg	– 0.85

[a] For an assumed carbon fee of $2/tonne CO_2

power-based hydrogenation of CO_2 to methanol. Methanol synthesis from water, renewable electricity, and carbon may lead to chemical storage of renewable energy, carbon recycle, fixation of carbon in chemical feedstock, as well as extended market potential for electrolysis. Methanol synthesis can fix the CO_2 when it is used as feedstock for producing various other chemicals and be used as transportation fuel and hence recycles CO_2 [25].

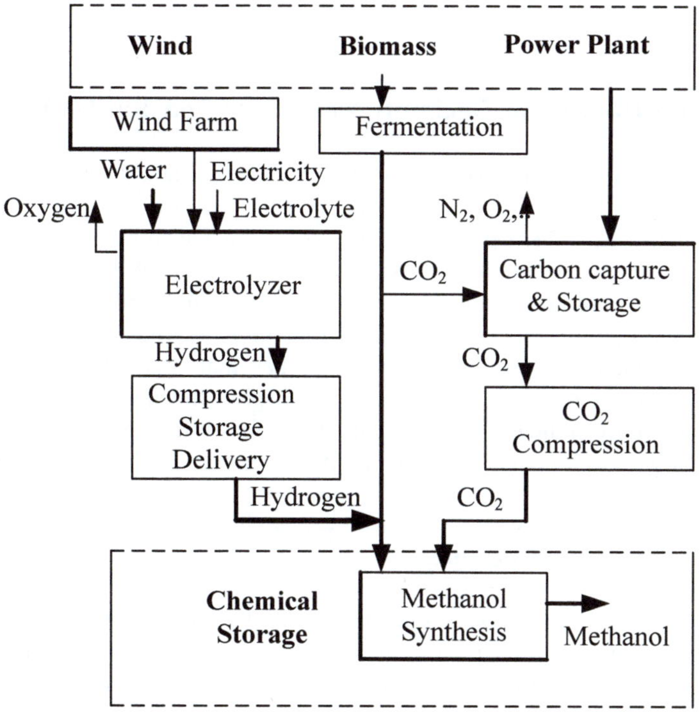

Fig. 11.6 Schematic of methanol production using renewable hydrogen and CO_2 [25, 26]

Fixation of carbon reduces the carbon dioxide equivalent emission by around − 0.85 kg CO_2e/kg methanol. Complete combustion increases the emission by around + 0.53 kg CO_2e/kg methanol. The electrolytic hydrogen production cost is the largest contributor to the economics of the integral plant. Table 11.13 presents the sustainability metrics, which show that the methanol facility requires 1.39 mt CO_2/mt methanol. The environmental impact metrics show that the integral methanol facility reduces emission by around − 0.84 kg CO_2/kg methanol when utilizing it as a chemical feedstock and recycles 0.53 kg CO_2/kg methanol after its complete combustion [25].

Sustainable Refinery Operation—Energy and environmental sustainability metrics for a crude oil refinery consisting three distillation columns can be assessed by the thermodynamic analysis and energy analyzer. Distillation columns consume considerable energy used to operate the plants in petrochemical and chemical process industries in North America. The objective is to explore the scope of reducing the thermal energy consumption and CO_2 emissions for a more sustainable refinery operation. Thermodynamic analysis is carried out by using the thermal analysis capability of "column targeting tool" to address the "energy intensity metrics," and the "energy analyzer" to design and improve the performance of the heat exchanger network system for process heat integration. Environmental pollution impact metrics are estimated from the "carbon tracking" options with a selected CO_2 emission data source of US-EPA-Rule-E9-5711 and using crude oil as a primary fuel source for the hot utilities [1].

11.6 Multi-criteria Decision Matrix for Feasibility Analysis

Beside the economics analysis, sustainability metrics should also be an integral part of the feasibility of energy sector. For this purpose, a multi-criteria Pugh decision matrix [23, 24] can be used to assess the feasibility of renewable and nonrenewable methanol production facilities (Table 11.14). With the weight factors adapted and the combined economic and sustainability indicators, the multi-criteria decision matrix shows that overall weighted score is around + 5.4 for the renewable integral methanol facility, which is higher than that of fossil fuel based methanol with a score of + 0.2. This may display the impact of multi-criteria with sustainability indicators on evaluating the feasibility of chemical processes requiring large investments and various types of energy resources [25, 28].

11.7 Life Cycle Analysis

Life cycle analysis (LCA) is a standardized technique that tracks all material, energy, and pollutant flows of a system from raw material extraction, manufacturing, transport, and construction to operation and end-of-life disposal. LCA

Table 11.14 Multi-criteria decision matrix for feasibility assessment of energy systems [25]

Economics and sustainability indicators	Weighting factor: 0–1	Fossil methanol	Renewable methanol
Economic indicators			
Net present value NPV	1	+	−
Payback period PBP	0.8	+	−
Rate of return ROR	0.8	+	−
Economic constraint EC	0.9	+	−
Impact on employment	1	+	+
Impact on customers	1	+	+
Impact on economy	1	+	+
Impact on utility	0.7	−	+
Sustainability indicators			
Material intensity	0.7	−	+
Energy intensity	0.8	+	−
Environmental impact: GHG in production	0.8	−	+
Environmental impact: GHG in utilization	0.8	−	−
Toxic/waste material emissions-	1	−	+
Potential for technological improvements	0.8	−	+
Security/reliability	0.9	−	+
Political stability and legitimacy	0.8	−	+
Quality of life	0.8	−	+
Total positive score		8	11
Total minus score		− 9	− 6
Net score (positive-minus)		− 1	+5
Weighted total score		+0.2	+5.4

is used for product and process development, improvement, and comparison of technologies. LCA can lead to quantified assessment to [6, 26]:

- Minimize the impact of pollution by most effective technique.
- Conserve nonrenewable energy and material resources.
- Develop and utilize cleaner energy technologies.
- Maximize recycling and optimize waste material production and treatment.

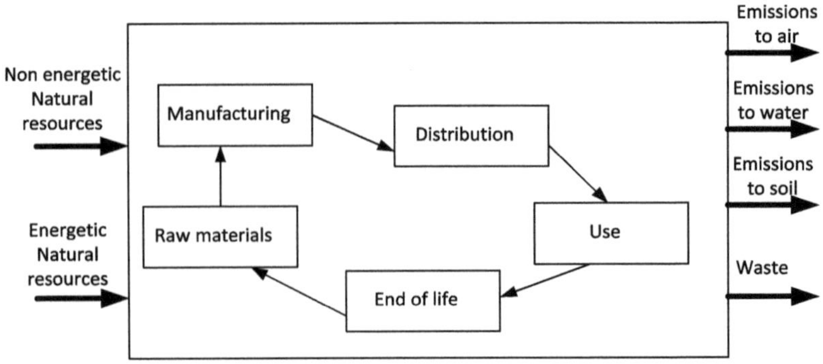

Fig. 11.7 Principles of life cycle assessment for cradle-to-grave boundary

LCA can help determine environmental impacts from "cradle-to-grave" or "cradle-to-gate" scenarios and can compare energy technologies (Fig. 11.7). Cradle-to-gate scenario estimates energy use (natural gas and electricity), GHG emissions, for feedstock production and conversion processes. LCA starts from the extraction of raw materials, production of product(s), and use of the product to reuse, recycling and final disposal of wastes. Applying LCA provides a broader view of a product's environmental impact through the value chain, not just at the final manufacturing stage. LCA assesses the environmental aspects and potential impacts associated with a product, process, or service, by:

- Compiling an inventory of relevant energy and material inputs and environmental releases.
- Evaluating the potential environmental impacts associated with identified inputs and releases.
- Interpreting the results to make a more informed decision.

LCA examines the environmental impacts of a product by considering the major stages of a product's life, which are [6]:

- Raw material harvesting and transportation to manufacturing site.
- Raw materials processing and transportation to production sites.
- Manufacturing product and assembly, packaging, and transportation to final distribution site.
- Energy and emissions during normal product life, required maintenance, and product reuse (refurbishing, material reuse).
- Waste management/end of life, including recycling and landfills.

11.7.1 Life Cycle Assessment Principles

International standards in the ISO 14040 series define the principles of LCA. These standards are compiled within the ISO 14044 standard [19]. The LCA technique can be narrowed down to four main steps which address one or more of the product's life stages at a time:

1. The definition and scope are determined along with information needs, data specificity, and collection methods and data presentation (ISO 14041).
2. The life cycle inventory (LCI) is completed through process diagrams, data collection, and evaluation of the data (ISO 14041). *Life cycle inventory* consists of emissions data, the raw materials, and energy used during the life cycle of a product.
3. The life cycle impact assessment (LCIA) is determined with impact categories and their weights, as well as any subsequent results (ISO 14042). *Life cycle impact assessment* measures the impact of these emissions and nonrenewable raw materials depletion on environment
4. The final report should include significant data, data evaluation and interpretation, final conclusions, and recommendations (ISO 14043). *Life cycle interpretation* consists of the results of the impact assessment in order to suggest improvements.

The first three steps of a life cycle analysis are related to one another. More importantly, however, data interpretation is an integral part of all three steps and should be done after each of the subanalyses is completed.

1. *Raw material acquisition.* This stage includes the removal of raw materials and energy sources from the earth, such as the extraction of crude oil and coal. Raw materials from the point of acquisition to the gate of processing are considered part of this stage.
2. *Manufacturing.* The manufacturing stage produces the product from the raw materials and delivers it to consumers:
   - Material manufacture stage converts raw materials into a form that can be used to fabricate a finished product. For example, the crude oil is refined to its various transportation liquid fuels.
   - Process stage includes the conversion of raw materials to a new product and makes it ready to be stored.
   - Filling, packaging, and distribution include all processes and transportation required to store, package, and distribute a finished product. Energy and environmental outputs caused by transporting the product to the consumer are accounted for a product's life cycle.
3. *Use.* The actual use, reuse, and maintenance of the product are considered with energy requirements and environmental wastes associated with product storage and consumption.

4. *Recycling and waste management*. Energy requirements and environmental wastes associated with product disposal are included in this stage, as well as post-consumer waste management options, such as recycling, composting, and incineration [6].

The LCA model aims to cover all processes related to an energy system within a long time scale and with all effects anywhere in the world, and all relevant substances and environmental themes. An effective LCA allows analysts to:

- Identify and estimate the positive or negative environmental impact of a process or product.
- Find opportunities for process and product improvement.
- Compare and analyze several processes based on their environmental impacts.
- Quantitatively justify a change in a process or product.

Within the life cycle, inputs and outputs of material, energy and emissions of each process are quantified. These input and output flows are associated to various environmental impacts such as nonrenewable resource (material and/or energy) depletion, global warming, and toxicities [6]. The assessments of input and output flows are called life cycle inventories (LCIs).

11.7.2 Benefits of International Organization for Standardization

ISO 50001 is an International Standard developed by the International Organization for Standardization (ISO) to provide organizations a framework to manage and improve their energy systems [19]. The improvements include operational efficiencies, decreasing energy intensity, and reducing environmental impacts. The standard addresses the following:

- Energy use and consumption
- Measurement, documentation, and reporting of energy use and consumption
- Design and procurement practices for energy using equipment, systems, and processes
- Development of an energy management plan and other factors affecting energy performance that can be monitored and influenced by the organization

ISO 50001 requires continuous energy performance improvement, but it does not include prescriptive energy performance improvement goals. Rather, it provides a framework through which each organization can set and pursue its own goals for improving energy performance.

The International Organization for Standardization (ISO) established a protocol for performing an LCA study. While simple in concept, the conduct of an

LCA can be complicated, mainly due to the large amount of data needed. However, the increasing availability of LCA databases and software programs makes it easier to conduct an LCA. In addition, The ISO 50001 energy management system standard is an international framework that helps industrial plants, commercial facilities, and organizations to manage energy, reduce costs, and improve environmental performance. Some of the specific benefits include:

- Assists organizations in optimizing their existing energy consumptions.
- Measures, documents, and reports energy intensity improvements and their projected impact on greenhouse gas emissions.
- Promotes energy management best practices and reinforces better energy management behaviors.
- Helps facilities evaluate and prioritize the implementation of new energy-efficient technologies.
- Provides a framework to promote energy efficiency throughout the supply chain.

11.7.3 Life Cycle Analysis Stages

Performing a life cycle analysis requires the following stages:

1. *The purpose of the LCA*: Eco-design,
   - Environmental impact (positioning products with respect to others in terms of their environmental balance).
   - Comparative LCA (positioning product by comparing with existing nonrenewable equivalents).
2. *The purpose of the LCA*: Eco-design,
   - *Cradle to gate.* This scope does not consider the end of product life scenarios, such as removal to landfill site, incineration, and recycling. If only energy consumption and GHG emission indicators are considered, the end of life impact would be less significant (see Fig. 11.8).
   - Cradle to grave. If the purpose is related to eco-design, then cradle-to-grave scope should be performed. When impact indicators such as eutrophication, air acidification, and human toxicity are assessed, it is vital that the end of life impact stage should be considered. The system boundaries describe which processes of the entire life cycle are included in the assessment [26]. Figure 11.7 shows the two common scopes for LCA.
3. *The functional unit*

A function unit such as kg of product or other relevant unit such as energy unit MJ should be specified.

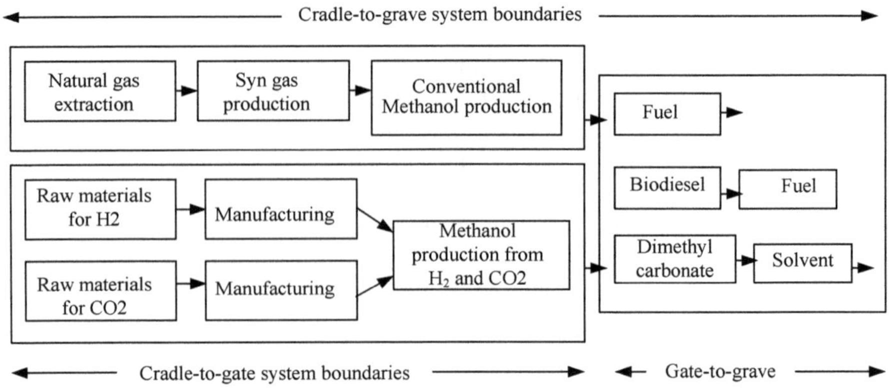

Fig. 11.8 Cradle-to-gate and cradle-to-grave system boundary definitions for methanol production using renewable and nonrenewable resources

4. *The indicators*

Impact indicators consist of input and output flows generated by a product by using emission factors for each identified flow in order to represent the following impact indicators of this product:

- Change in GHG emissions.
- Change in toxicity.
- Change in nonrenewable energy resources.
- Change in nonrenewable material resources.
- Change in water usage.

One should assess all the possible impacts of the products by at least one of the indicators (see Table 11.3). Nonrenewable energy consumption and global warming potential are the best-assessed indicators [5, 6, 25].

Best practice in LCA is to include all the by-products within the system boundaries and in the functional units. Figure 11.9 shows a biofuel life cycle analysis with cradle-to-grave scope. The environmental impacts are estimated for the entire system and associated to all products, e.g., kg CO_2eq/kg product. This is called system expansion in LCA, which is recommended by the ISO standards to avoid ambiguous choices in allocating inputs and emissions [6].

11.7.4 Life Cycle Assessment

A life cycle sustainability assessment (LCSA) is: LCSA = LCA + LCC + SLCA, where LCC is the life cycle costing and SLCA is social LCA. Here, the quantification of SLCA is a major challenge [6]. The comprehensive sustainability methods incorporate technoeconomic analysis and life cycle analysis. Both the methods are

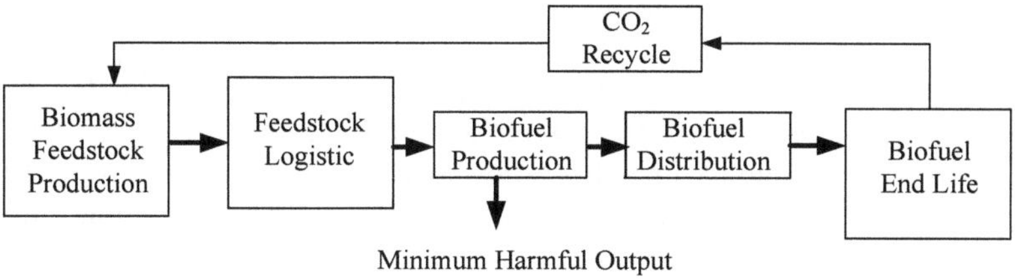

Fig. 11.9 Schematic of biofuel life cycle analysis with cradle-to-grave scope

very broad, information and data intensive. A comparative assessment method for various processes at design stage would be an essential step toward sustainability. Such a method compares an innovative new process with a conventional process based on these selected parameters [6]:

1. Economic constraint (EC)
2. Environmental impact of raw materials (EI)
3. Process cost and environmental impact (PCEI)
4. Environment, health and safety index (EHSI)
5. Risk aspects (RA)

1. Economic constraint (EC)

Economic feasibility is critical for economic sustainability. One of the definitions for EC is:

$$\mathrm{EC} = \frac{\text{Average discounted annual cost of product}}{\text{Average discounted annual revenue}} \tag{11.10}$$

For raw material case and for a single reaction, EC may be [6]

$$\mathrm{EC} = \frac{\sum_i (c_i f_i)_{\mathrm{RM}}}{\sum_i (c_i f_i)_{\mathrm{Prod}}} \tag{11.11}$$

where c and f are the cost and flow rates, respectively, RM and Prod stand for raw materials and products, respectively. For R number of multiple reactions, we have

$$\mathrm{EC} = \sum_i^R \mathrm{EC}_i \tag{11.12}$$

With these definitions, one needs $\mathrm{EC} \leq 1$ for economic sustainability. Fluctuations in prices and overall economic data affect the values of EC, and hence, an average over a time interval should be considered.

2. *Environmental impact of raw materials* (EI)

Environmental impact of raw materials is required for the production of a unit mass of product. Cumulative energy (renewable and nonrenewable) demand (CED) and GHG emission associated with all the raw materials from cradle to the relevant system boundary indicates the Environmental Impact (EI). Economic allocation factor for the main product (mp) is

$$A_{\mathrm{mp}} = \frac{c_{\mathrm{mp}} f_{\mathrm{mp}}}{\sum_i (c_i f_i)} \tag{11.13}$$

$\mathrm{CED}_{\mathrm{mp}}$ (process energy + energy content of main product) is

$$\mathrm{CED}_{\mathrm{mp}} = \left(\frac{A_{\mathrm{mp}}}{f_{\mathrm{mp}}} \sum_i f_i (\mathrm{CED}_i - E_i) \right) + E_{\mathrm{mp}} \tag{11.14}$$

$\mathrm{GHG}_{\mathrm{mp}}$ is

$$\mathrm{GHG}_{\mathrm{mp}} = \left(\frac{A_{\mathrm{mp}}}{f_{\mathrm{mp}}} \sum_i f_i \mathrm{GHG}_i \right) + \mathrm{FC}_{\mathrm{mp}} \tag{11.15}$$

where E is the calorific value and FC is the embedded fossil carbon, and f is the flow rate. CED and GHG emissions are good indicators for environmental impact. Depending on the availability, other data such as water and land use can be incorporated into the method. Limitation usually comes from toxicity inclusion.

3. *Process cost and environmental impact* (PCEI)

Energy loss in the reaction and separation sections of processing sequence can be used as an indicator for the expected cost and environmental impact based on energy loss index. Some possible indicators are:

- Presence of water at the reactor outlet and the difference in the boiling temp of water and products.
- Product concentration (conversion rate, yield).
- Boiling point differences between the main product and other components.
- Total mass of all the components to mass of the product indicating the level of mass loss.
- Reaction duty and possibility of energy recovery.
- Number of coproducts for understanding the separation needs.
- Pretreatment of feedstock.

4. *Environment, health, and safety index (EHSI)*

This index considers the safety, health and environment such as ecological toxicity aspects of a process. Each chemical present within the process is associated with persistency that is half-life in water or air. Heat is associated with irritation and chronic toxicity. Safety is associated with mobility (pressure, temperature), fire/explosion (flash point), reaction/decomposition, and acute toxicity. The property parameters and hazard classification of each chemical are taken into account to assign an index value.

$$\mathrm{EHSI} = A_{\mathrm{mp}}[Ew_{\mathrm{E}} + Hw_{\mathrm{H}} + Sw_{\mathrm{S}}] \tag{11.16}$$

where w is the weights for each category and A is the allocation factor. For multiple reaction steps, the methodology may be applied separately for each step.

5. *Risk analysis*

Risk analysis may be based on external aspects such as environmental, economic, and technical for commodity chemical and fuels with allocated weight factors for a target process or industry such as biofuel production [6, 22]. Some of the aspects include:

- Environmental
- Feedstock supply
- Economics
- Social
- Technical aspects of application
- Operational safety
- Chemicals (functional groups, retention of raw material functionality)
- Fuels (energy content, energy compatibility, storage and transportation

Index ratio (IR) displays the potential benefit associated with the proposed novel process:

$$\mathrm{IR} = \text{total weighted score of novel process/ total weighted score of conventional process} \tag{11.17}$$

A lower index (< 1) shows that the new process can provide certain benefits compared with the conventional process.

There may be some uncertainty associated with the data input, selected aspects and their weighting. The effect of these uncertainties can be analyzed using the Monte Carlo analysis in terms of probability distributions based on deviations within the weight factor ranges shown in Table 11.15. A sensitivity analysis can

Table 11.15 Distribution of deviations in weight factors [6]

	Weight	Minimum weight	Maximum weight
EC	0.3	0.25	0.6
PCEI	0.2	0.15	0.35
EI-CED	0.1	0.05	0.3
EI-GHG	0.1	0.05	0.3
EHSI	0.2	0.05	0.3
RA	0.1	0.05	0.25

be performed by estimating the impact of yield on the index ratio. This method combines the aspects of technoeconomic analysis, life cycle assessment, and green chemistry [6].

11.7.5 LCA Impact Categories

There are several environmental impacts and impact indicators [6] that are subject to LCA and shown previously in Table 11.3:

1. *Abiotic resource depletion potential*

Abiotic resource depletion potential (ARDP) includes depletion on nonrenewable resources such as fossil fuels, metals, and minerals. The total ARDP is estimated by

$$\mathrm{ARDP} = \sum_{i}^{n} \frac{r_i}{\mathrm{twr}_i} \tag{11.18}$$

where r_i is the quantity of resource i used per functional unit of product, n is the total number of resources, and twr is the total world reserves of the resource i. Burdens resource depletion (world reserves) are: coal reserves 8.72×10^{13} tons, oil reserves 1.24×10^{11} tons, and gas reserves 1.09×10^{14} m^3 [6, 17]

2. *Global warming potential*

Global warming potential (GWP) is the sum of emissions of the GHGs such as CO_2, N_2O, CH_4, and VOCs, multiplied by their respective GWP factors pf_i. GWP is estimated by

$$\mathrm{GWP} = \sum_{i}^{n} e_i(\mathrm{pf}_i) \tag{11.19}$$

where e_i is the emission rate of GHG i and pf$i$ is the potency factor. The values of GWP depend on the time horizon for assessment of GWP; the short-term effect is for 20 and 50 years and long-term effect for 100 and 500 years. Some values for potency factors of GWP are for chlorinated hydrocarbons: 400, CFCs: 5000, and other VOCs: 11 [17].

3. *Ozone depletion potential*

Ozone depletion potential (ODP) indicates the potential emissions of chlorofluorohydrocarbons (CFCs) and the chlorinated hydrocarbons (HCs) for depleting ozone layer and is estimated by

$$\mathrm{ODP} = \sum_{i}^{n} e_i(\mathrm{pf}_i) \tag{11.20}$$

where e_i is the emission rate of ozone depleting gas i and pf_i is the ODP factor for gas i expressed relative to the ozone depleting potential of CFC-11. Some values of ODP are chlorinated hydrocarbons: 0.5, CFCs: 0.4, and other VOCs: 0.005 [17].

4. *Acidification potential*

Acidification potential (AP) indicates the contribution of SO_2, NOx, HCL, NH_3, and HF to the potential acid formation and deposition such as their potential to form H + ions, AP is estimated by

$$\mathrm{AP} = \sum_{i}^{n} e_i(\mathrm{pf}_i) \tag{11.21}$$

The potency factor pf_i for gas i expressed relative to the AP of SO_2. Some values for AP are: NOx: 0.7, SO_2: 1, HCl: 0.88, HF: 1.6, NH_3: 1.88 [17].

5. *Eutrophication potential*

Eutrophication potential (EP) is the over-fertilization of water and soil resulting in increased growth of biomass. Eutrophication is characterized by excessive plant and algal growth due to the increased availability of one or more limiting growth factors needed, such as sunlight, carbon dioxide, and nutrient fertilizers. However, human activities have accelerated the rate and extent of eutrophication through both point-source discharges and nonpoint loadings of limiting nutrients, such as nitrogen and phosphorus, into aquatic ecosystems, with negative consequences for drinking water sources, fisheries, and recreational water bodies [17]. EP is estimated by

$$\mathrm{EP} = \sum_{i}^{n} e_i(\mathrm{pf}_i) \tag{11.22}$$

The value of EP is relative to the PO_4^{-3} [17].

6. *Photochemical oxidants creation potential*

Photochemical oxidants creation potential (POCP) or photochemical smog is usually expressed relative to the POCP factors of ethylene and calculated by

$$\mathrm{POCP} = \sum_{i}^{n} e_i(\mathrm{pf}_i) \tag{11.23}$$

Mainly, the values of volatile organic chemicals (VOCs) are classified into the following categories: alkanes, halogenated hydrocarbons, alcohols, ketones, esters, ethers, olefins, acetylenes, aromatics, and aldehydes. The values of pf_i are the respective POCP factors relative to ethylene [6, 17].

7. *Human toxicity potential*

Human toxicity potential (HTP) is estimated by adding the releases, which are toxic to humans, air, water, and soil

$$\mathrm{HTP} = \sum_{i}^{n} (e_i\mathrm{pf}_i)_\mathrm{A} + \sum_{i}^{n} (e_i\mathrm{pf}_i)_\mathrm{W} + \sum_{i}^{n} (e_i\mathrm{pf}_i)_\mathrm{S} \tag{11.24}$$

where each term shows the emissions of different toxic substances and their toxicological classification factors for the effect of the toxic emissions to air, water, and soil, respectively. The factors are estimated using the acceptable or tolerable daily intake of the toxic substances. These factors are still at their developing stages so that HTP can only be taken as an indication and not as an absolute measure of the toxic potential. Some values of HTP are: CO: 0.012, NOx: 0.78, SO2: 1.2, HC (excluding CH_4): 1.7, chlorinated hydrocarbons: 0.98, CFCs: 0.022, As: 4700, Hg: 120, F2: 0.48, HF: 0.48, NH_3: 0.02, Cr: 0.57, Cu: 0.02, Fe: 0.0036, Ni: 0.057, Pb: 0.79, Zn: 0.0029, fluorides: 0.041, nitrates: 0.00078, phosphates: 0.00004, pesticides: 0.14, phenols: 0.048 [17].

8. *Aquatic toxicity potential*

Aquatic toxicity potential (ATP) is estimated by

$$\mathrm{ATP} = \sum_{i}^{n} e_i(\mathrm{pf}_i) \tag{11.25}$$

The values of ATP are based on the maximum tolerable concentration of different toxic substances in water by aquatic organisms. Like HTP, classification factors for ATP are still developing so they can only be used as an indication of potential toxicity. Some values for ATP are Cr: 9.07×10^8, Cu: 1.81×10^9, Ni: 2.99×10^8, Pb: 1.81×10^9, Zn: 3.45×10^8, pesticides: 1.18×10^9, phenols: 5.45×10^9 [17].

11.8 Life Cycle Analysis of Energy Systems

LCA can provide a comprehensive framework to compare renewable energy technologies with fossil-based and nuclear energy technologies. Life cycle GHG emissions from renewable electricity generation technologies are generally less than from those from fossil fuel-based technologies [5, 6]. Comparisons also show that the proportion of GHG emissions from each life cycle stage differs by technology:

- Fossil fuel combustion emits the vast majority of GHGs.
- For nuclear and renewable energy technologies, the majority of GHG emissions are upstream of operation.
- Most emissions for biopower are generated during feedstock production, where agricultural practices play an important role.
- For nuclear power, fuel processing stages are most important, and a significant share of GHG emissions is associated with construction and decommissioning.
- For other renewable technologies (solar, wind, hydropower, ocean and geothermal), most life cycle GHG emissions stem from component manufacturing and, to a lesser extent, facility construction.

Comparison of published results for each specific electricity generation technology is challenging, due to variation of methods and assumptions and needs building on experience.

11.9 Economic Input–Output Life Cycle Assessment (EIO-LCA)

EIO-LCA method estimates the materials and energy resources required for processes and the environmental emissions resulting from energy production. Results from the EIO-LCA may provide guidance on the relative impacts of various products or industries with respect to resource use and emissions throughout the supply chain. Thus, the effect of producing a product would include not only the impacts at the final assembly facility, but also the impact from mining raw materials, transportation, storing, and others. LCA and EIO-LCA methods focus for the green design. The environmental impacts covered include global warming, acidification, energy use, nonrenewable ores consumption, eutrophication, conventional pollutant emissions and toxic releases to the environment. The EIO-LCA model is comprised of national economic input–output models and publicly available resource use and emissions data [6, 8].

11.10 Environment and Exergy

Since exergy is a measure of the departure of the state of the system from that of the environment, it relates the system to the environment. When a system is in thermal, mechanical, and chemical equilibrium with the environment, there are no processes taking place and the system is as the dead state. At dead state, the system has no motion and elevation relative to coordinates in the environment. Only after specifying the environment can we estimate a value for exergy. Specifying environment usually refer to some portion of a system's surroundings. For example, in estimating exergy values, the temperature and pressure of the environment are usually the standard state values, such as 298.15 K and 101.31 kPa.

The environment is composed of large numbers of common species within the earth's atmosphere, ocean, and crust. The species exist naturally. They are in their stable forms and do not take part in any chemical or physical work interactions between different parts of the environment. We mainly assume that the intensive properties of the environment are unchanging, while the extensive properties can change because of interactions with other systems. In the natural environment, however, there are components of states differing in their composition or thermal parameters from thermodynamic equilibrium state. These components can undergo thermal and chemical processes. Therefore, they are natural resources with positive exergy [8, 11].

11.10.1 Resource Depletion and Exergy

Resource depletion may cause environmental change. By reducing resource depletion, we can reduce ongoing environmental transformation. We may quantify the resource depletion by the *Depletion number Dp*, which is related to a nondimensional indicator Ex_{Dp} per unit consumption Ex_C

$$\mathrm{Dp} = \frac{\dot{E}x_{\mathrm{Dp}}}{\dot{E}x_{\mathrm{C}}} \tag{11.26}$$

Depletion number provides a measure of system progress or maturity. It is a useful basis for studying the evolution of resource depletion patterns and the implementation of resource conservation strategies. The depletion number is a function of the following three indicators showing the level of implementation of resource conservation strategies [8, 11, 13, 17]:

- The *exergy cycling fraction* ψ is a measure of recycling that accounts for both the throughput and quality change aspects of resource consumption and upgrading.
- The *exergy efficiency* η is a universal measure of process efficiency that accounts for the available energy.
- The *renewable exergy fraction* Ω is a measure of the extent to which resources supplied to an industrial system are derived from renewable sources. Industrial

systems consume resources by supporting processes associated with supplying and removing resources.

The definition of depletion number defined in Eq. (11.26) is

$$\mathrm{Dp} = \frac{\dot{E}x_{\mathrm{Dp}}}{\dot{E}x_{\mathrm{C}}} = 1 + \frac{\dot{E}x_{\mathrm{Dsl}}}{\dot{E}x_{\mathrm{C}}} + \psi\left[\frac{(1-\Omega_{\mathrm{RU}})}{\eta_{\mathrm{RU}}} - 1\right] + \frac{\dot{E}x_{\mathrm{TV}}}{\dot{E}x_{\mathrm{C}}}\left[\frac{(1-\Omega_{\mathrm{VU}})}{\eta_{\mathrm{VU}}} - 1\right] \tag{11.27}$$

where $\dot{E}x_{\mathrm{Dsl}}$ is the exergy dissipation rate, Ω_{RU} and η_{RU} are the renewable exergy fraction and transfer efficiency for the recovered resource upgrade process, respectively, $\dot{E}x_{\mathrm{TV}}$ is the exergy transfer rate to the nonrenewable source, and Ω_{VU} and η_{VU} are the renewed exergy fraction and transfer efficiency, respectively, for the nonrenewable resource upgrade process. Two structural constants $\alpha_{\mathrm{Ds\text{-}C}}$ and $\alpha_{\mathrm{V\text{-}C}}$ are defined by

$$\alpha_{\mathrm{Ds\text{-}C}} = \frac{\dot{E}x_{\mathrm{Dsl}}}{\dot{E}x_{\mathrm{C}}(1-\psi)},\ \alpha_{\mathrm{V\text{-}C}} = \frac{\dot{E}x_{\mathrm{TV}}}{\dot{E}x_{\mathrm{C}}(1-\psi)} \tag{11.28}$$

With these definitions, Eq. (11.27) becomes

$$\mathrm{Dp} = 1 + \psi\left[\frac{(1-\Omega_{\mathrm{RU}})}{\eta_{\mathrm{RU}}} - 1\right] + (1-\psi)\left\{\alpha_{\mathrm{Ds\text{-}C}} + \alpha_{\mathrm{V\text{-}C}}\left[\frac{(1-\Omega_{\mathrm{VU}})}{\eta_{\mathrm{VU}}} - 1\right]\right\} \tag{11.29}$$

The depletion number is a function of a system's structural constants, the exergy efficiency, and renewed exergy fraction of the individual resource upgrade processes, and the extent of resource cycling.

The generalized depletion number may result from numerous consumption processes, such as incomplete cycling or partial upgrading, and the direct reuse of resources without upgrading. Recycling may reduce the need for resources and the exergy requirements of manufacturing processes. Generally, increasing resource cycling reduces depletion due to less exergy transfer from other sources. For example, producing one-ton aluminum from bauxite requires 27,400 MJ of exergy transfer, while converting the recycled aluminum to feedstock requires far less exergy transfer [8, 17].

11.10.2 Extended Exergy Analysis

The attribute "extended" refers to the additional inclusion in the exergetic balance of previously neglected terms, such as labor and to environmental remediation expenditures. Extended exergetics is defined by its raw state exergy, augmented by the sum of all the net exergetics inputs received, directly or indirectly, in various processes pertaining to its extraction, preparation, transportation, and pretreatment, including the exergetic equivalents of labor, capital, and environmental costs.

Therefore, extended exergy accounting may be capable of properly addressing environmental issues.

An extended representation of exergy flow diagrams constitutes a substantial generalization of Szargut's cumulative exergy consumption procedure and may provide a coherent and consistent framework for including nonenergetic quantities like capital cost, labor cost, and environmental impact into an engineering optimization procedure. In this sense, the environmental approach represents already a significant extension of thermoeconomics, though its "environmental penalty" functions suffer from their direct dependence on monetary cost. Exergetic approach for the calculation of environmental costs may establish exergy as one of the proper measures of environmental impact. Extended exergy accounting can be considered a further development of the preexisting theories and methods of "engineering cost analysis."

A natural extension of life cycle assessment may be the *exergetic life cycle analysis*, which performs a lifelong analysis of a plant or process using exergy as a quantifier. The "exergetic life cycle analysis" includes labor, environmental damage, and recycle of energy and matter. Some of the efforts might be overstating the consequences of the laws of thermodynamics [8] as it tries to combine the thermodynamic imperfections with the profitability of a product as well as the environmental and social issues [8, 11, 17].

11.11 Ecological Planning

Ecological planning for sustainable development should consider the uncontrollable waste exergy emission in the environment. There are environmental impacts associated with the production and transmission of electricity. Emissions that result from the combustion of the fossil fuels include carbon dioxide (CO_2), carbon monoxide (CO), sulfur dioxide (SO_2), nitrogen oxides (NO_x), particulate matter, and heavy metals such as mercury. Nearly all combustion by-products may have negative impacts on the environment and human health:

- CO_2 is a greenhouse gas and a source of global warming. Power plants that burn fossil fuels and materials made from fossil fuels and some geothermal power plants are the sources of about 40% of total US carbon dioxide emissions.
- SO_2 causes acid rain, which is harmful to plants and to animals that live in water, and it worsens or causes respiratory illnesses and heart diseases, particularly in children and the elderly. SO_2 emissions are controlled by wet and dry scrubbers, which involve mixing lime in the fuel (coal) or by spraying a lime solution into the combustion gases. Fluidized bed combustion can also be used to control SO_2.
- NO_x contributes to ground level ozone, which irritates and damages the lungs. NO_x emissions can be controlled by several different techniques and technologies, such as low NO_x burners during the combustion phase or selective catalytic and noncatalytic converters during the post combustion phase.

- Particulate matter results in hazy conditions in cites and scenic areas, and, along with ozone, contributes to asthma and chronic bronchitis, especially in children and the elderly. Very small, or "fine particulate matter" is also thought to cause emphysema and lung cancer. Heavy metals such as mercury can be hazardous to human and animal health. Particulate matter emissions are controlled with devices that clean the combustion gases that exit the power plant, such as "Baghouses" use large filters, electrostatic precipitators use charged plates, and wet scrubbers use a liquid solution [18].

11.12 Chemical Looping Combustion for Sustainability

Chemical-looping combustion (CLC) is a novel technology in which power production and CO_2 capture are intrinsically combined using an oxygen carrier that transfers oxygen from the air to the fuel preventing direct contact between them. Well-designed and operated systems of chemical-looping combustion of fossil fuel/biomass offer scalable, diverse, economical, and environmentally sustainable energy pathways with inherited carbon capture. The fuel may be coal, natural gas, and biomass. The oxygen carrier can be alternately oxidized and reduced. The product gas contains mainly CO_2 and water undiluted with nitrogen, and without the production of nitrogen oxides (NO_X) as the high temperatures associated with the use of flame is avoided (Fig. 11.10a). The oxidation of the oxygen carrier is strongly exothermic and hence can be used to heat airflow to high temperatures (1000–1200 °C) and can drive a gas turbine [10].

In a new reactor concept, the oxygen carrier is not circulated between air and fuel reactors (see Fig. 11.10b) but kept inside a packed bed reactor and is alternately exposed to oxidizing and reducing conditions by periodic switching of the feed streams of fuel and air.

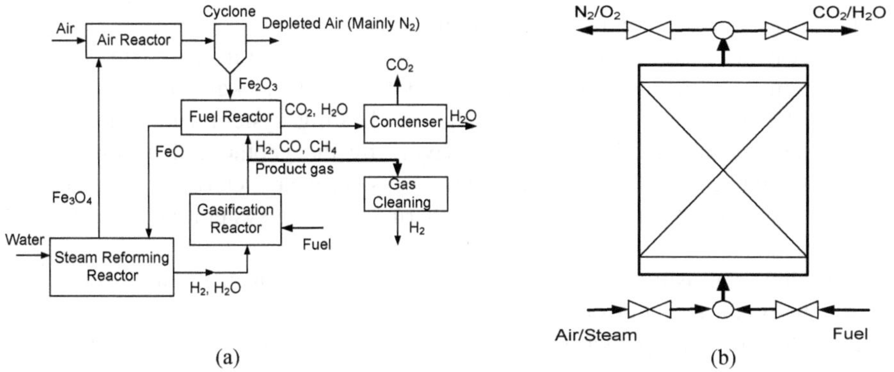

Fig. 11.10 Reactor configurations for chemical-looping technology: **a** Schematic of the chemical-looping combustion technology system, **b** periodically operated chemical-looping technology in packed bed system [10]

Thermochemical conversions of fuel/biomass can be analyzed by using the systems of chemical-looping combustion, chemical-looping reforming for producing hydrogen, and chemical-looping steam gasification for producing liquid transportation fuels of gasoline and diesel by the Fischer–Tropsch synthesis. The outcomes of such analyses would be:

(1) The lowest possible energy and economic costs for the fuel/biomass conversion systems without adverse environmental/societal consequences.
(2) A reduction of carbon intensity from energy conversion and use.
(3) Interactions of systems and patterns at the local/regional scale with systems/patterns at the global scale.

Problems

11.1. What are the advantages and disadvantages of using coal?
11.2. Does an electric car reduce the use of fossil fuels?
11.3. Is a fuel oil heater or an electric resistance heater the best for the environment?
11.4. Is a natural gas heater or a geothermal heating system the best for the house?
11.5. A car's daily traveling distance is about 80 miles/day. A car has a city-mileage of 20 miles/gal. If the car is replaced with a new car with a city-mileage of 30 miles/gal and the average cost of gasoline is $4.50/gal, estimate (a) the amount of fuel, energy, and money conserved with the new car per year, (b) reduction in CO_2 emission.
11.6. A car's daily traveling distance is about 80 miles/day. A car has a city-mileage of 10 miles/gal. If the car is replaced with a new car with a city-mileage of 32 miles/gal and the average cost of gasoline is $4.50/gal, estimate (a) the amount of fuel, energy, and money conserved with the new car per year, (b) reduction in CO_2 emission.
11.7. How can you control your carbon footprint?
11.8. A district uses natural gas for heating. Assume that average NO_x and CO_2 emissions from a gas furnace are 0.0045 and 6.4 kg/therm, respectively. The district wants to replace the gas heating system with a geothermal heating system. The projected saving by the geothermal heating system would be 40×10^6 therms of natural gas per year. Determine the amount of NO_x and CO_2 emissions the geothermal heating system would save every year.
11.9. How can using wind power help reduce pollution in the atmosphere and help conserve our oil supplies?
11.10. How can using thermal heat storage help reduce pollution in the atmosphere and help conserve our oil supplies?

References

1. Alhajji M, Demirel Y (2015) Energy and environmental sustainability assessment of crude oil refinery by thermodynamic analysis. Int J Energy Res 39:1925–1941
2. Alhajji M, Demirel Y (2016) Energy intensity and environmental impact metrics of the back-end separation of ethylene plant by thermodynamic analysis. Int J Energy Environ Eng 7:45–59
3. Allen JW, Unlu S, Demirel Y, Black P, Riekhof W (2018) Integration of biology, ecology and engineering for sustainable algal based biofuel and bioproduct biorefinery. Bioresources and Bioprocessing 5(47):1–28
4. Armstrong K, Styring P (2015) Assessing the potential of utilization and storage strategies for post-combustion CO_2 emission reduction. Front Energy Res 3:1–9
5. Cuellar-Franca RM, Azapagic A (2015) Carbon capture, storage and utilization technologies: a critical analysis and comparison of their life cycle environmental impacts. J CO_2 Util 9:82–102
6. Curran MA (2015) Life cycle assessment: a systems approach to environmental management and sustainability. CEP October
7. Demirel Y (2015) Sustainability and economic analysis of propylene carbonate and polypropylene carbonate production process using CO_2 and propylene oxide. Chem Eng Process Technol 6:236
8. Demirel Y, Gerbaud V (2025) Nonequilibrium thermodynamics: transport and rate processes in physical, chemical and biological systems, 5th edn. Elsevier, Amsterdam
9. Demirel Y (2013) Sustainable distillation column operations. Chem Eng Process Techniques 1005:1–15
10. Demirel Y, Matzen M, Winters C, Gao X (2015) Capturing and using CO_2 as feedstock with chemical-looping and hydrothermal technologies and sustainability metrics. Int J Energy Res 39:1011–1047
11. Demirel Y (2013) Thermodynamics analysis. Arab J Sci Eng 38:221–249
12. Demirel Y (2017) Lignin for sustainable bioproducts and biofuels. J Biochem Eng Bioprocess Technol 1:1–3
13. Demirel Y (2025) Energy conservation. In: Dincer I (ed) Comprehensive energy systems, 2nd edn. Elsevier, Amsterdam, Vol 9, pp 1–51
14. Demirel Y (2025) Biofuels. In I. Dincer (ed.), Comprehensive energy systems, 2nd edn. Elsevier, Amsterdam, Vol. 2, pp. 344–384
15. Dincer I, Ratlamwala TAH (2013) Development of novel renewable energy-based hydrogen production systems: a comparative study. Int J Hydrogen Energy 72:77–87
16. Hall CAS, Lambert JG, Balogh SB, Gupta A, Arnold M (2014b) EROI and quality of life. Energy Policy 64153
17. IChemE-Institution of Chemical Engineers (2002) The sustainability metrics, institution of chemical engineers sustainable development progress metrics recommended for use in the process industries
18. IEA (2023) International Energy Agency. World Energy Outlook. https://www.iea.org/reports/world-energy-outlook-2023
19. International Organization for Standardization (2006) Environmental management—Life cycle assessment—Principles and framework, 2006, International Organization for Standardization: Geneva, Switzerland
20. IRENA (2023) International Renewable Energy Agency, Renewable Power Generation Costs in 2023. https://www.irena.org/Publications/2024/Sep/Renewable-Power-Generation-Costs-in-2023
21. IRENA (2024) International Renewable Energy Agency, World Energy Transition Outlook 2024. https://www.irena.org/Publications/2024/Nov/World-Energy-Transitions-Outlook-2024
22. Kaylee A, Demirel Y (2024) Measuring risk of renewable diesel production processes using a multi-criteria decision strategy. Chemosphere 354:141695
23. Kutscher CF (ed) (2007) Tackling climate change in the U.S. American Solar Energy Society. www.ases.org/climatechange. Accessed in May 2014

24. Madden S, Alles K, Demirel Y (2021) Measuring sustainability of renewable diesel productions using multi-criteria decision-matrix. Bioproducts & Biorefining, Biofpr 15:1621–1637
25. Matzen M, Alhajji M, Demirel Y (2015) Techno economics and sustainability of renewable methanol and ammonia productions using wind power-based hydrogen. Adv Chem Eng 5:128
26. Matzen M, Demirel Y (2016) Methanol and dimethyl ether from renewable hydrogen and carbon dioxide: alternative fuels production and life-cycle assessment. J Clean Prod 139:1068–1077
27. UNESCO and Earth Charter International Secretariat (2007). https://earthcharter.org/library/good-practices-using-the-earth-charter-2/
28. Wang X, Demirel Y (2018) Feasibility of power and methanol production by an entrained-flow coal gasification system. Energy Fuels 32:7595–7610

Renewable Energy

12

Introduction and Learning Objectives: Energy demand continues to grow worldwide while the extraction of fossil fuels becomes more difficult and expensive. The ways we produce, convert, store, and use energy are changing Earth's climate and effecting the environment hence the ways of human's life as well as the next generation's future. As the industrial revolution progressed increased the use of electricity and the importance of liquid transportation fuels. Since the supply depended largely on fossil fuels plus some hydro power and then nuclear energy, slowly concerns appeared for Greenhouse Gas (GHG) emissions contributing to possible global warming. After advances, global efforts concentrated in replacing energy dense fossil fuels with the relatively dilute nature of wind and solar energy, which is very materials intensive. With the global consensus to utilize wind and solar energy, their costs have fallen and become comparable with the conventional sources as the increased costs of fossil fuel technologies, especially with decarbonization charges. A new focus may be achieving a balance between supply and demand of renewables.

The learning objectives of this chapter are:

- Solar and wind energy
- Bioenergy
- Renewable energy and ecology.

12.1 Renewable Energy

Renewable energy from natural resources are naturally replenished. Major renewable energy sources are hydroelectric, solar energy, biomass, wind, geothermal heat, ocean. In its various forms, renewable energy comes directly from the sun,

Y. Demirel, *Energy*, Green Energy and Technology,
https://doi.org/10.1007/978-3-031-97111-2_12

or from heat generated deep within the earth. For each of these resources, the IRENA (2020) reported the capacity factor from 2010 to 2019 as well as long term projections [12, 13].

Solar and wind energy have properties different from conventional forms of power generation: (1) their maximum output fluctuates according to the real-time and location. (2) The fluctuations can be predicted accurately only a short time, (3) they use converters to connect to the grid, (4) they are more modular and can be deployed in a distributed way, and (5) they cannot be transported and frequently are located at a distance from load centers, requiring connection costs [10].

- There is global support for using solar and wind energy, which provides electricity from widespread resources without GHG emissions.
- Renewable electricity depends on the cost and efficiency of the wind and solar technology, which are improving and thus reducing costs at the source.
- Utilizing renewable electricity from solar and wind in a grid becomes challenging mainly due to the mismatch between supply and demand.
- Utility storage capacity is required due to the intermittent nature of solar and wind.
- In many countries, policy settings exist to support renewables with subsidies.

In an age where sustainability is an ethical norm, the focus on the environmental implications of sources and the cost imposed on carbon dioxide changed the economic scope of clean energy sources. This creates a unique challenge in generating technologies, which efficiently supply renewable power to the consumer at a competitive price.

12.1.1 Solar Energy

Photosynthesis as solar energy's main human application in agriculture and forestry has been extended in many ways. Technologies for solar energy capture, conversion, and distribution may be either passive or active. *Active solar techniques* use of solar thermal collectors to capture solar energy in the form of solar radiation into a heat transfer fluid, which may be water, air, or others for water *space heating/cooling.* A typical water heating system includes a storage tank holding the hot liquid to manage the demand. *Passive solar systems* rely on gravity and the tendency for fluid to naturally circulate as it is heated.

Solar-based electricity relies on photovoltaics and heat engines that use parabolic trough collectors to collect the solar energy to generate steam to drive a conventional steam turbine [10]. The parabolic mirrors automatically track the sun throughout the day. The sunlight is directed to a central tube carrying a heat transfer medium like synthetic oil, which heats around 400°C. *Photovoltaic solar cell* is a solid-state electrical device that converts the energy of light directly into electricity using various semiconductors crystalline silicon, thin-film, and concentrator [14]. Photovoltaic systems produce direct current, which must be converted to alternating current to be used in grid. A major goal is to increase solar photovoltaic efficiency and decrease costs [10–12].

Example 12.1 **Estimation of emissions because of lost work**

A turbine discharges steam from 6 MPa and 400°C to saturated vapor at 360.15 K while producing 500 kW of shaft work. The temperature of surroundings is 290 K. Determine maximum possible production of power in kW, the amount of work lost, and emissions based on coal because of the waste energy.

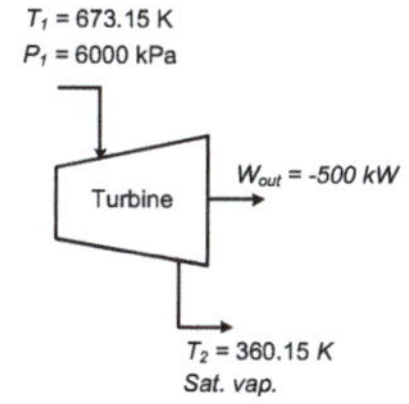

Solution:

Assume: The turbine operates at steady state. Kinetic and potential energy changes are negligible.

Basis: 1 kg/s steam flow rate

Inlet: superheated steam: $H_1 = 3180.1\,\text{kJ/kg},\ S_1 = 6.5462\,\text{kJ/kg K}$ (steam tables)

Outlet: saturated steam: $H_2 = 2655.3\,\text{kJ/kg},\ S_2 = 7.5189\,\text{kJ/kg K}$ (steam tables)

$$\dot{W}_{\text{out}} = -500\,\text{kW},\ T_{\text{o}} = 290\,\text{K}$$
$$\Delta H = H_2 - H_1 = (2655.3 - 3180.1)\,\text{kJ/kg} = -524.8\,\text{kJ/kg}$$

The amount of heat transfer: $q = -W + (H_2 - H_1) = -24.8\,\text{kJ/kg}$

We can determine the entropy production from an entropy balance on the turbine operating at steady state that exchanges heat only with the surroundings:

$$\dot{S}_{\text{prod}} = \dot{m}\left[S_2 - S_1 - \frac{q}{T_o}\right] =$$
$$(1\,\text{kg/s})[(7.5189 - 6.5462)(\text{kJ/kg K}) + 24.8\,\text{kJ/kg}/(290)\,\text{K}] = 1.06\,\text{kJ/s K}$$

Lost work: $\dot{W}_{\text{lost}} = T_o\dot{S}_{\text{prod}} = (290\,\text{K})(1.06\,\text{kJ/s K}) = 307.4\,\text{kW}$

Maximum work output: $\dot{W}_{\text{ideal}} = \dot{W}_{\text{out}} - \dot{W}_{\text{lost}} = -500 - 307.4 = -807.4\,\text{kJ/kg}$

Annual Emission for 1 kg/s steam flow rate:

$$(0.37\,\text{kg CO}_2/\text{kWh})(307.4\,\text{kW})(24\,\text{h/day})(365\,\text{day/year}) = 996{,}344.8\,\text{kg CO}_2/\text{year}.$$

Example 12.2 **Avoided emissions in a compressor operated with minimum power**

A compressor receives air at 15 psia and 80°F with a flow rate of 1.0 lb/s. The air exits at 40 psia and 300°F. Estimate the minimum power input to the compressor and avoided emissions if it is operated with solar concentrated power instead of power from natural gas-fired plant. The surroundings are at 520 R.

Solution:

Assume that potential energy effects are negligible, and steady process.

Basis: airflow rate = $\dot{m}$ = 1 lb/s. The surroundings are at 520 R

The properties of air from Table D1 after conversions from SI units:

State 1: $P_1 = 15\,\text{psia}, \quad T_1 = 540\,\text{R}, \quad H_1 = 129.0\,\text{Btu/lb}, \quad S_1 = 0.6008\,\text{Btu/lb R}$

State 2: $P_2 = 40\,\text{psia}, \quad T_2 = 760\,\text{R}, \quad H_2 = 182.0\,\text{Btu/lb}, \quad S_2 = 0.6831\,\text{Btu/lb R}$

Compressor work:
$\dot{W}_s = \dot{m}(H_2 - H_1) = (1\,\text{lb/s})(182.08 - 129.06)\,\text{Btu/lb} = 53.0\,\text{Btu/s}$

The entropy production: $\dot{S}_{\text{prod}} = \dot{m}(S_2 - S_1) = 0.0823\,\text{Btu/s R}$

Lost work: $\dot{W}_{\text{lost}} = T_o\dot{S}_{\text{prod}} = 42.8\,\text{Btu/s}$

Minimum work required: $= (53.0 - 42.8)\,\text{Btu/s} = 10.2\,\text{Btu/s} = 10.76\,\text{kW}$

Annual emission for electricity from natural gas-fired power by replacing with solar power

Use Table 11.2: 0.23 kg CO_2/kWh

$$(0.23\,\text{kg CO}_2/\text{kWh})(10.76\,\text{kW})(24\,\text{h/day})(365\,\text{day/year}) = 21{,}679.2.8\,\text{kg CO}_2/\text{year}.$$

12.1.2 Wind Energy

Despite the low-capacity factor (30–40% utilization) and intermittency, wind power receives widespread installations and investments (594 GWe onshore and 28 GWe offshore in 2019) [10–12]. According to Betz's law [2], no technological device can capture more than 59.5% of the kinetic energy in the wind or water. Utility-scale advanced wind turbines can achieve up to 80% of Betz's limit [2].

The differential heating drives a global atmospheric convection system reaching from the earth's surface to the stratosphere. Most of the energy stored in these wind movements can be found at high altitudes. Wind power is a totally renewable energy source with no greenhouse gas emissions, but due to its unpredictability, has problems integrating with national grids. To mitigate the intermittent wind power is to store it in the form of hydrogen by electrolysis and feed it into the gas grid and eventually electricity grid. The potential for wind power is considerable and integrating with existing transmission capacity helps large-scale deployment and reduces the cost [10, 11]. As technology has improved, the capacity factors increased for wind power generation [12].

Example 12.3 **Avoided emissions by windmill power generation**

A farm of windmills supplies a power output of 1 MW for a community. Each windmill has blades 10 m in diameter. At the location of the windmills, the average velocity of the wind is 11 m/s and the average temperature is 20°C. Estimate the minimum number of windmills to be installed.

Solution:

Air is the ideal gas and the pressure is atmospheric.

Inlet: $v = 11\,\text{m/s},\ R = 8.314\,\text{kPa m}^3/\text{kmole K},\ T = 293\,\text{K},$
$D = 10\,\text{m},\ MW_{\text{air}} = 29\,\text{kg/kmole}$

$$\text{Power output} = 1\,\text{MW}$$

Density of air: $\rho = (MW)\frac{P}{RT} = \frac{(29\,\text{kg/kmole})\,101.3\,\text{kPa}}{(8.314\,\text{kPa m}^3/\text{kmole K})(293\,\text{K})} = 1.2\,\text{m}^3/\text{kg}$

Air mass flow rate: $\dot{m} = \rho A v = \rho \pi \frac{D^2}{4} v = 1036.2\,\text{kg/s}$

Power from each windmill:
$KE = -\dot{m}\frac{v^2}{2} = (1036.2\,\text{kg/s})\frac{(11\,\text{m/s})^2}{2}\left(\frac{\text{kJ/kg}}{1000\,\text{m}^2/\text{s}^2}\right) = -62.7\,\text{kW}$

The minimum number of windmills to be installed:

$$1000\,\text{kW}/|62.7|\text{kW} = 16\,\text{windmills}.$$

12.1.3 Hydropower

Hydroelectric energy from the potential energy of rivers is the best-established electricity generation from renewable sources. The largest power plants in the world use dams on rivers. Different from wind and solar energy, hydro generating has significant mechanical inertia and is synchronous that supplies reliable energy to the grid. Hydro energy can be applied to peak-load demand easily as the individual turbines can be controlled from zero to full power in about ten minutes. Therefore, hydro energy can complement the wind and solar power in a grid system.

Pollution is minimum from hydroelectricity as it is produced without fuel. Hydroelectric plants can operate for a long time and with less operating and maintenance costs. The sale of electricity from the station may cover the construction costs after 5 to 8 years of full operation.

In the transforming energy scenario (TES) [11], hydropower capacity would need to increase 25% by 2030, and 60% by 2050, while pumped hydro storage capacity would need to double not only by new construction but by upgrading the existing turbines. Hydropower provides cost effectiveness to counteract the variability of wind and solar generation, and seasonal complementarities in resource patterns besides its benefits in regulating river flows and reducing flooding. New hydropower investments need to consider local environmental impacts and engage in discussions with the stakeholders.

Example 12.4 **Avoided emissions by hydroelectric power generation**

A hydroelectric plant operates by water falling from a 200 ft height. The turbine in the plant converts potential energy into electrical energy, which is lost by about 5% through the power transmission so the available power is 95%. If the mass flow rate of the water is 396 lb/s, estimate the power output of the hydro plant and the avoided emissions compared with coal-fired power plant.

Solution:

Equation: $\dot{W} = \frac{\dot{m} g \Delta z}{g_c}$

Data: $\Delta z = 200$ ft, water flow rate 396 lb/s; transmission loss = 5%

$$\dot{W} = -\frac{(396\,\mathrm{lb/s})(32.2\,\mathrm{ft^2/s})(200\,\mathrm{ft})}{32.2\,\mathrm{ft\,lb/lbf\,s^2}}\left(\frac{1.055\,\mathrm{kW}}{778\,\mathrm{lbf\,ft/s}}\right)$$
$$= -107.4\,\mathrm{kW}\ (\text{with the sign convention})$$

With the transmission loss of 5% of the available power:
$-107.4\,\mathrm{kW}\,(1 - 0.05) = -102.0\,\mathrm{kW}$

Annual avoided emission using Table 11.2:

$$(0.37\,\mathrm{kg\,CO_2/kWh})(102.0\,\mathrm{kW})(24\,\mathrm{h/day})(365\,\mathrm{day/year}) = 330.602.4\,\mathrm{kg\,CO_2/year}.$$

12.1.4 Geothermal Energy

Geothermal energy is the heat originating from radioactive decay of minerals, from volcanic activity, and from solar energy absorbed at the Earth's surface. The hot water and steam from reservoirs can be used to drive generators and produce electricity. Geothermal power is cost effective, reliable, sustainable yet has declining capacity factor [12].

Example 12.5 **Avoided emission by using a geothermal energy source**

A steam power plant uses a geothermal energy source. The geothermal source is available at 220°C and 2320 kPa with a flow rate of 200 kg/s. The hot water goes through a valve and a flash drum. Steam from the flash drum enters the turbine at 550 kPa and 428.62 K. The discharged steam from the turbine has a quality of $x_4 = 0.96$. The condenser operates at 10 kPa. The water is a saturated liquid after passing through the condenser. Determine the work output of turbine and emissions avoided compared with coal-fired and natural gas-fired power productions.

Solution:

Assume: The kinetic and potential energy changes are negligible, and this is a steady state process.

$\dot{m}_1 = 200\,\text{kg/s}$. From steam tables

$$T_1 = 493.15\,\text{K},\;\; P_1 = 2319.8\,\text{kPa},\;\; H_1 = H_2 = 943.7\,\text{kJ/kg},\;\; S_1 = 2.517\,\text{kJ/kg K}$$
$$T_3 = 428.62\,\text{K},\;\; P_3 = 550\,\text{kPa},\;\; H_3 = 2751.7\,\text{kJ/kg},\;\; S_3 = 6.787\,\text{kJ/kg K (saturated)}$$
$$H_{3\text{sat vap}} = 2551.7\,\text{kJ/kg},\;\; H_{3\text{sat liq}} = 655.80\,\text{kJ/kg},\;\; S_{3\text{sat vap}} = 6.787\,\text{kJ/kg K},$$
$$S_{3\text{sat liq}} = 1.897\,\text{kJ/kg K},\;\; P_4 = 10\,\text{kPa},\;\; H_{4\text{sat vap}} = 2584.8\,\text{kJ/kg},$$
$$H_{4\text{sat liq}} = 191.8\,\text{kJ/kg}$$

In this geothermal power plant, the hot water is flushed, and steam is produced and used in the turbine.

The rate of vapor is estimated from the quality at state 2. The fraction of steam after flashing is:

$$x_2 = \frac{943.7 - 655.8}{2751.7 - 655.8} = 0.159$$
$$S_2 = (1 - 0.159)1.897 + 0.159(6.787) = 2.6756\,\text{kJ/kg K}$$

The steam flow rate is: $\dot{m}_3 = x_2(\dot{m}_1) = 0.159(200) = 31.84\,\text{kg/s}$

From the mass balance around the flash drum, we have $\dot{m}_6 = \dot{m}_1 - \dot{m}_3 = 168.15\,\text{kg/s}$

The discharged steam has the quality of: $x_4 = 0.96$

$$H_4 = (1 - 0.96)H_{4\text{ sat liq}} + (0.96)H_{4\text{ sat vap}} = 2489.08\,\text{kJ/kg}$$

From the flash drum at state 6, we have: $\dot{W}_{net} = \dot{m}_3(H_4 - H_3) = -1993.82\,\text{kW}$

Table 11.2: (0.37 kg CO_2/kWh)

Annual avoided mission:

$$(0.37\,\text{kg}\,CO_2/\text{kWh})(1993.82\,\text{kW})(24\,\text{h/day})(365\,\text{day/year}) \\ = 6{,}462{,}369.4\,\text{kg}\,CO_2/\text{year}.$$

12.1.5 Ocean Energy

Ocean energy falls into three categories that are tidal, wave, and temperature gradient. Ocean thermal energy conversion uses the temperature difference that exists between deep and shallow waters to run a heat engine system to harvest electrical power from ocean waves is a viable technology. The generator is powerful, and the turbine has minimal environmental impact as the rotors pose little or no danger to wildlife as they turn quite slowly.

12.2 Bioenergy

Bioenergy is derived from biological sources, to be used for heat, electricity, or vehicle fuel. Biofuel derived from plant materials is among the most rapidly growing renewable energy. Bioenergy currently makes up a large share of renewable energy use and would remain a significant source for power and heat generation in industry and as a fuel in the transportation sector. Sustainable bioenergy is a necessary pre-condition on existing farmlands, grasslands, and residues without encroaching upon rainforests and threatening food production. In the Transforming Energy Scenario (TES), bioenergy plays an important role in shipping, aviation, and in industrial sectors. Bioenergy is increasing its share, consequently, traditional uses of bioenergy would be phased out and replaced with cleaner options, such as modern bioenergy and other renewables [12].

12.2.1 Biomass Resources

Biomass feedstock originates from diverse resources including naturally growing terrestrial and aquatic plants, and man-made wastes, Biomass contains carbon, oxygen, and hydrogen in a variety of compounds including carbohydrate, lipids,

Table 12.1 Biomass resources [21]

Generation	Type	Source	Examples
First	Food crops	Starch crops	Corn, wheat
		Sugar crops	Sugar cane, sugar beet, sweet sorghum
		Feed	Grass
Second	Lignocellulosic crops	Woody	Short-rotation crops, willow poplar
		Herbaceous	Miscanthus, switch grass. corn stover, bagasse
Third	Aquatic	Microalgae	*Chlamydomonas rheinhardii, chlorella, spirulina,*
		Macro algae	Seaweed
		Water plants	Salt marches, sea grass
Wastes	Natural	Agricultural	Animal manure, crop residues
		Forest	Logging residues, tree wastes
	Man-made	Municipal	Solid waste, sewage sludge, waste oil
		Industrial	Pulp and paper industry, sludge

and protein [14]. It has high moisture content and low energy density. Based on this diversity, biomass feedstock is classified as the first, second and third generations, as well as wastes as seen in Table 12.1. The first generation is food-based biomass such as corn, sugar cane, plant, vegetable oils and fats. The second generation refers to nonfood type biomass that includes lignocellulose, while the third-generation biomass mainly refers to algae and cyanobacteria. Cellulose, hemicellulose, and lignin are the structural components of lignocellulosic biomass. All these types of biomasses can be converted to biofuels and bioproducts by using biochemical, thermochemical, and hydrothermal processes [3, 4]. Biofuel supply chain can promote sustainable energy and development [5].

Table 12.2 shows the proximate and ultimate analyses of various biomass that contains between 40 and 53% of carbon, 20 and 43% oxygen, 5 and 6% of hydrogen, and very little sulfur, while the percentage of ash varies between 0.25 and 12%. Lignocellulosic biomass contains cellulose (38–50%), hemicellulose (23–32%), and lignin (15–25%) [21].

12.2.2 Energy from Solid Waste

Garbage, often called municipal solid waste contains biomass like paper, cardboard, food scraps, grass clippings, leaves, wood, and leather products, and other non-biomass combustible materials, mainly plastics and other synthetic materials made from petroleum. Recycling and composting programs may reduce the share of biomass in municipal solid waste that is landfilled or burned. Solid waste can be burned in special waste-to-energy plants, which produce heat to produce steam or

Table 12.2 Proximate and ultimate analyses of the second generation of biomass in wt% and dry base [21]

Biomass type	Fixed carbon	Volatiles	Ash	C	H	O	N	S	HHV[a] MJ/kg
Redwood	16.10	83.50	0.40	53.50	5.90	40.30	0.10	0.00	21.03
Wheat straw	19.80	71.30	8.90	43.20	5.00	39.40	0.61	0.11	17.51
Corn stover	19.25	75.17	5.58	43.65	5.56	43.31	0.61	0.01	17.65
Bagasse	14.95	73.78	11.27	44.80	5.35	39.55	0.38	0.01	17.33
Sawdust	14.33	76.53	0.25	40.00	5.98	44.75	0.01	0.01	19.95
Manure and sludge			38.8	29.9	4.3	22.40	2.90	0.62	13.1
Municipal waste			27.1	40.4	4.9	25.30	0.91	0.35	16.5

[a] HHV: higher heating values

produce electricity. Such plants help reduce the amount of solid waste to be buried in landfills. There also are solid waste incinerators that simply burn the solid waste without electricity production [1].

12.3 Biofuels

Biofuels are biomass feedstock-based fuels and include hydrogen, ethanol, butanol, methanol, biooil, biogas, and biodiesel, while renewable electricity is called biopower. Bioethanol and biodiesel have by far the largest share of the global biofuels markets. The biofuel supply chain involves the growing/production, harvesting, collecting, storing, and transporting of biomass to the biorefinery where it is converted to biofuel, bioproducts, heat and power. Some agricultural products grown for biofuel production include corn, soybeans, willow switchgrass, rapeseed, wheat, sugar beet, palm oil, miscanthus, sorghum, cassava, and jatropha. Biomass and biodegradable outputs from industry, agriculture, forestry, and households can be used for biofuel production using: (i) fermentation to produce ethanol, (ii) transesterification for biodiesel production (iii) anaerobic digestion to produce biogas, (iv) gasification to produce bio syngas (H_2, CO, CO_2), or (vi) direct combustion. The use of biomass, therefore, contributes to waste management as well as fuel.

Biomass can be fermented directly to produce hydrogen, ethanol, and other high-value chemicals. Certain photosynthetic microbes produce hydrogen from water in their metabolic activities using light energy. Photo biological technology holds great promise, but because oxygen is produced along with hydrogen, the technology must overcome the limitation of oxygen sensitivity of the hydrogen-evolving enzyme systems. A new system is also being developed that uses a metabolic switch (sulfur deprivation) to cycle algal cells between a photosynthetic growth phase and a hydrogen production phase [5, 15].

12.3.1 Biorefinery Systems

Biomass can be converted to biofuels, heat, and power in a biorefinery concept. Since biomass comes from diverse resources and requires various conversion processes, biorefinery is beneficial in converting various biomass feedstocks into various valuable fuels. Conversion processes include biochemical, chemical, thermochemical, and hydrothermal. Biochemical processes refer to fermentation, anaerobic and aerobic digestions. Chemical processes refer to transesterification of lipids, and Fischer-Tropsch (F-T) synthesis of bio syngas (mainly CO_2, CO and H_2) into various liquid transportation fuels. Thermochemical processes refer gasification, pyrolysis, reforming, and hydrothermal processes. Biorefineries are built in locations where an adequate supply of biomass exists to maintain their operations [3–5].

12.3.2 Ethanol Production

Ethanol is a clear, colorless alcohol fuel made from the sugars found in grains, such as corn, sorghum, barley, sugar cane, and sugar beets. The most common processes to produce ethanol today use yeast to ferment the sugars (glucose). Fermentation is a natural microbiological process where sugars are converted to alcohol and carbon dioxide by yeast (*Saccharomyces cerevisiae*) in about 24–36 h. The overall reaction within the yeast may be represented by

$$\begin{aligned} C_6H_{12}O_6 &\rightarrow 2(CH_3CH_2OH) + 2CO_2 \\ \text{Sugar} &\rightarrow \text{Alcohol} + \text{Carbon dioxide} \end{aligned} \tag{12.1}$$

Wet and dry milling routes are used to produce bioethanol from corn. Dry milling requires less investment and produces dried distiller's grain with solubles (DDGS) beside bioethanol, while the wet milling produces oil and animal feed beside the bioethanol. Figure 12.1 shows the basic steps of converting starch into bioethanol by biochemical process using 6-carbon sugars sources. Most corn is ground to a meal, and then the starch from the grain is hydrolyzed by enzymes to glucose (dry mill). The 6-carbon sugars are then fermented to ethanol by natural yeast and bacteria. The fermented mash is separated into ethanol and residue by distillation. Since water-ethanol mixture forms azeotropic mixture, molecular sieves produce fuel grade ethanol (0.4 vol% water) [5]. The average yield of converting corn starch to ethanol is around 100 gallons bioethanol per dry ton corn [21]. About one-third of every kilogram of corn grain is converted to ethanol, one-third to DDGS, and one-third to CO_2. Ethanol is produced at ASTM D4806 standards and shipped to the refiner or distributor for blending with conventional gasoline.

Lignocellulosic biomass and other nonfood biomass resources produce second generation bioethanol using corn stover, corn cobs, sorghum stalks, wheat straw, cotton residue, alfalfa stems, wood, fast growing plants such as grass, bagasse of

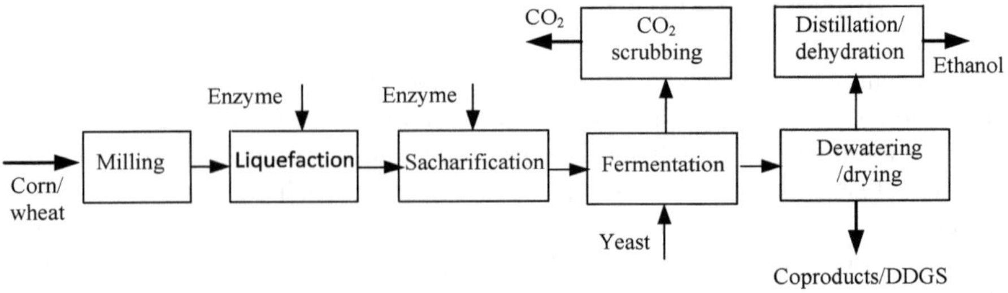

Fig. 12.1 Main steps for the first-generation biofuel production process producing bioethanol, Dried Distiller's Grain Solubles (DDGS), and CO_2

sugarcane, and sorghum stalks. As cellulose is protected by lignin and hemicellulose, pretreatment is required to hydrolyze hemicellulose sugars using enzymes. After pretreatment, cellulolytic enzymes are used to hydrolyze the cellulose polymers to C5 and C6 sugars (xylose and glucose). Unlike glucose, xylose is not readily fermented to ethanol. Genetically modified or metabolically engineered yeasts or bacteria are used to ferment both glucose and xylose to enhance the yield of ethanol from lignocellulose [5, 7, 8].

12.3.3 Biodiesel

Chemical process in biodiesel production is the transesterification of triglycerides of fatty acids into biodiesel (fatty acid methyl ester) using mainly methanol or ethanol and a catalyst (mainly acids and alkali-NaOH) [17] and producing glycerin as a byproduct. General transesterification reaction using methanol is represented by

$$\begin{gathered}\mathrm{RCOOR'} + \mathrm{CH_3OH} = \mathrm{RCOOCH_3} + \mathrm{R'OH} \\ \text{Triglyceride} + \text{Methanol} = \text{Fatty acid methyl esters (FAME)} + \text{Glycerin}\end{gathered} \tag{12.2}$$

Triglycerides are esters of glycerol and present in oilseed crops including soybean, rapeseed, palm, sunflower, and animal fat. Table 12.3 shows the typical oil content, and the number of carbon atoms. During the transesterification, methyl or ethyl esters of fatty acids are produced. Waste cooking oil contains free fatty acid and needs to be esterified with alkali or acid first before being converted to biodiesel by transesterification [5, 17].

The type of oil affects the quality of biodiesel. Soybean and palm oils are the two largest oil seed crops. On a dry-weight basis, soybean contains around 41% protein, 21% oil, and 29% carbohydrate, on average. Triacylglycerol (triglycerides-$C_{55}H_{98}O_6$) (94%) are the primary component in the soybean oil, while phospholipids content is around 3.7%. Other vegetable oils and animal

Table 12.3 Fatty acid composition (wt%) of various oils and fats [21]

Oil or fat	Typical oil content %	Number of carbon atoms: number of double bonds						
		14:0	16:0	18:0	18:1	18:2	18:3	20:1
Rapeseed	39–43	–	4.3	1.3	59.9	21.1	13.2	
Soybean	16–18	–	6–10	2–5	20–30	50–60	5–11	
Sunflower	40–50	–	7.2	4.1	16.2	72.5	–	
Jatropha	28–38	–	11.3	17.0	12.8	47.3	–	1.8
Tallow	24–32	3–6	24–32	20–25	27–43	2–3	–	

fats such as canola, camelina and jatropha constitute a small fraction of biodiesel feedstock [5, 21].

12.3.4 Biodiesel from Algae

Oil-rich microalgae strains can produce biodiesel with (1) algae growth in open ponds or in photobioreactors, (2) algae harvesting and dewatering, (3) lipid extraction, and 4) transesterification of lipid to biodiesel. Algae require water, sunlight, carbon, and nutrients like nitrogen and phosphorus to grow. Many strains of algae can grow nearly in any water resource. The residue of lipid extraction can be fed into anaerobic digestion to produce biogas and bio power, or used as a feedstock for fermentation to alcohols, or used as fertilizer [1]. Table 12.4 compares the properties of biofuels and diesel; fuel properties of biodiesel from soybean oil and conventional diesel are comparable and biodiesel contains less sulfur and higher Cetane numbers [5].

Table 12.4 Properties of biodiesel from soybean and conventional diesel [5, 21]

Property	Green diesel	Biodiesel	Petroleum diesel
Density g/cm^3	0.78	0.88	0.84
Cetane number	70–90	50–65	40
Heating value, MJ/kg	44	38	43
Sulfur, ppm	< 1	< 1	< 10
Oxygen, %	0	11	0
Cold flow properties	Poor	Poor	Good
Oxidative stability	Good	Marginal	Good
Cloud point, °C	− 20 to + 20	− 5 to + 15	− 5
NO_x	− 10 to 0	+ 10	Baseline

12.3.5 Green Diesel

Triglycerides contain long, linear aliphatic hydrocarbon chains, which are partially unsaturated and have a carbon number range similar to the molecules found in petroleum diesel fuels. *Green diesel* production requires large volumes of H_2 and a catalyst to hydrogenate triglycerides into a high-cetane diesel fuel by removing all the oxygen from the triglyceride and saturating all the olefinic bonds in the fatty acids. The primary products from this hydrogenation are water, CO_2, propane, and a mixture of normal paraffin. Green diesel is fully compatible with petroleum-based diesel and is a drop-in-fuel and has a heating value equal to conventional diesel (see Table 12.4). Feedstocks rich in saturated fats, such as palm and tallow oil, require less hydrogen than feedstocks higher in olefin content, such as soybean or rapeseed oil. Diesel yield depends on both feedstock type and the level of hydro isomerization required to achieve product cloud point specification [5].

Biodiesel comes with several shortcomings that include NO_x and SO_x emissions, a cloud point with adverse effects, filter clogging, marginal stability, reduced energy content, and need to be blended with conventional petroleum diesel [4, 5]. Biodiesel has around 11% oxygen, whereas petroleum-based diesel and green diesel have no oxygen. Petroleum diesel has around 10 ppm sulfur and biodiesel and green diesel have less than 1 ppm sulfur.

12.3.6 Biogas and Renewable Natural Gas

Anaerobic digestion of wet biomass feedstock (< 15% solid), such as animal manure, agricultural residue, and sewage sludge from municipal wastewater treatment produce methane rich biogas using bacteria (methanogenic), archaea, and fungi at 37–55°C. The chemical reactions in biochemical processes undergo at lower temperatures as well as at lower conversion rates in nonpolluting natural processes requiring low energy and other chemicals. Biogas contains around 60% methane and 40% carbon dioxide and hence a lower heating value of 20–25 MJ/m^3 at normal conditions Biogas can be used for heating, steam and consequently electricity production. It can also be used as renewable natural gas after desulfurization and methane enrichment by removing 90% of carbon dioxide [5, 21].

12.3.7 Bio Synthesis Gas

A biomass, represented by C_nH_{2m}, may be oxidized to CO_2 and water and releases heat of combustion, which can be used to produce steam and electricity

$$C_nH_{2m}\,(n + 0.5m)O_2 \rightarrow mH_2O + nCO_2 \qquad (12.3)$$

In a conventional gasification process, biomass (or other carbon-containing feedstock) reacts with limited oxygen (or air), CO_2, at high temperatures (750 –

1100 °C) to produce synthesis (syngas) containing mainly H_2, CO, CO_2, methane besides tars, and a solid char [4, 17]. The following reaction represents the steam gasification (reforming) as seen in Fig. 12.2

$$C_nH_m + H_2O = nCO + (m/2 + n)H_2 \tag{12.4}$$

The water-gas shift reaction can increase the hydrogen content from 6 to 6.5 % in the initial synthesis gas to 30–50 vol% in the final stage

$$CO + H_2O = CO_2 + H_2 \tag{12.5}$$

Syngas is cooled and sent to a gas cleaning unit to remove the CO_2 and H_2S. Purification of the syngas accounts for 60–70% of the total capital cost. The gasification process requires proper utilization of heat integration to reduce the operating costs. Various simulation and modeling studies of biomass (corn stover and distiller grain) gasifier can predict the flowrate and composition of product from given biomass composition and gasifier operating conditions. *Fischer-Tropsch* (F-T) synthesis can produce various liquid biofuels from syngas. A representative Fischer-Tropsch (F-T) reaction is

$$(2n + 1)H_2 + nCO \rightarrow C_nH_{2n+2} + nH_2O - 170\,\text{kJ mole}^{-1}\left(\text{at } 250\,^\circ\text{C and } 15\,\text{atm}\right) \tag{12.6}$$

In the production of diesel fuel "*n*" can be in the range of 12–25; therefore, a H_2 to CO molar ratio of close to 2 is required. An iron-based catalyst and an operating temperature of 350°C will produce mostly gasoline, while a cobalt base and an operating temperature of 200°C will produce mostly diesel fuel. The crude biooil produced in the F-T synthesis may be distilled to naphtha, distillate, and wax [4].

Chemical-looping gasification and reforming produces syngas (CO and H_2) or a mixture of CO, H_2, CO_2, and CH_4 (gasification), or a relatively pure stream of H_2 with inherited carbon capture [4, 17].

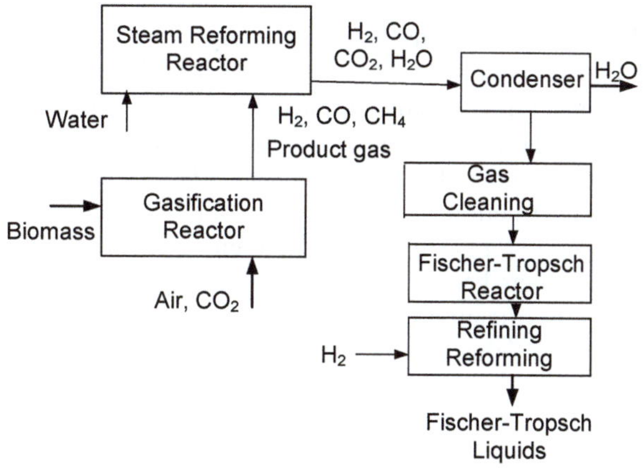

Fig. 12.2 Biomass steam gasification with production of liquid transportation fuels by Fischer-Tropsch synthesis [4, 5].

12.3.8 Biooil

Pyrolysis uses fast heating (700 – 800 °C) under anaerobic conditions to break down biomass into a volatile mixture of hydrocarbons and char. This mixture of hot gases is condensed into a biooil that is a rich mixture of hydrocarbons, some of which can be converted into biofuels.

Hydrothermal liquefaction of biomass also produces biooil (200–400°C, 5–40 MPa). This liquefaction system can work with high-moisture biomass feedstocks and municipal waste streams such as sewage sludge and wet algae at much lower temperatures compared with the gasification and pyrolysis processes. Hydrogen required for hydrogenation of carbon comes from water [4].

12.3.9 Butanol

Sugars can be converted into butyric, lactic, and acetic acids by fermentation process. Butyric acid is converted by fermentation into butanol. Corn starch also can be converted to biobutanol via the acetone-butanol-ethanol fermentation pathway [5]. Coproducts include alcohols with lower molecular weight than butanol and acetone. Butanol's toxicity to the microorganisms that ferment sugar creates obstacles. If corn grain is the source of sugar for fermentation, a residue similar to dried distillers' grain is produced. This might require additional processing to remove any toxic biobutanol and acetone residue before it could be used as an animal feed. Gas stripping can be used to extract the butanol.

Iso-butanol can be produced by anaerobic process using *E. coli* strains and continuous vacuum stripping for butanol fermentation. This process is still in the developing stage. The sugar to iso-butanol conversion yield is around 85%. As the yield increases butanol purification improves considerably. Both butanol costs, water usage, and direct CO_2 emissions are higher than that of cellulosic bioethanol [5], while butanol is far superior to ethanol in energy efficiency [5].

12.3.10 Methanol

Renewable methanol is produced by using hydrogen from wind power in the electrolytic system and carbon dioxide from bioethanol plants (Fig. 12.3). The energy efficiency for the concentrated CO_2 and H_2 based methanol is around 46% [15].

Sewage sludge is a residue of a municipal wastewater treatment plant. Mechanically dewatered sludge contains 12–25% solid (LHV $= 12.0\,\mathrm{MJ/kg}$) with an organic fraction of 56% of dry solids rich in carbon (50%), hydrogen (7%) and oxygen (31%). This may lead to energy recovery by producing biofuels as well as bioproducts beside the incineration commonly used. One way for energy recovery is the gasification of the sludge and conversion of syngas produced into methanol as well as other chemicals. This will help minimize waste as well as recover energy

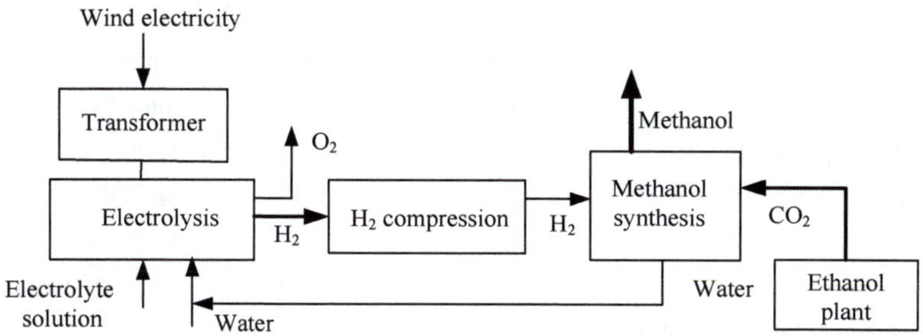

Fig. 12.3 The integral methanol production facility based on the feedstock of renewable hydrogen and CO_2 [15].

Table 12.5 Properties of bioethanol, biomethanol, biobutanol, bioisobutanol, and conventional gasoline [5, 21]

Properties	Methanol	Ethanol	Butanol	Isobutanol	Gasoline
C %	37.5	52.1	64.8	64.8	85–88
H %	12.6	13.1	13.5	13.5	12–15
O %	49.9	34.7	21.6	21.6	–
Density (20°C) kg/m	791	789	809	802	690–800
Normal boiling point °C	65.0	78.5	117.7	107.9	27–225
Motor octane number	91	92	84	90	80–88
Higher Heating value (20°C) MJ/kg	22.3	29.8	37.3	37.2	47.2
Stoichiometric air/fuel ratio kg/kg	6.4	8.9	11.2	11.2	14.7
Flash point °C	12	13	37	28	− 43
Autoignition temperature °C	470	363	340	415	250–300
Energy density MJ/L	16	21.4	26.9	26.6	30–33
CO_2 production MJ/kg fuel	15	13	15	15	14

[6, 20]. Table 12.5 Compares the properties of methanol, biobutanol, bioisobutanol, and conventional gasoline. As seen from the table, the energy densities of biobutanol, bioisobutanol close to conventional gasoline.

12.3.11 Dimethyl Ether

Dimethyl ether (DME) is the simplest ether (CH_3OCH_3) and can be produced by catalytic methanol dehydration

$$2CH_3OH \rightarrow CH_3OCH_3 + H_2O \tag{12.7}$$

Table 12.6 Environmental impacts for methanol or dimethyl ether [5, 21]

Indicator	Methanol	DME	Unit/mt product
Global warming potential	0.30	0.53	mton CO_2 eq
Acidification potential	0.67	0.97	kg SO_2 eq
Photochemical oxidant formation	0.69	1.17	kg NMVOC eq[a]
Particulate matter (PM) Formation	0.29	0.44	kg PM eq
Human toxicity	0.10	8.18	kg 1,4-DB eq[b]

[a] NMVOC: non-methane volatile organic compound
[b] 1,4-DM: 1,4 dichlorobenzene

DME is a colorless, non-toxic, highly flammable gas in ambient conditions, but can be handled as a liquid under slight pressure (0.5 MPa). The properties of DME are similar to those of liquefied petroleum gas with a lower heating value of 28.4 MJ/kg and a density of 0.67 kg/liter. It has a conventional diesel fuel equivalency of 0.59 as the lower heating value of diesel is 43.1 MJ/kg. DME is not greenhouse gas and can be used as a substitute for diesel fuel and domestic gas [16]. Table 12.6 compares methanol with dimethyl ether. Methanol and DME from CO_2 hydrogenation may outperform conventional petroleum-based fuels, reducing greenhouse gas emissions 82–86%, minimizing other pollutants (SO_x, NO_x) and reducing fossil fuel depletion by 82–91% [16].

12.4 Hydrogen Production

Hydrogen is a zero-emission fuel with combustion products of water and a trace amount of NO_x. Steam methane reforming is a current and economic process of producing hydrogen in large scales with around 86% energy efficiency. Hydrogen can be produced mainly from the second generation of biomass feedstock by using: thermochemical processes of gasification (steam reforming). Figure 12.4 shows a schematic of conventional hydrogen production by gasification of a biomass feedstock. Energy efficiency for biomass gasification for H_2 production is around 55–65%.

Figure 12.5 shows the schematic of wind energy-based renewable hydrogen production. The system includes a transformer, thyristor, electrolyzer unit, feed water

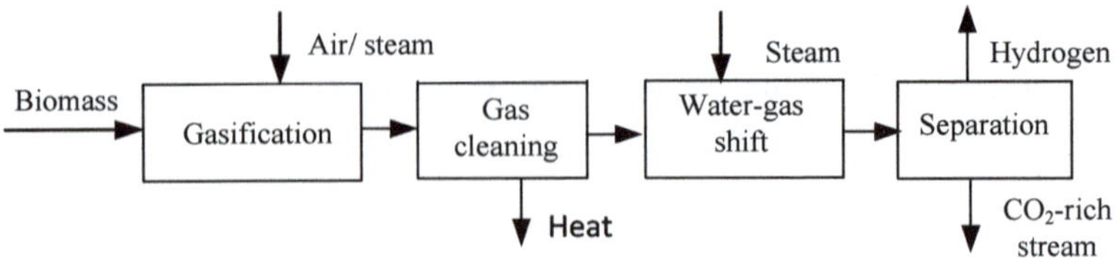

Fig. 12.4 Block flow diagram of biohydrogen production by gasification of biomass feedstock.

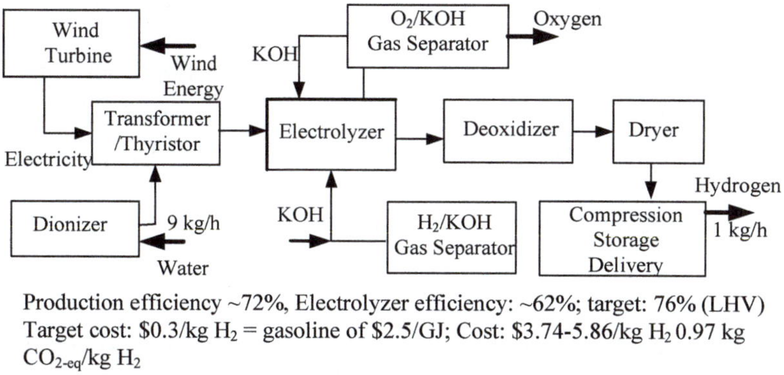

Fig. 12.5 Schematic for alkaline electrolysis of water for green hydrogen production with compression, storage, and delivery [15].

demineralizer, hydrogen scrubber, gas holder, two compressor units, deoxidizer, and twin tower dryer. For producing one kg H_2, approximately 26.7 kg water is necessary. The total greenhouse gas emission is around 0.97 kg CO_2e/kg H_2 [9, 15]. The hydrogen production cost largely depends on the electricity price. Therefore, electrolysis plants take advantage of low electricity prices at off-peak hours.

Declining of coal-fired and nuclear electricity generation capacity may lead to gain in electricity generation by natural gas and renewables [9]. Currently, 96% of H_2 is produced directly from steam reforming of natural gas, coal gasification, and partial oxidation of hydrocarbons such as biomass. If the carbon is captured in the process, then *blue hydrogen* is produced. Figure 12.6 shows some commercial processes for H_2 production from syngas carbon capture and storage.

Hydrogen can offer solutions supplied by electricity. Certain energy-intensive industries, such as iron making, and ammonia may relocate to areas with good renewable energy resources to produce cheap green hydrogen. Table 12.7 shows that by 2050, there would be 160 Mt (19 EJ) of green hydrogen produced annually

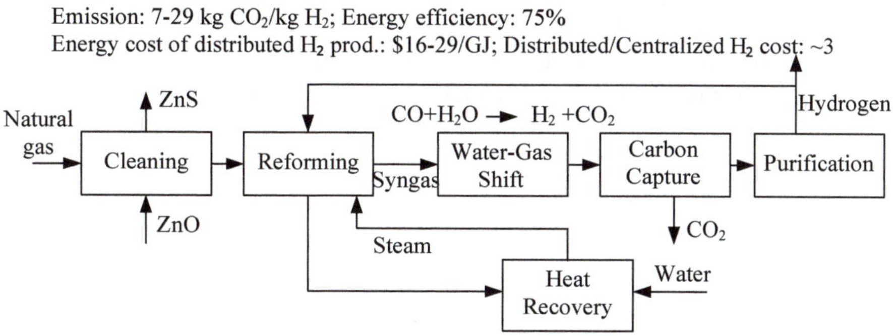

Fig. 12.6 Blue hydrogen production by steam reforming of natural gas with carbon capture [5, 15].

Table 12.7 Renewable energy share in power generation transforming energy scenario (TES) [11]

	TES (2030)	TES (2050)
Blue hydrogen production tonne	30	80
Green hydrogen production tonne	25	160
Electrolyzers GW	270	1700
Electricity demand to produce hydrogen from renewables, TWh*:	1200	7500

* $T = 10^{12}$

in the Transforming Energy Scenario (TES). Considerable scale-up of electrolyzers with increased efficiency would help produce between 50 GW and 60 GW per year of new capacity from now until 2050 [12].

12.5 Renewable Fuel Standards

The U.S. Congress created the Renewable Fuel Standard (RFS) program to reduce the GHG emissions and expand the Nation's renewable fuels sector while reducing reliance on imported oil. The RFS program was authorized under the Energy Policy Act of 2005 and expanded under the Energy Independence and Security Act of 2007. This new renewable fuel standard is known as RFS2.The Clean Air Act requires EPA to set the RFS volume requirements annually [18].

The RFS program includes four categories of renewable fuel, each with specific fuel pathway requirements and RIN D-Codes [18]:

- Advanced biofuels (D5) are produced from any type of renewable biomass except corn starch ethanol. Required lifecycle GHG emissions reduction is at least 50% compared to the petroleum baseline.
- Biomass-based Diesel (D4) includes biodiesel and renewable diesel and required lifecycle GHG emissions reduction is at least 50% compared to the diesel baseline.
- Cellulosic biofuel (D3 or D7) produced from cellulose or hemicellulose of biomass and required lifecycle GHG emissions reduction is at least 60% compared to the petroleum baseline.
- Conventional renewable biofuel (D6) includes ethanol derived from corn starch, or any other qualifying renewable fuel with lifecycle GHG emissions reduction is at least 20% .13compared to the average petroleum baseline.

12.6 Renewable Energy and Ecology

Energy is conserved in ecosystems. Carbon is linked to energy and water flow because of synthesis of C–H bonds in organic material through photosynthesis, which is the main source of energy in most ecosystems. Understanding how this link affects other elements and nutrients in native habitat will be important to protect ecosystems.

Extreme weather, flooding, rising sea levels and other effects of climate change in ecosystems will have adverse effects on the quality of life, health, economy, and ecology. Therefore, governments, investors, and stakeholders should avoid new investments into the new fossil fuel facilities but improve the existing ones with promising carbon capture and utilization technologies. The Task Force on Climate Related Financial Disclosure encourages voluntary action on climate-related financial risks disclosures [13].

Eco-friendly energy systems need policy makers who would establish long-term targets and adapt policies and regulations accordingly. Energy scenarios including traditional fuels, power sector, and industrial sector can facilitate open dialogue among stakeholders and other agencies to reach a consensus of target and challenges for the eco-friendly energy transition.

Climate change is a global threat to ecology and the transforming energy sector may need different approaches for different countries depending on the priorities. Construction of some large solar projects, wind farms, and hydroelectric dams may harm habitat including water quality, nutrient, and species flow representing critical natural balances in biodiversity in native ecosystems. Growing energy crops, such as corn, soybean, and oil palms may affect tropical rainforests adversely through excessive land and water usage leaving less resources for growing food and for native species [6, 8].

Example 12.6 **Avoided emissions in green energy production**

A superheated steam (stream 1) expands in a turbine from 5000 kPa and 325 °C to 150 kPa and 200 °C. The steam flow rate is 10.5 kg/s. If the turbine generates 1.1 MW of power, estimate the avoided emissions if the wind power instead of coal is used in the boiler to produce the steam.

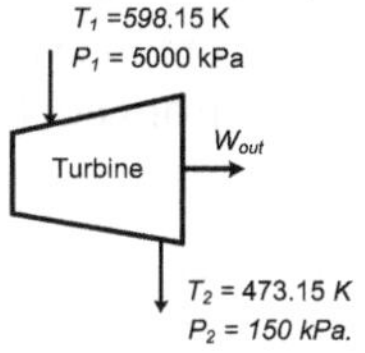

Solution:

Assume that the kinetic and potential energy effects are negligible; this is a steady process.

The properties of steam from the steam tables:

Stream 1: Superheated steam:

$$P_1 = 5000\,\text{kPa},\ T_1 = 325\,^{\circ}\text{C}\,(598.15\,\text{K}),\ H_1 = 3001.8\,\text{kJ/kg}$$

Stream 2: Superheated steam:

$$P_2 = 150\,\text{kPa},\ T_2 = 200\,^{\circ}\text{C}(473.15\,\text{K}),\ H_2 = 2872.9\,\text{kJ/kg}$$
$$\dot{W}_{\text{out}} = -1100\,\text{kW} = -1100\,\text{kJ/s}$$

The mass and energy balances for the turbine are:

$$\text{Mass balance}: \dot{m}_{\text{out}} = \dot{m}_{\text{in}}$$
$$\text{Energy balance}: \dot{E}_{\text{out}} = \dot{E}_{\text{in}} \quad \rightarrow \quad \dot{m}_1(H_2 - H_1) = \dot{q}_{\text{out}} + \dot{W}_{\text{out}}$$

Heat loss from the energy balance

$$\dot{q}_{\text{out}} = -\dot{W}_{\text{out}} + \dot{m}_1(H_2 - H_1)$$
$$= 1100\,\text{kJ/s} + (10.5\,\text{kg/s})(2872.9 - 3001.8)\,\text{kJ/kg} = -253.45\,\text{kW}$$

The sign is negative as the heat is lost from the system.

Table 11.2: (0.37 kg CO_2/kWh)

Annual avoided emission from 1.1 MW power production:

$$(0.37\,\text{kg}\,CO_2/\text{kWh})(1100\,\text{kW})(24\,\text{h/day})\,(365\,\text{day/year}) = 3{,}565{,}320\,\text{kg}\,CO_2/\text{year}.$$

Example 12.7 **Avoided emissions in compression of air**

Air enters a compressor at 100 kPa, 300 K, and a velocity of 2 m/s through a feed line with a cross-sectional area of 1.0 m^2. The effluent is at 500 kPa and 400 K. Heat is lost from the compressor at a rate of 5.2 kW. If the air behaves as an ideal gas, estimate the power requirement of the compressor in kW and the avoided emissions if the wind power instead of electricity from coal is used.

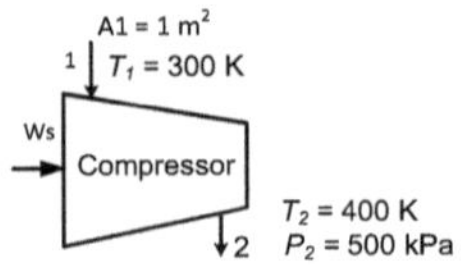

Solution:

Assume: The compressor is at steady state, air is ideal gas, and the change in the potential and kinetic energy of the fluid from the inlet to the outlet is negligible.

$$R = 8.314\,\text{kPa m}^3/\text{kmole K}, \quad MW = 29\,\text{kg/kmole}; \quad v_1 = 2\,\text{m/s}, \quad A_1 = 1.0\,\text{m}^2$$
$$P_1 = 100\,\text{kPa}, \; T_1 = 300\,\text{K}, \quad P_2 = 500\,\text{kPa}, \quad T_2 = 400\,\text{K}, \quad \dot{q}_{\text{loss}} = -5.2\,\text{kW}$$

For ideal gas: $V_1 = \frac{RT_1}{P_1} = 0.025\,\text{m}^3/\text{mole}$, the specific volume $(1/\rho)_{\text{air}} = \frac{V_1}{\text{MW}_{\text{air}}} = 0.86\,\text{m}^3/\text{kg}$

The mass flow rate of air: $\dot{m} = \frac{v_1 A_1}{(1/\rho)} = 1.72\,\text{kg/s}$

From Table D1 for ideal gas of air: $H_2 = 400.9\,\text{kJ/kg}, \; H_1 = 300.2\,\text{kJ/kg}$

Energy Balance: $\dot{m}(H_2 - H_1) - \dot{q} = \dot{W}$

$$(1.72\,\text{kg/s})((400.9 - 300.2)\,\text{kJ/kg}) + 5.2\,\text{kW} = \dot{W} = 178.4\,\text{kW}$$

This is the work needed by the compressor.

Table 11.2: (0.37 kg CO_2/kWh)

Annual avoided emission:

$$(0.37\,\text{kg CO}_2/\text{kWh})(178.4\,\text{kW})(24\,\text{h/day})(365\,\text{day/year}) = 578{,}230.0\,\text{kg CO}_2/\text{ year}.$$

Example 12.8 **Avoided emission in pumping**

A pump increases the pressure in liquid water from 100 kPa at 25°C to 4000 kPa. Estimate the minimum horsepower motor required and the avoided emissions if the wind power instead of electricity from coal is used in the pump for a flow rate of 0.5 m^3/s.

Solution:

Assume that the pump operates adiabatically and nearly isothermally, changes in potential and kinetic energy are negligible, and water is incompressible.

$$P_1 = 100\,\text{kPa}, \quad T_1 = 298.15\,\text{K}, \; P_2 = 4000\,\text{kPa}, \; Q = 0.01\,\text{m}^3/\text{s}$$

Energy Balance: $\dot{q} + \dot{W}_{\text{pump}} = \dot{m}\Delta H$

For an adiabatic operation ($\dot{q} = 0$): $\dot{W}_{\text{pump}} = \dot{m}\Delta H$

Since the water is incompressible the internal energy is only the function of temperature.

$\Delta U = 0$ as the system is isothermal.

$$\Delta H = \Delta U + \Delta(PV) = V\Delta P$$

The specific volume of saturated liquid water at 25°C: $V_1 = 0.001003\,\text{m}^3/\text{kg}$ (steam tables)

The mass flow rate: $\dot{m} = \frac{Q}{V_1} = 9.97\,\text{kg/s}$

$$\begin{aligned}\dot{W}_{\text{pump}} &= \dot{m}V_1(P_2 - P_1)\\ &= (9.97\,\text{kg/s})\big(0.001003\,\text{m}^3/\text{kg}\big)(4000 - 100)\,\text{kPa}\big(\text{kJ/kPa m}^3\big)\\ &= 39\,\text{kJ/s} = 39\,\text{kW} = 52.3\,\text{hP}(1\,\text{hp} = 745.7\,\text{W}).\end{aligned}$$

Table 11.2: for $(0.37\,\text{kg}\,CO_2/\text{kWh})$ annual avoided emission:

$$\begin{aligned}&(0.37\,\text{kg}\,CO_2/\text{kWh})(39\,\text{kW})(24\,\text{h/day})(365\,\text{day/year})\\ &\qquad = 126{,}406.8\,\text{kg}\,CO_2/\text{year}.\end{aligned}$$

12.7 Impact of Biofuels on Sustainability

Greater demand for biomass from the biofuel industry may increase the dead zones that have overload of nitrogen and phosphorous which can harm the aquatic flora and fauna, and marine life. To clean the dead zones, one needs to purify and oxygenate the existent waterways by removing the excess fertilizer run-offs and fully restore the flora and fauna of the natural habitat.

It is a future challenge to increase the biofuel production from the second and third generations, and waste biomass feedstock resources. Improving the technologies of multi production would be essential for the economic feasibility of the biofuel sector [3, 4]. Within the biorefinery and multi generation settings the unit cost of biofuels can be reduced because the current cost of biofuels is not competitive enough with those of fossil fuels. Distributed versus centralized biorefinery concepts should also be fully analyzed for an optimum process setting for sustainable biofuel production, regional economic development, and environmental protection [8].

The contribution of various biofuels to the transport sector keeps increasing. Some additional incentives in investments and electrification particularly in the transportation sector help increase the shares of renewables in power generation. This trend of increase is also projected in the 'Sustainable Development Scenario' (SDS) worldwide, complemented mainly by nuclear power and carbon capture technologies. By 2040, contributions from wind and solar PV will exceed hydropower if investments in renewables technologies accelerate [10–12].

Higher energy efficiency guarantees secure, sustainable, and inclusive economic growth and reduces the environmental footprint of the energy systems. Half of these energy efficiency savings come from industry, with major contributions also from transport and buildings. The rate for primary energy demand is slower than in the past (1% vs. 2.7%) as energy consumption and economic growth continue to decouple by 2040 [12]. The SDS depends upon fundamental changes to how energy is produced and consumed with rapid energy sector transformation. Because of its cost effectiveness and attractive payback, thermal efficiency improvements are critical elements that bring the world closer to the SDS. Increases in the electrification of economies and the share of renewables can cause fluctuations in the carbon footprint of electricity use. Greater emphasis on circular economy with higher materials efficiency may reverse the trend of growing emissions from the industrial sector, such as steel and cement sectors.

Example 12.9 **Avoided emissions due to biofuel in a car**

An average car consumes 50 gallons gasoline per month. Estimate the amount of energy and the avoided emissions if 85% ethanol blended gasoline is used instead of 100% gasoline consumed by the car per year.

Solution:

Assume that gasoline has an average density of $0.72\,\text{g/cm}^3$ and the heating value of $47.3\,\text{MJ/kg}$ (Table 2.6). Cost of gasoline: \$3/gallon

Data: $V = 50$ gal/month $=$ 189.25 l/month, 2271.0 l/year (3.785l $=$ 1 gal)

$$\rho_{\text{gas}} = 0.72\,\text{g/cm}^3 = 0.72\,\text{kg/l}$$

Mass of gasoline: $m_{\text{gas}} = \rho V = 1635.1$ kg/year

Energy consumed per year: $E_{\text{gas}} = 1635.1\,\text{kg/year}\,(47{,}300\,\text{kJ/kg})$
$= 77{,}340{,}230\,\text{kJ/year} = 77{,}340.2\,\text{MJ/year}.$

Cost per year: C (gasoline): (\$3/gallon) (50 gallon)(12) $=$ \$1800/ year .

Example 12.10 **Energy consumed by low and a high-mileage cars**

Assume that the average daily traveling distance is about 40 miles/day. A car has a city-mileage of 20 miles/gal. If the car is replaced with a new car with a city-mileage of 30 miles/gal and the average cost of gasoline is \$3.50/gal, estimate the amount of fuels.

Solution:

Assume: The gasoline is incompressible with $\rho_{av} = 0.75\,\text{kg/l}$.

Lower heating value (LHV) $= 44{,}000\,\text{kJ/kg}$; 44,000 kJ of heat is released when 1 kg of gasoline is completely burned, and the produced water is in vapor state.

Fuel needed for the old car: (40 miles/day)/(20 miles/gal) = 2 gal/day

Fuel needed for the new car: (40 miles/day)/(30 miles/gal) = 1.34 gal/day

Old car: 20 miles/gal

Mass of gasoline: $m_{gas} = \rho_{av}$ (Volume) =
(0.75 kg/1)(2.0 gal/day)(3.785 1/gal) = 5.7 kg/day

$$\text{Energy of gasoline:} \dot{E}_{\text{gas}}(\text{LHV}) = (5.7\,\text{kg/day})(44{,}000\,\text{kJ/kg})$$
$$= 250{,}800\,\text{kJ/day}(365\,\text{day/year})$$
$$= 91{,}542{,}000\,\text{kJ/year} = 91{,}542\,\text{MJ/year}.$$

Cost: (\$3.50/gal)(2 gal/day)(365 day/year) = \$2555/ year

New car: 30 miles/gal

Mass of gasoline:
$m_{\text{gas}} = \rho_{\text{av}\cdot}$ (Volume) = (0.75 kg/1)(1.34 gal/day)(3.7851/gal) = 3.8 kg/day

$$\text{Energy of gasoline: } \dot{E}_{\text{gas}}(\text{LHV}) = (3.8\,\text{kg/day})(44{,}000\,\text{kJ/kg})$$
$$= 167{,}200\,\text{kJ/day}\,(365\,\text{day/year})$$
$$= 61{,}028{,}000\,\text{kJ/year} = 61{,}028\,\text{MJ/year}.$$

Cost: (\$3.50/gal)(1.34 gal/day)(365 day/year) = \$1712/ year .

The new car reduces fuel consumption by around 33%, which is significant.

Example 12.11 Avoided emissions by the biogas use

A city consumes natural gas at a rate of 500 × 10^6 ft^3/day. The volumetric flow is at standard conditions of 60°F and 1 atm = 14.7 psia. If natural gas is costing \$6/GJ of higher heating value estimate the yearly energy required and the emission avoided if biogas instead of natural gas used in the city.

Solution:

Q = 500 × 10^6 ft^3/day at 60°F and 1 atm = 14.7 psia.

The higher heating value is the heat of combustion of natural gas when the water product is at liquid state. From Table 2.6, the value of HHV is: 50,000 kJ/kg ~ 1030 Btu/ft^3 (Table 2.7)

Heating value: $\left(1030\,\text{Btu/ft}^3\right)\left(500 \times 10^6\,\text{ft}^3/\text{day}\right) = 515.0 \times 10^9\,\text{Btu/day}$

$$\left(515.0 \times 10^9\text{Btu/day}\right)(1055\,\text{J/Btu}) = 543{,}325\,\text{GJ/day} = 6.28\,\text{GW} = 6{,}288{,}483\,\text{kW}$$

Yearly cost: (543,325 GJ/day) (\$6/GJ) (365 days/year) = \$11.9 × 10^8/year

Table 11.2: for ($0.23\,kg\,CO_2/kWh$), Annual avoided emission:

$$(0.23\,\text{kg}\,CO_2/\text{kWh})(6{,}280{,}000\,\text{kW})(24\,\text{h/day})(365\,\text{day/year}) \\ = 1.267 \times 10^{10}\,\text{kg}\,CO_2/\text{year}.$$

Example 12.12 **Avoided emission by a car**

An average car consumes about 2 gallons (US gallon = 3.785 liter) a day, and the capacity of the fuel tank is about 15 gallons. Therefore, a car needs to be refueled once every week. The density of gasoline ranges from 0.72 to 0.78 kg/l (Table 2.6). The lower heating value of gasoline is about 44,000 kJ/kg. Assume that the average density of gasoline is 0.75 kg/l. If the car was able to use 85% ethanol instead of gasoline estimate the avoided emissions.

Solution: $\dot{q}_{\text{process}} = \dot{m}_7(H_3 - H_7)$

Assume: The gasoline is incompressible with $\rho_{av} = 0.75\,\text{kg/l}$.

Cost of gasoline $3.5/gallon

Lower heating value (LHV) = 44,000 kJ/kg; 44,000 kJ of heat is released when 1 kg of gasoline is completely burned, and the produced water is in vapor state.

Mass of gasoline per day:

$m_{\text{gas}} = \rho_{av}\,V = (0.75\,\text{kg/l})(2\,\text{gal/day})(3.785\,\text{l/gal}) = 5.67\,\text{kg/day}$

Energy of gasoline per day:

$\dot{E}_{\text{gas}} = (5.67\,\text{kg/day})(44{,}000\,\text{kJ/kg}) = 249{,}480\,\text{kJ/day} = 2.88\,\text{kW}$

Energy of E85 per day:

$\dot{E}_{\text{gas}} = \dot{m}_{\text{gas}}(\text{LHV}) = (5.67\,\text{kg/day})(31{,}816\,\text{kJ/kg}) = 180{,}396\,\text{kJ/day} = 2.08\,\text{kW}$

Avoided energy: $(2.88 - 2.08)\,\text{kW} = 0.8\,\text{kW}$

Avoided emission Table 11.2: ($0.27\,kg\,CO_2/kWh$)

Annual avoided emission:

$$(0.27\,\text{kg}\,CO_2/\text{kWh})(0.8\,\text{kW})(24\,\text{h/day})(365\,\text{day/year}) \\ = 1892.1\,\text{kg}\,CO_2/\text{year}.$$

Example 12.13 **Thermal efficiency impact on emissions of an actual Brayton cycle**

A power plant is operating on an ideal Brayton cycle with a pressure ratio of $r_p = 9$. The fresh air temperature is 295 K at the compressor inlet and 1300 K at the end of the compressor and at the inlet of the turbine. Assume the gas-turbine cycle operates with a compressor efficiency of 85% and a turbine efficiency of 85%. Determine the thermal efficiency of the cycle and impact of efficiency on emissions.

Solution:

Assume that the cycle is at steady-state flow and the changes in kinetic and potential energy are negligible. Heat capacity of air is temperature dependent, and the air is an ideal gas. The standard-air assumptions are applicable

Basis: 1 kg air.

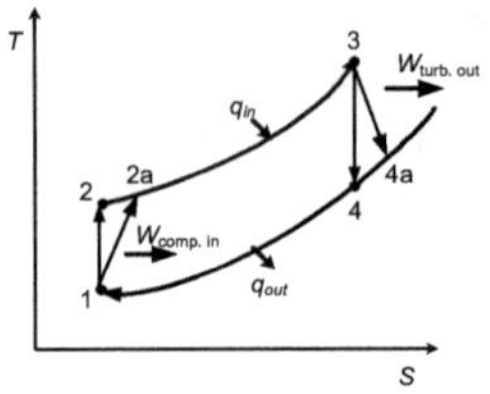

At $T_1 = 295\,\text{K}$, $H_1 = 295.17\,\text{kJ/kg}$; $P_{r1} = 1.3068$; (Table D1)

P_r is the relative pressure (Eq. 7.36).

$$\frac{P_{r2}}{P_{r1}} = \frac{P_2}{P_1} = P_r \rightarrow P_{r2} = (9)(1.3068) = 11.76$$

Approximate values from Table D1 for compressor exit at

$P_{r2} = 11.76$: $T_2 = 550\,\text{K}$, $H_2 = 555.74$ kJ/kg

Process 3–4 isentropic expansion in the turbine as seen on the *TS* diagram above

$T_3 = 1300\,\text{K}$, $H_3 = 1395.97\,\text{kJ/kg}$; $P_{r3} = 330.9$ (From Table D1)

$$\frac{P_{r4}}{P_{r3}} = \frac{P_4}{P_3} \rightarrow P_{r4} = \left(\frac{1}{9}\right)(330.9) = 36.76$$

Approximate values from Table D1 at the exit of turbine at 36.76:

$T_4 = 745$ K and $H_4 = 761.87$ kJ/kg

The work input to the compressor for a mass flow rate of 1 kg/s:

$$\dot{W}_{\text{comp.in}} = \dot{m}(H_2 - H_1) = (555.74 - 295.17)\text{kW} = 260.6\,\text{kW}$$

Work output of the turbine:

$\dot{W}_{\text{turb.out}} = \dot{m}(H_4 - H_3) = (761.87 - 1395.97)\,\text{kW} = -634.10\,\text{kW}$

From the efficiency definitions for compressor and turbine, we have:

$\eta_C = \frac{\dot{W}_{Cs}}{\dot{W}_{Ca}} \rightarrow \dot{W}_{Ca} = \frac{\dot{W}_{Cs}}{\eta_C} = \frac{260.6\,\text{kJ/kg}}{0.85} = 306.6\,\text{kW}$ (Actual compression work)

Work output turbine with 85% efficiency:

$$\eta_T = \frac{\dot{W}_{Ta}}{\dot{W}_{Ts}} \rightarrow W_{Ta} = \eta_T(\dot{W}_{Ts}) = 0.85(-634.1\,\text{kW}) = -539.0\,\text{kW}$$

The net work out: $\dot{W}_{net} = \dot{W}_{out} - \dot{W}_{in} = (-53.90 + 306.6) = -232.4$ kW.

The back-work ratio r_{bw} becomes: $r_{bw} = \left|\frac{W_{comp.in}}{W_{turb.out}}\right| = \left|\frac{306.6}{539.0}\right| = 0.568.$

This shows that the compressor now is consuming 56.8% of the turbine output. The value of back work ratio increased from 41 to 56.8% because of friction, heat losses and other nonideal conditions in the cycle.

Enthalpy at the exit of compressor:

$$W_{Ca} = H_{2a} - H_1 \rightarrow H_{2a} = W_{Ca} + H_1 = (306.6 + 295.2)\text{kJ/kg} = 601.8\,\text{kJ/kg}$$

Heat added: $q_{in} = H_3 - H_{2a} = 1395.97 - 601.8 = 794.2\,\text{kJ/kg}$

Thermal efficiency: $\eta_{th} = \frac{W_{net}}{q_{in}} = \frac{|232.4\ \text{kJ/kg}|}{794.2\ \text{kJ/kg}} = 0.292$ (or 29.2%).

The actual Brayton-gas cycle thermal efficiency drops to 0.292 from 0.444. The efficiency of the compressor and turbine affects the performance of the cycle. Therefore, for a better cycle thermal efficiency, significant improvements are necessary for the compressor and turbine operations.

Example 12.14 **Avoided emission in a cogeneration plant**

A cogeneration plant uses steam at 8200 kPa and 773.15 K (see Fig. 6.13). One-fourth of the steam is extracted at 700 kPa from the turbine for cogeneration. After it is used for process heat, the extracted steam is condensed and mixed with the water output of the condenser. The rest of the steam expands from 8200 kPa to the condenser pressure of 10 kPa. The steam flow rate produced in the boiler is 60 kg/s. Determine the work output and process heat produced and estimate the avoided emissions if this cogeneration plant is replaced by a concentrated solar system producing power and heat.

Solution:

Assume that the kinetic and potential energy changes are negligible, and this is a steady state process.

From steam tables

$\dot{m} = 60$ kg/s, $z = 0.25$; basis: mass flow rate = 1 kg/s

$$P_1 = P_8 = 10\,\text{kPa},\ H_{1\text{sat vap}} = 2584.8\,\text{kJ/kg},\ H_{1\text{sat liq}} = 191.83\,\text{kJ/kg},$$

$$V_1 = 0.00101\,\text{m}^3/\text{kg},\ S_{1\text{sat vap}} = 8.1511\,\text{kJ/kg K},\ S_{1\text{sat liq}} = 0.6493\,\text{kJ/kg K},$$
$$P_3 = P_7 = P_2 = P_4 = 700\,\text{kPa},\ H_3 = 697.06\,\text{kJ/kg},\ S_3 = 1.9918\,\text{kJ/kg K},\ z = 0.25$$
$$P_6 = 8200\,\text{kPa},\ T_6 = 773.15\,\text{K},\ H_6 = 3396.4\,\text{kJ/kg},\ S_6 = 6.7124\,\text{kJ/kg K}$$

In this cogeneration cycle, the steam extracted from the turbine is used as process heat. The liquid condensate from the process heat is combined with the output of the condenser.

$$W_{\text{p1}} = V_1(P_2 - P_1) = 0.00101(700 - 10)\left(\frac{1\,\text{kJ}}{1\,\text{kPa m}^3}\right) = 0.697\,\text{kJ/kg}$$
$$H_2 = H_1 + W_{\text{p1}} = 191.83 + 0.697 = 192.53\,\text{kJ/kg}$$

From the energy balance around the mixer, we have $\dot{m}_3/\dot{m}_6 = 0.25$

$$\dot{m}_6 = \dot{m}_4 = 60\,\text{kg/s},\ \dot{m}_3 = \dot{m}_7 = 15\,\text{kg/s},\ \dot{m}_8 = \dot{m}_1 = 0.75(60) = 45.0\,\text{kg/s}$$
$$\dot{m}_4 H_4 = \dot{m}_2 H_2 + \dot{m}_3 H_3 \rightarrow H_4 = [45(192.53) + 15(697.06)]/60 = 318.66\,\text{kJ/kg}$$
$$T_4 = 349.15\,\text{K},\ V_4 = 0.001027\,\text{kg/m}^3$$
$$W_{p2} = V_4(P_5 - P_4) = 0.001027(8200 - 700)\left(\frac{1\,\text{kJ}}{1\,\text{kPa m}^3}\right) = 7.70\,\text{kJ/kg}$$
$$H_5 = H_4 + W_{\text{p2}} = 326.36\,\text{kJ/kg}$$

Isentropic processes: $S_6 = S_7 = S_8 = 6.7124$ and $P_7 = 700$ kPa, $H_7 = 2765.68$ kJ/kg.
We estimate the quality of the discharged wet steam at state 8: $x_8 = \frac{6.7124-0.6493}{8.1511-0.6493} = 0.808$

$$H_8 = 191.83(1 - 0.808) + 2584.80(0.808) = 2125.87\,\text{kJ/kg}$$

The energy balance yields the fraction of steam extracted

$$\dot{W}_{\text{total}} = \dot{m}_6(H_7 - H_6) + \dot{m}_8(H_8 - H_7) = -66634.44\,\text{kW}$$
$$\sum \dot{W}_{\text{pi}} = \dot{m}_1 W_{p1} + \dot{m}_4 W_{p2} = 493.51\,\text{kW}$$

The net work output: $\dot{W}_{\text{net}} = \dot{W}_{\text{total}} - \sum \dot{W}_{\text{pi}} = -66140.93$ kW (work output of the turbine)

$\dot{q}_{\text{process}} = \dot{m}_7(H_3 - H_7) = -31029.3$ kW (heat discharged from the steam)

Table 11.2: for (0.37 kg CO_2/kWh) for coal-based power, annual avoided emission:

$$(0.37\,\text{kg}\,CO_2/\text{kWh})(66140.93\,\text{kW})(24\,\text{h/day})(365\,\text{day/year}) = 214{,}375{,}982\,\text{kg}\,CO_2/\text{year}.$$

Problems

12.1 The primary energy source for cells is the aerobic oxidation of sugar called glucose ($C_6H_{12}O_6$). Estimate the energy released from the oxidation 500 g of glucose:

$$C_6H_{12}O_6(s) + 6O_2(g) \rightarrow 6CO_2(g) + 6H_2O(g)$$

12.2 Synthesis gas is a mixture of H_2 and CO. One of the uses of synthesis gas is to produce methanol from the following reaction: $CO(g) + 2H_2(g) \rightarrow CH_3OH(g)$. Estimate the energy released after burning 100 gallons of methanol.

12.3 Estimate the energy released by the combustion of 500 ft^3 propane:

$$C_3H_8(g) + 5O_2(g) \rightarrow 3CO_2(g) + 4H_2O(g)$$

12.4 Estimate the energy released by the combustion of 1000 ft^3 ethane:

$$C_2H_6 + \frac{7}{2}O_2 \rightarrow 2CO_2(g) + 3H_2O(g)$$

12.5 Estimate the energy released from the combustion of 2000 ft^3 methane:

$$CH_4 + 2O_2 \rightarrow CO_2(g) + 2H_2O(g)$$

12.6 What is the energy released by the combustion at 25°C of 500 gallons of CH_3OH when the reaction products are $CO_2(g)$ and $H_2O(g)$? The reaction is: $6\,CH_3OH(g) + 9\,O_2 \rightarrow 6\,CO_2(g) + 12\,H_2O(g)$.

12.7 Estimate the energy released by the combustion of n-pentane gas at 25°C when the combustion products are $CO_2(g)$ and $H_2O(l)$: $C_5H_{12}(g) + 8O_2(g) \rightarrow 5CO_2(g) + 6H_2O(l)$.

12.8 A heating oil with an average chemical composition of $C_{10}H_{18}$ is burned with oxygen completely in a bomb calorimeter. The heat released is measured as 43,900 J/g at 25°C. The combustion on the calorimeter takes place at constant volume and produces liquid water. Estimate the energy released by the oil at 25°C when the products are (g) H_2O and $CO_2(g)$.

12.9 A superheated steam (stream 1) expands in a turbine from 5500 kPa and 325°C to 100 kPa and 100°C. The steam flow rate is 12 kg/s. If the turbine generates 1.1 MW of power, estimate the total amount of avoided emissions if the wind power instead of coal is used in the boiler to produce the steam.

12.10 An air compressor compresses air from 100 to 600 kPa. The airflow rate is 900 kg/h, and specific volumes of air at the inlet and outlet are 0.95 and 0.19 m^3/kg, respectively. The jacket cooling water removes 100 kW of heat generated due to the compression. Estimate the power input to the compressor and the total amount of avoided emissions if the wind power instead of electricity from coal is used

12.11 An air compressor compresses air from 100 to 1000 kPa. The airflow rate is 400 kg/h, and specific volumes of air at the inlet and outlet are 0.95 and 0.19 m^3/kg, respectively. The jacket cooling water removes 100 kW of heat generated due to the compression. Estimate the power input to the compressor and the avoided emissions if the wind power instead of electricity from coal is used

12.12 A centrifugal pump delivers 1.8 m^3/h water to an overhead tank 20 m above the eye of the pump impeller. Estimate the work required by the pump and the avoided emissions if the wind power instead of electricity from natural gas is used.

12.13 A pump increases the pressure in liquid water from 100 kPa at 25°C to 2000 kPa. Estimate the minimum horsepower motor required and the total amount of avoided emissions if the wind power instead of electricity from coal is used in the pump for a flow rate of 0.4 m^3/s.

12.14 A centrifugal pump delivers 1.8 m^3/h water to an overhead tank 20 m above the eye of the pump impeller. Estimate the work required by the pump and the total amount of avoided emissions if the wind power instead of electricity from natural gas is used.

12.15 An average car consumes about 2 gallons (US gallon = 3.785 l) a day, and the capacity of the fuel tank is about 15 gallons. Therefore, a car needs to be refueled once every week. The density of gasoline ranges from 0.72 to 0.78 kg/l (Table 2.6). The lower heating value of gasoline is about 44,000 kJ/kg. Assume that the average density of gasoline is 0.75 kg/l. If the car was able to use 85% ethanol instead of gasoline estimate the total amount of avoided emissions.

12.16 A city consumes natural gas at a rate of 500×10^6 ft^3/day. The volumetric flow is at standard conditions of 60°F and 1 atm = 14.7 psia. If natural gas is costing $6/GJ of higher heating value estimate the yearly energy required and the total amount of emission avoided if biogas instead of natural gas used in the city.

12.17 An average car consumes 50 gallons gasoline per month. Estimate the amount of energy and the total amount of avoided emissions if 85% ethanol blended gasoline is used instead of 100% gasoline consumed by the car per year.

12.18 For a car consuming 2000 gallons gasoline per year, estimate the amount of energy and the avoided emissions if 85% ethanol blended gasoline is used instead of 100% gasoline consumed by the car.

12.19 The overall efficiencies are about 25–28% for gasoline car engines, 34–38% for diesel engines, and 40–60% for large power plants [3]. A car engine with a power output of 120 hp has a thermal efficiency of 24%. Determine the fuel consumption of the car and emissions avoided if the gasoline is mixed with 85% ethanol and if the gasoline has a higher heating value of 20400 Btu/lb.

12.20 A steam power plant is using a geothermal energy source. The geothermal source is available at 220°C and 2320 kPa with a flow rate of 150 kg/s. The hot water goes through a valve and a flash drum. Steam from the flash drum enters the turbine at 550 kPa and 428.62 K. The discharged steam from the turbine has a quality of $x_4 = 0.90$. The condenser operates at 10 kPa. The water is a saturated liquid after passing through the condenser. Determine the work output of turbine and the total amount of avoided emissions compared with coal-fired power productions.

12.21 A steam power plant is using a geothermal energy source. The geothermal source is available at 180°C and 2000 kPa with a flow rate of 100 kg/s. The hot water goes through a valve and a flash drum. Steam from the flash drum enters the turbine at 550 kPa and 428.62 K. The discharged steam from the turbine has a quality of $x_4 = 0.96$. The condenser operates at 10 kPa. The water is a saturated liquid after passing through the condenser. Determine the work output of turbine and the total amount of avoided emissions compared with coal-fired and natural gas-fired power productions.

12.22 A cogeneration plant uses steam at 900 psia and 1000°F to produce power and process heat. The steam flow rate from the boiler is 16 lb/s. The process requires steam at 70 psia at a rate of 3.2 lb/s supplied by the expanding steam in the turbine with a value of $z = 0.2$. The extracted steam is condensed and mixed with the water output of the condenser. The remaining steam expands from 70 psia to the condenser pressure of 3.2 psia. Determine the process heat and power generated and estimate the total amount of avoided emissions if this plant is replaced with a concentrated solar power system.

12.23 A cogeneration plant uses steam at 900 psia and 1000°F to produce power and process heat. The steam flow rate from the boiler is 25 lb/s. The process requires steam at 70 psia at a rate of 10.5 lb/s supplied by the expanding steam in the turbine with a value of $z = 0.2$. The extracted steam is condensed and mixed with the water output of the condenser. The remaining steam expands from 70 psia to the condenser pressure of 9.5 psia. Determine the process heat and power generated and estimate the total amount of avoided emissions if this plant is replaced with a concentrated solar power system.

12.24 A hydroelectric plant operates by water falling from a 150 ft height. The turbine in the plant converts potential energy into electrical energy, which is lost by about 8% through the power transmission so the available power is 95%. If the mass flow rate of the water is 600 lb/s, estimate the power output of the hydro plant and the total amount of avoided emissions compared with coal-fired power plant.

12.25 A hydroelectric plant operates by water falling from a 90 ft height. The turbine in the plant converts potential energy into electrical energy, which is lost by about 8% through the power transmission so the available power is

90%. If the mass flow rate of the water is 350 lb/s, estimate the power output of the hydro plant and the total amount of avoided emissions compared with coal-fired power plant.

12.26 In a hydropower plant, a hydro turbine operates with a head of 46 m of water. Inlet and outlet conduits are 1.80 m in diameter. If the outlet velocity of the water is 5.5 m/s estimate the power produced by the turbine and the total amount of avoided emissions compared with natural gas-fired power plant.

12.27 A farm of windmills supplies a power output of 0.8 MW for a community. Each windmill has the blades with 10 m in diameter. At the location of the windmills, the average velocity of the wind is 14 m/s and the average temperature is 20°C. Estimate the minimum number of windmills to be installed and the total amount of avoided emissions compared with natural gas-fired power plants.

12.28 A wind turbine with 10 m rotor diameter is to be installed on a hilltop where the wind blows steadily at an average speed of 15 m/s. Determine the maximum power that can be generated by the turbine operated with a 35% efficiency and the total amount of avoided emissions compared with natural gas-fired power plant.

12.29 A farm of windmills supplies a power output of 3 MW for a community. Each windmill has blades 11 m in diameter. At the location of the windmills, the average velocity of the wind is 14 m/s and the average temperature is 20°C. Estimate the minimum number of windmills to be installed and the total amount of avoided emissions compared with natural gas-fired power plants.

12.30 A farm of windmills supplies a power output of 4.2 MW for a community. Each windmill has blades 10 m in diameter. At the location of the windmills, the average velocity of the wind is 15 m/s and the average temperature is 20°C. Estimate the minimum number of windmills to be installed and the total amount of avoided emissions compared with natural coal-fired power plants.

12.31 A refrigerator using tetrafluoroethane (R-134a) as a refrigerant operates with a capacity of 250 Btu/s. Cooling water enters the condenser at 70°F. The evaporator is at 10°F, and the condenser is at 80°F. The refrigerated space is at 20°F. Determine the ideal and actual power necessary and compare the avoided emissions if the wind power used instead of power from coal-fired plant. Assume that the compressor efficiency is 80% and the kinetic and potential energy changes are negligible.

12.32 A refrigerator using tetrafluoroethane (R-134a) as a refrigerant operates with a capacity of 2500 kW. Cooling water enters the condenser at 280 K. Evaporator is at 271.92 K, and the condenser is at 299.87 K. The refrigerated space is at 280 K. Determine the ideal and actual power necessary and compare the avoided emissions if the wind power used instead of power from coal-fired plant. Assume that the compressor efficiency is 80% and the kinetic and potential energy changes are negligible.

References

1. Allen JW, Unlu S, Demirel Y, Black P, Riekhof W (2018) Integration of biology, ecology and engineering for sustainable algal based biofuel and bioproduct biorefinery. Bioresour Bioprocess 5(47):1–28
2. Betz A (1966) Introduction to the theory of flow machines (Randall DG, Trans.). Pergamon Press, Oxford
3. Cherubini F (2010) The biorefinery concept: Using biomass instead of oil for producing energy and chemicals. Energy Convers Manage 51:1412–1421
4. Demirel Y, Matzen M, Winters C, Gao X (2015) Capturing and using CO_2 as feedstock with chemical-looping and hydrothermal technologies. Int J Energy Res 39:1011–1047
5. Demirel Y (2025) Biofuels. In: Dincer I (ed) Comprehensive energy systems, 2nd edn. Elsevier, Amsterdam, vol 2, pp 344–384
6. Demirel Y (2025) Energy conservation. In: Dincer I (ed) Comprehensive energy systems, 2nd edn. Elsevier, Amsterdam, vol 9, pp 1–51
7. Demirel Y (2018) Sugar versus lipid for sustainable biofuels. Int J Energy Res 42:881–884. https://doi.org/10.1002/er.3914
8. Demirel Y (2017) Lignin for sustainable bioproducts and biofuels. J Biochem Eng Bioprocess Technol 1:1–3
9. Dincer I, Ratlamwala T (2013) Development of novel renewable energy-based hydrogen production systems: a comparative study. Energy Convers Manage 72(2013):77–87
10. IEA (2023) International energy agency. World Energy Outlook. https://www.iea.org/reports/world-energy-outlook-2023
11. IRENA (2025) International Renewable Energy Agency. Record-breaking annual growth in renewable power capacity. https://www.irena.org/News/pressreleases/2025/Mar/Record-Breaking-Annual-Growth-in-Renewable-Power-Capacity
12. IRENA (2023) International Renewable Energy Agency. Renewable power generation costs in 2023. https://www.irena.org/Publications/2024/Sep/Renewable-Power-Generation-Costs-in-2023
13. IRENA (2024) International Renewable Energy Agency. World energy transition outlook 2024. https://www.irena.org/Publications/2024/Nov/World-Energy-Transitions-Outlook-2024
14. Jaffe RL, Taylor W (2018) The physics of energy. Cambridge University Press, Cambridge
15. Küçük K, Tevatia R, Sorgüven E, Demirel Y, Özilgen M (2015) Bioenergetics of growth and lipid production in *Chlamydomonas reinhardtii*. Energy 83:503–510
16. Matzen M, Demirel Y (2016) Methanol and dimethyl ether from renewable hydrogen and carbon dioxide: alternative fuels production and life-cycle assessment. J Cleaner Production 139:1068–1077
17. Moghtaderi B (2012) Review of the recent chemical looping process developments for novel energy and fuel applications. Energy Fuels 26(2012):15–40
18. National Research Council (2011) Renewable fuel standard: potential economic and environmental effects of U.S. biofuel policy. The National Academies Press, Washington, DC
19. Nguyen N, Demirel Y (2011) A novel biodiesel and glycerol carbonate production plant. Int J Chem Reactor Eng 9:1–25
20. Olah GA, Goeppert A, Surya GK (2009) Chemical recycling of carbon dioxide to methanol and dimethyl ether: from greenhouse gas to renewable, environmentally carbon neutral fuels and synthetic hydrocarbons. J Org Chem 74:487–498
21. Ptasinski KJ (2016) Efficiency of biomass energy. In: An exergy approach to biofuels, power, and biorefineries. Wiley, New York

Energy Management and Economics 13

13.1 Introduction

Energy management refers to energy production, distribution and storage toward controlling and optimizing energy uses and conservation. Energy management programs provide organizations with information, tools, and assistance to reduce energy waste, cost, greenhouse gas emissions, and increase energy efficiency. The program provides the organizations with annual performance data, energy savings performance contracts, utility energy service contracts, and support with energy efficient products. There are several basic steps that are commitment, data analysis, analysis and implementation of energy conservation options and continued feedback and analysis. In a comprehensive feasibility analysis, techno economic analysis is being complemented with sustainability metrics to account current socioeconomics, socio-ecology, and eco-economics changes in large energy production projects.

Learning objectives of this chapter are:

- Techno economic analysis of various fuels
- Economics of energy production and conversion
- Bioeconomy
- Circular economy.

13.2 Energy Usage and Management

Energy is the main driver of technological development for the prosperity of modern society by being able to produce all goods and services. A 50% increase in global energy use and despite the increased use of renewable energy, fossil fuels will continue to supply global energy through 2050. Therefore, GHG emissions

Y. Demirel, *Energy*, Green Energy and Technology,
https://doi.org/10.1007/978-3-031-97111-2_13

Table 13.1 Global energy consumption by end-use sector [16, 19]

End-use sector	2050	
	Quads	Share %
Residential	90.6	15
Commercial	48.7	8
Industrial	316.7	51
Transportation	167.2	27
Total delivered energy	623.2	100

will continue to rise [16]. Table 13.1 shows the forecast of global energy consumption by end-use sector for 2050 in comparison to the current results of 2018. The growth of energy use in the industrial sector is driven by increasing product demand.

By 2050 the demand for energy will nearly double with water and food increasing by over 50% globally. The interlinkage (Fig. 13.1) between the food-energy and water supply systems is a major consideration in sustainable development strategies. The complex interconnections show that:

(i) water supply is influenced by demands from energy and food sectors,
(ii) food production depends both on water and energy, and
(iii) energy production and operations require water.

Understanding the connections fully between food-energy and water can change day-to-day practices of resource managers and policymakers [11].

Figure 13.2 shows that industrial and transportation sectors will keep being the largest energy consumers in an increasing trend. Energy use in residential and commercial sectors will grow gradually. Liquid transportation fuels, natural gas,

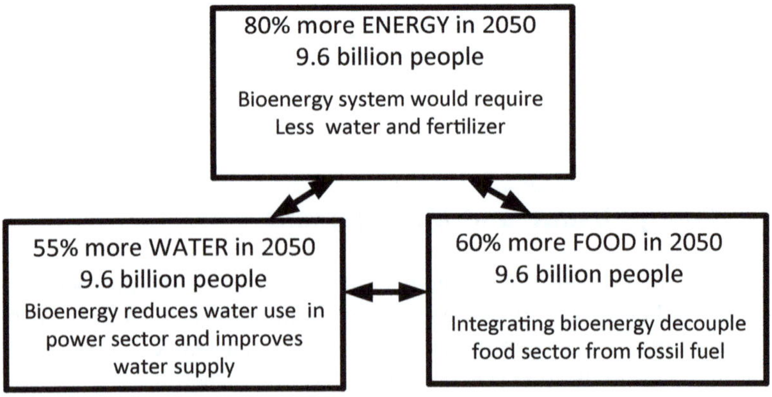

Fig. 13.1 Food, energy, water interdependency; sustainable bioenergy can positively affect water, energy, and food security, and slows climate change [11]

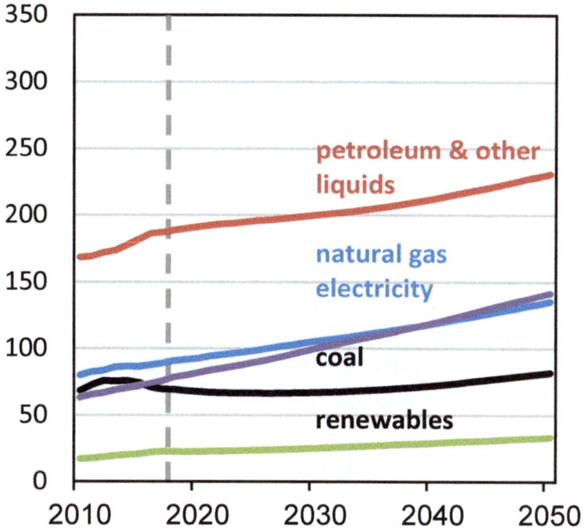

Fig. 13.2 World end use energy consumption by various fuels: quadrillion British thermal units [19]

and electricity will increasingly dominate the world's end use of energy, while coal will remain stagnant, and renewables will have the lowest share. Industrial, residential, and commercial sectors are the main users of electricity, while the transportation sector has the lowest share as electricity users. The projections show that the residential sector merges with the industrial sector by 2050 [18, 19].

Advanced, efficient, and low-emission energy technologies utilizing renewable and nonrenewable resources are vital for sustainable energy technology, which plays a major role in global climate change as well as in international politics and trade. Still the fossil fuel-based energy would play important roles in global energy use and global energy management policies [26].

Green certificates are increasingly common and get awarded related to the units of green electricity that they produce. The value of a green certificate is not fixed and can vary depending on market conditions. The *emission trading system* (ETS) requires big companies to acquire *emission certificates* for the GHG emissions produced. If a company can switch partly from fossil fuels to bioenergy and the avoided cost for buying certificates could be an additional benefit of the bioenergy plant [23].

For effective energy management, there are several basic steps that are commitment, data analysis, analysis and implementation of energy conservation options and continued feedback and analysis. Energy management programs provide organizations with information, tools, and assistance to reduce energy, water, nonrenewable energy use, and greenhouse gas emissions. The program provides the organizations with annual performance data, energy savings performance contracts, utility energy service contracts, and support with energy efficient products. Within this framework organizations may prepare sustainability/energy score cards. The program offers expertise regarding all levels of project and policy implementation to reduce the energy intensity of operations. For example, heat pumps efficiency gains ranging from two to four times higher than conventional heating systems may lead to an increase of ninefold by 2050 [8, 18, 26].

13.2.1 Energy Assessment

Country and industry-specific energy assessment can improve energy-efficiency considerably in a cost-effective way in the industrial sector. Several government programs have been established around the world to encourage, facilitate, or mandate industrial facilities to undertake energy assessment. Energy assessment team prepares "Assessment Report" to help identify ways to save energy, reduce waste, and improve productivity with the following activities.

- In pre-assessment information gathering stage, the client is sent a pre-assessment form that includes the size of a plant and plant layout, industry type and process description, production levels, units and costs, operating hours, history of utility bills, list of major energy consuming equipment, and key plant managers of plant, energy, environment, and maintenance.
- The pre-assessment analysis involves an analysis of the manufacturing process, analysis of chart and graph of utility bills, establishing the unit cost of energy, plant profile, identify key energy systems, review design and other technical documentations, identify possible energy saving potential recommendations and develop assessment strategy.
- Day of the assessment involves the description of manufacturing, utility bills, and waste materials, develop a list of potential energy saving opportunities with potential recommendations, discuss findings with management, preliminary estimate of potential savings, and prioritize recommendations of analysis.
- Post assessment activities conduct engineering and financial analysis including implementation cost and delivering the report to the client. A follow-up report involves calling the client in 6–9 months for implementation data [8, 27].

13.2.2 Energy Policy

The relation between economic growth and energy consumption is the main factor for energy policy implication since the Gross Domestic Product (GDP) (one of the main indicators of economic growth) increases with the increasing usage of energy [18]. This is directly related to international policies to cut nonrenewable energy to reduce emissions. Therefore, governments and energy policymakers design energy policies based on data relating to GDP to energy usage toward conservation hypothesis and growth hypothesis. According to the conservation hypothesis, in some developed nations economic growth has less reliance on energy usage, while the growth hypothesis predicts that in some developing nations economy is more dependent on energy and economic growth and energy usage are directly related to each other. Based on the hypotheses, policymakers may decide to reduce energy usage and hence emissions.

Energy policy recognizes the importance of energy efficiency and the necessary transition to renewables to tackle the risks and challenges of fossil fuel use. One effort called the "Green Growth Strategy" helps policymakers and stakeholders to increase the renewables in energy development. In developed countries, growth in energy consumption is relatively due to slower population and economic growth, improvements in energy efficiency, and less growth in energy-intensive industries with the help of energy conservation policies [18, 27]. Major progress in energy conservation is needed to achieve a sustainable world with the collaboration of organizations and policymakers around the globe.

Higher energy efficiency guarantees secure, sustainable, and inclusive economic growth and reduces the environmental footprint of the energy system. Half of these energy efficiency savings come from industry, and major contributions also from the transport and building sectors [18, 19].

13.3 Power Production

Power production technologies from fossil fuels and renewable sources are the main part of energy management. The sustainability of existing fossil-fuel fired power plants, especially with the decarbonizing constraints, is decreasing, while the renewables-based power production is gaining global support to replace conventional power production (Fig. 13.3). Renewable power has become affordable. The global weighted-average levelized cost of electricity (LCOE) from renewable resources have all been within the range of fossil fuel-fired power generation costs [17, 18]. Need for biofuel and abundant supply of fossil fuels may lead to these objectives: (1) Sequester carbon to reduce GHG emissions, (2) Achieve greater energy efficiency, (3) Integrate rural programs into increasing energy security, (4) Stimulate economic growth and development, (5) Obtain feasible conversion technologies [23]. A proper assessment would be needed to identify the proportion of the global coal power risk by analyzing existing coal plants and new renewable costs at a country level [28].

Below is the summary of the efforts in both fossil fuel-based and renewables-based power production [26, 27]:

Coal-fired power plants are required to meet standards that limit the amounts of some of the pollutants that these plants release into the air. These plants must use coal that is low in sulfur content and ash. Coal can also be pre-treated and processed to reduce the types and amounts of undesirable compounds in combustion gases.

Integrated coal gasification combined cycles (IGCC) offer power production with steam turbine and gas turbine with heat recovery steam generation. The gasification unit produces synthesis (syngas), which can be a feedstock to produce fuels and

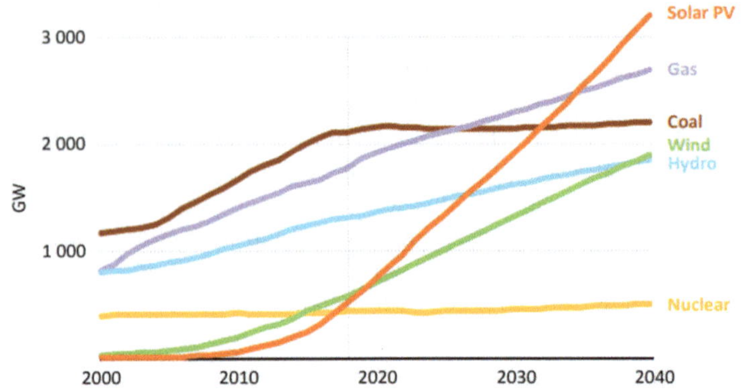

Fig. 13.3 Global power capacity of renewables and fossil fuels with projections. The projections display sharp increases in power generation by solar photovoltaic (PV) and wind, as well as by natural gas, while stagnation in power generation by coal and nuclear continues. Solar PV capacity catches wind power by 2023 and hydropower before 2030 and increases sharply to dominate as a source of power by 2040. However, coal and natural gas will keep their considerable contributions by 2040 [16, 19]

chemicals like methanol [29]. Power from IGCC is economical, offers high efficiency, and decreases emissions of sulfur dioxide, nitrogen oxides, and particulate matters.

Natural gas power plants are replacing the coal power plants because of their lower capital costs and relatively low GHG emissions.

Chemical-looping combustion with inherent CO_2 capture, if properly developed, can produce water and CO_2 stream, and helps reduce carbon capture and storage costs [6].

Advanced biofuels refer to the lignocellulosic-based biofuel model that contains all unit operations to transform biomass such as corn stover into ethanol using dilute acid pretreatment and co-fermentation of C5 and C6 sugars. [7, 9, 21, 25]. Biofuels involve Renewable Fuel Standards (RFS), waiver credits, renewable volume obligations, renewable identification numbers, energy costs, and third-party relationships [24].

Solar energy device provides a direct conversion of solar radiation into electrical power. Photovoltaic (PV) solar cells play an important role in worldwide power generation. *Solar Insolation* in units of energy per unit area is around 1366 W/m^2, which is also called solar constant. Solar radiation is mainly concentrated around the wavelength of visible light [20]. Commercial solar PV cells have an efficiency range of up to 26% conversion and can be connected to electricity grids. Efficiency is increased using concentrating solar systems. To mitigate grid integration problems with solar power, battery systems, and other means are being developed [19].

Wind power involves *onshore and offshore* wind farms that come with high costs of access to the site and perform maintenance on towers and cabling. The use of a high level of wind and solar energy will require costly enhancement of system integration and management that includes flexible power sources, demand-side measures, electricity storage, strong and smart transmission, and distribution grids.

Hydro power offers affordable power with low maintenance cost, but it has environmental concerns [19].

Geothermal energy may offer regional opportunities of green energy, but it needs operational and maintenance cost.

Organic Rankine cycle (ORC) operates between 50 and 150°C using organic chemicals such as isopentane as the working fluid and the turbine operates in the superheater vapor phase. They present sustainable energy solutions in rural areas. However, their efficiency is low [18].

Bioenergy may be produced from any type of organic materials including trees, crops, grass, woody material, animal waste, and agricultural waste. Fermentation, transesterification and anaerobic digestion processes produce ethanol, biodiesel, and biogas that are among the most used bioenergy globally. Biomass and waste biodegraded by bacteria in landfill produce biogas mainly methane and carbon dioxide, which can be used for heat and power production [7, 10].

The technologies above will have their unique impacts in managing energy production in terms of capital cost, operating cost, environmental and social implications. The evolution of their technological maturity will be assessed by industry and policy makers toward their managements of selected energy mix with regional constraints.

13.3.1 Assessment of Environmental Impact

Fossil fuels are well known to have many detrimental environmental effects such as air pollution, greenhouse gas emissions, and natural habitat disruption associated with drilling practices. To better compare the environmental effects of fossil fuels with biofuels, the life cycle assessments of the fuels may be helpful. For example, there are numerous studies comparing corn ethanol with gasoline and the results vary. The variance in results can be attributed to the differences in the studies, the process by which the study was carried out, and the scope of the study.

The methods to assess the impact of biofuels over the duration of their lifecycle focus on different aspects of the biofuel lifecycle are very sensitive to variable inputs. This may cause a large margin of error that can be difficult to pinpoint at the end of the assessment. For example, the potential changes in the use of land to grow crops can influence the carbon content contained in the soil. Based on the previous vegetation, large amounts of carbon can be released into the atmosphere or stored in the soil. Once the crops are grown, they need transportation to biorefineries operating with necessary utilities which release emissions that must be accounted for when attributing greenhouse gas emissions to biofuels or fossil fuels.

Life cycle Analysis of biofuels may be challenging because of the sheer number of variables that need to be accounted for with reasonable assumptions and minimum errors possible. Many of these unknowns are from the agricultural practices adopted by producers. Soil nutrients vary drastically across regions and so do water and moisture levels. One of the keys for future policy decisions on bioenergy may be the comparison of greenhouse gas emissions from biofuel production and use with the emissions from fossil fuels [3, 5, 25, 27, 28].

Environment impact assessment identifies the short- and long-term impacts of energy production on the environment within the liability of local and federal government agencies. Such impacts include natural physical resources, natural biological/resources, and quality-of-life values [15]. Without careful assessment of the mass and energy balances, and ecological impacts, energy projects may be unsustainable. Dependence on non-renewables poses risks and challenges including:

- Depletion of fossil fuel resources
- Decrease in useful energy output
- Uneven distribution of fossil fuels around the globe
- Price volatility and increases in international energy markets
- Security of energy supply
- Damaging effects of fossil fuel use on the environment and society.

13.3.2 Gas Resource Planning

Gas utilities use a wide range of factors when planning resource allocation, which makes up their Integrated Resource Planning (IRP). This includes regional market prices, weather data, historical usage data, new regulations, and improving technology. The IRP process is broken down into four main stages starting with understanding the demands of the market in a large scope and using previous data to characterize specific areas. The next is to break the data into smaller sections and look for trends based on previous forecasts and current demand. The third is defining and predicting variance due to theoretical risk and event occurrences. The final step of the process is then to develop a portfolio that can handle the daily energy demands and has contingencies in place for emergencies [26].

In business as usual, gas utilities are the energy source that strives to meet policy goals, but there is pushback from environmentalists to decarbonize the energy industry. The advancement of clean energy technologies is becoming a global goal that is in line with how investments made in the energy sector grow based on policy. As decarbonization efforts are increased and more heavily enforced, gas utility companies are challenged to come up with near-term solutions and long-term goals, such as limits to the amount of allowable GHG emissions [16].

13.3.3 Grid Resilience

Resilience in utilities focuses on how quickly the system can become operational after failure. The degree of resilience increase is directly tied to investment costs and management. Resiliency of a system directly involves the response first responders, government agencies in the event of failure. Increasing resiliency is a goal for many utilities by undertaking and adopting affordable measures that provide incentives based on increased economic benefit and reliability in different capacities. Changing to a higher portion of renewable energy may create challenges for the reliability of renewable energy sources, their storage, and connections to the existing power grid [16, 19].

13.3.4 Optimum Energy Production: Distributed Versus Centralized

Underutilization of capacity reduces profits, which may be a survival issue for technology. Projections of the demand–supply gap in the market should be made based on technology, equipment, public policy, and finances. When energy production technology is mature, large capacity produces more energy per unit and more revenue can be earned with the availability of transportation and other infrastructure. A large part of the land area should be reserved for service roads, storm water mains, railways, overhead gas, steam, and air pipelines, and water reservoirs.

Centralized plants need a reliable supply of a large amount of feedstock throughout the year. This may be the case for primary energy production in large scale operations, which helps lower the unit cost of energy generated and create considerable economic activity. For renewable fuels from biomass, however, large-scale operation may adversely affect the economics because of the nature of biomass with low density and widely scattered supply chains around the plant. Biomass may contain water dissolved organic and inorganic compounds, and solid particulates of various sizes. Distributed process technology may offer economic solutions to regions where the large amounts of biomass feedstock are not readily transportable, and end users would be far from the production site.

13.3.5 Risk Assessment

Risk analysis consists of financial risk, environmental risk, technical risk, and social risk. For a standalone plant, such as heat production in sawmill for drying wood, availability and failure risk will be of minor importance. However, for emerging technologies, some risk of unexpected failure or production breakdown should be taken into consideration. If the bioenergy plant is integrated into industrial production, for example, energy supply of a pulp and paper mill, the availability of the feedstock is high (~ 99.5%) and the risk of unexpected failure is low. Most developers will carry out some form of risk assessment as part of their project activities to identify key constraints of finance, environment, technical, and social in construction and operating energy producing plants [4, 7].

13.4 Source Conservation and Energy Management

The cultivation of different feedstocks for biofuel production utilizes different amounts of resources depending upon the chosen type. The use of short-rotation woody crops can also have different greenhouse gas emissions in comparison to annual crops such as corn or soybeans. The way in which the crops used as feedstock are grown and cultivated also may have adverse effects on the environment. Feedstock crops, like all agricultural crops, must be planted. The farm management practices used during this stage of the process can also have varying effects on the total environmental effects of the biofuel feedstock. There are numerous factors producers must consider when growing crops, including maximizing yields and profits. These decisions play a part in the total impact biofuels have on the environment during their lifetime, from plant to product. Crops require water and nutrients to grow, machinery to harvest and transport, and industrial processes to refine. By minimizing the impact at each stage, biofuels provide a renewable alternative to fossil fuels.

Energy conservation focuses on reducing energy consumption and increasing efficiency in energy usage for the same useful energy output. For example, using less fuel or reducing the demand on a limited supply such as fossil fuels and replacing it with renewable energy would lead to source conservation. Energy recovery also may be a part of energy conservation through captured and reduced waste energy. Energy conservation may lead to increased security, financial gain, and environmental protection. For example, electric motors consume a considerable amount of electrical energy and operate at efficiencies between 70 and 90%. Therefore, using an electric motor operating with higher efficiency will conserve energy throughout its useful life [8, 23].

Reducing energy consumption through efficiency and conservation requires a combination of increased research and development on energy efficiency and policies. It will also require structural and behavior changes in residential and transportation systems such as replacing fossil fuels with renewable energy.

Since power at peak demands is very expensive residential energy efficiency and conservation programs focus on space heating and cooling as well as water usage. The Industrial sector is a good candidate for energy conservation, improving energy efficiency, and reducing GHG emissions. Energy recovery may lead to the use of available energy from a subsystem to be used in another part of the same system. There is a large potential for energy recovery in industries and utilities leading to reduced use of fossil fuels and hence less GHG emission. Some examples of energy recovery practices are [8]:

- Energy and water recycling
- Installing heat recovery steam generator (HRSG)

- Using heat regenerative cyclone engine
- Using thermal diode and thermoelectric modules.

Regenerative brake is used in electric cars and trains, where the part of kinetic energy is recovered and stored as chemical energy in a battery. Besides, active pressure reduction systems conserve energy when the differential pressure in a pressurized fluid flow is recovered rather than converted to heat in a pressure reduction valve.

13.4.1 Scenarios

The *Technical Potential Scenarios* estimate the level of energy consumption that would occur if all industrial processes were upgraded with Energy Conservation Management (ECM) leading to technically feasible solutions that are:

- *Economic potential* scenario 1 estimates the level of energy consumption that would occur if all industrial processes were upgraded with ECMs that are economically feasible with a 2-year simple payback period.
- The e*conomic potential* scenario 2 predicts a 5-year simple payback period at a predefined uptake rate and trend. The projections for economic and technical potential are up to 2050 on a 5-year increment [8].

13.4.2 Energy Efficiency and Energy Conservation

Energy efficiency and ECMs often focus on a situation where internal and external perspectives converge. Internal perspective may consist of many under evaluated behavioral elements. By overlaying the energy saving potentials with the associated economic, organizational, and technical issues, the following potential measures encourage ECMs effectiveness [8]:

1. Improving energy efficiency through *energy management*, instead of replacing equipment
2. Mandatory implementation of energy management systems for large energy-intensive sectors
3. Mandatory sub-metering requirement for significant energy consuming equipment
4. Facilitate the development of insurance products for energy savings guarantee
5. Promoting and facilitating further potential for resource sharing among industrial sectors.

Example 13.1 **Energy cost of a car**

An average car consumes 50 gallons gasoline per month. Estimate the amount and cost of energy consumed by the car per year.

Solution:

Assume that gasoline has an average density of 0.72 g/cm^3 and the heating value of 47.3 MJ/kg (Table 2.6). Cost of gasoline: $3/gallon.

Data: $V = 50\,\text{gal/month} = 89.251/\text{month}, 2271.01/\text{year}\ (3.7851 = 1\,\text{gal})$; $\rho_{gas} = 0.72\,\text{g/cm}^3 = 0.72\,\text{kg/l}$.

Mass of gasoline: $m_{gas} = \rho V = 1635.1\,\text{kg/year}$.

Energy consumed per year:

$E_{gas} = 1635.1\,\text{kg/year}\ (47{,}300\,\text{kJ/kg}) = 77{,}340{,}230\,\text{kJ/year}$.

Cost per year: C(gasoline) : ($3/gallon)(50 gallon)(12)=$1800/year.

Example 13.2 **Fuel consumption by a low and a high-mileage car**

Assume that the average daily traveling distance is about 40 miles/day. A car has a city-mileage of 20 miles/gal. If the car is replaced with a new car with a city-mileage of 30 miles/gal and the average cost of gasoline is $3.50/gal, estimate the amount of fuel, energy, and money conserved with the new car per year.

Assume: The gasoline is incompressible with $\rho_{av} = 0.75$ kg/l.

Lower heating value (LHV) = 44,000 kJ/kg; 44,000 kJ of heat is released when 1 kg of gasoline is completely burned, and the produced water is in vapor state.

Fuel needed for the old car: $(40\,\text{miles /day})/(20\,\text{miles /gal}) = 2\,\text{gal/day}$.

Fuel needed for the new car: $(40\,\text{miles/day})/(30\,\text{miles/gal}) = 1.34\,\text{gal/day}$.

Old car: 20 miles/gal

$$\begin{aligned}\text{Mass of gasoline: } m_{gas} &= \rho_{av}(\text{ Volume })\\ &= (0.75\,\text{kg/l})(2.0\,\text{gal/day})(3.785\,\text{l/gal}) = 5.7\,\text{kg/day}\end{aligned}$$

$$\begin{aligned}\text{Energy of gasoline: } E_{gas}\ (\text{LHV}) &= (5.7\,\text{kg/day})(44{,}000\,\text{kJ/kg})\\ &= 250{,}800\,\text{kJ/day}\ (365\,\text{day /year})\\ &= 91{,}542{,}000\,\text{kJ/year} = 91{,}542\,\text{MJ/year}\end{aligned}$$

$$\text{Cost: } (\$3.50/\text{gal})(2\text{gal/day })(365\text{ day /year}) = \$2555/\text{year}$$

New car: 30 miles/gal

$$\begin{aligned}\text{Mass of gasoline: } m_{\text{gas}} &= \rho_{\text{av}}(\text{volume}) \\ &= (0.75\,\text{kg/l})(1.34\,\text{gal/day}) \\ &\quad (3.785\,1/\text{gal}) = 3.8\,\text{kg/day}\end{aligned}$$

$$\begin{aligned}\text{Energy of gasoline: } E_{\text{gas}}\ (\text{LHV}) &= (5.7\,\text{kg/day})(44{,}000\,\text{kJ/kg}) \\ &= 250{,}800\,\text{kJ/day}(365\,\text{day/year}) \\ &= 91{,}542{,}000\,\text{kJ/year} = 91{,}542\,\text{MJ/year}\end{aligned}$$

Cost: ($3.50/gal)(1.34 gal/day)(365 day/year) = $1712/year.

Annual fuel savings: 30,514 MJ/year = (91,542 − 61,028) MJ/year.

Annual cost savings: $843/year = ($2555 − $1712) MJ/year.

Approximate annual emission reduction:

$$\left.\begin{aligned}(5.7 - 3.8)\text{kg/day } (365\,\text{day/year})&\left(3.3\,\text{kgCO}_2/\text{kg gasoline}\right) \\ &= 2288.5\,\text{kgCO}_2/\text{year}\end{aligned}\right\} (\text{Table 11.2})$$

The new car reduces fuel consumption by around 33%, which is significant.

Example 13.3 **Yearly consumption of natural gas by a city**

A city consumes natural gas at a rate of 500×10^6 ft^3/day. The volumetric flow is at standard conditions of 60°F and 1 atm = 14.7 psia. If the natural gas is costing $6/GJ of higher heating value what is the yearly cost of the gas for the city.

Solution:

$$\text{Q} = 500 \times 10^6\,\text{ft}^3/\text{day at } 60°\text{F and } 1\,\text{atm} = 14.7\,\text{psia}.$$

The higher heating value is the heat of combustion of the natural gas when the water product is at liquid state. From Table 2.6, the value of HHV is:

50,000 kJ/kg ~ 1030 Btu/ft^3 (Table 2.7)

$$\text{Heating value: } \left(1030\,\text{Btu/ft}^3\right)\left(500 \times 10^6\,\text{ft}^3/\text{day}\right) = 515.0 \times 10^9\,\text{Btu/day}$$

$$\left(515.0 \times 10^9\,\text{Btu/day}\right)(1055\,\text{J/Btu}) = 543{,}325\,\text{GJ/day}$$

$$\text{Yearly cost: } (543{,}325\,\text{GJ/day})(\$6/\text{GJ})(365\,\text{days/year}) = \$11.9 \times 10^8/\text{year}.$$

***Example 13.4* Avoided fossil fuel by a car**

An average car consumes about 2 gallons (US gallon = 3.785 L) a day, and the capacity of the fuel tank is about 15 gallons. Therefore, a car needs to be refueled once every week. The density of gasoline ranges from 0.72 to 0.78 kg/l (Table 2.6). The lower heating value of gasoline is about 44,000 kJ/kg. Assume that the average density of gasoline is 0.75 kg/l. If the car was able to use 0.2 kg of nuclear fuel of uranium-235, estimate the time in years for refueling and avoided cost of fossil fuel of gasoline.

Solution:

Assume: The gasoline is incompressible with $\rho_{av} = 0.75$ kg/l.
Cost of gasoline $3.5/gallon.

Lower heating value (LHV) = 44,000 kJ/kg; 44,000 kJ of heat is released when 1 kg of gasoline is completely burned, and the produced water is in vapor state.

Complete fission energy of U-235 = 6.73×10^{10} kJ/kg

Mass of gasoline per day:
$m_{gas} = \rho_{av} V = (0.75\,\text{kg/l})(2\,\text{gal/day})(3.785\,\text{l/gal}) = 5.67\,\text{kg/day}$

Energy of gasoline per day: $E_{gas} = m_{gas}\,(\text{LHV})$
$= (5.67\,\text{kg/day})(44{,}000\,\text{kJ/kg}) = 249{,}480\,\text{kJ/day}$

Energy released by the complete fission of 0.2 kg U-235:

$E_{U\text{-}235} = (6.73 \times 10^{10}\ \text{kJ/kg})(0.2\ \text{kg}) = 1.346 \times 10^{10}\ \text{kJ}.$

Time for refueling: $(1.346 \times 10^{10}$ kJ)/(249,480 kJ/day) = 53,952 days = 148 years.

Therefore, the car will not need refueling for about 148 years.

Avoided cost of gasoline:

($3.5/gallon) (2 gallon/day) (365 days/year) (148 year) = $378,140.

***Example 13.5* Energy densities and emissions of ethanol and gasoline**

Assuming complete combustion of ethanol and gasoline (assuming 100% octane) estimate the HHVs by using the oxidations and carbon dioxide intensity in $kgCO_2/MJ$.

Solution:

$$C_2H_5OH + 3O_2 \rightarrow 2CO_2(g) + 3H_2O(l) \quad \Delta H_c = -1368\,\text{kJ/mole}$$
$$\text{MW} = 46\,\text{g/mole} \quad E_c = 29.7\,\text{MJ/kg}$$

$$C_8H_{18} + 25/2O_2 \rightarrow 8CO_2(g) + 9H_2O(l) \quad \Delta H_c = -5476\,\text{kJ/mole}$$
$$\text{MW} = 114\,\text{g/mole} \quad E_c = 47.5\,\text{MJ/kg}$$

We estimate the specific energy densities (E_c) by dividing the heats of combustions by their molecular weights (MW). Comparisons of the specific energy densities show that ethanol produces 63% (29/7/47.5) of energy per unit mass of octane.

Combustion of 46 g of ethanol produces two moles of carbon dioxide that is 88 g (2 × 44 g), while combustion of 114 g of octane produces eight moles that is 352 g (8 × 44 g). If we use the CO_2 intensity defined as amount of CO_2 released per unit of energy, we have

Amount of CO_2 produced/kg of ethanol = 44 g/mole x (number of moles of CO_2 produced by 1000 g of ethanol) = (44 g/mole) 2x (1000 g/46 g = 43.47)
= 1.913 kg CO_2/kg ethanol.

Amount of CO_2 produced/kg of octane = 44 g/mole x (number of moles of CO_2 produced by 1000 g of octane) = (44 g/mole) 8x (1000 g/114 g = 70.17)
= 3087.7 kg CO_2/kg octane

$$\mathrm{CO_2\ intensity\ of\ ethanol} = 1913\,\mathrm{g}/29.7 = 64.4\,\mathrm{gCO_2/MJ}$$

$$\mathrm{CO_2\ intensity\ of\ octane} = 3087\,\mathrm{g}/47.5 = 65.0\,\mathrm{gCO_2/MJ}$$

The CO_2 intensities based on burning are almost the same for these two fuels, except the processes required for producing them would be different. This shows the estimation of environmental impact of fuels is involved in their feedstock, conversion, as well as their use.

Example 13.6 Cost estimations of electricity

A superheated steam (stream 1) expands in a turbine from 5000 kPa and 325°C to 150 kPa and 200°C. The steam flow rate is 10.5 kg/s. If the turbine generates 1.1 MW of power, estimate the emissions caused to produce the steam from natural gas and revenue of electricity.

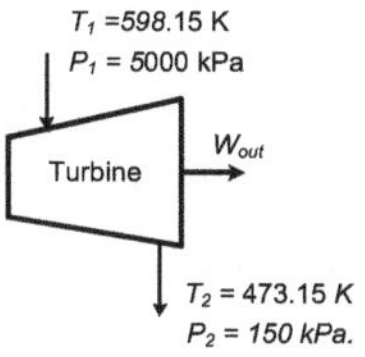

Solution:

Assume that the kinetic and potential energy effects are negligible; this is a steady process.

The properties of steam from the steam tables:

Stream 1: Superheated steam: $P_1 = 5000\,\text{kPa}$, $T_1 = 325°\text{C}(598.15\,\text{K})$, $H_1 = 3001.8\,\text{kJ/kg}$.

Stream 2: Superheated steam: $P_2 = 150\,\text{kPa}$, $T_2 = 200°\text{C}(473.15\,\text{K})$, $H_2 = 2872.9\,\text{kJ/kg}$

$$\dot{W}_{\text{out}} = -1100\,\text{kW} = -1100\,\text{kJ/s}$$

The mass and energy balances for the turbine are: Mass balance: $\dot{m}_{\text{out}} = \dot{m}_{\text{in}}$

Energy balance: $\dot{E}_{\text{out}} = \dot{E}_{\text{in}} \rightarrow \dot{m}_1(H_2 - H_1) = \dot{q}_{\text{out}} + W_{\text{out}}$.

Heat loss from the energy balance

$$\begin{aligned}\dot{q}_{\text{out}} &= -\dot{W}_{\text{out}} + \dot{m}_1(H_2 - H_1)\\ &= 1100\text{ kJ/s} + (10.5\text{ kg/s})\\ &\quad (2872.9 - 3001.8)\text{kJ/kg} = -253.45\,\text{kJ/s}\end{aligned}$$

The sign is negative as the heat is lost from the system.

Enthalpy If water entering the boiler is at 98°C and 94 kPa

$H = 410.6$ kJ/kg (saturated steam table)

Energy to produce the steam: (10.5 kg/s) (3001.8–410.6) kJ/kg

= 27,207.6 kW = 27,207 kJ/s.

From Table 2.8: 53.6 MJ/kg natural gas = 53,600 kJ/kg NG.

Amount of natural gas needed:

(27,207 kJ/s)/(53,600 kJ/kg) = 0.507 kg/s = 1827.3 kg/h.

Every 1 kg natural gas produces 2.8 kg CO_2 (Table 11.2).

Emissions for every hour:

(1827.3 kg/h) (2.8 kg CO_2 /kg NG) = 5116.4 kg CO_2/h.

1.1 MW power plant = 1100 kWh each hour operating at full power

and 26,400 kWh/day and 9,636,000 kWh per year.

For \$0.14/kWh the total revenue from electricity would be:

(\$0.14/kWh) (9,636,000 kWh/year) = \$1,349,040/year.

Example 13.7 **Cost of compression work**

Air enters a compressor at 100 kPa, 300 K, and a velocity of 2 m/s through a feed line with a cross-sectional area of 1.0 m^2. The effluent is at 500 kPa and 400 K. Heat is lost from the compressor at a rate of 5.2 kW. If the air behaves as an ideal gas, estimate the power requirement of the compressor in kW and cost of power.

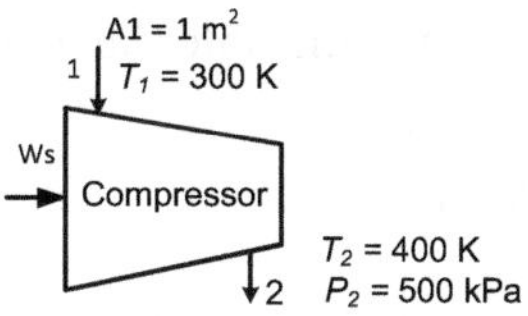

Solution:

Assume: The compressor is at steady state, air is ideal gas, and the change in the potential and kinetic energy of the fluid from the inlet to the outlet is negligible.

$$R = 8.314\,\text{m}^3\text{kPa/kmoleK},\ \text{MW} = 29\,\text{kg/kmole};\ v_1 = 2\,\text{m/s},\ A_1 = 1.0\,\text{m}^2$$

$$P_1 = 100\,\text{kPa},\ T_1 = 300\,\text{K},\ P_2 = 500\,\text{kPa},\ T_2 = 400\,\text{K},\ q_{\text{loss}} = -5.2\,\text{kW}$$

For ideal gas: $V_1 = \frac{RT_1}{P_1} = 0.025\ \text{m}^3\text{/mol}$,

the specific volume $(1/\rho)_{\text{air}} = \frac{V_1}{\text{MW}_{\text{air}}} = 0.86\ \text{m}^3\text{/kg}$

The mass flow rate of air: $\dot{m} = \frac{v_1 A_1}{(1/\rho)} = 1.72\ \text{kg/s}$

From Table D1 for ideal gas of air: $H_2 = 400.9$ kJ/kg, $H_1 = 300.2$ kJ/kg.

Energy Balance: $\dot{m}(H_2 - H_1) - \dot{q} = \dot{W}$

$$(1.72\,\text{kg/s})((400.9 - 300.2)\,\text{kJ/kg}) + 5.2\,\text{kW} = \dot{W} = 178.4\,\text{kW}$$

This is the work needed by the compressor.

Daily electricity: (178 kW) (24 h/day) = 4281.6 kWh/day.

For \$0.14/kWh electricity.

Annual cost = \$218.939.6/year = (\$0.14/kWh)(4281.6 kWh/day)(365 day/year).

Example 13.8 **Cost of pumping**

A pump increases the pressure in liquid water from 100 kPa at 25°C to 4000 kPa. Estimate the minimum horsepower motor required and its cost to drive the pump for a flow rate of 0.01 m^3/s.

Solution:

Assume that the pump operates adiabatically and nearly isothermally, changes in potential and kinetic energy are negligible, and water is incompressible.

$$P_1 = 100\,\text{kPa},\ T_1 = 298.15\,\text{K},\ P_2 = 4000\,\text{kPa},\ Q = 0.01\,\text{m}^3/\text{s}$$

Energy Balance: $\dot{q} + \dot{W}_{\text{pump}} = \dot{m}\Delta H$

For an adiabatic operation ($\dot{q} = 0$): $\dot{W}_{\text{pump}} = \dot{m}\Delta H$.

Since the water is incompressible the internal energy is only the function of temperature.

$\Delta U = 0$ as the system is isothermal.

$$\Delta H = \Delta U + \Delta(PV) = V\Delta P$$

The specific volume of saturated liquid water at 25°C: $V_1 = 0.001003$ m^3/kg (steam tables).

The mass flow rate: $\dot{m} = \frac{Q}{V_1} = 9.97$ kg/s

$$\begin{aligned}\dot{W}_{\text{pump}} &= \dot{m}V_1(P_2 - P_1)\\ &= (9.97\,\text{kg/s})\left(0.001003\,\text{m}^3/\text{kg}\right)(4000 - 100)\text{ kPa}\left(\text{kJ/kPa m}^3\right)\\ &= 39\,\text{kJ/s} = 39\,\text{kW} = 52.3\,\text{hP}\,(1\text{ hp} = 745.7\,\text{W}).\end{aligned}$$

For \$0.14/kWh electricity

Annual cost = \$48.829.6/year = (\$0.14/kWh)(39 kW) (24 h/day)(365 day/year).

Example 13.9 **Cost and emissions of pump work**

A centrifugal pump delivers 1.8 m^3/h water to an overhead tank 20 m above the eye of the pump impeller. Estimate the work required by the pump cost of power used and emissions from that power production using natural gas.

Solution:

Assume that the pump operates adiabatically and nearly isothermally, changes in potential and kinetic energy are negligible, and water is incompressible and steady frictionless flow.

Data: $Q = 1.8$ m^3/h and density of water: $\rho = 1000$ m^3/h.

Energy balance: $\Delta\left(H + \frac{1}{2}v^2 + zg\right)\dot{m} = \dot{q} + \dot{W}_s$.

For an adiabatic operation ($\dot{q} = 0$) with no kinetic and potential energy effects:

$$\begin{aligned}\dot{W}_{\text{pump}} = \dot{m}g\Delta z &= \left(\frac{1.8\ \text{m}^3/\text{h}}{3600\,\text{s/h}}\right)(1000)\,\text{kg/m}^3\left(9.81\,\text{m/s}^2\right)(30\,\text{m})\\ &= 147.15\,\text{Nm/s} = 147.15\,\text{W}\end{aligned}$$

From Table 2.8: 53.6 MJ/kg natural gas = 53,600 kJ/kg NG.

Amount of natural gas needed:

(147.15 kJ/s)/(53,600 kJ/kg) = 0.00274 kg/s = 9.88 kg/h.

Every 1 kg natural gas produces 2.8 kg CO_2 (Table 11.2).

Emissions for every hour: (9.88 kg/h) (2.8 kg CO_2/kg NG) = 27.66 kg CO_2/h.

Example 13.10 **Saving of fuel cost by regeneration in a Brayton cycle**
A power plant is operating on an ideal Brayton cycle with a pressure ratio of $r_p = 9$. The fresh air temperature at the compressor inlet is 295 K. The air temperature at the inlet of the turbine is 1300 K. The cycle operates with a compressor efficiency of 80% and a turbine efficiency of 80%. The unit cost of fuel is \$0.14/kWh. The cycle operates 360 days per year.

(a) Using the standard-air assumptions, determine the thermal efficiency of the cycle.
(b) If the power plant operates with a regenerator with an effectiveness of 0.78, determine the thermal efficiency of the cycle and annual saving of fuel cost.

Solution:

Assume that the cycle is at steady-state flow and the changes in kinetic and potential energy are negligible. Heat capacity of the air is temperature dependent, and the air is an ideal gas.

(a) Basis: 1 kg/s air. Data given: $\eta_{turb} = 0.8$, $\eta_{comp} = 0.8$, $r_p = P_2/P_1 = 9$

Process 1-2: isentropic compression

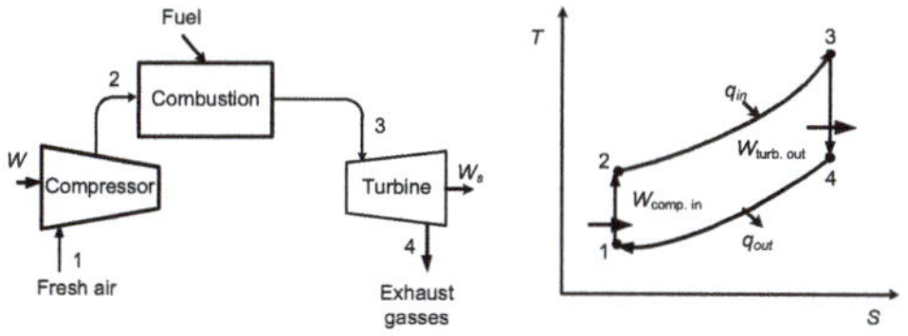

Data from (Table D1): $T_1 = 295$ K, $H_1 = 295.17$ kJ/kg, $P_{r1} = 1.3068$.

P_r shows the relative pressure defined in Eq. (7.36).

$$\frac{P_{r2}}{P_{r1}} = \frac{P_2}{P_1} = r_p \rightarrow P_{r2} = (9)(1.3068) = 11.76$$

Approximate values from Table D1 for the compressor exit: $P_{r2} = 11.76$,

$T_2 = 550$ K, and $H_2 = 555.74$ kJ/kg.

Process 3–4: isentropic expansion in the turbine as seen on the *TS* diagram above

$$T_3 = 1300\,\text{K},\ H_3 = 1395.97\,\text{kJ/kg},\ P_4/P_3 = 1/r_p = 1/9,\ P_{r3} = 330.9$$

$$\frac{P_{r4}}{P_{r3}} = \frac{P_4}{P_3} \rightarrow P_{r4} = \left(\frac{1}{9}\right)(330.9) = 36.76$$

Approximate values from Table D1 at the exit of turbine: at $P_{r2} = 11.76, T_2 = 550\text{K}$, and $H_2 = 555.74\text{kJ/kg}$

The work input to the compressor:

$W_{\text{comp.in}} = \frac{H_2 - H_1}{\eta_{\text{comp}}} = \frac{(335.74 - 295.17)\ \text{kJ/kg}}{0.8} = 325.7\ \text{kJ/kg}.$

The work output of the turbine:

$W_{\text{turb.out}} = \eta_{\text{turb}}(H_4 - H_3) = 0.8(761.87 - 1395.97) = -507.3\ \text{kJ/kg}.$

The back-work ratio r_{bw}: $r_{\text{bw}} = \frac{W_{\text{comp.in}}}{|W_{\text{turb.out}}|} = \frac{325.7}{507.3} = 0.64.$

This shows that 64% of the turbine output has been used in the compressor.

Work output: $W_{\text{net}} = W_{\text{out}} - W_{\text{in}} = -181.6\ \text{kJ/kg}.$

Actual values of air enthalpy at the compressor outlet:

$$H_{2\text{a}} = H_1 + W_{\text{comp.in}} = (295.17 + 325.70)\ \text{kJ/kg} = 620.87\ \text{kJ/kg}$$

The amount of heat added: $q_{\text{in}} = H_3 - H_{2a} = 1395.97 - 620.87 = 775.1\ \text{kJ/kg}$

The thermal efficiency: $\eta_{\text{th}} = \frac{|W_{\text{net}}|}{q_{\text{in}}} = \frac{181.6}{775.1} = 0.234$ or 23.4%

The temperature of exhaust air, T_4, and the actual enthalpy are estimated from the energy balance: $W_{\text{turb out}} = H_{4a} - H_3$, $H_{4\text{a}} = 1395.97 - 507.3 = 888.67\ \text{kJ/kg} \rightarrow T_4 = 860\ \text{K}$ (From Table D1).

Advances in the compressor and turbine designs with minimal losses increase the efficiency of these components. In turn, a significant increase in the thermal efficiency of the cycle is possible.

(b) $T_4 = 860$ K and $T_2 = 550$ K, since $T_4 > T_2$ regeneration is possible.

Regeneration with effectiveness of $\varepsilon = 0.78$:

$$\varepsilon = \left(\frac{H_5 - H_2}{H_4 - H_2}\right) = \frac{H_5 - 620.9}{888.67 - 620.9} = 0.78 \rightarrow H_5 = 829.80\ \text{kJ/kg}$$

$$q_{\text{in}} = H_3 - H_5 = 1395.97 - 829.8 = 566.2\ \text{kJ/kg}$$

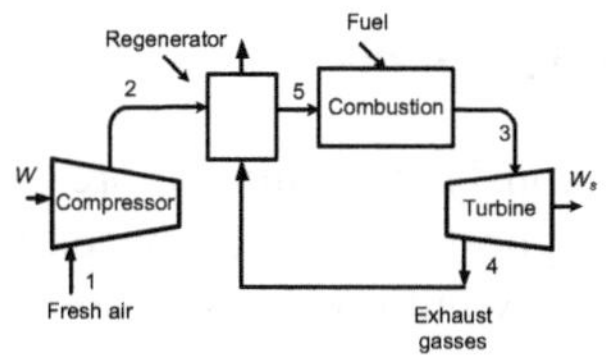

This represents a conservation of 840.2 – 566.2 = 274.0 kJ/kg from the fuel required.

The thermal efficiency: $\eta_{th} = \frac{|W_{\text{net}}|}{q_{\text{in}}} = \frac{181.6}{566.2} = 0.32$ or 32%

Unit cost of fuel = \$0.14/kWh.

Days of operation 360, hours of operation per year = 360(24) = 8640 h/year.

Conserved fuel: (274 kJ/kg)(1 kg/s) (8640 h/year) = 2,367,360 kWh/year.

Saved money: (2,367,360 kWh/year) (\$0.14/kWh) = \$331,430/year.

After the regeneration, the thermal efficiency has increased from 23 to 32% in the actual Brayton cycle operation. The addition of a regenerator (operating without thermal losses) does not affect the network output of the cycle. Savings for fuel and costs are considerable.

13.5 Energy Economics

Expenditures on solar, wind and storage are very close to each other, while the hydro expenditure is relatively low. Annual operating and maintenance costs are high and similar for wind and storage, while lower for solar and much lower for hydro energy. Energy economies should be handled with extreme care due to its impact in the ecosystem [20]. Accurate estimation of the cost of plants for generating energy is possible with accurate assumptions, scale-down/scale-ups, competitive utility costs, and waste treatment. Risk analysis and sensitivity analysis help large investments ~~to~~ operate successfully.

The mix of wind and solar-based renewable energy has the optimum cost of replacing fossil fuels worldwide. Solar power may be used during the day and wind power during the night. In addition, reliable grid structure, utility-scale energy storage, and smaller wind turbines help increase the contribution of renewable energy without nuclear power. Battery storage costs would be reduced by about 60% if around 20% of energy is supplied from clean natural gas generation with carbon capture. Cheap and abundant natural gas can be an important back-up fuel for intermittent renewable energy besides the mixture of hydroelectricity and stored energy [19, 20].

13.5.1 Renewable Energy Cost

Renewable power is continuously becoming more competitive each year in the order of affordability from highest to lowest: concentrated solar power, bioenergy, solar photovoltaic and onshore wind power, hydropower, geothermal, and offshore wind power. This trend also follows low-cost decarbonization, which is encouraging in sustainability concerns. Especially onshore wind and solar power are becoming less expensive than any fossil-fuel options proving that renewable power is cost-effective and more evenly distributed worldwide compared with fossil fuels [19].

13.5.2 Techno Economic Analysis

Techno economic analysis (TEA) undergoes pre-investment, investment, and operation phases. The pre-investment phase involves the identification of investment opportunities, preliminary selection of the project ideas, project design, and final evaluation based on the expectations and requirements of various stakeholders. The investment phase involves several inter-disciplinary tasks including negotiation & contracting for project plant, construction, detailed project design and cost estimates, feasibility, and startup of trial operation. The operation phase involves day to day operation and marketing the products [26, 29].

TEA produces discounted cash flow diagrams (DCFD) prepared for 20 to 25 years of operation using the current economic data. Based on the equipment list from the process flow diagrams, fixed capital investments (FCI), operating and maintenance costs, and cost of manufacturing are estimated. Chemical Engineering Plant Cost Index (CEPCI-2014) (= 595.8) can be used to estimate and update the costs and capacity to the present date by [26, 28]

$$\mathrm{Cost}_{\mathrm{New}} = \mathrm{Cost}_{\mathrm{New}} \frac{\mathrm{CEPCI}_{\mathrm{New}}}{\mathrm{CEPCI}_{\mathrm{Old}}} \left(\frac{\mathrm{Capacity}_{\mathrm{New}}}{\mathrm{Capacity}_{\mathrm{Old}}} \right)^{x} \tag{13.1}$$

where x is the factor, which is usually assumed to be 0.6. A possible depreciation method is the Maximum Accelerated Cost Recovery System. After estimating the revenue DCFDs are prepared to generate the three economic feasibility criteria that are Net Present Value, Payback Period, and Rate of Return. At least two out of three criteria should be favorable for the operation to be feasible. Plant capacity affects feasibility.

The sudden growth of renewables markets can lead to high energy prices in the short term if the supply does not match the demand. It is also possible that excess supply causes the energy prices to fall below the production costs. This can make analyzing the cost of renewable power generation technologies unreliable in given locations and times [19].

Stochastic Analysis—The estimated profitability criteria using discounted cash flow diagram may be affected by the changes in the economic data and market conditions including fixed costs, raw material prices, revenue, interest rates, and tax rate. Therefore, a stochastic model such as Monte-Carlo method helps account for the viability of the economic data due to the uncertainties. The stochastic model produces probability density distributions of important profitability criteria [1].

13.5.3 Levelized Cost of Electricity

Levelized cost of electricity (LCOE) is the price at which electricity must be produced from a specific source to break even. It is an economic assessment tool and includes initial investment, cost of operations and maintenance, cost of fuel, and cost of capital [18]

$$\mathrm{LCOE} = \frac{\sum_i^n \frac{I_i + M_i + F_i}{(1+r)^i}}{\sum_i^n \frac{E_i}{(1+r)^i}} = \frac{\text{Total costs}}{\text{Total electricity produced}} \tag{13.2}$$

where LCOE is the average lifetime levelized electricity generation cost, I the Investment expenditures, M the operations and maintenance expenditures, F the fuel expenditures, E the electricity production, i the interest rate, and n the useful life of operation. Typically, levelized energy costs are calculated over 20- to 40-year operational lifetimes and are given in the units of $/kWh or $/MWh. The simple levelized cost of energy is calculated by:

$$\begin{aligned}\mathrm{LCOE} = \{&(\text{capital cost} * \text{capital recovery factor} + \text{fixed and variable O\&M cost})/\\&(8760 * \text{capacity factor})\}\end{aligned} \tag{13.3}$$

where capital cost is in $/kW, while fixed operation and maintenance (O&M) costs in $/kW-yr and variable O&M costs in $/kWh. In denominator 8760 is the number of hours in a year and capacity factor is a fraction between 0 and 1 representing the portion of a year that the power plant is generating power. *Fixed* O&M *costs* include labor, scheduled maintenance, routine component/ equipment replacement (for boilers, gasifiers, feedstock handling equipment), insurance, and others. The fixed O&M costs of larger plants are lower per kW due to economies of scale. *Variable* O&M *costs* include ash disposal, unplanned maintenance, equipment replacement and incremental serving costs. Capital recovery factor (CRF) is the ratio of a constant annuity to the present value of receiving that annuity for a given length of time. Using an interest rate i, the capital recovery factor is

$$\mathrm{CRF} = \frac{i(1+i)^n}{(1+i)^n - 1} \tag{13.4}$$

Fuel cost is expressed in ($/MMBtu) and heat rate is measured in Btu/kWh. Fuel cost is optional since some generating technologies like solar and wind do not have fuel costs. The LCOE indicates the average cost per unit of electricity generated at the actual plant and allows the recovery of all costs over the lifetime of the plant. The LCOE of renewable energy technologies varies by technology, location, and project, based on the renewable energy resource, capital and operating costs, and the efficiency/performance of the technology [18, 19].

13.5.4 Economic Assessment of Biofuels

The biochemical process of fermentation converts the first-generation biomass including corn, sugarcane, and wheat into bioethanol, while biochemical and thermochemical processes convert mostly the lignocellulosic and algal biomass into bioethanol, biodiesel, green diesel, and other fuels and chemicals. Conversion of lignocellulosic feedstock requires first the complex process of conversion of cellulose and hemicellulose into sugars. The cost of the conversion processes of biomass increases in this direction: triglycerides → starch → lignocellulosic, while the cost of biomass increases in this order: lignocellulosic → starch → triglycerides. Bioethanol producers have adopted various technologies such as high tolerance yeasts, continuous ethanol fermentation, co-generation of steam and electricity, and molecular sieve driers to reduce ethanol production costs [7, 25].

Second generation biomass resources are geographically more evenly distributed than fossil fuel resources, which may lead to energy security. Biofuels might create local economic activity and employment when they are produced sustainably by integrating biology, ecology, and engineering [3]. Most agricultural biomass production, except for forest products, is seasonal and results in a large volume of feedstock that needs to be stored and transported to a biorefinery. Regional preprocessing infrastructure can be set up to clean, sort, chop or grind, control moisture, densify, and package the feedstocks before transporting them to biorefineries. In contrast, forest products are available year-round so long-term storage might not be necessary [9, 10].

Table 13.2 shows the large amounts of water requirements of bioethanol and biodiesel productions from various biomass feedstocks. The effects of biofuel production on the world-wide trade of grains, livestock, biomass, and crude oil are a part of economic assessment. The biofuel industry has also some economic effects related to the national budget spending such as tax credits, subsidies, incentives, and other policy matters. The diversion of land to corn and soybean production and a greater demand for biofuels may coincide with an increase in the price of the wheat, corn, and soybean [3, 25].

13.5.5 Cost Trends in Renewable Energy

The International Renewable Energy Agency (IRENA) reported data showing how renewable energy generation has produced lower-cost electricity and continued to decrease over the past ten years. Using this data, predictions can be made as to

Table 13.2 Comparison of water requirements for ethanol and biodiesel productions [25]

Biofuel crop	Water use m^3 water/kg crop	Biofuel conversion liter fuel/kg crop	Crop water use m^3 water/kg fuel	Crop water uses per unit energy m^3 water/GJ
Ethanol from				
Corn (grain)	833	409	2580	97
Sugarcane	154	334	580	22
Corn stover	634	326	2465	92
Switchgrass	525	336	1980	74
Grain sorghum	2672	358	9460	354
Sweet sorghum	175	238	931	35
Biodiesel from				
Soybean	1818	211	9791	259
Canola	1798	415	4923	130

what the future costs may be based upon the trends. As technology has improved in the past years, the cost of installing new onshore wind facilities and turbines has decreased. Electricity costs passed to the customer are lowered. Offshore wind is a more volatile class on a year-to-year basis due to a more dispersed market. Advances in solar power technology have increased the power generation of photovoltaic cells at a lower cost and industrial level installation projects are cheaper than smaller scale installations. The levelized cost of electricity is expected to continue to decrease [19].

13.5.6 Bio Break Model

The purchase price for feedstocks should be obtained by surveying biorefineries, and the costs of producing and delivering biomass feedstocks to a biorefinery should be based on observed production practices. The Bio break model represents the regional feedstock supply system and biofuel biorefinery economics. The Bio break model estimates willingness to pay (*WTP*), willingness to accept (*WTA*) and the price gap (*PG*) between them [7]:

(1) *WTP* is the maximum price that a biorefinery pays for a dry ton of biomass delivered at the gate. *WTP* is a function of the price of bioethanol, the conversion yield, and the cost of processing biomass.
(2) *WTA* is the minimum price that a biomass producer would accept for a dry ton of biomass delivered at the gate of a biorefinery. *WTA* depends on the biomass opportunity cost, production, and delivery costs.

Assumptions used in the bio break model are:

- Producer minimizes costs on the long-run average cost curve
- A yield distribution for biomass crops is based on the expected mean yield
- A transportation cost is based on the average hauling distance for a defined circular capture region
- Biorefinery has an annual capacity of 89 million liter or more to be competitive in the market
- Each biorefinery uses a single feedstock with no market disruptions
- Energy price uncertainty on biofuel investment is neglected

Willingness to pay (WTP) for lignocellulosic bioethanol production—Eq. (13.5) shows the estimation of WTP for 1 dry ton of cellulosic material delivered to a biorefinery

$$WTP = \left(P_{\mathrm{gas}}E_{\mathrm{V}} + T + V_{\mathrm{CP}} + V_{\mathrm{O}} - C_{\mathrm{I}} - C_{\mathrm{O}}\right)Y_{\mathrm{E}} \tag{13.5}$$

The market price of bioethanol is estimated as the energy equivalent price of gasoline, where the P_{gas} denotes per gallon price of gasoline and E_{V} the energy equivalent factor of gasoline to ethanol. Based on historical data gasoline and crude oil are related by $P_{\mathrm{gas}} = 0.13087 + 0.023917P_{\mathrm{oil}}$. Beyond direct ethanol sales, the ethanol processor also receives revenues from the tax credits T, the coproduct production V_{CP}, and the octane benefits V_{O} per gallon of processed bioethanol. Biorefinery costs are the investment costs C_{I} and the operating C_{O} costs per gallon. A conversion ratio is used for gallons of ethanol produced per dry ton of biomass Y_{E}. Therefore, Eq. (13.5) provides the maximum amount the processor can pay per dry ton of biomass delivered to the biorefinery and still break even. The values of the variables in Eq. (13.5) are based on the following assumptions [7]:

- Energy equivalent factor (E_{V}) for ethanol to gasoline and the value of octane benefits (V_{O}) would be fixed based on the current economic data per gallon.
- Coproduct value (V_{CP}) is the excess energy as the only coproduct
- Conversion ratio (Y_{E}) is assumed as a mean value of 70 gallons per dry ton feedstock.
- Investment costs (C_{I}) are estimated for an optimized *n*th biorefinery for biochemical conversion of corn stover to ethanol.
- Operating costs (C_{O}) are separated into two components: enzyme costs and non-enzyme operating costs including salaries, maintenance, overhead, insurance, taxes, and other conversion costs.
- Biofuel production incentives and tax credits (T) for cellulosic ethanol producers would be determined locally.

Willingness to accept (WTA) for lignocellulosic bioethanol production—Eq. (13.6) shows the estimation of *WTA* for 1 dry ton of cellulosic material delivered to the biorefinery [28]

$$WTA = \{(C_{\mathrm{ES}} + C_{\mathrm{Opp}})/Y_{\mathrm{B}} + C_{\mathrm{HM}} + SF + C_{NR} + C_{\mathrm{S}} + DFC + DVC^{*}D\} - G \tag{13.6}$$

The value of *WTA* is equal to the total production costs less than the government incentives G (tax credits and production subsidies). Costs include the establishment and seeding C_{ES} per acre land and biomass opportunity costs C_{Opp} per acre, the harvest and maintenance C_{HM}, the stumpage fees *SF*, the nutrient replacement C_{NR}, the biomass storage C_{S}, the transportation fixed costs *DFC*, and the variable transportation costs calculated per mile *DVC* multiplied by the average hauling distance to the biorefinery. Therefore, the biomass yield per acre Y_{B} is used to convert the per acre costs into per dry ton costs. Equation (13.6) is used with the following assumptions:

- Nutrient replacement costs (C_{NR}) are based on the added value by the uncollected cellulosic material to the soil through enrichment and protection against rain, wind, and radiation, thereby limiting the loss of vital soil nutrients such as nitrogen, phosphorus, and potassium.
- Harvest and maintenance costs (C_{HM}) and stumpage fees (*SF*) are adjusted by the current economic data for short-rotation woody crops.
- Transportation costs (*DVC, DFC*, and *D*)—One-way transportation distance *D* has been evaluated up to around 140 miles for woody biomass and between 5 and 75 miles for all other feedstocks. The average hauling distance is between 13 and 53 miles.
- Biomass storage costs (C_{S}) depend on the feedstock, harvest technique, and storage area.
- Establishment and seeding costs (C_{ES}) are assumed to not incur for corn stover, wheat straw, and forest residue suppliers, whereas all other feedstock suppliers would have to be compensated for their establishment and seeding costs.
- Opportunity costs (C_{Opp}) of using biomass for ethanol production are assumed to be a mean opportunity cost per acre of switchgrass and *Miscanthus.*
- Biomass yield (Y_{B}) depends on the feedstock and the selected process.
- Biomass supplier government incentives (*G*) are the dollar/dollar matching payments for collecting, harvesting, storing, and transporting (CHST), which is a temporary (2-year) payment.

Bio break model estimates the price gap (*PG*): $PG = WTA - WTP$; if the PG is negative or zero, a biomass market is economically feasible, otherwise the biomass market is not [14]. Policy incentives for carbon emissions could also affect the *PG* for a possible interaction with biofuel policy [28]. Biomass supplier government incentives are crucial to maintaining a moderate *WTA* cost. They can subsidize the delivery costs significantly, encouraging buyers to purchase more biomass for biorefinery purposes. They can be adjusted accordingly to manipulate the supply and demand of biomass in the market to ensure that both the suppliers' and buyers' benefit—maintains a reasonable market value for biomass. *WTA* does not take into consideration external changes in cost independent of energy costs, e.g., drought, pest infestations, agricultural disease—this prevents it from being used as an accurate overall measure of the viability of biorefinery. *WTA* can be used as an approximation/indicator of profitability in the industry instead—high *WTA* suggests less chance of the process being profitable in the long-term. Harvest and maintenance cost (C_{HM}) and the Opportunity cost (C_{Opp}) are the main contributors to *WTA* and need to be minimized to reduce the *WTA* substantially.

13.5.7 Thermoeconomics

Thermoeconomics assigns costs to exergy-related variables by using the *exergy cost theory* and *exergy cost balances. Extended exergy* accounts for the environmental impact in a more systematic way by estimating the resource-based value of a commodity. Some concerns in thermoeconomics evaluations are: (i) costs of fuel and equipment change with time and location, (ii) optimization of an individual process does not guarantee an overall optimum for the system, and (iii) for the whole system, often several design variables should be considered and optimized simultaneously [9].

For any process or subsystem *i*, the specific cost of exergy *c* in \$/kW-unit time for a stream is

$$c = \dot{C}/\dot{E}x \tag{13.7}$$

where $\dot{C}$ and $\dot{E}x$ are the cost rate and the rate of exergy transfer for a stream, respectively. However, the cost of a product and other exiting streams would include the fixed capital investment $\dot{C}_{FCI}$ and the annual operating cost of process $\dot{C}_{OP}$. This will be called the total cost of process

$$\dot{C}_P = \dot{C}_{OP} + \dot{C}_{FCI} \tag{13.8}$$

Then the cost rate balance for a single process is

$$\left(\sum_i c_i \dot{E}x_i\right)_{out} = \left(\sum_i c_i \dot{E}x_i\right)_{in} + \dot{C}_P \tag{13.9}$$

For example, as seen in Example 3.11 the cost rate balance for a boiler (control volume 1) relates the total cost of producing high-pressure steam to the total cost of the entering streams plus the cost of the boiler $\dot{C}_B$, and from Eq. (13.8) we have

$$c_{HP}\dot{E}x_{HP} + c_{EG}\dot{E}x_{EG} = c_F\dot{E}x_F + c_A\dot{E}x_A + c_W\dot{E}x_W + \dot{C}_B \tag{13.10}$$

where the symbol HP denotes the high-pressure steam, EG the exhaust gas, while F, A, and W are the fuel, air, and water, respectively. All the cost estimations are based on exergy as a measure of the true values of work, heat, and other interactions between a system and its surroundings. By neglecting the costs of air and water, and assuming that the combustion products are discharged directly into the surroundings with negligible cost, Eqs. (13.7) and (13.9) yield the specific cost of high-pressure steam

$$c_{\mathrm{HP}} = c_F\left(\frac{\dot{E}x_F}{\dot{E}x_{\mathrm{HP}}}\right) + \frac{\dot{C}_B}{\dot{E}x_{\mathrm{HP}}} \tag{13.11}$$

The ratio $\left(\dot{E}x_F/\dot{E}x_{\mathrm{HP}}\right) > 1$ due to inevitable exergy loss is in the boiler, and hence $c_{HP} > c_F$.

Similarly, the cost rate balance for the turbine (control volume 2) is

$$c_E\dot{W}_E + c_{\mathrm{LP}}\dot{E}x_{\mathrm{LP}} = c_{\mathrm{HP}}\dot{E}x_{\mathrm{HP}} + \dot{C}_T \tag{13.12}$$

where c_E, c_{LP}, and $\dot{C}_T$ are the specific costs of electricity, low-pressure steam, and the total cost of the turbine, respectively; $\dot{W}_E$ and $\dot{E}x_{\mathrm{LP}}$ are the work produced by the turbine and the exergy transfer rate of low-pressure steam, respectively. Assuming that the specific costs of low and high-pressure steams are the same $c_{\mathrm{LP}} = c_{\mathrm{HP}}$, we have

$$c_E = c_{\mathrm{HP}}\left(\frac{\dot{E}x_{\mathrm{HP}} - \dot{E}x_{\mathrm{LP}}}{\dot{W}_E}\right) + \frac{\dot{C}_T}{\dot{W}_E} \tag{13.13}$$

Using the exergetic efficiency of turbine $\eta_t = \dot{W}_E/\left(\dot{E}x_{\mathrm{HP}} - \dot{E}x_{\mathrm{LP}}\right)$, Eq. (13.13) becomes

$$c_E = \frac{c_{\mathrm{HP}}}{\eta_t} + \frac{\dot{C}_T}{\dot{W}_E} \tag{13.14}$$

As $\eta_{\mathrm{t}} < 1$, the specific cost of electricity (product) will be higher than that of high-pressure steam. Example 13.11 illustrates the cost calculations in power production.

Example 13.11 **Approximate thermoeconomics of power generation**

Consider the exergy cost on a boiler and turbine system shown below:

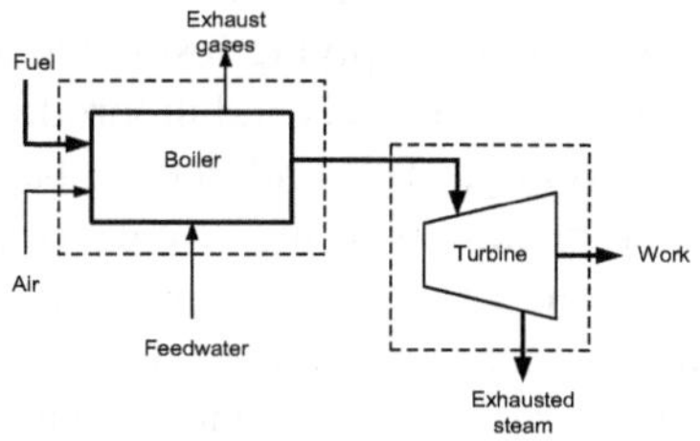

A turbine produces 30 MW of electricity per year. The average cost of steam is \$0.017/(kWh) of exergy (fuel). The total cost of the unit (fixed capital investment and operating costs) is $\$1.1 \times 10^5$. If the turbine exergetic efficiency increases from 84 to 89%, after an increase of 2% in the total cost of the unit, evaluate the change of the unit cost of electricity [12].

Solution:

Assume that heat transfer effects between the turbine and surroundings are negligible. Also, kinetic and potential energy effects are disregarded. From Eq. (13.13), we have

$$c_E(84\%) = \frac{c_{\text{HP}}}{\eta_t} + \frac{\dot{C}_T}{\dot{W}_E} = \frac{0.017}{0.84} + \frac{1.1 \times 10^5}{30 \times 10^6} = \$0.0239/(\text{kW h})$$

$$c_E(89\%) = \frac{0.017}{0.89} + \frac{(1.02)1.1 \times 10^5}{30 \times 10^6} = \$0.0228/(\text{kW h})$$

The reduction in the unit cost of electricity after the increase in efficiency is about 4.4%. This simple example shows the positive effect of exergetic efficiency on the unit cost of electricity.

13.5.8 Algae-Based Energy Economy

The third-generation biofuels and bioproducts derived from algae continue to be a promising new low carbon economy for several reasons. Algae can be produced on non-arable areas such as lakes, oceans, or deserts, thus reducing competition with the food supply chain. High photosynthetic carbon sequestration efficiencies and carbon capture percentages of 90% paired with the ability to harvest and use the totality of the biomass makes algae very well suited as a source for biofuels and bioproducts [3].

Integration of engineering, ecology, and biology is essential since neither of them alone is likely to yield the desired solution. Biology will mostly involve the understanding of the capabilities of strain and/or communities of strain, cultivation with optimized growth and stability, understanding the lipid pathway, maintenance, and carbon capture, as well as the protection of algae culture from contamination. The engineering part will involve in supplying resources of CO_2, light, and energy, extraction and conversion of lipid into fuels and bioproducts, waste and co-product processing with possible recycling, mass and energy balances, cost analysis, and sustainability assessment. Applications of the science of aquatic ecology can play an important role: (i) in optimum nutrients supplied to algal cultivation systems, (ii) to design and construct biotic communities that will help to maximize algal biomass yields and minimize grazing losses, and biomass losses to infectious disease, (iii) to help guide the magnitude and frequency of algal crop harvests and

(iv) to create biologically adaptive algal biomass production systems that are both resistant and resilient to future climate change [3].

In general, the use of photosynthetic organisms as a feedstock will aid in mitigating ever-increasing anthropogenic CO_2 emissions. Microalgae can fix CO_2 10 – 50 times more efficiently than other energy plants. Algae based biofuels would have a positive impact on energy and environment and can improve energy security and reduce GHG emissions from the transportation sector provided: (i) research, development, and demonstration of algal strain selection, (ii) useful energy output that is comparable to other transportation fuels, and (iii) the use of wastewater and recycling nutrients for cultivating algae.

However, despite their excellent potential and the well-studied cultivation options, microalgae are not yet commercially viable feedstock for biofuels and bioproducts. The major technological constraints include the cost-effective harvesting of algae and the extraction of lipids. A national assessment of land requirements for algae cultivation that accounts climatic conditions, fresh water, inland and coastal saline water, wastewater resources, and CO_2 sources would help estimate the potential of algal biofuels that could be produced economically.

13.5.9 Optimum Cost of Algae Biomass

Sustainable development of algal biofuels requires favorable energy efficiency and emissions, water use, supply of nitrogen, phosphorus, CO_2, and appropriate land resources. The quantity of algal biomass (M_{AB}, tons) representing the energy equivalent of a barrel of crude petroleum is [3]

$$M_{AB} = \frac{E_{petroleum}}{Q(1 - w)E_{biogas} + \mathrm{YwE}_{biodiesel}} \tag{13.15}$$

where $E_{petroleum}$ (~ 6100 MJ/brl) is the energy contained in a barrel of petroleum, Q ($m^{3/}$ton) is the biogas volume produced by anaerobic digestion (400 m^3/ton), E_{biogas} (MJ/m^3) is the energy content of biogas (~ 2.4 MJ/m^3), Y is the yield of biodiesel from algal oil (80% by weight), $E_{biodiesel}$ is the average energy content of biodiesel (37,800 MJ/ton), and w is the oil content of algae biomass. If a barrel of crude oil has the same energy of M tons of algae, acceptable cost of algae C_{Algae} becomes

$$C_{Algae} = \frac{C_{petroleum}}{M_{AB}} = \frac{C_{petroleum}}{E_{petroleum}}\left(Q(1 - w)E_{biogas} + YwE_{biodiesel}\right) \tag{13.16}$$

Figure 13.4 shows the acceptable cost al algae biomass with respect to crude oil prices. When the cost of petroleum is \$100/brl biodiesel produced from algae oil costing \$2.61/gal is likely to be competitive with petroleum diesel.

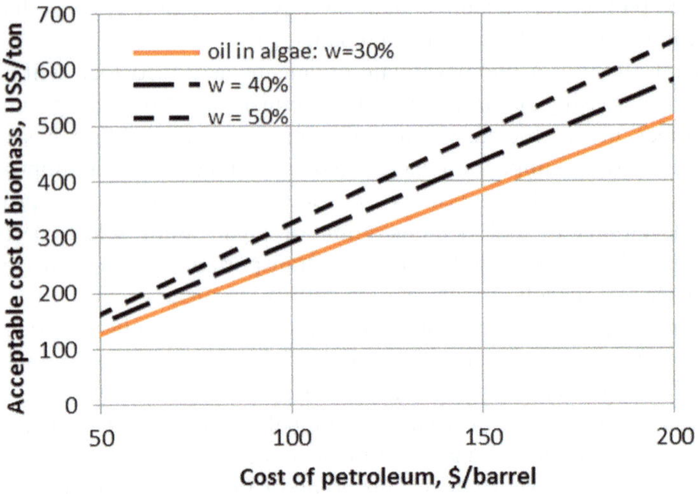

Fig. 13.4 Acceptable cost of biomass with oil percentages of $w = 30\%$, $w = 40\%$, and $w = 50\%$ [3]

13.5.10 Energy Return on Investment (EROI)

Society is aware of the contribution/ limitations of type of energy on technological development and economic activity. The ratio of useful energy to input energy required may be called the energy return on investment (EROI) that helps assess the energy cost on society and its impact on development. Most renewable energy resources have lower values of EROI compared with conventional fossil fuels. This means that more energy output must be available to society with respect to energy needed to produce and hence funds can be redirected toward societal needs, such as health and educational expenses.

Some examples of EROI are: EROI (sugarcane ethanol) ~ 8 and EROI (corn ethanol) ~ 1.2–1.6 (excluding the energy content of dried distillers' grains with solubles) that shows a better overall economy for bioethanol production from sugarcane [22].

$$\mathrm{EROI_{Society}} = \frac{\text{Energy returned to society}}{\text{Energy invested to get this energy}} = \frac{\mathrm{ER}}{\mathrm{EI}} \qquad (13.17)$$

In the case of $\mathrm{EROI_{Society}}$, the numerator (ER) is composed of a nation's Gross Domestic Product (in US$) multiplied by the MJ per unit of energy used in the generation of that GDP. The denominator (EI), the energy invested to produce the energy output, is composed of the total energy consumed by that nation each year (in MJ) multiplied by dollars per unit spent in the acquisition of that fuel.

The estimated minimum EROI of 3 is required for an energy source to be beneficial to society [14.15]. There is a strong correlation between EROI and societal well-being [16, 17]. Indicators of quality of life may be percent children under weight, health expenditures, gender inequality index, literacy rate, and access to improved water. As the value of EROI increases, so does the quality of life. According to this trend, a minimum EROI of 14 might be necessary to maintain the qualities of an advanced society. The values of EROI are around 0.8 to 1.7 for bioethanol, and 1.3 for biodiesel [14].

The energy from the traditionally high EROI deposits may need to be supplemented or rapidly replaced by alternative energy sources to avoid future energy constraints and the potential effects of climate change. These "new" energy sources must be sufficiently abundant and have a large enough EROI value to power society [17]. If the EROI values of traditional fossil fuel energy sources (e.g., oil) continue to decline and renewable energy resources fail to provide enough high EROI alternatives, then we may be moving toward the "net energy cliff." The human development index (HDI) is a commonly used composite index of well-being and is calculated using four measures of societal well-being: life expectancy at birth, adult literacy, combined educational enrollment, and per capita GDP [18].

Table 13.3 shows the energy ratio that is the biofuel energy output to fossil energy inputs in the production process with different feedstock. The energy ratios for all the biodiesels from various feedstocks are higher than 1 suggesting that biofuels are renewable energy with positive net energy outputs and reductions in GHG emissions.

13.5.11 Circular Economy

The linear economy is characterized by "take, make, and dispose," while the circular economy closes the loop by collecting and recycling waste, and using it to manufacture new products and protect natural resources. Circular economy describes strategies for waste prevention and resource efficiency toward regenerative manufacturing models. This is achieved through design, maintenance, repair, and recycling (Fig. 13.5). In this bottom-up approach, the shareholders and investors are raising concerns about natural resource depletion, climate change, sustainability, energy consumption, water consumption, and regulatory compliance. Pressure to reform business practices is from many sectors including energy sector such as oil and gas and others [12]. Some focused sustainability and circular initiatives with new models enable companies to:

- Reduce nonrenewable material/energy and water use such as replacing fossil fuels with renewable energy and wastewater treatment
- Reduce emissions such as CO_2 and NO_x and emphasize decarbonization with feasibility assessment
- Improve process thermal efficiency with processes intensification and product value chain integration

Table 13.3 Energy ratio for bioethanol and biodiesel production processes from various feedstocks and change in life cycle GHG emissions per kilometer traveled by replacing diesel with 100% biodiesel fuel; the cost ratio (price of biodiesel/price of fossil fuel) for biodiesel is around 1.2 [7, 25]

Feedstock	Energy ratio[a]	EROI	GHG emissions change[b] %
Sugarcane	≈ 8	0.8–10	− 87 to − 96
Sugar beets	≈ 2		− 35 to − 56
Sweet sorghum	≈ 1		
Corn	≈ 1.5	0.84–1.65	− 21 to − 38
Wheat	≈ 2		− 19 to − 47
Lignocellulosic	≈ 2–36	0.69–6.61	− 37 to − 82
Rapeseed	≈ 2.5	1.0–1.5	− 21 to − 51
Soybeans	≈ 3	0.7–2.0	− 63 to − 78
Sunflower	≈ 3	0.4–1.2	
Castor	≈ 2.5		
Palm oil	≈ 2.5–9		
Jatropha	≈ 1.4		
Waste vegetable oil	≈ 5–6		− 92

[a] Energy from biofuel/fossil energy used in production of biofuel

[b] Approximate avoided GHG emissions because of the biomass feedstock used in bioethanol production

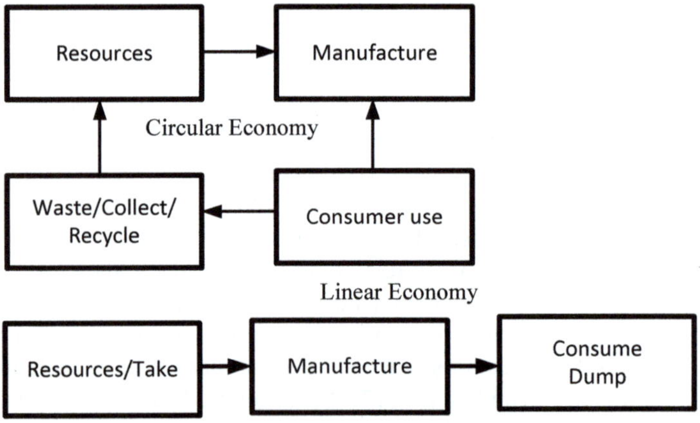

Fig. 13.5 Circular economy versus linear economy

- Develop innovative products and process solutions including recycling/reusing
- Safe and reliable operation with predictive maintenance solutions.

Sustainability-related objectives are expanding and becoming more specific process metrics for emissions, resource use, waste reduction, and exploring new energy sources with lower carbon footprints. The circular economy forces companies to rethink how to design and manufacture products, as well as to interact with customers. Here, the customer's role is no longer just consumption, but the use of a function. This demands the business community to build long-term relationships with the consumers by retaining the value and integrity of the product.

13.5.12 Circular Economy and Sustainability

The Circular Economy is viewed as a condition for sustainability. Both notions emphasize regeneration commitments, shared responsibilities motivated by environmental hazards and problems on a global scale. Both concepts integrate non-economic aspects into development through system design and innovations, and cooperation between stakeholders. Multiple and coexisting pathways of development are essential for both concepts.

However, there are different goals associated with the circular economy and sustainability. Sustainability focuses on benefiting the environment, the economy, and society at large, while the Circular Economy emphasizes the efficient use of resources with less waste and emission benefiting the economic actors that implement the system. Society benefits also from environmental improvements and more economic activity with. In the sustainability debate, responsibilities are not clearly defined but loosely shared with the focus on interest alignment between stakeholders, while private businesses, regulators, and policymakers are responsible for the transition to a circular economy [12, 13].

13.5.13 Bioeconomy

Bioeconomy is based on all economic activity originating from invention, development, production, and use of biological products and processes in the production of food, energy, and materials [21, 22]. Bioeconomy responses to diminishing fossil-based resources, climate change, growing world population, resource-efficient fuels. and processes for the well-being of societies.

Production of biofuels and bioproducts depends on the availability of biomass feedstock and biotechnology. The costs of different types of biomasses for producers as well as the converting processes are key for the bioeconomy. The intersecting of newly emerging bioenergy market with established markets in agriculture, forestry, water, and energy is causing substantial impacts on the prices of agricultural commodities, food, feedstuffs, forest products, fossil fuel energy, and land values.

Large-scale bioenergy generation plants (> 50 MW) are unattractive compared to fossil fuel plants because of the logistical costs of transporting low-density feedstock from far production sites. In addition, large-scale productions of biofuels may have the adverse effect of land available for food crops to be allocated for energy crops. Biofuels are promoted worldwide by tax credits and subsidies to reduce petroleum imports and GHG emissions from vehicles.

Energy efficiency of a conversion process for producing biofuels from a biomass affects the bioeconomy. Energy efficiency for biomass to biofuel can be estimated by

$$\begin{pmatrix}\text{Energy efficiency}\\ \text{of biomass to biofuel}\\ \text{conversion}\end{pmatrix} = \frac{\left\{\begin{pmatrix}\text{LHV of biofuel}\\ \text{produced per 1 kg}\\ \text{of biomass}\end{pmatrix} - \begin{pmatrix}\text{External energy}\\ \text{used in 1 kg biomass}\\ \text{to biofuel conversion}\end{pmatrix}\right\}}{\begin{pmatrix}\text{LHV of the 1 kg biomass}\\ \text{used in the convesion}\end{pmatrix}} \quad (13.18)$$

Here LHV is the lower heating value.

13.5.14 Bioeconomy and Circular Economy

Sustainability-related objectives are expanding and becoming more specific process metrics for emissions, resource use, waste, and discharge reduction from production units as well as exploring new energy sources with lower carbon footprints. Therefore, the benefits of the circular economy overlap with the elements of sustainability and align with the sustainable development policy. The inclusion of bioenergy into the circular economy needs the combination of interests and concerns of policymakers, researchers, technology developers, project developers, and society. In addition, new and innovative technologies are necessary to produce and consume without depleting natural resources, which will not be possible without society rethinking ecological balances in production-consumption chain. This shows that bioenergy and circular economy are closely linked with multiple elements overlapping, in which mature circular bioeconomy can merge urban and rural communities, as well as cross-cutting policies, such as environmental protection, waste management, eradication of poverty, and manufacturing strategies [13, 25].

13.5.15 Agricultural Economic Models

Three of the models FAPRI, FASOM, and POLYSYS are partial-equilibrium (PE) models, which means that not all sectors of the economy are included in the model. These PE models mainly focus on the agricultural sector. General-equilibrium (GE) models, such as GTAP, cover all sectors of the economy and all regions of the world. Thus, GE models capture the interactions among sectors and between product and factor markets. However, GE models, especially global ones like GTAP, cannot model the interactions among detailed sectors of agriculture or regions as PE models can. PE are appropriate for responding to policy questions regarding the sector(s) of interest regarding cropping practices, land quality and use, and regional variations, and they can permit more in-depth analysis of sector-specific policies. However, the models implicitly assume that the agricultural sector can be analyzed without worrying explicitly about what happens in the rest of the economy. GTAP is more heavily focused on the trade dimension.

FASOM is a forward-looking model and solves all years simultaneously, whereas FAPRI and POLYSYS are recursive dynamic models. GTAP can be run either as a comparative static or dynamic model. Comparative static models compute the market equilibrium under one set of conditions; when conditions change, the models compute the new equilibrium without worrying about the path from one equilibrium to the other. The three PE models originally had a heavy focus on agricultural policy, although FASOM was designed to examine competition between forestry and agricultural sectors for land from its early stages. GTAP originally was developed to evaluate the effects of alternative trade policies in international trade negotiations and regional and bilateral trade agreements [7, 28].

13.5.16 Methanol Economy

An affordable and viable alternative to hydrogen can be methanol, which has more hydrogen by mass in the same volume (Fig. 13.6). Therefore, methanol may be a safe carrier fuel for hydrogen. The reported octane number of methanol is 108.7 as compared to that of gasoline, which is 91. This suggests that the methanol-air mixture can be compressed more before it is ignited. This boosts the efficiency of the engine and ensures cleaner emissions. Unlike gasoline and diesel, methanol does not emit any soot or smoke on combustion. Also, methanol can acquire the same infrastructure as used for gas and diesel and thus can relieve infrastructure and dispensing costs.

Methanol Economy may solve many issues of climate change and oil shortage due to the following:

(i) methanol is produced from synthesis gas (syngas) obtained from coal and natural gas, (ii) methanol can also be produced by reductive hydrogenation by utilizing CO_2. (iii) methanol has a higher "flame speed" which enables faster and more complete fuel combustion in the cylinders, (iv) methanol has a higher-octane number than gasoline, which increases the efficiency of the internal combustion engines [13].

13.5.17 Ecological Cost

Ecological cost analysis may help minimize the excessive depletion of natural resources. Determining the extraction of raw materials and fossil fuels from natural resources is not sufficient in fully understanding the ecological impact of

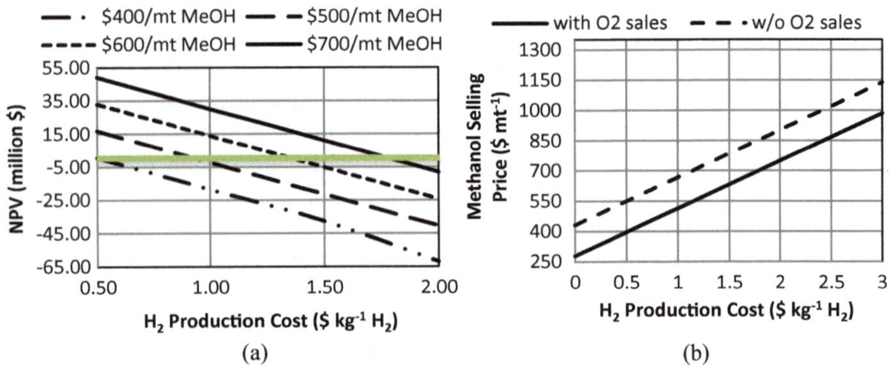

Fig. 13.6 **a** Methanol economy with hydrogen cost: **b** Selling price of methanol for NPV = 0 with and without selling O_2 byproduct at $100/tonne [7]

production processes. The production, conversion, and utilization of energy may lead to environmental problems, such as air and water pollution, impact on the use of land and rivers, thermal pollution due to mismanagement of waste heat, and global climate change. The influence of waste products discharged into the environment should also be considered. The waste products may be harmful to agriculture, plant life, human health, and industrial activity.

Several options exist for reducing CO_2 emissions, such as nonrenewable energy conservation, energy efficiency improvement, increasing reliance on nuclear and renewable energy, and carbon capture and storage (CCS) systems [2]. CCS is a process in which CO_2 is separated from effluent streams and injected into geologic formations, avoiding its release into the atmosphere. Some significant challenges to CSS are the cost of building and operating capture-ready industrial facilities, the feasibility of permanently storing CO_2 underground, and the difficulty of constructing infrastructure to transport CO_2 to injection sites [7, 9].

Example 13.12 Cost of solar heat storage

A typical square two-story home with a roof surface area of 1260 ft^2 and a wall surface area of 2400 ft^2 is to be heated with solar energy storage using a salt hydrate as phase change material. It presently has insulation of 6 inches on the roof and 1 inch on the walls. Inside the temperature will be held at 70°F and an expected outside low temperature is 10°F. Average solar radiation is 650 Btu/ft^2 and the cost of the solar air collector is $1.1/ft^2. The salt hydrate costs around $0.15/lb. Estimate the costs of salt hydrate and solar air collectors.

Solution:

Assume that the approximate thermal conductivity of the walls and roof is $k = 0.025$ Btu/h °F ft^2.

Average solar radiation is 650 Btu/ft^2 and the cost of solar air collector is \$1.1/ft^2. The salt hydrate costs around \$0.15/lb and its heat of melting is 145 Btu/lb.

Heating requirement of building with present insulation:
Heat loss from roof ($\Delta x = 6/12 = 0.5$ ft); $\Delta T = (10 - 70) = -60$°C

$$q_{\text{loss, roof}} = -kA\frac{\Delta T}{\Delta x} = -\left(0.025\,\text{Btu/h°F ft}^2\right)$$
$$\left(1260\,\text{ft}^2\right)(-60°\text{F}/0.5\,\text{ft}) = 3780\,\text{Btu/h}$$

Heat loss from walls ($\Delta x = 1$ in $= 0.083$ ft)

$$q_{\text{loss, wall}} - kA\frac{\Delta T}{\Delta x} = -\left(0.025\,\text{Btu/h°F ft}^2\right)\left(2400\,\text{ft}^2\right)$$
$$(-60°\text{F}/0.083\,\text{ft}) = 43{,}373\,\text{Btu/h}$$

Total heat loss = Total heating requirement →

$q_{\text{total loss}} = 3780$ Btu/h + 43,373 Btu/h = 47,153 Btu/h.

Total heat storage material needed:

m_s = (47,153 Btu/h) (24 h)/145 Btu/lb = 7804.0 lb.

Cost of salt hydrate = (7804.0 lb) (\$0.15/lb) = \$1170.6

Size and cost of collector to meet present heating requirements:

Area of collectors = (47,153 Btu/h) (24 h)/(650 Btu/ft^2) = 1740 ft^2

Cost of collectors: (1740 ft^2)(\$1.1/ft^2) = \$1914.0

Total cost of solar energy capture and its storage:

Cost total = \$1170.6 + \$1914.0 = \$3084.6
The annual cost of heating may be optimized by compromising between the two opposing effects of capital costs of thermal energy storage system and fuel as shown below.

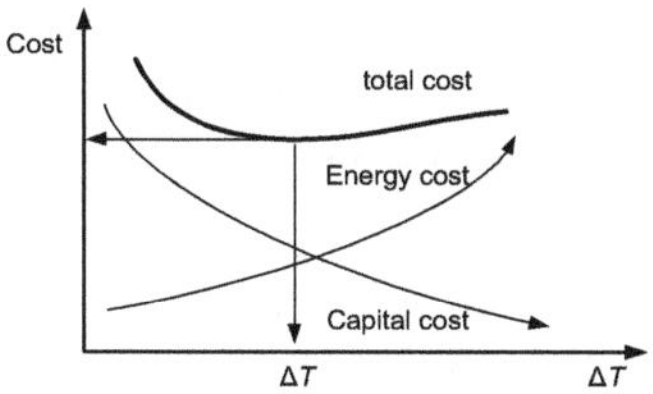

As the temperature difference between hot and cold space increases, the capital cost may decrease while the energy (fuel) cost increases. The total cost should be optimized for the required value for ΔT.

Example 13.13 **Cost saving from heat pump**

A heat pump provides 60 MJ/h to a house. If the compressor requires an electrical energy input of 5 kW, calculate the COP. If electricity costs \$0.08 per kWh and the heat pump operates 100 h per month, how much money does the homeowner save by using the heat pump instead of an electrical resistance heater?

Solution:

The heat pump operates at a steady state.

COP for a heat pump with a heat supply of 60 MJ/h = 16.66 kW:

and $W_{HP} = 5$ kW

$$\mathrm{COP_{HP}} = \frac{q_{out}}{W_{HP}} = 3.33$$

An electrical resistance heater converts all the electrical work supplied W_e into heat q_H. Therefore, to get 16.66 kW into your home, you must buy 16.66 kW of electrical power.

Cost of resistance heater:

Power = 16.66 kW

$$\begin{aligned} &\mathrm{Cost}(\$/\mathrm{month}) \\ &= \mathrm{Electricity}(\$0.08/\mathrm{kWh})\ \mathrm{Power}(16.66\ \mathrm{kW}) \\ &\mathrm{Time}(100\ \mathrm{h/month}) = 133.3\ \$/\mathrm{month} \end{aligned}$$

Cost of heat pump with a power of 5 kW:

$$\begin{aligned} &\mathrm{Cost}(\$/\mathrm{month}) \\ &= \mathrm{Electricity}(\$0.08/\mathrm{kWh})\ \mathrm{Power}(5.0\ \mathrm{kW}) \\ &\mathrm{Time}(100\ \mathrm{h/month}) = 40.0\$/\mathrm{month} \end{aligned}$$

Therefore, monthly savings are (133.3 − 40.0) \$/month = 93.3 \$/month.

Electrical resistance heaters are not very popular, especially in cold climates. The thermal efficiency of a heat pump drops significantly as the outside temperature falls. When the outside temperature drops far enough that the $\mathrm{COP_{HP}} \sim 1$, it becomes more practical to use the resistance heater.

Example 13.14 **Increasing the efficiency of a Rankine cycle by reducing the condenser pressure**

A steam power plant is operating on the simple ideal Rankine cycle. The steam mass flow rate is 20 kg/s. The steam enters the turbine at 3500 kPa and 400°C. Discharge pressure of the steam from the turbine is 78.5 kPa.

(a) Determine the thermal efficiency of the cycle.
(b) If the pressure of the discharge steam is reduced to 15 kPa determine the thermal efficiency.
(c) Determine the annual cost saving if the unit cost of electricity is \$0.10/kWh and avoid emission from a coal-fired plant.

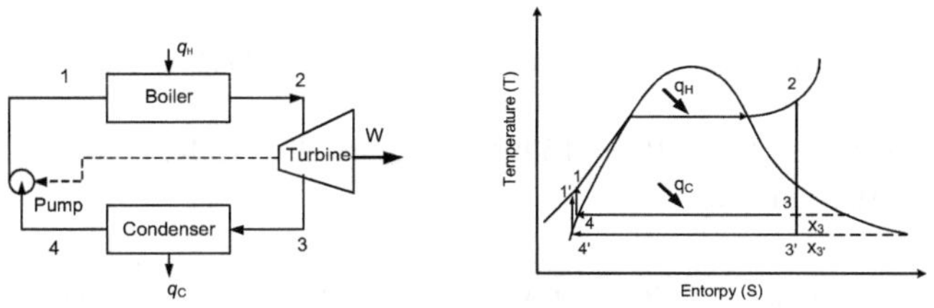

Solution:

Assume that the cycle is at steady-state flow and the changes in kinetic and potential energy are negligible. Efficiency of pump and turbine is 100%.

(a) $\dot{m}_s = 20.0$ kg/s

Using the data from the steam tables

Superheated steam: $P_2 = P_1 = 3500\,\text{kPa},\ H_2 = 3224.2\,\text{kJ/kg}$,

$S_2 = 6.8443\,\text{kJ/kg K},\ T_2 = 400°\text{C}$.

Saturated steam: $P_3 = P_4 = 78.5\,\text{kPa}(T_{\text{sat}} = 366.15\,\text{K}), V_4 = 0.001038\,\text{m}^3/\text{kg}$

$H_{3\ \text{sat vap}} = 2665.0\,\text{kJ/kg},\ H_4 = H_{3\ \text{sat liq}} = 389.6\,\text{kJ/kg},$

$S_{3\ \text{sat vap}} = 7.4416\,\text{kJ/kg K},\quad S_{3\ \text{sat liq}} = 1.2271\,\text{kJ/kg K}$

Basis: 1 kg/s steam. With a pump efficiency of $\eta_{\text{pump}} = 100\%$

$$W_{\text{p,in}} = V_1(P_1 - P_4) = (0.001038)(3500 - 78.5)\left(\frac{1\,\text{kJ}}{1\,\text{kPa}\,\text{m}^3}\right) = 3.55\,\text{kJ/kg}$$

$$H_1 = H_4 + W_{\text{p,in}} = 393.1\,\text{kJ/kg}$$

Isentropic process $S_1 = S_4$ and $S_3 = S_2$.

The quality of the discharged wet steam ($S_2 < S_{3\text{sat vap}}$): $6.8443 < 7.4416$

$$x_3 = (6.8463 - 1.2271)/(7.4416 - 1.2271) = 0.90$$

$$H_3 = 389.6(1 - 0.9) + 2665.0 \times 0.9 = 2437.5\,\text{kJ/kg}$$

Heat interactions: $q_{\text{in}} = H_2 - H_1 = 3224.2 - 393.1 = 2831.1$ kJ/kg
$q_{\text{out}} = -(H_4 - H_3) = -(389.6 - 2437.5) = 2048.0$ kJ/kg (heat received by the cooling medium that is the system in the condenser)

The thermodynamic efficiency of the cycle: $\eta_{\text{th}} = 1 - \frac{|q_{\text{out}}|}{q_{\text{in}}} = 0.276$ or 27.6%

Therefore, the plant uses only 27.6% of the heat it received in the boiler.

Turbine work output: $W_{\text{out}} = H_3 - H_2 = 3224.2 - 2437.5 = -786.7$ kJ/kg.

(b) Steam properties: $P_3 = P_4 = 15\,\text{kPa}, \quad T_{\text{sat}} = 327.15\,\text{K},$
$V_4 = 0.001014\,\text{m}^3/\text{kg}$

$$H_{3\text{sat vap}} = 2599.2\,\text{kJ/kg}, \quad H_4 = H_{3satliq} = 226.0\,\text{kJ/kg},$$

$$S_{3\text{ sat vap}} = 8.0093\,\text{kJ/kg K}, \quad S_{3\text{ sat liq}} = 0.7550\,\text{kJ/kg K}$$

With a pump efficiency of $\eta_{\text{pump}} = 100\%$

$$W_{\text{p,in}} = V_1(P_1 - P_4) = (0.001014)(3500 - 15)\left(\frac{1\,\text{kJ}}{1\,\text{kPa m}^3}\right) = 3.53\,\text{kJ/kg}$$

$$H_1 = H_4 + W_{\text{p,in}} = 226.0 + 3.53 = 229.5\,\text{kJ/kg}$$

Isentropic process $S_1 = S_4$ and $S_3 = S_2$.

The quality of the discharged wet steam ($S_3 < S_{3\text{satvap}}$): 6.8443 < 8.0093

$$x_3' = (6.8443 - 0.7550)/(8.0093 - 0.7550) = 0.84$$

$$H_3' = 226.0(1 - 0.84) + 2599.2 \times 0.84 = 2219.5\,\text{kJ/kg}$$

Heat interactions: $q_{\text{in}} = H_2 - H_1 = 3224.2 - 229.5 = 2994.7$ kJ/kg

$q_{\text{out}} = -(H_4 - H_{3'}) = -(226.0 - 2219.5.5) = 1993.5$ kJ/kg (heat received by the cooling medium that is the system in the condenser)

The thermal efficiency of the cycle: $\eta_{\text{th}} = 1 - \frac{|q_{\text{out}}|}{q_{\text{in}}} = 0.334\, or\, 33.4\%$

Therefore, the plant uses only 33.4% of the heat it received in the boiler:

Turbine work output: $W_{\text{out}} = H_{3'} - H_2 = 2219.5 - 3224.2 = -1004.7$ kJ/kg

(heat produced by the turbine).

Cycle work out: $W_{\text{net}} = (q_{\text{out}} - q_{\text{in}}) = (1993.5 - 2994.7) = -1001.2$ kJ/kg.

Energy gain: (20 kg/s)(1001.2 − 783.1) kJ/kg = 4362 kW.

For 360 days of operation = (360)(24) = 8640 h/year.
The annual saving in the net power output:

(20 kg/s)(1001.2 − 783.1) kJ/kg (8640) h/year = 37,687,680 kWh/year.

For a unit selling price of electricity of \$0.1/kWh, annual saving = \$3,768,768/year.

Avoided CO_2 emissions from coal-fired plant with 0.37 kg CO_2/kWh from Table 11.2.

(0.37 kg CO_2/kWh) (37,687,680 kWh/year) = 13,944,441 kg CO_2/year.

The thermal efficiency increased from 0.276 to 0.334 by reducing the condenser pressure from 78.5 to 15.0 kPa. However, the quality of the discharged steam decreased from 0.9 to 0.84, which is not desirable for the blades of the turbine. Savings are considerable.

Example 13.15 **Economic and ecological gain of a Rankine cycle by increasing the boiler pressure**

A steam power plant is operating on the simple ideal Rankine cycle. The steam mass flow rate is 20 kg/s. The steam enters the turbine at 3500 kPa and 400°C. Discharge pressure of the steam from the turbine is 78.5 kPa.

(a) If the pressure of the boiler is increased to 9800 kPa while maintaining the turbine inlet temperature at 400°C, determine thermal efficiency.
(b) Determine the annual saving if the unit cost of electricity is \$0.10/kWh and avoided emission from a natural gas-powered plant.

Solution:

Assume that the cycle is at steady-state flow and the changes in kinetic and potential energy are negligible. Pump efficiency of $\eta_{pump} = 1$.

Basis: 1 kg/s steam. Using the data from the steam tables.
From Example 9.2 part a: $\eta_{th} = 0.276$ and $W_{net} = 783.1$ kJ/kg at 3500 kPa.

(a) Superheated steam:

$$P_2 = P_1 = 9800\,\text{kPa};\ H_2 = 3104.2\,\text{kJ/kg},\ S_2 = 6.2325\,\text{kJ/kg K},\ T_2 = 400°\text{C}$$

Saturated steam: $P_3 = P_4 = 78.5\text{kPa},\quad T_{sat} = 366.15\text{K},\ V_4 = 0.001038\text{m}^3/\text{kg}$

$$H_{3\ sat\ vap} = 2665.0\,\text{kJ/kg},\ H_4 = H_{3\ sat\ liq} = 389.6\,\text{kJ/kg},$$
$$S_{3\ sat\ vap} = 7.4416\,\text{kJ/kg K},\ S_{3\ sat\ liq} = 1.2271\,\text{kJ/kg K}$$

$$W_{p,in} = V_1(P_1 - P_4) = (0.001038)(9800 - 78.5)\left(\frac{1\,\text{kJ}}{1\,\text{kPa}\,\text{m}^3}\right) = 10.1\,\text{kJ/kg}$$

$$H_1 = H_4 + W_{p,in} = 389.6 + 10.1 = 399.7\,\text{kJ/kg}$$

Isentropic process $S_1 = S_4$ and $S_3 = S_2$.

The quality of the discharged wet steam x_3: ($S_3 < S_{3\text{sat vap}}$): $6.2325 < 7.4416$

$$x_3 = (6.2325 - 1.2271)/(7.4416 - 1.2271) = 0.80$$

$$H_3 = 389.6(1 - 0.8) + 2665.0 \times 0.8 = 2210.0\,\text{kJ/kg}$$

Heat interactions:

$$q_{in} = H_2 - H_1 = 3104.2 - 399.7 = 2704.5\,\text{kJ/kg}$$

$q_{out} = -(H_4 - H_3) = -(389.6 - 2210.0) = 1820.4\,\text{kJ/kg}$ (heat received by the cooling medium in condenser)

The thermodynamic efficiency of the cycle: $\eta_{th} = 1 - \frac{q_{out}}{q_{in}} = 0.326$ or 32.6%

Therefore, the plant uses only 32.6% of the heat it receives in the boiler.

The thermal efficiency increased from 0.276 to 0.326 by increasing the boiler pressure from 3500 to 9800 kPa. However, the quality of the discharged steam decreased from 0.9 to 0.80, which is not desirable for the blades of the turbine.

Turbine work output: $W_{out} = H_3 - H_2 = (2210.0 - 3104.2)\,\text{kJ/kg} = -894.2\,\text{kJ/kg}$.

Cycle work out: $W_{net} = (q_{out} - q_{in}) = (1820.4 - 2704.5)\,\text{kJ/kg} = -884.1\,\text{kJ/kg}$.

(b) $\dot{m}_s = 20.0$ kg/s, and for 360 days of operation = (360)(24) = 8640 h/year

Annual gain power output:

$(20\,\text{kg/s})(884.1 - 783.1)\,\text{kJ/kg}\ (8640)\,\text{h/year} = 17{,}452{,}800\,\text{kWh/year}$.

For a unit selling price of electricity of \$0.1/kWh: annual saving = \$1,745,280.

Avoided CO_2 emissions from natural gas-fired plant with 0.23 kg CO_2/kWh from Table 11.2.

(0.23 kg CO_2/kWh) (17,452,800 kWh/year) = 4,014,144 kg CO_2/year.

Example 13.16 **Economic and ecological gains from a Rankine cycle by increasing the boiler temperature**

A steam power plant is operating on the simple ideal Rankine cycle. The steam mass flow rate is 20 kg/s. The steam enters the turbine at 3500 kPa and 400°C. Discharge pressure of the steam from the turbine is 78.5 kPa.

(a) If the temperature of the boiler is increased to 525°C while maintaining the pressure at 3500 kPa, determine the thermal efficiency.
(b) Determine the annual saving if the unit cost of electricity is $0.10/kWh and the avoided emission from a natural-gas powered plant.

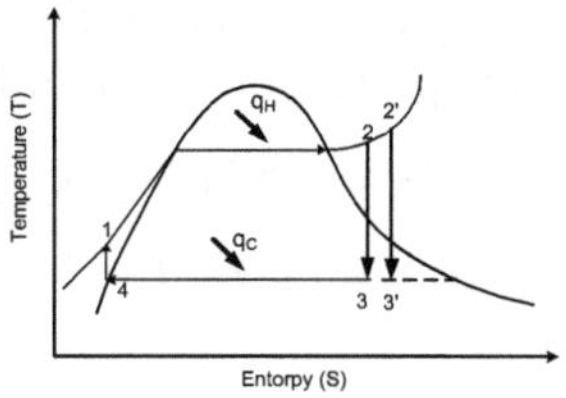

Solution:

Assume that the cycle is at steady-state flow and the changes in kinetic and potential energy are negligible.

(a) Basis: 1 kg/s steam. Using the data from the steam tables:

From Example 9.2 part a: $\eta_{th} = 0.276$ and $W_{net} = 783.1$ kJ/kg at 400°C.

Superheated steam: $P_2 = P_1 = 3500\,\text{kPa}$; $H_2 = 3506.9\,\text{kJ/kg}$;

$S_2 = 7.2297\,\text{kJ/kg K}, \quad T_2 = 525°\text{C}$.

Saturated steam: $P_3 = P_4 = 78.5\text{kPa}, \; T_{sat} = 366.15\,\text{K}, \; V_4 = 0.001038\,\text{m}^3/\text{kg}$

$$H_{3\text{ sat vap}} = 2665.0\,\text{kJ/kg}, H_{3\text{ sat liq}} = 389.6\,\text{kJ/kg},$$
$$S_{3\text{ sat vap}} = 7.4416\,\text{kJ/kg K}, S_{3\text{ sat liq}} = 1.2271\,\text{kJ/kg K}$$

$$W_{p,in} = V_1(P_1 - P_4) = (0.001038)(3500 - 78.5)\left(\frac{1\,\text{kJ}}{1\,\text{kPa m}^3}\right) = 3.55\,\text{kJ/kg}$$

$$H_1 = H_4 + W_{p,in} = 389.6 + 3.55 = 393.15\,\text{kJ/kg}$$

Isentropic process $S_1 = S_4$ and $S_3 = S_2$.

The quality of the discharged wet steam ($S_2 < S_{3\text{sat vap}}$): 7.2297 < 7.4416

$$x_{3'} = (7.2297 - 1.2271)/(7.4416 - 1.2271) = 0.96$$

$$H_{3'} = 389.6(1 - 0.96) + 2665.0 \times 0.96 = 2574.0\,\text{kJ/kg}$$

Heat interactions: $q_{\text{in}} = H_2 - H_1 = 3506.9 - 393.1 = 3113.8$ kJ/kg

$q_{\text{out}} = -(H_4 - H_{3'}) = -(389.6 - 2574.0) = 2184.4$ kJ/kg

(heat received by the cooling medium)

The thermodynamic efficiency of the cycle is $\eta_{th} = 1 - \frac{q_{\text{out}}}{q_{\text{in}}} = 0.298$ or 29.8%

Therefore, the plant uses only 29.8% of the heat it receives in the boiler.

The thermal efficiency increased from 0.276 to 0.298 by increasing the boiler temperature from 400 to 525°C. The quality of the discharged steam increased from 0.9 to 0.96, which is desirable for the protection of the turbine blades.

Turbine work out: $W_{\text{out}} = H_{3'} - H_2 = (2574.0 - 3506.9) = -932.9$ kJ/kg.

Cycle work out: $W_{\text{net}} = (q_{\text{out}} - q_{\text{in}}) = (2184.4 - 3113.8) = -929.4$ kJ/kg.

(b) $\dot{m}_s = 20.0$ kg/s and for 360 days of operation = (360)(24) = 8640 h/year

Annual increase in the power output:

(20 kg/s)(929.4 – 783.1) kJ/kg (8640) h/year = 25.280,640 kWh/year.

With a unit selling price of electricity of \$0.1/kWh: Annual saving = \$2,528,064.

Avoided CO_2 emissions from natural gas-fired plant with 0.23 kg CO_2/kWh from Table 11.2

(0.23 kg CO_2/kWh) (25.280,640 kWh/year) = 5,814,547 kg CO_2/year.

Example 13.17 **Economic and ecological gains in a two-stage compression work by intercooling**

Air with a flow rate of 2 kg/s is compressed in a steady state and reversible process from an inlet state of 100 kPa and 300 K to an exit pressure of 1000 kPa. Estimate the work for (a) polytropic compression with $\gamma = 1.3$, and (b) ideal two stage polytropic compression with intercooling using the same polytropic exponent of $\gamma = 1.3$, (c) estimate conserved compression work by intercooling and electricity per year if the unit cost of electricity is \$0.15/kWh and the compressor is operated 360 days per year. Estimate the avoided emissions from a natural gas fired plant.

Solution:

Assumptions: steady state operation; air is ideal gas; kinetic and potential energies are negligible.

$$P_2 = 1000\,\text{kPa},\ P_1 = 100\,\text{kPa},\ \ T_1 = 300\,\text{K},\ \ \gamma = 1.3,\ \text{MW}_{\text{air}} = 29\,\text{kg/kmol}$$

Basis 1 kg/s airflow rate.

(a) Work needed for polytropic compression with $\gamma = 1.3$

$$W_{\text{comp}} = \frac{\gamma \text{RT}_1}{\text{MW}(\gamma - 1)}\left[\left(\frac{P_2}{P_1}\right)^{(\gamma-1)/\gamma}\right]$$

$$= \frac{1.3(8.314\,\text{kJ/kmol K})(300\,\text{K})}{29\,\text{kg/kmol}\,(1.3-1)}\left[\left(\frac{1000}{300}\right)^{(1.3-1)/1.3} - 1\right]$$

$$= 261.3\,\text{kJ/kg}$$

(b) Ideal two stage polytropic compression with intercooling ($\gamma = 1.3$)

$$P_x = (P_1 P_2)^{1/2} = (100 \times 1000)^{1/2} = 316.2\,\text{kPa}$$

$$W_{\text{comp}} = \frac{2\gamma \text{RT}_1}{\text{MW}(\gamma - 1)}\left[\left(\frac{P_x}{P_1}\right)^{(\gamma-1)/\gamma}\right]$$

$$= \frac{2(1.3)(8.314\,\text{kJ/kmol K})(300\,\text{K})}{29\,\text{kg/kmol}\,(1.3-1)}\left[\left(\frac{316.2}{100}\right)^{(1.3-1)/1.3} - 1\right]$$

$$= 226.8\,\text{kJ/kg}$$

Recovered energy = (261.3 – 226.8) kJ/kg = 34.5 kJ/kg

Reduction in energy use: $\frac{261.3-226.8}{261.3} = 0.13$ or 13%

(c) Conservation of compression work = (2 kg/s) (261.3 – 226.8) kJ/kg = 69 kW

Yearly conserved work: (69 kW) (8640 h/year) = 596,160 kWh/year.

Saving in electricity (2,385,745 kWh/year) ($0.15/kWh) = = $89,424/year.
The compression work has been reduced by 13% when two stages of polytropic compression are used instead of single polytropic compression and conserved 596,160 kWh/year.

Avoided CO_2 emissions from natural gas-fired plant with 0.23 kg CO_2/kWh from Table 11.2

(0.23 kg CO_2/kWh) (596,160 kWh/year) = 137,116 kg CO_2/year.

Example 13.18 **Economic and ecological gains from compression work**

Ten adiabatic compressors are used to compress air from 100 kPa and 290 K to a pressure of 900 kPa at a steady-state operation. The isentropic efficiency of the compressor is 80%. The airflow rate is 0.4 kg/s. Determine the minimum and actual power needed by the compressor and the avoided emissions from a coal-fired plant.

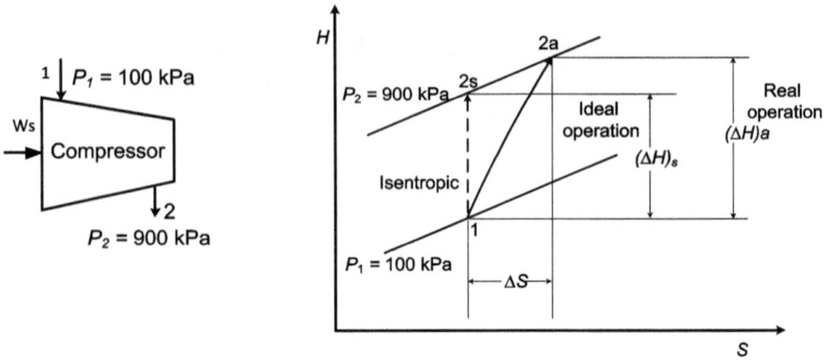

Solution:

Assume: Steady-state adiabatic operation. Air is an ideal gas. The changes in kinetic and potential energies are negligible.

Enthalpies of ideal gas depend on temperature only.

The air mass flow rate = 0.4 kg/s, $\eta_C = 0.8$

Inlet conditions (Table D1): $P_1 = 100\,\text{kPa},\ T_1 = 290\,\text{K},$
$H_1 = 290.16\,\text{kJ/kg},\ \text{Pr}_1 = 1.2311.$

Exit conditions (Table D1): $P_2 = 900$ kPa

$$\text{Pr}_2 = \text{Pr}_1 \frac{P_2}{P_1} = 1.2311\left(\frac{900\,\text{kPa}}{100\,\text{kPa}}\right) = 11.1$$

From Table D1: for $\text{Pr}_2 = 11.1$, we find$T_2 = 540$ K and $H_{2\text{s}} = 544.35$ kJ/kg

$\eta_C = \frac{H_{2s}-H_1}{H_{2a}-H_1} = 0.8 \rightarrow H_{2\text{a}}H_{2\text{a}} = 607.9\,\text{kJ/kg}$ and $T_2 = 600\,\text{K}$ (Table D1) (Approximate)

As seen from the *TS* diagram above $H_{2\text{a}}$ is the actual enthalpy at the exit

Actual power:

$\dot{m}\Delta H_a = \dot{W}_{\text{net,in}} = \dot{m}(H_{2a} - H_1) = 0.4$ kg/s(607.9 – 290.16) kJ/kg $= 127.1\,\text{kW}.$

Minimum power:

$\dot{m}\Delta H_s = \dot{W}_{\text{net,in}} = \dot{m}(H_{2s} - H_1) = 0.4$ kg/s(544.3 – 290.1) kJ/kg $= 101.7\,\text{kW}.$

If the operation was ideal the rate of conserved energy would be 2540 kW for ten compressors.

Yearly gain: $2540\,\text{kW}(24\,\text{h/day})(365\,\text{day/year}) = 22{,}250{,}400\,\text{kWh/year}$.

Avoided CO_2 emissions from coal-fired plant with 0.37 kg CO_2/kWh from Table 11.2

(0.37 kg CO_2/kWh) (22,250,400 kWh/year) = 8,232,648 kg CO_2/year.

Example 13.19 **Economic and ecological gain from heating a house by heat pump**

A heat pump is used to heat a house and maintain it at 18°C. On a day where the outside temperature is – 2°C, the house is losing heat at a rate of 75,000 kJ/h. The heat pump operates with a coefficient of performance (COP) of 2.8. Determine:

(a) Power needed by the heat pump,
(b) The rate of heat absorbed from to the surrounding cold air
(c) Cost saving and the avoided emissions from a natural-gas powered plant.

Solution:

Assume: Steady-state operation.

(a) House is maintained at 18°C.

Heat pump must supply the same amount of lost heat from the cold source:

$q_H = 75{,}000$ kJ/h = 20.83 kW.

Power required by the heat pump:

$$\text{COP}_\text{HP} = \frac{\dot{q}_H}{\dot{W}_\text{net,in}} \rightarrow \dot{W}_\text{net,in} = \frac{\dot{q}_H}{\text{COP}_\text{HP}} = \frac{20.83\ \text{kW}}{2.8} = 7.44\ \text{kW}.$$

(b) Energy balance: $\dot{q}_H - \dot{q}_C = \dot{W}_\text{net,in} = 7.44$ kW

$$\dot{q}_C = \dot{q}_H - \dot{W}_\text{net,in} = (20.83 - 7.44)\,\text{kW} = 13.39\,\text{kW}$$

13.39 kW is extracted from the outside. The house is paying only for the energy of 7.44 kW that is supplied as electrical work to the heat pump.

If we have to use electricity in a resistance heater, we have to supply 20.83 kW.

Energy conserved by using heat pump instead of electric heater is:

13.39/20.83 = 0.64 (or 64%).

Energy conserved by using heat pump instead of electric heater is:

(20.83–13.39) kW = 7.44 kW.

Yearly gain: 7.44 kW (24 h/day)(365 day/year) = 65,174.4 kWh/year.

Avoided CO_2 emissions from natural gas-fired plant with 0.23 kg CO_2/kWh from Table 11.2

(0.37 kg CO_2/kWh) (65,174.4 kWh/year) = 8,232,648 kg CO_2/year.

Example 13.20 **Economic and ecological gains from air conditioner**

Estimate the cost of electricity saving for a 5000 Btu/h (1500 W) air-conditioning unit operating, with a SEER of 10 Btu/Wh. The unit is used for a total of 1500 h during an annual cooling season and the unit cost of electricity is $0.14/kWh. Also, estimate the total cost of electricity.

Solution:

Air conditioner sizes are often given as "tons" of cooling where 1 ton of cooling is equivalent to 12,000 Btu/h (3500 W).

The unit considered is a small unit and the annual total cooling output would be:

$$(5000\,\text{Btu/h}) \times (1500\,\text{h/year}) = 7{,}500{,}000\,\text{Btu/year}$$

With a seasonal energy efficiency ratio (SEER) of 10, the annual electrical energy usage:

$$(7{,}500{,}000\,\text{Btu/year})/(10\,\text{Btu/Wh}) = 750{,}000\,\text{Wh/year} = 750\,\text{kWh/year}$$

With a unit cost of electricity of $0.14/kWh, the annual cost:

(750 kWh/year)($0.14/kWh) = $105/year.

The average power usage may also be calculated more simply by:

Average power = (Btu/h)/(SEER, Btu/Wh) = 5000/10 = 500 W = 0.5 kW.

With the electricity cost of $0.14/kWh, the cost per hour:

(0.5 kW) ($0.14)/kWh = $0.07/h.

For 1500 h/year, the total cost: ($0.07/h) (1500 h/year) = $105.

Example 13.21 **Economic and ecological gains of cooling with a unit at a higher SEER rating**

Estimate the annual cost of electric power consumed by a 6-ton air conditioning unit operating for 2000 h per year with a SEER rating of 10 and a power cost of $0.16/kWh.

Solution:

Air conditioner sizes are often given as "tons" of cooling:

1 ton of cooling = 12,000 Btu/h (3500 W).

This is approximately the power required to melt one ton of ice in 24 h.

The annual cost of electric power consumption:

(6)(12,000 Btu/h) = 72,000 Btu/h

$$\text{Cost} = \frac{(\text{Size, Btu/h})(\text{time, h/year})\left(\text{Cost of energy, \$/kWh}\right)}{(\text{SEER, Btu/Wh})(1000\ \text{W/kW})}$$

(72,000 Btu/h)(2000 h/year)($0.16/kWh)/[(10 Btu/Wh)(1000 W/kW)]

= $2304/year.

For the temperatures of hot and cold sources:

$$T_H = 273.15 + 20\,^{\circ}\text{C} = 293.15\,\text{K},\ T_C = 273.15 - 4\,^{\circ}\text{C} = 269.15\,\text{K},\ \text{and}$$
$$T_H - T_C = 24\,\text{K}$$

$$\text{EER}_{\text{Carnot}} = 3.41\left(\frac{T_C}{T_H - T_C}\right) = 3.41(269.13/24) = 38.24$$

The maximum EER decreases as the difference between the inside and outside air temperature increases.

***Example 13.22* Cost of coal in a steam power plant**

An adiabatic turbine is used to produce electricity by expanding a superheated steam at 4100 kPa and 350°C. The power output is 50 MW. The steam leaves the turbine at 40 kPa and 100°C. If the combustion efficiency is 0.75 and the generator efficiency is 0.9, determine the yearly cost of coal and amount of emissions.

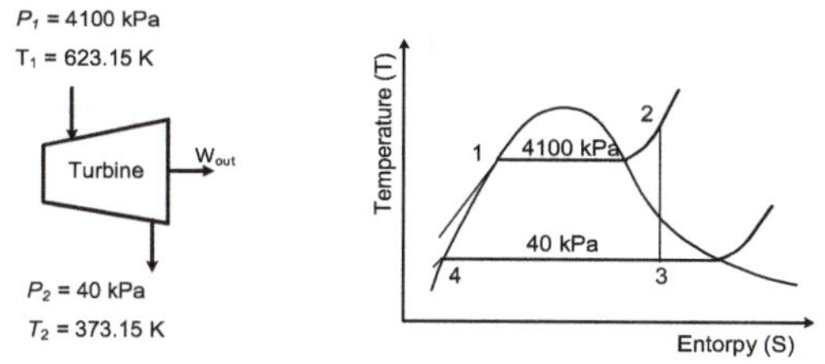

Solution:

Assume: Steady-state adiabatic operation. The changes in kinetic and potential energies are negligible.

Basis: stems flow rate = 1 kg/s.

Combustion efficiency $\eta_{\text{comb}} = 0.75$, Generator efficiency $\eta_{\text{gen}} = 0.9$

Power output = 50 MW = 50,000 kW.

From steam tables. Turbine inlet conditions:

$$P_2 = 4100\,\text{kPa},\ T_2 = 623.15\,\text{K},\ H_2 = 3092.8\,\text{kJ/kg},\ S_2 = 6.5727\,\text{kJ/kg K}$$

Turbine outlet conditions:

$P_4 = P_3 = 40$ kPa, $T_3 = 373.15$ K, $V_4 = 1.027\ \text{cm}^3/\text{g}\ = 0.001027\ \text{m}^3/\text{kg}$

$$S_{4\text{sat vap}} = 7.6709\,\text{kJ/kg K},\ S_{4\text{sat liq}} = 1.2026\,\text{kJ/kg K},$$
$$H_{4\text{sat vap}} = 2636.9\,\text{kJ/kg},\ H_{4\text{sat liq}} = 317.6\,\text{kJ/kg}\ (40\,\text{kPa})$$

$$W_{\text{p,in}} = V_4(P_1 - P_4) = \left(0.001027\,\text{m}^3/\text{kg}\right)$$
$$(4100 - 40)\,\text{kPa}\,\left(1\,\text{kJ}/1\,\text{kPa m}^3\right) = 4.2\,\text{kJ/kg}$$

$$H_1 = H_4 + W_{\text{p,in}} = 317.6 + 4.2 = 321.8\,\text{kJ/kg}$$

For the isentropic operation $S_3 = S_2 = 6.5727$ kJ/kg K

Since $S_2 < S_{3\text{sat vap}}$: 6.5727 < 7.6709 the steam at the exit is saturated liquid–vapor mixture.

Quality of that mixture: $x_3 = \frac{S_2 - S_{2\text{sat liq}}}{S_{2\text{sat vap}} - S_{2\text{sat liq}}} = \frac{6.5727 - 1.2026}{7.6709 - 1.2026} = 0.83$

$$H_3 = (1 - x_3)H_{3\text{sat liq}} + x_3 H_{3\text{sat vap}} = 2243.1\ \text{kJ/kg}$$

Heat interactions: $\dot{q}_{\text{in}} = \dot{m}(H_2 - H_1) = (3092.8 - 321.8)\ \text{kJ/kg}\ = 2771.0\,\text{kJ/s}$

$\dot{q}_{\text{out}} = -\dot{m}(H_4 - H_3) = -(317.6 - 2243.1)\ \text{kJ/kg}\ = 1925.5\ \text{kJ/s}$

(heat received by the cooling medium that is the system in the condenser)

Thermal efficiency: $\eta_{th} = 1 - \frac{q_{\text{out}}}{q_{\text{in}}} = 0.305$ or 30.5%

Overall plant efficiency:

$\eta_{\text{overall}} = \eta_{\text{th}}\eta_{\text{comb}}\eta_{\text{gen}} = (0.305)(0.75)(0.9) = 0.205$ or 20.5%

The rate of coal energy required: $\dot{E}_{\text{coal}} = \frac{|\dot{W}_{\text{net}}|}{\eta_{\text{overall}}} = \frac{50{,}000\ \text{kW}}{0.205} = 243{,}900\ \text{kJ/s}.$

Energy of coal (bituminous) = 24,000 kJ/kg (Table 2.6)

$$\dot{m}_{\text{coal}} = \frac{\dot{E}_{\text{coal}}}{\eta_{\text{overall}}} = \frac{243{,}900}{24{,}000}\left(\frac{1\text{tonne}}{1000\text{kg}}\right) = 0.0101/\text{s} \;= 36.6\,\text{tonne/h}$$

Annual average price of bituminous coal = \$60/tonne (2018).

Annual consumption of coal: (36.6 tonne/h) (24 h/day) (365 day/year)

= 320,616 ton/year.

Annual cost of coal: (320,616 tonne/year) (\$60 tonne) = \$19,236,960/year.

Emission from coal: 2.3 kg CO_2/kg coal (Table 11.2); short ton = 2000 lb = 907.1 kg.

Annual emission from coal:

(320,616 ton/year) (907.1 kg/ton) (2.3 kg CO_2/kg coal) = 668,910,779 kg CO_2/year.

Example 13.23 **Economic and ecological gains with compact fluorescent bulbs**

Assume that the average residential rate of electricity is \$0.14/kWh and a household consumes about 10,000 kWh per year. If the lighting is provided by compact fluorescent bulbs only, estimate the yearly cost saving and avoided emissions from a natural gas = fired power plant.

Solution:

Assume that 11% of the energy budget of households is for lighting, and compact fluorescent bulbs consume about 75% less energy than standard lighting.

Annual energy use and cost with standard lighting:

(10,000 kWh/year) (0.11) = 1100 kW/year.

(1100 kW/year)(\$0.14/kWh) = \$154/year.

Annual energy use and cost with compact fluorescent bulbs:

(1100 kW/year) (1– 0.75) = 275 kWh/year.

(275 kWh/year) (\$0.14/kWh) = \$38.5/year.

Annual energy and cost savings:

(1100 – 275) kWh = 825 kWh.

(154 – 38.5)\$/year = \$115 5/year.

Avoided emission from coal-fired plant:

(825 kWh)(24 h/day)(365 day/year) (0.37 kgCO_2/kg coal) = 2,673,990 kg CO_2.

Conservation of energy and cost with compact fluorescent bulbs are considerable.

Problems

13.1. An overused car may consume around 150 gallons gasoline per month. Estimate the cost of energy consumed by the car per year.

13.2. A city consumes natural gas at a rate of 500×10^6 ft^3/day. The volumetric flow is at standard conditions of 60°F and 1 atm = 14.7 psia. If the natural is costing $12/GJ of higher heating value what is the daily cost of the gas for the city.

13.3. A city consumes natural gas at a rate of 800×10^6 ft^3/day. The volumetric flow is at standard conditions of 60°F and 1 atm = 14.7 psia. If the natural is costing $10/GJ of higher heating value what is the daily cost of the gas for the city.

13.4. A car consumes about 6 gallons a day, and the capacity of a full tank is about 15 gallons. The density of gasoline ranges from 0.72 to 0.78 kg/l (Table 2.2). The lower heating value of gasoline is about 44,000 kJ/kg. Assume that the average density of gasoline is 0.75 kg/l. If the car was able to use 0.2 kg of nuclear fuel of uranium-235, estimate the time in years for refueling and avoided cost of gasoline.

13.5. A car consumes about 3 gallons a day, and the capacity of the full tank is about 11 gallons. The density of gasoline ranges from 0.72 to 0.78 kg/l (Table 2.2). The lower heating value of gasoline is about 44,000 kJ/kg. Assume that the average density of gasoline is 0.75 kg/l. If the car was able to use 0.1 kg of nuclear fuel of uranium-235, estimate the time in years for refueling and avoided cost of gasoline.

13.6. When a hydrocarbon fuel is burned, almost all the carbon in the fuel burns completely to form CO_2 (carbon dioxide), which is the principal gas causing the greenhouse effect and thus global climate change. On average, 0.59 kg of CO_2 is produced for each kWh of electricity generated from a power plant that burns natural gas. A typical new household uses about 7000 kWh of electricity per year. Determine the amount of CO_2 production and carbon fee in a city with 100,000 households.

13.7. When a hydrocarbon fuel is burned, almost all the carbon in the fuel burns completely to form CO_2 (carbon dioxide), which is the principal gas causing the greenhouse effect and thus global climate change. On average, 0.59 kg of CO_2 is produced for each kWh of electricity generated from a power plant that burns natural gas. A typical new household uses about 10,000 kWh of electricity per year. Determine the amount of CO_2 production and carbon fee in a city with 250,000 households.

13.8. A large public computer lab operates Monday through Saturday. There the computers are either being used constantly or remain on until the next user comes. Each computer needs around 240W. If the computer lab contains 53 computers and each is on for 12 h a day, during the year how much CO_2 will the local coal power plant have to release and carbon fee to be paid to keep these computers running?

13.9. The average university will have a large public computer lab open Monday through Saturday. There the computers are either being used constantly or remain on until the next user comes. Each computer needs around 13.10 W. If the computer lab contains 53 computers and each is on for 12 h a day, during the year how much coal will the local coal power plant have to consume to keep these computers running?

13.10. A large public computer lab runs six days per week from Monday through Saturday. Each computer uses a power of around 120 W. If the computer lab contains 45 computers and each is on for 12 h a day, during the year how much CO_2 will the local coal power plant have to release to the atmosphere in gram moles to keep these computers running?

13.11. If a car consumes 60 gallons of gasoline per month. Estimate the energy consumed by the car per year.

13.12. A car having an average 22 miles/gal is used 32 miles every day. If the cost of a gallon fuel is \$3.8 estimate the yearly cost of fuel.

13.13. A car having an average of 22 miles/gal is used 32 miles every day. Estimate yearly energy usage.

13.14. A 150-Watt electric light bulb is used on average 10 h per day. A new bulb costs \$2.0 and lasts about 5000 h. If electricity cost is \$0.15/kWh, estimate the yearly cost of the bulb.

13.15. A laptop consuming 90 Watt is used on average 5 h per day. If a laptop costs \$500 and will be used for four years estimate the total electricity cost in four years for the laptop. Electricity cost is \$0.15/kWh.

13.16. A laptop consuming 90 Watt is used on average 7 h per day. If a laptop costs \$500 and will be used for four years estimate the total electricity cost in four years for the laptop. Electricity cost is \$0.10/kWh.

13.17. A 20-hP electric motor is used to pump ground water into a storage tank for four hours every day. Estimate the work done by the pump in kW every year and the cost of electricity every year. Assume that the electricity unit cost is \$0.1/kWh.

13.18. A city consumes natural gas at a rate of 250×10^6 ft^3/day. The volumetric flow is at standard conditions of 60°F and 1 atm = 14.7 psia. If natural gas costs \$6/GJ of higher heating value what is the daily cost of the gas for the city.

13.19. A home consumes natural gas at a rate of 4.3ft^3/day to heat the home. The volumetric flow is at standard conditions of 60°F and 1 atm = 14.7 psia. If natural gas is costing \$0.67/MJ of higher heating value what is the daily cost of the gas for the home?

13.20. A water heater consumes propane, which provides 80% of the standard heat of combustion when the water produced after combustion is vapor. If the price of propane is \$2.2/gal measured at 25°C. What is the heating cost in \$ per million Btu and in \$ per MJ?

13.21. An average video games system consumes 170 W of power during gameplay. If a person were to play an hour a day for 80% of the year how many liters of gasoline would the person have burned and what is the cost? (Evaluated at HHV).

13.22. Assuming complete combustion of methanol and gasoline (assuming 100% octane) estimate the HHVs by using the oxidations and carbon dioxide intensity in $kgCO_2/MJ$.

13.23. Assuming complete combustion of ethanol and diesel (assuming 100% dodecane) estimate the HHVs by using the oxidation and carbon dioxide intensity in $kgCO_2/MJ$.

13.24. Assuming complete combustion of butanol and natural gas estimate the HHVs by using the oxidation and carbon dioxide intensity in $kgCO_2/MJ$.

13.25. Assuming complete combustion of ethanol and bituminous coal estimate the HHVs by using the oxidations and carbon dioxide intensity in $kgCO_2/MJ$.

13.26. Consider the exergy cost on a boiler and turbine system shown below

A turbine produces 100 MW of electricity per year. The average cost of the steam is \$0.05/(kWh) of exergy (fuel). The total cost of the unit (fixed capital investment and operating costs) is $\$2.1 \times 10^5$. If the turbine exergetic efficiency increases from 80 to 85%, after an increase of 3% in the total cost of the unit, evaluate the change of the unit cost of electricity.

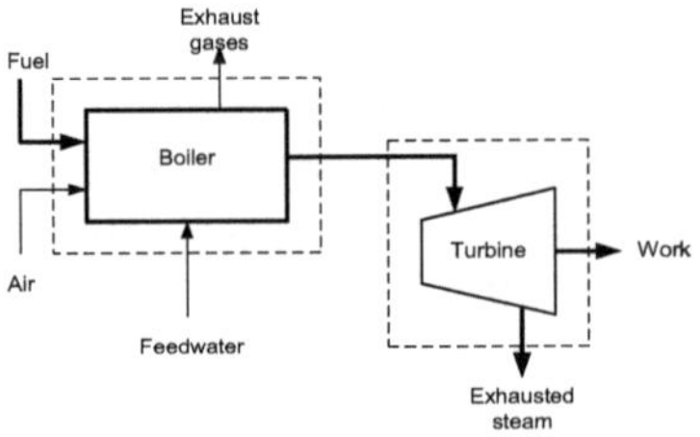

13.27. Consider the exergy cost on a boiler and turbine system shown in Pr. 13.26. A turbine produces 80 MW of electricity per year. The average cost of the steam is \$0.05/(kWh) of exergy (fuel). The total cost of the unit (fixed capital investment and operating costs) is $\$1.51 \times 10^5$. If the turbine exergetic efficiency increases from 75 to 87%, after an increase of 2% in the total cost of the unit, evaluate the change of the unit cost of electricity.

13.28. Consider the exergy cost on a boiler and turbine system shown in Pr. 13.26. A turbine produces 200 MW of electricity per year. The average cost of the steam is \$0.05/(kWh) of exergy (fuel). The total cost of the unit (fixed capital investment and operating costs) is $\$5.5 \times 10^5$. If the turbine exergetic efficiency increases from 80 to 88%, after an increase of 2% in the total cost of the unit, evaluate the change of the unit cost of electricity.

13.29. A typical square two-story home with a roof surface area of 1260 ft^2 and a wall surface area of 2400 ft^2 is to be heated with solar energy storage using a salt hydrate as phase change material. It presently has insulation of 6 inches on the roof and 1 inch in the walls. Inside temperature will be held at 70°F and expected outside low temperature is 10°F. Average solar radiation is 650 Btu/ft^2 and the cost of solar air collector is \$3/ft^2. The salt hydrate costs around \$0.15/lb. Estimate the costs of salt hydrate and solar air collectors.

13.30. A large building with a roof surface area of 3700 ft^2 and a wall surface area of 5000 ft^2 is to be heated with solar energy storage using a salt hydrate as phase change material. It presently has insulation thickness of 5 inches in the roof and 1 inch on the walls. Inside temperature will be held at 70°F and expected outside low temperature is 20°F. Average solar radiation is 700 Btu/ft^2 and the cost of solar air collector is \$5.1/ft^2. The salt hydrate costs around \$0.15/lb. Estimate the costs of salt hydrate and solar air collectors.

13.31. A typical square two-story home with a roof surface area of 1260 ft^2 and a wall surface area of 2400 ft^2 is to be heated with solar energy storage using paraffin as phase change material. It presently has insulation of 6 inches on the roof and 1 inch in the walls. Inside temperature will be held at 70°F and an expected outside low temperature is 10°F. Average solar radiation is 300 Btu/ft^2 and the cost of solar air collector is \$4.5/ft^2. The paraffin costs around \$0.4/lb. Estimate the costs of salt hydrate and solar air collectors.

13.32. A large building with a roof surface area of 3200 ft^2 and a wall surface area of 4200 ft^2 is to be heated with solar energy storage using octadecane as phase change material. It presently has insulation of 6 inches on the roof and 1 inch in the walls. Inside temperature will be held at 70°F and expected outside low temperature is 10°F. Average solar radiation is 350 Btu/ft^2 and the cost of solar air collector is \$3.4/ft^2. The octadecane costs around \$0.45/lb. Estimate the costs of salt hydrate and solar air collectors.

13.33. A power plant is operating on an ideal Brayton cycle with a pressure ratio of $r_p = 9$. The fresh air temperature at the compressor inlet is 295 K. The air temperature at the inlet of the turbine is 1200 K. The cycle operates with a compressor efficiency of 75% and a turbine efficiency of 80%. The unit cost of fuel is \$0.10/kWh. The cycle operates 360 days per year.
   (a) Using the standard-air assumptions, determine the thermal efficiency of the cycle.
   (b) If the power plant operates with a regenerator with an effectiveness of 0.78, determine the thermal efficiency of the cycle and annual saving of fuel cost.

13.34. A steam power plant is operating on the simple ideal Rankine cycle. The steam mass flow rate is 25 kg/s. The steam enters the turbine at 3500 kPa and 400°C. The discharge pressure of the steam from the turbine is 90 kPa.
   (a) Determine the thermal efficiency of the cycle.
   (b) If the pressure of the discharge steam is reduced to 20 kPa determine the thermal efficiency.
   (c) Determine the annual cost saving if the unit cost of electricity is \$0.10/kWh and avoid emission from a coal-fired plant.

13.35. A steam power plant is operating on the simple ideal Rankine cycle. The steam mass flow rate is 40 kg/s. The steam enters the turbine at 3500 kPa and 400°C. The discharge pressure of the steam from the turbine is 85 kPa.

(a) Determine the thermal efficiency of the cycle.

(b) If the pressure of the discharge steam is reduced to 25 kPa, determine the thermal efficiency.

(c) Determine the annual cost saving if the unit cost of electricity is \$0.10/kWh and avoid emission from a coal-fired plant.

13.36. A steam power plant is operating on the simple ideal Rankine cycle. The steam mass flow rate is 25 kg/s. The steam enters the turbine at 3500 kPa and 400°C. Discharge pressure of the steam from the turbine is 80 kPa.

(a) If the pressure of the boiler is increased to 9000 kPa while maintaining the turbine inlet temperature at 400°C, determine thermal efficiency.

(b) Determine the annual saving if the unit cost of electricity is \$0.10/kWh and avoided emission from a natural gas-powered plant.

13.37. A steam power plant is operating on the simple ideal Rankine cycle. The steam mass flow rate is 20 kg/s. The steam enters the turbine at 3500 kPa and 400°C. Discharge pressure of the steam from the turbine is 78.5 kPa.

(a) If the temperature of the boiler is increased to 500°C while maintaining the pressure at 3500 kPa, determine the thermal efficiency.

(b) Determine the annual saving if the unit cost of electricity is \$0.12/kWh and the avoided emission from a natural-gas powered plant.

13.38. Ten adiabatic compressors are used to compress air from 100 kPa and 290 K to a pressure of 1000 kPa at a steady-state operation. The isentropic efficiency of the compressor is 80%. The airflow rate is 0.5 kg/s. Determine the minimum and actual power needed by the compressor and the avoided emissions from a coal-fired plant.

13.39. A heat pump system is used to heat a district with 1200 houses and maintain them at 18°C. On a day where the outside temperature is – 2 °C, the house is losing heat at a rate of 80,000 kJ/h. The heat pump operates with a coefficient of performance (COP) of 3.0. Determine:

(a) Power needed by the heat pump,

(b) The rate of heat absorbed from the surrounding cold air

(c) Cost saving and avoided emissions from a natural-gas powered plant.

13.40. A heat pump system is used to heat a district with 1200 houses and maintain them at 18°C. On a day where the outside temperature is – 2°C, the house is losing heat at a rate of 80,000 kJ/h. The heat pump operates with a coefficient of performance (COP) of 3.0. Determine:

(a) Power needed by the heat pump,

(b) The rate of heat absorbed from the surrounding cold air

(c) Cost saving and avoided emissions from a natural-gas powered plant.

13.41. An air conditioner with a 5000 Btu/h (1500 W) operates with a SEER of 10 Btu/Wh. The unit is used for a total of 1700 h during an annual cooling season and the unit cost of electricity is \$0.10/kWh. Estimate the cost of electricity saving and the avoided emissions from a coal-fired plant if a district uses 1500 units of the air conditioner.

13.42. An air conditioner with a 5000 Btu/h (1500 W) operating with a SEER of 22 Btu/Wh. The unit is used for a total of 2000 h during an annual cooling season and the unit cost of electricity is $0.10/kWh. Estimate the cost of electricity saving and the avoided emissions from a coal-fired plant if a district uses 1500 units of the air conditioner.

13.43. (a) Estimate the annual cost of electric power consumed by a 5-ton air conditioning unit operating for 2000 h per year with a SEER rating of 10 and a power cost of $0.10/kWh. (b) Estimate the emission avoided after replacing natural-gas powered plant with solar concentrated power plant.

13.44. (a) Estimate the annual cost of electric power consumed by a 6-ton air conditioning unit operating for 3000 h per year with a SEER rating of 22 and a power cost of $0.12/kWh. (b) Estimate the emission avoided after replacing natural-gas powered plant with solar concentrated power plant.

13.45. A 5 tons of current residential air conditioner operates with a seasonal energy efficiency ratio (SEER) rating of 10. This unit will be replaced with a newer unit operating with a SEER rating of 22. The unit operates 130 days with an average of 10 h per day. Average inside and outside temperatures are 21 and 3°C, respectively. The unit cost of energy is $0.16/kWh. Estimate the saving in the cost of electricity and the maximum energy efficiency ratio, as well as the avoided emission if you replace a natural gas fired plant with wind power.

13.46. An adiabatic turbine is used to produce electricity by expanding superheated steam at 4100 kPa and 350°C. The power output is 75 MW. The steam leaves the turbine at 90 kPa and 100°C. If the combustion efficiency is 0.85 and the generator efficiency is 0.9, determine the yearly cost of coal and amount of emissions.

13.47. An adiabatic turbine is used to produce electricity by expanding superheated steam at 4800 kPa and 350°C. The power output is 100 MW. The steam leaves the turbine at 80 kPa and 100°C. If the combustion efficiency is 0.70 and the generator efficiency is 0.9, determine the yearly cost of coal and amount of emissions.

13.48. An adiabatic turbine is used to produce electricity by expanding superheated steam at 4500 kPa and 350°C. The steam flow rate is 42 kg/s. The steam leaves the turbine at 40 kPa and 100°C. If the combustion efficiency is 0.77 and the generator efficiency is 0.95, determine the yearly cost of coal and amount of emissions.

13.49. Assume that an average residential rate of electricity is $0.10/kWh and a household consumes about 12,000 kWh per year. If the lighting is provided by compact fluorescent bulbs only, estimate the yearly cost saving and avoided emissions from a natural gas-fired power plant.

13.50. Assume that the average residential rate of electricity is $0.12/kWh and a district with 2000 households consumes about 20,000,0000 kWh per year.

If the lighting is provided by led bulbs (a light emitting diode) only, estimate the yearly cost saving and avoided emissions from a natural gas-fired power plant.

References

1. Alhajji M, Demirel Y (2015) Energy and environmental sustainability assessment of crude oil refinery by thermodynamic analysis. International Journal Energy Research 39:1925–1941
2. Adamu A, Russo-Abegão F, Boodhoo K (2020) Process intensification technologies for CO_2 capture and conversion—a review. BMC Chem Eng 2:2
3. Allen JW, Unlu S, Demirel Y, Black P, Riekhof W (2018) Integration of biology, ecology and engineering carbon capture for sustainable algal based biofuel and bioproduct biorefinery. Bioresources Bioprocessing 5(47):1–28
4. Alles K, Demirel Y (2024) Measuring risk of renewable diesel production processes using a multi-criteria decision strategy. Chemosphere 354:141695
5. Cuellar-Franca RM, Azapagic A (2015) Carbon capture, storage and utilisation technologies: a critical analysis and comparison of their life cycle environmental impacts. J CO_2 Util 9:82–102
6. Demirel Y, Matzen M, Winters C, Gao X (2015) Capturing and using CO_2 as feedstock with chemical-looping and hydrothermal technologies. Int J Energy Research 39:1011–1047
7. Demirel Y (2025) Biofuels. In: Dincer I (ed) Comprehensive energy systems, 2nd edn. Elsevier, Amsterdam, Vol. 2, pp. 344–384
8. Demirel Y (2025) Energy conservation. In: Dincer I (ed) Comprehensive energy systems, 2nd edn. Elsevier, Amsterdam, Vol. 9, pp. 1–51
9. Demirel Y, Gerbaud V (2025) Nonequilibrium thermodynamics: transport and rate processes in physical, chemical and biological systems, 5th edn. Elsevier, Amsterdam
10. Demirel Y (2017) Lignin for sustainable bioproducts and biofuels. J Biochem Eng Bioproc Technol 1:1–3
11. Ferroukhi R, Nagpal D, Lopez-Peña A, Hodges T, Mohtar RH, Daher B, Keulertz M (2015) Renewable energy in the water, energy & food nexus. IRENA, Abu Dhabi
12. Geissdoerfer M, Savaget P, Bocken NMP, Hultink EJ (2017) The circular economy 'a new sustainability paradigm? J Cleaner Prod 143:757–768
13. Goeppert A et al (2018) Chapter 3.26—toward a sustainable carbon cycle: the methanol economy. In: Török B (ed) Timothy dransfield, green chemistry, pp 919–962
14. Hall CAS, Lambert JG, Balogh SB (2014) EROI of different fuels and the implications for society. Energy Policy 64:141–152
15. Hall CAS, Lambert JG, Balogh SB, Gupta A, Arnold M (2014b) EROI and quality of life. Energy Policy 64153
16. IEA (2023) International Energy Agency. World Energy Outlook. https://www.iea.org/reports/world-energy-outlook-2023
17. IRENA (2025) International renewable energy agency. In: Record-breaking annual growth in renewable power capacity. https://www.irena.org/News/pressreleases/2025/Mar/Record-Breaking-Annual-Growth-in-Renewable-Power-Capacity
18. IRENA (2023) International Renewable Energy Agency, Renewable Power Generation Costs in 2023. https://www.irena.org/Publications/2024/Sep/Renewable-Power-Generation-Costs-in-2023
19. IRENA (2024) International Renewable Energy Agency, World Energy transition outlook 2024. https://www.irena.org/Publications/2024/Nov/World-Energy-Transitions-Outlook-2024.
20. Jaffe RL, Taylor W (2018) The physics of energy. Cambridge Univ. Press, Cambridge

21. Kumar A, Demirel Y, Jones DD, Hanna MA (2010) Optimization and economic evaluation of industrial gas production and combined heat and power generation from gasification of cornstover and distillers grains. Biores Technol 101:3696–3701
22. Lago C, Herrera I, Caldés N, Lechón Y (2019) Nexus bioenergy bioeconomy. In: Lago C, Caldés N, Lechón Y (eds) The role of bioenergy in the emerging bioeconomy. Amsterdam, Elsevier
23. McCardell SB (2014) Effective energy use: rewards and excitement. In: Anwar S (ed) Encyclopedia of energy engineering and technology, 2nd edn. CRC Press, Boca Raton
24. National Research Council (2011) Renewable fuel standard: potential economic and environmental effects of U.S. biofuel policy. The National Academies Press, Washington, DC
25. Ptasinski KJ (2016) Efficiency of biomass energy. In: An exergy approach to biofuels, power, and biorefineries. Wiley, New York
26. Price L, Lu H (2011) Industrial energy auditing and assessments: a survey of programs around the world. European Council for an Energy Efficient Economy (ECEEE)
27. Theising TR (2016) Preparing for a successful energy assessment. AIChE
28. Theis J (2019) Quality guidelines for energy systems studies: cost estimation methodology for NETL assessments of power plant performance. In *Quality guidelines for energy systems studies: cost estimation methodology for NETL assessments of power plant performance (technical report)* | OSTI.GOV
29. Wang X, Demirel Y (2018) Feasibility of power and methanol production by an entrained-flow coal gasification system. Energy Fuels 32:7595–7610

Decarbonization 14

Introduction and Learning Objectives: Decarbonization is the process to reduce and eventually eliminate greenhouse gas (GHG) emissions with innovations to overcome challenges in clean energy in fields where reducing the emissions are difficult and expensive such as industry and transportation sectors. There is global consensus in reducing GHG emissions and pledges to achieve energy and climate goals in a timely joint effort with policies and scenarios. Carbon capture, utilization and storage in power, industrial, and transportation sectors play key role in decarbonization. The primary innovations for decarbonization include improvements in energy efficiency, environmental protection, sustainable technologies, health, employment, and public engagement. This effort may need substantial undertaking in how we produce and use energy globally. This chapter will summarize some major strategies and technologies to replace fossil fuels with renewable energy toward reducing and eliminating greenhouse gas emissions globally to mitigate the adverse effects of climate change and global warming.

The learning objectives of this chapter are:

- Greenhouse gases and their impacts
- Carbon capture, storage, and utilization
- Emerging green fuels
- Decarbonization efforts in industrial sector and transportation sector
- Decarbonization and artificial intelligence.

Y. Demirel, *Energy*, Green Energy and Technology,
https://doi.org/10.1007/978-3-031-97111-2_14

14.1 Decarbonization

Decarbonization is the process to reduce and eliminate greenhouse gas (GHG) emissions with innovations to overcome challenges in clean energy. The innovations include energy efficiency, environmental protection, sustainable technologies, health, employment, and public engagement in eight sectors: electricity, buildings, land use, transportation, industry, finance, fossil fuels, and non-federal actors [64]. Innovations must focus on reducing costs, development of infrastructure supporting low-carbon solutions, and supply chain engagement. Some recommendations for decarbonization are:

- a broadened policy portfolio,
- rigorous and transparent analysis and reporting for adaptive management,
- ensuring procedural equity in planning new infrastructure and programs,
- ensuring justice, health, and fairness of impacts,
- reforms for interstate transmission,
- tightened targets for the buildings and industrial sectors and transport sector,
- managing the future of the fossil fuel sector,
- building the needed workforce and capacity,
- reforming financial markets,
- research, development, and demonstration needs.

Possible risks in achieving targeted decarbonization may be:

- execution risk,
- technological risk,
- political, judicial, and societal polarization risk,
- risk from events outside the energy systems.

The three pillars of decarbonization are energy efficiency, low-carbon electricity, and fuel switching, including in the transportation sector. In a low-carbon systems with the end uses like electric vehicles or more efficient boilers lead to reduced fossil fuel purchases. Decarbonization strategies are:

1. improve energy and materials efficiency along with circular economy practices,
2. direct use of clean electricity,
3. direct use of renewable heat and biomass,
4. indirect use of clean electricity via synthetic fuels and feedstocks,
5. use of CO_2 removal and measures including carbon capture, utilization, and/or storage (CCUS).

As more consumers pursue sustainable products and services, a shift toward a circular economy would gain momentum with reusable elements toward zero or little waste. This may be determined by the following key performance indicators (KPI-https://www.bluebite.com/brand-strategy/sustainability-kpi-measures) [27, 66]:

1. Reduce supply chain waste
2. Increase recycling rates
3. Reduce GHG gas emissions/carbon footprint
4. Reduce energy consumption
5. Manage load density
6. Manage water use
7. Watch ecological systems impact
8. Saving levels and improvements
9. Improve sustainability assessment scorecards
10. Track products at the item specifics

14.1.1 Greenhouse Gases

Greenhouse gases (GHGs) include carbon dioxide (CO_2), methane (CH_4), nitrous oxide (N_2O), chlorofluorocarbons (CFCs), ozone (O_3), perfluorinated compounds (PFCs), and hydrofluorocarbons (HFCs). GHG emissions are typically reported in carbon dioxide equivalents (CO_2e) for consistent reporting across various gas types. GHG emissions may be used with other data to report intensity, expressed as emissions per square foot or per unit of energy produced to compare normalized data [17, 18, 23, 24]. Major carbon sources are:

- Biogenic carbon sources include (1) waste wood and forest wood (soft and hard), (2) sewage sludge, manure, industry waste, and household garbage, (3) agriculture residues.
- Nonbiogenic carbon sources include (1) coal, natural gas, light fuel oil, diesel, gasoline, and jet-fuel (kerosene), (2) domestic waste, (3) construction emission for technologies and direct cement manufacturing emissions.

14.1.2 Methane

Methane is responsible for around 30% of the rise in GWP and cutting methane emissions is one of the most effective near-term ways to limit global warming and improve air quality. Methane is the second most important GHG contributing to human-driven warming, behind carbon dioxide (CO_2). Methane in the free atmosphere and naturally converts to carbon dioxide via oxidation [64]. Main sources of methane include natural and anthropogenic methane sources. Agriculture is the main contributor to anthropogenic sources. Targeted research on hard-to-abate sources would inform potential applications of atmospheric methane removal technologies.

These are the key differences of methane from carbon dioxide:

- CO_2 has a lifetime of hundreds of years, whereas methane is short lived, with an average lifetime of 9–12 years.
- Methane is naturally oxidized in the atmosphere to form CO_2. Atmospheric methane removal technologies would generally accelerate this process by reducing the lifetime of methane in the atmosphere.
- The main natural process that oxidizes methane in the atmosphere is reaction with the hydroxyl radical (OH).

Engagement with relevant parties and communities will be important for all atmospheric methane removal technologies, especially open system technologies, prior to consideration of deployment and in tandem with ongoing small-scale research. Technology categories for methane removal are:

- Methane reactors are operating down to ~ 1000 ppm methane today, while they need to heat bulk air.
- Methane concentrators need efficiently concentrate methane to enable methane reactor technologies.
- Surface treatments remove other reactive species.
- Ecosystem uptake enhancement involves enhancing soil services with different microbial communities. But unintended consequences for nutrient cycling, biodiversity, ecosystem services, and other attributes are not well understood.

14.1.3 Greenhouse Gas Emissions

Greenhouse gas (GHG) emissions may come from fossil fuel combustion in boilers, vehicles and furnaces, utilities of electricity, steam, heat or cooling. GHGs absorb infrared radiation in the atmosphere, leading to global warming by absorbing energy and slowing the rate at which the energy escapes to space. GHGs act like a blanket insulating the Earth. Total demand for coal declines as it has been replaced mostly by natural gas. Quantifying biogenic and nonbiogenic carbon emissions and the circular can help policy makers for decarbonization.

GHG emissions are classified into three categories, or scopes shown in Table 14.1. The GHG inventory development process [23] consists of three key steps as shown in Fig. 14.1:

1. Reviewing accounting standards, methods, and scope
2. Quantify GHG emissions
3. Develop a GHG inventory management plan to formalize data collection procedures
4. Set GHG emission reduction target

Table 14.1 Types of greenhouse gas emissions

	Scope 1 (direct emissions)	Scope 2 (indirect emissions)	Scope 3 (indirect emissions)
Cycle	During production	Upstream	Both upstream and downstream
Activities	Emissions released/ changed when manufacturing chemicals and fuels or providing a service	Emissions from off-site utility providers: Steam, electricity, Water, fuel, refrigeration	Upstream: Emissions released from the activities before the production process. Downstream: Emissions from use or disposal of a product or service

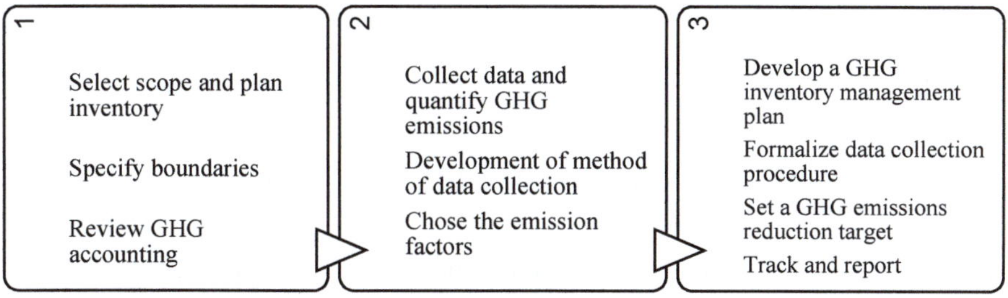

Fig. 14.1 Steps of GHG emissions inventory and reporting

The GHG Protocol defines 15 categories of scope 3 emissions, though not every category will be relevant to all organizations. Figure 14.2 shows scope 3 category that includes upstream and downstream emissions initiated from process-based and end-of-life products.

Basic organizational phases of scope 3 engagement should establish a relevant and accurate scope 3 emissions inventory with determining relevant scope 3 categories, estimating emissions, and improving and expanding emissions estimates [23]. Figure 14.2 overviews of accounting Scope 3 emissions that can be based on three steps.

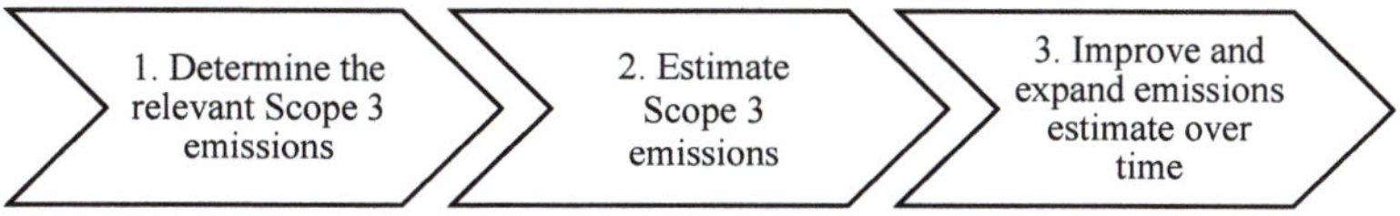

Fig. 14.2. Estimating and reporting stages for scope 3 emissions

14.1.4 Carbon Intensity

Carbon intensity (CI) typically refers to the amount of carbon dioxide equivalent emissions per unit of energy, for example: gCO2e/MJ. Carbon intensity in power generation is the amount of carbon dioxide released per kWh power generated. Table 14.2 shows the CO_2 emissions coefficients for various fuels in kg CO_2 per million Btu [24: https://www.eia.gov/environment/emissions/CO2_vol_mass.php]. Targets for CI reduction are often dependent on technologies including carbon capture and sequestration (CCS), biomass-based heat, renewable electricity, and farm practices. CCS and carbon capture and utilization (CCU) options explore pipelines and executing on-site projects based on life cycle-analysis and technoeconomic analysis for successful implementations. Some strategies to reduce the CI include:

- Energy efficiency increase
- Optimization of process
- Renewable energy use

Table 14.2 Carbon dioxide emissions coefficients by fuel

Fuel	kg CO_2/Million Btu
Propane	62.88
Diesel and heating fuel	74.14
Kerosene	73.19
Coal (all types)	95.99
Natural gas	52.91
Finished motor gasoline	70.66
Motor gasoline[a]	76.11
Residual heating fuel	75.09
Jet fuel	72.23
Aviation gas	69.15
Petroleum coke	102.12
Asphalt and road oil	75.35
Lubricants	74.07
Naphthas	68.02
Waxes	72.60
Anthracite coal	103.69
Bituminous coal	93.24
Subbituminous coal	97.13
Lignite coal	98.27
Coke	113.67

Source https://www.eia.gov/environment/emissions/CO2_vol_mass.php

- Supply chain improvements
- Work force availability.

14.2 Global Warming Potential

Starting in 1990, the Intergovernmental Panel on Climate Change (IPCC) used the global warming potential (GWP) to allow comparisons of the global warming impacts of different gases. GWPs in units of carbon dioxide equivalent (CO_2e) include GWPs for methane (CH_4) and nitrous oxide (N_2O) and is a measure of how much energy the emission of 1 ton of gas will absorb over a given time period, relative to the emission of 1 ton of carbon dioxide (CO_2). Different GHGs can have different effects on Earth's warming [26, 45, 46, 48]. Two keys in which these gases differ from each other are:

(1) their ability to absorb energy, and
(2) how long they stay in the atmosphere known as their lifetime.

The lifetime for GWPs is usually 100 years. GWPs provide a common unit of measure, which allows analysts to add up emissions estimates of different gases and allows policymakers to compare emissions reduction opportunities across sectors and gases. Differences for some GHGs are:

- CO_2 is used as the reference and has a GWP of 1 and remains in the climate system for a very long time.
- Methane (CH_4) is estimated to have a GWP of 27–30 and absorbs much more energy than CO_2 but lasts about a decade on average. Methane is also a precursor to the ozone, which is also a GHG.
- Nitrous Oxide (N_2O) has a GWP 273 times that of CO_2 for a 100-year time scale.
- Chlorofluorocarbons (CFCs), hydrofluorocarbons (HFCs), hydrochlorofluorocarbons (HCFCs), perfluorocarbons (PFCs), and sulfur hexafluoride (SF_6) are sometimes called high-GWP gases because, for a given amount of mass, they trap substantially more heat than CO_2.

14.3 Impact of Climate Change

GHG emissions cause climate-induced disasters including floods, droughts, and fires. Negative impacts of climate change will be increasing over time and varying across regions due to differing climate vulnerabilities. Socio-economic outcomes can be improved through appropriate national policy measures. Limiting global warming, two methods are generally considered [45, 54]:

- Mitigation of carbon sources by the massive penetration of renewable resources, and efficiency improvement for carbon-intensive areas,
- Reinforcement of carbon sinks in either natural way, such as afforestation and reforestation, or artificial way, typically the deployment of carbon capture utilization and sequestration (CCUS) technologies,
- The clean energy economy is gaining ground, but greater efforts are needed including CCUS set to expand rapidly with reduced costs.

14.4 Reduction of Greenhouse Gas Emissions

Major technological tools for reducing emissions include use of renewable energy, increase energy efficiency, electrification, and carbon removals [48]. Scopes 1 and 2 emissions are the main concerns for organizations to reduce their carbon footprints. Determining and reducing Scope 3 emissions create challenges as they require collaboration with customers, suppliers, and other stakeholders [23, 24]. *Green logistics* shifts toward greener supply practices and may contribute to reducing Scope 3 emissions. Some Green logistics practices include decarbonizing warehouses, shifting or mixing transport modes to achieve cost and carbon reductions. Biogenic resources are explicitly indicated as carbon neutral. Carbon layers with specifications on the principal carbon sources and sinks can be analyzed with optimization algorithms that may solve the carbon loop issue in highly interconnected energy systems. Two categories of carbon sinks are considered:

- Natural carbon sinks: Biomass is regarded as a carbon sink in the process of biogenic formation.
- Artificial carbon sinks: Carbon capture utilization and storage CCUS, where the three processes, CC (carbon capture), CCS (carbon capture sequestration), and CCU (carbon capture utilization) are treated separately [38, 40].

Atmospheric methane removal, mainly a sociotechnical problem, involves the conversion of methane into a less radiatively potent form or to physically remove methane from the atmosphere and store it elsewhere. This requires human intervention to reduce methane emissions at the source, typically anthropogenic in origin and offers the most cost-effective means of mitigation. Due to the short atmospheric lifetime of methane, sustained removal efforts could potentially address hard-to-abate methane emissions [64].

Table 14.3. Estimated reduction of CO_2 emissions in various sectors under 1.5°C Scenario between 2023 and 2050 ($GtCO_2$: Giga tonnes CO_2) [48]

	Industry	Power and heat	Transport	Buildings	Other
CO_2 emissions $GtCO_2$/year	– 12.7	– 9.8	– 7.9	– 2.5	– 1.3
Emissions %	37	28	23	7	5

14.4.1 Scenarios for Reducing Emissions

Some major scenarios [43, 48] for reducing emissions to control global warming are:

- IRENA's 1.5°C *Scenario* considers a pathway with the average global temperature increase by the end of the present century to 1.5°C, relative to pre-industrial levels (Table 14.3). Hydrogen would play a key role in the decarbonization of end-uses and flexibility of the power system with efficiency improvements, deployment of renewables, and changes in consumer behavior.
- *Net Zero Emissions* (NZE) *Scenario* is a pathway for the energy sector to help limit the global temperature rise and to meet the key energy-related UN Sustainable Development Goals (SDGs). The NZE model requires a three-fold increase in global renewables power generation and reduction in GHG emissions.
- *Announced Pledges Scenario* (APS) assumes that governments, businesses and other stakeholders will meet their commitments, including longer term net zero emissions targets and pledges, as well as commitments in energy access.
- *Stated Policies Scenario* (STEPS) is designed to provide a sense of the prevailing direction of energy system progression, based on a detailed review of the current policy landscape considering the pledged energy targets and objectives across the energy economy.

14.4.2 Policies in Reducing Emissions

Some major policies [44, 48] for reducing emissions are:

- *Energy and climate policies* consider annual energy outlook reviews and assess new policies, regulatory measures, targets and announcements. This may affect the evolution of the energy system, including decisions to proceed with new oil and gas licensing permits, GHG emission reduction action plans, and initiatives and strategies for energy access.
- *Industrial policies* refer to specific policies in place to encourage clean energy manufacturing, costs and targets with deployment trends.
- *Policies for aviation* consider the International Civil Aviation Organization to adopt two global aspirational goals for the international aviation sector, which include 2% annual fuel efficiency improvement and carbon-neutral growth.

Table 14.4. Sample of factors impacting the carbon intensity of ethanol (with permission from ACE American Coalition for Ethanol 2025: https://ethanol.org/carbon-calculator, U.S. Department of Energy GREET Model and the GREET 2020 Corn Feedstock Carbon Intensity Calculator)

Factors	User specified CI score
Corn yield (bushels/acre)	166 (Farm average corn yield)
N_2O Emissions from fertilizer	Current: 13.2 gCO_2/MJ
Penalty for land use change	19.8 gCO_2/MJ
Credit for climate-smart farming practices	GREET: 0 gCO_2/MJ
Ethanol yield (gallons/bushel)	(Gall/bu): 2.9
Natural gas use by ethanol plant	GREET: 28408 Btu/gallon
Credit for capture and sequestration of CO_2	GREET: 0 g CO_2/MJ)

- *Policies for shipping* involve the International Maritime Organization that can set sectoral decarbonization targets and strategies and promoting its wide adoption by industries and countries.

14.4.3 Carbon Intensity in Bioethanol Industry

For the corn-based ethanol industry, emissions come from growing corn, including the carbon intensity (CI) from parts of the process such as fertilizer, indirect land use changes, and transportation (Table 14.4). The use of renewable energy such as wind, solar or renewable natural gas, must ensure that those energy sources are tied into the plant with an assessment of the local energy source's current CI. This will help producers decide if the reduction with a specific renewable energy source is worth the investment. Some coproduct credit value in the CI score should be protected. For example, the use of biogas as process energy can lower your CI, but in theory, you may be losing your coproduct credit value. CI score may become less favorable when producing biogas utilizing coproducts at the ethanol plant. CCS has the largest potential benefit to CI reduction from a cycle of development from conception to competition when implementing their chosen decarbonization strategy (https://ethanol.org/calculate-ci).

14.4.4 Climate-Smart Agriculture

Climate-smart agriculture is a new strategy to reduce Scope 3 emissions. With the emissions associated with farm practices constituting roughly 28 CI points of ethanol's total score and the practices that could reduce corn's CI include applying low-carbon fertilizers, planting cover crops, implementing no-till or strip-till, and controlling nutrient runoff. This would involve working with a data aggregator that collects data on agricultural practices, machines and farmers for verification [10].

14.4.5 Net Zero Emissions

Net zero may be achieved when GHG emissions are balanced by removals over a specified period toward carbon neutrality. Achieving net-zero CO_2 emissions is essential to stabilizing climate. Net zero energy and net zero carbon can encourage different strategies, such as on-site generation or carbon offsets. The geographical variation in decarbonization efforts is characterized by several factors:

- Variations in the carbon intensity from the emissions impact of electricity
- Policy differences
- Types of data collected, and geographic challenges

For example, building decarbonization efforts are not evenly distributed across communities, following other patterns of income inequality in affordable housing and commercial buildings, affecting businesses, economic opportunity, and climate resilience in these communities [44, 48, 75].

14.4.6 Role of Renewable Energy in Decarbonization

Reducing GHG emissions from power generation is essential for decarbonization. Electrification of buildings, vehicles, and other uses of energy can support decarbonization only if their supply of electricity is renewable [43, 46]. Decarbonization in the private sector and civil society requires:

(1) Public engagement and decision-making support for the new transmission facilities for electricity generation with carbon-free resources.
(2) Regulatory support for investment in distribution-system infrastructure, controls, and interconnection policies.
(3) Increased energy efficiency and low-emissions energy and fuels meeting an increasing share of the rising energy need.

14.4.7 Transformative Decarbonization

Climate concerns besides economic inequality require a comprehensive global economic transformation with holistic solutions in which renewable energy plays a key role. The decarbonization of the global economy calls for large-scale policy interventions and substantial resource allocations. Market mechanisms alone may not bring the desired emissions reductions.

Enabling policies include industrial policies, labor-market interventions, educational and skill development, and social protection measures. At the international level, the Climate Investment Platform, the International Renewable Energy Agency, the United Nations Development Program, the multi-partner Sustainable Energy for All initiative, and the Green Climate Fund aims to mobilize energy-transition investments on a scale in line with the climate goals. Investment decisions should be evaluated to the extent at which they accelerate the shift toward an inclusive low-carbon economy [44].

14.5 Carbon Capture Strategies

GHG emissions from combustion of fossil fuels are the main concern toward a right strategy for decarbonization. The two approaches for carbon capture are pre-combustion and past combustion carbon captures. A solvent-based CO_2 capture process is an absorption process that can be used in both pre- and post-combustion power plant applications for CO_2 capture. There are two types of solvents: chemical and physical. A chemical solvent relies on the chemical reaction between CO_2 and the solvent to enhance absorption, whereas a physical solvent absorbs molecular CO_2 via van der Waals interactions. Chemical absorption involves the formation of chemical bonds between CO_2 and solvent molecules and thus can be highly selective and exhibits saturation at a certain level:

- *Pre-combustion* CO_2 *capture* involves gasification process. In gasification/reforming processes the fuel reacts with air or oxygen, or steam undergoes to generate synthesis gas containing mainly H_2 and carbon monoxide (CO) under high pressure and temperature involving the following equations.

$$C_nH_m + n2O_2 \rightarrow 2H_2 + nCO \qquad \text{Gasification}$$
$$C_nH_m + nH_2O \rightarrow (m+n)2H_2 + nCO \ \text{Reforming}$$

After removing the impurities, syngas passes through the water-gas shift reactor (WGSR) where the carbon monoxide CO in the existence of water is converted into CO_2 and H_2

$$CO + H_2O \rightarrow CO_2 + H_2$$

It becomes energy efficient and less costly to capture the concentrated CO_2 leaving hydrogen as a fuel for combustion. When integrated with catalytic WGSR process can provide high CO conversion at low steam to carbon ratios and hence shows promising for modular scale pre-combustion CO_2 capture [8, 66].

- *Post-combustion carbon capture* involves oxidation of fossil fuels with excessive air causing GHG emissions including acidic gases. More environmentally combustion processes include oxy combustion and chemical looping combustion. CO_2 is captured from the flue gas at the temperature range of 120–180°C and is considered a more feasible process for existing coal plants. The low volumetric concentration of CO_2 in the flue gas ($< 20\%$) is the main issue and needs a right solvent, leading to a higher regeneration energy requirement. Post-combustion is comparatively economical but less efficient than pre-combustion technology due to low concentration of CO_2 [36].

- *Oxy-fuel combustion* (OFC) uses pure oxygen provided from an air separation unit and CO_2 can be separated by condensation with ~75% mass flow reduction due to little or no nitrogen and nitrogen oxides (NOx) in the flue gas. The chemical equation that evolves during the combustion process can be written as

$$C_xH_y + (x + y4)O_2 \rightarrow xCO_2 + y2H_2O$$

The large quantity of oxygen needed makes this technology costlier with around 7.0–11% energy penalty. Cryogenic air separation is considered the most feasible method but contributes to 3–4% of the energy penalty and 14% of the overall cost of the OFC plant. The oxy-fuel combustion technology is suited to coal and gas-fired plants to capture CO_2. The use of biomass with a supercritical CO_2 cycle makes this method economical with a carbon tax [77, 78].

- *Chemical looping combustion* (CLC) uses low-cost metallic oxides as an oxygen carrier and provides the oxygen needed for combustion by oxidizing the fuel to CO_2 and water, while deoxidizing to metal in the fuel reactor. The deoxidized metal is re-cycled back in the air reactor for re-oxidation, involving the following reactions during oxidation and reduction steps. CO_2 from flue gas can be separated by condensation of vapor, thus eliminating the separation steps. The process has a superior performance and is cost-effective compared to the pre-combustion process. However, the mechanical and chemical stabilities at high temperatures ($\sim 800 - 1200$ °C) are operational issues [15].
- *Direct air carbon capture* (DAC) offers two technological approaches, (a) passing air through solutions, such as hydroxide solution/amine/amino acid and (b) use of solid sorbent filters to reduce the cost. In the presence of vacuum, when the filters are heated, the CO_2 gets released for storage or other applications. The challenge is capturing CO_2 from low-concentrated air and need for a large flow of air in the plant, which requires large thermal energy for CO_2 separation, making this process costlier.
- *Membrane systems* for the post combustion capture should be highly CO_2 permeable, CO_2/N_2 selective, chemically and thermally stable, low cost, and durable.
- *Bioenergy with carbon capture and sequestration* (BECCS) involves the capturing CO_2 during energy generation from biomass. Therefore, the carbon captured during photosynthesis by the plants would not be recycled back to the atmosphere but captured and sequestered underground or converted into other chemicals permanently fixing the carbon that may lead to negative emission.

14.6 Carbon Capture Technologies

Carbon capture storage and utilization (CCSU) technologies help the economic, environmental, and social pillars of sustainability by significantly mitigating the carbon emission [21, 38, 60, 68]. The use of a single technique for CO_2 capture may be either energy-intensive or costlier. Some common and mature technologies as well as new emerging technologies are briefly introduced in the following sections.

14.6.1 Post Combustion Carbon Capture Technologies

The carbon capture cost contributes to $\sim 70 - 80\%$ of the total cost decarbonization efforts. The dissociation of concentrated $CO_2(> 20\%)$ from the mixture of CO_2/H_2 is much more compatible compared to a mixture of CO_2/N_2 because of different molecular weight and kinetic parameters. Chemical and physical absorption, adsorption, membrane separation, cryogenic processing, and chemical looping combustion are the principal technologies to capture CO_2 from flue gas (Fig. 14.3). Hybrid technologies, such as a combination of the above, are rapidly evolving as alternative options to reduce costs and energy requirements. Other advances for CO capture from exhaust streams include ionic liquids, nanofluids, microencapsulation, phenoxide salts, enzymes, membranes and sorbents [21, 55]. Figure 14.3 shows various carbon capture technologies.

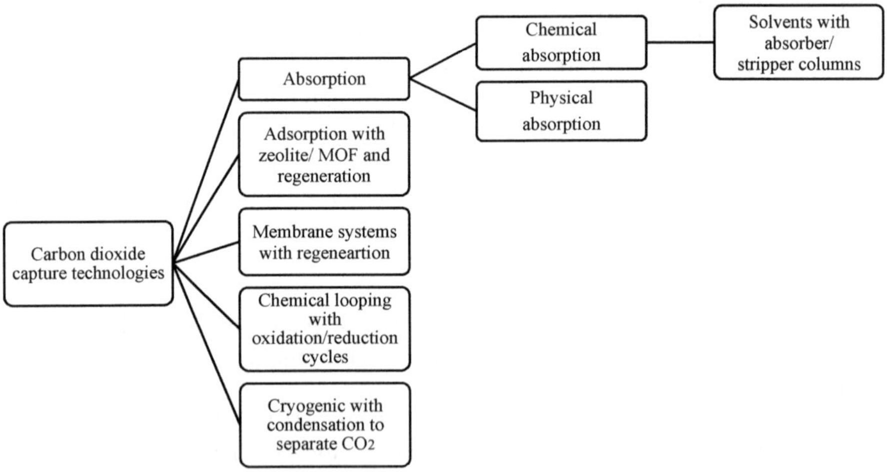

Fig. 14.3. Some major post-combustion carbon capture technologies: (MOF: Metallic organic frameworks)

14.6.1.1 Chemical Absorption

Two columns for absorption and releasing CO_2 by stripping are associated with the use of amines and other solvents. Based on hydrogen atoms in the structure, amines are classified as primary, secondary, and tertiary with varying absorption capacity (mol CO_2/mol absorbent). The technique uses absorbents of higher selectivity for CO_2 than nitrogen to capture the CO_2 from flue gases containing excessive nitrogen using some of the following aqueous absorbents:

- AMP (2-amino-2-methyl-1-propanol) from a gas mixture of N_2 and CO_2
- DEA (Diethanolamine) from a gas mixture of CH_4, C_2H_6, C_3H_8, N_2, CO_2 and H_2S.
- Mixed DEA and MDEA (methyl-diethanolamine) from a gas mixture of CO_2, CH_4, C_2H_6, C_3H_8, $n-C_4H_{10}$, $i-C_4H_{10}$, $n-C_5H_{12}$ and N_2.
- DGA (Diglycolamine) from a gas mixture of N_2, O_2, CO_2 and H_2O.
- K_2CO_3 (Potassium Carbonate) from a gas mixture of N_2, H_2O, CO_2, and H_2S.
- MDEA from a gas mixture of CH_4, CO_2, H_2S and H_2O.
- MEA (monoethanolamine) from a gas mixture of N_2, O_2, CO_2 and H_2O.
- Mixed MEA and MDEA aqueous solution from a gas mixture of CO_2 and N_2.
- NaOH (sodium hydroxide) from a gas mixture of N_2, O_2, H_2O and CO_2.
- NH_3 (Ammonia).
- Piperazine (PZ) (cyclic amine) solution.
- Mixed piperazine (PZ) and monoethanolamine (MEA).
- Mixed solvent composed of sulfolane, DIPA (diisopropanolamine) and water.
- Mixed solvent composed of sulfolane, MDEA and water.
- TEA (Triethanolamine) solution from a gas mixture of H_2 and CO_2.

Table 14.5 shows the absorption capacity and regeneration energy required of some of the common absorbents. Monoethanolamine (MEA) is the most common solvent and comes with high regeneration energy requirement. The faster reaction, low solubility, and high reactive heat are associated with primary and secondary amines, while higher absorptivity of CO_2 using tertiary amines. Therefore, the blends of various aqueous amines have been used to improve performance and found that the combination of MEA with aminomethyl propanol (AMP) compared to MEA, piperazine (PZ), has faster CO_2 absorption rate (~two times of MEA), low degradation rate, and is suitable at low concentration (<10%) as an additive to increase the absorption using an aqueous solution of tertiary amine. Adding PZ with MEA reduces the energy requirement due to the improved absorption efficiency of the mixture [9, 21, 60, 65].

Aqueous potassium carbonate, calcium carbonate, and sodium carbonate are used for CO_2 absorption due to their low capturing cost, high solubility, low toxicity, high efficiency and low degradability compared to amines. Potassium carbonate

Table 14.5 Absorption capacity a of some solvents [9]

Aqueous solvent	Boiling point K	Absorption capacity mole CO_2/mole absorbent	Energy required for regeneration MJ/kg CO_2
MEA (30–40 wt%)	443.0	0.35–56	3.8–4.2
DEA (30%)	544.2	0.50	
DGA (8–27 wt%)		88–90	
MDEA (30 wt%)	520.2	1	0.61–1.34
K_2CO_3 (20–40 wt%)		0.7–1.05	1.98–325
NaOH (1–5 wt%)		0.5–1	4.09–5.07
PZ (30 wt%)		0.775	

(K_2CO_3) has a higher capture capacity and involves the following exothermic reaction

$$CO_2 + H_2O + K_2CO_3 = 2KHCO_3$$

Negative impacts include limited CO_2 loading, corrosiveness, high energy consumption, evaporation loss, and solvent degradation. The capture of CO_2 using alkaline solutions requires significant thermal energy to release CO_2 from a (bi)carbonate-enriched solution.

A bicarbonate electrolyzer at lower pH forms acid equivalents to convert a (bi)carbonate-enriched and hence can reduce the energy required for desorption of CO_2 as well as an upgraded carbon product such as CO and ethylene. In electrochemical CO_2 reduction reaction, the CO_2 single-pass conversion efficiency is a key performance metric. The porous solid electrolyte reactor may achieve high CO_2 conversion efficiency to high-purity formic acid (HCOOH) while preserving the cell voltage and HCOOH Faradaic efficiency. Flue gas may be converted to high-purity formic acid with conversion efficiency of more than 80% through an electrochemical sequential CO_2 capture and reduction system [3, 49, 51, 52].

14.6.1.2 Physical Absorption

Physical absorption of CO_2 is based on Henry's law without any chemical reaction and can be separated out at low pressure in a mild thermal regeneration. The process is most suited for pre-combustion CO_2 capture. The low vapor pressure, low toxicity, less corrosiveness, and less energy requirement are the main advantages [3]. Physical absorption is a mature technology, however, many aspects such as energy requirement and operating cost can be improved by discovering new absorbents.

14.6.1.3 Emerging Carbon Capture Technologies

Emerging technologies involve advanced materials and processes to improve absorption capacity, energy penalty, stability, cost, society, and climate [9, 34]. Some of the emerging technologies are:

- Metal-organic frameworks (MOFs) are porous crystalline solids with large dimensions for gases with fine dispersal properties.
- Ionic liquids (ILs) are synthesized using organic cations and organic/inorganic anions with varying compositions with the aim of increasing the CO_2 solubility. ILs sufficiently absorbs CO_2 in the presence of philic groups at room temperatures. However, they suffer from undesired high viscosity, cost, and lower absorption performance than amines. A nano metal-organic framework along with ILs is reported to be an effective adsorbent for CO_2 over nitrogen and methane gas.
- Hybrid membrane may provide an increase in CO_2 capture when ILs were combined with the membranes due to the production of carbonate and bicarbonate, which can be regenerated easily.
- Molecular sieves are extremely permeable materials suited for the hole sizes of gas molecules to separate a specific gas easily. The addition of IL with a sieve improves the performance of CO_2 separation.
- Cryogenic carbon capture involves a physical separation process based on the differences between the boiling points and the desublimation properties of the components in the gas mixture. This technology is less sensitive to impurities and has high CO_2 capture but requires large amounts of energy for CO_2 separation during refrigeration cycle.
- Nanofluids addition to a solvent may lead to significant improvement in CO_2 absorption, stability, and energy efficiency, depending on the nanofluid, particle size, temperature, pressure, and CO_2 concentration.
- The surface area of the core absorbents increases with microencapsulation by housing the absorbents in semi-permeable microcapsules, which allow CO_2 and water to flow, achieving the advantages of both solid sorbents and liquid solvents.

14.6.2 Challenges in Carbon Capture

The use of even mature techniques for CO_2 capture may be energy-intensive and hence costly. Before implementation on the industrial scale, each technique must overcome the associated challenges of cost, scalability, and environmental impact. The major challenges include the effect of impurities, low partial pressure of CO_2 absorption capacity, absorption kinetics, regeneration energy requirement, cost, and retrofitting to existing plants (Table 14.6). The next generation technologies such as amine blends, membranes, MOFs, and cryogenic systems are at early-stage [55, 79].

At modular scale, the capital cost of the carbon capture equipment was most important for reducing the levelized cost of electricity (LCOE). A range of physical solvents are suitable for pre-combustion CO_2 capture to produce electricity, hydrogen, and ammonia. Power plants can balance their capture requirements and costs by optimizing the size of absorption towers deployed. DAC contactors have

Table 14.6. Summarized of various absorbents [9, 55]

Chemicals		Advantage	Disadvantages
Amines	MEA	High absorption rate	High energy consumption
	MDEA	High absorption rate	Low absorption rate
	DEA	Mature	High energy consumption, degradation
	AMP	Mature	Costly
	PZ	Mature	Costly, toxic, and corrosive
ILs (ionic liquids)	Conventional		Medium absorption rate
Ammonia		Easy regeneration	High volatility and pollution
Carbonate		High solubility Low cost, toxicity	High corrosiveness Low reactivity
Cryogenic		High capture efficiency	Solid carbon formation, high cost
Amino acid salt		Low oxidation	High regeneration energy cost
MOF		Low toxicity	Impurities affect the adsorption
Enzymes		Low vapor pressure	Low CO_2 selectivity
Hydrate based		Low Easy recycling	Immature technology, high pressure

very different footprints from absorption towers and are expensive. Three fundamental transformations need to occur simultaneously to achieve ambitious GHG emission reduction targets:

- Electrification of end-use sectors,
- Decarbonization of electricity generating supply, and
- Efficiency improvements.

14.6.3 Case Study: CO_2 Capture from Syngas Using K_2CO_3

Carbon capture, utilization and storage (CCUS) is a part of decarbonization in sustainable development scenario. Primary focus of CCUS is on retrofitting existing fossil fuel-based power and industrial plants to reduce emissions. This capture technology is being actively investigated with rigorous models necessary for the design, scale up, integration, optimization and operation (Fig. 14.4). This example from Aspen Plus design package demonstrates how to model CO_2 capture from syngas by using potassium carbonate (K_2CO_3) (www.aspentech.com).

Process Description: The syngas with CO_2 is sent to the absorber column to remove CO_2. The absorber column has packing inside it and is divided into 10 stages in the simulation. The sour gas enters the column at the bottom and the lean

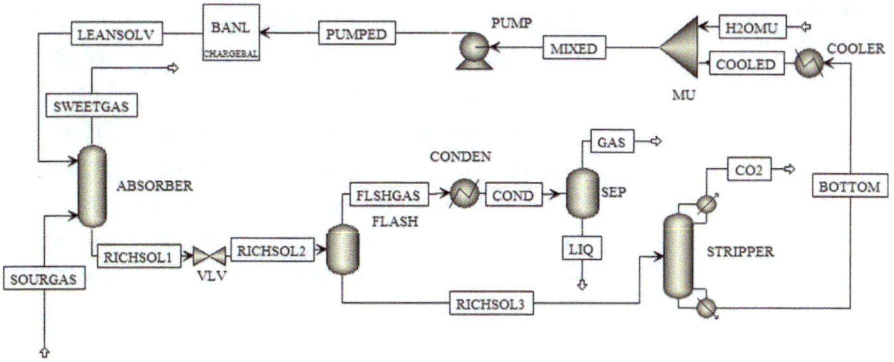

Fig. 14.4 Process flow diagram for CO_2 capture from syngas using K_2CO_3. (www.aspentech.com)

K_2CO_3 solvent enters the column on the top. Lean gas comes out from the top of the absorber and the rich solvent leaves the absorber column from the bottom and is sent to the stripper column. The stripper is divided into 12 stages (including the condenser and reboiler) in the simulation. Rich solvent enters stripper column on stage 2. CO_2 is stripped from the rich solvent in the column and the K_2CO_3 solvent is regenerated. The lean solvent from the bottom of the stripper is recycled with makeup back to the absorber. The greenhouse gases in streams GAS and CO_2 are captured and sent to the next process (www.aspentech.com).

Model Description: The Electrolyte Non-Random Two- Liquid (Electrolyte NRTL) model is used for electrolyte thermodynamics and Redlich-Kwong Equation of State for vapor phase properties with aqueous-phase equilibrium and kinetics reactions required for rigorous calculations. The absorber and stripper columns are simulated with the rate-based model, which utilizes the mass and heat-transfer correlations, and interfacial area estimations based on transfer properties and tray/packing geometry. This makes it more accurate over a wider range of operating conditions compared with the equilibrium-stage model. The key simulation results of the absorber are shown in Table 14.7. The specifications and results of the stripper are summarized in Tables 14.8 and 14.9.

Table 14.7 Simulation results of absorber

Property	LEANSOLV	SOURGAS	SWEETGAS	RICHSOL1
Temperature K	368.67	373.15	369.29	377.29
Pressure atm	21.41	21.41	21.12	21.12
Mass flow kg/h	750000	115000	89300	776000
Composition (mole fraction of apparent components)				
CO_2	0.0193	0.12	0.0087	0.0417
K_2CO_3	0.0680	0	0	0.0668

Table 14.8 Simulation results of stripper

Property	RICHSOL3	CO_2	BOTTOM
Temperature K	376.75	317.37	400.69
Pressure atm	2.02	1.67	1.67
Molar flow kmole/h	31200	67200	31200
Mass flow kg/h	775000	28600	747000
Composition (mole fraction of apparent components)			
CO_2	0.0412	0.9448	0.0195
K_2CO_3	0.0669	0	0.0685

Table 14.9 Results of CO_2 emissions

	Feed	Products			Utility	Net CO_2 emissions
Stream	Sour gas	Sweet gas	Gas	CO_2		
CO_2 emission (tonne/year)	496000	2.45000	6090	245000	74500	−1.77000

Aspen Plus supports five well known standards for Global Warming Potential, including IPCC AR6 (2021). Finally, the process CO_2 emissions including both material streams and utilities are shown in Table 14.9.

Stream CO_2 and GAS represent that about 0.25 million tonne CO_2/year is captured in the process, so the net CO_2 emission is negative (www.aspentech.com).

14.7 Carbon Dioxide Sequestration and Storage

Carbon dioxide storage/sequestration technologies commonly include geological, mineral carbonation, terrestrial and ocean storage options as analyzed in Table 14.10. In carbon sequestration, the CO_2 is buried into underground formation. CO_2 transport costs may depend on pressure and onshore and offshore pipelines.

Two examples of technologies that will need to be deployed widely in deep decarbonization pathways are carbon capture and storage (CCS) and renewable energy generation. CCS technology needs to be matured to be widely deployed throughout the industrial sector, transportation sector, and buildings. Besides carbon capture and utilization (CCU), CCS should move to bioenergy with carbon capture and storage (BECCS). With CCS/CCU, BECCS the action items may be in the following order shown in Figure 14.5.

Still fossil fuel-based carbon capture and utilization (CCU) and storage are vital for GHG reduction in cement, iron, steel and chemicals industries [76].

Table 14.10. Comparison and summary of different carbon storage technologies [55]

Technology	Advantages	Current stage	Outlook
Geological storage	Large storage capacity	Mature	Growing for larger scales
Mineral carbonation	Use in construction	Early stage	Needs advances in technology
Terrestrial storage	Benefits ecosystems, biodiversity	Mature	Need support from governments, industries and communities
Ocean storage	Large storage capacity	Early stage	Uncertain

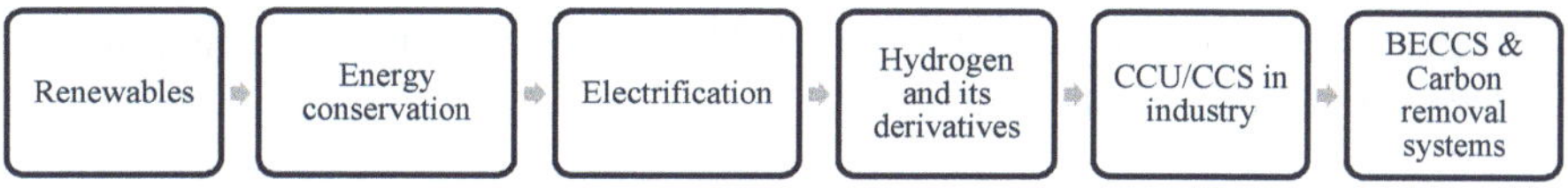

Fig. 14.5. Bioenergy with carbon capture stages

14.8 CO_2 Utilization and Conversion

Carbon dioxide utilization requires extensive infrastructure for capture, purification, transportation, and conversion. A regional or national CO_2 pipeline system would enable low-cost transportation. Carbon utilization infrastructure focuses on preventing GHG emissions or using as feedstock through circular processes. In a net zero scenario, CO_2 can be used to sustainably create products that fixed carbon permanently rather than recycle it. Successful carbon utilization addresses environmental, societal, and economic challenges by demonstrating carbon utilization pathways, techno-economic, environmental, and social life cycle assessments, and workforce development with community engagement [33, 37, 49, 61, 62, 77].

Chemical, biochemical, or biological conversions of CO_2 can produce:

- Short-lived products, including chemicals, pharmaceuticals, aviation fuels and food and animal feed, which can participate in a circular carbon economy.
- Long-lived products, including concrete, aggregates, construction materials, and carbon fiber.
- Elemental carbon materials, which durably store carbon.
- Sustainable carbon-based materials for construction, electronics, and others that do not result in near-term CO_2 emissions.

Pathways toward CO_2 utilization research, development, and demonstration include:

(1) Mineralization of CO_2 into inorganic carbonates,
(2) Conversion of CO_2 into elemental carbon materials, and

(3) Chemical and biological pathways for CO_2 conversion into organic chemicals and fuels.

CO_2 can be used to increase the efficiency of photosynthesis for algae converted into biofuels and other bio-products and crop cultivations. The biological CO_2 conversion and genetic engineering techniques can improve the metabolic pathways.

Biogas and biomethane purification may be possible by biomass-derived activated carbon modified with metal oxides for the removal of hydrogen sulfide (H_2S) and CO_2. Some technological innovations in biogas upgrading include water scrubbing, pressure swing adsorption, and membrane separation have their operational efficiency, environmental impact, and cost-effectiveness [52].

In the energy system decarbonization, CO_2 can be treated as a feedstock to produce synthetic fuels and chemicals:

- Mineralization of CO_2 to inorganic carbonates results in materials with durable carbon storage potential.
- Chemical conversion of CO_2 to elemental carbon materials yields materials high conductivity and mechanical strength
- CO_2 can be reduced to elemental carbon materials through thermochemical, electrochemical, photochemical, or plasma chemical reaction pathways.
- Chemical and biological conversion of CO_2 leads to fuels, including methanol and methane
- Methanation is a catalytic conversion process through which CO_2 reacts with H_2 to produce synthetic methane referred to as the Sabatier process by the following reactions:

$$CO_2 + 4H_2 = CH_4 + 2H_2O$$
$$CO + 3H_2 = CH_4 + H_2O$$

Methanation requires moderate temperatures (in the range of 250–350 °C) and pressures as high as 20–25 bars. As the reaction makes use of a catalyst to aid with the conversion CO_2-to-diesel is a blend of mainly linear hydrocarbons, diesel, jet fuel, can be mainly produced through Fischer–Tropsch synthesis using a carbon source and hydrogen.

- CO_2-to-methanol-to-X where methanol is an intermediate chemical, solvent. It is produced from CO_2 hydrogenation [61]. Low-carbon chain olefins such as ethylene and propylene can be produced by catalytic conversion of methanol via dimethyl ether (DME) in a process known as methanol to olefins (MTO), which follows the following reaction scheme:

$$2CH_3OH \rightarrow CH_3OCH_3 + 2H_2O$$

$$CH_3OCH_3 \rightarrow C_2H_4 + H_2O$$
$$3CH_3OCH_3 \rightarrow 2C_3H_6 + 2H_2O$$

Methanol can also be transformed into a mixture of aromatic and biogenic carbon inputs lead to at maximal net-zero emissions, and negative emissions may be possible in the presence of bioenergy with carbon capture and storage (BECCS).

- Nonbiogenic carbon inputs result in at least net-zero emission.

14.8.1 Some Carbon Conversion Technologies

Some of the carbon conversion technologies are [9, 67]:

- *Pyrolysis* is mainly thermal cracking of wood, or solid biomass in general, in the absence of oxidizing conditions producing gas, liquid, and solid products,
- *Gasification* is used to convert woody biomass to a gaseous fuel stream comprised mainly of CO and H_2 (syngas) and to a smaller extent CO_2 and CH_4. This thermochemical process employs biomass oxidation in sub stoichiometric conditions to favor the production of CO instead of CO_2.
- *Fischer–Tropsch* (FT) *synthesis* is a catalytic process to produce synthetic biofuels from gasification-derived syngas. The chemical reaction is described by a simple representation of the form

$$(2n + 1)H_2 + nCO \rightarrow C_nH_{2n+2} + nH_2O$$

where n is the number of carbon atoms in the products. The conversion conditions (around 200–300°C and pressure in the range of >10 bars) are used to promote long-chain hydrocarbons with 10–20 carbon atoms. The FT chemical process uses a Fe/Co catalyst to convert syngas containing H_2 and CO to a mixture of straight chain hydrocarbons. The syngas has strict limitations of the amount of sulfur, tars, and phenols to avoid processing issues including catalyst poisoning. The H_2:CO ratio varies depending on which catalyst is used, but water gas shift reactions will be required.

- *Anaerobic digestion* is a biochemical process that is used to break the biogenic carbon of wet biomass and produce biogas, a gaseous mixture of CH_4 and CO_2
- *Hydrothermal gasification* converts wet organic streams into methane using high pressure to reach supercritical water conditions, while avoiding the energy-intensive drying step.
- *Biomass to alcohols* conversion process produces alcohols, such as methanol and ethanol that are used as fuels and as intermediate feedstocks to produce other fuels and chemicals.

14.8.2 Gasification

Integrated gasification combined cycle (IGCC) power generation can provide low-cost, low-emission energy. Combining the applications of power, heat, and chemicals production by modular coal gasification may help decarbonization with potential to increase the types of feedstocks and products [4, 7, 70, 72, 73]. Figure 14.6 shows the stages to produce some major fuels and chemicals from the gasification process using coal, natural gas, or biomass.

To integrate carbon capture technology with novel gasification processes, it is important to assess the syngas properties for different applications:

1. Besides decarbonization, pre-combustion CO_2 capture supports polygeneration gasification applications with H_2, fertilizer, chemicals, and liquid fuels productions.
2. Modular scale capture processes need to integrate modular scale gasifiers with automated controls to vary flow rates to optimize production of electricity vs polygeneration products with water gas shift reaction (WGSR) temperatures to reduce heating/cooling requirements of syngas.
3. Hybrid processes, such as integrated "H_2 membrane WGSR followed by physical solvent separation", can provide compact operation while simultaneously allowing varying flow to power generation and polygeneration processes.
4. Local production of fertilizers for agricultural use could provide a sizable revenue stream from modular scale gasification with polygeneration cases in which ammonia sulfate, potassium nitrate, gypsum, and/or other agricultural fertilizers are generated at the modular scale gasifier for local use.
5. H_2 separating membranes are best suited for polygeneration cases where pure hydrogen or ammonia are co-products.
6. Air blown gasification processes are not suitable at modular scale for non-power applications due to the large amount of nitrogen in the syngas.

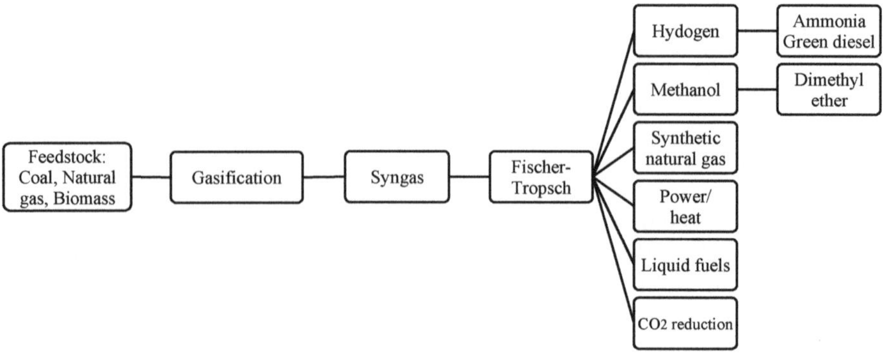

Fig. 14.6 Some major fuels and chemicals from gasification process using coal, natural gas, or biomass

Gasification systems require syngas cleaning processing including CO_2 capture. SNG consists of ~ 95% methane and is produced by passing cleaned syngas over a metal catalyst to react H_2 with CO_2 and CO to form CH_4. The syngas requires a CO: H_2 ratio of ~ 1:3 which requires CO_2 removal. Under these syngas conditions, a chemical solvent such as MDEA capturing the CO_2 and H_2S. Global demand for natural gas makes production of SNG from coal gasification an attractive option. Some uses of syngas options are:

- Hydrogen production
- Electricity generation
- Ammonia/Urea production
- F-T Chemicals and fuels
- Methanol production

14.8.2.1 Hydrogasification

Hydrogasification occurs in hydrogen-rich environments, starting from coal or other gasifying feedstocks, gas clean up, methanation, water removal, and syngas production. The hydrogasification of coal and biomass produces methane (CH_4) called synthetic natural gas and fuels such as methanol. For that the reactor configurations should promote the intensification of gas-solid contact to reduce by-product formation form biomass feedstocks [4, 5, 70].

Hydrolysis of coal and hydrogasification to convert biomass and/or coal to fuels offers to reduce the harmful effects of coal. Hydrogen and/or pre-treated biomass are firstly fed to the hydrogasifier reactor, the produced stream is fed to a clean-up unit working at high temperatures for contaminant removal. Methanation reactors are used to convert residual carbon monoxide as seen in Fig. 14.7.

Hydrogasification offers:

- Lower operative temperatures and higher thermal efficiency,
- low production of harmful chemicals compared with typical gasification units,
- Higher fraction of final product methane and lower carbon monoxide yield,
- Eliminating the residual CO_2 through its combination with H_2 and further methane production.

Feeding low-purity CO_2 feedstock, including waste plastics both to coal and biomass to the gasifier can increase the CH_4 yield. Due to the slow kinetics of the hydrogasification reaction, the choice of proper catalysts is desired. The capromoted iron-group metals-based catalysts are the most promising samples for the char hydrogasification of both coal and biomass [70].

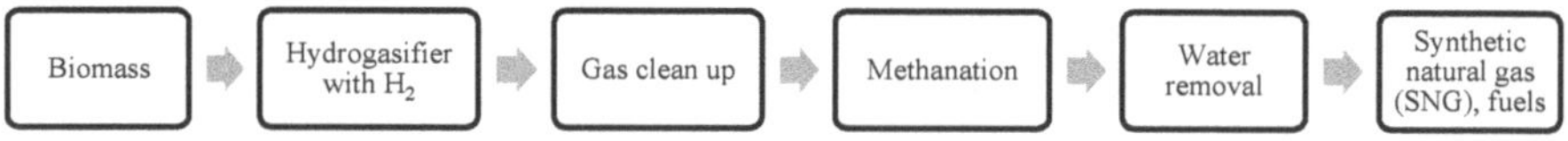

Fig. 14.7 Hydrogasification stages for synthetic natural gas (SNG) and other fuels

14.8.3 Electrochemical Reduction of Carbon Dioxide

Electrochemical CO_2 reduction using catalysts such as copper and zinc incorporated copper can produce multiple chemicals and fuels, including ethylene and ethanol with the increased efficiency of reduction reactions. Table 14.11 and Fig. 14.8 show some valuable chemicals and fuels produced by electrochemical reduction of CO_2 and conversion of CO [56, 65, 80]. This leads to:

- Decarbonization by converting captured CO_2 into fuels and chemicals.
- Storage of renewable electricity in chemicals and fuels.

14.9 Hydrogen and Decarbonization

Hydrogen has a dual role as a fuel and as a chemical agent to enhance energy efficiency, reduce GHG emissions, and the energy need for energy storage. In the steel sector, hydrogen can also be used as a co-fuel in steel furnaces integrating with both the electrical grid and energy-intensive systems that show hydrogen's critical role in industrial decarbonization. Therefore, renewables-based hydrogen can help both industrial and transportation decarbonization [30, 40–42].

Table 14.11. Electrochemical reactions for carbon dioxide reduction to chemical and fuels [56, 65, 80]

Reaction	Product
$xCO_2 + nH^+ + ne^- \rightarrow \text{product} + yH_2O$	CO_2 reduction,
$CO_2 + 2H^+ + 2e^- \rightarrow HCOOH$	Formic acid
$CO_2 + 2H^+ + 2e^- \rightarrow CO + H_2O$	Carbon monoxide (CO)
$CO_2 + 6H^+ + 6e^- \rightarrow CH_3OH + H_2O$	Methanol
$CO_2 + 8H^+ + 8e^- \rightarrow CH_4 + 2H_2O$	Methane
$2CO_2 + 2H^+ + 2e^- \rightarrow (COOH)_2$	Oxalic acid
$2CO_2 + 8H^+ + 8e^- \rightarrow CH_3COOH + 2H_2O$	Acetic acid
$2CO_2 + 10H^+ + 10e^- \rightarrow CH_3CHO + 3H_2O$	Acetaldehyde
$2CO_2 + 12H^+ + 12e^- \rightarrow C_2H_5OH + 3H_2O$	Ethanol
$2CO_2 + 12H^+ + 12e^- \rightarrow C_2H_4 + 4H_2O$	Ethylene
$2CO_2 + 14H^+ + 14e^- \rightarrow C_2H_6 + 4H_2O$	Ethane
$3CO_2 + 16H^+ + 16e^- \rightarrow C_2H_5CHO + 5H_2O$	Propionaldehyde
$3CO_2 + 18H^+ + 18e^- \rightarrow C_3H_7OH + 5H_2O$	Propanol
$CO + 6H^+ + 6e^- \rightarrow CH_4 + H_2O$	Methane
$2CO + 8H^+ + 8e^- \rightarrow CH_3CH_2OH + H_2O$	Ethanol
$2CO + 8H^+ + 8e^- \rightarrow C_2H_{4(g)} + 2H_2O$	Ethylene

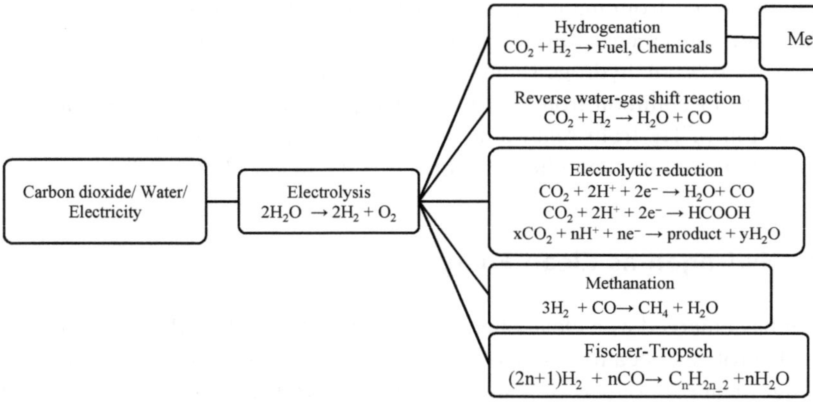

Fig. 14.8. Decarbonization pathways using CO_2, water and renewable electricity to produce chemical and fuels, as well as store electricity in chemical and fuels [56, 65, 80]

14.9.1 Hydrogen Production and Storage

Several low-carbon processes can produce hydrogen using nuclear and renewable electricity, biomass, and fossil fuels with CCUS. Steam methane reforming (SMR) is a mature and economical process for hydrogen production. SMR heats methane and steam in the presence of a catalyst to produce a synthetic gas (syngas) consisting of mainly carbon monoxide (CO) and hydrogen. Because SMR and coal gasification account for most of hydrogen production today, CO_2 capture retrofits are an attractive option to decarbonize existing production capacity.

Production of hydrogen is a mature process where steam and oxygen react with coal under high temperatures and pressures to produce syngas, which then undergoes a water-gas shift reaction to produce CO_2 and more hydrogen (Fig. 14.9). On average, this process is nearly twice as CO_2-intensive as SMR [42]. Electrolysis splits water into hydrogen and oxygen using electricity in an electrolyzer, which consists of an anode and a cathode that are separated by an electrolyte material [28, 29, 58, 74]. The low volumetric energy density of hydrogen complicates and increases the cost of storage and transport relative to conventional fossil fuels, as larger volumes and compression or liquefaction are needed.

Production of zero-emission hydrogen involves the addition of a carbon capture system and the electrolysis of water powered by renewable energy sources. Blue

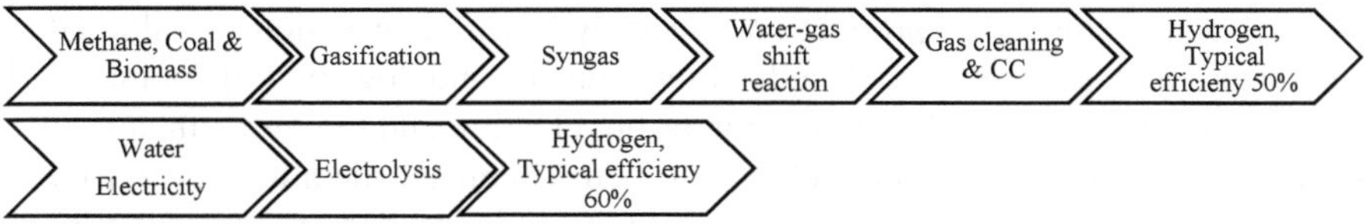

Fig. 14.9 Major hydrogen productions pathways based on feedstock: CC: carbon capture

hydrogen, produced from natural gas with carbon capture, can serve as an interim solution, while the green hydrogen production from renewable electricity scales grows. Up to 10% of hydrogen can be mixed with methane and can be transported directly in common iron-based methane pipes with minimal modification. The production of H_2 from renewable sources derived from agricultural or other waste streams increases flexibility and improves the cost [12, 20, 40].

14.9.2 Hydrogen as Clean Energy

Hydrogen has emerged as a clean energy, however, hydrogen production processes such as electrolysis, steam methane reforming (SMR), and biomass gasification require significant amounts of water. Besides, hydrogen must be pure enough to protect the pipelines, meet the regulations, and earn environmental tax benefits. Integrating hydrogen with industrial processes leads to energy storage and more efficient process solutions. Specifically, in the iron-steel sector, hydrogen can replace carbon as a reductant in the direct reduced iron (DRI) process, which accounts for around 7% of global steel production. Hydrogen can be a co-fuel in steel furnaces. "Virtual" power plants integrating with both the electrical grid and energy-intensive systems are proposed highlighting hydrogen's critical role in industrial decarbonization [17, 30–32]. Hydrogen, energy efficiency and CCUS projects are growing with an accelerated scale-up of this energy efficiency and carbon reduction technology. Examples are:

(1) carbon capture with old hydrocarbon reservoirs that can store CO_2 and have the technical ability to develop joint solutions with industries—such as steel—that are difficult to decarbonize,
(2) biomass to supply refineries can increase the renewable content of their end products.

Direct electrification may be increasingly prominent with green hydrogen, which can be co-fuel and can be blended with natural gas in existing furnaces. Mixtures of up to 20% hydrogen with natural gas are already being used, resulting in measurable reductions in CO_2 emissions [12, 32, 40, 41, 42].

14.9.3 Utilization of Hydrogen

Hydrogen has unique energy content to weight ratio and the high conversion efficiencies achievable in its production (electrolysis), its conversion to value added chemicals, and its use in fuel cells. Low density and low boiling point of H_2 increase the energy cost of compression or liquefaction and the investment costs of storage. Hydrogen uses include chemical, petrochemical, metallurgy, food industry, and fuel. Hydrogen is used in the refining industry for upgrading fuels and reducing their sulfur content, in the production of ammonia, SNG, liquid fuels and other chemicals, in hydrogen powered vehicles, and for electricity generation via Integrated Gasification Combined Cycle (IGCC) or fuel cells [4].

The CO_2 hydrogenation process produces methanol and methane directly from hydrogen and CO_2. Methanol can then also be catalytically converted to produce synthetic liquid fuels or undergo a dehydration reaction to produce dimethyl ether (DME). Methanol, like ammonia, is an important feedstock for the chemical industry. Green methanol produced from electrolytic hydrogen or biomass feedstocks could play an important role in decarbonization of the plastic production industry. DME is a gas at ambient conditions but can be liquefied under pressure [40, 42, 61, 62]. Figure 14.10 shows possible usage of hydrogen in the industrial sector and other applications.

Gas grid injection—methane produced from electrolytic hydrogen can be injected into the natural gas grid, at desired quantities for use in home heating, Compressed Natural Gas (CNG) vehicle fueling, and other traditional natural gas applications [44].

Shipping—Methanol can be used in marine diesel engines with minor system modifications using a small amount of pilot fuel for ignition. Because methanol is a liquid, its associated distribution, handling, and bunkering approaches are like those of conventional marine fuels. However, methanol's energy density is approximately one-half that of conventional marine fuels [17].

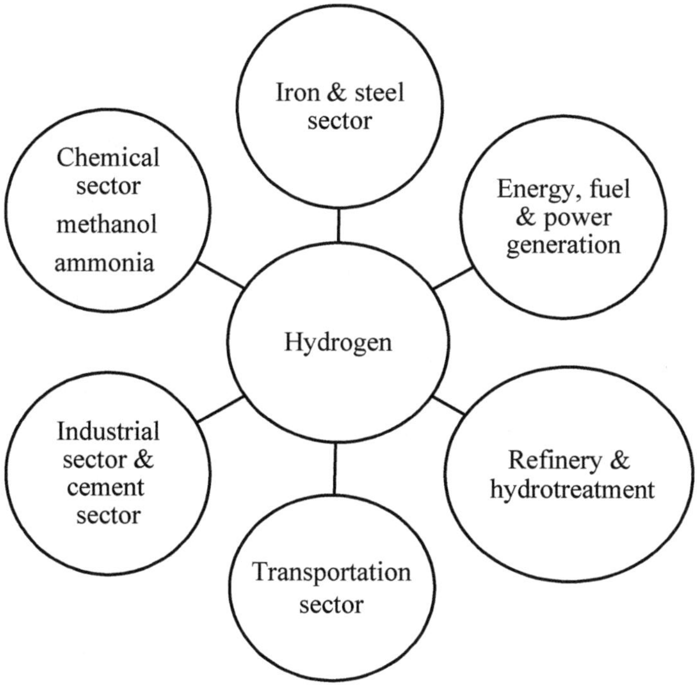

Fig. 14.10 Hydrogen impact on decarbonization in various sectors

Use gaseous hydrogen as an energy carrier requires a complete technical and organizational infrastructure for storage, transport, distribution, and use. Hydrogen can be fed into the existing natural gas grid network, which contains up to 5% hydrogen. Hydrogen embrittlement may lead to failures in vessels containing hydrogen, such as pipelines and storage tanks.

- *Hydrogenation of* CO_2 *to methanol*: Costs of methanol produced from natural gas may be the lowest compared with those from coal, biomass, and CO_2 captured from flue gas. Conversion of gaseous hydrogen to liquid methanol may result in a more convenient alternative energy carrier that can be handled by the existing infrastructure. Methanol can be used as a hydrogen carrier, a fuel, or a feedstock for chemical synthesis. Methanol is specially favored in fuel cell-powered vehicles as an onboard hydrogen storage medium [61, 62].
- *Hydrogenation of coal and coal tar*: The Bergius process is a method of production of liquid hydrocarbons for use as synthetic fuel by hydrogenation at high temperature and pressure. and dried in a stream of hot gas. Coal tar hydrogenation leads to high quality naphtha fractions and diesel with a high economic return program for environment protection and for development comprehensive utilization of coal at present.
- *Biogas upgrading*: After removal of CO_2, biogas can be upgraded to natural gas quality, giving more utilization possibilities, such as utilization as auto gas, or distant utilization by using the existing natural gas grid. Biogas may be upgraded into a separate biogas reactor, containing enriched hydrogenotrophic methanogens and fed with biogas and hydrogen [39].

14.9.4 Emerging Fuels: Clean Hydrogen and Its Derivatives

Hydrogen and its derivative fuels, including ammonia, methanol and kerosene, emerge as key components of the energy mix. These fuels may control emissions for industry and transport sectors that are hard to decarbonize through direct electrification. Direct electrification using hydrogen is preferable based on the efficiency of the gas' conversion to useful energy. About two to three times more electricity is needed to deliver the same service via hydrogen as direct electricity, due to conversion losses. Economic efficiency depends on the availability and cost of primary energy sources, the cost of electricity generation and storage technologies [40, 41].

14.9.4.1 Hydrogen and Decarbonization

Petroleum refining, ammonia production, and methanol production dominate hydrogen use and transitioning these applications to low-carbon hydrogen sources offers substantial GHG emissions reduction. Besides, low-carbon hydrogen's emerging applications include use as a fuel for power generation, transportation applications, or heat provision for industry and buildings. Hydrogen can be used as a chemical building block for other hydrogen-based molecules, such as ammonia, methanol, dimethyl ether, methane, and other hydrocarbons, leading to further decarbonization (Fig. 14.11). Delivering hydrogen blended with natural gas via existing gas pipeline networks could play an important role in certain regions where building space and water heating would help reduce GHG emissions [32, 40–42].

Impacts of green/low-carbon hydrogen with carbon capture utilization and storage include:

- Remote/decentralized hydrogen-based heat and power productions
- Hydrogen derivative fuels and chemicals
- Low-cost renewable energy
- Decarbonized feedstocks
- Energy storage in chemicals
- Novel technologies in electrolysis, membranes, compressors, transportation pipelines, and distributions
- Synthetic natural gas
- Green diesel
- Sustainable aviation fuels.

Technical benefits include:

- Considerable help for hard-to-decarbonize industries
- Impact on low-carbon liquid transportation fuels for land, aviation, and shipping
- Renewable energy integration with industrial sector and buildings
- Compatibility with existing natural gas infrastructure with storage, transportation, and distribution
- Energy technology and efficiency improvements
- Energy transition
- Energy security and access.

Barriers of hydrogen impact on decarbonization include:

- Policy
- Financing
- Regulatory issues
- Modeling
- Certification
- Technology
- Market
- Workforce
- Public acceptance.

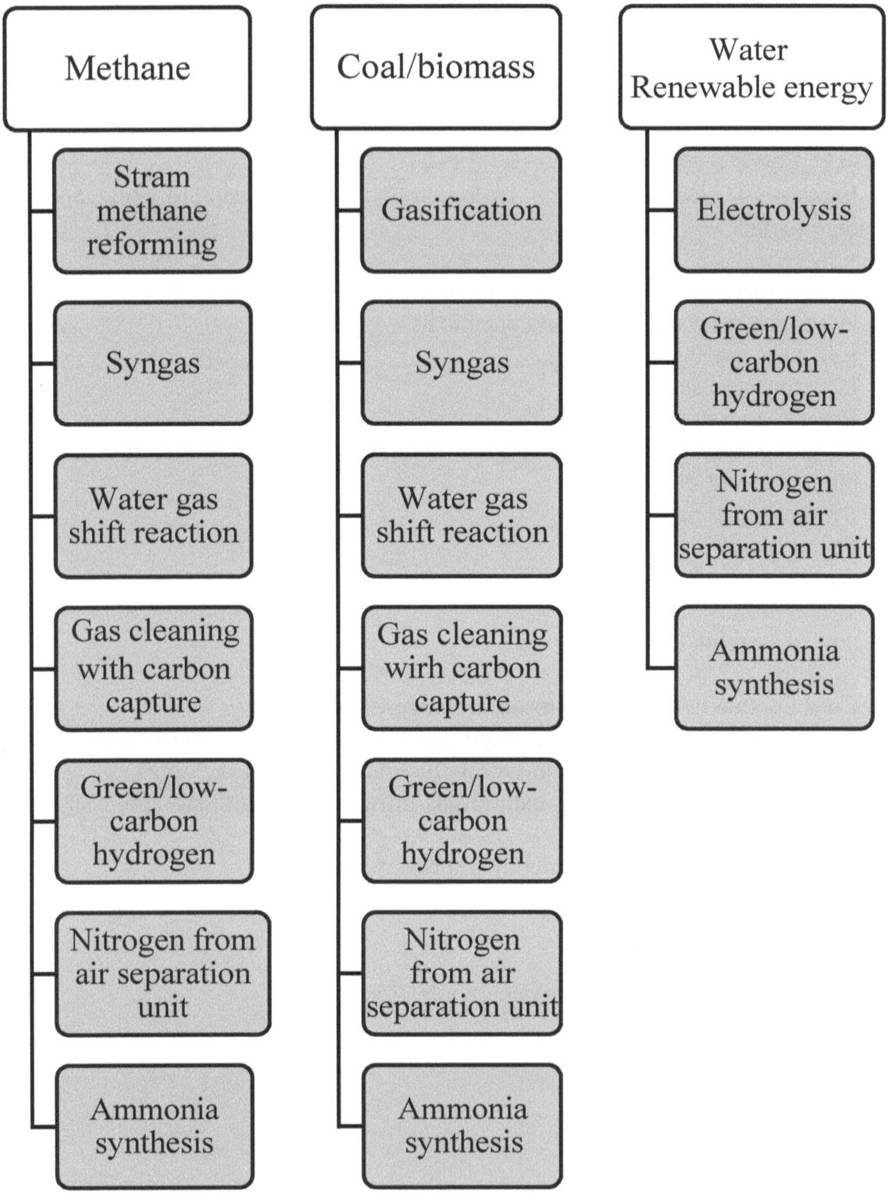

Fig. 14.11 Green ammonia production pathways based on feedstock

14.9.4.2 Ammonia for Decarbonization

Ammonia can be synthesized using hydrogen and nitrogen with reaction

$$3H_2 + N_2 \rightarrow 2NH_3$$

Hydrogen may be low-carbon or green hydrogen produced by using natural gas, coal/biomass, renewable electricity, and water as shown in Fig. 14.11. Ammonia, which is a gas at room temperature, liquifies at a much higher temperature and lower pressure than hydrogen, which reduces the cost of its transport and storage. As an energy storage medium, liquid ammonia has a potential advantage in that its volumetric energy density is nearly three times higher than compressed hydrogen [61].

Pure nitrogen and hydrogen are produced with no direct CO_2 emissions when ammonia is combusted. Processes for low-carbon ammonia include pairing conventional fossil-based production routes with CCUS for hydrogen produced from the hydrocarbon feedstocks. By-product CO_2 from ammonia plants is commonly used as a chemical feedstock for production of urea, which is an ammonia-based fertilizer that accounts for more than one-half of global ammonia demand. CO_2 utilization in urea production is by far the largest carbon capture and utilization application in fertilizer sector [25, 71].

- Liquid ammonia is often transported by truck and pipeline, which would be less costly than new hydrogen pipelines.
- Ammonia may be a "hydrogen carrier" because of its higher volumetric energy density and greater ease of compression and liquefaction relative to hydrogen. This means hydrogen would be converted to ammonia for transport and storage before being converted back to hydrogen known as "ammonia cracking" which causes efficiency losses of approximately 15% near the point of use. Transporting hydrogen as ammonia would likely be the most cost-effective option.
- Although combustion of ammonia produces no CO_2, emissions of nitrogen oxides (NOx), an air pollutant, must be mitigated. Emerging uses of ammonia include its application as a fuel to power engines, boilers, turbines, and solid oxide fuel cells.
- Fuel switching to ammonia is an option being explored for the decarbonization of the maritime sector. Manufacturers are developing 100% ammonia-fueled maritime engines, dual-fuel configurations (ammonia blended with fuel oil) may offer nearer-term commercial viability.
- Like hydrogen, ammonia can be used as a fuel for power generation: (a) ammonia can be used in blends for coal-fired boilers, (b) ammonia can fuel existing reciprocating engines with relatively minor modifications, and (c) in addition to hydrogen, ammonia blended with natural gas is another potential fuel for gas turbines.

14.9.5 Low-Carbon Fuels for Decarbonization

When produced with zero or low GHG emissions from clean electricity, renewable feedstocks, or fossil resources with carbon capture, utilization, and storage (CCUS), the low-carbon fuels could provide effective pathways to decarbonization. The global energy system is transforming with a momentum in low-carbon fuels, such as hydrogen, ammonia, methanol, and synthetic drop-in fuels that could provide solutions for the difficult-to-decarbonize sectors with the following end-use options [25]:

- Maritime shipping and aviation
- Long haul and heavy-duty transport
- Providing high-temperature heat for industry
- Providing heat for certain segments of the building heating markets.

14.9.5.1 Biofuels

There are various types of biofuels, including conventional and drop-in fuels based on sugar, lipid, or lignin feedstocks [13, 14, 17, 25, 39]:

- *Conventional biofuels* refer to biomass-derived liquid fuel and are often used for transport applications. Such as ethanol and biodiesel.
- *Advanced biofuels* typically refer to biomass-derived fuels produced from lignocellulosic feedstock including organic waste streams, agricultural and forestry residues, and dedicated energy crops in some cases that may provide net GHG emissions relative to the fuels they replace. Many advanced biofuels can function as "drop-in" and can be blended in high concentrations or used in conventional end-use engines, appliances without modifications.
- *Drop-in fuels* such as synthetic methane, gasoline, diesel, and jet fuel are chemically and functionally identical to their petroleum-derived counterparts. They are fully compatible with the existing transport and storage infrastructure for the transport sector. Power-to-gas projects are producing synthetic methane for application in compressed natural gas vehicles. Methanol and dimethyl can be blended with gasoline and diesel, respectively, for use in the transport sector to decarbonize.
- *Renewable diesel* is a class of advanced biofuel drop-in diesel replacements produced through various processes (including hydrotreatment, Fischer-Tropsch synthesis, and pyrolysis). Renewable diesel production through hydrotreatment is commercially mature.

There are various primary conversion processes for biofuels from biomass as seen in Figure 14.12:

- *Pyrolysis* Dry biomass feedstocks are heated in the absence of oxygen, producing pyrolysis oil, solid charcoal, and by-product gases. The output of pyrolysis

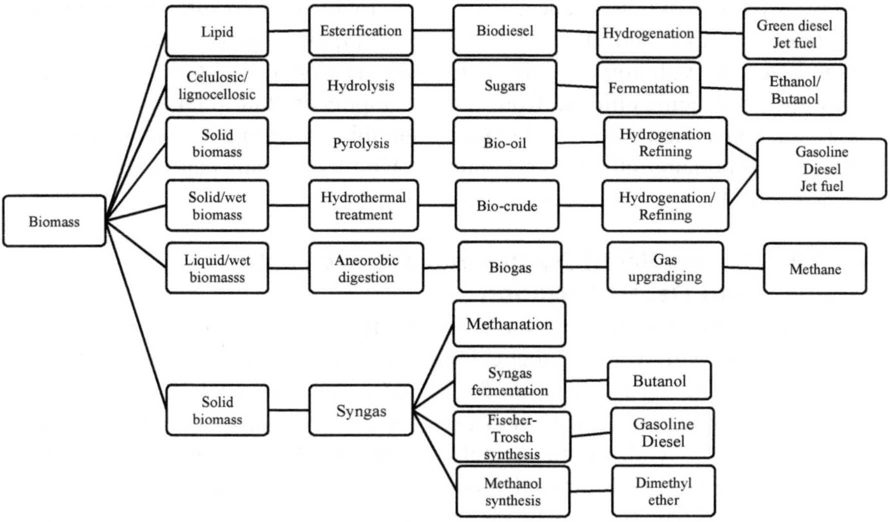

Fig. 14.12 Various pathways for producing biofuels from biomass for decarbonization by controlling the emissions,

can be upgraded by hydro-treating to produce high-quality liquid fuels that can substitute gasoline, diesel, or jet fuel.

- *Fischer-Tropsch synthesis* converts syngas to intermediate hydrocarbons in a catalytic process, which are subsequently upgraded to diesel, kerosene, or gasoline.
- *Hydrothermal upgrading* is a process in which biomass with high water content is heated and pressurized to convert it to liquid "bio-crude", which can act as a substitute for heavy fuel oil or be upgraded to gasoline and diesel via hydrotreating.
- *Hydrotreatment* involves reaction of lipid-based feedstocks, such as waste oils, with added hydrogen under high temperatures and pressures in the presence of a catalyst producing jet fuel, diesel, or renewable diesel, and gasoline.
- *Methanol synthesis* results from syngas reacting with CO_2 over a catalyst and is refined. The resulting methanol can be catalytically converted and refined to produce gasoline (via the methanol-to-gasoline process) or be dehydrated to produce DME.
- *Methanation* results from syngas undergoing a catalytic reaction before being upgraded to biomethane known as biosynthetic natural gas.
- *Aqueous phase reforming* with a biomass feedstock first undergoes pretreatment and hydrolysis to produce an aqueous solution of sugars, which is later converted via a multi-step catalytic reforming process that requires hydrogen as an input to produce liquid biofuels.
- *Anaerobic digestion* involves the biological decomposition of biomass materials in the absence of air and produces biogas, which is a mixture of methane and

CO_2. Biogas avoids methane emissions and can be used directly or upgraded by removing CO_2 and impurities to produce pipeline-quality biomethane.

- *Fermentation* with lignocellulosic biomass is first pretreated to separate the cellulose and hemicellulose from other components that then undergo hydrolysis for conversion into simple sugars suitable for fermentation with yeast or bacteria, producing ethanol and butanol. Microorganisms can also ferment the syngas, producing ethanol and butanol.

14.9.5.2 Green Electricity-Based Synthetic Fuels and Decarbonization

Hydrogen produced from renewable electricity can be combined with CO_2 to produce a wide variety of synthetic fuels, such as methane, methanol, dimethyl ether, gasoline, diesel, and aviation fuels. These synthetic fuels can be more easily handled, transported, stored, and used than hydrogen. Potential CO_2 sources for production of synthetic fuels include captured CO_2 from fossil-fuel usage for power generation or industrial processes. Synthetic fuel production is energy intensive as each conversion step involves sizeable energy losses, which account for approximately 45-60% of the electricity used to produce synthetic fuels [6, 47]. Figure 14.13 shows some primary pathways to produce synthetic fuels from electricity, water and carbon dioxide for decarbonization.

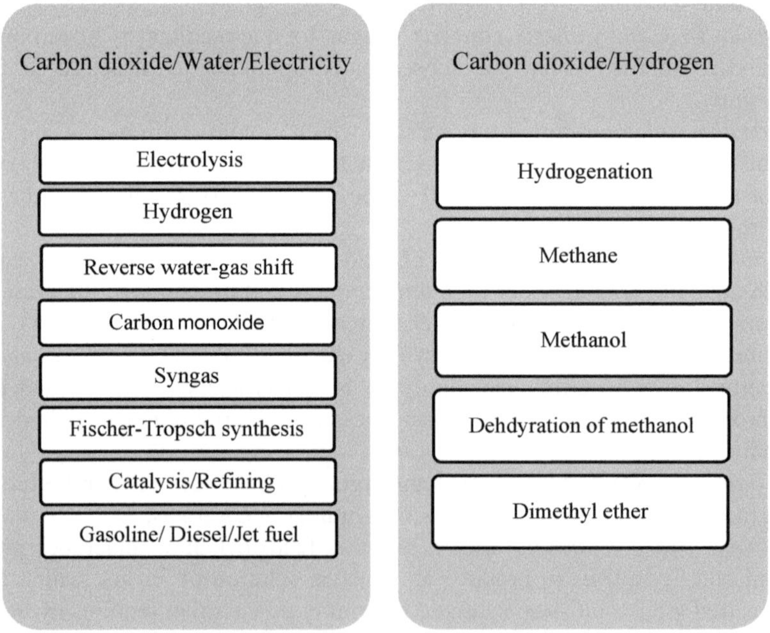

Fig. 14.13 Electricity-based synthetic fuel production pathways

14.9.5.3 Bioenergy's Potential for Emissions Reductions

Pairing biofuel production with CCS, which removes CO_2 from the atmosphere and sequesters it below ground may lead to "reduced emissions" but not zero emissions as the possible GHG changes due to land-use and market implications including supply chain.

Butanol: Bio-derived butanol can be blended with gasoline, similarly to ethanol.

Pyrolysis oil and biocrude: Pyrolysis oil and biocrude have higher water and oxygen content and tend to be more corrosive and they need upgrading.

Biomethane: Biogas containing methane and CO_2 can be upgraded to pipeline-quality biomethane that is similar to conventional natural gas and is typically stored similarly to conventional natural gas in a compressed or liquified form. Biomethane can be injected into existing natural gas pipeline networks.

Power and heat: Most producers use biogas for on-site electricity, heat, or combined heat and power (CHP) production.

14.10 Decarbonization in Industrial Sector

Major Industrial sectors include:

- Iron and steel manufacture of basic metals
- Chemicals manufacture of chemicals and chemical products
- Cement and lime manufacture of other non-metallic mineral products
- Food and drink manufacture of food products and beverages
- Pulp and paper manufacture of paper and paper products
- Glass manufacture of other non-metallic mineral products
- Aluminum manufacture of basic metals
- Refining manufacture of coke and refined petroleum products
- Ceramics manufacture of other non-metallic mineral products.

The industry is undergoing a transformation to produce more sustainable chemicals and increase green feedstocks just like green hydrogen, while staying competitive. Companies, therefore, must look beyond fossil fuels and find new, renewable ways of electrification and automation, and focus on sustainable use of resources through innovative technology. Energy, raw material costs, and rise in product quality standards require high efficiency and productivity for the industry to stay competitive globally. A range of mature decarbonization options based on CCUS, fuel switching to hydrogen, electrification, and biomass-based fuel can reduce emissions in industrial sectors. However, many industrial decarbonization technologies require high capital and operational costs [69].

Best practices for energy-efficient industrial system design are to improve energy and materials efficiency, implement beneficial electrification, and use low-carbon energy sources and feedstocks (Fig. 14.14). The decarbonization opportunity may be divided into the following design layers:

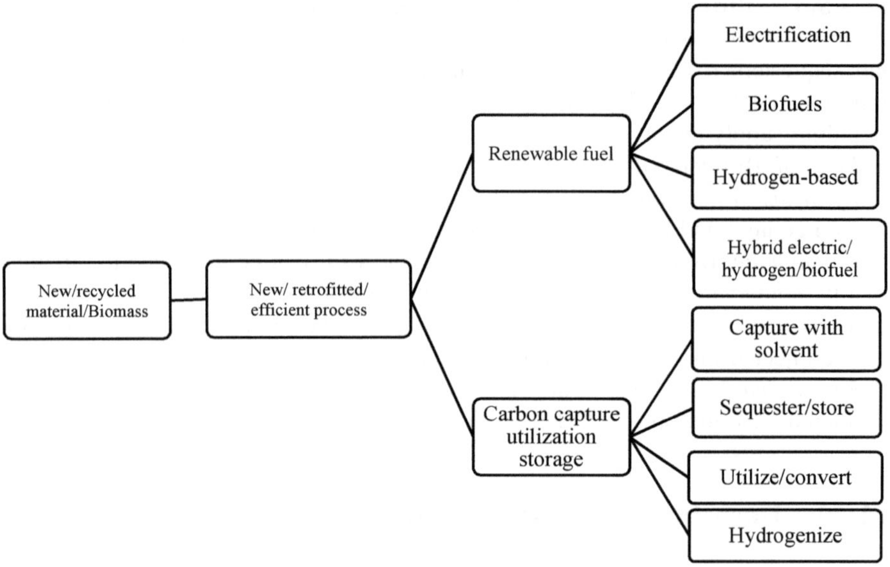

Fig. 14.14. Generalized energy intensive industry decarbonization

1. Optimize the core process system for material and energy efficiency
2. Design an efficient distribution system
3. Select the right size and efficient equipment
4. Control the system for efficient operation.

The industrial sector accounts for considerable global energy consumption and direct GHG emissions and needs improvements in energy efficiency, carbon capture utilization and storage, electrification, and green hydrogen use as a heat source and feedstock [31]. The roadmap toward decarbonization identifies four pillars required:

- policy,
- climate leadership and movements,
- finance, and
- exponential innovation in technology.

Industrial decarbonization requires new and upgraded supporting infrastructure, including the electricity network, reduced electricity demand with flexible innovative technologies. Since industrial sites have different capacities, transport and distribution distances, and storage conditions, CO_2 and hydrogen infrastructure size varies significantly. Biomass switching for industrial processes may create several barriers, such as its availability as an energy source and the impact on land use, the alignment of biomass uses with international sustainability standards, and trade rules.

14.10.1 Circular Economy and Industrial Decarbonization

Circular economy decouples economic activity from the consumption of finite resources to reduce production costs, GHG emissions, waste and other pollutants, improve competitiveness, increase stability and resilience in the production [45]. Using renewable material, energy and reducing toxic chemicals, the design of industrial products can be improved so they can be easily retrofitted, reused, and recycled to provide energy or raw material for new products [13, 18]. For example, aluminum can be recycled many times without losing its original properties and can reduce emissions by 95% relative to primary aluminum.

The circular economy has three key principles:

- eliminating waste and pollution,
- circulating products and materials at their highest value, and
- regenerating nature.

Buildings are becoming less carbon-intensive and more energy efficient in large part to the green building movement with renewable energy. However, decarbonization progress and grid conditions are not distributed evenly. Public engagement for deep decarbonization is necessary to buy and use new technologies and to transform energy systems. The development, economics, and deployment of new energy technologies and electrification in vehicles, power plants, buildings, and industrial applications will have a positive impact on decarbonization [30, 35. 47, 69].

14.10.2 Bioeconomy and Decarbonization

Bioeconomy is the production, utilization, conservation and regeneration of biological resources, with related know-how, science, technology and innovation. Bioeconomy policies have roots in agriculture, aquaculture, livestock and forestry, sustainable use of biomass, biodiversity, and circular bioeconomy. Bioeconomy can offer bio-based materials like mycelium (fungus) and hempcrete (hemp-based concrete) that are sustainable alternatives to traditional building materials. These materials are biodegradable and provide improved insulation properties and reduce the energy demand for heating and cooling buildings. Combining bioeconomy growth with ecosystem conservation and restoration may lead to the use of biotechnology in cost reduction and hence encourage governments and private investors commitment to the bioeconomy's growth [11, 18].

Biogenic resources are explicitly indicated as carbon-neutral from Life Cycle Assessment (LCA) perspective by the IPCC. Quantifying biogenic and nonbiogenic carbon footprints and optimizing the circular economy help policy makers for sustainable decarbonization in the long term.

14.10.2.1 Biotechnology

Biotechnology is biology-based technology and involves cellular and biomolecular processes to develop technologies and products that help improve the health of society. As more data such as electronic health records, genomics, metabolomics, and even life-style information, further insights and opportunities for biotechnology will become apparent.

Bioprocessing and utilizing robust microbial strains can be used to make biofuels, biochemicals and bioplastics, to optimize fermentation processes and scalability, reduce costs and environmental impacts. Engineering biology can deliver high-impact commercial solutions that are less carbon intensive and resource saving by design to replace the processes that are carbon-intensive. This contributes significantly to global decarbonization efforts. Challenges include immature regulations, lengthy product development timelines, and high costs and risks to commercialize [1, 6, 59].

14.10.2.2 Synthetic and Generative Biology

Generative biology helps restore ecosystems and biodiversity, healthcare solutions, sustainable manufacturing of fuels, bioproducts with advancements in information technology and next-generation biology-based computing. Genetic biotechnologies can deliver groundbreaking innovations for biofuels. Synthetic biology has significant potential to decarbonize as it could be applied to what we eat (alternative and cultured proteins), the clothes we wear (textile fibers originating from engineered microbes), how we build our homes and workplaces (e.g., mycelium-based construction materials with enhanced carbon capture technologies), and even the way we travel (e.g., mobility powered by sustainable biobased alternatives) [18].

14.10.3 Thermal Process Intensification

Thermal process Intensification (TPI) involves transformative technologies and strategies to substantially improve the performance of thermal processing systems in the industrial sector, including energy productivity, energy conservation, thermal efficiency, reduced GHG emissions, and reduced number of thermal process steps [18, 22]. Thermal processing is the largest contributor of GHG emissions from combustion of fossil fuels, such as in the case of cement and lime production as the established industrial base still depends mainly on carbon-based processes and energy sources. Industries in the following four groups have similar thermal processes:

- High-temperature metal processing (iron and steel industry, alumina-aluminum industry),
- High-temperature nonmetal and mineral processing (cement and glass industry),
- Medium- to low-temperature thermal processing (food processing and pulp and paper industry as part of forest products sector), and
- Hydrocarbon processing (petroleum refining and chemical industry).

Different types of thermal process intensification technologies are [22]:

1. Low–thermal budget transformative technologies (mid- to long-term) that may use alternative energy sources while offering disruptive changes in the current production methods (e.g., electrolysis and electrodialysis)
2. Alternative thermal processing (near- to mid-term) technologies may use an alternative source of energy while maintaining the current production methods to gain more flexibility/control (e.g., induction and resistance furnaces)
3. Transformative supplemental technologies (near-term) with emerging energy efficiency and supplemental technologies that reduce thermal demand (e.g., smart Internet of Things devices for system optimization)
4. Waste heat management technologies (near-term) with emerging waste heat reduction, recycling, and recovery technologies (e.g., thermoelectric devices and heat pipes).

Some major research needs for thermal process intensification include [22]:

- *Low–thermal budget technologies*: There is a need to develop non-carbon or low-carbon technologies for oxygen removal (reduction) from natural raw materials like iron ore, bauxite, calcium carbonite to produce commodities. Alternative methods include reducing agents such as hydrogen or the electrolysis process. Non-heat–based technologies that rely on mechanical vibrations (contact and non-contact ultrasound), advanced membrane (reverse osmosis and ultrafiltration) may be beneficial in drying and separation processes. Besides, electromagnetic radiation (microwave and radio frequency) interacts with materials in unique ways and may lead to energy and quality improvements in annealing, sintering, drying, and electrolytic processes applications in different industries.
- *Low-carbon fuels and electric heating applications*: The use of renewable-based electrical heating systems including induction, resistance can lead to higher productivity and reduce losses given the improved controls and carbon-free operations. Increased use of renewable fuels such as biogas, hydrogen, and solar energy can be more broadly adopted.
- *Smart manufacturing technologies*: Smarts sensors with AI and deep learning algorithms and systems can be used for online continuous measurement of critical process parameters (oxygen, temperature) and can be used for improving consistency of performance of process heating equipment. Digital twins of virtual replicas of thermal equipment and processes can be used to optimize operations and improve product quality [18].
- *Waste heat management technologies*: Design and material improvements with energy conservation and recovery of waste heat generated from thermal processes would be beneficial when matching between available and required heats is enabled through pinch analysis and heat exchanger network systems [16, 18].
- *Advanced materials*: Substitution or reduced use of certain basic materials, including cement, lime, glass that require very high ($> 1600\,°F$) temperatures

can reduce thermal demand and cut down on a large amount of the CO_2 emitted by the raw material. Similarly, the use renewable resources to replace fossil fuel–based raw materials in the production of chemicals and other products would help reduce emissions.

14.10.4 Challenges for Using Biomass to Decarbonize Industry

Three potential applications of biomass in industry require additional research and development to be implemented on a commercial scale. First, use of biomass as a direct replacement for fossil fuels might be limited by its lower calorific value compared to some fuel types (e.g., bituminous coal) or by the fuel quality requirements and standards in some applications. Second, bio-oils derived from pyrolysis technologies are often unstable and acidic, and current stabilization processes add cost. Third, while lignin is a promising source of aromatics, its integration into cellulose and tendency to condense into oligomers present challenges for extracting and cracking it in a manner that produces clean aromatic monomers [1, 59].

To use biomass for industrial decarbonization depends on policies and incentives, competition between biomass use cases, and consumer demand. For example, food waste can be anaerobically digested to generate organic acids, which can serve as feedstocks in chemical syntheses. Alternatively, the same food waste could be converted to renewable natural gas, reformed into hydrogen, which is more attractive in the current policy environment. Changing consumer preferences could also influence industrial biomass use; consumers may be willing to pay more for sustainable biochemicals derived from a biomass feedstock [1].

Adding carbon capture to a mix of low-carbon generation technologies can reduce the costs of deep decarbonization. One option may be bioenergy with carbon capture for net-zero electric sector emissions targets. There are questions about the speed, scale, cost, and acceptability of different carbon capture options such as bioenergy with carbon capture and sequestration (BECCS), direct air capture (DAC), and afforestation. Policy makers and resource planners should incorporate these options into the technology choice for their deep decarbonization and net-zero targets. Carbon capture availability can impact power sector decarbonization pathways and help to lower costs for higher mitigation levels. Carbon capture can also provide flexibility in meeting zero targets avoiding over dependence on any single technology. Without carbon reduction, electric sector abatement costs increase sharply, even with significant cost reductions for renewables and energy storage [18].

Energy market projections are inherently uncertain because many of the events that will shape future energy markets including regional developments in policy, technology, demographics, and resources:

1. Increasing population and income may compete with declining energy and carbon intensity on emissions.
2. The shift to renewable electricity demand is driven by regional resources, technology costs, and policy.

3. Energy security concerns may limit transition from fossil fuels in some countries, while drive increased fossil fuel consumption in others.

The relative prices between using electricity compared to sustainable biomass or CCS need to change for making decarbonization a competitive option. This implies a need for very high carbon prices currently presents a large barrier to implementation. Many of the basic electro-thermal technologies already exist but to scale up, some new key technology building blocks including electrolysis, the Sabatier process, and electrowinning [43].

Downstream industries do not respond to technology shocks directly, but they do respond to price shocks. A 5% upstream price increase is followed by a 4% increase in capital investments. Transitioning energy-and natural resource intensive industries from fossil-based to fossil-free production methods is a technological, economic and political challenge. Although this is technically possible, the productions costs are likely to increase and hence decrease economic welfare. Utilities tend to have relatively long investment cycles compared with manufacturing firms. Still utilities should invest in new capital and enhance their productivity. Higher levels of competition in the manufacturing sector prevent manufacturing firms from pushing on the cost increase to their consumers. Instead, the firms innovate and invest to make themselves even more productive when the material costs increase. Utilities and construction firms commonly face less competition and can pass on the cost increase to their consumers without losing customers. The third potential factor is the ability to innovate. The products of utilities, and to some extent construction firms, are more standardized than those of manufacturing firms, and the scope for updating the product to increase its value is more limited than for manufacturing firms [2, 18].

Three policy options to encourage the decarbonization of the electricity sector may be:

1. A carbon price or carbon tax
2. Long-term contracts for renewables
3. Using markets to achieve renewable goals

A typical refinery consisting of separations, purifications, and chemical reactions requires large energy inputs that include a combination of heat, electricity, and hydrogen. Each type of energy input may create a unique challenge for decarbonization. Heat energy includes either steam or direct firing to drive chemical processes, while electricity is used to power mechanical components. The emissions targets will require both efficiency gains as well as lower carbon energy sources, both of which require innovation and deployment of new technologies, including utilizing co-generation of power and heat from the same unit, low energy-based separations, process intensification, lower carbon energy and fuel sources, CCS, low-carbon production of hydrogen, and biomass fuels [18, 38].

14.10.5 Iron and Steel Industry

Worldwide steel production cost could increase by 15% by 2050 if low-carbon technologies are used. Over 70% of the global steel is produced using blast furnaces/basic-oxygen furnaces, which rely mostly on metallurgical coal as the chemical reducing agent. There are several options to decarbonize the iron and steel sector:

(1) Circular economy practices along with increased use of recycled scrap in electric arc furnace–based steel production will play a key role. Global steel production is projected to increase by around 40% by 2050, while approximately one-third of it would involve scrap-based electric arc furnaces.

(2) A structural shift in iron and steel production is needed, with renewables displacing fossil fuels as both energy consumption and reducing agents. The steel sector would see a growth in renewables' share (bioenergy, other renewables, supply in electricity and green hydrogen production) from 6% in 2020 to 20% in 2030 and 61% in 2050. Key innovations such as hydrogen-based direct iron reduction with electric arc furnaces represent a promising, efficient and low-emission technological pathway integrated with Industry 4.0 practices [2, 18, 47].

14.10.5.1 Case Study: Energy Efficiency in an Iron and Steel Coke Plant

This case study uses the coke plant section of an iron and steel plant to calculate the net power generated and CO_2 emissions to produce one ton of crude steel. The sector employs high temperature furnaces for refining and production with mostly from coal to melt the burden and to produce coke which is the main reducing agent for removing oxygen in the iron ore. In this case study, an integrated simulation of coke production, COG cooling, coke quenching and flue gas waste heat recovery in a coke plant in the steel industry is carried out

Process Description- A large amount of heat in the coke production process is lost from the discharge of flue gas, quenching of incandescent coke, and raw coke oven gas (COG). Coke production, which heats coking coals at 1100°C in the coke oven to drive off volatile compounds, is a high energy consuming process. The coke oven is composed of a combustion chamber and a coking chamber. Clean COG is combusted in the combustion chamber, following which heat is transferred across the bricks to the coking chamber, where the coking process takes place. After the coking is finished, the incandescent coke is pushed to the quenching system. The raw coke oven gas leaves the coking chamber and is cooled to collect the by-products. Waste heat can be captured from incandescent coke, hot raw COG, and flue gas to increase the efficiency of the coking plant and thus decrease CO_2 emissions. The incandescent coke is quenched by inert gas, while the raw COG is quenched by water and enters the byproduct recovery section to recover valuable byproduct chemicals such as coal tar, sulfur, and crude benzenes.

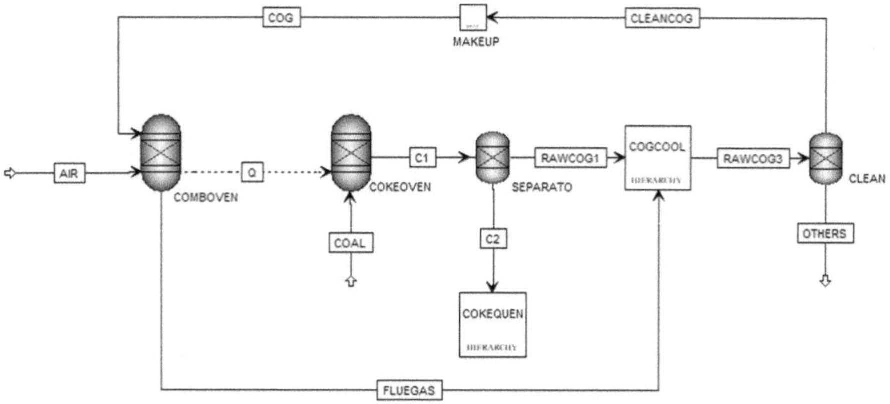

Fig. 14.15. Coking plant consists of coking, coke quenching, COG quenching, flue gas cooling and COG cleaning and combustion

At the combustion chamber, heat is also recovered from the flue gas and high temperature raw COG.

Model Description—As shown in Fig. 14.15, the whole coking plant consists of coking, coke quenching, COG quenching, flue gas cooling and COG cleaning and combustion.

At about 1100°C, the coal undergoes the coking process to decompose to coke, COG, coal tar, crude benzenes, sulfur, ammonia, and water. The raw COG is cleaned by the separator, which has mostly hydrogen and methane, giving COG a higher heating value. Energy for heating the coking stream by burning a mixture of air and recycled COG in combustion chambers was simulated by the RGibbs reactor for the combustion process. The heat of the flue gas emitted from the combustion oven is recycled in the flue gas cooling unit to be combined with COG quenching unit.

Modeling of the COG quenching unit—During the process of coal dry-distillation, the volatile matter in the coal is released to form raw COG. The gas is further processed in the COG quenching unit where COG is cooled and cleaned recovering tar, benzene, and sulfur. The COG cooling is combined with flue gas cooling section, while the flue gas from the combustion chamber is used as an economizer to preheat the feed water. Then, the excess heat of the flue gas is recovered by generating low pressure steam. To recycle waste heat in the high temperature raw COG, preheated feed water is fed to the COG cooling unit and medium pressure steam (MPS) produced through an evaporator and superheater is used in a steam turbine to generate electricity.

Modeling of the coke quenching unit—The coke product from the coke oven is sent to a dry quenching tower where inert gas is used to cool down the red-hot coke. A large amount of thermal energy is recovered during the coke dry quenching (CDQ) process. The inert gas is recirculated by passing through a series of heat exchangers which are used to heat feed water, which is then sent to the

Table 14.12 Performance of coke oven gas (COG) and coke dry quenching (CDQ) process

	COG	CDQ	Total
Quenched COG/coke kg/h	120.3	542.1	
MPS generated (4.8 MPa, 450°C) kg/h	470.0	273.0	743.0
Power generated kW	101.37	58.88	160.26
Power consumed kW	1.19	0.75	1.94
Net power generated kWh	100.19	58.14	158.32
CO_2 emissions kg/h			551.14

medium pressure steam (MPS) cycle to generate electricity. The results include MPS generation, power generation and consumption, CO_2 emissions are listed in Table 14.12.

For generating one ton of hot metal, 158.3 kWh of electricity can be generated, while causing only 0.552 tons of CO_2 emissions. Since the model is only part of the iron and steel process, for further energy conservation and decarbonization, the whole iron and steel production process should be simulated and optimized based on the proposed model. [Energy Efficiency in an Iron and Steel Coke Plant with Aspen Plus. 2024 by Aspen Technology; Wang R (2022); Energy saving technologies and optimization of energy use for decarbonized iron and steel industry, Durham theses, Durham University.]

14.10.6 Transport Sector

Transportation emissions represent nearly one-third of GHG emissions, the majority of which will be reduced through vehicle electrification with cost-effective electrification of port and airport operations, cost-effective state and local policies to reduce vehicle emissions. Transport sector decarbonization will require scaling up the adoption of renewable fuels, the production of sustainable liquid biofuels [17, 44, 47]:

- *Land transport*-Population and economic growth in developing economies is expected to boost road transport activity. Road transport would see hydrogen demand coming mostly from heavy-duty trucks. International shipping would see a diverse mix of low-carbon fuels for decarbonization, with ammonia, methanol and hydrogen. Accelerating electric vehicle adoption for road transport, along with green power supply, is an important approach for transport sector decarbonization. Electrification is also crucial for deeper decarbonization of the rail sector.
- *Aviation*-Aviation demand is expected to recover to pre-pandemic levels by 2023. The introduction of novel hydrogen and electric aircraft (or hybrids) from 2035 for short-haul flights would radically transform the aviation sector.

synthetic kerosene use in total energy consumption would grow up. Synthetic drop-in jet fuels produced via Fischer-Tropsch synthesis from renewable electricity can lead to long-term decarbonization of the aviation sector. Sustainable aviation fuel (SAF) is a key scalable in-sector solution for decarbonizing aviation contributing to reducing emissions from air transport [19, 48].

- *Shipping*-The international shipping sector would see a diverse mix of low-carbon fuels to decarbonize the sector, with ammonia, methanol and hydrogen composing almost 61% of the fuel mix by 2050. The maritime sector is already adopting dual-fuel engines for both new vessels and retrofits, with methanol-fueled options [48, 50]. Advanced liquid biofuels could represent the most cost competitive low-carbon option compared to alternatives, such as ammonia and hydrogen for the shipping industry. Biofuel intermediates, such as bio-crude from hydrothermal upgrading and pyrolysis oil, are also being explored as substitute fuels for marine engines that use low-quality heavy fuel oil. While the existing liquefied natural gas (LNG) fleet size is relatively small but growing, biomethane can act as a drop-in fuel for LNG vessels. Methanol and dimethyl ether are also being explored as marine fuels [39].

Emerging technologies for the e-mobility sector-Electric vehicles (EVs) are emerging as solutions for decarbonizing the transport sector through direct electrification using high shares of renewables integrated into power systems. Ensuring universal access to charging would entail the possibility to charge at any shared space such as those on streets, at workplaces, in commercial establishments, public spaces. With increasing EV sales, battery demand is rising, while battery prices are decreasing realizing the targets on zero-emission vehicles [43, 47].

14.10.7 Cement Sector

Decarbonized cement and paper production cost might increase by nearly 30% and 8%–15% in 2050, respectively. To successfully transform the cement sector:

1. Reduce the use of cement with material having the same strength while using less cement in concrete and less concrete in infrastructure and buildings.
2. Enhance the use of additives and fillers to substitute for clinkers, complemented by circular economy practices. The most common type of cement is "Portland cement", which contains clinker along with smaller quantities of gypsum and ground limestone. Conventional clinker can be partially substituted with. blast furnace slag, coal fly ash, red mud, and calcined clay), most of which are waste from other materials production processes.
3. Promote efficiency measures in cement production with defined minimum standards and promote the application of the best available technologies to also improve the efficiency of existing units.

4. Switch to renewable energy and alternative fuels. The overall share of renewables (bioenergy, other renewables, electricity) would increase from 4% in 2020 to 55% by 2050, dominated by bioenergy.
5. Apply CO_2 removal measures such as CCS-CCUS and BECCS technologies, particularly to capture residual process emissions, with total capture reaching the order of 2.1 $GtCO_2$/year in 2050. Bioenergy derivatives are ideal in energy transitions in transport, power and heat production, industrial processes, and clean cooking as they are versatile, storable, and dispatchable. For example, biofuels are the most feasible options in aviation and shipping, which require energy-dense carriers [2, 47].

14.10.8 Chemical Industry

The chemical industry consumes high quantities of water for cooling purposes or to convert water into steam, hydrogen or oxygen. With global demand for freshwater expected to exceed supply by 40% by 2030, operators face regulatory pressures to better control how chemical plants manage water and minimize wastewater production. High water use also translates to higher costs for the plant, through increased energy consumption during production processes. Therefore, plant operators are exploring smarter water usage with artificial intelligence (AI) technologies to help reduce excessive water use and create high-quality products with less waste. To reduce water usage, new chemical plants must invest in smart sensors and meters in place to optimize water intake, precisely control and reduce water use during production, and maximize water recycling. This may lead to a considerable reduction in water costs by implementing recycling systems, while better monitoring and control of boiler feed water can decrease water usage by 20–30%. Producing olefins and aromatics via a decarbonized steam cracking route or methanol to olefins process can increase their cost by 50%–20% by 2050. Green ammonia production is likely to increase its cost by 13–41% by 2030. [6, 18, 47].

14.10.9 Wastewater: Electrodialysis

Wastewater and by-product treatments are a part of eco-efficient and circular solutions by limiting the use of new raw materials and reducing waste. Electrodialysis (ED) processes such as conventional, selective, bipolar membranes, and filtration membranes are processes that have demonstrated their potential and eco-efficiency. There are valorization opportunities for industrial sectors to manage waste or by-product resources through electrodialysis to improve industrial decarbonization. ED processes can be optimized to decrease energy consumption and costs, and to increase efficiency; however, eco-efficiency scores should be determined to compare electrodialysis with conventional processes [11].

14.10.10 Built Environment

The planning, design, construction, maintenance, and disposition of the built environment creates considerable greenhouse gas (GHG) emissions. The overall reduction of total energy demand in the built environment through energy-efficiency improvements and advanced energy management is critical of carbon-based energy directly used by them through building electrification and property-level renewable energy installations. Most of these actions are well within the current technological capacity and knowledge of the industries.

Buildings would be smart, inter-connected, highly energy efficient, and powered and heated or cooled predominantly by renewable energy with the following recommendations:

- Enhance the energy efficiency of buildings with efficient lighting and appliances, and renewable energy-based heating and cooking technologies. Heat pumps can achieve energy efficiency levels three to five times greater than fossil fuel- based boilers and can be powered by renewable electricity.
- Prepare for a rise in electricity demand.
- Replace traditional, unsustainable bioenergy sources with sustainable biomass, biogas and electricity.
- Promote renewables such as biofuels, biomethane and solar thermal for heating.
- Couple heating systems with thermal and seasonal storage for flexibility.
- Promote systemic innovation, including new technologies, materials, designs and business models
- A considerable reduction in commercial building industry emissions can be achieved through a combination of grid decarbonization, growth of heat pump (see Sect. 7.15.1) installations modeled to the trajectory of solar, and net zero carbon government buildings.

14.10.10.1 Energy Use Intensity

Energy use intensity (EUI) is a metric that expresses a building's energy use as a function of its size, usually expressed as energy consumed in kBtu per square foot per year. Source EUI represents the amount of fuel required to operate the building, incorporating all transmission, delivery, and production losses. Site EUI represents the amount of heat and electricity consumed by a building, as reflected in utility bills.

Electrification-In addition to building efficiency measures that reduce a building's EUI, a building's source of energy is key to decarbonization. Electrification refers to the process of replacing technologies that use fossil fuels (coal, oil, and natural gas) with technologies that use electricity as a source of energy. As electric grid regions across the country switch from fossil fuel energy sources to renewable sources, buildings can reduce their emissions by utilizing electrification to access this increase in grid renewable energy generation. Zero Energy Design

Designation-It is critical that building professionals are prepared to design and construct high-performing, affordable buildings. Objectives are based on the Building Science Education.

14.10.10.2 Net-Zero Buildings

Smart energy management systems in buildings and digitalization, such as the Internet of Things, will change how buildings consume energy and even allow them to provide grid services through enhanced demand flexibility. A parallel shift to EVs will enable home charging, while decentralized energy supply will enable more local generation of buildings' electricity demand through solar energy systems and storage, both electric and thermal [48].

14.11 Decarbonization and Digital Technologies

Inter-Governmental Panel on Climate Change (IPCC) stages project development into three parts: project preparation, construction and commercial operation. Recent developments in artificial intelligence may accelerate project proposal development, at the asset level (wind, solar and transmission), at national levels and globally, to reduce time and costs to deployment, and hence accelerate global decarbonization efforts. Decarbonization solutions must be cost-competitive, for example, industrial heat pumps can improve efficiency and dramatically cut emissions and costa from industrial sector thermal processes [10, 18].

14.11.1 Distributed Renewable Energies and Artificial Intelligence

Digitalization can enhance supply and demand-side flexibility by integrating distributed renewable energy and flexibility assets. The right digital equipment can identify the optimal placement of distributed assets allowing operators to monitor and control them in real time and reinforces power system security and reliability and flexibility. This reduces costs and maximizes the utilization of renewable resources, pairs distributed renewable energy assets with optimized electric end-use technologies, such as electric vehicles and heat pumps, and integrates demand-response mechanisms and applications (Fig. 14.16).

14.11.2 Artificial Intelligence in Decarbonization

Data is the new feedstock set to optimize and improve decarbonization with data analytics for collecting, standardizing, analyzing, and infusing data into operation. AI-powered analytics on real-time and stored data can improve products yields by providing operators with suggestions to increase plant performance. AI creates the

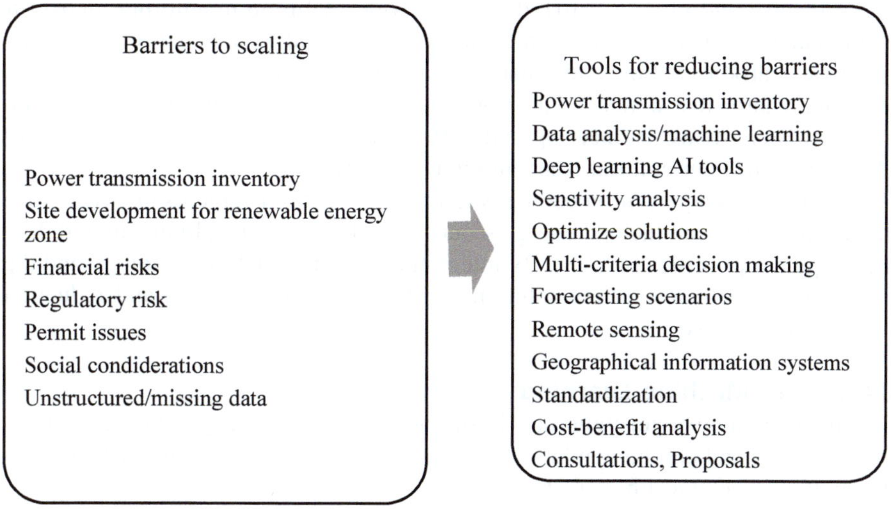

Fig. 14.16. AI impact on managing the barriers to improve and scale wind and solar power sites and transmission

potential to identify and map the clean energy infrastructure to support optimization of infrastructure for wind, solar and transmission. The criteria for identifying sites are based on delivering a minimum viable product (MVP) with initial and final site assessment and return on investment. AI, particularly generative artificial intelligence, offer powerful new tools to support asset-level project initiation of large-scale wind and solar infrastructure by taking account of environmental and social trade-offs [10, 18].

- The Global Energy Monitor (GEM) provides selected location data for large installations, while Multi-criteria Analysis for Planning Renewable Energy (MapRE) provides a framework for the systematic identification and valuation of areas for renewable energy development.
- The Renewable Energy Zoning (REZoning) tool combines spatial analysis and economic calculations to give users insights into the technical and economic potential of renewable energy resources for all countries.
- Geographic Information System (GIS) explores spatial data, such as maps and satellite images, weather patterns, land use and topography, which can be processed using machine learning algorithms to identify potential renewable energy sites [6, 10, 18].

14.11.3 Data and AI

Digital technologies like big data, AI, and the internet of things (IoT) can further improve the bioeconomy by analyzing large datasets, making bioprocessing more precise and effective. IoT devices provide real-time monitoring, delivering comprehensive data on environmental conditions and system performance. For example, carbon capture and smart climate agriculture involve working with a data aggregator that collects data from satellites, machines, and farmers for verification. Biosystems engineering and optimizing the use of machine learning and AI help control agricultural base emissions [53, 57].

Bioenergy as a part of bioeconomy helps achieve improvements in environmental, economic, social factors, energy security, biodiversity, and alignment with local development priorities [47, 48]. By integrating advances in biotechnology and digital tools, bioeconomy offers solutions to drive economic growth and reductions of GHG emissions.

14.11.3.1 Machine Learning

Machine learning algorithms can be trained on labeled geographic information systems data and can be applied to data to identify suitable sites for wind or solar farms [63]. The algorithms that can be used are:

- *Evolutionary algorithms* are complex optimization algorithms inspired by the process of natural selection.
- *Deep-learning* AI tools can help enable the accomplishment of most of the decarbonization.
- *Large language models and generative* AI (GAI)-could support decision-making in renewable energy projects. GAI for example can bring together fragmented information to provide a rapid assessment and synthesis of information to gain data and insights from across academic literature and policy reports, across multiple languages, and from unstructured data sources. This could support p*roject generation* scenarios, workflows, project plans financial assessments for each project and location [53].
- *Reinforcement learning* as a type of ML algorithm where an agent learns to make decisions by iteratively taking actions in an environment to maximize some notion of cumulative reward.
- *Decision tree* works like a flowchart where each branch represents a decision rule leading to an outcome at leaf nodes.
- *Randomforests* is a collection of decision trees as it combines the output of multiple decision trees to arrive at a single result with classification and regression problems.

Support vector machines are a type of supervised learning model used for:

- *Permitting* drafts and review contracts to speed up permitting processes.
- *Environmental assessments* are detailed analysis of the impacts of infrastructure on ecology.
- *Auctions* help governments create scenarios that simulate different formats for the best outcomes.

- *Consultations* help answer common questions and gather important information from stakeholders.
- *Financing* generates predictions about future energy prices, and costs of materials that impact the financial viability of renewable energy projects.
- *Regulatory compliance* generates reports or checklists to ensure compliance with the regulations.

The project development tool needs to be designed using various AI systems to adapt to different needs and factors [27]:

1. *Environmental factors* include ecosystems, biodiversity, land use, and water availability
2. Social factors include population, health and safety factors, historical/cultural factors
3. Economic factors include energy costs, transmission access, economic trade-offs, and supply chain access
4. Regulatory and political factors include regulatory environment, local community support, political and financial stability

The best design teams will embrace the most advanced digital technologies, including hybrid models that incorporate artificial intelligence, through rigorous design tools and advanced approaches to achieve a low-risk, repeatable, modular design. Some top projects are:

- energy efficiency project services,
- hydrogen asset design,
- carbon capture storage and utilization design,
- energy optimization,
- solar/wind generation assets,
- materials recycling, and
- bio-feedstocks.

The opportunity to play a leading role in reducing the GHGs would be digitalization of the capital projects to achieve global geopolitical, financial, and social goals.

14.11.4 Case Study: Optimizing Carbon Capture, Utilization, and Storage

Global oil and gas, petrochemical and chemical companies announced ambitious and broad carbon mitigation, plastics re-use and water conservation targets. But many are finding it challenging to identify the right solutions that deliver the level of promised carbon capture and storage (CCS) and carbon capture and utilization

(CCU) are key pathways being developed today to help companies realistically reach their goals. The viable, energy efficient alternatives are [6]:

1. Continuous innovation and development of technologies to improve efficiency
2. Economics that scales efficiently to drive wider adoption of the technology
3. Accelerated pace of implementation
4. Complete confidence in subsurface storage and sequestration

Digital technologies will be crucial enablers in each of these four areas. Examples of recent innovations that are delivering results today include system-level risk modeling technology that ranks the most practical and economically feasible options for carbon reduction, capture, storage and re-use.

Approaching the pursuit of effective solutions below ground, industry leaders are using:

- Subsurface geophysical and geological modeling software to quickly determine the best reservoir targets for reliable, long-term carbon storage.
- Digital grid management software is another effective solution, optimizing the use of distributed energy sources.
- Separately, these digital solution areas impact CCS and CCU projects; combined in a more holistic, collaborative approach, the possibilities are even greater.
- The powerful combination of breakthroughs in process simulation, subsurface geophysical and geological modeling, AI-powered hybrid modeling, process optimization software and digital grid management can deliver results at scale—both economically and at an accelerated pace to meet the requirements of industrial carbon mitigation.
- Achieving end-to-end optimization through digital technology certainly, digitalization will play a central role in ensuring the rapid scaling and adoption of CCS and CCU.

Two technology areas in particular are (i) process modeling and optimization and (ii) subsurface characterization and modeling. They present organizations with a unique opportunity to rapidly screen projects, look at multiple options and design the most effective, economic solutions to address carbon mitigation requirements.

Identifying the best carbon removal approach at the earliest stages of a CCS or CCU initiative, the project's leaders typically evaluate:

- long-term impacts,
- risks and overall economics,
- assessing the entire lifecycle of capturing carbon,
- generating and storing power to run the capture facility,
- transporting carbon,
- injecting and sequestering carbon in geological formations, or
- converting carbon to usable fuels and chemicals

Today's process modeling and system-level risk modeling capabilities enable these efforts by looking at the flow, capacities, efficiencies, technology risks and external stochastic factors that will affect the system.

Many renewable energy sources for carbon capture are intermittent or seasonal. Power and hydrogen storage facilities are critical aspects enabling successful operation of CCUS systems. Additionally, the carbon value chain is generating new business models like carbon capture and storage and looking for companies to optimize capture and storage together:

- *Feasibility.* Geophysical interpretation and reservoir modeling help to effectively analyze all extant data and rapidly screen candidate subsurface reservoirs and aquifers for suitability and capacity.
- *Funding.* Simulation and economic modeling tools rapidly optimize designs and technology selections to match the processing system with the target storage.
- *Implementation.* Subsurface characterization and well engineering help with injections, well planning, and drilling. Process and subsurface modeling can be optimized together, employing reduced order modeling and AI, to achieve the most secure and effective execution plan.
- *Storage.* Geophysical interpretation, reservoir modeling and simulations will form the basis for converting monitor data into auditable visibility and decision support, helping to secure storage of the injected CO_2 in the target reservoir.

14.11.5 Predictive Maintenance

Predictive maintenance can benefit industrial asset reliability and sustainability as a part of optimizing industrial operations, maintenance, environmental impacts. Unplanned disruptions cause increased emissions, water consumption, and transportation needs, adding to an organization's ecological footprint. In sectors like oil and gas, for example, asset failure leads to inefficiencies caused by underperforming assets or the ripple effects of production interruptions on waste generation [6, 18].

In the mining industry where operations often take place in remote and environmentally sensitive areas, recovery and clean-up can become an ecological burden, negatively impacting company profitability, reputation, and future investment. When industrial organizations use AI and machine learning for proactive maintenance may increase asset efficiency, reduce emissions, fuel consumption, water usage, and waste leading to more sustainable operational models. Two steps to accelerate sustainable decarbonization are:

1. *Unlock efficiencies*—By leveraging real-time data in simulation and AI-based models, teams easily perform trouble shooting and improve efficiencies. Efficiency core uses predictive asset optimization and monitor energy efficiency in real time.

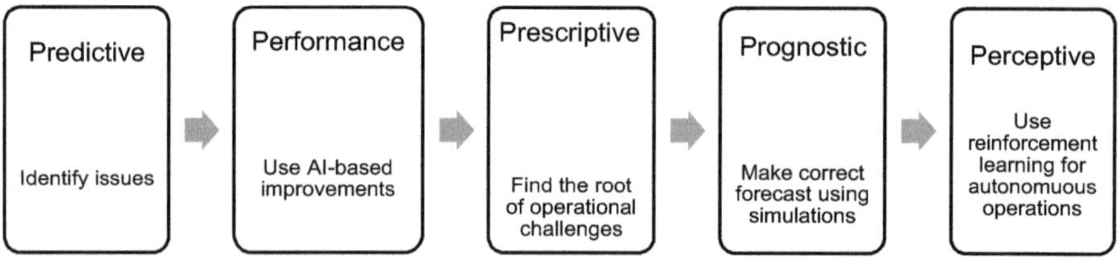

Fig. 14.17 Digital twins stages from predictive to perceptive

2. *Embrace green evolution*—Growth and sustainability depend on greener feedstocks, renewable energy, recyclable products, low-carbon chemical processes, and improve decarbonization.

14.11.6 Digital Twins

Integrating ideal and current operations into reliable digital twins enables understanding and optimizing operations with the incorporating AI-powered simulations in the following steps (Fig. 14.17).

Investment funds for the energy transition and accelerated decarbonization projects may implement the right digital twins design tool for success. The energy transition needs energy security, efficiency, and ensuring enough conventional supply [6, 18].

14.11.7 Challenges

One of the challenges is balancing energy with energy efficiency, bio-feedstocks, carbon capture, electrification, hydrogen, and advanced recycling. To meet decarbonization targets, energy efficiency, and energy transition capital project contracts must be supported by new project execution strategies supported by new digital execution approaches. Economic feasibility impacts future carbon taxes, credits and offsets, renewable energy incentives, bio-feedstock incentives and the manufacturability risks associated with new low-carbon processes. Digital solutions integrated workflows between process design and cost engineering enable risk assessment, the optimal flow of information and avoid duplication of work. Digital solutions can optimize early design phases by reducing overall engineering and construction costs, schedules and risks. Digital execution approaches can assist project leaders with regulatory ambiguities, the skills gap, and help answer questions including what specific initiatives are driving capital spending momentum. For example, AI algorithms and machine learning tools helped synthetic biology mature toward "generative biology" that can produce tailored bioproducts and biofuels from waste carbon sources. All these are the results of advances in data generation, laboratory automation, computing power, and AI innovations [18, 46, 47, 48, 75].

References

1. Alles K, Demirel Y (2024) Measuring risk of renewable diesel production processes using a multi-criteria decision strategy. Chemosphere 354:141695
2. Andersson FNG (2020) Effects on the manufacturing, utility and construction industries of decarbonization of the energy-intensive and natural resource-based industries. Sustain Production Consumpt 21:1–13
3. Ashkanani HE et al (2020) Levelized cost of CO2 captured using five physical solvents in pre-combustion applications. Int J Greenh Gas Control 101:103135
4. Barbuzza E et al (2019) Gasification of wood biomass with renewable hydrogen for the production of synthetic natural gas. Fuel 242:520–531
5. Bargiacchi E et al (2021) Life cycle assessment of substitute natural gas production from biomass and electrolytic hydrogen. Int J Hydrogen Energy 46:35974–35984
6. Beck R, Plougoulen A, Munoz G (2022) Optimizing carbon capture, utilization and storage to meet ambitious sustainability goals White Paper www.aspentech.com
7. Burchart D et al (2023) Life cycle assessment of hydrogen production from coal gasification as an alternative transport fuel. Energies 16:383
8. Cao M et al (2020) A carbon molecular sieve membrane-based reactive separation process for pre-combustion CO_2 capture Membr. Sci 605:118028
9. Chai SYW, Ngu LH, How BS (2022) Review of carbon capture absorbents for CO_2 utilization, greenhouse gas. Sci Technol 12:294–427
10. Chen L et al (2023) Artificial intelligence-based solutions for climate change: a review. Environ Chem Lett 21:2525–2557
11. Cournoyer A, Bazinet L (2023) Electrodialysis processes an answer to industrial sustainability: toward the concept of eco-circular economy?—A review. Membranes 13:205
12. Dash SK, Chakraborty S, Elangovan D (2023) A brief review of hydrogen production methods and their challenges. Energies 16:1141
13. Demirel Y (2017) Lignin for sustainable bioproducts and biofuels. J Biochem Eng Bioprocess Technol 1:1–3
14. Demirel Y (2018) Sugar versus lipid for sustainable biofuels. Int J Energy Res 42:881–884
15. Demirel Y, Matzen M, Winters C, Gao X (2015) Capturing and using CO2 as feedstock with chemical-looping and hydrothermal technologies and sustainability metrics. Int J Energy Res 39:1011–1047
16. Demirel Y (2025) Energy conservation. In: Dincer I (ed) Comprehensive energy systems, 2nd ed. Elsevier, Amsterdam, vol 9, pp 1–51
17. Demirel Y (2025) Biofuels. In: Dincer I (ed) Comprehensive energy systems, 2nd ed. Elsevier, Amsterdam, vol 2, pp 344–384
18. Demirel Y, Rosen MA (2023) Sustainable engineering. process intensification, energy analysis, artificial intelligence, CRC Press, Taylor and Francis
19. DOE (2024) U.S. Department of Energy. Sustainable Aviation Fuel (SAF) Grand Challenge: Building Supply Chains. https://www.energy.gov/eere/bioenergy/articles/sustainable-aviation-fuel-saf-grand-challenge-building-supply-chains
20. 11.DOE (2024b) Heat pump systems. https://www.energy.gov/energysaver/heat-pump-systems
21. Dubey A, Arora A (2022) Advancements in carbon capture technologies: a review. J Cleaner Prod 373:133932
22. EERE (2022) Advanced manufacturing office thermal process intensification: transforming the way industry uses thermal process energy may 2022. Thermal Process Intensification: Transforming the Way Industry Uses Thermal Process Energy | Department of Energy

23. EPA (2024a) Environmental protection agency, simplified guide to greenhouse gas management for organizations https://www.epa.gov/climateleadership
24. EPA (2024b) Environmental protection agency, carbon dioxide emissions coefficients www.epa.gov.
25. EPRI (2020) Low-carbon fuels primer. https://www.epri.com/research/products/000000003002031268
26. Fawzy S et al (2020) Strategies for mitigation of climate change: a review. Environ Chem Lett 18:2069–2094
27. Forster PM et al (2023) Indicators of Global Climate Change 2022: annual update of large-scale indicators of the state of the climate system and human influence. Earth Syst Sci Data 15:2295–2327
28. Franco A, Giovannini C (2023) Routes for hydrogen introduction in the industrial hard-to-abate sectors for promoting energy transition. Energies 16:6098
29. Franco A, Giovannini C (2023) Recent and future advances in water electrolysis for green hydrogen generation: critical analysis and perspectives. Sustainability 15:16917
30. Franco A, Rocca M (2024) Renewable electricity and green hydrogen integration for decarbonization of "Hard-to-Abate" industrial sectors. Electricity 5:471–490
31. Gailani A et al (2024) Assessing the potential of decarbonization options for industrial sectors. Joule 8:576–603
32. Griffiths S et al (2021) Review Industrial decarbonization via hydrogen: a critical and systematic review of developments, socio-technical systems and policy options. Energy Res Soc Sci 80:102208
33. Gür TM (2022) Carbon dioxide emissions, capture, storage and utilization: review of materials, processes and technologies progress energy and combustion. Science 89:100965
34. Hasan S, Abbas AJ, Nasr GG (2021) Improving the carbon capture efficiency for gas power plants through amine-based absorbents. Sustainability 13(1):72
35. Holmes KJ et al (2021) Scaling deep decarbonization technologies. Earth's Future 9:e2021EF002399
36. Jiang K et al (2020) Achieving zero/negative-emissions coal-fired power plants using amine-based post combustion CO capture technology and biomass co-combustion. Environ Sci Technol 54(4):2429–2438
37. IEA (2019a) Putting CO2 to Use. International Energy Agency, Paris: 2019. https://www.iea.org/reports/putting-co2-to-use
38. IEA (2020a) Flagship report, CCUS in Clean Energy Transitions. https://www.iea.org/reports/ccus-in-clean-energy-transitions
39. IEA (2020b) Outlook for Biogas and Biomethane: prospects for organic growth. International Energy Agency, Paris. https://www.iea.org/reports/outlook-for-biogas-and-biomethane-prospects-for-organic-growth
40. IEA (2019b) The future of hydrogen. international energy agency, Paris. https://www.iea.org/reports/the-future-of-hydrogen
41. IEA (2020c) Hydrogen projects database. international energy agency, Paris. https://www.iea.org/reports/hydrogenprojects-database
42. IEA (2024b) International energy agency. Global hydrogen review. https://iea.blob.core.windows.net/assets/89c1e382-dc59-46ca-aa47-9f7d41531ab5/GlobalHydrogenReview2024.pdf
43. IEA (2023) International energy agency. renewable energy market update—Analysis. https://www.iea.org/reports/renewable-energy-market-update-june-2023
44. IEA (2024a) International energy agency | World Energy Outlook 2023, France
45. IPPC (2022) Intergovernmental panel on climate change. AR6 Climate Change 2022: Mitigation of Climate Change. https://www.ipcc.ch/report/sixth-assessment-report-working-group-3/
46. IRENA (2023c) International renewable energy agency. Global landscape of renewable energy finance 2023. https://www.irena.org/Publications/2023/Feb/Global-landscape-of-renewable-energy-finance-2023

47. IRENA (2024) Decarbonizing hard-to-abate sectors with renewables: perspectives for the G7; international renewable energy agency: Abu Dhabi, United Arab Emirates
48. IRENA (2023a) International renewable energy agency. World Energy Transitions Outlook 2023: 1.5 °C Pathway. https://www.irena.org/Publications/2023/Jun/World-Energy-Transitions-Outlook-2023
49. Kim Y et al (2024) Economic viability of integrated CO_2 capture and conversion, ACS Energy Letters. https://doi.org/10.1021/acsenergylett.4c02852, 403-409
50. Kirstein L, Halim R, Merk O (2018) Decarbonizing maritime transport: pathways to zero-carbon shipping by 2035. Case-specific policy analysis series. international transport forum—Organisation for economic co-operation and development: March 2018. https://www.itf-oecd.org/decarbonising-maritime-transport
51. Kopac T, Demirel Y (2024) Impact of thermodynamics and kinetics on the carbon capture performance of the amine-based CO_2 capture system. Environ Sci Pollut Res 31:39350–39371
52. Kopac T, Demirel Y (2025) Optimizing CO_2 capture from biogas: a comparative study of amine-based solvents through aspen plus simulations. Biomass Convers Biorefinery. https://doi.org/10.1007/s13399-025-06668-5
53. Larosa F et al (2023) Halting generative AI advancements may slow down progress in climate research. Nat Clim Change 13:497–499
54. Lenton TM et al (2023) Quantifying the human cost of global warming. Nat Sustain https://www.nature.com/articles/s41893-023-01132-6
55. Leonzio G, Shah N (2024) Recent advancements and challenges in carbon capture, utilization and storage. Current Opinion Green and Sustainab Chem 46:100895
56. Leonzio G, Hankin A, Shah N (2024) CO_2 electrochemical reduction: a state-of-the-art review with economic and environmental analyses. Chem Eng Res Des 208:934–955
57. Li H et al (2019) Prediction of CO_2 absorption by physical solvents using a chemoinformatics-based machine learning model. Environ Chem Lett 17(3):1397–1404
58. Li Q et al (2024) Hydrogen impact: a review on diffusibility, embrittlement mechanisms, and characterization. Materials 17:965
59. Madden S, Alles K, Demirel Y (2021) Measuring sustainability of renewable diesel productions using multi-criteria decision-matrix, Bioproducts & Biorefining. Biofpr. 15:1621–1637
60. Madejski P et al (2022) Methods and techniques for CO_2 capture: review of potential solutions and applications in modern energy technologies. Energies 15(887):1–21
61. Matzen M, Alhajji M, Demirel Y (2015) Technoeconomics and sustainability of renewable methanol and ammonia productions using wind power-based hydrogen'. Adv Chem Eng 5:128
62. Matzen M, Alhajji M, Demirel Y (2015) Chemical storage of wind energy by renewable methanol production: feasibility analysis using a multi-criteria decision matrix. Energy 93:343–353
63. Milojevic-Dupont N, Creutzig F (2021) Machine learning for geographically differentiated climate change mitigation in urban areas. Sustain Cities Soc 64:102526
64. NASEM (2024) National academies of sciences, engineering, and medicine. Accelerating Decarbonization in the United States: technology, policy, and societal dimensions. The National Academies Press, Washington, DC. https://doi.org/10.17226/25931
65. Nitopi S et al (2019) Progress and perspectives of electrochemical CO_2 reduction on copper in aqueous electrolyte. Chem Rev 119(12):7610–7672
66. Olabi AG et al (2022) Assessment of the pre-combustion carbon capture contribution into sustainable development goals SDGs using novel indicators. Renew Sustain Energy Rev 153:111710
67. Owebor K, Diemuodeke EO, Briggs TA (2022) Thermo-economic and environmental analysis of integrated powerplant with carbon capture and storage technology. Energy 240:122748
68. Patil SC (2024) Electrothermal carbon capture and utilization review. ACS Energy Lett. https://doi.org/10.1021/acsenergylett.4c02851:371-392
69. Rissman J et al (2020) Technologies and policies to decarbonize global industry: review and assessment of mitigation drivers through 2070. Appl Energy 266:114848

70. Saraceno E, Ruocco C, Palma V (2023) A review of coal and biomass hydrogasification: process layouts, hydrogasifiers, and catalysts. Catalysts 13:417
71. Scarlett E (2020) Could ammonia be the next key player in energy storage? Power Technology, https://www.power-technology.com/features/ammonia-next-key-player-energy-storage
72. Smith K, Chen S, Siefert N (2020) Modular CO_2 capture processes for integration with modular scale gasification technologies: literature review and gap analysis for future R&D; DOE/NETL-2020.2149; NETL: Pittsburgh, PA. 136
73. Tosti S et al (2019) Process analysis of refuse derived fuel hydrogasification for producing SNG. Int J Hydrogen Energy 44:21470–21480
74. Vidas L, Castro R (2021) Recent developments on hydrogen production technologies: state-of-the-art review with a focus on green electrolysis. Appl Sci 11:11363
75. WEF (2024) World economic forum, net-zero industry tracker 2024: Key insights. https://www.weforum.org/publications/netzero-industry-tracker-2024/key-insights/
76. Wang Z et al (2022) Life cycle assessment of combustion-based electricity generation technologies integrated with carbon capture and storage: a review. Environ Res 207:112219
77. Wei X, Manovic V, Hanak DP (2020) Techno-economic assessment of coal- or biomass-fired oxy-combustion power plants with supercritical carbon dioxide cycle. Energy Convers Manag 221:113143
78. Yadav S, Mondal SS (2022) A review on the progress and prospects of oxy-fuel carbon capture and sequestration (CCS) technology. Fuel 308:122057
79. Yun X et al (2024) Advancements in CO_2 capture by absorption and adsorption: a comprehensive review. J CO_2 Utilization 81:102727
80. Zhang J et al (2023) Accelerating electrochemical CO_2 reduction to multi-carbon products via asymmetric intermediate binding at confined nanointerfaces. Nat Commun 14:1298

Energy Transition

15

Introduction and Learning Objectives: Energy transition (ET) refers to technological developments toward mowing renewable energy from fossil fuels. ET would dramatically decrease fossil fuel's share globally with a shift that not only reduces carbon emissions but designs an energy system that supports a resilient and sustainable global energy supply. ET requires the creation of innovative and energy efficient technologies before the results are widely available in other regions. Besides the technological advances, the ET requires policy, regulations, business models, and public experience with actual deployment. An array of technological capabilities can accelerate transition to clean-energy sourcing and allocation enabled by smart-grid networks and software-enabled renewable energy-storage systems to optimize supply and demand forecasting. This chapter will summarize some major strategies and technologies to replace fossil fuels with renewable energy toward energy transition. The learning objectives of this chapter are:

- Energy transition
- Emerging fuels for transition
- Transforming industry
- Energy transition and digitalization

Supplementary Information The online version contains supplementary material available at https://doi.org/10.1007/978-3-031-97111-2_15.

Y. Demirel, *Energy*, Green Energy and Technology,
https://doi.org/10.1007/978-3-031-97111-2_15

15.1 Energy Transition

The policy and regulatory systems remain shaped around fossil fuels. However, coal is being replaced by natural gas alongside promoting electrification with advanced technologies and financing. This promotion with energy efficiency improvement aims at policy development to overcome the cost advantage of fossil fuels. Energy transition (ET) would dramatically decrease fossil fuel's share with a strategic shift that not only reduces carbon emissions but designs an energy system that supports a resilient and inclusive global economy [15, 20].

ET consists of both temporal and spatial processes. Spatially refers to the creation of innovative technologies before the results are widely available in other regions. Temporally, deployment occurs in three phases: an innovation phase, a growth phase, and a maturity phase. Besides the technological advances, the ET requires policy, regulations, business models, and public experience with actual deployment. ETs can be possible with careful design of technology push and demand-pull, coupled with transformed social structure supported by design of learning cycles between generations and deploying the technologies [20]. Three types of deployments of ETs are:

1. A societal shift in fuel sources toward clean energy,
2. A transition in fuel dependence of an economic system with low carbon energy, and
3. Electrification of the sectors difficult to decarbonize including transportation sector and industrial sector

The world energy transitions outlook envisages three pillars:

1. Building the necessary infrastructure and sources at scale in grids, land and sea routes, new production locations, trade patterns and demand centers.
2. Advancing policy and regulatory structure towards realigning resources, investments, skills and capabilities with matching the clean energy systems to create.

ET refers not only to technological developments, but also socio-economic aspects with the world moving from fossil fuels to renewables. The changes in socio-economic patterns could slow down the GHG emission growth and the mitigation requirement would lead to more efficient and renewable energy structure changes including electrification [16, 20, 30].

With phasing out fossil fuels and scaling up renewables it would be possible to achieve ET with electrification, smart-grid technology, upgraded transmission lines, and energy storage in a society where everyone can afford clean energy. To transition at the necessary pace may also require decarbonize electricity use of biomass and emissions removal technologies.

Many countries with around 80% of people live in countries that import fossil fuels. Those who would benefit from a transition vastly outnumber those who benefit from fossil fuels. ET with the drop in renewable energy costs and other technological improvements includes [15]:

- Efficiency gains have already curbed energy demand growth factor,
- Renewable electricity costs less than fossil fuel electricity in two-thirds of the world,
- Renewable energy integration technologies keep increasing with variable renewables mix,
- Policy makers may afford more space and opportunity to act.

15.1.1 Global Energy Transition Indicators

The IEA's *Global Energy Transitions indicators* collect the data and analysis on the ET, including energy sector GHG emissions, technology developments, energy sector financing, energy access and energy employment. These indicators may track global progress of ET [18]:

- Clean energy sectors continue to fuel energy-related job growth worldwide
- In a complex global energy markets the emergence of a new clean energy economy provides hope for reducing polluted air pollution
- Momentum around low-emission hydrogen keeps growing, but progress is still hampered by cost challenges and uncertainty around demand
- Progress on electricity access has remained slow in 2023, hobbled by financial strains
- The clean energy economy is gaining ground, but greater efforts are needed
- Clean energy investments in developing countries need to grow steeply to reach ET goals
- Clean energy investment reaches record high
- Renewable capacity additions increase
- Public investment in energy-based research and development increase
- Clean energy supply chains expand substantially
- Global car market share of EVs keeps increasing
- Global heat pump sales are increasing
- Growth in emissions is decreasing
- Carbon Capture, Utilization and Storage (CCUS) expands with realizing all planned projects
- Half the emission reductions come from the technologies that are not yet on the market.

15.2 Physical Infrastructure

Planning, modernization development, storage, distribution, transmission, and using renewables are necessary in physical infrastructure to facilitate national, regional and global strategies for new supply–demand dynamics as follows [20]:

- Plan integrated cross-sector infrastructure for the energy transition with targets for expansion.
- Provide incentives for infrastructure investments.
- Streamline permitting procedures with addressing environmental and social impact assessments.
- Set mandatory targets for new or renovated buildings.
- Provide finance for the development of the infrastructure.
- Design of policy and regulatory frameworks for deployment and trade of renewables-based energy.
- Encourage awareness and capacity-building institutions to drive and sustain ET.

15.3 Socio-economic Impacts of the Energy Transition

Energy transition because of renewable energy technologies affects employment, GDP, and human welfare based on a model that integrates energy into sustainability. This is directly related to energy system planning and economic policy making. In the TES, jobs in the overall energy sector of renewable energy, fossil fuels, and nuclear power could reach 72% more than the total present employment by 2050 [17]. New jobs in solar PV, wind, and bioenergy would compensate job losses in fossil fuels and nuclear energy. This expanded renewable energy workforce will require growing investments, as well as focused education, training, and social policies such as just and inclusive transition policies to limit the job losses. GDP gains can result by changes in consumer spending and other indirect and induced factors including increased investment and the prospects of economic growth. The primary goal of the transition would be to improve people's overall welfare through secure, clean energy supply, economic and social development, and mitigation of climate change [20, 25].

Affordability is always a concern for consumers central to clean energy transitions. Many clean energy technologies are already the most affordable ones, especially when lifetime costs are considered. Cost reductions have mainly been driven by a virtuous circle of innovation, accelerated deployment, economies of scale and policy support, with examples extending beyond renewable power to light emitting diode (LED) lighting and electric cars. Efficient appliances, such as air conditioning units and heat pumps (see Sect. 7.15) creating affordable openings for everything from utility-scale projects to home solar systems, enhanced with their cheaper batteries [16, 20, 34].

15.3.1 Welfare Index

Welfare, envisaged with the sustainability elements of economics, social, and environment, would improve 13.5% by 2050 under the TES [20]. The economic element is measured via household consumption and a composite of total investment and employment. Jobs gains would be linked to the transitions in renewable energy, energy efficiency, power grids, and energy flexibility under TES. The social element reflects spending on education and health. The environmental element entails lower greenhouse gas emissions and the consumption of nonrenewable materials. Overall, health benefits and emission reductions influence welfare gain besides substantial socio-economic benefits globally.

However, owing to variations in regional and country-specific socio-economic structures, the welfare gains could vary at the country and regional levels and local solutions would be necessary. Regional socio-economic structures, resource endowment, industrial productive capacity and support policies, trade structures and domestic supply chains would determine the opportunities offered by the energy transition. A Successful transition requires that energy policies must be compatible with regional economic, industrial, labor, educational, and social policies [18, 20].

Human welfare and security must remain at the center to successful energy transition. Policy makers mainly focus on the technological, institutional, regulatory phases of ET, with less attention to its socio-economic implications. ET would create more employment throughout the economy. To attract talent to the sector, jobs must offer decent wages and opportunities evenly distributed across regions. Achieving this transition may depend on three fundamental pillars:

1. physical infrastructure,
2. policy and regulatory enablers, and
3. skills and capacities.

These elements are interconnected and engage in continuous feedback loops with both the economy and societal structures. The energy transition presents a historic opportunity to adopt domestically available resources, boost socio-economic benefits and spread them broadly. Solar and wind power could run factories with affordable power and can be chemically stored to replace the use of fossil fuels. Programs could be established to create demand when energy is available and cheap, producing hydrogen for later use, or pulling carbon dioxide out of the air to sequester it underground.

Energy security, economic and social resilience offer new opportunities for improved human welfare index analyzes the socio-economic of the ET with five dimensions (Table 15.1): economic, social, environmental, distributional and access. The index's dimensions provide insights into how socio-economic outcomes can be improved through appropriate policy measures [20].

Table 15.1 Welfare index dimensions [20]

Environment	Social	Economic	Distribution	Access
• GHG emissions • Resourse depletion	• Social expenditure • health impact	• Consumption & investment • job creation	• Within contry/region • Across countries/regions	• Basic energy access • Sufficiency

The technological aspects of the ET have both direct and indirect impacts on overall welfare, including the positive impact of reduced GHG emissions, creating jobs, increased consumer spending in health and education. Enhancing overall welfare requires policy action to ensure a level of human well-being at lower levels of energy and materials intensity.

15.4 Scaling Energy Transition

The following factors may help scale ET [21]:

Policy—Strong climate policies to scale a just transformation include:

- Introducing national targets to reduce GHG emissions
- Developing national sectoral road maps and support policies and strategies
- Pricing carbon emissions at a sufficient level.
- Establishing strong energy efficiency and emissions standards and policies
- Integrating climate action into technology strategies and digital, circular and sharing economies.
- Incentivizing behavior towards ET.

Climate leadership—Engagement in climate changes is demanding climate leadership to support 100% renewable electricity and explore investing in their own renewable energy production. Social movements are demanding large-scale scientific progress to address climate change in recognition of the record-breaking flooding and drought events.

Finance—Investors are recognizing climate risk and financial and fiscal incentives, such as tax rebates, tax credits, subsidies and grants to mitigate the costs from low-carbon ET. Sector-specific policies and actions can accelerate the transformation of high emitters, including the cement, iron and steel, and chemical industries. Cement decarbonization could benefit from support for the research, development and demonstration of innovative solutions for low-carbon and renewable-based clinker alternatives and cement in operational plants.

Exponential technology—Digitalization can improve the global economy to reduce energy and materials across and enable the emissions reductions. Policy makers need to explore how more energy efficient technology, the digital industry could help achieving societal, economic, and environmental goals.

15.5 Challenges and Opportunities for Energy Transitions

A definition of environmental and social risks around investing in energy investments is needed for the ET required enhancements in power grids to scale up renewables. The existing grids have been designed to accommodate large, centralized dispatchable power plants and the integration of renewable sources into the power grid presents a challenge. Investment in management and control systems is necessary to accommodate distributed power generation capacity for faster electrification [20, 29, 34]. Table 15.2 shows some important barriers and possible solutions for ET.

15.5.1 Decarbonization

Most of the technologies needed to decarbonize the energy system are already available, including those for massive electrification of most applications. In the meantime, fossil fuels and expanded use of low-carbon fuels like biofuels and

Table 15.2 Some important barriers and possible solutions for ET

	Enabling infrastructure	Policy and regulations	Skills and instituional capacity
Barriers	*Insufficient infrastructure to connect renewable energy mix, energy storage, and grid integration *Lack of distribution infrastructure for electricity and fuels *Lack of end-use sector facilities to replace fossil fuels	*Policy and regulatory frameworks shaped around fossil fuels * lack of integrated planning for energy production and use *Insufficient policy toward socio-economic needs	*Lack of alignment between fossil fuel jobs and renewable energy jobs that are skills, spatial, temporal, and sectorial related *Skills gabs and lack of awareness of opportunities * Lack of work force and appropriate training
Solutions	*Planning, expansion, and modernizatition of supporting infrastructure *Developing energy storage, and distribution renewable energy mix *National and global strategies for resillient supply-demand dynamics	*Design policy and reulatory frameworks that help deployment, integration and trade of renewable energy mix *Improve socio-economic gains and environmental protection * Decrease the cost of basic energy	*Awareness and capacity building of intstitutions, and individuals to gain necessary skills and knowledge to drive and sustain the energy transition *Coordination between universities and industry * Support social and institutional dialogue

hydrogen can be used for applications that are difficult to electrify, such as transportation and manufacturing industry [17, 22, 27, 28].

Engineers need to develop models of future energy consumption globally that move away from coal, oil, and natural gas toward wind, solar, nuclear, and other low-carbon energy sources and hydrogen fuels for those applications that are difficult to electrify. Besides, those models must make the ET less expensive with lower costs of carbon capture, solar power, or bringing nuclear fusion to modular commercial-scale applications. Some performance indicators to monitor progress towards controlling global warming with ET involve:

- Increase the use of renewables to generate electricity globally
- Direct uses of renewables in total final energy consumption
- increased quantity of bioenergy used
- Improvements in energy intensity
- The electrification of end-use sectors
- Production and supply of clean fuels with carbon capture, utilize, and store (CCUS)

15.5.2 Behavioral Changes

Behavioral changes include actions that consumers consider reducing wasteful energy consumption, or using differently, or using less, and cut energy intensity. These changes depend in part on individual choices and evolving socio-cultural norms, as well as by targeted and well-designed policy interventions. ET can benefit from overcoming existing barriers across infrastructure, policy, workforces and institutions that hamper progress and impede inclusivity by [16, 20]:

- Upgrading physical infrastructure to increase resilience and flexibility for an interconnected energy system.
- Accommodating transmission and distribution for highly localized, decentralized renewables mix.
- Planning for interconnectors to enable electricity and fuel trade to promote the diversification and resilience of energy systems.
- Enabling widespread energy storage solutions with geo-economic impacts in mind.
- Securing public acceptance through project transparency.
- Investing in a well-skilled workforce development.

15.5.3 International Co-operation

A common consensus across all countries is the need to electrify heating and transportation using renewable electricity, efficiency measures, and the direct use of renewables. The dynamism of energy sectors and geopolitical developments necessitate international co-operation in ET and achieving greater resilience. A diverse energy market would reduce supply chain risks, improve energy security, and ensure local value creation for commodity producers. Access to technology, training, capacity building, and affordable finance enhance the full potential of countries' contributions to global ET. Specific technologies can drive ET in end-use sectors and can improve supply-side and transformational processes, depending on resource availability and infrastructure [16, 18].

15.5.4 Workforce

ET needs a workforce with the required skills by integrating renewable energy into educational curricula and training opportunities, including [6, 17]:

- Efforts to anticipate future occupational needs in each renewable energy sector and work with industry associations and training institutions to align their planning.
- Develop pathways for fossil fuel industry workers to retrain and recertify careers in renewable energy.

15.6 Energy Transition Fuels

Biofuels in the transportation sector increased global awareness of their impact on ET. Fuels such as hydrogen and its derivatives (e.g. ammonia and methanol), and bioenergy will play a unique role in the energy transition, especially for industrial processes and certain transport modes. This would require rapid expansion of both renewable power and electrolyzer capacity [8].

15.6.1 Renewable Energy Mix

Power sector transformation would see a vast scale-up of all renewable mixes, including wind, solar, and hydrothermal power. Renewable energy policies have evolved as technologies have matured into sustainable power production systems. Power systems are integrated to support ET with increasing variable renewable energy (VRE) and flexibility to be achieved [15, 17, 20]. This helps facilitate VRE investments to lower the cost of renewable power generation with capacity expansion, investment in energy efficiency, electrification in buildings and industry. Besides, the transport sector, driven by electric vehicles and battery research, development and investment could have a positive impact on ET. Improvements in production and distribution of renewable energy-based power include:

- Adoption of a power system with variable renewable energy with socio-economic benefits.

- Streamlining permitting procedures for renewable power projects.
- Synchronizing power grid expansion and other infrastructure developments.
- Design renewable energy technologies with least supply chain disruptions.

Advanced efficiency and electrification based on renewables are cost-effective strategies for the power sector, as well as for the transport and building sectors. The cost of solar and wind energies declined so clean solar energy that people could generate themselves to charge electric vehicles and to serve as batteries at night. ET requires a rapid expansion of affordable and renewable energy-based electricity generation. Wind and solar power must be supported with advanced power electronics such as solid-state transformers and electronics converters to manage the flow of electricity between renewable sources and the grid.

Targeted policies can contribute towards increasing global clean power capacity, energy efficiency, and reducing GHG emissions with accelerating the ET away from fossil fuels. Besides solar and wind energy, hydropower, bioenergy and other renewables should be supported with policies [15]. Some common challenges to deploy clean power capacity include waiting times for permits, inadequate investment in grid infrastructure, and financing costs of integrating variable renewables. Countries can take to address these obstacles to:

- streamline permitting with simplifying rules, procedures and administrative structures,
- invest in spatial planning to streaming zoning,
- involving local communities throughout the permitting process,
- incentivize power system flexibility,
- develop greater energy storage capacity, and
- leverage digitalization to enable greater demand response.

A clean energy system is more efficient and cheaper to run as a more electrified, renewable-rich and efficient system brings important gains for affordability, alongside the clear environmental benefits. As the running costs of decarbonizing energy systems are much lower, this is raising awareness globally. For example, the combination of increased clean electricity generation, electrified transport, phasing out of polluting fuels and progress with access to clean cooking fuels leads to a significant improvement in health outcomes from decreasing air pollution.

15.6.2 Possible Risks

Many power systems are vulnerable to an increase in extreme weather events and cyberattacks, adding a cost on investments in resilience and digital security. Readiness to manage risks needs design of support mechanisms with [6, 16, 26]:

1. The transition to a clean energy system is participated by all parts of society.
2. Make the most of clean technologies that are cost-competitive, bringing costs down.
3. Recognize and value the wider benefits of clean energy transitions.
4. Know your consumers and their needs.
5. Create the conditions for large-scale investment and finance from the private sector.
6. Minimize the chances of adverse price shocks.
7. Be clear on long-term objectives and on possible short-term risks.

15.6.3 Bioenergy and Energy Transition

Bioenergy plays a key role in ET as bioenergy's share in the primary energy supply increase considerably in various industries, including transport and buildings. To ensure the environmental, social and economic benefits of bioenergy are maximized, countries would have to implement regulations and certificates and promote partnerships to ensure sustainability of biomass feedstock and the entire supply chain [8, 15].

Wood pellets, biodiesel and bioethanol are major commodities. Biofuels contribute the largest share of renewables utilization worldwide, and growth in production and use of modern bioenergy will be critical for global ET with low carbon to net zero emissions. It requires dispatchable electricity to systems with a high share of variable renewable energy, or combined heat and power (CHP) generation, and bioenergy with carbon capture and storage (BECCS) [33].

For policy makers, bioenergy involves a wide range of stakeholders and issues as it interacts with agriculture, forestry, environmental protection and waste management, and the supply chain [21]. Ensuring the sustainability of bioenergy along the supply chain, including most notably biomass feedstocks, is the most fundamental element of bioenergy policy making. Besides, international bioenergy trade further increases the complexity. Cross-cutting barriers to bioenergy deployment include:

- Political and institutional,
- Financial and economic,
- Technical and infrastructure-related,
- Supply chain related.

15.6.4 Bioenergy and Power Generation

The efficiency of steam turbines in biomass power generation-only plants can be between 15 and 30%, which is lower than fossil fuel options that can be over 50% for modern integrated gasification combined cycle (IGCC) plants [15, 21]. Bioenergy has been increasingly used for power generation, mainly based

Table 15.3 Biomass for power generation

Feedstock			Biomass to power pathways with carbon capture		
Forestry and agricultural residues	Municipal solid waste (processing waste)	Organic waste (food waste, livestock waste)	Power generation with biomass gasification	Biomass and coal co-firing	Combined heat and power generation

on pellets, biogas, municipal solid waste, and agricultural and forestry residues (Table 15.3). Weak supply chains and high operation costs are among the major barriers. Renewable obligations and blending mandates for coal plants can secure deployment.

Biomass has been widely used in the pulp and paper industry and sugar mills to provide industrial heat and cogeneration to provide process heat for the chemical and petrochemical sectors, and the cement, steel and iron industries. Bioenergy, including bioethanol, biodiesel and biomethane, can replace fossil fuels and decarbonize road transport and shipping [8].

15.6.5 Sustainability of Bioenergy

Bioenergy can help reduce GHG emissions in power generation, heating, transport and industry, as well as, it brings environmental and socio-economic benefits such as land restoration and job creation and improved health from clean cooking. However, there is also scope for potential benefits by using bioenergy by-products to improve soil quality, plantations for phytoremediation to improve water quality and agroforestry to increase biodiversity. Achieving sustainably produced bioenergy would play an important role in medium and high-temperature heat applications in industry [6]. Fuel efficiency improvements and the increased number of electric vehicles on the road with increasing biofuel blending ratios could help reduce emissions from road transport (Fig. 15.1).

However, bioenergy supply chains can cause negative environmental, social or economic impacts due to their interactions with eco systems including agriculture, forestry, rural development and waste management. Increased bioenergy supply could cause land-use changes from the ecological service to energy use, raising issues such as food security or biodiversity loss. Table 15.4 shows the potential benefits and negative impacts of bioenergy.

15.6.6 Transforming Traditional Bioenergy Systems

Inefficient, traditional use of bioenergy remains the largest component of global bioenergy consumption. Carbon debt is usually linked to the conversion of forests

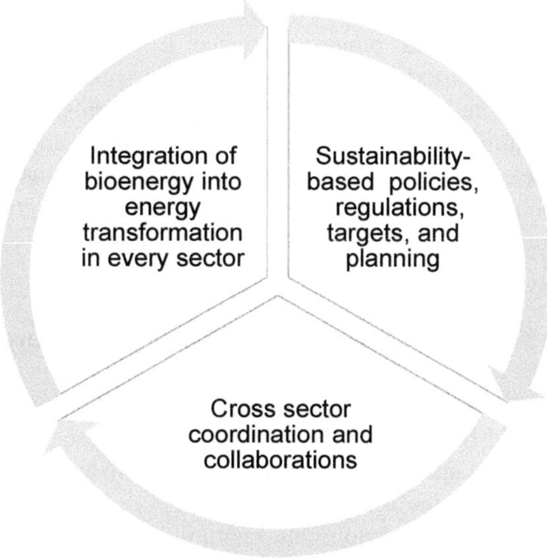

Fig. 15.1 Sustainable bioenergy development framework towards energy transformation

Table 15.4 Bioenergy sustainability aspects

Bioenergy sustainability aspects		
Possible benefits		Negative impacts
*Emission reductions *Net zero emissions with BECCS *Reduced methane emissions	1.GHG gas emissions	*Emissions from land use changes and supply chain
*Land restoration *Improved water quality and soil quality and waste management	2.Environmental aspects	*Biodiversity loss *increased water consumption *Reduced water and soil quality
*Improved public health, energy security, and job creation	3.Socio-economic aspect	*Competiotion with food supply chain *Expensive BECCS applications

to biofuel production, which causes significant one-off emissions due to deforestation and would take a long time to "pay off". Similar to other agriculture and tree planting activities, bioenergy production will increase the volume of water consumption, which must be considered for the potential competition with food production. Bioenergy policy needs to be coordinated with environment, forestry and agriculture policies [6, 7].

Bioenergy generated from waste outputs has extra benefits in emissions reduction, such as biogas produced from methane released from landfills, animal manure, food waste and other organic residues. A biogas digester can turn a waste issue into a revenue opportunity for farms, waste treatment industries and rural communities by selling biogas to utilities and gas companies or saving on their fuel costs. Additionally, the digestate can be used or sold as a fertilizer. *Low Carbon Fuel Standard* incentivizes biogas production from livestock manure which may increase the financial gains in many parts of the world [1, 8]

15.6.7 Bioenergy with Carbon Capture

Biomass use combined with CCS in the power sector and some industrial sectors would be critical in achieving ET. Processes using biomass, CCS, bioenergy with carbon capture and storage (BECCS) are assumed to capture and store considerable carbon in the industrial sector.

BECCS technologies capture carbon emissions from bioenergy conversion processes, such as biomass-based power generation and CHP, and digestion or fermentation of biomass to produce biogas and liquid biofuels, and to permanently store the CO_2 in geological formations. Integrating BECCS into bioenergy production and industrial processes (e.g. chemicals and cement production) is a possible option to further reduce emissions. Pathways would rely on CCS technologies including BECCS, which is constrained by sustainable bioenergy potential. Concerns exist about their low efficiency, life-cycle emissions and sustainability. Therefore, stimulating the deployment of BECCS requires policy support such as sustainable governance and financial incentives [15, 35].

15.7 Emerging Fuels for Energy Transition

Global economies aim to become carbon neutral and competitive using emerging fuels, including hydrogen and its derivatives like ammonia, methanol and kerosene, they offer broad base decarbonization solutions [27, 28].

15.7.1 Hydrogen and Its Derivatives

Hydrogen from renewable sources can accelerate ET in energy intensive sectors, including steel, petrochemicals, glass, cement, and paper. Hydrogen currently is produced from fossil fuels and therefore, new technologies are needed to expand the production of clean hydrogen, which requires updated policy and technology in production, storage, and transportation. Besides, improved efficiency and certification schemes would help ensure the sustainability of low-carbon hydrogen production and minimize cost of hydrogen processing. Key areas to consider for research and development include (i) electrolysis, (ii) carbon capture and

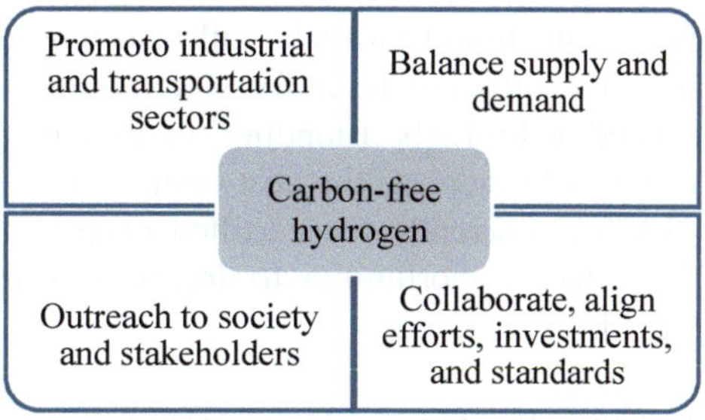

Fig. 15.2 Management of hydrogen production and use

sequestration (CCS), (iii) advanced materials, and (iv) fuel cell technologies. Blue hydrogen, produced from natural gas with CCS, can serve as an interim solution while green hydrogen production from renewable electricity scales up continues to mature [28].

It is important that industries including steel and fertilizer, shipping, and aviation should be given priority in clean hydrogen use with these recommendations as shown in Fig. 15.2.

The production of clean hydrogen and its related compounds, like ammonia, methanol and kerosene should increase with investments in electrolysis, fuel cells, pipelines, and storage. It is vital to increase hydrogen applications in end-use sectors with carbon reduction goals. Policy makers can facilitate the promotion of hydrogen electrolyzers, sustainable aviation fuels, and carbon capture utilization and storage (CCUS) by balancing the costs and benefits of these programs [14, 15].

15.7.2 Green Hydrogen and Methane for Upscaling Renewables

Green hydrogen could be key in mitigating emissions from harder-to-decarbonize sectors such as aviation, shipping and heavy industry sectors. Green hydrogen and biomethane will require appropriate infrastructure to scale up of its use. While biomethane can rely on existing natural gas infrastructure, the transport of hydrogen via pipeline requires upgraded infrastructure and costly investments that may limit their viability:

- Aviation and shipping require port and bunkering infrastructure for hydrogen.
- Electric vehicle charging and hydrogen refueling infrastructure are required for road transport with the introduction of electric and hydrogen vehicles [8].

15.7.3 Aviation Fuel

Aviation fuels based on hydro-processed esters of fatty acids or hydrogenated vegetable oil can also have unique roles in decarbonizing aviation. Replacing fossil fuels with liquid biofuels will also require necessary fuel transportation and distribution infrastructures for which sustainability risks have been the main barrier related to biofuels. Blending regulations and certificates, grants and loan guarantees would increase the cost-competitiveness of biofuels. Sustainable aviation fuel (SAF) is currently the certified drop-in fuel. SAF together with other synthetic fuels open opportunities to decarbonize the sector in a cost-effective scale-up [3, 9, 10].

- Currently most SAF is produced via the hydrotreatment of fats, oils and waste cooking oils. Small volumes of alcohol-to-jet are currently available and this technology, as well as others such as gasification and Fischer–Tropsch synthesis, will provide greater volumes in the longer term.
- The SAF fraction is one component of the fuel output of biorefineries.
- With limited investment and with the right policy drivers in place, nearly all existing renewable diesel facilities could be incentivized to produce SAF as well as renewable diesel.
- In the future, due to the finite availability of lipid/oleochemical feedstocks, to significantly scale up production, alternative feedstocks such as lignocellulosic biomass, and technologies such as gasification with Fischer–Tropsch and/or pyrolysis or hydrothermal liquefaction will be required.
- Supply chain innovation and the process optimization will play a key role in future cost improvements of SAF production.

15.7.4 Sustainability of Aviation Fuels

Two certification bodies in the assessment of the sustainability of SAF fuels are the Roundtable on Sustainable Biomaterials (RSB) and International Sustainability and Carbon Certification (ISCC). As the emission reduction potential of SAF fuel is a central part of its sustainability, it is typically determined via the life-cycle assessment (LCA) of a SAF fuel supply chain as compared to the emissions from using conventional jet fuel. Several factors will affect the actual carbon intensity of SAF fuel. These include the nature of biomass feedstock, agricultural practices used and harvest yields, as well as technological factors such as the source of electricity/hydrogen, total product yields, amount and nature of the co-products and wastewater treatment [3, 22].

15.7.5 Carbon Intensity of Aviation Fuels

The specific carbon intensity of the fuel will have an impact on the incentives available through policy measures. An alternative way of looking at cost is calculating the specific cost of carbon abatement, rather than just the minimum fuel selling price, as this considers the price of the fuel and the price relative to the emission reductions offered.

Carbon intensity of SAF can be further reduced with the following:

- The energy source can be changed from fossil based to renewable energy.
- Changing the source of hydrogen from natural gas to renewable sources such as biogas or produced with electrolysis of water from renewable electricity.
- Increasing the number of co-products to allocate carbon emissions to all products.
- Reducing emissions in the feedstock supply chain.
- Including carbon capture and storage processes at the biofuel facility.

Production of conventional biofuels (e.g. bioethanol/biodiesel) involves relatively mature technologies that have been primarily promoted by market policies. However, a similar policy approach is unlikely to be as effective for SAFs as many of the technologies are not yet mature, with initial capital investment proving to be a major barrier. Thus, a greater focus on technology push policies, including capital investment financing, will be important. The future production and use of SAFs will be primarily influenced by the three challenges [6, 8, 26]:

(i) feedstock,
(ii) echnology and policy, and
(iii) an effective combination of SAF policies.

Case Study 1: Retrofit Improves Process Design Accuracy by Debottlenecking Gas Processes
The right retrofitting of an existing operation can increase capacity by 20% by rigorous heat exchanger modeling with the broader process simulation to determine the most cost-effective pathway for the target increase. This exercise found that it was possible the project goal could be met by replacing three of the existing exchangers with more appropriately sized models. This revamp allowed the site to meet 20% increased capacity requirements, with limited capital expenses. Cost estimation tools can effectively determine the best configuration of heat exchanger trains and compressors by assessing benefits based on thermal efficiency, capacity, and availability, as well as costs based on capital, maintenance, and energy requirements for equipment re-vamps. [Davidson J, Potdar S Enabling affordable and reliable delivery of low-carbon fuel for the energy transition. Aspen Tech, 2024 (www.aspentech.com)].

Case Study 2: Dynamic simulation of natural gas liquefaction technology-Liquefied Natural Gas Is a Strategic Component of the Global Energy Supply
The dual mixed refrigerant model uses two refrigeration cycles with two mixed refrigerants and plate fin heat exchangers. This increases the adaptability of LNG operations to a wide range of feed stream and environmental conditions. This case study analyzed the feasibility of increasing the capacity of two LNG plants with

0.5 million mton/yr and 0.6 million mton/yr. The case study evaluated the feasibility of scaling this technology to 3–5 million Mt/yr using dynamic simulation models to ensure reliability, safety, and profitability. This was accomplished by modifying the existing designs with the results from dynamic base case models that were in line with the design data by doing an interference test and confirming the model recovered back to steady state operation. [Davidson J, Potdar S Enabling affordable and reliable delivery of low-carbon fuel for the energy transition. Aspen Tech, 2024 (www.aspentech.com); https://www.shell.com/energy-and-innovation/natural-gas/liquefied-natural-gas-lng/lng-outlook-2023.html 3 2023 World LNG Report | IGU; Bassoini, G.; Harald, K. Liquefaction of Natural Gas and Simulated Process Optimization—A Review, Ain Shams Engineering Journal, 2024, 15, 102431. https://doi.org/10.1016/j.asej.2023.102431].

15.8 Energy Storage

Seasonal disconnects between energy demand and energy generation may require some form of seasonal energy storage, such as batteries, pumped hydropower, hydrogen, or reliance on nuclear energy. Storage systems include not only the underlying mechanisms such as batteries but also power conversion systems, energy management systems, and the integration of these and other components. The question involves the relative cost and ability to install transmission versus the relative cost and ability to install storage. Innovations in infrastructure and investment should be prioritized, as the costs need to be further reduced. Grid-scale storage is available through the deployment of lithium-ion batteries, but research and development are also needed on new battery chemistries with new materials beside lithium.

As the deployment of variable renewable technologies increases, the need for storage technologies to complement renewable electricity rises rapidly. Whereas batteries can provide time-related flexibility by addressing an eventual mismatch between the demand and production. Renewable energy can be stored in producing green fuels such as methanol synthesis by using electrolytic hydrogen and carbon dioxide. Power system flexibility can be achieved through short- and long-term energy storage and demand response, which can couple the electricity sector to the provision of heating, charging electric vehicles, and the production of green hydrogen [28].

15.8.1 Chemical Storage of Renewable Electricity

Chemical storage may involve the production of hydrogen with renewable electricity in an electrolytic system. Later this hydrogen can be used to produce synthetic natural gas, and chemicals including methanol and ammonia. A combination of several storage applications together may help electricity storage to be more feasible and flexible. The initial investment requires a cost per unit of power and a cost per unit of energy capacity, which are technology dependent [7, 19]. The economics of electricity storage are influenced by the type of storage technology, electricity price, the requirements of each application, the frequency of charging and discharging cycles, and the system in which the storage facility is located.

Multi-chemical production from biorefineries can produce methane, methanol, ethanol, dimethyl ether and others using renewable electricity and hence store it in various chemicals. For example, renewable power-based hydrogenation of carbon dioxide to methanol may lead to chemical storage of renewable energy. The produced methanol can be further converted to dimethyl ether, which is a valuable fuel and chemical feedstock [16]. The required carbon may come from many sources including biomass fermentation and effluents of fossil-fuel-fired power plants. Hydrogen can come from biological processes or from the electrolysis of water using renewable energy. This may also help fix the carbon in chemical feedstock and extend the market potential for electrolysis. The variability of hydrogen sources allows for a continuous process to be achieved despite fluctuating conditions of hydrogen production because of peak electricity demands and fluctuating wind and solar power.

15.9 Energy Efficiency

Much improved energy efficiency, structural and behavioral changes are needed in the power sector with clear efficiency standards for buildings, appliances and cars to reduce not only emissions globally, but also energy bills. Energy efficiency is one of the most cost-effective solutions for expanding access to clean energy. Retrofits in energy efficiency should be funded toward more efficient heating and cooling packages to support accelerated ET globally. Energy efficiency increase is one of the most effective ways of increasing energy security and reducing GHG emissions and energy cost. Energy efficiency helps control global energy intensity that is the amount of energy necessary to produce one unit of Gros Domestic Product (GDP).

Electrification and energy conservation bring important efficiency improvements, including more efficient electric motors than internal combustion engines and heat pumps more efficient than gas-fired boilers. Energy conservation may be the most affordable approach toward reducing the required energy and hence increasing energy efficiency. Upgrading technical efficiency gains from electrification, less use of solid biomass, and using energy and materials more efficiently [7, 12, 19, 20].

Energy efficiency standard (EES) produces techno-economic and sustainability analyses. These analyses create standards that achieve maximum improvement in energy efficiency that are feasible and lead to considerable energy savings.

Thermal efficiencies of residential furnaces and boilers are measured by annual fuel utilization efficiency, which is the ratio of heat output of the furnace or boiler to the total energy consumed by them over a typical year. he Energy Efficiency

Ratio of a cooling device is the ratio of output cooling (in Btu/ h) to input electrical power (in Watt) at a given operating point [7, 8].

15.9.1 Comparison of Energy-Efficiency Standards

The EES for residential appliances, equipment and lighting have been adopted globally and is contributing considerably to achieving energy conservation. Many countries have mandatory Minimum Energy Efficiency Standards (MEES) and labeling programs beside voluntary standards [33, 35]. EES for equipment and appliances are the MEES in most countries.

Energy Star programs are used in many countries to help industries, businesses, and consumers to adopt energy saving products and practices, reduce energy waste and GHG emissions. Energy Star program: (1) establishes specifications, testing procedures and verifications requirements for various consumer appliances and products, (2) combines research into residential energy use to promote energy-efficient homes, and (3) develops commercial building energy asset rating program to assess energy usage accurately.

15.9.2 Rebound Effect and Energy Efficiency

Increased energy efficiency reduces the cost of energy, so consumers may use it more. An example is the energy-efficient light bulbs reduce electricity bills, savings are used for other goods and services, while energy use increases slightly. The direct rebound is the amount of energy savings from increased efficiency that are offset by increased use of energy. This is acknowledged in energy economics and interpreted as allocating the savings to other beneficial uses. The indirect rebound results from the way the consumers use their savings; for example, savings can be used in clean energy resources. Overall, the rebound effect may redirect the benefits, but it does not cause the gains of efficiency to be lost [18].

15.10 Electrification

A high rate of electrification in sectors such as transport and buildings would require a considerable increase in renewable electricity capacity, may require existing infrastructure to be modernized with grid reinforcement and expansion. ET features a synergy between increasingly affordable renewable power technologies and the wider adoption of electric technologies for end-use applications, especially in transport and heat.

Renewable energy deployment, improvements in energy efficiency and the electrification of end-use sectors will lead to electricity becoming the main energy carrier. Energy transitions rely on electrification and technologies like wind, solar PV and batteries, and push for electricity security and diversified supply chain for clean technologies and critical minerals up in the policy agenda. The acceleration of renewable energy deployment is being coupled with increasing electrification of industrial sector and in buildings. Clean energy is the most dynamic aspect of global energy investment [14, 32, 33].

Considerable acceleration is needed in deeper end-use electrification of transport and heat, direct renewable use, energy efficiency and infrastructure additions:

- The share of renewable energy in the global energy mix must increase considerably.
- Primary energy supply would be stable with increased energy efficiency and growth of renewables.
- Renewables would increase across all end-use sectors.
- Biomass and hydrogen would both play more significant roles.
- Hydrogen must be renewables-based.
- Besides the investments in solar PV and wind power, biofuels, hydropower and geothermal energy. Must also be supported.

Bioenergy—Bioenergy in various forms, including solid biomass, biogas, biomethane and liquid biofuels would increase considerably in the transport sector. In addition, drop-in fuels and green hydrogen or hydrogen with carbon capture both play significant roles.

Power systems—Power system flexibility is essential for a successful energy transition because it enables the decarbonization of the electricity sector with some of the lowest-cost power sources available today. This is crucial in effectively achieving climate goals by integrating solar and wind energy integration with the balance between supply and demand and supporting the system's stability.

Transmission network—Grids should not be a bottleneck for ET as low-carbon fuel or green hydrogen-based electricity production may occur in locations away from demand centers. Robust national transmission networks dispatch power generation where it is most needed. Transmission lines connect the new and planned offshore wind farms and solar power generation centers to the mainland. Cross-border power exchange facilities share flexibility across an entire region.

15.10.1 Sustainable Bioenergy for Electricity Production

Bioenergy-based electricity generation is increasing with major co-firing technologies, including [32, 33, 35]:

- Direct co-firing is the simplest, cheapest and most common option. Biomass can either be milled jointly with coal, typically less than 5% in terms of energy content, or pre-milled and then fed separately into the same boiler with common or separate burners.

- Indirect co-firing is expensive and a less common process in which a gasifier converts the solid biomass into a fuel gas that is then burned with coal in the same boiler. Gas cleaning and filtering are needed to remove gas impurities before burning, and the ashes of the two fuels remain separate.
- Parallel co-firing requires a separate biomass boiler that supplies steam and allows for high biomass percentages and is frequently used in pulp and paper industrial facilities.

15.10.2 Electrification and Hydrogen

Direct electrification, where possible, is preferable to using hydrogen, based on the efficiency of the gas' conversion to useful energy due to conversion losses. Economic efficiency depends on a range of factors, including the availability and cost of primary energy sources, the cost of electricity generation and storage technologies, and the cost of hydrogen production, storage, and transport. Therefore, the use of hydrogen needs to be carefully considered when there are no practical alternatives as the electricity needed for hydrogen production could be considerable. Integrated planning is needed to ensure a pipeline of renewable power projects to meet this growing demand for hydrogen as well as electrolyzers [27, 28].

15.10.3 Critical Materials

Increasing electrification and energy transitions increase demand for critical materials, including copper, rare earth elements, silicon and various battery metals, notably lithium.

- Copper is used extensively in electricity transmission and distribution grids, but its conductive properties also make it an essential component for low emissions power generation technologies such as solar photo voltaic modules, wind turbines and batteries.
- Rare earth elements (REEs) are used to manufacture permanent magnets for the motors of direct drive and hybrid wind turbines.
- Silicon is used to manufacture solar panels.
- Lithium-ion batteries dominate EVs and are the fastest growing electricity storage technology in the world, making lithium indispensable for electrification [14, 15].

Large investments and strong policy support are needed to ensure that the supply of some key materials grows sustainably to meet rapidly growing demand for ET. Some key actions needed include:

- Building new mines and expanding the existing supply of materials quickly
- Building a more diverse and secure supply

Table 15.5 Some key material for energy transition and clean energy systems: H: high requirement; M: Medium requirement [14]: https://www.energy-transitions.org/publications/material-and-resource-energy-transition/)

	Solar	Wind	Grids	Batteries and EVs	Electrolyzer	Nuclear power	Hydropower
Aluminum	H	H	H		M		M
Cobalt				H			
Copper	H	H	H	H	H	M	M
Graphite				H			
Lithium				H			
Neodymium		H		H	M		
Nickel		M		H	H	M	M
Platinum					H		
Palladium					H		
Siver	H						
Steel	H	H	H		H	H	H
Uranium						H	
Polysilicon	H						
Iridium				H	H		
Tin				H			

- Driving sustainable and responsible material production
- Boosting innovation and recycling to reduce pressure on primary supply

Some rare earth elements (REEs) are critical raw materials for global ET and economy. An electrokinetic mining technique could enable green and selective extraction of REEs from ores by addressing challenges related to its mining efficiency, environmental footprint and economic performance [29]. The Energy Transitions Commission (ETC) has developed a series of material factsheets for key materials, including cobalt, copper, graphite, lithium, neodymium and nickel (Table 15.5). Iridium is important for the current generation of electrolyzers, while tin is used in electrical connections, cobalt for batteries and polysilicon for solar panels. Industrial materials such as steel, copper and aluminum are required across most of the clean energy technologies [14].

15.10.4 Recycling of Critical Materials

Without strong action to improve materials efficiency, increase recycling or increase mined supply, there could be significant supply gaps for six key ET materials: lithium, nickel, graphite, cobalt, neodymium and copper. Efficiency and recycling can reduce primary material requirements since over 50% of the demand could be met by recycled supply for three key battery materials: cobalt, graphite and lithium. In the case of wind turbines, up to 90% of a wind turbine's mass can be recycled, and there are established recycling systems for the foundation, tower and parts of the nacelle. The mass of solar panel materials should ideally re-used or recycled. Close to total recycling is technically feasible, and the high cost of primary minerals creates strong economic incentives for recycling to be deployed. In the case of nickel-manganese-cobalt batteries (which include cobalt and nickel as well as lithium) extensive recycling would occur even without regulation [14].

15.10.5 Prospects for the Grid

The technological entity ties together the components of the electricity grid consisting of substations, transformers, and power lines. In addition, the use of renewable energy will continue to increase more rapidly than the overall demand for energy, but renewable energy sources are not always "on," which means that the grid will have to incorporate massive amounts of energy storage from pumped hydropower to batteries to hydrogen systems. The power systems need to integrate very large shares of generation from variable renewables, which creates a considerable challenge in supply and demand dynamics. Linking this expansion with battery capacity would enable smoother integration of these sources over time as all power sector components must combine to deliver an effective, reliable, highly renewable and decarbonized power system.

In addition, land-based wind and solar sites are often located in remote locations far from where the electricity is consumed, so new transmission lines will be needed to transport electricity from where it is generated. To plan, permit, and build a high-voltage transmission line requires time and investments due to the costs of steel, insulators, and conductors. Grid may look like a collection of smaller grids that are interconnected so that they can be powered up, turned off, and isolated, creating a reliability only with advanced power electronics reacting faster when necessary [17, 18].

15.10.6 Power Demand

The electrification will increase power demand, which needs to be met with increased efficiency and with renewables integrated into the grid. According to the TES, the number of electric vehicles (EVs) will increase from around 8 million in 2019 to over 1100 million in 2050. Despite falling renewable energy subsidies, renewable power generation technologies are increasing new capacity while lowering the cost. In the TES, electricity is projected as the main energy carrier, growing from 20% to almost 50% share by 2050 doubling the gross electricity consumption. Table 15.6 shows that the share of renewable power must increase 86% by 2050 [20].

15.10.7 Power Flexibility

Flexibility in power systems is a key to enabling the increased shares of renewable power in a decentralized and digitalized system at a share of over 30% or higher on an annual basis (see Table 15.7). In the TES 73% of the installed capacity and over 60% of all power generation would come from solar PV and wind. Power storage is essential for flexibility, and the amount of storage would need to expand from around 230 gigawatt-hours (GWh) today to over 23,000 GWh by 2050. Flexibility will also be achieved by grid expansion, smart charging of EVs, demand-side flexibility, and sector coupling. This would require considerable investment worldwide [20].

Although a variable source, offshore wind has the technical potential to meet today's electricity demand as it offers considerably higher capacity factors than solar PV and onshore wind with ever-larger turbines more reliable wind speeds. Increasingly cost-competitive offshore wind projects can attract higher investment toward 2040. In the Sustainable Development Scenario, offshore wind rivals its onshore counterpart and helps achieve full decarbonization and the production of low-carbon hydrogen [6, 15, 16].

In the TES, half of the energy demand could be supplied by electricity by 2050. Of this, one third is already supplied by end-used renewable sources, with the remaining two thirds by fossil fuels. Direct use of bioenergy, solar thermal, geothermal heat, energy efficiency and structural changes can reduce energy demand, leading to deeper electrification. However, more is needed in shipping, aviation, and heavy industry to reduce emissions [20].

Industry, such as iron and steel, cement, and petrochemicals, is the dominant energy consumption sector in many countries, where the sector consumes around half of the final energy. Innovation is needed to find low-to-zero GHG emission

Table 15.6 Electrification projections by IRENA in the Transforming Energy Scenario (TES) [20]

	2030	2050
Renewable share in electricity production %	57	86
Electrification share of final energy %	29	49
Renewable energy share in total final energy consumption %	28	66
Energy intensity improvement rate %/year	3.6	3.2

Table 15.7 Renewable energy share in power generation by the Transforming Energy Scenario (TES) [12]

	2030	2050
Variable renewable energy share in generation %	35	61
Solar power generation GW	32,279	8828
Wind power generation GW	2526	6044

solutions for the industrial process and to address transport modes of aviation and shipping sectors that are hard to electrify [24].

15.11 Transforming Industry

A radical shift in the industry sector value chain is required as key materials are an essential enabler of modern economies. Industrial production processes are carbon intensive as most of the industrial energy uses fossil fuels. Reducing emissions across industrial sectors will require radical shifts in how materials are produced, consumed and disposed. The three main industry sectors of chemical, cement and steel are the most challenging to electrify [16, 20, 30]. Some solutions for ET may be:

- For the chemical industry, electric crackers are in the pilot phase.
- The iron and steel sector needs additional financial support for the research and development of innovative solutions, including hydrogen-based direct iron reduction, infrastructure for CCS and the retrofit of any remaining fossil fuel-based installations.
- The steel sector is considering electrolytic reduction processes.
- The cement industry is working on new kilns where heat is provided via plasma generators.
- Programs to electrify steam cracking furnaces helps decarbonize one of the most carbon-intensive aspects of petrochemical manufacturing.
- Stronger energy and emissions efficiency standards can drive dramatically higher material efficiency, particularly with steel, cement, plastics and aluminum production.
- Developing a circular economy with reuse of materials, scrap collection, and recycling programs would be helpful.

The cross-sector and crosscutting policies include:

- Introduce fiscal policy measures for fossil fuels and carbon prices.
- Increase public finance and strategic plan
- Define 'risk' in a more comprehensive way to include broader environmental and social risks.
- Enhance international collaboration in sustainability, energy, technology and innovation, regional power grids, and green fuel development.
- Develop national bioenergy and/or hydrogen strategies.
- Incentivize/mandate a circular economy approach.

The energy intensity of industry sector with the increase of electrification rate and the decrease in coal consumption would be much required improvement, although adopting ET while maintaining competitiveness is a challenge. Many energy-intensive products such as steel or chemicals are traded internationally in competitive markets with margins that are too slim to absorb elevated production costs [16]. Hard-to-decarbonize industry sectors require green hydrogen, bioenergy, direct electrification, and the integration of CCS and BECCS to control residual emissions, as well as energy efficiency and circular economy principles.

Ultimately, consumers need to pay for the investments that are required to transform the energy system, whether directly through bills or indirectly through taxation or the cost of goods and services over time. Well-designed policies can bring clean energy technologies to households and communities with carbon pricing schemes and the associated revenues be used for social aspects of energy transitions. For example, expanding the scope of carbon pricing instruments to cover sectors such as transport and buildings may force lower income household pay to a larger proportion of overall CO_2 costs, if they cannot switch away from fossil fuels [26].

15.11.1 Power Sector

Electricity markets must consider that ET should occur at a price consumers can afford as ET is largely an economic rather than an engineering challenge. Renewables-based power generation is the global effort as the decline in renewable electricity costs has made it the most economical ET. Besides these direct cost savings, renewables' deployment brings substantial economic benefits due to reduced GHG emissions and local air pollutants. Renewable energy mix will dominate the power capacity increase globally and transform electricity systems coupled with electricity storage [17].

15.11.1.1 Modular Nuclear Reactors

A way to generate power without producing carbon dioxide may be possible using small modular nuclear reactors that produce less than 300 megawatts of power. Designing a nuclear reactor for the modern grid requires safety and auto shut down. The design must also be approved for off-grid operation, which is a first for nuclear power. The reactors can thus go right next to end users that need power for carbon-free hydrogen production, ammonia production, or other applications or stand-alone desalination systems as the economics of water production proves. Fission reactions today produce considerable electricity globally [22].

15.11.2 Transport Sector

In transport sector, direct electrification would dominate road and rail transport, whereas green ammonia and methanol would be key in shipping. Adoption of sustainable renewable fuels, including biofuels, hydrogen, synthetic fuels, and a mix of synthetic fuels and biofuels would be needed in aviation, besides the phasing out of internal combustion engine for low-emission vehicles [3]. Fuel economy standards have proven effective in enhancing vehicle efficiency with policies and standards.

Electric vehicles (EVs) emerge as solutions through direct electrification using high shares of renewables integrated into power generation. Charging infrastructure development is key for deploying EVs. Accelerating EV adoption for road transport, along with power supply decarbonization, for transport sector ET. Evolution of batteries has greatly improved the economic and the scope of application is quickly expanding to another road vehicle segments and service types. Electrical and plug-in hybrid vehicles continue to grow. Heavy transportation can be also optimized, retrofitted, electrified and use renewable fuels. Long-haul transport requires ambitious technology roadmaps with the right policy incentives and collaboration across the maritime industry and stakeholders.

Electrification is also crucial for deeper decarbonization of the rail sector. Electric train technology is mature, and electrification is technically possible and can be cost-effective. High-speed train routes reduce the demand for short-haul aviation. International shipping would see a diverse mix of low-carbon fuels for decarbonization, with ammonia, methanol and hydrogen [27, 28].

15.11.2.1 Green Balance Mechanism

Green Balance Mechanism to incentivize the maritime sector's energy transition with the benefits ranging from infrastructure deployment across different regions to enabling policies. This innovative approach incorporates a GHG fuel-intensity standard and establishes a fee on fossil fuels, redistributing the collected funds to support clean fuels. This creates a strong incentive for fuel providers to invest in clean energy, accelerating economies of scale that will lower production costs and drive down prices. What's needed is a mechanism that bridges the green premium, rewarding early adopters by compensating truly green vessels for their emissions reductions [21, 29].

15.11.3 Cement Sector

Cement manufacture involves different kiln designs. Clinker production is of two basic types ("dry" and "wet") depending on the raw materials' moisture content (IEA 2018). Currently, dry-process kilns are the most widely deployed technology for cement production, since they require less energy than wet-process kilns [20]. The substitution of clinker, circular economy practices and the increase of renewables will be needed to transform the cement sector.

To successfully transform the cement sector, while maintaining its competitiveness, several strategies will have to be pursued with the complete cement cycle:

- Reduce the use of cement by maintaining quality and material strength.
- Enhance the use of additives and fillers to substitute for clinker, complemented by circular economy practices to provide use of waste (blast furnace slag, coal

fly ash, red mud, calcined clay) from other materials production processes [17]. The availability of these alternative materials and national standards for cement properties should be considered.
- Promote efficiency measures in cement production.
- Switch to renewable energy and alternative fuels.
- Apply carbon removal measures such as CCS-CCUS and BECCS technologies, particularly to capture residual process emissions.
- Emerging solutions to decarbonize heating processes in the cement sector include the use of electric arc calciners and solar kilns.

15.11.4 Iron and Steel Sectors

Key innovations based on hydrogen, complemented by electrification and recycling, would enable low-emission iron and steel production. Global steel is produced using blast furnaces/basic-oxygen furnaces, which rely mostly on metallurgical coal as the chemical reducing agent. Steel can be produced via two routes: primary steelmaking from iron ore and secondary steelmaking from scrap steel. Primary steelmaking is highly energy and emission intensive. However, decarbonizing the steel industry would require deploying low-emission primary steelmaking alternatives. These include mainly using hydrogen, along with carbon capture or direct electrification [1, 15]. There are several options to decarbonize the iron and steel sector:

- Circular economy practices along with increased use of recycled scrap in electric arc furnace–based steel production will play a key role.
- A structural shift in iron and steel production is needed, with renewables displacing fossil fuels as both energy consumption and reducing agents. The steel sector would see a growth in renewables' share (bioenergy, other renewables, supply in electricity and green hydrogen production).
- Key innovations such as hydrogen-based direct iron reduction with electric arc furnaces represent a promising, efficient and low-emission technological pathway, complemented by process digitalization with the integration of Industry 4.0 practices.
- CCS is an alternative to further reduce emissions in cases of continuation of usage of fossil fuel-based processes.
- Relocation of industries with new value and supply chains created while achieving emission reductions if iron ore mining and green ironmaking are coupled with abundant and low-cost renewable resources.
- Decoupling the ironmaking and steelmaking processes in countries relying heavily on fossil fuels.
- Decarbonize heating processes in the iron and steel sector include the use of high- or very-high-temperature heat pumps, electric boilers and iron ore electrolysis

To accelerate the transition to a decarbonized steel sector:

- Top steelmaking countries must implement robust common definitions and standards for low carbon and near-zero-carbon steel.
- Steps must be taken urgently to create a demand for low-carbon and near-zero-carbon steel to overcome cost premiums.
- Commercial-scale projects will require a far-sighted approach that includes developing electricity transmission and CO_2 and hydrogen storage and transportation.
- As technology and policy pathways become apparent, public and private sectors must establish medium- and long-term strategies with milestones for deep decarbonization.

15.11.5 Chemicals Sector

The chemical industry is among the harder-to-decarbonize sectors and can therefore benefit from the deployment of high-efficiency heat pumps and other energy-efficiency solutions. The chemical industry is undergoing a big transformation to produce more sustainable chemicals and increase green feedstocks just like green hydrogen. Therefore, efficiency and productivity are essential for the chemical industry to stay competitive. Energy efficiency, circular economy measures and use of clean energy resources are necessary for the ET of the chemical sector. The total global demand for plastics would increase and meeting this demand would require increasing the sector's process energy. Circular economy strategies help reuse, and accelerated recycling coupled with substantial reduction of plastic demand. Achieving these objectives would require multiple actions as part of an energy transition adapted to the challenges of the chemical sector:

- Transfer to alternative fuels like renewable energy and increased energy efficiency
- Consider process energy generation with CCS and BECCS
- Consider hydrogen and biomass as feedstock as well as renewable energy
- Introducing circular economy and bioeconomy
- Promote green chemistry and process intensification

In addition to these measures, the chemical sector could achieve heating process decarbonization using the emerging technology of electric cracking furnaces, and by considering high- or very-high-temperature heat pumps and other technologies [12, 15, 24].

15.11.6 Building Sector

In the buildings sector, energy efficiency is the main enabler of the energy transition. Efficient appliances and heat pumps will play an important role in electrification space and water heating and making space cooling more efficient [17, 20]. Cooking, which relies heavily on fossil fuels and traditional biomass globally, would need to adopt electricity-powered efficient stoves and sustainable biomass.

Bioenergy, natural gas, and direct use of fuels for heating consume the most energy in the sector. Buildings would be smart, inter-connected, highly energy efficient, and powered and heated or cooled by renewable energy. It also offers to be part of distributed clean energy resources, by providing energy storage to the power system (both with batteries and with EVs).

Electrifying the residential heating sector will significantly increase the electricity load. Energy efficiency together with flexible, smart electrification strategies seek to enhance renewable integration, reduce peak loads and decrease grid congestion. This would require considering some measures including:

- efficient lighting and appliances, renewable and electricity-based heating and cooking technologies.
- apply heat pumps for electricity-based heating and cooling.
- replace traditional, unsustainable bioenergy sources causing indoor air by sustainable biomass, biogas and electricity.
- promote renewables such as biofuels, biomethane and solar thermal for heating.
- couple heating systems with thermal and seasonal storage to provide flexibility.
- Promote smart energy management systems in buildings and digitalization.
- shift to electric vehicles to enable home charging.
- decentralized energy supply through solar energy systems and storage, both electric and thermal.

15.11.6.1 Heat Pumps

Heat pumps need a compressor, indoors and outdoors coils, and an expansion valve to exchange heat either in or out of the building regulated by the thermostat providing both heating and cooling [2]. When an air heat pump is in heating mode, heat is collected from the exterior air by the exterior coil, which acts as the evaporator because the liquid is changing into gas as the refrigerant heats. The compressor compresses the refrigerant, which causes the refrigerant to become hotter. The hot, gaseous refrigerant circulates through the interior coil acting as condenser, while a fan helps with direct heating a space by blowing air through the interior coil. The expansion valve allows the refrigerant to expand, causing the refrigerant to become very cold before it moves through the exterior coil. The cold refrigerant is then warmed up by the outdoor air before it repeats the cycle again [11] (see Sect. 7.15).

15.12 Energy Transition and Digitalization

Digital innovations such as advanced data analytics, artificial intelligence and machine learning can make system operation more effective and enhance flexibility [5, 20]. Deep-learning tools can support ET and hence create significant disruptions and improve productivity rapidly.

The energy market needed greater confidence in artificial intelligence (AI) decision-making. Secure, and resilient energy future requires an upgrade and transform energy systems from how we generate and store energy to how we deliver it to consumers. AI is an essential tool to navigate the complexities of this transition, accelerating innovation and improving efficiency and reliability. Some major steps include [13]:

- AI-powered predictive tools can help modernize the grid with the integration of renewable resources to anticipate and mitigate grid disruptions caused by extreme weather or cyberattacks, improving resilience and ensuring a consistent power supply. AI is also optimizing load forecasting for cost-effectiveness and minimizing the impact of variability in renewable energy generation.
- Accelerating Energy Innovation such as advanced solar photovoltaics, next-generation batteries, and sustainable biofuels, is crucial for energy transition. AI accelerates the discovery and development of innovative materials and designs, optimizing manufacturing processes, and reducing the time and cost, co-designing of materials and components, considering system-level requirements for grid integration and lifecycle performance.
- AI-driven technology is improving energy efficiency in buildings, transportation, and industrial processes by using smart control systems to optimize energy consumption and for more efficient vehicles and engines.
- AI tools help improve the site location and permits for energy projects for faster energy infrastructure deployment.

15.12.1 Transforming the Global Energy System and AI

Although each country has its own energy mix and demand pattern, interconnecting them may maximize the compatibility of production and consumption patterns, share storage resources and the surplus of energy produced to meet demand. This requires improving flexibility with investment costs, regulatory barriers, system integration, data and modelling, and public acceptance.

The digital technology and energy storage are reaching the power industry. For example, hybrid projects, coupling storage with solar cells may create a computer-controlled power plant. How equipment interacts with the grid and leverage our increased connectivity is changing to optimize the response and behavior. This change in the electrical grid helps transition from thermal and mechanical controls to electronic and digital controls, from grid following to grid forming, from valuing capacity to valuing flexibility and balancing, and from top-down controls to distributed, intelligent agents. High voltage direct current transmission lines would be a useful tool in decarbonizing the electrical system [34].

Digital tools enable asset intensive industries to navigate the challenges of ET with optimization, control technologies, planning, scheduling, and fundamental science. The growth of electrification and the increasing deployment of renewable energy incorporate higher renewable capacity into the grid with digital grids management for rapid energy transition [5, 6, 23].

Industrial artificial intelligence created using a combination of first principles and AI ensure the technology is applied safely. The Internet of Things allows greater traceability of materials and products, while artificial intelligence can optimize processes for efficiency. In asset intensive industries with complex processes AI can recommend a set point or decisions within the context of engineering analysis, maintenance, and safety [6, 30, 34].

15.12.2 Artificial Intelligence's Impact on the Electrical Grid

Modern and decarbonized grids will be vital toward a low carbon or net-zero emissions ET. AI can help overcome barriers to integrate renewable energy into existing grids. The variability in renewable energy production often results in overproduction and underproduction depending on the demand, leading to wasteful energy consumption and grid instability. Analyzing available datasets, from weather patterns to energy consumption trends, AI could monitor load shifting to make sure consumers use energy when electricity from renewable energy sources is available to ensure optimal grid stability and efficiency. This will require powerful analytical tools of AI [23].

Power systems are becoming complex as demand for electricity grows and they must be able to work with rising number of grid-connected devices, from electric vehicle (EV) charging stations to residential solar installations, the transportation, building and industrial sectors. The result is a need for information processing tools to plan and operate power systems. The capabilities of AI applications with advanced machine learning models can reliably analyze data and inputs and generate responses. For example, smart meters produce and send data to utilities than their analogue predecessors. The global operations of wind turbines are estimated to produce more than 400 billion data points per year [31].

Machine learning helps match variable supplies with rising and falling demands and hence maximizes the economic value of renewable energy and allows it to be integrated more easily into the grid with:

- Facilitating demand response uses a range of processes such as forecasting electricity prices, scheduling and controlling response loads, and setting dynamic pricing.
- Providing improved consumer services using machine learning processes online improves customers' billing experience.

15.12.2.1 Grid Security and AI

AI can perform predictive maintenance, monitor and analyze the performance of energy assets to identify potential faults ahead of time. For this, a range of utilities are developing AI-enabled schemes to help monitor physical assets and use past data on performance and outages to predict when intervention is required. Potential uses for AI across power systems are likely to increase for better forecasting of energy production and predictive maintenance of physical assets to manage and control grids [23, 31].

15.12.3 Digitalization and Risk

Risks associated with AI include:

- threats to cybersecurity and privacy,
- the influence of biases in data, and miscorrelations due to insufficient training, data or coding mistakes,
- increased use of automated and self-learning software raises questions about the reliability of the outcomes of these systems,
- AI uses more energy than other forms of computing systems, which is a consideration in the efficient energy systems.

Governments will also need to develop secure mechanisms for data sharing, storage, and governance to expedite the energy transition while reducing its costs. AI can also help increase the energy efficiency of buildings by optimizing the performance of heating and air conditioning, by predictive maintenance [31].

15.13 Industrial Clusters for Energy Transition

Industrial clusters can build successful clean energy infrastructure deployment by integrating multi-fuel, multi-modal ecosystem. These clusters can share risks, aggregate and optimize demand, secure capital for production, distribution and consumption of clean energy with facilitated global logistics and connections between international markets and regional industries (Fig. 15.3). It presents three solutions within clean energy value chain [4]:

- Develop a common vision at cluster level
- Expedite the scaling of cluster-level clean energy initiatives
- Strengthen the collaboration across clusters and regions

Successfully transitioning industrial clusters requires four core components:

(1) a balanced focus on economic, social and environmental values,
(2) adoption of technologies and processes that support GHG emissions reduction,

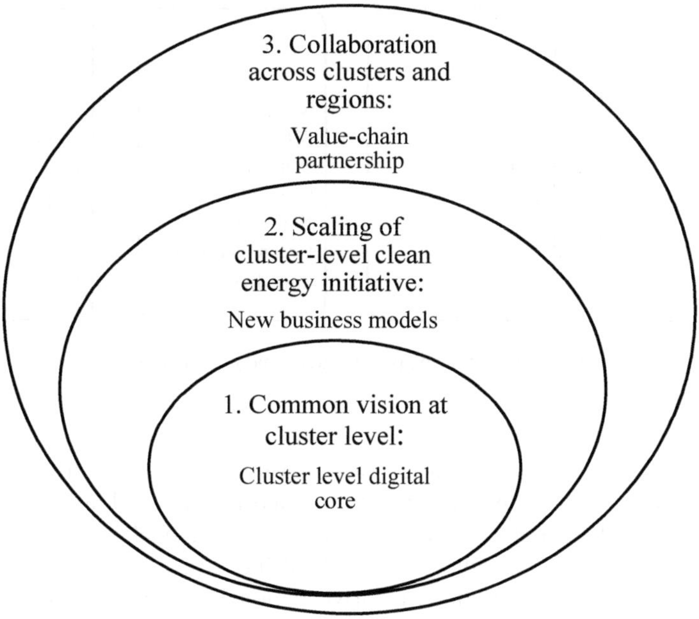

Fig. 15.3 Cluster development stages

(3) cross-sector collaboration among industry, government, financers, labor and communities, and
(4) coordinated, agile strategies across partnerships, policy, and technology deployment.

Together, these elements form the foundation for a sustainable and scalable energy transition for hard-to-abate sectors [15].

Industrial clusters create a base for accelerating the ET as infrastructure supports new energy flows between clusters. This may involve repurpose of existing assets, operations, feedstocks for low carbon energies. Connecting clusters into a global network supports the expansion of cluster networks to establish a global clean energy infrastructure model and facilitate a more interconnected world. A regional fuel study may help a cluster identify its primary value propositions and connecting clusters into a global network toward a more interconnected operation.

15.13.1 Clusters for Low-Carbon Fuels

To accelerate the energy transition, industry leaders and policy makers can harness the power of industrial clusters and streamline collaboration across the clean energy value chain [4]. Mobilizing co-located companies may help optimize opportunities for scale, share risk/resources, aggregation and optimize demand and

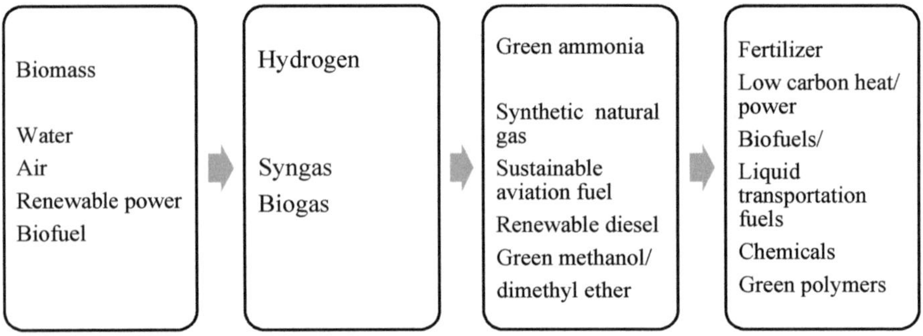

Fig. 15.4 Hydrogen hub in clean energy, power, fuels, and chemical in a cluster platform [4, 27, 28]

resources. The vision for the cluster encompasses the region's diverse industrial sectors, such as energy, steel, manufacturing and transport, alongside its existing infrastructure and natural resources. For example, regional clean hydrogen hub industrial cluster brings together a network of industries, businesses, community groups, and academic institutions to drive regional decarbonization (Fig. 15.4). With an established ethanol industry and the potential to use renewable power for electrolysis, the cluster shows strong promises for locally produced renewable fuels, which would support regional aviation, maritime and energy needs, hence provide benefits including CO_2 emissions reduction, increase in jobs, GDP, and energy security [4].

15.13.2 Clusters with Digitalization

Data collaboration, underpinned by a strong digital core, is a critical foundation to enable clusters to unlock system value benefits for their constituents and local stakeholders. Trust among players in a cluster is key to enabling transparency and data collaboration via a robust cluster-level digital core in a secure environment. In turn, digital technologies can collect and integrate complex information to produce valuable insights at the cluster level. Building upon the core, digital applications enable operational and commercial benefits and unlock related system value benefits (e.g. energy security). These digital tools not only facilitate secure data collaboration but also set the foundation of resilience and reinvention to help clusters innovate and incorporate new digital technologies such as AI [5, 6].

Clusters achieve the greatest impact when digital technology providers are deeply integrated in the collaborative process that unlock significant value, driving performance, sustainability and innovation. Digital technologies enable low-carbon industrial clusters, underpinning the energy transition. As ET accelerates, digital solutions can manage complex infrastructure and low carbon products and services from industrial clusters. In CCUS, digital platforms enable end-to-end

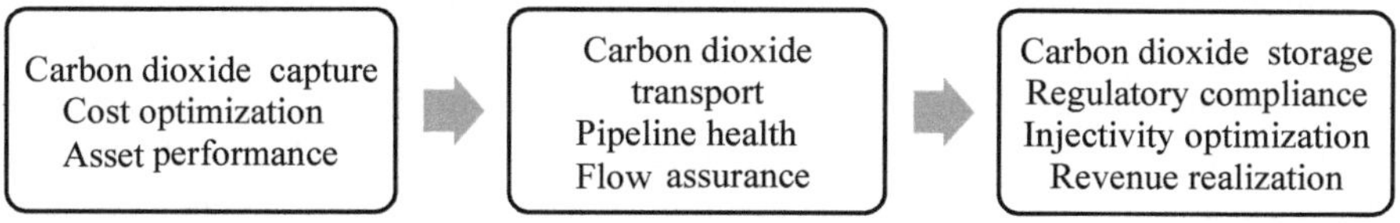

Fig. 15.5 Stages in the digital solutions platform

monitoring, reporting and verification across the capture, transport, utilization and storage stages. These platforms ensure seamless data integration, operational alignment, and performance assurance for both process and products. For instance, a low carbon ammonia project, digital tools and solutions, enable the measurement, monitoring and verification of CO_2 with real-time data, tracking and management capabilities across multiple facilities, ensuring compliance and verification for permanent underground storage. Digital solutions enable existing and new infrastructure to reach their full potential with end-to-end digital solutions with the stages shown in Fig. 15.5.

15.13.3 Digital Twins for Decarbonization

By digitally modelling the hydrogen value chain with r4elevant data, digital twins (DTs) encompass the production of blue and green hydrogen, their storage, transport and consumption within and beyond the cluster. The DT will facilitate the modelling of decarbonization pathways and scenarios, analyze risk, carbon abatement potential, hydrogen market evolution, and the development of hydrogen supply chain. As the clusters cooperate, DT technology can be a decision-supporting tool to facilitate the envisioning, shaping and building of clusters from policy makers to investors and operators.

The project's output demonstrates the potential of digital models in solving complex supply chain problems in industrial clusters. The product-level DT showcases the benefits of connectivity between hydrogen producers, industrial emitters and various cross-sector actors. It also includes financial modelling of different supply and demand scenarios, providing investors with economic return on investment [6, 30].

Vertical integration to drive down costs and enhance the efficiency of the production of green hydrogen and its derivatives by managing the supply chain from renewable energy equipment manufacturing and generation to electrolyzer manufacturing. CO_2 captured from cement production is combined with green hydrogen to produce low-carbon methanol. In addition, the proximity of the industrial cluster to the country's largest commercial port facilitates the transport and export of green hydrogen and its derivatives, further lowering logistics costs. By combining large-scale production capabilities with the adjacent industrial ecosystem, decarbonization within the cluster is supported, while enabling the trade and export of green hydrogen and its associated derivatives [31].

Digital twins are highly accurate replicas of the physical plant at both a molecular and sitewide level. Therefore, they can enable optimized production and drive higher production performance. Offline models can run simulations, identify and troubleshoot bottlenecks, and optimize asset performance with "what-if" analysis toward accurately modeling critical equipment such as compressors and cryogenic heat exchanger trains for the impact of external constraints on efficiency. Digital twins can make data-driven decisions by combining first principles models with real-time plant operational data to optimize the assets online [13].

References

1. Andersson FNG (2020) Effects on the manufacturing, utility and construction industries of decarbonization of the energy-intensive and natural resource-based industries. Sustain Prod Consump 21:1–13
2. Arpagaus C et al (2018) High temperature heat pumps: market overview, state of the art, research status, refrigerants, and application potentials. Energy 152:985–1010
3. Bergero C et al (2023) Pathways to net-zero emissions from aviation. Nat Sustain 6:404–414
4. Bocca R, Espinoza N, Jamison S (2025) Unleashing the full potential of industrial clusters: infrastructure solutions for clean energies white paper. World Economic Forum (WEF)
5. Chen L et al (2023) Artificial intelligence-based solutions for climate change: a review. Environ Chem Lett 21:2525–2557
6. Demirel Y, Rosen MA (2023) Sustainable engineering: process intensification, energy analysis, artificial intelligence. CRC Press, Taylor & Francis
7. Demirel Y (2025) Energy conservation. In: Dincer I (ed) Comprehensive energy systems, 2nd edn. Elsevier, Amsterdam, vol 9, pp 1–51
8. Demirel Y (2025) Biofuels. In: Dincer I (ed) Comprehensive energy systems, 2nd edn. Elsevier, Amsterdam, vol 2, pp 344–384
9. DOE (2024) U.S. Department of Energy. Sustainable aviation fuel (SAF) grand challenge: building supply chains. https://www.energy.gov/eere/bioenergy/articles/sustainable-aviation-fuel-saf-grand-challenge-building-supply-chains
10. Doliente S et al (2020) Bio-aviation fuel: a comprehensive review and analysis of the supply chain components. Front Energy Res. https://doi.org/10.3389/fenrg.2020.00110
11. DOE (2024) Heat pump systems. https://www.energy.gov/energysaver/heat-pump-systems
12. EERE (2022) Advanced manufacturing office thermal process intensification: transforming the way industry uses thermal process energy. Department of Energy
13. EERE (2024) Advanced research directions on AI for energy report on winter 2023 workshops. In: Artificial intelligence for energy. Department of Energy; AI for Energy Report 2024 | Argonne National Laboratory
14. ETC (2023) Material and resource requirements for energy transition. https://www.energy-transitions.org/publications/material-and-resource-energy-transition/
15. Franco A, Rocca M (2024) Renewable electricity and green hydrogen integration for decarbonization of "hard-to-abate" industrial sectors. Electricity 5:471–490
16. IEA (2020) Flagship report, CCUS in clean energy transitions. https://www.iea.org/reports/ccus-in-clean-energy-transitions
17. IEA (2024) International energy agency. Special Report. Strategies for affordable and fair clean energy transitions
18. IEA (2024) Global energy-transition stocktate. https://www.iea.org/topics/global-energy-transitions-stocktake
19. IEA (2024) Energy efficiency. International Energy Agency. https://www.iea.org/reports/energy-efficiency

20. IRENA (2023) International renewable energy agency. World energy transitions outlook 2023: 1.5°C pathway. https://www.irena.org/Publications/2023/Jun/World-Energy-Transitions-Outlook-2023
21. IRENA (2022) Scaling up biomass for the energy transition: untapped opportunities in Southeast Asia. International Renewable Energy Agency, Abu Dhabi
22. IRENA (2021) Reaching zero with renewables: SAF fuels. International Renewable Energy Agency, Abu Dhabi. ISBN 978-92-9260-350-2 (efficiency-2024)
23. Kemene E, Valkhof B, Tladi T (2024) AI and energy: will AI help reduce emissions or increase demand? Here's what to know. World Economic Forum. https://www.weforum.org/stories/2024/07/generative-ai-energy-emissions/
24. Lechtenböhmer S et al (2016) Decarbonizing the energy intensive basic materials industry through electrification—implications for future EU electricity demand. Energy 115:1623–1631
25. Lenton TM et al (2023) Quantifying the human cost of global warming. Nat Sustain 6:1237–1247
26. Madden S, Alles K, Demirel Y (2021) Measuring sustainability of renewable diesel productions using multi-criteria decision-matrix. Bioprod Bioref Biofpr 15:1621–1637
27. Matzen M, Demirel Y (2016) Methanol and dimethyl ether from renewable hydrogen and carbon dioxide: alternative fuels production and life-cycle assessment. J Clean Prod 139:1068–1077
28. Matzen M, Alhajji M, Demirel Y (2015) Technoeconomics and sustainability of renewable methanol and ammonia productions using wind power-based hydrogen'. Adv Chem Eng 5:128
29. NAE (2024) National Academy of Engineering. Transitioning to net-zero carbon: engineering challenges and opportunities: proceedings of a forum. The National Academies Press, Washington, DC. https://doi.org/10.17226/28542
30. Pietri A (2024) The future of the energy transition., AspenTech (www.aspentech.com)
31. Rozite V, Miller J, Oh S (2023) Why AI and energy are the new power couple-(EIA 2023). https://www.iea.org/commentaries/why-ai-and-energy-are-the-new-power-couple
32. Wang Z et al (2022) Life cycle assessment of combustion-based electricity generation technologies integrated with carbon capture and storage: a review. Environ Res 207:112219
33. Wang X, Demirel Y (2018) Feasibility of power and methanol production by an entrained-flow coal gasification system. Energy Fuels 32:7595–7610
34. WEF (2023) World economic forum, 2023. https://www.weforum.org/stories/2023/09/steps-businesses-demystify-just-energy-transition/
35. Wei X, Manovic V, Hanak DP (2020) Techno-economic assessment of coal- or biomass-fired oxy-combustion power plants with supercritical carbon dioxide cycle. Energy Convers Manag 221:113143

Physical and Critical Properties

A

See Tables A1 and A2.

Y. Demirel, *Energy*, Green Energy and Technology,
https://doi.org/10.1007/978-3-031-97111-2

Table A1 Physical properties of various organic and inorganic substances

Compound	Formula	*MW* g/mole	Sp Gr	T_m K	T_b K	ΔH_v kJ/kg	ΔH_m kJ/kg
Air		28.97					
Ammonia	NH_3	17.03	0.817	195.4	239.7	1374.0	322.4
Benzene	C_6H_6	78.11	0.879	278.7	353.3	394.3	126.0
Benzyl alcohol	C_7H_8O	108.13	1.045	257.8	478.4		
Butane	n-C_4H_{10}	58.12	0.579	134.8	272.7	383.6	80.3
Iso-Butane	i-C_4H_{10}	58.12	0.557	113.6	261.4	366.4	105.7
Carbon dioxide	CO_2	44.01	1.530	217.0			
Carbon disulfide	CS_2	76.14	1.261	161.1	319.4	351.9	
Carbon monoxide	CO	28.01	0.968	68.1	81.7	214.2	
Cyclohexane	C_6H_{12}	84.16	0.779	279.8	353.9	357.6	
Cyclopentane	C_3H_{10}	70.13	0.745	179.7	322.4	389.2	
Diethyl ether	$(C_2H_5)_2O$	74.12	0.708	156.9	307.8	352.1	
Ethane	C_2H_6	30.07	1.049	89.9	184.5	488.8	
Ethanol	C_2H_6O	46.07	0.789	158.6	351.7	837.8	109.0
Ethylene glycol	$C_2H_6O_2$	62.07	1.113	260.0	470.4	916.7	181.1
Gycerol	$C_3H_8O_3$	92.09	1.260	291.4	563.2		200.6
Heptane	C_7H_{16}	100.20	0.684	182.6	371.6	316.3	
Hexane	C_6H_{14}	86.17	0.659	177.8	341.9	335.3	
Hydrogen	H_2	2.02	0.069	14.0	20.4	445.5	59.5
Hydrogen chloride	HCI	36.47	1.268	158.9	188.1	444.2	
Hydrogen sulfide	H_2S	34.08	1.189	187.6	212.8	548.7	
Mercury	Hg	200.61	13.546				
Methane	CH_4	16.04	0.554	90.7	111.7	511.2	58.4
Methanol	CH_3OH	32.04	0.792	175.3	337.9	1101.7	99.2
Nitric acid	HNO_3	63.02	1.502	231.6	359.0	480.7	
Nitrogen	N_2	28.02	12.500	63.2	77.3	199.8	25.3
Nitrogen dioxide	NO_2	46.01	1.448	263.9	294.5	319.4	
Nitrogen oxide	NO	30.01	1.037	109.5	121.4	459.8	
Nitrogen pentoxide	N_2O_5	108.02	1.630	303.0	320.0		
Nitrous oxide	N_2O	44.02	1.226	182.1	184.4		
Oxygen	O_2	32.00	1.105	54.4		212.5	13.7
n-Pentane	C_5H_{12}	72.15	0.630	143.5	309.2	357.5	
Iso-pentane	i-C_5H_{12}	72.15	0.621	113.1	300.9		
Propane	C_3H_8	44.09	1.562	85.5	18.8		80.0
Propylene	C_3H_6	1500	7.792	1.637	22.706	−6.915	
n-Propyl alcohol	C_3H_8O	60.09	0.804	146.0			

(continued)

Table A1 (continued)

Compound	Formula	*MW* g/mole	Sp Gr	T_m K	T_b K	ΔH_v kJ/kg	ΔH_m kJ/kg
Iso-Propyl alcohol	C_3H_8O	60.09	0.785	183.5			
n-Propyl benzene	C_9H_{12}	120.19	0.862	173.7	38.2		
Sodium hydroxide	NaOH	40.00	2.130	592.0	1663.0		
Sulfur dioxide	SO_2	64.07	2.264	197.7	263.1	388.6	
Sulfur trioxide	SO_3	80.07	2.750	290.0	316.5	522.0	
Sulfuric acid	H_2SO_4	98.08	1.834	283.5			
Toluene	$C_6H_5CH_3$	92.13	0.866	178.2	383.8	363.6	
Water	H_2O	18.02	1.000	273.2	373.2	2253.0	333.7

Table A2 Critical properties

Species	MW g/gmole	ω	T_c K	P_c atm	V_c cm^3/gmole
Air	28.97	0.035	132.5	37.2	88.3
H_2	2.02	− 0.216	33.3	12.8	65.0
Air	28.97	0.035	132.0	36.4	86.6
N_2	28.01	0.038	126.2	33.5	90.1
O_2	32.00	0.022	154.4	49.7	74.4
CO	28.01	0.048	132.9	34.5	93.1
CO_2	44.01	0.224	304.2	72.8	94.1
NO	30.10	0.583	180.0	64.0	57.0
N_2O	44.01	0.141	309.7	71.7	96.3
SO_2	64.06	0.245	430.7	77.8	122.0
Cl_2	70.91	0.069	417.0	76.1	124.0
CH_4	16.04	0.012	191.1	45.8	98.7
C_2H_6	30.07	0.100	305.4	48.2	148.0
C_3H_6	42.08	0.140	365.0	45.5	181.0
C_3H_8	44.10	0.152	369.8	41.9	200.0
n-C_4H_{10}	58.12	0.200	425.2	37.5	255.0
i-C_4H_{10}	58.12	0.181	408.1	36.0	263.0
n-C_5H_{12}	72.15	0.252	469.5	33.2	311.0
n-C_6H_{14}	86.18	0.301	507.3	29.7	370.0
n-C_7H_{16}	100.20	0.350	540.1	27.0	432.0
n-C_8H_{18}	114.23	0.400	568.7	24.5	492.0
n-C_9H_{20}	128.26	0.444	594.6	22.6	548.0
Cyclohexane	84.16	0.210	553.0	40.0	308.0
Benzene	78.11	0.210	562.6	48.6	260.0

Heat Capacities

B

See Tables B1, B2, B3 and B4.

Y. Demirel, *Energy*, Green Energy and Technology,
https://doi.org/10.1007/978-3-031-97111-2

Table B1 Heat capacities in the ideal-gas state: $C_p^{ig}/R = A + BT + CT^2$ $T = 298$ K to T_{max} K

Chemical species	Formula	T_{max} K	C_p^{ig}/R 298 K	A	10^3 B	10^6 C
Methane	CH_4	1500	4.217	1.702	9.081	−2.164
Ethane	C_2H_6	1500	6.369	1.131	19.225	−5.561
Propane	C_3H_8	1500	9.011	1.213	28.785	−8.824
Acetylene	C_2H_2	26.04	0.906	191.7	191.7	672.0
Ethylene	C_2H_4	28.05	0.975	104.0	169.5	481.2
n-Butane	C_4H_{10}	1500	11.928	1.935	36.915	−11.402
n-Pentane	C_5H_{12}	1500	14.731	2.464	45.351	−14.111
n-Hexane	C_6H_{14}	1500	17.550	3.025	53.722	−16.791
n-Heptane	C_7H_{16}	1500	20.361	3.570	62.127	−19.486
n-Octane	C_8H_{18}	1500	23.174	4.108	70.567	−22.208
Benzene	C_6H_6	1500	10.259	−0.206	39.064	−13.301
Cyclohexane	C_6H_{12}	1500	13.121	−3.876	63.249	−20.928
Ethanol	C_2H_6O	1500	8.948	3.518	20.001	−6.002
Methanol	CH_4O	1500	5.547	2.211	12.216	−3.450
Toluene	C_7H_8	1500	12.922	0.290	47.052	−15.716
Air		2000	3.509	3.355	0.575	–
Ammonia	NH_3	1800	4.269	3.578	3.020	–
Carbon monoxide	CO	2500	3.507	3.376	0.557	–
Carbon dioxide	CO_2	2000	4.467	5.457	1.045	–
Chlorine	Cl_2	3000	4.082	4.442	0.089	–
Hydrogen	H_2	3000	3.468	3.249	0.422	–
Nitrogen	N_2	2000	3.502	3.280	0.593	–
Nitrogen dioxide	NO_2	2000	4.447	4.982	1.195	–
Oxygen	O_2	2000	3.535	3.639	0.506	–
Sulfur dioxide	SO_2	2000	4.796	5.699	0.801	–
Sulfur trioxide	SO_3	2000	6.094	8.060	1.056	–
Water	H_2O	2000	4.038	3.470	1.450	–

Table B2 Heat capacities of liquids Constants for the equation $C_p/R = A + BT + CT^2$ T (K); $T = 298$ K to T_{max} K

Chemical species	$C_{p,\,298\,K}/R$	A	10^3 B	10^6 C
Ammonia	9.718	22.626	−100.75	192.71
Aniline	23.070	15.819	29.03	−15.80
Benzene	16.157	−0.747	67.96	−37.78
Carbon tetrachloride	15.751	21.155	−48.28	101.14
Chlorobenzene	18.240	11.278	32.86	−31.90
Chloroform	13.806	19.215	−42.89	83.01
Cyclohexane	18.737	−9.048	141.38	−161.62
Ethanol	13.444	33.866	−172.60	349.17
Ethylene oxide	10.590	21.039	− 86.41	172.28
Methanol	9.798	13.431	−51.28	131.13
n-Propanol	16.921	41.653	−210.32	427.20
Toluene	18.611	15.133	6.79	16.35
Water	9.069	8.712	1.25	−0.18

Table B3 Heat capacities of solids+ Constants for the equation $C_p/R = A+BT+CT^{-2}$: $T = 298$ K to T_{max} K

Chemical species	T_{max}, K	$C_{p,\,298}/R$	A	10^3 B	10^{-5} C
CaO	2000	5.058	6.104	0.443	−1.047
$CaCO_3$	1200	9.848	12.572	2.637	−3.120
$CaCl_2$	1055	8.762	8.646	1.530	−0.302
C(graphite)	2000	1.026	1.771	0.771	−0.867
Cu	1357	2.959	2.677	0.815	0.035
CuO	1400	5.087	5.780	0.973	−0.874
Fe	1043	3.005	−0.111	6.111	1.150
FeS	411	6.573	2.612	13.286	
NH_4Cl	458	10.741	5.939	16.105	
NaOH	566	7.177	0.121	16.316	1.948
SiO_2 (quartz)	847	5.345	4.871	5.365	−1.001

Table B4 Ideal-gas specific heats of various common gases

T K	C_p kJ/kg.K	C_v kJ/kg.K	γ	C_p kJ/kg.K	C_v kJ/kg.K	γ	C_p kJ/kg.K	C_v kJ/kg.K	γ
	Air			Carbon dioxide CO_2			Carbon monoxide CO		
250	1.003	0.716	1.401	0.791	0.602	1.314	1.039	0.743	1.400
300	1.005	0.718	1.400	0.846	0.657	1.288	1.040	0.744	1.399
350	1.008	0.721	1.398	0.895	0.706	1.268	1.043	0.746	1.398
400	1.013	0.726	1.395	0.939	0.750	1.252	1.047	0.751	1.395
450	1.020	0.733	1.391	0.978	0.790	1.239	1.054	0.757	1.392
500	1.029	0.742	1.387	1.014	0.825	1.229	1.063	0.767	1.387
550	1.040	0.753	1.381	1.046	0.857	1.220	1.075	0.778	1.382
600	1.051	0.764	1.376	1.075	0.886	1.213	1.087	0.790	1.376
650	1.063	0.776	1.370	1.102	0.913	1.207	1.100	0.803	1.370
700	1.075	0.788	1.364	1.126	0.937	1.202	1.113	0.816	1.364
750	1.087	0.800	1.359	1.148	0.959	1.197	1.126	0.829	1.358
800	1.099	0.812	1.354	1.169	0.980	1.193	1.139	0.842	1.353
900	1.121	0.834	1.344	1.204	1.015	1.186	1.163	0.866	1.343
1000	1.142	0.855	1.336	1.234	1.045	1.181	1.185	0.888	1.335
Hydrogen, H_2				Nitrogen, N_2			Oxygen, O_2		
250	14.051	9.927	1.416	1.039	0.742	1.400	0.913	0.653	1.398
300	14.307	10.183	1.405	1.039	0.743	1.400	0.918	0.658	1.395
350	14.427	10.302	1.400	1.041	0.744	1.399	0.928	0.668	1.389
400	14.476	10.352	1.398	1.044	0.747	1.397	0.941	0.681	1.382
450	14.501	10.377	1.398	1.049	0.752	1.395	0.956	0.696	1.373
500	14.513	10.389	1.397	1.056	0.759	1.391	0.972	0.712	1.365
550	14.530	10.405	1.396	1.065	0.768	1.387	0.988	0.728	1.358
600	14.546	10.422	1.396	1.075	0.778	1.382	1.003	0.743	1.350
650	14.571	10.447	1.395	1.086	0.789	1.376	1.017	0.758	1.343
700	14.604	10.480	1.394	1.098	0.801	1.371	1.031	0.771	1.337
750	14.645	10.521	1.392	1.110	0.813	1.365	1.043	0.783	1.332
800	14.695	10.570	1.390	1.121	0.825	1.360	1.054	0.794	1.327
900	14.822	10.698	1.385	1.145	0.849	1.349	1.074	0.814	1.319
1000	14.983	10.859	1.380	1.167	0.870	1.341	1.090	0.830	1.313

Enthalpies and Gibbs Energies of Formation

C

See Table C1.

Y. Demirel, *Energy*, Green Energy and Technology,
https://doi.org/10.1007/978-3-031-97111-2

Table C1 Standard enthalpies and Gibbs energies of formation at 298.15 K

Chemical species	Formula	State	ΔH_{f298} kJ/mole	ΔG_{f298} kJ/mole
Methane	CH_4	(g)	−74.520	−50.460
Ethane	C_2H_6	(g)	−83.820	−31.855
Propane	C_3H_8	(g)	−104.680	−24.290
n-Butane	C_4H_{10}	(g)	−125.790	−16.570
n-Pentane	C_5H_{12}	(g)	−146.760	−8.650
Benzene	C_6H_6	(g)	82.930	129.665
Benzene	C_6H_6	(l)	49.080	124.520
Cyclohexane	C_6H_{12}	(g)	−123.140	31.920
Cyclohexane	C_6H_{12}	(l)	−156.230	26.850
Ethanol	C_2H_6O	(g)	−235.100	−168.490
Ethanol	C_2H_6O	(l)	−277.690	−174.780
Methanol	CH_4O	(g)	−200.660	−161.960
Methanol	CH_4O	(l)	−238.660	−166.270
Toluene	C_7H_8	(g)	50.170	122.050
Toluene	C_7H_8	(l)	12.180	113.630
Ammonia	NH_3	(g)	−46.110	−16.450
Calcium carbonate	$CaCO_3$	(s)	−1206.920	−1128.790
Calcium oxide	CaO	(s)	−635.090	−604.030
Carbon dioxide	CO_2	(g)	−393.509	−394.359
Carbon monoxide	CO	(g)	−110.525	−137.169
Hydrochloric acid	HCI	(g)	−92.307	−95.299
Hydrogen sulfide	$H\ _2S$	(g)	−20.630	−33.560
Iron oxide	FeO	(s)	−272.000	
Nitric acid	HNO_3	(l)	−174.100	−80.710
Nitrogen oxides	NO	(g)	90.250	86.550
	NO_2	(g)	33.180	51.310
Sodium carbonate	Na_2CO_3	(s)	−1130.680	−1044.440
Sodium chloride	NaCl	(s)	−411.153	−384.138
Sodium hydroxide	NaOH	(s)	−425.609	−379.494
Sulfur dioxide	SO_2	(g)	−296.830	−300.194
Sulfur trioxide	SO_3	(g)	−395.720	−371.060
Sulfur trioxide	SO_3	(l)	−441.040	
Water	H_2O	(g)	−241.818	−228.572
Water	H_2O	(I)	−285.830	−237.129

Ideal Gas Properties of Some Common Gases

D

See Table D1.

Y. Demirel, *Energy*, Green Energy and Technology,
https://doi.org/10.1007/978-3-031-97111-2

Table D1 Ideal-gas properties of air

T, K	H kJ/kg	P_r	U kJ/kg	V_r	S kJ/kg K
270	270.11	0.9590	192.60	808.0	1.596
280	280.13	1.0889	199.75	783.0	1.632
285	285.14	1.1584	203.33	706.1	1.650
290	290.16	1.2311	206.91	676.1	1.668
295	295.17	1.3068	210.49	647.9	1.685
300	300.19	1.3860	214.07	621.2	1.702
305	305.22	1.4686	217.67	596.0	1.718
310	310.24	1.5546	221.25	572.3	1.734
315	315.27	1.6442	224.85	549.8	1.751
320	320.29	1.7375	228.42	528.6	1.766
325	325.31	1.8345	232.02	508.4	1.782
330	330.34	1.9352	235.61	489.4	1.797
340	340.42	2.149	242.82	454.1	1.827
350	350.49	2.379	250.02	422.2	1.857
360	360.58	2.626	257.24	393.4	1.885
370	370.67	2.892	264.46	367.2	1.913
380	380.77	3.176	271.69	343.4	1.940
390	390.88	3.481	278.93	321.5	1.966
400	400.98	3.806	286.16	301.6	1.991
410	411.12	4.153	293.43	283.3	2.016
420	421.26	4.522	300.69	266.6	2.041
430	431.43	4.915	307.99	251.1	2.065
440	441.61	5.332	315.30	236.8	2.088
450	451.80	5.775	322.62	223.6	2.111
460	462.02	6.245	329.97	211.4	2.134
470	472.24	6.742	337.32	200.1	2.156
480	482.49	7.268	344.70	189.5	2.177
490	492.74	7.824	352.08	179.7	2.198
500	503.02	8.411	359.49	170.6	2.219
510	513.32	9.031	366.92	162.1	2.239
520	523.63	9.684	374.36	154.1	2.259
530	533.98	10.37	381.84	146.7	2.279
540	544.35	11.10	389.34	139.7	2.299
550	555.74	11.86	396.86	133.1	2.318
560	565.17	12.66	404.42	127.0	2.336

(continued)

Table D1 (continued)

T, K	H kJ/kg	P_r	U kJ/kg	V_r	S kJ/kg K
570	575.59	13.50	411.97	121.2	2.355
580	586.04	14.38	419.55	115.7	2.373
590	596.52	15.31	427.15	110.6	2.391
600	607.02	16.28	434.78	105.8	2.409
610	617.53	17.30	442.42	101.2	2.426
620	628.07	18.36	450.09	96.92	2.443
630	683.63	19.84	457.78	92.84	2.460
640	649.22	20.64	465.50	88.99	2.477
650	659.84	21.86	473.25	85.34	2.493
660	670.47	23.13	481.01	81.89	2.509
670	681.14	24.46	488.81	78.61	2.525
680	691.82	25.85	496.62	75.50	2.541
690	702.52	27.29	504.45	72.56	2.557
700	713.27	28.80	512.33	69.76	2.572
710	724.04	30.38	520.23	67.07	2.588
720	734.82	32.02	528.14	64.53	2.603
730	745.62	33.72	536.07	62.13	2.618
740	756.44	35.50	544.02	59.82	2.632
750	767.29	37.35	551.99	57.63	2.647
760	778.18	39.27	560.01	55.54	2.661
780	800.03	43.35	576.12	51.64	2.690
800	821.95	47.75	592.30	48.08	2.717
820	843.98	52.59	608.59	44.84	2.745
840	866.08	57.60	624.95	41.85	2.771
860	888.27	63.09	641.40	39.12	2.797
880	910.56	68.98	657.95	36.61	2.823
900	932.93	75.29	674.58	34.31	2.848
920	955.38	82.05	691.28	32.18	2.873
940	977.92	89.28	708.08	30.22	2.897
960	1000.55	97.00	725.02	28.40	2.921
980	1023.25	105.2	741.98	26.73	2.944
1000	1046.04	114.0	758.94	25.17	2.967
1020	1068.89	123.4	776.10	23.72	2.990
1040	1091.85	133.3	793.36	23.29	3.012
1060	1114.86	143.9	810.62	21.14	3.034

(continued)

Table D1 (continued)

T, K	H kJ/kg	P_r	U kJ/kg	V_r	S kJ/kg K
1080	1137.89	155.2	827.88	19.98	3.056
1100	1161.07	167.1	845.33	18.896	3.077
1120	1184.28	179.7	862.79	17.886	3.098
1140	1207.57	193.1	880.35	16.946	3.118
1160	1230.92	207.2	897.91	16.064	3.139
1180	1254.34	222.2	915.57	15.241	3.159
1200	1277.79	238.0	933.33	14.470	3.178
1220	1301.31	254.7	951.09	13.747	3.198
1240	1324.93	272.3	968.95	13.069	3.217
1260	1348.55	290.8	986.90	12.435	3.236
1280	1372.24	310.4	1004.76	11.835	3.255
1300	1395.97	330.9	1022.82	11.275	3.273
1320	1419.76	352.5	1040.88	10.747	3.291
1340	1443.60	375.3	1058.94	10.247	3.309
1360	1467.49	399.1	1077.10	9.780	3.327
1400	1515.42	450.5	1113.52	8.919	3.362
1460	1587.63	537.1	1168.49	7.801	3.412
1500	1635.97	601.9	1205.41	7.152	3.445
1560	1708.82	710.5	1260.99	6.301	3.492
1600	1757.57	791.2	1298.30	5.804	3.523
1640	1806.46	878.9	1335.72	5.355	3.553
1700	1880.1	1025	1392.7	4.761	3.597
1750	1941.6	1161	1439.8	4.328	3.633
1800	2003.3	1310	1487.2	3.994	3.668
1850	2065.3	1475	1534.9	3.601	3.702
1900	2127.4	1655	1582.6	3.295	3.735
1950	2189.7	1852	1630.6	3.022	3.767
2000	2252.1	2068	1678.8	2.776	3.799
2100	2377.7	2559	1775.3	2.356	3.860

Index

Y. Demirel, *Energy*, Green Energy and Technology,
https://doi.org/10.1007/978-3-031-97111-2

MIX
Papier aus verantwortungsvollen Quellen
Paper from responsible sources
FSC® C105338
FSC
www.fsc.org

If you have any concerns about our products,
you can contact us on
ProductSafety@springernature.com

In case Publisher is established outside the EU,
the EU authorized representative is:
Springer Nature Customer Service Center GmbH
Europaplatz 3, 69115 Heidelberg, Germany

Printed by Libri Plureos GmbH
in Hamburg, Germany